Equipment Commonly Used in the Organic Chemistry Laboratory

Filter flask

Büchner funnel

Hirsch funnel

Separatory funnel with ground-glass joints

Conical vial

Claisen adapter

Air condenser

Reflux condenser

Hickman stillhead with port

Drying tube

Pasteur pipet

West condenser

Hempel column

Round-bottom flask

Claisen adapter

Stillhead

Thermometer adapter with Neoprene fitting

Vacuum adapter with ground-glass joints

Rubber septum

Experimental Organic Chemistry

a miniscale and microscale approach
second edition

Experimental Organic Chemistry

a miniscale and microscale approach
second edition

John C. Gilbert
Stephen F. Martin
The University of Texas at Austin

Saunders Golden Sunburst Series

SAUNDERS COLLEGE PUBLISHING
Harcourt Brace College Publishers

Fort Worth   Philadelphia   San Diego   New York   Orlando   Austin
San Antonio   Toronto   Montreal   London   Sydney   Tokyo

Vice President/Publisher: Emily Barrosse
Vice President/Publisher: John Vondeling
Senior Developmental Editor: Elizabeth C. Rosato
Project Editor: Linda Boyle
Production Manager: Alicia Jackson
Art Director: Carol Bleistine

Cover Credit: Cholesteryl Acetate photographed by Rae Nadler-Olenick

---

WARNINGS ABOUT SAFETY PRECAUTIONS

Some of the experiments contained in this Laboratory Manual involve a degree of risk on the part of the instructor and student. Although performing the experiments is generally safe for the college laboratory, unanticipated and potentially dangerous reactions are possible for a number of reasons, such as improper measurement or handling of chemicals, improper use of laboratory equipment, failure to follow laboratory safety procedures, and other causes. Neither the Publisher nor the Authors can accept any responsibility for personal injury or property damage resulting from the use of this publication.

Printed in the United States of America

EXPERIMENTAL ORGANIC CHEMISTRY: A MINISCALE AND MICROSCALE APPROACH, second edition

0-03-020632-4

Library of Congress Catalog Card Number: 97-67176

890123456 032 10 9876543

Contents Overview

Preface

The laboratory course in organic chemistry has undergone significant changes during the past decade. The motivating forces behind these changes derive largely from concerns about the effects of chemicals on the environment, the increasing costs of the chemicals that are used in the laboratory, and the costs associated with their safe disposal. In response to these concerns, teachers have turned to reducing the scale at which the experiments are performed. The two pedagogically distinct approaches that have thus emerged may be termed **"miniscale"** and **"microscale"**. We have revised this edition of our textbook to encompass *both* approaches, each of which is meritorious with respect to financial cost and environmental impact. Both miniscale and microscale procedures have been written for most of the experiments included in the textbook, and it remains for individual instructors to decide which format or combination of the two is better suited to the needs of their students.

Theoretical Discussions

The philosophy we adopted in developing the discussions preceding the Experimental Procedures was to provide sufficient theoretical background for the individual experiments so that the textbook is self-contained. However, to minimize unnecessary duplication of material found in lecture texts, the discussion is focused upon the experiments themselves. Thus, the discussions are intended to give the experimentalist the basic information needed for complete understanding of the mechanistic and practical aspects of the specific reactions and procedures being performed.

Experimental Procedures

The *miniscale approach* will appeal to those who believe in the importance of performing experiments on a scale that allows isolation and characterization of products using conventional laboratory glassware, while reducing the costs of purchasing and disposing of chemicals. The quantities of starting materials used in these experiments are less than 5 grams, usually in the range of 1 to 3 grams, so the technical challenges that attend working with much smaller amounts of material are avoided. Normally you will be able to develop the techniques required to purify your products and fully characterize them by comparing their melting or boiling points with literature values. You may then further characterize these products by IR and NMR spectroscopy so the spectral properties of starting materials and products may be compared. In short, you are able to experience the real world of organic chemistry where usable quantities of compounds must be made.

On the other hand, the *microscale approach* is especially attractive when the cost of purchasing and disposing of chemicals is a major concern. The specialized glassware that is required for performing experiments on such small scales is now readily available.

Indeed, many of the components found in a microscale kit are also found in the advanced organic laboratory where trained researchers often must work with minute amounts of material. The weights of reactants and reagents that are used in these procedures are typically less than one gram and are often only 100–200 mg. Because the quantities of materials that must be handled are small, you must be particularly meticulous in order to isolate your products. Purifying small quantities of materials by distillation or recrystallization is tedious, so it will frequently be impractical to characterize pure products. Nevertheless, the experiments selected for performing on the microscale should provide tangible quantities of material so that you can verify that the product was formed by a technique other than the ''sniff test,'' which is generally unwise.

The experiments we have selected are intended not only to reinforce concepts given in the lecture component of organic chemistry but also to familiarize you with the techniques that organic chemists routinely use to test mechanistic hypotheses and to synthesize compounds in the laboratory. Because certain fundamental laboratory operations must be mastered prior to actually conducting experiments, the procedures in Chapters 3–6 are designed to introduce you to the different techniques for distillation, liquid-liquid and liquid-solid extraction, thin-layer, column, and gas-liquid chromatography; the basic principles for these techniques are also described in the respective chapters. Experiments that illustrate concepts such as kinetic and thermodynamic control of reactions (Chapter 13), selectivity of free-radical substitution (Chapter 9), and the stereochemistry and regiochemistry of addition reactions (Chapters 10, 11, 12, and 17) are designed to provide a better understanding of these important subjects. Other experiments illustrate specific chemical transformations such as the generation, reactions, and rearrangements of carbocations (Chapters 10 and 15), electrophilic aromatic and nucleophilic substitution processes (Chapters 15 and 14, respectively), eliminations (Chapters 10 and 11), oxidations and reductions (Chapters 16 and 17, respectively), nucleophilic additions to carbonyl compounds and imines (Chapters 17 and 18), and the generation and reactions of Grignard reagents (Chapter 19). The value of enzymes for effecting enantioselective reactions is illustrated in Chapter 17. Because the current practice of organic chemistry in industry frequently involves multi-step transformations, examples of multi-step synthesis are contained in Chapter 20. A rational approach to solving the structures of unknown compounds with and without the aid of spectroscopic data is given in Chapter 23. In summary, the experimental procedures cover a broad range of topics and provide a sufficient number of options so that the instructor may focus the course in whatever manner is desired.

Spectroscopic Techniques

The importance of spectroscopy in modern experimental organic chemistry cannot be overemphasized. Consequently, a thorough discussion of the theory and practical techniques for infrared and nuclear magnetic resonance, including ^{1}H and ^{13}C NMR is presented in Chapter 8. To help reinforce the basic principles set forth in this chapter, the infrared and nuclear magnetic spectra of all of the organic starting materials and products are provided in this textbook. Moreover, in the CD-ROM that accompanies this text, it is possible to perform simple manipulations of the ^{1}H NMR spectra to measure integrals and coupling constants directly. In several instances, raw spectral data are provided so that it is possible for you to process the data to obtain the final spectra. This ''hands on'' experience has proven an invaluable aid in teaching the basics of interpreting ^{1}H NMR spectra.

It also gives you a more realistic sense for the modern organic laboratory, especially when such instrumentation is not readily available.

Safety and the Environment

With each procedure we incorporate sections entitled ''Safety Alert'' and ''Finishing Touches.'' The first section is designed to alert you and your instructor to possible hazards associated with the experimental operations being performed. Because of the flammable nature of the solvents and the chemicals that are handled in the laboratory, the use of *flameless* heating sources is emphasized. Careful adherence to these recommendations will make the laboratory a safer workplace. The guidelines and methods in the Finishing Touches section are provided to assist you in the proper disposal of the spent chemicals and other by-products of the experimental procedures. Use of these recommended methods will protect the environment and lessen the costs associated with the necessary disposal of these materials.

Essays

Several ''Historical Highlights'' are included as part of the chapters to which they are related. Some of these brief essays are biographical to familiarize you with the lives of the chemical pioneers who have advanced our science. Other accounts are intended to provide you with a sense of the excitement and insight of the individuals whose observations evolved into the experiments that you perform in the laboratory. As you will see, luck has played a major role in some very important scientific discoveries!

Pre-Lab and Post-Lab Exercises

Each experiment in this textbook is accompanied by a set of questions. The ''Pre-Lab Exercises'' are intended to be worked as part of preparing for the laboratory and are contained in a separate book. They are designed to test your understanding of the basic concepts underlying the experiments being performed and the techniques required to conduct the procedures in a safe and successful manner. Your instructor may require that answers to these questions be submitted *before* you may begin work in the laboratory. Even if submission of answers is not required, working the exercises will help you derive the greatest educational benefit from the laboratory experience.

The ''Post-Lab Exercises'' are found under the heading, ''Exercises,'' at the end of each Experimental Procedure. These questions are written to reinforce the principles that are illustrated by the experiments and to determine whether you understand the observations you made or the particular laboratory operations you performed. Questions on spectroscopy will help you develop the skills required to interpret IR and ^{1}H and ^{13}C NMR spectra.

Instructor's Manual

Teachers who adopt this textbook may obtain an Instructor's Manual. Its contents include answers to all the Pre- and Post-Lab exercises and tables of compounds and their derivatives to support the qualitative organic analysis of Chapter 23.

Saunders College Publishing may provide complimentary instructional aids and supplements or supplement packages to those adopters qualified under our adoption policy. Please contact your sales representative for more information. If as an adopter or potential user you receive supplements you do not need, please return them to your sales representative or send them to

Attn: Returns Department
Troy Warehouse
465 South Lincoln Drive
Troy, MO 63379

Significant Changes from the First Edition

A number of new features have been incorporated in the second edition of this textbook. The *addition of microscale procedures* to accompany most of the experiments is one of the most significant changes, as it maximizes the flexibility of the instructor in tailoring the content of the course to address local pedagogical, environmental, and economic issues. Chapter 2 has thus been modified extensively to incorporate the discussion of techniques and glassware used for microscale experiments.

All of the *IR and ^{1}H NMR spectra are new* and have been measured using modern spectrometers so that you are exposed to state-of-the-art spectroscopic data. Nearly all the IR spectra of solids have been obtained using IR cards, which greatly facilitate sample preparation. A *CD-ROM version of the ^{1}H NMR spectra has been added*. This is a unique ancillary instructional tool that for the first time allows you to analyze the NMR spectra of the compounds you use in the organic laboratory by directly measuring integrations and coupling constants. A series of questions on spectroscopy have been added to aid you in learning to interpret IR and NMR spectra.

In addition to these important changes, a number of other enhancements have been added that render the textbook more readable and helpful to the student. The *background and discussion sections* that precede the experiments *have been rewritten* to focus only on those concepts essential to understanding the experiment. The *experimental procedures have been extensively reorganized* to assist you in understanding the purpose for doing the experiment, in reviewing the theory and techniques involved, in determining the apparatus that is needed, and in understanding the various stages of the procedures. To make it easier to assemble the apparatus required for many experiments, *icon artwork for these set-ups has been added* in the margins.

Finally, there have been a number of *revisions to the experimental procedures*. Several procedures that either proved troublesome or were considered too hazardous have been deleted or modified to avoid the problem. Commercially available starting materials are used, even in those cases where the preparation of one of the requisite starting materials was incorporated as part of the procedure in the previous edition. A *new multi-step synthesis* has been added that may be performed on either a miniscale or a microscale; this new procedure increases your instructor's flexibility for introducing students you to this important aspect of organic chemistry.

We would appreciate your comments, criticisms, and suggestions regarding our textbook. Our e-mail and ''snail-mail'' addresses are provided below, and we look forward to hearing from you. We promise to respond to any messages that you send.

Acknowledgments

We wish to thank the following reviewers who reviewed the first edition of EXPERI-MENTAL ORGANIC CHEMISTRY: A Miniscale Approach and offered many suggestions for creating this new edition:

Sally Mallory	University of Pennsylvania
Nancy Mills	Trinity University, TX
Denise Guinn	Regis College, CO
Miroslav Krumpolc	University of Illinois at Chicago

The following individuals responded to a detailed survey we created:

Larry S. Helwick	Cedarville College, OH
David E. Minter	Texas Christian University
Terry Kruger	Ball State University
Woodrow Elias	Moorpark College, CA
Don England	Harding University, AR

A number of individuals made important contributions to this textbook. These include Brian Barnes, Laura Breksa, Ingrid Fellows, Jason Fellows, David Flaxbart, Duen-ren Hou, Connie Kiang, Darin Laird, Ann Lockwood, Janet Macdonald, Timothy Marquart, Sarah Mathews, Ben Shoulders, Steven Sorey, and Teresa Winterringer (The University of Texas at Austin), and Arthur Meyer (Austin Community College). Prof. M. Robert Willcott (The University of Texas Medical Branch at Galveston) kindly provided the MRI plots accompanying the Historical Highlight in Chapter 8. In addition, we thank the many individuals at Saunders College Publishing for cajoling, encouraging, and empathically responding to the stresses we encountered during the course of this exciting project. We particularly acknowledge the efforts of Carol Bleistine, Linda Boyle, and Beth Rosato for making this textbook ''happen.'' Finally, we thank 3M for its generous gift of the cards used for obtaining IR spectra of solids, Jennifer Andrews (The Perkin-Elmer Corporation) for providing the software for digital manipulation of IR data, and Virginia Miner and Woodrow Conover (Acorn NMR Inc.) for their contributions to developing the CD-ROM containing the ^{1}H NMR spectra presented in our textbook.

In Memoriam

Professor Royston Roberts, whose commitment to a meaningful and educational experience in the undergraduate organic laboratory led to the development of the original version of this textbook, died in 1996. Much of the philosophy on which our writing is based reflects his own approach to the subject of experimental organic chemistry, and we have an everlasting debt of gratitude to him for his vision and mentorship.

jgilbert@mail.utexas.edu   **John C. Gilbert**

sfmartin@mail.utexas.edu   **Stephen F. Martin**

Department of Chemistry and Biochemistry
The University of Texas at Austin
Austin, TX 78712

February 1997

Contents

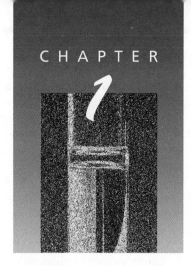

CHAPTER 1

Introduction, Record Keeping, and Laboratory Safety

The laboratory component of a course in organic chemistry has an important role in developing and augmenting your understanding of the subject matter. The theoretical concepts, functional groups, and reactions presented in the lecture part of the course may be abstract, but should become more understandable as a result of the experiments you perform. The ''hands-on'' experience gained in the laboratory, as you gather and interpret data from a variety of reactions, provides a sense of organic chemistry that is nearly impossible to communicate in lectures. For example, it is one thing to be told that the addition of bromine (Br_2) across the π-bond of most alkenes is a rapid process at room temperature but quite another to personally observe the *immediate* decoloration of a reddish solution of bromine in dichloromethane (Br_2/CH_2Cl_2) as a few drops of it are added to cyclohexene. The principles developed in the lectures will help you to predict what reaction(s) should occur when various reagents are combined in experimental procedures and to understand the mechanistic course of the process(es). Performing reactions allows you to verify what you have been told in lecture.

Of course, the laboratory experience in organic chemistry has another important function beyond reinforcing the concepts presented in lecture—and that is to introduce you to the techniques and procedures that are important to the successful practice of experimental organic chemistry. You will learn how to handle a variety of chemicals safely and how to manipulate apparatus properly, talents that are critical to your success as a student of the chemical sciences. Along with becoming more skilled in the technical aspects of laboratory work, you should also develop a proper scientific approach to the execution and interpretation of experiments. By reading and understanding this chapter, you will be better able to achieve these valuable goals.

1.2 Preparing for the Laboratory

One of the most common misconceptions students have about performing experiments is that it is much like cooking; that is, one merely follows the directions given—the ''recipe''—and the desired product or data will result. Such students enter the laboratory prepared to follow the provided experimental procedure in a more or less rote manner. This is an unfortunate attitude that can lead to inefficiencies, accidents, and minimal educational benefit and enjoyment from the laboratory experience.

The correct approach to being successful in the laboratory is *never* to begin any experiment until you understand its overall purpose and the reasons for each operation that you are to do. This means that you must *study*, not *just read*, the entire experiment *prior* to arriving at the laboratory. Rarely, if ever, can you complete the necessary preparation in five or ten minutes, which means that you should not wait until just before the laboratory period begins to do the study, thinking, and writing that are required. *Planning* how to spend your time in the laboratory is the key to efficient completion of the required experiments. Your performance in the laboratory will benefit enormously with proper advance work and so will your grade!

The specific details of what you should do prior to coming to the laboratory will be provided by your instructor. However, to help you prepare in advance, we have developed a set of ''Pre-Lab Exercises'' for each of the experimental procedures in this textbook. Your instructor may require you to submit answers to the exercises for approval before authorizing you to do the assigned experiments. Even if you are not asked to complete the exercises, you will find that working them *prior* to the laboratory period will be a valuable educational tool to self-assess your understanding of the experiments to be done.

You will undoubtedly be required to maintain a laboratory notebook, which will serve as a complete, accurate, and neat record of the experimental work that you do. Once more, your instructor will provide an outline of what specific information should appear in this notebook, but part of what is prescribed will probably necessitate advance preparation, which will further enhance your ability to complete the experiments successfully. The laboratory notebook is a *permanent record* of your accomplishments in the course, and you should take pride in the quality and completeness of its contents!

1.3   Working in the Laboratory

You should be aware that experimental organic chemistry is *potentially* dangerous, because many of the chemicals used are toxic and/or highly flammable, and most of the procedures require the use of glassware that is easily broken. Careless handling of these chemicals and sloppy assembly of apparatus are sources of danger not only to yourself but also to those working near you. You should *not* be afraid of the chemicals and equipment that you will be using, but you *should* treat them with the respect and care associated with safe experimental practice. To facilitate this, there is an emphasis on the proper handling of chemicals and apparatus throughout the textbook, and the importance of paying particular attention to these subjects *cannot* be overemphasized. In a sense, laboratory safety is analogous to a chain that is only as strong as its weakest link: the possibility that an accident will occur is only as great as the extent to which unsafe practices are followed. In other words, if you and your labmates adhere to proper laboratory procedures, the risk of an accident will be minimized.

It is important that you closely follow the experimental procedures in this textbook. There is a good reason why each operation should be performed as it is described, although that reason may not be immediately obvious to you. Suffice it to say that it is *dangerous* for a beginning experimentalist to be ''creative'' when it comes to modifying the steps and approaches that we have specified! As you become more experienced in the organic laboratory, you may wish to develop alternate procedures for performing a reaction or purifying a desired product, but *always* check with your instructor *before* trying any modifications.

Note that rather detailed experimental procedures are given early in the textbook, whereas somewhat less detailed instructions are provided later on. This is because many of the basic laboratory operations will have become familiar to you and need not be spelled out. It is hoped that this approach to the design of procedures will decrease your tendency to think that you are essentially following a recipe in a cookbook. Moreover, many of the experimental procedures given in the literature of organic chemistry are relatively brief and require the chemist to "fill in the blanks," so it is valuable to gain some initial experience in figuring out some details on your own.

Most of your previous experience in a chemistry laboratory has probably required that you precisely measure quantities, using analytical balances, burets, pipets, and other precise measuring devices (Secs. 2.5 and 2.6). Indeed, if you have done any quantitative inorganic analysis, you know that it is often necessary to measure weights to the third or fourth decimal place and volumes to at least the first. Experiments in organic chemistry that are performed at the **microscale** level, that is, experiments in which less than about 1 mL of the principal reagents is used and the amounts of solvents are less than 2 or 3 mL, also require relatively precise measuring of quantities. For example, if you are to use 0.1 g of a reagent and your measuring device only allows measuring to the nearest 0.1 g, you could easily have as much as about 0.15 g or as little as 0.05 g of the reagent. Such deviations from the desired quantity represent significant *percentage* errors in measurement and can result in serious errors in the proportions of reagents involved in the reaction. Consequently, weights should be accurate to within about 0.01 g and volumes to within about 0.1 mL. This requires the use of appropriate analytical balances and graduated pipets.

Experiments being performed at the **miniscale** level, which we define as involving 1 to 5 g of reagents and usually less than about 25 mL of solvent, normally do not require such precise measuring. Weighing reagents to the nearest tenth of a gram is usually satisfactory, as is measuring out liquids in graduated cylinders, which are accurate to $\pm 10\%$. For example, if you are directed to use 20 mL of diethyl ether as solvent for a reaction, the volume need *not* be 20.0 mL. In fact, it probably will make little difference to the success of the reaction whether anywhere from 15 to 25 mL of the solvent is added. This is not to say that care need not be exercised in measuring out the amounts of materials that you use. Rather, it means that valuable time need not be wasted in making these measurements highly precise.

Markers, in the form of **stars** (★), have been inserted within many of the experimental procedures in this textbook. These indicate places where the procedure can be interrupted without affecting the final outcome of the experiment. These markers are designed to help you make the most efficient use of the time you have in the laboratory. For example, you may be able to start a procedure at a point in the period when there is insufficient time to complete it, but enough to be able to work through to the location of a star (★); you can then safely store the reaction mixture and finish the sequence during the next laboratory period. Stars have *not* been inserted at every possible stopping point, only at those where it is not necessarily obvious that interruption of the procedure will have no effect on the experimental results. Consult your instructor if in doubt about whether a proper stopping point has been reached.

As noted above, a *carefully* written **notebook** and *proper* **safety procedures** are important components of an experimental laboratory course. These aspects are further discussed in the following two sections.

1.4 The Laboratory Notebook

One of the most important characteristics of successful scientists is the habit of keeping a complete and understandable record of the experimental work that has been done. Did a precipitate form? Was there a color change during the course of the reaction? At what temperature was the reaction performed and for how long did the reaction proceed? Was the reaction mixture homogeneous or heterogeneous? On what date(s) was the work performed? These are observations and data that may seem insignificant at the time but may later prove critical to the interpretation of an experimental result or to the ability of another person to reproduce your work. All of them belong in a properly kept laboratory notebook. Suggestions for such a document follow. Your instructor may specify other items to be included, but the list we give is representative of a good notebook.

1.5 General Protocol for the Laboratory Notebook

1. Use a *bound* notebook for your permanent laboratory record to minimize the possibility that pages will be lost. If a number has not been printed on each page, do so manually. Some laboratory notebooks are designed with pairs of identically numbered pages so that a carbon copy of all entries can be made. The duplicate page can then be removed and turned in to your instructor or put in a separate place for safekeeping. Many professional scientists use this type of notebook.
2. Reserve the first page of the notebook for use as a title page, and leave several additional pages blank for a Table of Contents.
3. As the main criterion for what should be entered in the notebook, adopt the rule that the record should be sufficiently complete so that anyone who reads it will know exactly what you did and will be able to repeat the work in precisely the same way as it was done originally.
4. Record all experimental observations and data in the notebook as they are obtained. Include the *date* and, if appropriate, the *time* when you did the work. In a legal sense, the information entered into the notebook *at the time of performance* constitutes the primary record of the work, and it is important for you to follow this principle. Many patent cases have been determined on the basis of dates and times recorded in a laboratory notebook. One such example is described in the Historical Highlight at the end of this chapter.
5. Make all entries in ink, and *do not delete anything you have written* in the notebook. If you make a mistake, cross it out and record the correct information. Using erasers or correction fluid to modify entries in your notebook is unacceptable scientific practice!

   Do not scribble notes on odd bits of paper with the intention of later recording the information in your notebook. Such bad habits only lead to problems, since the scraps of paper are easily lost or mixed up. They are also inefficient, since transcribing the information to your notebook means that you must write it a second time. This procedure can also result in errors if you miscopy the data.

   Finally, do not trust your memory with respect to observations that you have made. When the time comes to write down the information, you may have forgotten a key observation that is critical to the success of the experiment.
6. Unless instructed to do otherwise, do not copy *detailed* experimental procedures provided elsewhere in your notebook; this consumes valuable time. Rather, provide

a specific reference to the source of the detailed procedure and enter a *synopsis* of the written procedure that contains enough information that (1) you need not refer to the source while performing the procedure and (2) another chemist will be able to *duplicate* what you did. For example, when performing an experiment from this textbook, a reference should be given to the page number on which the procedure appears, and any *variations* made in the procedure should be detailed along with the reason(s) for doing so.

7. Start the description of each experiment on a new page titled with the name of the experiment. The recording of data and observations from several different procedures on the same page can lead to confusion, both for yourself and for others who may read your notebook.

1.6 Types of Organic Experiments and Notebook Formats

This textbook contains two general classes of experiments: **investigative experiments** and **preparative experiments**. Investigative experiments normally involve making observations and learning techniques that are common to laboratory work in organic chemistry but do not entail conversion of one compound into another. Some examples are solubility tests, distillation, recrystallization, and qualitative organic analysis. Preparative experiments, in contrast, involve interconversion of different compounds. Most of the procedures described in this textbook fall into the latter category.

The format of the laboratory notebook is usually different for these two types of experiments. Once again, your instructor may have a particular style that is recommended, but we provide suggested formats in the sections that follow.

NOTEBOOK FORMAT FOR INVESTIGATIVE EXPERIMENTS

1. **Heading.** Use a new page of the notebook to start the entries for the experiment. Provide information that includes your name, the date, the title of the experiment, and a reference to the place in the laboratory textbook or other source where the procedure may be found.

2. **Introduction.** Give a brief introduction to the experiment in which you clearly state the purpose(s) of the procedure. This should require no more than one-fourth of a page.

3. **Synopsis of and Notes on Experimental Procedure—Results.** Enter a one- or two-line statement for each part of an experiment. Reserve sufficient room to record results as they are obtained. As noted in Part (6) of ''Notebook Format for Preparative Experiments,'' do *not* copy the experimental procedure from the textbook, but provide a synopsis of it.

   Much of this section of the write-up can be completed before coming to the laboratory to ensure that you understand the experiment and that you will perform all parts of it.

4. **Interpretation of Instrumental Data.** If instructed to do so, discuss any instrumental data, such as gas-liquid chromatographic analyses and spectral data you have obtained or that are provided in the textbook.

5. **Conclusions.** Record the conclusions that can be reached, based on the results you have obtained in the experiment. If the procedure has involved identifying an unknown compound, summarize your findings in this section.

6. **Answers to Exercises.** Enter answers to any exercises for the experiment that have been assigned from the textbook.

A sample write-up of an investigative experiment is given in Figure 1.1.

NOTEBOOK FORMAT FOR PREPARATIVE EXPERIMENTS

1. **Heading.** Use a new page of the notebook to start the entries for the experiment. Provide information that includes your name, the date, the title of the experiment, and a reference to the place in the laboratory textbook or other source where the procedure may be found.

2. **Introduction.** Give a brief introduction to the experiment in which you clearly state the purpose(s) of the procedure. This should require no more than one-fourth of a page.

3. **Main Reaction(s) and Mechanism(s).** Write *balanced* equations giving the main reaction(s) for conversion of starting material(s) to product(s). The reason for balancing the equations is discussed in Part (4) below. Whenever possible, include the detailed mechanisms for the reactions that you have written.

4. **Table of Reactants and Products.** Set up a Table of Reactants and Products as an aid in summarizing the amounts and properties of reagents and catalysts being used and the product(s) being formed. Only those reactants, catalysts, and products that appear in the main reaction(s) should be listed in the table; many other reagents may be used in the work-up and purification of the reaction mixture, but these should *not* be entered in the table.

   Your instructor will have specific recommendations about what should appear in the table, but the following items are illustrative.

   a. The name and/or structure of each reactant, catalyst, and product.
   b. The molecular weight of each compound.
   c. The weight used, in grams, of each reactant and the volume of any liquid reactant. We recommend that the weight and/or volume of any catalysts used be entered for purposes of completeness.
   d. The molar amount of each reactant used; this can be calculated from the data in Parts **b** and **c**.
   e. The theoretical mole ratio, expressed in whole numbers, for the reactants and products; this ratio is determined by the *balanced* equation for the reaction, as given in Part **3**.
   f. Physical properties of the reactants and products; this entry might include data such as boiling and/or melting point, density, solubility, color, and odor.

5. **Yield Data.** Compute the maximum possible amount of product that can be formed, called the **theoretical yield**. This can easily be calculated from the data in the Table of Reactants and Products as follows. First determine which of the reactants corresponds to the **limiting reagent**. This is the reagent that is used in the *least* molar amount relative to what is required theoretically. In other words, the reaction will stop once this reactant is consumed, so its molar quantity will define the maximum quantity of product that can be produced. From the number of moles of limiting reagent involved and the balanced equation for the reaction, determine the theoretical yield, in moles (mol), of product. This value can then be converted into the theoretical yield in grams, based on the molecular weight of the product.

   Once the isolation of the desired product(s) has been completed, you should also calculate the **percent yield**, which is a convenient way to express the overall efficiency of the reaction. This is done by obtaining the **actual yield** of product(s)

1.   **Your Name**
                                                                             Date

Separation of Green Leaf Pigments by TLC

Reference: *Experimental Organic Chemistry: A Miniscale and Microscale Approach,*
by Gilbert and Martin, Section 6.3.

2. INTRODUCTION

The pigments in green leaves are to be extracted into an organic solvent, and the extract is to be analyzed by thin-layer chromatography (TLC). The presence of multiple spots on the developed TLC plate will indicate that more than a single pigment is contained in the leaves.

3. SYNOPSIS OF AND NOTES ON EXPERIMENTAL PROCEDURE—RESULTS

Procedure: Grind five stemless spinach leaves in mortar and pestle with 5 mL of 2:1 pet. ether and EtOH. Swirl soln. with 3×2-mL portions H_2O in sep. funnel; dry org. soln. for few min over anhyd. Na_2SO_4 in Erlenmeyer. Decant and concentrate soln. if not dark-colored. Spot 10-cm $\times$ 2-cm TLC plate about 1.5 mm from end with dried extract; spot should be less than 2-mm diam. Develop plate with $CHCl_3$. *Variances and observations:* Procedure followed exactly as described in reference. Org. soln. was dark green in color; aq. extracts were yellowish. Half of org. layer lost. TLC plate had five spots having colors and R_f-values shown on the drawing below.

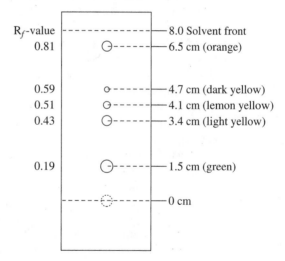

4. INTERPRETATION OF INSTRUMENTAL DATA

No data provided for this experiment.

5. CONCLUSIONS

Based on TLC analysis, the procedure used allows the extraction of at least five different pigments from the spinach leaves. Judging from colors, one of these is a carotene, three are xanthophylls, and the last is chlorophyll *b*.

6. ANSWERS TO EXERCISES

(Answers omitted intentionally)

FIGURE 1.1   Sample notebook format for investigative experiments.

in grams, and then applying the expression in Equation 1.1. Generally, the calculated value of percent yield is rounded to the nearest whole number. As points of reference, most organic chemists consider yields of 90% or greater as being "excellent," and those below 20% as "poor."

$$\text{Percent yield} = \frac{\text{actual yield (g)}}{\text{theoretical yield (g)}} \times 100 \qquad \textbf{(1.1)}$$

6. **Synopsis of and Notes on Experimental Procedure.** Provide an outline of the experimental procedure that contains enough detail so that you do not have to refer to the textbook repeatedly while performing the experiment. Note any variations that you use, as compared to the referenced procedure, and observations that you make while carrying out the formation and isolation of the product(s).

7. **Observed Properties of Product.** Record the physical properties of the product that you have isolated in the experiment. Appropriate data under this heading might include boiling and/or melting point, odor, color, and crystalline form, if the product is a solid. Compare your observations with those available on the compound in various reference books (for example, the *CRC Handbook of Chemistry and Physics* or Lange's *Handbook of Chemistry*).

8. **Side Reactions.** If instructed to do so, list possible **side reactions** (those reactions leading to undesired products) that are likely to occur in the experiment. It is important to consider such processes because the by-products that are formed must be removed by the process used to purify the desired product. You may need to consult your lecture notes and textbook in order to predict what side reactions might be occurring.

9. **Other Methods of Preparation.** If instructed to do so, suggest alternate methods for preparing the desired compound. Such methods may involve using entirely different reagents and reaction conditions. Your lecture notes and textbook can serve as valuable resources for providing possible entries for this section.

10. **Method of Purification.** Develop a flow chart that summarizes the sequence of operations that will be used to purify the desired product. The chart will show at what stages of the work-up procedure unchanged starting materials and unwanted by-products are removed. By understanding the logic of the purification process, you will know why each of the various operations specified in the purification process is performed.

    Purifying the final product of a reaction can be the most challenging part of an experimental procedure. Professional organic chemists are constantly required to develop work-up sequences that allow isolation of a pure product, free from starting materials and other contaminants. They do this by considering the chemical and physical properties of both the desired and undesired substances, and it is important for you to gain experience in devising such schemes as well.

11. **Interpretation of Instrumental Data.** If instructed to do so, discuss any instrumental data, such as gas-liquid chromatographic analyses and spectral data you have obtained or that are provided in the textbook.

12. **Answers to Exercises.** Enter answers to any exercises for the experiment that have been assigned from the textbook.

A detailed example of the write-up for a preparative experiment involving the dehydration of cyclohexanol (Sec. 10.3) is given in Figure 1.2. You may not actually perform this reaction; nevertheless, you should carefully study the example in order to see how to prepare specific entries for the first eight items listed. The various entries in Figure 1.2 are labeled with circled, **boldface** numbers, and are discussed further in

(text continued on page 12)

1.　　　　　　　　　　　　　　　　　　　　　　　　　**Your Name**
　　　　　　　　　　　　　　　　　　　　　　　　　　　Date

Dehydration of Cyclohexanol

Reference: *Experimental Organic Chemistry: A Miniscale and Microscale Approach,*
by Gilbert and Martin, Section **10.2.**

2. INTRODUCTION

Cyclohexene is to be prepared by the acid-catalyzed dehydration of cyclohexanol.

① **3.** MAIN REACTION(S) AND MECHANISM(S)

$$\text{—OH} \xrightarrow[\text{H}_2\text{SO}_4]{\Delta} \quad + \text{H}_2\text{O}$$

(Catalytic
amount)

　　　1　　　　　　　　　　2

(Mechanism omitted intentionally)

② **4.** TABLE OF REACTANTS AND PRODUCTS

③ Compound	④ M.W.	⑤a Volume used (mL)	⑤b Weight used (g)	⑥ Moles used	⑦ Moles required	⑧ Other data
Cyclohexanol	100.2	5.2	5	0.05	1	bp 161 °C (760 Torr), mp 25.1 °C, d 0.962 g/mL, colorless
Sulfuric acid (9 *M*)	98.1	2.5	#	#	0	d 1.84 (18 *M* H$_2$SO$_4$)
Cyclohexene	82.2	*	*	*	1	bp 83 °C (760 Torr), mp 103.5 °C, d 0.810, colorless

#Entry left blank because this row is for the catalyst.
*Entry left blank because this row is for the product.

⑨ LIMITING REAGENT: Cyclohexanol

⑩ **5.** YIELD DATA

Theoretical yield of cyclohexene = moles of limiting reagent (cyclohexanol) × M.W. of cyclohexene
= 0.05 mol × 82.2 g/mol
= 4.1 g

Actual yield = 2.7 g　　　　　　　　　　　　　　　　　　　*(Continued)*

FIGURE 1.2　Sample notebook format for preparative experiments.

Percent yield = [Actual yield (g)/theoretical yield (g)] $\times$ 100

$\qquad$ = [2.7/4.1] $\times$ 100 = 66%

(11)   6. SYNOPSIS OF AND NOTES ON EXPERIMENTAL PROCEDURE—RESULTS

Procedure: Put alcohol in 25-mL rb flask and add H_2SO_4. Mix, add stirbar, attach to fractional dist. apparatus. Heat with oil bath; heating rate such that head temp. stays below 90 °C. Stop when 2.5 mL remain in rxn. flask. Put distillate in 25-mL Erlenmeyer and add 1–2 g K_2CO_3.★ Occasionally swirl mix. for 15 min and transfer liquid to 10-mL rb by decantation or pipet. Add stirbar and do simple distillation (no flames!); receiver must be close to drip tip of adapter to minimize losses by evaporation. Collect product at 80–85 °C (760 Torr). *Variances and observations:* Procedure followed exactly as described in reference. Distillate cloudy throughout dehydration step; formed two layers in receiver. Head temperature never exceeded 77 °C. Liquid in stillpot darkened as reaction proceeded. Addition of carbonate (1 g) to distillate caused evolution of a few bubbles of gas (CO_2?) Had to add about 0.5 g more of carbonate to get rid of cloudiness. Left solution over drying agent for one week (next lab period). Used pipet to transfer dried liquid to distillation flask. Collected cyclohexene in ice-cooled 10-mL rb flask attached to vacuum adapter protected with $CaCl_2$ tube. Stopped distillation when about 1 mL of yellowish liquid remained in stillpot.

(12)   7. OBSERVED PROPERTIES OF PRODUCT

bp 80–84 °C (760 Torr); colorless liquid; insoluble in water; decolorizes Br_2/CH_2Cl_2 solution and produces brown precipitate upon treatment with $KMnO_4/H_2O$.

(13)   8. SIDE REACTIONS

(Continued)

FIGURE 1.2   *(Continued)*

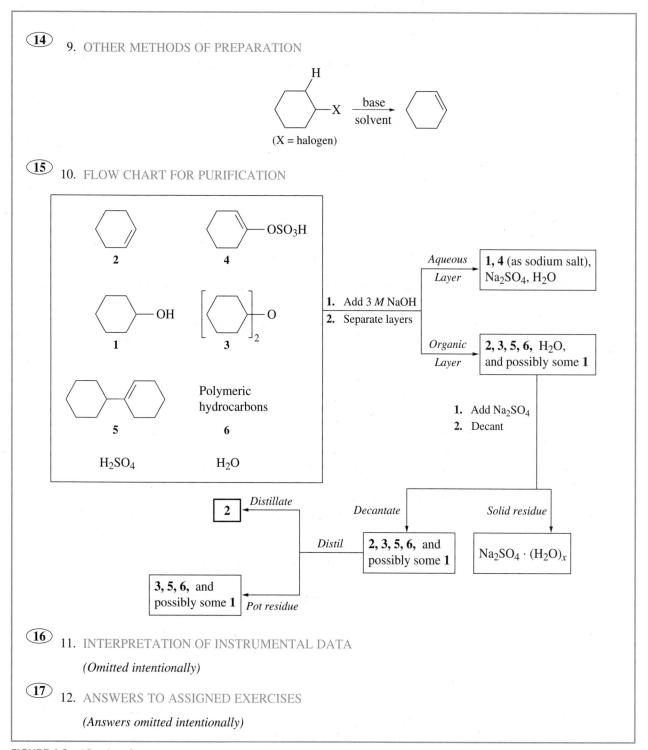

FIGURE 1.2 *(Continued)*

the following paragraphs. It is assumed for illustrative purposes that an actual yield of 2.7 g is obtained.

(1) Use a new page of the notebook to start the entries for the experiment. Provide information that includes your name, the date, the title of the experiment, and a reference to the place in the laboratory textbook or other source where the procedure can be found.

(2) Self-explanatory.

(3) There is only a single reaction in our example, but in many cases more than one step is involved; write equations for *all* of the main reactions. A mechanism for the reaction is intentionally omitted in our example.

(4) Use the illustrated format for the Table of Reactants and Products unless instructed to do otherwise.

(5) Enter the name or structure of each reactant catalyst, if any, and desired product.

(6) Record the molecular weight (MW) of each reactant and desired product. For completeness, make an entry for any catalyst used, although this may be optional.

(7a) Give the volume, in milliliters (mL) or microliters (μL), of each *liquid* reactant and catalyst.

(7b) Record the weight, in grams (g) or milligrams (mg), of each reactant. This entry is optional for liquid catalysts, but should be provided for reference purposes.

(8) Calculate the moles (mol) used of each reactant. For completeness, a value for the catalyst is computed in our example.

(9) Obtain the **theoretical ratio** for reactant(s) and product(s) by referring to the *balanced* main equation(s) for the reaction.

(10) List selected physical properties of reactant(s) and product(s). The information needed is generally available in reference books.

(11) Determine the **limiting reagent** in the following way. Compare the actual ratio of reactants used to that theoretically required. The reagent that is used in the least molar amount, relative to the theoretical amount, is the limiting reagent. In our example, there is only a single reactant, cyclohexanol, so it obviously must be the limiting reagent.

(12) Calculate the **theoretical yield** of the desired product both in moles and in grams. Knowing in our case that the limiting reagent is cyclohexanol and, from the main equation, that 1 mole of alcohol yields 1 mole of cyclohexene, it is clear that no more than 0.05 mole of the alkene can be formed.

Assuming that you were able to isolate 2.7 g of pure cyclohexene in the experiment, the **percent yield** would be calculated according to Equation 1.1.

(13) Self-explanatory.

(14) Self-explanatory.

(15) Self-explanatory.

(16) Self-explanatory.

(17) Develop this diagram by considering what components, in addition to the desired product, may be present in the reaction mixture *after* the main reaction is complete. The chart shows how and where each of the inorganic and organic contaminants of the product is removed by the various steps of the work-up procedure. Ideally, pure cyclohexene results.

(18) Self-explanatory.

(19) Self-explanatory.

Students frequently have difficulty in setting up Tables of Reactants and Products and calculating theoretical yields, so two hypothetical examples are provided for your reference.

EXAMPLE 1

Problem • Consider the reaction shown in Equation 1.2. Assume that you are to use 5 g (7.8 mL) of 1-pentene and 25 mL of concentrated HBr solution. Prepare a Table of Reactants and Products, determine the limiting reagent, and calculate the theoretical yield for the reaction.

$$CH_2 =\!\!=\!CHCH_2CH_2CH_3 + HBr \longrightarrow CH_3CH(Br)CH_2CH_2CH_3 \qquad (1.2)$$

Answer • First of all, note that the equation is balanced, because the "1" that signifies that 1 mole of each reactant will react to produce 1 mole of product is omitted by convention. Because an aqueous solution of HBr, rather than the pure acid, is being used, the amount of HBr present must be determined. Concentrated HBr is 47% by weight in the acid, and its density, d, is 1.49 g/mL, a value that would be recorded in the column headed "Other Data." Consequently, 25 mL of this solution contains 17.5 g of HBr (25 mL × 1.49 g/mL × 0.47). The needed data can then be entered into Table 1.1.

The limiting reagent is seen to be the 1-pentene, because theory requires that it and HBr react in a 1:1 molar ratio, yet they have been used in a ratio of 0.07:0.22. This means that no more than 0.07 mole of product can be formed, since theory dictates that the ratio between 2-bromopentane and 1-pentene also be 1:1. The calculation of the theoretical yield is then straightforward.

Table 1.1   **Table of Reactants and Products for Preparation of 2-Bromopentane**

Compound	M.W.	Volume Used (mL)	Weight Used (g)	Moles Used	Moles Required	Other Data
1-Pentene	70.14	7.8	5	0.07	1	*
HBr	80.91	25	17.5	0.22	1	*
2-Bromopentane	151.05	†	†	†	1	*

* These entries have been intentionally omitted in this example.
† These entries are left blank because this line is for the product.
Limiting reagent: 1-Pentene
Theoretical yield: 151.05 g/mol × 0.07 mol = 10.5 g

You may find it convenient to use units of milligrams (mg), microliters (μL), and millimoles (mmol) instead of grams, milliliters, and moles, respectively, in performing measurements and calculations when small quantities of reagents are used, as is the case for microscale reactions. For example, consider how Table 1.1 would be modified

Table 1.2	Table of Reactants and Products for Preparation of 2-Bromopentane					
Compound	M.W.	Volume Used (mL)	Weight Used (mg)	Mmols Used	Mmols Required	Other Data
1-Pentene	70.14	0.16	100	1.4	1	*
HBr	80.91	0.5	358	4.4	1	*
2-Bromopentane	151.05	†	†	†	1	*

* These entries have been intentionally omitted in this example.
† These entries are left blank because this line is for the product.
Limiting Reagent: 1-Pentene
Theoretical yield: 151.05 mg/mmol $\times$ 1.4 mmol = 211 mg

if 0.1 g of 1-pentene and 0.5 mL of concentrated HBr solution were used. If the calculations were done in grams and moles, the entries under "Moles Used" would be 0.0014 and 0.0044, respectively. Errors can arise when making such entries because a zero may inadvertently be added or dropped.

This potential problem is less likely if you enter the data as milligrams and millimoles. By recognizing that 0.1 g of alkene is 100 mg, and 0.5 mL of HBr solution contains 358 mg of HBr (0.5 mL $\times$ 1490 mg/mL $\times$ 0.48), the entries would be those shown in Table 1.2. Determining the limiting reagent and calculating the theoretical yield are done as in Example 1. Note that the necessary cancellation of units occurs when the molecular weight is expressed in mg/mmol.

Although the volume of 1-pentene to be used is expressed in milliliters, you may be measuring out this amount with a device that is calibrated in microliters (μL, 1 μL = 10^{-3} mL). Thus, in the present example, you would be using 160 μL of 1-pentene (0.16 mL $\times$ 10^3 μL/mL).

EXAMPLE 2

Problem • Now consider the transformation illustrated in Equation 1.3. Assume that you are to use 7 mL of ethanol and 0.1 mL of concentrated H_2SO_4 as the catalyst. Prepare a Table of Reactants and Products, determine the limiting reagent, and calculate the theoretical yield for the reaction.

$$2 \text{ CH}_3\text{CH}_2\text{OH} \xrightarrow[\text{(Catalytic amount)}]{\text{H}_2\text{SO}_4} \text{CH}_3\text{CH}_2\text{OCH}_2\text{CH}_3 + \text{H}_2\text{O} \qquad (1.3)$$

Answer • As in the previous example, a volumetric measurement must first be converted to a weight. The density of ethanol is 0.789 g/mL, information that would be entered under the column headed "Other Data," so that means that 5.5 g is being used. Table 1.3 can then be completed. Note that the catalyst, although recorded in the table, is not used in any of the calculations because, by definition, it is not consumed during the reaction. Its inclusion should help remind the experimentalist that it is indeed required to make the reaction work!

Calculation of the theoretical yield is performed as in Example 1, with the important exception that a factor of 0.5 is incorporated to adjust for the fact that only one-half mole of diethyl ether would be produced for each mole of ethanol that is used.

Table 1.3	Table of Reactants and Products for Preparation of Diethyl Ether					
Compound	M.W.	Volume Used (mL)	Weight Used (g)	Moles Used	Moles Required	Other Data
Ethanol	46.07	7	5.5	0.12	2	*
Sulfuric Acid	†	0.1	†	†	†	*
Diethyl ether	74.12	‡	‡	‡	1	*

* These entries have been intentionally omitted in this example.
† These entries are left blank for reactants that serve only as catalysts.
‡ These entries are left blank because this line is for the product.

Limiting reagent: Ethanol
Theoretical yield: 74.12 g/mol $\times$ 0.12 mol $\cdot$ 0.5 = 4.4 g

1.8 Safe Laboratory Practice: Overview

There is little question that one of the most important abilities that you, the aspiring organic chemist, can bring to the laboratory is a sound knowledge of how to perform experimental work in a safe manner. But just knowing *how* to work safely is insufficient! You must also make a *serious* commitment to follow standard safety protocols. In other words, having the knowledge about safety is useless if you do not put that knowledge into practice. What you actually do in the laboratory will determine whether you and your labmates are working in a safe environment.

Chemistry laboratories are potentially dangerous because they commonly house flammable liquids, fragile glassware, toxic chemicals, and equipment that may be under vacuum or at pressures above atmospheric. They may also contain gas cylinders that are under high pressure. The gases themselves may or may not be hazardous—for example, nitrogen is not whereas hydrogen certainly is—but the fact that their containers are under pressure makes them so. Imagine what might happen if a cylinder of nitrogen fell and ruptured: You could have a veritable rocket on your hands, and, as in the case of hydrogen, the ''rocket'' might even come equipped with a fiery tail! This is another way of saying *all* substances are hazardous under specified conditions.

Fortunately the laboratory need be no more dangerous than a kitchen or bathroom, but this depends on you *and* your labmates practicing safety as you work. Should you observe someone else doing anything that is unsafe, let them know about it in a friendly manner. Everyone will benefit from your action. We shall be alerting you repeatedly to the possible dangers associated with the chemicals and apparatus that you will use so that you can become well-trained in safe laboratory practice. Mastery of the proper procedures is just as important in the course as obtaining high yields of pure products, and careful reading of our suggestions will assist you in achieving this goal. Some safety information will be contained in the text describing a particular experiment or in the experimental procedure itself. It will also appear in highlighted sections titled ''**Safety Alert**.'' These are designed to draw your special attention to aspects of safety that are of particular importance. Read these sections carefully and follow the guidelines in them carefully.

1.9   Safety: General Discussion

We highlight here, in the form of a **Safety Alert**, some general aspects regarding safe practices in the laboratory.

Safety Alert

Personal Attire

1. *Do not wear shorts or sandals in the laboratory.* Proper clothing gives protection against chemicals that may be spilled accidentally. It is advisable to wear a laboratory coat, but in any case, the more skin that is protected by clothing the better.

2. *Always wear safety glasses or goggles in the laboratory.* This applies even when you are writing in your laboratory notebook or washing glassware, since workers nearby may have an accident. Unless your laboratory instructor says otherwise, do *not* wear contact lenses in the laboratory. Even if you are wearing eye protection, chemicals may get into your eyes, and you may not be able to get the contact lenses out before damage has occurred. Should you have to wear corrective glasses while working in the laboratory, make certain that the lenses are shatter-proof. Wearing goggles over such glasses is recommended because the goggles give additional protection from chemicals entering your eyes from the side of the lenses.

3. *Wear latex gloves when working with particularly hazardous chemicals.* Some reagents are particularly hazardous if they come into contact with your skin. The ones you are most likely to encounter in the organic laboratory are concentrated acids and bases, and bromine and its solutions. Check with your instructor whenever you are uncertain whether you should be wearing gloves when handling reagents.

General Considerations

1. *Become familiar with the layout of the laboratory room.* Locate the exits from the room and the fire extinguishers, fire blankets, eye-wash fountains, safety showers, and first-aid kits in and near your workspace. Consult with your instructor regarding the operation and purpose of each of the safety-related devices.

2. *Find the nearest exits from your laboratory room to the outside of the building.* Should evacuation of the building be necessary, use stairways rather than elevators to exit. Remain calm during the evacuation, and walk rather than run to the exit.

3. *Become knowledgeable about basic first-aid procedures.* The damage from accidents will be minimized if first aid is applied promptly. Read the section, "First Aid in Case of An Accident" on the inside front cover of this book.

4. *Never work alone in the laboratory.* In the event of an accident, you may need the immediate help of a co-worker. Should you have to work in the laboratory outside of the regularly scheduled periods, do so only with the express permission of your instructor and in the presence of at least one other person.

Apparatus and Chemicals

1. *Always check carefully for imperfections in the glassware that you will be using.* This should be done not only when checking into the laboratory for the first time but also when setting up the apparatus needed for each experiment. Look for cracks, chips, or other imperfections in the glass that weaken it. Use care in storing your glassware so that it is not damaged upon opening or closing your locker or drawer.

   Pay particular attention to the condition of round-bottom flasks and condensers. The flasks often have "star" cracks (multiple cracks emanating from a central point) as a result of being banged against a hard surface. Heating or cooling a flask having this type of flaw may cause the flask to rupture with loss of its contents. This could result in a serious fire, not just loss of the desired product. To detect such cracks, hold the flask up to the light and look at all its surfaces closely. With respect to condensers, their most vulnerable points are the ring seals (the points where the inner tube and the water jacket of the condenser are joined). Special care must be given to examining these seals for defects because if cracks are present, water might leak into your apparatus and cause violent reactions.

   If you detect imperfections in your glassware, consult with your teacher immediately regarding replacement. Cracked or seriously chipped apparatus should always be replaced, but glassware with slight chips may still be safe to use.

2. *Dispose of glassware properly.* The laboratory should be equipped with a properly labeled special container for broken glassware and disposable glass items such as Pasteur pipets and melting point capillaries. **It is not appropriate to throw such items in the regular trash containers because maintenance personnel may injure themselves while removing the trash.** Broken thermometers are a special problem because they usually contain residual mercury, which is toxic and relatively volatile. There should be a separate, closed container for disposal of thermometers. If mercury has spilled as a result of the breakage, it should be cleaned up immediately. Consult with your instructor about appropriate procedures for doing so.

3. *Know the properties of the chemicals used in the experiments.* Understanding the properties of the chemicals that you will be using helps you to take the proper precautions when handling them and to minimize danger in case of an accident. **Handle all chemicals with care.**

   Refer to Material Safety Data Sheets (Sec. 1.10) to learn about toxicity and other potential hazards associated with the chemicals you use. Most chemicals are at least slightly toxic, and many are *very* toxic and irritating if inhaled or allowed to come in contact with the skin. It is a good laboratory practice to wear plastic or latex gloves when handling chemicals, and there may be times when it is imperative to do so. Your instructor will advise you on the need for gloves.

Should chemicals come in contact with your skin, they can usually be removed by a thorough and *immediate* washing of the affected area with soap and water. Do *not* use organic solvents like ethanol or acetone to rinse chemicals from your skin, as these solvents may actually assist the absorption of the substances into your skin.

4. *Avoid the use of flames as much as possible.* Most organic substances are flammable, and some are highly volatile as well, which increases their potential for being ignited accidentally. Examples of these are diethyl ether, commonly used as a solvent in the organic laboratory, and acetone. Occasionally, open flames, such as that from a Bunsen burner, may be used for flame-drying an apparatus or distilling a high-boiling liquid. In such cases, a "**Safety Alert**" section will give special precautions for their use. Some general guidelines follow.

   a. *Never use an open flame without the permission of your instructor.*
   b. *Never use a flame to heat a flammable liquid in an open container.* Use a water or steam bath, hot plate, aluminum block, or similar electrical heat device instead. If a flammable liquid is to be heated with an open flame, equip the container holding the liquid with a *tightly* fitting reflux condenser.

      Information about the flammability of many commonly used organic solvents is provided in Table 3.1. Do *not* assume, however, that a solvent is not flammable just because it is not listed in the table. In such cases, refer to the Material Safety Data Sheets (see discussion in Sec. 1.10) or other sources to determine flammability.
   c. *Do not pour flammable liquids when there are open flames within several feet.* The act of transferring the liquid from one container to another will release its vapors into the laboratory, and these could be ignited by a flame some distance away.
   d. *Do not pour flammable water-insoluble organic solvents into drains or sinks.* First of all, this is an environmentally unsound way to dispose of waste solvents, and second, the solvents may be carried to locations where there are open flames which could ignite them. Water-soluble solvents can be flushed down the drain if local regulations permit; consult with your instructor about this.

5. *Avoid inhaling vapors of organic and inorganic compounds.* Although many of the odors you encounter in everyday life are organic in nature, it is prudent not to expose yourself to such vapors in the laboratory. Work at a fume hood when handling particularly noxious chemicals, such as bromine or acetic anhydride, and, if possible, when performing reactions that produce toxic gases.

6. *Never taste anything in the laboratory unless specifically instructed to do so.* You should also never eat or drink in the laboratory, as your food may become contaminated by the chemicals that are being used.

7. *Minimize the amounts of chemicals you use and dispose of chemicals properly.* This aspect of laboratory practice is so important that we have devoted a portion of Section 1.10 to it. Read the relevant paragraphs *carefully* and consult with your instructor if there are any questions about the procedures.

1.10 Safety: Material Safety Data Sheets

The variety and potential danger of chemicals used in the organic chemistry laboratory probably exceed that of any laboratory course you have had. It is imperative to understand the nature of the substances with which you are working. Fortunately, the increased emphasis on the proper handling of chemicals has led to a number of different types of publications containing key information about the chemical, physical, and toxicological properties of the majority of organic and inorganic compounds used in the experiments in this textbook. The most comprehensive of the sources is *The Sigma-Aldrich Library of Chemical Safety Data* (Reference 8), and it or similar compilations should be available in your library or some other central location. Alternatively, the information may be available to you on a computer-accessible database; consult with your instructor about this possibility. The data provided by such references are basically a summary of the information contained in the Material Safety Data Sheets (MSDSs) published by the supplier of the chemical of interest. Your instructor may also be able to provide these sheets because, by Federal regulation, an MSDS must be delivered to the buyer each time a chemical is purchased.

The information contained in an MSDS can be overwhelming. For example, the official MSDS for sodium bicarbonate is currently *seven* pages although it was originally only two pages in length! Fortunately, the data provided in most compilations are summaries containing the type of information shown in Figure 1.3; the specific data given are those for diethyl ether. Entries regarding the structure and physical properties of the compound, including melting point (mp), boiling point (bp), and flash point (fp), are included along with its CAS (Chemical Abstracts Service) Registry number, which is unique for each different chemical substance (see Chap. 24 for a further discussion of CAS registry numbers), and RTECS (Registry of Toxic Effects of Chemical Substances) number. Further data are provided concerning its toxicity, the permissible levels set by OSHA for exposure to it in the air you breathe (time-weighted average of 400 ppm), and possible health consequences resulting from contact with the compound. For diethyl ether, the entry for "Toxicity Data" represents the *lowest* recorded *lethal* concentration for ingestion of the chemical. Valuable information is also given regarding first-aid procedures, classes of substances with which diethyl ether reacts, and thus is "incompatible," products of its decomposition, and materials suitable for extinguishing fires involving ether. Finally, protocols for safe handling and storage are included, along with procedures for disposal of diethyl ether and cleaning up spills of it.

Remember that the information in an MSDS is targeted toward dealing with much larger quantities of the chemical than would be used in the laboratory. For example, it is unnecessary to wear a respirator when working with the amounts of diethyl ether normally encountered in a teaching or research laboratory! Moreover, some statements describe needs that are standard to the properly equipped laboratory (for example, safety goggles, shower, and eye-wash fountain) and to routine laboratory practice—such as

Name	Ether	Reviews and Standards	OSHA standard-air: TWA 400 ppm
Other Names	Diethyl ether	Health Hazards	May be harmful by inhalation, ingestion, or skin absorption. Vapor or mist is irritating to the eyes, mucous membranes, and upper respiratory tract. Causes skin irritation. Exposure can cause coughing, chest pains, difficulty in breathing and nausea, headache, and vomiting.
CAS Registry No.	60-29-7	First Aid	In case of contact, immediately flush eyes or skin with copious amounts of water for at least 15 min while removing contaminated clothing and shoes. If inhaled, remove to fresh air. If not breathing, give artificial respiration; if breathing is difficult, give oxygen. If ingested, wash out mouth with water. Call a physician.
Structure	$(CH_3CH_2)_2O$	Incompatibilities	Oxidizing agents and heat.
MP	$-116\ ^\circ C$	Extinguishing Media	Carbon dioxide, dry chemical powder, or alcohol or polymer foam
BP	$34.6\ ^\circ C$ (760 Torr)	Decomposition Products	Toxic fumes of carbon monoxide, carbon dioxide
FP	$-40\ ^\circ C$	Handling and Storage	Wear appropriate respirator, chemical-resistant gloves, safety goggles, other protective clothing. Safety shower and eye bath. Do not breathe vapor. Avoid contact with eyes, skin, and clothing. Wash thoroughly after handling. Irritant. Keep tightly closed. Keep away from heat, sparks, and open flame. Forms explosive peroxides on prolonged storage. Refrigerate. Extremely flammable. Vapor may travel considerable distance to source ignition. Container explosion may occur under fire conditions. *Danger:* tends to form explosive peroxides especially when anhydrous. Inhibited with 0.0001% BHT
Appearance	Colorless liquid	Spillage	Shut off all sources of ignition. Cover with activated carbon adsorbent, place in closed containers, and take outdoors.
Irritation Data	Human eye 100 ppm	Disposal	Store in clearly labeled containers until container is given to approved contractor for disposal in accordance with local regulations.
Toxicity Data	Man, oral LDL_0 260 mg/kg		

FIGURE 1.3   Summary of MSDS for diethyl ether.

not breathing the vapor and avoiding contact with skin and clothing. However, some comments deserve particular attention—such as those regarding the flammability of diethyl ether and its vapors, and the tendency of this substance to form explosive peroxides upon prolonged storage.

You may be required to show your knowledge of safety procedures associated with handling chemicals by incorporating some of the MSDS data into your laboratory notebook. One possible format for doing so is shown in Figure 1.4. The first time that a particular chemical is used in the course, data like those shown can be recorded at the end of your laboratory notebook on pages reserved for such information. Whenever this same substance is used in later experiments, a reference to the page of the notebook

Compound	Health hazards, First Aid, Incompatibilities, Extinguishing Media, and Handling
Diethyl Ether	May be harmful by inhalation, ingestion, or skin absorption. Avoid contact with eyes, skin, and clothing. In case of contact, immediately flush eyes or skin with copious amounts of water. Incompatible with oxidizing agents and heat. Keep tightly closed when not in use. Keep away from heat, sparks, and open flames. Extremely flammable. Vapor may travel considerable distance to source of ignition. If spilled, shut off all sources of ignition. Extinguish fire with carbon dioxide or dry chemical extinguisher.

FIGURE 1.4   Abstract of MSDS for diethyl ether.

containing these data is all that would be needed. *However*, you should reread the information that has originally been recorded so that you can continue to handle the chemical properly.

To summarize, you may think that reading about and recording data like those contained in Figures 1.3 and 1.4 is not a good investment in time. This is absolutely *wrong*! By knowing more about the chemicals that are used in the laboratory, you will be able to work safely and to deal with accidents, should they occur. The end result will be that you should accomplish a greater amount of laboratory work and have a more valuable educational experience.

1.11 Safety: Disposal of Chemicals

The proper disposal of inorganic and organic chemicals is one of the biggest responsibilities that you have in the organic laboratory. Your actions, and those of your labmates, can minimize the environmental impact and even financial cost to your school for handling the waste chemicals that are necessarily produced in the experiments you do.

The experimental procedures in this textbook have been designed at a scale that should allow you to isolate an amount of product sufficient to see and manipulate, but they also involve the use of minimal quantities of reactants, solvents, and drying agents. Bear in mind, however, that minimizing the amounts of chemicals that are used is only the *first* part of an experimental design that results in the production of the least possible quantity of waste. The *second* part is to reduce the amounts of materials that you, the experimentalist, *define* as waste, thereby making the material subject to regulations for its disposal. From a legal standpoint, the laboratory worker is empowered to declare material as waste; that is, unneeded materials are not waste until you say they are! Consequently, a part of most of the experimental procedures in this textbook is reduction of the quantity of residual material that eventually must be consigned to waste. This means some additional time will be required for completion of the experiment, but the benefits, educational, environmental, and economic in nature, fully justify your efforts. The recommended procedures that should be followed are described under the heading, "FINISHING TOUCHES."

How do you properly dispose of spent chemicals at the end of an experiment? In some cases this involves simply flushing chemicals down the drain with the aid of large volumes of water. As an example, solutions of sulfuric acid can be neutralized with base, and the aqueous sodium sulfate that is formed can safely be washed into the sanitary sewer system. However, the environmental regulations that apply in your particular community may require use of alternative procedures. *Be certain to check with your instructor before flushing any chemicals down the drain!*

For water-insoluble substances, and even for certain water-soluble ones, this option is not permissible under *any* circumstances, and other procedures must be followed. The laboratory should be equipped with various containers for disposal of both liquid and solid chemicals; the latter should not be thrown in a trash can because (1) this exposes maintenance personnel to potential danger, and (2) it is environmentally unsound. The containers must be properly labeled as to what can be put in them, because it is very important for safety and environmental reasons that different categories of spent chemicals be segregated from one another. Thus, you are likely to find the following types of containers in the organic laboratory: hazardous solids, nonhazardous solids, halogenated organic liquids, hydrocarbons, and oxygenated organic liquids. Each student must assume the responsibility for seeing that his or her spent chemicals go into the appropriate container; otherwise dangerous combinations of chemicals might result and/or a much more expensive method of disposal be required.

Historical Highlight

The Importance of Record Keeping

Alexander Graham Bell is a household name whereas that of Elisha Gray, a formidable competitor of Bell's in the race to invent the telephone, is not. Therein lies a tale of the importance of accurate record keeping.

Both Bell and Gray were originally interested in inventions associated with transmitting information electrically, and Gray's accomplishments in this area prior to 1875 far exceeded those of Bell. Indeed, Gray's considerable inventive abilities had resulted in his being awarded a number of patents associated with telegraphy by 1870, and in 1872 he became a co-founder of what was to become the Western Electric Company. As early as the winter of 1866, Gray had been intrigued with the possibility of a "harmonic multiple telegraph system," which is the basis of telephonic communication because it permits transmitting tone or pitch in the form of electrical signals. However, he focused his attention on other telegraphic devices and managing the early growth of his company. It was not until 1874 that he could return to his research endeavors.

Meanwhile, Bell's lifelong interest in educating the deaf had led him toward studying the transmission of speech electrically and he, like Gray, became interested in the harmonic telegraph. His work, although not hindered by his having to manage an emerging company, was impeded by the demands made by his duties as a professor, tutor, and promoter of methods for teaching the deaf. Nonetheless, because of his research in telegraphy on November 23, 1874, he was able to write the following momentous words in a letter to his parents: "Please keep this paper as a *record of the conception of the idea* [emphasis our own] in case any one else should at a future time discover that the vibrations of a permanent magnet will induce a vibrating current of electricity in the coils of an electromagnet."

This concept of a "harp apparatus" to induce a variable current comprises a fundamental basis of telephonic speech, and Bell's foresight in instructing his parents to keep the *dated* letter as a record of his idea proved to be the crucial element in the legal wrangling that eventually developed between Bell and Gray as to who had precedent for this invention. Gray may have conceived of the principles for a harmonic multiple telegraph as early as the spring of 1874, but he failed to include his ideas in patent

HISTORICAL HIGHLIGHT
The Importance of Record Keeping *(continued)*

renewals he filed in January, 1875. However, he filed a patent application on February 23, 1876, that did outline his concept, predating Bell's own application by *two* days. Determining to whom the patent would be awarded, though, depended on the respective dates of conception by the two inventors. It was the written record, contained in Bell's letter to his parents, that led to his being awarded the first key patent for the harmonic multiple telegraph. Unfortunately for Gray, his claim of conceiving the same idea at an even earlier date had no corresponding documentation.

It was yet a second *dated* letter written by Bell that established his priority in conceiving the final fundamental principle for the modern telephone: variable resistance. This permits the *amplitude* as well as the pitch of a sound to be modulated electrically. Although he did not file a patent for this concept until February 14, 1876, an action again contested by Gray, Bell only three weeks later was awarded patent No. 174,465, which is generally considered the key patent for the invention of the telephone. The rest is history!

A fascinating description of the path that led Alexander Graham Bell to inventing the telephone is found on the World-wide Web (htpp://lists.village.virginia.edu/albell/homepage.html), and the definitive biography of his life is provided by the following reference: Robert V. Bruce, "Alexander Graham Bell and the Conquest of Solitude," Ithaca: Cornell University Press, 1973, 564 pp.

Relationship of Historical Highlight to Experiments

The importance of completely documenting the experiments you perform in the laboratory cannot be overemphasized. Writing down the observations you make and the date on which you make them may not be the key to winning litigation over patent rights, as was the case with Bell's inventions, but there is always the possibility that they will. A more likely outcome is that such documentation will demonstrate your competence in the laboratory and may result in winning the contest for the top mark in the class!

REFERENCES

1. Lunn, G.; Sansone, E. B. *Destruction of Hazardous Chemicals in the Laboratory*, 2nd ed., John Wiley & Sons, New York, 1994. A handbook providing procedures for decomposition of materials or classes of materials commonly used in the laboratory.

2. Committee on Hazardous Substances in the Laboratory. *Prudent Practices for Disposal of Chemicals from Laboratories*, National Academy Press, Washington, D.C., 1995. An excellent reference containing information for the minimization of waste generated in the laboratory and for the proper handling and disposal of waste chemicals, both organic and inorganic.

3. Young, J. A., ed. *Improving Safety in the Chemical Laboratory: A Practical Guide*, 2nd ed., John Wiley & Sons, New York, 1991. A book containing thorough discussions of the full range of safe practices in the laboratory.

4. Mahn, W. J. *Fundamentals of Laboratory Safety. Physical Hazards in the Academic Laboratory*, Van Nostrand Reinhold, New York, 1991.

5. Lide, D. A., ed. *CRC Handbook of Chemistry and Physics*, annual editions, CRC Press, Boca Raton, FL.

6. Dean, J. A., ed. *Lange's Handbook of Chemistry*, 14th ed., McGraw-Hill, New York, 1992.

7. Bretherich, L., ed. *Hazards in the Chemical Laboratory*, 4th ed., The Royal Society of Chemistry, London, 1986.

8. Lenga, R. E., ed. *The Sigma-Aldrich Library of Chemical Safety Data*, 2nd ed., Sigma-Aldrich, Milwaukee, WI, 1988.

9. Budavari, S., ed. *Merck Index of Chemicals and Drugs*, 12th ed., Merck and Co., Rahway, N.J., 1996.

Techniques and Apparatus

This chapter introduces the basic experimental techniques and associated glassware and apparatus that are commonly used in the organic chemistry laboratory. In some instances, only the practical aspects of a particular technique are discussed in this chapter, whereas the theoretical principles underlying it are presented in later chapters.

It is important to skim the contents of this entire chapter, but we do not recommend that you do a detailed reading of all parts of it at this time. We make reference to specific techniques and/or apparatus as part of the procedures for many experiments; thus, when you are preparing for an experiment, *carefully* read the appropriate sections of this chapter. This should be done *prior* to entering the laboratory to perform the experimental procedure!

2.1 Glassware: Precautions and Cleaning

Laboratory experiments in organic chemistry are commonly conducted in specialized glassware, which is usually expensive. Since you are responsible for the maintenance of your glassware, you should follow proper procedures for safely handling and cleaning it. Failure to do so is likely to result in injury to yourself and broken or dirty glassware that is difficult to clean.

The cardinal rule in handling and using laboratory glassware is *never apply undue pressure or strain to any piece of glassware*. Strained glassware may break at the moment the strain is induced, when it is heated, or even upon standing for a period of time. When setting up a glassware apparatus for a particular experiment, be sure that the glassware is properly positioned and supported so that strain does not develop (Sec. 2.4).

Sometimes it is necessary to insert thermometers or glass tubes into rubber or cork stoppers or rubber tubing. *If you have to force it, do not do it!* Either make the hole slightly larger or use a smaller piece of glass. A useful aid for minimizing the hazard of this procedure is to lubricate the glass tube with a little water containing soap or glycerol prior to insertion into stoppers or tubing. *Always* grasp the glass piece as close as possible to the rubber or cork part when trying to insert it. It is also wise to wrap a towel around the glass tube and the rubber or cork stopper while inserting the tube. This usually prevents a serious cut in the event the glass happens to break.

Glassware should be thoroughly cleaned *immediately* after use. Residues from chemical reactions may attack the surface of the glass, and cleaning becomes more difficult the longer you wait. Before washing glassware, it is good practice to wipe off any lubricant or grease from standard-taper ground-glass joints (Sec. 2.2) with a towel or tissue moistened with a solvent such as hexane or dichloromethane. This prevents the grease from being transferred during washing to inner surfaces of the glassware, where it may be difficult to remove. Most chemical residues can be removed by washing the glassware using a brush, special laboratory soap, and water. Acetone, which is miscible with water, dissolves most organic residues and thus is commonly used to clean glassware; use as little solvent as possible to do the job. Acetone should not be used to clean equipment that contains residual amounts of bromine, however, since a powerful lachrymator, bromoacetone, may form. A **lachrymator** is a chemical that adversely affects the eyes and causes crying, and may also affect the lungs and produce a burning sensation; tear gas is a lachrymator. After use, the spent wash acetone should be transferred to an appropriately marked container for disposal.

Stubborn residues may sometimes remain in your glassware. Often these may be removed by carefully scraping the glassware with a bent spatula in the presence of soap and water or acetone. If this technique fails, more powerful cleaning solutions may be required, but these must be used with great care as they are highly corrosive. *Do not allow these solutions to come into contact with your skin or clothing; they will cause severe burns and produce holes in your clothing.* Chromic acid, which is made from concentrated sulfuric acid and chromic anhydride or potassium dichromate, is sometimes an effective cleaning agent, but since it is a strong oxidizing acid, it must be used with great care. When handling chromic acid, always wear latex gloves and pour it *carefully* into the glassware to be cleaned. After the glassware is clean, pour the chromic acid solution into a specially designated bottle, *not* into the sink.

Another powerful cleaning solution is alcoholic potassium hydroxide. This is most conveniently prepared by putting some ethanol into the flask, adding a few pellets of solid potassium hydroxide, and warming the solution gently while swirling it around inside the flask. When the glassware is clean, pour the solution into a specially designated bottle, and rinse the glassware thoroughly with soap and water to complete the process. *Before using any cleaning solutions other than soap and water, consult your instructor for permission and directions concerning their safe handling and disposal.*

Brown stains of manganese dioxide that are left in the glassware can generally be removed by rinsing the apparatus with a 30% (4 M) aqueous solution of sodium bisulfite, $NaHSO_3$. If this fails, wash the equipment with water and then add a small amount of 6 M HCl. This must be done in a good hood, since chlorine gas is evolved; try this only after obtaining permission from your instructor.

The fastest method for drying your glassware is to rinse the residual droplets of water from the flask with a small volume (<5 mL) of acetone; this acetone may be recovered as wash acetone to remove organic residues in the future. Flasks and beakers should be inverted to allow the last traces of solvent to drain. Final drying is also aided by directing a *gentle* stream of dry compressed air into the flask or beaker or by placing the glassware in a drying oven.

2.2   Standard-Taper Glassware for Miniscale Procedures

Your laboratory kit probably contains **standard-taper glassware** with **ground-glass joints**; this equipment is safe and convenient to use. Before the advent of standard-taper

Female ground-glass joint

Thin even coating of lubricant

Male ground-glass joint

(a)   (b)

FIGURE 2.1
Standard-taper 𝕋 joints: (a) female, (b) male.

glassware, chemists had to bore corks or rubber stoppers to fit each piece of equipment, and the glass apparatus was assembled by connecting the individual pieces of equipment with glass tubing. This was not only time-consuming, but many chemicals reacted with and/or were absorbed by the cork or rubber stoppers. Although standard-taper glassware has greatly simplified the task of assembling the apparatus required for numerous routine laboratory operations, it is expensive, so handle it carefully.

A pair of standard-taper joints is depicted in Figure 2.1. Regardless of the manufacturer, a given size of a male standard-taper joint will fit a female joint of the same size. The joints are tapered to ensure a snug fit and a tight seal. Standard-taper joints come in a number of sizes, and are designated by the symbol 𝕋 followed by two sets of numbers separated by a slash as in 14/20, 19/22, 24/40. The first number is the diameter of the joint in millimeters at its widest point, and the second is the length of the joint in millimeters. A standard-taper joint that is designated as 𝕋 19/22 therefore has a widest diameter of 19 mm and a length of 22 mm.

When using glassware with standard-taper ground-glass joints, you must be sure that the joints are properly lubricated so that they do not freeze and become difficult, if not impossible, to separate. Lubrication is accomplished by spreading a *thin* layer of joint grease around the outside of the upper half of the male joint, mating the two joints, and then rotating them gently together to cover the surfaces of the joints with a thin coating of lubricant. *Applying the correct amount of grease to the joints is important.* If you use too much, the contents of the flask, including your product, may become contaminated; if too little lubricant is used, the joints may freeze. As soon as you have completed the experiment, disassemble the glassware to lessen the likelihood that the ground-glass joints will stick. If the pieces do not separate easily, the best way to pull them apart is to grasp the two pieces as close to the joint as possible and try to loosen the joint with a *slight* twisting motion.

West condenser

Hempel column

Round-bottom flask

Claisen adapter

Stillhead

Thermometer adapter with Neoprene fitting

Vacuum adapter

FIGURE 2.2   Miniscale standard-taper 𝕋 glassware.

Sometimes the pieces of glass will still not separate. In these cases there are a few other tricks that can be tried. These include the following options: (1) Tap the joint *gently* with the wooden handle of a spatula, and then try pulling the joint apart as described earlier. (2) Heat the joint in hot water or a steam bath before attempting to separate the pieces. (3) As a last resort, heat the joint *gently* in the yellow portion of the flame of a Bunsen burner. Heat the joint slowly and carefully until the outer joint breaks away from the inner section. Wrap a cloth towel around the hot joint to avoid burning yourself, and pull the joint apart as described earlier. Consult your instructor before attempting this final option, as it is tricky.

The common pieces of standard-taper glassware with ground-glass joints for miniscale procedures are shown in Figure 2.2, and apparatus involving them are pictured throughout the textbook. The proper techniques for erecting and supporting apparatus are discussed in Section 2.4.

2.3   Standard-Taper Glassware for Microscale Procedures

As with miniscale procedures (Sec. 2.2), **standard-taper glassware** with **ground-glass joints** is commonly used for performing reactions at the microscale level. The apparatus is smaller, however, so the joints themselves are shorter, typically being ⫪ 14/10, which corresponds to a length of only 10 mm (Fig. 2.3a).

When using glassware with standard-taper ground-glass joints, you must be sure that the joints are properly lubricated so that they do not freeze and become difficult, if not impossible, to separate. Lubrication is accomplished by spreading a *thin* layer of joint grease around the outside of the upper half of the male joint, mating the two joints, and then rotating them gently together to cover the surfaces of the joints with a thin coating of lubricant. *Applying the correct amount of grease to the joints is important.* If you use too much, the contents of the flask, including your product, may become contaminated; if too little lubricant is used, the joints may freeze. As soon as you have completed the experiment, disassemble the glassware to lessen the likelihood that the ground-glass joints will stick. If the pieces do not separate easily, the best way to pull them apart is to grasp the two pieces as close to the joint as possible and try to loosen the joint with a *slight* twisting motion.

Sometimes the pieces of glass will still not separate. In these cases there are a few other tricks that can be tried. These include the following options: (1) Tap the joint *gently* with the wooden handle of a spatula, and then try pulling the joint apart as described earlier. (2) Heat the joint in hot water or a steam bath before attempting to separate the pieces. (3) As a last resort, heat the joint *gently* in the yellow portion of

(a)   (b)   (c)

FIGURE 2.3   Standard-taper ⫪ 14/10 joints and greaseless assembly.

Conical vial   Round-bottom   Claisen adapter   Air condenser   Reflux condenser   Hickman stillhead   Drying tube
flask   with port

FIGURE 2.4   Microscale standard-taper ⊤ glassware.

the flame of a Bunsen burner. Heat the joint slowly and carefully until the outer joint breaks away from the inner section. Wrap a cloth towel around the hot joint to avoid burning yourself, and pull the joint apart as described earlier. Consult your instructor before attempting this final option, as it is tricky.

The design of some microscale apparatus allows for *greaseless* mating of joints, and your kit of glassware may permit this option. As shown in Figure 2.3a, a female joint may have threads on the outside, which permits use of Teflon®-lined screw-caps for joining pieces of glassware. The screw-cap contains a silicone rubber O-ring. When a male joint is inserted through the cap and O-ring (Fig. 2.3b) and then mated with the female joint, tightening the cap compresses the O-ring, producing a gas-tight seal that clamps the unit together (Fig. 2.3c). Grease is not used in this case, and this has obvious advantages with respect to preventing contamination of reaction mixtures and simplifying clean-up of the glassware. You must disconnect greaseless joints *immediately* after completing use of the apparatus, however; otherwise, the joint will become frozen.

Various pieces of microscale glassware you may be using are depicted in Figure 2.4. A number of different assemblies using this type of glassware are portrayed in this chapter as well as others. The proper techniques for erecting and supporting apparatus are discussed in Section 2.4.

2.4   Assembling Apparatus

To perform most of the experiments in the organic chemistry laboratory, you must set up the appropriate apparatus. In the sections that follow, you will be introduced to the various techniques that are needed. All experiments require certain items of glassware as well as various heating or cooling devices and a means to mix the contents of a reaction vessel. The assembly of the various pieces of glassware and equipment according to the diagrams will seem awkward at first, but with practice you will soon be able to set up your own apparatus with ease.

Providing adequate support for your apparatus is paramount to preventing breakage. In general you will accomplish this by clamping the apparatus to a ring stand or other vertical support rod using a rigid three-finger clamp (Fig. 2.5a) or a jaw clamp (Fig. 2.5b). Plastic Keck clips (Fig. 2.5c) are then used as needed to secure the union of the glass joints. It is best to begin assembling the apparatus with the reaction or distillation vessel and to use a single ring stand to minimize the manipulations that are required if the apparatus must be moved.

FIGURE 2.5 (a) Three-finger clamp. (b) Jaw clamp. (c) Keck clamp.

If the experimental technique requires heating or cooling the apparatus, then you should assemble the apparatus and support it about 6–10 in. above the base of the ring stand or bench top. You will then use a laboratory jack (Fig. 2.6) to raise and lower heating sources or cooling baths without moving the apparatus itself. In addition to providing this convenience, the laboratory jack allows you to quickly lower a heating source if an exothermic reaction begins to get out of control. If such lab jacks are not available, you should use an iron ring clamp to support the heating source or cooling bath; the ring clamp then may be lowered if necessary to remove the heat source or the cooling bath from the flask.

FIGURE 2.6
Laboratory jack.

MINISCALE SET-UP

As a rule, when you assemble the apparatus for a miniscale experiment, you should *tighten* only the clamp to the round-bottom flask that contains your reaction or distillation mixture. The proper way to tighten a rigid clamp is to place the neck of the flask or other piece of glassware being secured against the stationary side of the clamp, and then turn the wingnut or screw to bring the movable side of the clamp into contact with the glassware so that the piece of glassware fits snugly in the clamp. To avoid breaking the glassware, *do not overtighten the clamp.* Other clamps should be positioned to support the remainder of the apparatus and should be *loosely* tightened to avoid straining the glassware and risking breakage. Alternatively, plastic Keck clips (Fig. 2.5c) or rubber bands may be used in place of additional rigid clamps. Proper locations of the clamps for the various laboratory set-ups are provided in the diagrams that accompany the description of the technique.

MICROSCALE SET-UP

Because the glassware required for *microscale* experiments is substantially smaller than that used in miniscale experiments, a single rigid clamp is usually sufficient to secure the apparatus. To avoid breaking the glassware, *do not overtighten the clamp.* The plastic screw-caps hold the different pieces of glassware together so additional clamping is unnecessary. Generally you will clamp a reflux condenser or a stillhead by placing the glassware against the stationary side of the clamp and then turning the wingnut or screw to bring the movable side of the clamp into contact with the glassware so that the piece of glassware fits snugly in the clamp.

2.5 Measuring and Transferring Liquids

Proper handling of liquid reagents and products is an important technique. It is particularly critical when no more than 1 or 2 mL of a liquid are involved. Indeed, when working at the microscale level, quantities of liquids being used often are less than 0.5 mL (500 μL). Even when larger amounts of liquids are being measured, knowing how to transfer them safely and accurately is important.

GRADUATED CYLINDERS, BEAKERS, ERLENMEYER FLASKS, AND VIALS

FIGURE 2.7
Graduated cylinder and funnel.

Funnel

Plastic collar

Bottom of meniscus

Graduated cylinders (Fig. 2.7) are calibrated containers used to measure volumes in the range of 2 to 500 mL, depending on the size of the cylinder. The markings on the cylinder shaft represent the approximate volume contained when the bottom of the **meniscus** of the liquid is at the top of the line. The accuracy of calibration is only about $\pm10\%$, so graduated cylinders are not used for delivering precise quantities of liquids. As shown in Figure 2.7, the cylinder may be fitted with a plastic collar to minimize the chance of breakage if the cylinder is accidentally knocked over. You should not fill graduated cylinders by pouring liquids directly from very large containers such as 3-L bottles. In such cases, you should first pour the liquid into a beaker, trying to estimate how much you will need so as to minimize waste, and then transfer the liquid from the beaker into the graduated cylinder. You may use a stemmed funnel to help avoid spillage.

Beakers, **Erlenmeyer flasks**, and **conical vials** may also bear calibration marks. Keep in mind that, just as with graduated cylinders, volumetric measurements using these containers are only approximate, and other apparatus is used when more precise measurements are needed.

PASTEUR PIPET

1.5 mL

1.0 mL

0.5 mL

(a)   **(b)**

FIGURE 2.8
Pasteur pipets.

The **Pasteur pipet** is comprised of a glass barrel, drawn out at one end to form a tip through which liquid is pulled with the aid of a latex suction bulb (Fig. 2.8). These pipets are commonly available in two lengths. Liquid is drawn into the pipet by first compressing the bulb, and then inserting the tip of the pipet into the liquid to be transferred. To maximize the efficiency of the transfer, it is best not to compress the bulb more than is needed to draw the desired amount of liquid into the tip and barrel of the pipet. With practice, you will learn how much to compress the bulb to pull up a given volume of liquid.

Pasteur pipets are not used for quantitative measurements of volumes, but they can be used for qualitative measurements. This requires calibration of the pipet. Do this by first weighing, to the nearest 0.1 g, a specific amount of water (density 1.0 g/mL) into a test tube. You may also use a graduated cylinder to measure a given volume of liquid. Carefully draw the liquid into the pipet to be calibrated, and use a file to score (scratch) the pipet lightly at the level reached by the liquid (Fig. 2.8b). You may wish to place several calibration marks on the pipet, say at the 0.5-, 1-, and 1.5-mL levels. It is a good idea to calibrate several pipets at once since more than one may be needed in a given experimental procedure and because they are easily broken.

Although Pasteur pipets are often called ''disposable pipets,'' they can be reused many times if given proper care. This involves cleaning (Sec. 2.1) and drying the pipet as soon as possible after use. When the pipet is no longer cleanable or the tip has become chipped or broken, it should be discarded in a container labeled for broken glassware rather than in a wastebasket to prevent accidental injury to cleaning personnel.

MODIFIED PASTEUR PIPETS

The **Pasteur filter-tip pipet** (Fig. 2.9a) and **filtering pipet** (Fig. 2.9b) are two useful modifications of the Pasteur pipet. The filter-tip pipet is helpful for transferring liquids containing particularly volatile components like diethyl ether (bp 35 °C). The plug resists the backpressure that develops when the liquid is drawn into the pipet and volatilizes, thus tend to force the liquid back out of the pipet. The filtering pipet has its plug at the base of the barrel rather than in the tip (Fig. 2.9b). This modification is useful for operations such as removing solids from solutions during microscale recrystallizations (Sec. 2.17).

Cotton is the preferred material for the plug because it packs more tightly than glasswool, making removal of finely divided solids more effective. However, glasswool is used in certain instances, such as when particularly rapid transfer of a liquid is desired or the particles to be filtered from a liquid are relatively large. Moreover, strongly acidic solutions may react with cotton, so glasswool is preferred in these cases as well.

You may prepare a filter-tip pipet by first inserting a small piece of the packing material in the top of a Pasteur pipet and using a length of heavy-gauge wire or a narrow rod first to push it to the base of the barrel and then into the tip of the pipet. The second step is omitted when making a filtering pipet. For both types of pipet, the plug that results should neither be too large or packed too tightly because the flow of liquid into and out of the pipet may then be seriously impeded.

(a) **(b)**

FIGURE 2.9
Modified Pasteur pipets: (a) Pasteur filter-tip pipet; (b) Pasteur filtering pipet.

Plug of cotton or glasswool

GRADUATED PIPETS

Quantitative volumetric measurements are possible with **graduated pipets,** of which there are a number of styles and sizes. Two examples are shown in Figure 2.10. The graduated ''blow-out'' pipet (Fig. 2.10a), sometimes labeled as a ''to deliver'' (TD) pipet, is calibrated so that the full volume of liquid in the pipet is dispensed only when the liquid remaining in the tip is forced out using a pipet bulb. A drop, if any, remaining on the tip must also be added to the delivered liquid, and this is done by touching the drop to the inside of the container or to the surface of the liquid. Of course, you may measure volumes less than the maximum contained in the pipet by difference, in which case the entire contents of the pipet are not blown out.

The graduated pipet shown in Figure 2.10b is designed to deliver the full volume of liquid in the pipet without having to be blown out. Thus, a small amount of liquid will remain in the tip after dispensing. Proper use of this pipet requires that the bulb used to draw liquid into the pipet be removed as the contents are being delivered into a container. If there is a drop of liquid remaining on the tip, it too should be added to the dispensed liquid.

SYRINGES

Plastic or glass **syringes** (Fig. 2.11) are often used to deliver liquids into reaction mixtures. Syringes marked with gradations showing the amount of liquid contained in them normally have an accuracy of volumetric measurement on the order of ±5%, although the syringes originally designed for use in gas chromatography (Sec. 6.4) and having volumes of 500 μL or less are more accurate. The needles on syringes are either fixed (Fig. 2.11a) or demountable (Fig. 2.11b,c). With the latter style of syringe, needles having various gauges can be affixed to the barrel. Removing and attaching the needles

(a) **(b)**

FIGURE 2.10
Graduated pipets: (a) to deliver (TD); (b) full volume delivery.

FIGURE 2.11 Fixed and demountable syringes: (a) fixed needle; (b) demountable, locking; (c) demountable, non-locking.

often requires twisting the base of the needle to disengage it from or engage it to the barrel; this type of syringe is shown in Figure 2.11b. Otherwise, the needle simply slides on or off the plastic or ground-glass tip of the barrel (Fig. 2.11c).

Syringes are commonly used in the laboratory to transfer solutions or liquid reagents to reactions being performed under anhydrous conditions. The apparatus is usually fitted with a rubber septum, and the syringe needle is used to pierce the septum for delivery of the liquid. You must be careful when inserting the needle into the septum to avoid bending it. The danger of bending can be minimized by supporting the sides of the needle with two fingers as you insert it into the septum.

A syringe is a sharp-pointed instrument, and you must use it with appropriate care. *Never* point an assembled syringe at yourself or anyone else, and do not leave needles pointed upwards. You should clean a syringe immediately after use. This is done by pulling acetone or some other low-boiling solvent into the barrel and expelling it into an appropriate container. The rinsing should be repeated several times, after which the syringe should be dried by pumping the plunger to pull air through it or by attaching the barrel of the syringe to a water aspirator.

DISPENSING PUMPS AND AUTOMATIC PIPETS

FIGURE 2.12
Dispensing pump.

Dispensing pumps and **automatic pipets** are devices used for accurately and quickly dispensing liquids. Because of their costs, the use of these devices is normally limited to liquid transfers from containers, such as reagent bottles, that are being used by everyone in a laboratory.

Dispensing pumps (Fig. 2.12) are commonly used when quantities greater than about 0.5 mL must be dispensed. The plunger is made of Teflon, which allows using the unit with corrosive liquids and organic solvents, such as bromine in dichloromethane. The pump is loaded by pulling the plunger up as far as it will go, and the liquid that has been drawn into the glass reservoir is then expelled into the desired container by slowly depressing the plunger. Liquids of relatively low viscosity are usually expelled using only the weight of the plunger. More viscous liquids may require gently pushing the plunger down. If any liquid remains on the tip of the dispenser at the end of the transfer, the tip should be touched to the wall of the container to complete the operation.

The length the plunger travels, and thus the amount of liquid ultimately delivered, is adjustable, but this is usually done by your laboratory instructor. In cases where this liquid is the limiting reagent in a reaction, it should be transferred into a tared container, which is then reweighed to determine the precise amount of the reagent. To ensure maximum accuracy in measuring the volume of liquid delivered by the pump, be certain that an air bubble has not developed as the liquid is drawn into the reservoir and that the spout has no air voids. Should either of these conditions not be met, the proper volume of liquid will not be dispensed. You can correct these problems by raising and depressing the plunger until no air bubbles are visible; any liquid that is delivered by this operation should be collected and disposed of in the appropriate waste container.

Automatic pipets are used for accurately delivering 10- to 1000-μL (0.01- to 1-mL) quantities of liquid and are particularly valuable in microscale procedures. Although several styles of these pipets are available, the principles of their operation are basically the same. The pipet itself is comprised of a barrel, which contains a spring-loaded plunger whose length of travel may be variable, and a disposable plastic tip into which the liquid is drawn (Fig. 2.13a). There are two ''detent'' or stop positions of the plunger; these control the filling and dispensing stages of the liquid transfer. You *must* attach the plastic tip securely, and you should *not* totally immerse the tip in the liquid being transferred. These precautions are necessary because liquid should *never* come into contact with the barrel and plunger since this may damage the pipet. If liquid does accidentally touch these surfaces, notify your instructor immediately.

The operation of the pipet is straightforward. Holding the assembled pipet in the vertical position, depress the plunger to the *first* detent *before* inserting the tip in the liquid (Fig. 2.13a). Then insert the tip to a depth of 5–10 mm in the liquid and *slowly* release the plunger (Fig. 2.13b). If you allow the plunger to snap back to its original position, liquid may splash into the barrel of the pipet and damage the apparatus. Finally, dispense the liquid by placing the plastic tip, still held in a vertical position, against the inner wall of the tilted receiving vessel and slowly push the plunger down to the *first*

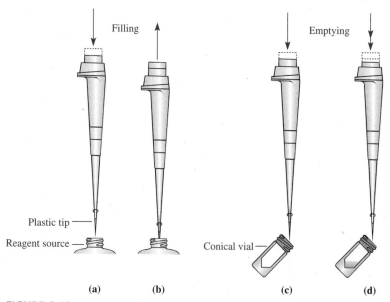

FIGURE 2.13 Automatic pipet: (a) depressing plunger to first detent; (b) releasing plunger to fill tip; (c) re-depressing plunger to first detent to deliver liquid; (d) depressing plunger to second detent to complete delivery.

detent (Fig. 2.13c); a second or two later, further depress the plunger to the *second* detent to expel the final drop of liquid from the plastic tip (Fig. 2.13d). Remove the pipet from the receiver and follow the directions of your instructor with respect to any further steps to be taken with the pipet and its tip.

2.6   Weighing Methods

FIGURE 2.14
A top-loading balance with draft shield.

It will be necessary to measure quantities of reagents and reactants for the reactions you will perform, and the techniques for measuring *volumes* of liquids are discussed in Section 2.5. The *weight* of a liquid whose density is known is often most conveniently measured by transferring a known volume of the liquid to the reaction flask using a graduated pipet or a syringe. This is especially true when the amount of a liquid to be weighed is less than 200–300 mg. However, for some liquids and all solids, weights are usually determined using a suitable **balance**. For quantities greater than 0.5 g (500 mg), a top-loading balance that reads accurately to the nearest 0.01 g (10 mg) is usually adequate. When performing reactions on the microscale, it is necessary to use a top-loading balance that has a draft shield and reads to the nearest 0.001 g (1 mg) (Fig. 2.14) or an analytical balance (Fig. 2.15) that reads to the nearest 0.0001 g (0.1 mg).

With modern electronic balances equipped with a taring device, it is possible to subtract automatically the weight of a piece of paper or other container from the combined weight to give the weight of the sample itself directly. For example, to weigh a liquid, tare a vial and remove it from the balance. Transfer the liquid from the reagent bottle to the vial with a pipet or a graduated pipet or syringe, and reweigh the vial. Be careful to avoid getting liquid on the outside of the vial or the balance pan. *Clean any spills promptly.*

To weigh a solid, place a piece of glazed weighing paper on the balance pan and press the tare device; the digital readout on the balance will indicate that the paper has 'zero' weight. You should *never* weigh solids directly onto the balance pan. Using a microspatula, transfer the solid to be measured from its original container to the weighing paper until the reading on the balance indicates the desired weight. Do not pour the solid directly from the original bottle because spills are more likely. Carefully transfer the solid to your reaction flask or other container. You may also weigh the solid directly into the reaction vessel by first taring the flask or conical vial. *Clean any spills promptly.*

FIGURE 2.15
An analytical balance with draft shield.

The laboratory practices and apparatus used to determine **melting points** of solids are discussed in this section, and the theory and use of melting points are described in Section 3.3. The task of determining a melting point of a compound simply involves heating a small amount of a solid and determining the temperature at which it melts. Many different types of heating devices can be used, but most equipment utilizes a capillary tube to contain the sample so that only a small amount of the sample is required.

CAPILLARY TUBES AND SAMPLE PREPARATION

The first step in determining a melting point is transferring the sample into a melting point capillary tube. Such tubes have one sealed end and are commercially available. The proper method for loading the sample into the capillary tube is as follows. Place a small amount of the solid on a clean watchglass and press the open end of the tube into the solid to force a small amount of solid (about 2–3 mm in height) into the tube (Fig. 2.16a); this operation should *not* be performed on filter paper rather than a watchglass because fibers of paper as well as the solid may be forced into the tube. Then take a piece of 6–8 mm tubing about 1-m long, hold it vertically on a hard surface such as the bench top or floor, and drop the capillary tube down the larger tubing several times with the sealed end *down* (Fig. 2.16b). This packs the solid sample at the closed end of the capillary tube.

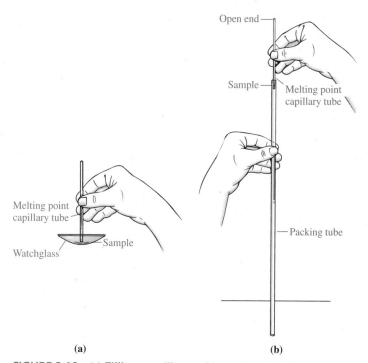

(a)   **(b)**

FIGURE 2.16   (a) Filling a capillary melting point tube. (b) Packing the sample at the bottom of the tube.

MELTING POINT DETERMINATION

The melting point of the solid is determined by heating the packed capillary tube until the solid melts. Some representative devices for measuring melting points are presented in Figures 2.17–2.19. The most reproducible and accurate results are obtained by heating the sample at the rate of about 1 to 2 °C/min to ensure that heat is transferred to the sample at the same rate as the temperature increases and that the mercury in the thermometer and the sample in the capillary tube are in thermal equilibrium.

Many organic compounds undergo a change in crystalline structure just before melting, perhaps as a consequence of release of the solvent of crystallization. The solid takes on a softer, ''wet'' appearance, which may also be accompanied by shrinkage of the sample in the capillary tube. These changes in the sample should *not* be interpreted as the beginning of the melting process. Wait for the first tiny drop of liquid to appear. Melting invariably occurs over a range of a degree or more, and the **melting point range** is defined as the temperature at which the first tiny drop of liquid appears and extends to and includes the temperature at which the solid has completely melted.

MELTING POINT APPARATUS

A simple type of melting point apparatus is the **Thiele tube**, shown in Figure 2.17a. This tube is shaped such that the heat applied to a heating liquid in the sidearm by a burner is distributed evenly to all parts of the vessel by convection currents, so stirring is not required. Temperature control is accomplished by adjusting the flame produced by the microburner; this may seem difficult at first but can be mastered with practice.

Proper use of the Thiele tube is required to obtain reliable melting points. Secure the capillary tube to the thermometer at the position indicated in Figure 2.17b using either a rubber band or a small segment of rubber tubing. Be sure that the band holding

FIGURE 2.17   (a) Thiele melting point apparatus. (b) Arrangement of sample and thermometer for determining melting point.

the capillary tube on the thermometer is as close to the top of the tube as possible. Now support the thermometer and the attached capillary tube containing the sample in the apparatus either with a cork, as shown in Figure 2.17a, or by carefully clamping the thermometer so that it is immersed in the oil. The thermometer and capillary tube must *not* contact the glass of the Thiele tube. Since the oil will expand on heating, make sure that the height of the heating fluid is approximately at the level indicated in Figure 2.17a and that the rubber band is in the position indicated. Otherwise, the hot oil will come in contact with the rubber, causing the band to expand and loosen; the sample tube may then fall into the oil. Heat the Thiele tube at the rate of 1–2 °C/min in order to determine the melting point. The maximum temperature to which the apparatus can be heated is dictated by the nature of the heating fluid, a topic that is discussed in Section 2.9.

The Thiele tube has been replaced in many laboratories by various **electric melting point devices**, which are more convenient to use. One common type of electric melting point apparatus is the Thomas-Hoover melting point unit, which is shown in Figure 2.18. This particular unit has a built-in vibrating device to pack the sample in the capillary tube, and it also allows for the determination of the melting points of up to five samples simultaneously. The oil bath in this unit is electrically heated and stirred. An electrical resistance heater is immersed in a container of silicone oil. The voltage across the heating element is varied by turning the large knob in the front of the apparatus so that the oil is heated at a slow, controlled rate. A motor drives a stirrer in the oil bath to ensure even heating; the rate of stirring is controlled by a knob at the bottom of the unit. Some models are equipped with a movable magnifying lens system that gives the user a better view of the thermometer and the sample in the capillary tube. The capillary tube containing the sample is inserted into the apparatus as illustrated in Figure 2.18.

The Mel-Temp® apparatus shown in Figure 2.19 is another electrical unit that utilizes a heated metal block rather than a liquid for transferring the heat to the capillary tube. A thermometer is inserted into a hole bored into the block, and the thermometer gives the temperature of the block and the capillary tube. Heating is accomplished by controlling the voltage applied to the heating element contained within the block.

FIGURE 2.18
Thomas-Hoover melting point apparatus. (Courtesy of Arthur H. Thomas Company.)

FIGURE 2.19
Mel-Temp® melting point apparatus. (Courtesy of Laboratory Devices.)

2.8   Boiling Point Methods and Apparatus

There are several techniques that may be used to determine the **boiling point** of a liquid, depending upon the amount of material available. When multi-gram quantities are available, the boiling point is typically determined by reading the thermometer during a simple distillation using the technique described in Section 2.13. However, for smaller amounts of liquid there is sometimes not enough sample to distil, so other techniques have been developed. Two of these are described here.

MINISCALE METHOD

FIGURE 2.20

Miniscale technique to determine the boiling point of a liquid.

An accurate boiling point may be determined with about 0.5–1.0 mL of liquid using the method illustrated in Figure 2.20. Place the liquid in a long, narrow Pyrex test tube, and add a small black carborundum boiling stone; do not use a white marble chip as bumping is more likely. Clamp the test tube and position the thermometer about 2 cm above the level of the liquid using a second clamp. Bring the liquid rapidly to a vigorous boil using a suitable heating device suggested by your instructor. You should see a reflux ring move up the test tube, and drops of the liquid should condense on the walls of the test tube. Do not heat so much that the liquid boils out of the test tube. Be sure the bulb of the thermometer is fully immersed in the vapor of the boiling liquid long enough to allow equilibration so a good temperature reading may be obtained.

MICROSCALE METHOD

To determine the boiling point of a small amount of liquid, a simple **micro boiling-point** apparatus may be easily constructed. First, prepare a capillary ebullition tube by taking a standard melting point capillary tube, which is already sealed at one end, and make a seal in it about 1 cm from the open end, using a hot flame (Fig. 2.21a). Alternatively, two melting point capillary tubes can be joined by heating the closed ends in a hot flame; a clean cut is then made about 1 cm below the point where the tubes have been joined (Fig. 2.21b). Second, seal a piece of 4–6-mm glass tubing at one end using a hot flame and cut it to a length about 1 cm longer than the capillary ebullition tube. These tubes are prepared most easily by using a glassblowing torch, and they may be provided by your instructor.

Attach the 4–6-mm tube to a thermometer with a rubber ring near the top of the tube. The bottom of the tube should be even with the mercury bulb of the thermometer. Put the capillary ebullition tube into the larger glass tube, and with a Pasteur pipet add the liquid whose boiling point you wish to determine until the level of the liquid is about 2 mm above the seal of the capillary tube (Fig. 2.21c).

Immerse the thermometer and the attached tubes in a heating bath. A Thiele tube (Fig. 2.17) is convenient for this purpose, but other heating baths can be used. *Be sure that the rubber ring is well above the level of the oil in the heating bath.* Heat the oil bath at the rate of about 5 °C/min until a *rapid* and *continuous* stream of bubbles comes out of the capillary ebullition tube. Before this occurs, you may see some bubbles form in an erratic fashion. This is due to expansion of air trapped in the capillary tube. You should see a marked change from the slow evolution of air bubbles to the rapid evolution of bubbles resulting from the boiling action of the liquid as the boiling point is reached. *However, this is **not** the boiling point!* Remove the heating source and allow the bath

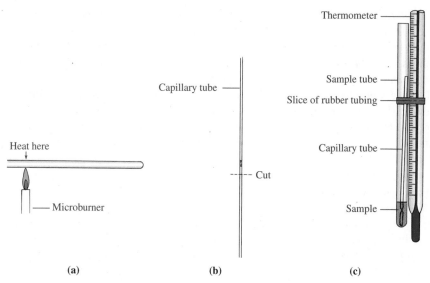

FIGURE 2.21 Micro boiling-point apparatus: (a) fusing a capillary tube; (b) joining capillary tubes and cutting off one end; (c) assembling micro boiling point apparatus with correct placement of ebullition and sample tubes, and thermometer.

to cool slowly. As the liquid starts to rise into the capillary tube, note the temperature measured by the thermometer; *this is the boiling point of the liquid.* If the liquid rises sufficiently slowly into the capillary tube, note the temperatures at which the liquid starts to rise and at which the capillary tube is full. This will be the **boiling point range** of the liquid. Remove the capillary ebullition tube and expel the liquid from the small end by gently shaking the tube. Replace it in the sample tube and repeat the determination of the boiling point by heating the oil bath at the rate of 1–2 °C/min when you are within 10–15 °C of the approximate boiling point as determined in the previous experiment. Observed boiling points may be reproduced to within 1 or 2 °C.

The physical basis of this technique is interesting. Before the liquid is heated, the capillary tube is filled with air. As the bath is heated, the air in the capillary tube is driven out and replaced with the vapor of the liquid. When the apparatus is heated until vigorous boiling of the liquid is observed, the actual boiling point of the liquid has been exceeded, and the air in the capillary tube has been completely replaced by the vapor of the liquid. On cooling, the vapor pressure of the liquid becomes equal to the external pressure, thus allowing the liquid to rise into the capillary tube. The temperature at which this occurs is, by definition, the boiling temperature of the liquid (Sec. 4.1).

2.9 Heating Methods

Heating is an important laboratory technique that serves a variety of purposes. For example, many chemical reactions require heating to proceed at a reasonable rate. Heating is also used to purify liquids by distillation, to remove volatile solvents during the work-up of a reaction, and to dissolve solids when purifying solid products by recrystallization.

Two general rules regarding heating are noted here. (1) Whatever device is being used to heat a liquid or solid, you must arrange the apparatus so that the heating source can be *rapidly* removed from the apparatus to prevent accidents. This normally means that the heating source is mounted on a ring clamp or a lab jack, either of which allows for quick removal of the device if necessary. (2) As a rule, the *safest* way to heat organic solvents is with a *flameless* heat source and in a hood; this minimizes the chance of fire and avoids filling the room with solvent vapors.

Heating with electrical devices is generally the method of choice in the organic laboratory, since it is much safer than using open flames. These devices are usually comprised of two essential components: (1) a resistance element that converts electrical into thermal energy; and (2) a variable transformer for controlling the voltage across the element. They differ, however, in the medium used to transfer heat from the element to the experimental apparatus. For example, air and oil are the heat transfer agents for a heat gun and an oil bath, respectively.

Although we describe below how burners are properly used for heating purposes, the experimental procedures in this textbook focus on *flameless* heating devices. Burners can be substituted for such devices in most cases, but not all. *Consult with your instructor if you have any questions regarding the suitability of substituting flames for other heating devices.*

BURNERS

Most chemistry laboratories are supplied with natural gas for use with various types of burners. A burner provides the convenience of a rapid and reasonably inexpensive source of heat. However, many organic substances, especially solvents such as ether and hexane, are highly flammable, and good judgment must always be exercised when considering the use of a burner for heating volatile organic compounds. *Before using a burner to heat anything in the laboratory, consult with your instructor for the proper precautions and directions.* If an alternate mode of heating is available, choose it in preference to a burner. *Never use a burner to heat flammable materials in open containers such as beakers or Erlenmeyer flasks.*

There are a number of situations in the laboratory where burners are appropriate heating sources. They may be used to heat Thiele tubes in the determination of melting or boiling points of organic substances and to heat a water bath to obtain and maintain temperatures from ambient to about 80 °C. When volatile solvents are not being used in the laboratory, burners may be safely substituted for heat guns to dry apparatus so a reaction may be conducted under anhydrous conditions (Sec. 2.26). On occasion you will need a burner to bend glass tubing or to fashion a piece of glass apparatus. Burners can also be used to heat aqueous solutions that do not contain flammable substances or to heat higher-boiling liquids that are completely contained in round-bottom flasks either fitted with a reflux condenser (Sec. 2.22) or equipped for distillation as discussed in Sections 2.13–2.16. In these instances, it is important to lubricate the joints of the apparatus with a hydrocarbon or silicone grease to avoid freezing of the joints or the leakage of vapors through them.

You must be aware of the type of work others are doing in the laboratory. Although you might be using a burner to perform a completely safe operation, someone else nearby may be working with a very volatile, flammable solvent, some of which can creep along the bench top for several feet! These vapors or those in the room may be ignited explosively by an open flame.

Two common types of laboratory burners are pictured in Figure 2.22. The classic **Bunsen burner**, named after its inventor, is shown in Figure 2.22a. The needle valve

FIGURE 2.22

Laboratory burners: (a) Bunsen burner; (b) microburner.

at the bottom of the burner serves as a fine adjustment of the gas flow, and turning the barrel of the burner regulates the air flow; adjustment of gas and air flow provides control of the flame. In the **microburner** in Figure 2.22b, the air flow is adjusted at the baffle at the bottom of the burner and the gas flow is adjusted at the gas valve on the laboratory bench.

Heating a flask with a burner may produce ''hot spots'' if most of the heat is supplied to a small area on the bottom of the flask. Hot spots can lead to severe bumping, since the heat must be dispersed throughout the liquid by convection or by means of the turbulence caused by boiling. Hot spots can be avoided by holding the burner and slowly moving the flame over the bottom of the flask. Alternatively, a piece of wire gauze, which diffuses the heat reaching the flask, may be placed between the flame and the flask; the gauze is supported with an iron ring.

HEATING MANTLE

A widely used device for heating round-bottom flasks is a **heating mantle**. The mantle is typically fashioned either from a blanket of spun fiberglass partially covered with a flexible or rigid cover (Fig. 2.23a) or from a ceramic core contained in an aluminum housing (Fig. 2.23b). In either case, heat is provided by an electrical resistance coil embedded in the fiberglass or ceramic core, so when using these devices it is important not to spill liquid on them. The mantle may be equipped with a thermocouple so that its internal temperature can be monitored, but this is seldom done in undergraduate laboratories.

Because heating mantles are constructed of nonferrous materials, they can be used in conjunction with magnetic stirring (Sec. 2.11), so that simultaneous heating and stirring of a reaction mixture is possible. A variable transformer connected to an electrical outlet provides the power for heating the mantle. The cord from the mantle itself must *always* be plugged into a transformer, *never* directly into a wall outlet!

Most heating mantles have a hemispherical cavity into which the round-bottom flask is fitted. This means in principle that a different size mantle is required for each size of flask, and this is indeed the case with those mantles having a spun fiberglass interior. However, if the hemisphere is a rigid ceramic material, it is possible to put a granular heat transfer agent such as sand in the cavity so that flasks of sizes smaller than the diameter of the cavity can be accommodated (see **Sand Bath,** p. 43).

There are some drawbacks to using heating mantles. They heat up rather slowly, and it is difficult to obtain a given temperature or maintain a constant temperature. Heating mantles have a high heat capacity, so if it becomes necessary to discontinue heating suddenly (for example, if a reaction begins to get out of control), *the heating mantle must be removed immediately from below the flask* to allow the flask to cool, either on its own or by means of a cooling bath; it is not sufficient simply to lower the voltage or turn off the electricity. After the mantle is removed, the electricity should be turned off at the transformer.

The amount of heat supplied by a heating mantle is moderated by the boiling liquid contained in the flask because the hot vapors of the liquid transfer heat away from the mantle. If the flask becomes dry or nearly so during a distillation, the mantle can become sufficiently hot to melt the resistance wire inside, thus causing the mantle to ''burn out.'' *Do not heat an empty flask with a heating mantle.* Moreover, most mantles are marked with a maximum voltage to be supplied, and this should not be exceeded.

Variable transformer

(a)

Thermowell

Variable transformer

(b)

FIGURE 2.23

(a) Woven-glass heating mantle. (b) Heating mantle with ceramic core.

OIL BATH

FIGURE 2.24
Electrically heated oil bath.

Electrically heated **oil baths**, which typically contain either mineral oil or silicone oil, are commonly employed in the laboratory. These baths are heated either by placing the bath container on a hot plate or by inserting a coil of Nichrome resistance wire in the bath. In the latter case, the resistance wire, which serves as the heating element, is attached to a transformer with an electrical cord and plug (Fig. 2.24). Thermal equilibration within the bath is maintained by placing a paper clip or stirbar in the bath and using a magnetic stirrer (Sec. 2.11).

Heating baths offer several important advantages over heating mantles. The temperature of the bath can easily be determined by inserting a thermometer in the liquid, and a given bath temperature may be obtained and accurately maintained by careful adjustment of the variable transformer. Although heat is transferred uniformly to the surface of the flask in the bath and there are no hot spots, there typically will be a temperature gradient of about 10 °C between the bath and the contents of the flask.

Some inconveniences are also encountered using oil baths. If the volume of heating liquid is fairly large, it may take a while to reach the desired bath temperature. The maximum temperature that may be safely attained in an oil bath is limited by the type of heating liquid being used. Silicone oils are more expensive but are generally preferable to mineral oils because they can be heated to 200 to 275 °C without reaching the **flash point**, the temperature at which a liquid can burst into flame, and without thickening through decomposition. Mineral oil should *not* be heated above about 200 °C because it will begin to smoke, and there is the potential danger of flash ignition of the vapors. Water must not be present in mineral and silicone oils, since at temperatures of about 100 °C, the water will boil, spattering hot oil. If water drops are present in the oil, the heating fluid should be changed, and the container should be cleaned and dried before refilling.

A minor nuisance associated with oil baths is removing the film of mineral or silicone oils, both of which are water-insoluble, from the outer surface of the flask. This is best done by wiping the flask using a *small* amount of hexane or dichloromethane on a paper towel prior to washing the flask with soap and water.

FIGURE 2.25
(a) Hot plate. (b) Stirring hot plate.

HOT PLATE

When flat-bottom containers such as beakers or Erlenmeyer flasks must be heated, **hot plates** (Fig. 2.25a) and **stirring hot plates** (Fig. 2.25b), which are hot plates with built-in magnetic stirrers, are convenient heat sources; round-bottom flasks cannot be heated effectively with hot plates. The flat upper surface of the hot plate is heated by electrical resistance coils to a temperature that is controlled by a built-in voltage regulator, which is varied by turning a knob on the front of the unit. A hot plate generally should be limited to heating liquids such as water, mineral or silicone oil, and nonflammable organic solvents such as chloroform and dichloromethane. *Under no circumstances should a hot plate be used to boil highly flammable organic solvents*, since the vapors may ignite as they billow onto the surface of the hot plate or the electrical resistance coils inside the hot plate. Furthermore, many hot plates use a relay that turns the electricity on and off to maintain the desired temperature, and these relays are often not explosion-proof and may produce sparks that can ignite fires.

ALUMINUM BLOCK

An **aluminum block** is a convenient accessory used in conjunction with hot plates for reactions being performed in a microscale apparatus. The block has indentations of various sizes and shapes to accommodate conical vials and flasks (Fig. 2.26). There may also be a receptacle for inserting the bulb of a glass thermometer, or the base of a metal dial thermometer, so the temperature of the block can be monitored. Although a glass thermometer may remain vertical without any further support, it should be held in place by a clamp or a wire to prevent breakage. This is not necessary with a metal dial thermometer.

The temperature of the block is controlled by varying the power to the hot plate, and it is helpful to have a correlation between the two. You may calibrate the block by selecting four or five evenly spaced settings on the dial controlling the power to the hot plate, determining the temperature of the aluminum block at each setting, and recording both the setting and the corresponding temperature in your notebook for future reference.

Thermometer —

Aluminum block

FIGURE 2.26   Aluminum block with conical vial in place.

HEAT GUN

A useful device for heating apparatus, but not reaction mixtures, is a **heat gun** (Fig. 2.27), which is essentially a high-powered hair dryer. It is particularly useful for drying assembled apparatus, as this eliminates the need to use a flame. The stream of hot air from the heat gun is directed at the apparatus to drive atmospheric moisture out through an open port, which, after the entire apparatus is heated, is then protected with a drying tube (Sec. 2.27) so that air is dried as it enters the cooling apparatus. Alternatively, a slow stream of dry nitrogen gas can be introduced during the heating process to help expel moisture from the apparatus; once again, a drying tube or some other device such as a bubbler containing mineral oil must be attached to the apparatus to prevent intrusion of atmospheric moisture as the apparatus cools and during the course of the reaction itself. For best results, you should start heating of the apparatus at a point most distant from the open port and work toward the opening.

FIGURE 2.27
Heat gun.

The temperature of the air stream emerging from a heat gun is commonly from 300–500 °C, far higher than that of the typical hand-held hair dryer. Consequently, you must use the heat gun prudently when drying apparatus held by plastic-containing devices such as Keck clamps or clamps with rubber- or vinyl-coated jaws (Fig. 2.5) because the plastic can melt.

You should never point a heat gun in the direction of another person. Moreover, when using this device, you should be certain that there are no light-weight objects in the vicinity that will become airborne on the powerful jet of air produced.

SAND BATH

Sand baths are convenient devices for heating small volumes of material in small flasks. These baths may be easily prepared by putting about 1–3 cm of sand in a Pyrex crystallizing or Petri dish and then placing the dish on a hot plate (Fig. 2.28). The temperature in the sand bath is controlled by varying the heat setting on the hot plate.

FIGURE 2.28
Sand bath heated by hot plate.

Alternatively, the sand may be placed in the ceramic well of a heating mantle, such as that shown in Figure 2.23b. The temperature of the sand bath in this case is controlled by the variable transformer attached to the mantle. The temperature may be monitored with a thermometer inserted into the bath to the same depth as the flask or conical vial being heated. Because sand is not a good conductor of heat, there is a temperature gradient in the bath, with the highest temperature being closest to the hot plate or heating mantle. This gradient can be exploited: The flask or vial can be deeply immersed in the sand bath for rapid heating; and once the mixture has begun to boil, the flask or vial can be raised to slow the rate of reflux or boiling.

Although sand is cleaner than mineral and silicone oil, there are some limitations to using sand baths. For example, sand baths in glass containers are normally not used to heat flasks to temperatures higher than about 200 °C, since the glass might break. Since sand is a poor heat conductor and it is not possible to stir the bath, temperature control is difficult. Furthermore, different devices may heat the sand bath to a different temperature gradient, so you must calibrate each heating device/sand bath combination for reproducible results. A procedure for doing this is discussed under **Aluminum Block** (p. 43).

WATER BATH

A **water bath** is a suitable alternative to an oil or sand bath when temperatures no higher than about 80 °C are desired. The water is contained in a wide-mouthed vessel such as a beaker or crystallizing dish, and preferably is heated with an electrical hot plate (Fig. 2.29a). Alternatively, a Bunsen burner can be used, in which case the bath is supported on a ring clamp bearing a wire mesh for diffusing the flame (Fig. 2.29b). You may find it helpful to cover the open portions of the bath with aluminum foil to minimize the loss of water through evaporation.

FIGURE 2.29 Water bath: (a) heated by hot plate; (b) heated by Bunsen burner.

STEAM BATH

Steam provides a useful and safe source of heat when temperatures only up to 100 °C are required. The steam outlet is connected to either a **steam bath** or a **steam cone** (Fig. 2.30), both of which have an outlet at the bottom to drain the condensed water. The tops of steam baths and cones are typically fitted with a series of overlapping concentric rings that may be removed in succession to provide openings of variable size. For example, if a rapid rate of heat transfer to a flask is desired, the rings are removed until up to one-half of the surface of the flask is immersed in the steam (Fig. 2.30a); cloth towels may be wrapped around the upper portion of the flask to insulate it. If a slower rate of heating is desired, the opening may be adjusted so less of the flask is in direct contact with the steam. To heat beakers and Erlenmeyer flasks, the opening should be small enough so that the container sits directly on top of the steam bath with only its lower surface exposed to steam.

When the steam valve is first opened, several minutes are usually required for the condensed water to drain out of the steam lines. Once the steam is issuing smoothly, the steam valve should be adjusted to provide a *slow*, steady flow of steam. There is no benefit to having a fast flow of steam, other than to fog your safety glasses and give everyone in the lab a sauna bath! Regardless of how fast the steam is flowing, the temperature will never exceed 100 °C. Another disadvantage of a fast flow of steam is that water will condense on your equipment and may even find its way into the reaction flask.

Steam baths are especially useful heat sources for the safe evaporation of volatile solvents during the course of working up a reaction and removing the solvent used for the reaction or extraction (Sec. 2.29) or when purifying a solid product by recrystallization (Sec. 3.2). As an alternative to working at a hood, an inverted funnel attached to a vacuum source may be placed over the top of the container (Fig. 2.30c) to remove vapors and keep them from entering the room.

FIGURE 2.30   (a) Steam bath being used to heat a reaction mixture. (b) Steam cone.
(c) Heating an Erlenmeyer flask on a steam cone, with a funnel attached to vacuum source for removing vapors.

2.10 Cooling Techniques

Sometimes it is necessary to cool a reaction mixture either to avoid undesired side reactions or to moderate the temperature of exothermic reactions that could become dangerously hot and uncontrollable. Cooling is also often used to maximize the recovery of solid products during purification by recrystallization (Sec. 3.2) or to lower the temperature of hot reaction mixtures so that work-up procedures can be performed.

The most common cooling medium in the undergraduate laboratory is an ice-water bath. Liquid water is a more efficient heat transfer medium than ice, because it covers the entire surface area of the vessel with which it is in contact. Consequently, when preparing this type of bath, do not use ice alone. Finally, stirring an ice-water bath is useful for maintaining a constant temperature throughout the bath.

An ice-water bath has an equilibrium temperature of 0 °C. For lower temperatures, an ice-salt bath can be prepared by mixing ice and sodium chloride in a proportion of about 3:1 to generate a temperature of approximately −20 °C. As the ice melts, more ice and salt must be added to maintain this temperature. Still lower temperatures are possible using combinations of organic liquids and either Dry Ice (solid carbon dioxide) or liquid nitrogen; directions for preparing these types of baths can be found in reference 5 of Chapter 1.

Immersing a vessel containing a warm reaction mixture in a container of tap water is useful for rapid cooling to room temperature so long as the glassware you are using is Pyrex®. Consult with your instructor to ensure that this is the type of glass you have!

2.11 Stirring Methods

Heterogeneous reaction mixtures must be stirred to distribute the reactants uniformly and facilitate the chemical reactions. Whenever the contents of a flask are being heated or cooled, stirring also ensures thermal equilibration. If a mixture is boiling, the associated turbulence is usually sufficient to provide reasonable mixing; however, stirring a boiling mixture is an alternative to using boiling stones to maintain smooth boiling action and avoid bumping. Stirring is most effectively achieved using mechanical or magnetic stirring devices, but often swirling is sufficient.

SWIRLING

The simplest means of mixing the contents of a flask is **swirling**, which is accomplished by manually rocking the flask with a circular motion. If a reaction mixture must be swirled, carefully loosen the clamp(s) that support the flask and attached apparatus, and swirl the contents periodically during the course of the reaction. If the entire apparatus is supported by clamps attached to a single ring stand, the clamp(s) attached to the flask do not have to be loosened. Make sure all the clamps are tight, pick up the ring stand, and gently move the entire assembly in a circular motion to swirl the contents of the flask.

MAGNETIC STIRRING

A very convenient mode for mixing contents of a flask is **magnetic stirring**. The equipment consists of a magnetic stirrer, which houses a large bar magnet that is spun by a variable-speed motor, and a stirbar or spinvane (Fig. 2.31) contained in a round-bottom flask or conical vial. The metallic core of the stirbar or spinvane is usually coated with a chemically inert substance such as Teflon, although glass is sometimes used for stirbars.

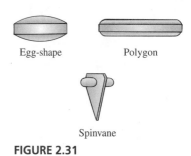

FIGURE 2.31
Stirbars and a spinvane.

The stirbar or spinvane normally is placed in the flask or vial *before* any other materials such as solvents or reagents. In the case of introducing stirbars into a flask, you should *not* simply drop the stirbar in—that may well crack or break the flask. Rather, tilt the flask and let the stirbar gently slide down the side.

A flat-bottomed container such as a beaker or Erlenmeyer flask may be placed directly on top of the stirrer (Fig. 2.32a), whereas a round-bottom flask or a conical vial must be clamped directly above the center of the stirrer (Fig. 2.32b). A flask containing a stirbar should be *centered* on the magnetic stirrer so that the stirbar rotates smoothly and does not wobble. As the stirrer motor turns, the stirbar rotates in phase with the motor-driven magnet. The stirring rate may be adjusted using the dial on the stirring motor, but excessive speed should be avoided because it often causes the stirbar to wobble rather than to rotate smoothly. Because the shapes of a conical vial and a spinvane are matched, wobbling is not a problem. Nonetheless, the stirring rate must still be adjusted to minimize splashing.

The use of magnetic stirring in conjunction with heating is illustrated in Figure 2.32b. This figure depicts apparatus in which the contents of a reaction vessel are simultaneously being heated and magnetically stirred, a common operation in the organic laboratory. A large stirbar or paper clip is used to stir the bath itself; this maintains

FIGURE 2.32   (a) Magnetically stirring contents of Erlenmeyer flask. (b) Using a heating source with magnetic stirring of reaction mixture.

a homogeneous temperature throughout the heating fluid. Other heating devices such as a sand bath, aluminum block, or heating mantle can be used as well, although monitoring of the heating source is more difficult with a mantle.

MECHANICAL STIRRING

Thick mixtures and large volumes of fluids are most efficiently mixed using a **mechanical stirrer**; a typical set-up is depicted in Figure 2.33. A variable-speed, explosion-proof, electric motor drives a stirring shaft and paddle that extend into the flask containing the mixture to be stirred. The motor should have high torque, so that it has sufficient power to turn the shaft and stir highly viscous mixtures. The stirrer shaft is usually constructed of glass, and the paddle, which agitates the contents of the flask, is constructed of an inert material such as stainless steel, Teflon, or glass. A glass paddle must be used to stir reaction mixtures containing active metals such as sodium or potassium. The paddle is easily removed from the shaft to facilitate cleaning, and different-sized paddles can be used according to the size of the flask. The glass shaft and the inner bore of the standard-taper bearing are ground to fit each other precisely. A cup at the top of the bearing is used to hold a few drops of silicone or mineral oil, which lubricates the shaft and provides an effective seal.

The stirrer shaft is connected to the motor with a short length of heavy-walled rubber tubing that is secured with twisted copper wire or a hose clamp. The motor and shaft *must* be carefully aligned to avoid wear on the glass surfaces of the shaft and bearing and to minimize vibration of the apparatus that could result in breakage. The bearing is held in place in the flask with either a rubber band or a clamp so that it does

FIGURE 2.33
Flask equipped with mechanical stirring.

not work loose while the motor is running. The rate of stirring is controlled by varying the speed of the motor with either a built-in or separate variable transformer.

Various operations can be performed while using mechanical stirring. For example, the flask in Figure 2.33 is a three-neck, standard-taper, round-bottom flask that is equipped with an addition funnel and a condenser. This apparatus could be used in cases where dropwise addition of a reagent to a stirred and heated reaction mixture is required.

2.12 Centrifugation

Centrifugation is a useful technique to facilitate the separation of two immiscible phases. The procedure simply involves spinning a sample in one or more tubes at high speed in a bench-top centrifuge (Fig. 2.34). It is important to balance the centrifuge before spinning to avoid vibrating the centrifuge during spinning. For example, if the sample to be centrifuged is placed in a single centrifuge tube, you should then fill a second tube with an equal volume of solvent and place this tube opposite the sample tube in the centrifuge. After the sample has been spun, the phases may be separated by decantation (Sec. 2.19) or by removing a liquid phase with a Pasteur or filter-tip pipet (Sec. 2.5).

Centrifugation is sometimes more effective than filtration for removing suspended solid impurities in a liquid, especially when the particles are so fine they would pass through a filter paper. It may also be used for Craig tube filtration during a microscale recrystallization (Sec. 2.17). Centrifugation will also aid in separating the organic and aqueous layers when performing an extraction (Sec. 2.21); this is particularly useful when emulsions are formed.

FIGURE 2.34
Centrifuge with centrifuge tube.

2.13 Simple Distillation

Simple distillation is a useful method for isolating a pure liquid from other substances that are not volatile. The experimental aspects of this technique are described in this section, whereas the theory and application of the method are discussed in detail in Section 4.2.

MINISCALE APPARATUS

Typical examples of laboratory apparatus for performing simple miniscale distillations are shown in Figure 2.35. The operation entails heating the liquid contained in the **stillpot**, also called the **distillation flask**, to its **boiling point**, which is defined as the temperature at which the total vapor pressure of the liquid is equal to the external pressure. The vapors then pass from the pot into a water- or air-cooled **condenser** where they condense to form the liquid phase that is collected in the **receiver**. The thermometer measures the temperature of the vapors. However, in order to obtain an accurate temperature reading, the thermometer must be carefully positioned so that the top of the mercury bulb is approximately even with the bottom of the sidearm outlet, as indicated in the inset in Figure 2.35a. To ensure an accurate temperature reading, a drop of condensate *must* adhere to the bottom of the mercury bulb. If a pure liquid is being distilled, the temperature read on the thermometer, which is termed the **head temperature**, will be identical to the temperature of the boiling liquid, which is called the **pot temperature**, provided that the liquid is not superheated. The head temperature corresponds to the boiling point of the liquid, and will remain constant throughout the distillation. If a mixture of volatile compounds is being distilled, the head and pot temperatures will be different.

(a)   **(b)**

FIGURE 2.35 (a) Typical apparatus for miniscale simple distillation at atmospheric pressure or under vacuum. Inset shows correct placement of thermometer in stillhead. (b) Typical apparatus for miniscale shortpath distillation with ice-water bath for cooling receiver.

To avoid possible breakage and spillage, certain techniques must be followed when assembling the apparatus shown in Figure 2.35. The apparatus in Figure 2.35a is typically used for volumes of distillate in excess of about 10 mL. If a condenser is not necessary, the modified apparatus shown in Figure 2.35b, comprised of a distillation head and a bent vacuum adapter, is well suited for distilling quantities in the range of 1–10 mL; the shorter path length in this set-up minimizes material losses. *You should thoroughly familiarize yourself with the following general guidelines before attempting to set up your apparatus.*

1. Assemble the equipment so that the distillation flask is elevated 15 cm (6 in.) or so above the bench to allow placement of a suitable heat source. Read Section 2.4 for a general discussion.

2. Review the various heating methods discussed in Section 2.9. Heating mantles may be used to heat the stillpot, but an oil bath is generally the best way to heat the flask, since the temperature of the bath, and hence the stillpot, is easier to control. The level of the liquid in the distillation flask should be below the level of the oil in the heating bath to minimize the risk of bumping. *Do not use a flame unless directed to do so by your instructor.* The flask should *never* be heated *directly* with a flame because this may produce hot spots and cause bumping; the flask may only be heated with a flame *if* wire gauze is placed under the flask to diffuse the heat.

3. When distilling materials that have boiling points in excess of 125 °C, wrap the stillhead with glasswool and then aluminum foil to prevent excessive heat loss; failure to do so usually means that it will be necessary to heat the distillation flask to higher temperature to drive over the last portion of product.

4. Place a stirbar in the distillation flask along with the liquid to be distilled *before* attaching the rest of the glassware; the flask should not be more than *half*-filled. *Start assembling the equipment by clamping the stillpot in position*, and attach the rest of the glassware by next putting the stillhead in place, then the condenser, and finally the vacuum adapter at the end of the condenser. Place a thin film of lubricant (Sec. 2.2) on each of the standard-taper glass joints before mating them. Joints must fit snugly so that flammable vapors do not leak from the apparatus into the room.

5. Note the location of the rigid clamps in Figure 2.35; *do not "overclamp" the apparatus*. An apparatus that is too rigidly clamped may be so stressed that it breaks during its assembly or the distillation. Align the jaws of the clamp *parallel* to the piece of glass being clamped so that the clamp may be tightened without twisting or torquing the glass and either breaking it or pulling a joint loose. Tighten the clamps only enough to hold each piece securely in place. *Do not tighten the clamp unless the piece of glassware is correctly positioned and the clamp is properly aligned.* Always remember the rule: *Do not apply undue pressure to the glassware.*

6. Use Keck clips to hold the condenser, vacuum adapter, and receiver in place. Additional support should be provided for flasks of 100-mL capacity or larger; smaller flasks may also require additional support if they become more than half-full during a distillation. Clamp the receiver. An iron ring holding a piece of wire gauze or a cork ring may be used to provide additional support underneath the receiving flask.

7. Note the location of the "water-in" and "water-out" hoses on the condenser. The tube carrying the incoming water is *always* attached to the lower point, which ensures that the condenser is filled with water at all times.

8. Adjust the water flow through the condenser to a *modest* flow rate. There is no benefit to a fast flow, and the increased pressure in the apparatus may cause a piece of rubber tubing to pop off, spraying water everywhere. Showers in the laboratory

should be restricted to the emergency shower! It is good practice to *wire* the hoses to the condenser and to the water faucet to minimize the danger that they will break loose.

MICROSCALE APPARATUS

The volume of distillate in microscale distillations is typically less than 500 μL (0.5 mL), so it is important to minimize the surface area of glass in which the distillate comes in contact to avoid unnecessary loss of material. This is accomplished by combining the stillhead and collector in a single piece called a **Hickman stillhead**, which has a reservoir for containing distillate that may have a port to facilitate removal of distillate (Fig. 2.36).

Because the volume of distillate is small, it is somewhat difficult to measure the distillation temperature (boiling point) accurately at the microscale level. An *approximate* method is to determine the temperature of the heat source at the point when distillation occurs. This temperature is about 20 °C or so above the distillation temperature, so it is necessary to make the corresponding correction in the boiling point recorded for the distillate. Alternatively, a thermometer can be inserted through the condenser attached to the stillhead so the thermometer bulb is in the lower neck of the stillhead. A more accurate measurement of the boiling-point of the distillate is possible using the technique of microscale boiling point determination (Sec. 2.8).

A Pasteur or filter-tip pipet (Sec. 2.5) is used to remove distillate from the Hickman stillhead. If the head has a port, the pipet can be inserted at this point. If not, the condenser must first be removed, and the pipet inserted through the top of the stillhead. Depending on the geometry of the stillhead, it may be necessary to use a Pasteur pipet having a curved tip to allow the tip to reach the bottom of the reservoir. A Bunsen or microburner (Sec. 2.9) may be used to produce the necessary bend.

Thermometer

Water out

Condenser

Water in

Clamp

Hickman stillhead

Conical reaction vial

Spinvane

Heating source

Magnetic stirrer

Laboratory jack

FIGURE 2.36
Microscale apparatus for distillation with external temperature monitoring.

2.14 Fractional Distillation

The technique of **fractional distillation** is a useful method for isolating the individual pure liquid components from a mixture containing two or more volatile substances. This technique is described in this section, whereas the theory and application of this method are discussed in detail in Section 4.32. An apparatus for fractional distillation at atmospheric pressure or vacuum is shown in Figure 2.37. The principal difference between an apparatus for fractional and simple distillation is the presence of a **fractional distillation column** (Fig. 2.38) between the stillpot and the stillhead. This column is similar to a condenser, the major difference being that a fractionating column has a large outside jacket and some indentations at the male end to hold the packing in place. Unpacked distillation columns such as the Hempel column shown in Figure 2.38a can also be used as condensers, but condensers lack the indentations and cannot serve as packed distillation columns.

Before assembling the apparatus, clean and dry the inner tube of the distillation column; it is not necessary to dry the water jacket, since traces of water in it will not

FIGURE 2.37
Miniscale apparatus for fractional distillation.

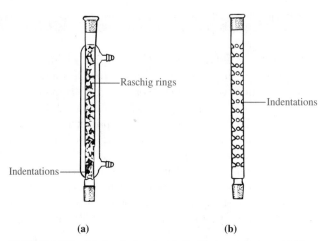

FIGURE 2.38   Miniscale fractional distillation columns: (a) Hempel column filled with packing material; (b) Vigreux column.

affect the distillation. Pack the fractional distillation column by adding the desired column packing, a small quantity at a time, through the top of the column while holding it vertical. The column packing is an inert material such as glass, ceramic, or metal pieces in a variety of shapes (helices, saddles, woven mesh, and so on); common packings include glass tubing sections, glass beads, glass helices, or a stainless steel sponge. Some column packings are sufficiently large that they will stop at the indentations. If the packing falls through the column, put a small piece of glasswool, wire sponge, or wire screen into the column just above the indentations by pushing it down the column with a wooden dowel or a piece of glass tubing. The column packing should extend to the top of the water jacket but should not be packed too tightly.

Assemble the apparatus using the general guidelines for simple distillation apparatus (Secs. 2.4 and 2.13). *Read those instructions carefully before proceeding with setting up the apparatus in Figure 2.37.* Start by putting the liquid to be distilled, along with a stirbar, in the distillation flask, then clamp the flask in place. Do not fill the flask more than *half*-full. Attach the distillation column and make sure that it is *vertical.* Lubricate and tighten all the joints after the stillhead and condenser are in place. Clamp only the distillation flask and condenser. *Do not run water through the jacket of the distillation column.*

2.15   Vacuum Distillation

It is most convenient to distil liquids at atmospheric pressure (760 Torr), but compounds that have high molecular weights or numerous functional groups may decompose, oxidize, or undergo molecular rearrangement at temperatures below their atmospheric boiling points. These problems may frequently be circumvented if the distillation is conducted at a lower temperature. Since a liquid boils when its total vapor pressure is equal to the external pressure, it is possible to lower the boiling point of the liquid by performing the distillation at a pressure *less* than one atmosphere. The technique involved in such distillations is termed **vacuum distillation**.

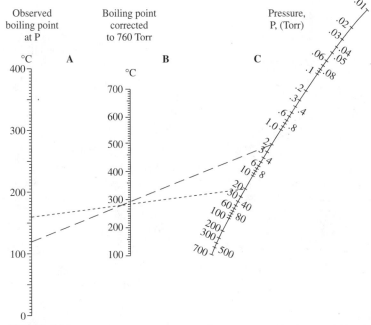

FIGURE 2.39 Pressure-temperature alignment nomograph (dashed lines added for illustrative purposes). How to use the nomograph: Assume a reported boiling point of 120 °C at 2 Torr. To determine the boiling point at 20 Torr, connect 120 °C (column **A**) to 2 Torr (column **C**) with a transparent plastic rule and observe where this line intersects column **B** (about 295 °C). This value would correspond to the normal boiling point. Next, connect 295 °C (column **B**) with 20 Torr (column **C**) and observe where this intersects column **A** (160 °C). The approximate boiling point will be 160 °C at 20 Torr.

Although accurate estimates of the effect of pressure upon the boiling point of a liquid may be made by use of charts or a nomograph (Fig. 2.39); two useful *approximations* of the effect of lowered pressure on boiling points are:

1. Reduction from atmospheric pressure to 25 Torr lowers the boiling point of a compound boiling at 250–300 °C at atmospheric pressure by about 100–125 °C.
2. Below 25 Torr, the boiling point is lowered by about 10 °C each time the pressure is reduced by one-half.

Reduced pressures may be obtained by connecting a **water aspirator pump** or a **mechanical vacuum pump** to a vacuum adapter that is fitted between the condenser and the receiving flask. The vacuum produced by a water aspirator is limited by the vapor pressure of the water at the ambient temperature and the condition of the aspirator pump. Pressures as low as 8–10 Torr may be obtained from a water aspirator with cold water, but pressures in the range of 15–25 Torr are more common. A good mechanical vacuum pump can evacuate the apparatus to less than 0.01 Torr; it is important to clean the oil periodically and maintain tight connections in the distillation apparatus to achieve the lowest possible pressures for a particular pump. Some laboratories are equipped with ''house'' vacuum lines that are connected to a large central vacuum pump, but such vacuum systems seldom reduce the pressure to less than 50 Torr. The pressure is measured by a device called a **manometer**.

FIGURE 2.40 (a) Miniscale apparatus for vacuum distillation. (b) Modified apparatus for miniscale shortpath vacuum distillation.

The apparatus for vacuum distillations is a modification of that used for simple distillations (Fig. 2.35), and two options are shown in Figure 2.40. The set-up depicted in Figure 2.40a is typically used for volumes of liquid in excess of about 10 mL. If a condenser is not required, the modified apparatus of Figure 2.40b can be used for distilling volumes of liquid in the range of 1–10 mL; the shorter path length of this apparatus minimizes loss of material.

The primary difference between the apparatus for simple and vacuum distillations is the attachment of the vacuum adapter to a vacuum source. Rather than being directly connected to the vacuum source, the adapter is normally linked to a heavy-walled, side-arm flask. The flask has several functions. For example, if a water aspirator is being used to produce a vacuum, the flask serves as a trap to prevent back-up of water into the apparatus, which may occur if there is sudden decrease in the water pressure in the aspirator. If it is equipped with a stopcock release valve, the flask also provides a means for connecting the apparatus and the manometer to the source of the vacuum and releasing the vacuum.

The apparatus shown in Figure 2.40 is assembled (Sec. 2.4) according to the same general guidelines that are used for setting up the apparatus for simple distillation (Sec. 2.13). *Read those instructions carefully before proceeding with assembling the apparatus in Figure 2.40.* Because magnetic stirring (Sec. 2.11) is the most convenient means of obtaining an even rate of vaporization during the distillation, that is the method described here. However, if this option is not available, consult your instructor regarding alternative set-ups for vacuum distillation.

Start the assembly of apparatus by tilting the stillpot to one side and sliding a stirbar down the side. Then add the liquid to the stillpot, making certain that the flask is no more than *half*-full. After lubricating all the ground-glass joints (Sec. 2.2), attach the stillhead and equip it with a thermometer adapter and thermometer. Continue the set-up as described in Section 2.13, clamping the condenser, if one is used, and the receiver. Make sure all of the joints are tightly mated. Finally, connect the apparatus to the side-arm flask, which in turn must be connected to the vacuum source and the manometer.

The volume and density of the vapor formed by volatilization of a given amount of liquid are highly pressure-dependent. For example, the volume of vapor formed from vaporization of a drop of liquid will be about 20 times as great at 38 Torr as it would be at 760 Torr. During a vacuum distillation, large quantities of vapor are produced in the distillation flask, and these vapors enter the condenser with high velocity prior to condensing. Since the density of the vapor is much lower at reduced pressure than at atmospheric pressure, it may be difficult to control the rate of vaporization and thereby minimize the difference in pressure between the distillation flask and the manometer where the pressure must be accurately measured.

The manometer measures the pressure at which the distillation is being conducted; this value is important and is reported with the boiling point. For example, benzaldehyde boils at 180 °C at atmospheric pressure and at 87 °C at 35 Torr, and these two boiling points are reported in the format: bp 180 °C (760 Torr) and 87 °C (35 Torr). To obtain an accurate measure of the pressure, however, you must carefully *control the rate of distillation.* The problem arises because a drop of condensate forms from a larger volume of vapor as the pressure is lowered, thereby causing higher vapor velocities to enter the condenser. This creates a back pressure so that the pressure in the distillation flask and stillhead is higher than that measured on the manometer, which is located beyond the condenser and receiver and is therefore insensitive to back pressure. The difference between the actual and measured pressure can be minimized by distilling the liquid at a slow but steady rate. Superheating the vapor can be avoided by maintaining the oil bath at a temperature no more than 25–35 °C higher than the head temperature.

The following paragraphs provide a general procedure for performing a vacuum distillation using the apparatus shown in Figure 2.40a. *Read these instructions carefully prior to executing a vacuum distillation.*

1. *Never use glassware with cracks, thin walls, or flat bottoms, such as Erlenmeyer flasks, in vacuum distillations.* Hundreds of pounds of pressure may be exerted on the *total* exterior surfaces of systems under reduced pressure, even if only a water aspirator pump is used. Weak or cracked glassware may *implode*, and the air rushing into the apparatus will shatter the glassware violently in a manner little different from that of an explosion. *Examine the glassware carefully.*

2. Lubricate and seal *all* glass joints carefully during assembly of the apparatus; this will help avoid air leaks and provide lower pressure. The rubber fittings holding the thermometer in place must be tight. The neoprene fittings normally used with the thermometer adapters may be replaced, if necessary, with short pieces of heavy-walled tubing. Do not use rubber stoppers elsewhere in the apparatus, since direct contact between the rubber and the hot vapors during distillation may cause contamination. The three-holed rubber stopper in the safety trap should fit snugly to the flask and the pieces of glass tubing. The safety trap must be made of heavy-walled glass and wrapped with electrical tape as protection in case it implodes. Heavy-walled vacuum tubing must be used for all vacuum connections. *Check the completely assembled apparatus to make sure that all joints and connections are tight.*

3. Now begin stirring the liquid and turn on the vacuum; if you are using a water aspirator pump, you should *fully open the water faucet.* The release valve on the safety flask should be *open*. *Slowly* close the release valve, but be prepared to reopen it if necessary. If the liquid contains small quantities of low-boiling solvents, as is often the case, foaming and bumping are likely to occur in the distillation flask. If this occurs, adjust the release valve until the foaming abates. This may have to be done several times until the solvent has been removed completely. When the surface of the liquid in the flask is relatively quiet, fully evacuate the system to the desired pressure. The release valve may have to be opened slightly until the desired pressure is obtained; a needle valve from the base of a Bunsen burner may be used for fine control of the pressure. Check the manometer for constancy of pressure. Note and record the pressure at which the distillation is being performed. This pressure should be monitored throughout the course of the distillation, even when fractions are not being collected. Begin heating the flask, and maintain the temperature of the bath so that the distillate is produced at the rate of 3 to 4 drops per minute. *Do not begin heating the flask until the system is fully evacuated and the vacuum is stable.* The best way to heat the stillpot is with an oil bath, since it is easier to control the temperature in the bath and distillation flask; however, other techniques (Sec. 2.9) may be used.

4. If it is necessary to use multiple receivers to collect fractions of different boiling ranges, the distillation must be interrupted to change flasks. Remove the heating source with caution and allow the stillpot to cool somewhat. *Slowly* open the vacuum release valve to re-admit air to the system. When atmospheric pressure is reached, change receivers, close the release valve, re-evacuate the system to the same pressure as previously, reapply heat, and continue distilling. The operation may result in a different pressure in the fully evacuated system. Periodically monitor and record the head temperature and the pressure, particularly just before and after changing receivers.

5. After the distillation is complete, discontinue heating, allow the pot to cool somewhat, slowly release the vacuum, and turn off the source of the vacuum.

One of the most inconvenient aspects of this procedure is the disruption of the distillation in order to change receivers. In order to eliminate this problem, you may wish to use a "cow" receiver, which may accommodate three or four round-bottom flasks (Fig. 2.41). These flasks are successively used as receivers by rotating them into the receiving position.

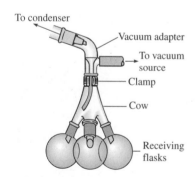

FIGURE 2.41
Multi-flask adapter for miniscale vacuum distillation.

To condenser

Vacuum adapter

To vacuum source

Clamp

Cow

Receiving flasks

2.16 Steam Distillation

Steam distillation is a useful and mild method for separating and purifying volatile liquid or solid organic compounds that are immiscible or insoluble in water. This technique is not applicable to substances that react with water, decompose on prolonged contact with steam or hot water, or have a vapor pressure of less than about 5 Torr at 100 °C. The practical features and the apparatus of this technique are described in this section; its theory and applications are discussed in detail in Section 4.4.

The two basic techniques commonly used to carry out a steam distillation in the laboratory are differentiated on the basis of whether the steam is introduced from an external source or generated internally. For larger scale reactions, the most common and most efficient method for conducting a steam distillation involves placing the organic compound(s) to be distilled in a round-bottom flask equipped with a Claisen adapter, a stillhead, and a water-cooled condenser, as depicted in Figure 2.42. The combination of introducing steam from an external source into the distillation flask via the inlet tube and of the turbulence associated with the boiling action tends to cause occasional violent splashing, and the Claisen adapter is necessary to prevent the mixture from splattering into the condenser. During the course of the distillation, water may condense in the distillation flask and fill it to undesirable levels. This problem can be avoided by gently heating the stillpot with an appropriate heat source.

Steam may be produced externally in a generator as shown in Figure 2.43. The round-bottom flask is initially half-filled with water, and boiling stones are added before heating. The safety tube relieves internal pressure if steam is generated at too high a rate or the outlet tube to the apparatus becomes clogged. Steam is also conveniently obtained from a laboratory steam line. A trap (Fig. 2.44) is then placed between the steam line and the distillation flask to permit removal of any water and/or impurities present in the steam.

If only a small amount of steam is necessary to separate a mixture completely, a simplified method involving *internal steam generation* may be employed. Water is added directly to the distillation flask together with the organic compounds to be

FIGURE 2.42 Apparatus for miniscale steam distillation. The steam tube is replaced with a stopper if steam is generated by direct heating.

FIGURE 2.43
Steam generator.

FIGURE 2.44   Water traps for use in steam distillations, (a) bent adapter trap and (b) separatory funnel trap.

separated. The flask is equipped for steam distillation by setting up the apparatus as shown in Figure 2.42, *except* that the steam inlet tube is replaced with a stopper. The flask is then heated with a Bunsen burner or heating mantle, and the distillate is collected. This technique is generally not applicable for distillations that require large amounts of steam, since it would be necessary to use an inappropriately large flask or to replenish the water in the flask frequently by using an addition funnel. However, this procedure is satisfactory and convenient for most miniscale reactions.

2.17   Filtration Apparatus and Techniques

Filtration is the primary technique used to separate solids from liquids. It is important to perform filtrations properly to avoid loss or contamination of the desired product, regardless of whether it is a solid or liquid.

Conceptually, the process of filtration is simple and involves passing a liquid containing a solid material through a barrier that ideally is permeable only to the liquid. The liquid is collected in an appropriate receiver, and the solid remains on the barrier. However, there are many nuances to how filtrations are performed, and some of them are discussed below.

FILTRATION MEDIA

Various types of materials are used in conjunction with funnels to produce the barrier that is the key to separating solids from liquids. The most common of these is **filter paper**, which is available in different sizes and porosities. The more porous the paper, the faster the rate of filtration but also the greater the possibility that solid particles will pass through the paper. **Glasswool** and **cotton** may serve as filters, but they are considerably more porous than paper and thus are used only when relatively large particles are to be separated from a liquid.

Inert finely divided solids such as silica gel and Celite®, both of which are forms of silicon dioxide, are sometimes used along with filter paper, glasswool, or cotton to assist filtration. These materials generally function by *adsorption* of solids and colloidal materials from the solution being filtered, thereby preventing their passing through the filter paper and/or clogging its pores. Applications of Celite as a ''filter-aid'' are described in Section 2.18.

For microscale filtrations, there is also a specialized technique that uses a Craig tube and depends on contact between the ground-glass strip on the inner shoulder of the glass tube and a glass or Teflon stopper (Fig. 2.45) to effect separation of liquid and solid. The ground-glass surface prevents complete contact with the glass or Teflon surface of the plug and thus allows for liquid to flow out of the tube upon centrifugation of the assembled unit. Very fine particles are not effectively removed by this technique.

FIGURE 2.45
Craig tube assembly.

GRAVITY FILTRATION

Gravity filtration is the filtering technique most commonly used to remove solids such as impurities, decolorizing carbon (Sec. 2.18), or drying agents (Sec. 2.24) from liquids prior to crystallization, evaporation, or distillation. In addition to filter paper of the appropriate size, only a stemmed funnel, ring clamp, and a receiver (Fig. 2.46) are needed for this operation. The filter paper must be folded to fit into the funnel, and there are two ways of doing this. The simplest involves folding the circular piece of paper in half, then in half again so that corners 1 and 2 of Figure 2.47a are coincident, inserting the paper in the funnel, and opening the folded paper so that the solution can be poured through it. Before starting the filtration, it is helpful first to wet the filter paper with a small amount of an appropriate solvent. This serves to form a seal between the paper and the funnel and to keep the paper from collapsing on itself and closing the opening you have made in it.

The rate and ease of filtration are increased if a piece of **''fluted'' filter paper** (Fig. 2.47g) is used, since fluting increases the surface area of paper contacted by the liquid being filtered. Although such paper is commercially available, it may also be easily prepared from a flat piece of round filter paper. One of several methods of fluting is shown in Figure 2.47 and involves the following sequence. Fold the paper in half, and then into quarters. Fold edge 2 onto 3 to form edge 4, and then 1 onto 3 to form 5

FIGURE 2.46
Apparatus for miniscale gravity filtration.

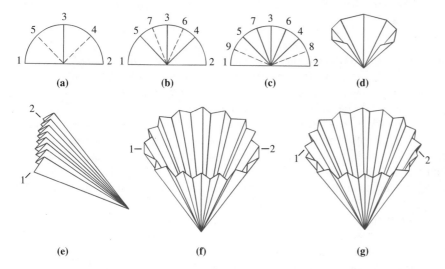

(a) (b) (c) (d)

(e) (f) (g)

FIGURE 2.47 Folding filter paper.

FIGURE 2.48
Executing a miniscale gravity filtration.

FIGURE 2.49
Gravity filtration with a Pasteur filtering pipet.

(Fig. 2.47a). Now fold edge 2 onto 5 to form 6, and 1 onto 4 to form 7 (Fig. 2.47b). Continue by folding edge 2 onto 4 to form 8, and 1 onto 5 to form 9 (Fig. 2.47c). The paper now appears as shown in Figure 2.47d. Do not crease the folds tightly at the center, because this might weaken the paper and cause it to tear during filtration. All folds thus far have been in the same direction. Now make folds in the *opposite* direction between edges 1 and 9, 9 and 5, 5 and 7, and so on, to produce the fanlike appearance shown in Figure 2.47e. Open the paper (Fig. 2.47f) and fold each of the sections 1 and 2 in half with reverse folds to form paper that is ready to use (Fig. 2.47g).

To perform a filtration, the folded or fluted filter paper is first placed in a funnel and *wetted* with the appropriate solvent. The solution containing the solid is then poured onto the filter paper or fluted filter paper with the aid of a glass stirring rod to minimize dripping down the sides of the flask. As the solution passes through the paper, the solid remains on the filter paper. A typical experimental set-up is shown in Figure 2.48. If you do not use a ring stand to support the funnel, you should insert a paper clip or a bent piece of wire between the funnel and the lip of the flask to avoid the formation of a solvent seal between the flask and the funnel. Such a seal will interfere with the filtration, since there will be no vent for the air displaced by the filtrate.

When volumes less than about 5 mL must be filtered, there is a danger of losing material on the filter paper because the paper absorbs a significant volume of liquid. In such cases a Pasteur filtering pipet (Sec. 2.5) should be used. The liquid mixture to be filtered is transferred by pipet into the filtering pipet as shown in Figure 2.49. The plug should be rinsed with 1–2 mL of solvent to maximize the recovery of the solute, since some of it may have crystallized on the cotton plug.

HOT FILTRATION

Sometimes it is necessary to remove solid impurities from an organic compound that is only sparingly soluble in the solvent at room temperature. In such cases, the mixture must then be heated, and **hot filtration** must be performed to remove the insoluble solids without leaving the desired material in the funnel, too. Such a situation commonly arises during purification of an organic solid by recrystallization when decolorizing carbon (Sec. 2.18) or other insoluble residues must be removed; a filter-aid may be required for a successful filtration.

Miniscale Apparatus

The apparatus for performing a miniscale hot filtration is shown in Figure 2.50. A small volume of the *pure* solvent is first placed in an Erlenmeyer flask containing a boiling stick or several boiling stones, and a short-stemmed or stemless filter funnel containing a piece of fluted filter paper is placed in the mouth of the flask. A small stick or piece of thick paper is inserted between the flask and the funnel to allow the vapors of solvent to rise around the outside of the funnel. The flask is heated with either a steam bath or hot plate until the solvent boils and the vapors rise up the sides of the flask and around the funnel; the solvent vapors heat the filter funnel to minimize problems of crystallization of the desired material in the funnel. A steam bath is preferred for flammable solvents that boil below 90 °C, whereas a hot plate can be used for all nonflammable solvents and, with proper care and in a hood, for flammable solvents boiling above 90 °C.

When the flask and funnel are hot, the hot solution is poured onto the filter paper. After all the hot solution has been poured onto the filter paper, several milliliters of hot

FIGURE 2.50 Miniscale hot filtration: (a) using steam cone for heating; (b) using hot plate for heating.

solvent are added to the flask that contained the original recrystallization solute and solvent. The additional hot solvent is poured onto the filter paper to ensure the complete transfer of material to the flask containing the filtered solution.

Microscale Apparatus

A microscale hot filtration is performed using a combination of a Pasteur filter-tip pipet (Fig. 2.9a) and a filtering pipet (Fig. 2.9b). The goal is filtering the hot solution containing solid contaminants as rapidly as possible so that none of the solute crystallizes before the filtrate reaches the receiver. This requires preheating both pipets by pulling hot solvent up into the barrel several times and then quickly transferring the hot solution into the top of the filtering pipet, as shown in Figure 2.51. As with miniscale gravity filtration, the filter should be rinsed with the solvent to maximize the recovery of the solute, since some of it may have crystallized on the cotton plug. However, no more than 0.2–0.5 mL of additional solvent should be used.

VACUUM FILTRATION

Vacuum filtration is a technique for collecting crystalline solids from solvents after recrystallization or precipitation. A typical apparatus is shown in Figure 2.52. Either a **Büchner** or a smaller **Hirsch funnel** is used, with the latter being better suited for isolating quantities of solid ranging from 100–500 mg. The funnel is fitted to a vacuum filter flask using a neoprene adapter or a rubber stopper. The sidearm of the filter flask is connected to an aspirator or house vacuum line through a **trap** using heavy-walled rubber tubing. The trap prevents water from the aspirator from backing up into the filter flask when there is a loss of water pressure. The trap should be a *heavy-walled* bottle or a second vacuum filter flask wrapped with electrical or duct tape. If a filter flask is used for the trap, it should be equipped with a two-holed stopper with its sidearm

FIGURE 2.51 Microscale hot filtration.

FIGURE 2.52   Apparatus for miniscale vacuum filtration; inset shows Hirsch funnel, used for small quantities of solids.

attached to the first filter flask with heavy-walled tubing. The glass tube that extends to within approximately 2–3 cm of the bottom of the filter trap is connected with heavy-walled tubing to the water aspirator pump so that any water that collects slowly in the trap is evacuated back through the aspirator pump and into the drain. If a sudden back-up occurs, the vacuum should be released immediately so that the trap does not fill completely with water. Any filtrate that may overflow the filter flask will also be collected in the trap. *The filter flask should be cleaned before doing the filtration*, since it may be necessary to save the filtrate.

A piece of filter paper is placed in the funnel and should lie flat on the funnel plate, covering all the small holes in the funnel. It should *not* extend up the sides of the funnel. A vacuum is applied to the system, and the filter paper is "wetted" with a small amount of pure solvent in order to form a seal with the funnel so that crystals do not pass around the edges of the filter paper and through the holes in the filter. The solution containing the crystals is then transferred to the funnel. The flask containing the crystals is swirled to suspend the crystals in the solvent, and the solution containing the crystals is poured slowly onto the funnel. A stirring rod or spatula may be used to aid the transfer. The last of the crystals may be transferred to the funnel by washing them from the flask with some of the filtrate, which is called the **mother liquor**. When all the solution has passed through the filter, the vacuum is released slowly by opening the screw clamp or stopcock on the trap. The crystals are washed to remove the mother liquor, which contains impurities, by adding a small amount of *cold*, *pure solvent* to the funnel to just cover the crystals. Vacuum is reapplied to remove the wash solvent, and the crystals are pressed as dry as possible with a clean spatula or a cork while the funnel is under vacuum.

Most of the solvent may be evaporated from the crystals by allowing the vacuum to pull air through the crystals on the funnel for a few minutes. A clean spatula is used to scrape the crystals gently from the filter paper and then to transfer them to a clean watchglass; care should be exercised to avoid contaminating the crystals with torn bits of filter paper. The crystals may be dried completely by allowing them to air-dry for a few hours or by leaving them loosely spread on a filter paper in a locker until the next class. The drying process may be accelerated by placing the crystals in an oven, but the temperature of the oven should be at least 20–30 °C below the melting point of the crystals. A vacuum desiccator also can be used to hasten the drying process. The vacuum desiccator should either be wrapped with electrical or duct tape or placed in a metal cage as protection should the desiccator implode. Specially designed desiccators permit heating samples under vacuum, but the temperature must be kept about 20–30 °C below the melting point of the solid. Heat or vacuum desiccators should not be used to dry crystals of compounds that sublime readily (Sec. 2.20).

Vacuum filtration can also be used to remove undesired solids from a solution. If the solid is finely divided or colloidal, a pad of a filter-aid such as Celite can be used to ensure complete removal of the solid. A pad of filter-aid can be formed on the filter funnel by first making a slurry of 0.5 to 1 g of filter-aid in a few milliliters of the solvent being used. The slurry is poured onto a filter paper in a Büchner or Hirsch funnel that is attached to a *clean* filter flask; vacuum is then slowly applied to draw the solvent through the filter paper, leaving a thin, even pad of the filter-aid; the pad should be about 2–3 mm thick. The solution containing the solid is then filtered as described above. Obviously this technique is useful for removing solid impurities *only* when the *solution* contains the substance of interest; a solid product is not collected in this manner as it would be necessary to separate the desired material from the filter-aid.

CRAIG TUBE FILTRATION

Craig tube filtration is the microscale analog to vacuum filtration for isolating solid products. However, rather than using a vacuum to pull solvent through a filter, this technique depends on centrifugal forces to force the solvent through the tiny openings between a ground-glass surface and a plug in contact with it.

The Craig tube itself is comprised of two components, a test tube-like glass vessel, the inner shoulder of which is ground-glass, and a plug made from glass or Teflon (Fig. 2.53a). In addition to the Craig tube, a centrifuge, centrifuge tubes, and a piece of wire to affix the Craig tube to a centrifuge tube are required.

The technique involves transferring a solution of the desired solid into the bottom portion of the Craig tube. After crystallization is complete, the top part of the Craig

FIGURE 2.53 Filtration using the Craig tube: (a) components of Craig tube, (b) components for Craig tube filtration, (c) assembled apparatus, (d) apparatus after centrifugation.

tube is attached, and a wire holder is affixed to the unit (Fig. 2.53b). The assembled unit is placed in an inverted centrifuge tube so that the top of the Teflon plug is at the bottom of the tube; the wire is looped over the lip of the tube (Fig. 2.53c). The entire assembly is then inverted carefully so that the two components of the Craig tube stay in close contact, thereby preventing solid from escaping past the plug into the centrifuge tube. After centrifugation, the solvent will be in the bottom of the centrifuge tube, whereas the desired solid remains in the Craig tube (Fig. 2.53d).

2.18   Decolorizing Carbon

When a solid is dissolved in a suitable solvent for recrystallization, a colored solution may result. The color can arise from the compound itself or from the presence of colored impurities in the solution. Similarly, an organic liquid may be colored because of impurities or because of the nature of the compound itself. If you are in doubt about the cause of the color, you may try to remove it by adding activated **decolorizing carbon** to the solution or liquid and then removing the carbon by filtration; sometimes heating is required. Decolorizing carbon, which is commercially available under trade names such as Norite® or Darco®, is finely divided, activated charcoal with an extremely large surface area. Polar, colored impurities bind preferentially to the surface of decolorizing carbon, but you should be aware that some of the desired compound may also be adsorbed, so excessive quantities of decolorizing carbon should *not* be used. After the decolorizing carbon is removed by filtration, the liquid or solution should be free of colored impurities; any remaining color is probably due to the compound itself.

MINISCALE TECHNIQUE

In a typical miniscale procedure, the liquid or solution, contained in an Erlenmeyer flask, is first brought to a temperature 10–20 °C below its boiling point. Decolorizing carbon is then added, and the hot solution is gently swirled. There are no firm rules concerning the amount of decolorizing carbon that should be used, but a good first approximation is to add about 0.5 to 1 g per 100 mL of solution. The decolorizing carbon is then removed by gravity filtration or hot gravity filtration, not vacuum filtration, using fluted filter paper; a filter-aid (Sec. 2.17) will facilitate complete removal of all of the carbon. Hot gravity filtration must be used whenever an organic solid is dissolved in a hot solvent, such as during a recrystallization. The solution or liquid must be refiltered if it contains any small black specks of carbon or any other solid material. Normally this extra step can be avoided if care is exercised when filtering the mixture and a filter-aid is used. For example, the filter paper should not contain any tears or holes, and the mixture should be carefully poured onto the filter paper so that none of it runs down the side of the funnel. If the compound is known to be colorless and if the decoloration process does not remove the color, the procedure should be repeated.

MICROSCALE TECHNIQUE

The microscale procedure for decolorizing solutions is analogous to that used at the miniscale level. A small amount of powdered decolorizing carbon (10–20 mg) is added to the solution contained in a conical vial, test tube or other suitable vessel and which is at a temperature 10–20 °C *below* its boiling point. After the mixture is gently swirled or stirred for several minutes, the decolorizing carbon is removed by hot filtration (Sec. 2.17). The particles of carbon are very fine, however, and some of them may pass through the cotton plug of the Pasteur filter pipet, in which case the filtrations should be repeated, preferably with a clean filter pipet.

Pelletized carbon may be available in your laboratory, and you may be instructed to use it instead of the powdered form. Because of its larger size (pelletized carbon is about 10 mesh, or 2 mm in diameter, whereas powdered carbon is typically 100 mesh or 0.14 mm in diameter), it is easier to separate from the solution. However, because its surface area is lower, pelletized carbon is less efficient at removing colored contaminants.

2.19   Decanting Solutions

Small amounts of dense solids are sometimes removed by simple **decantation**. This technique is a viable alternative to filtration for removing solid drying agents (Sec. 2.24), but it cannot be used to remove finely dispersed solids such as decolorizing carbon. To decant a liquid from a solid, the solid should first be allowed to settle to the bottom of the container. The container is then *carefully* tilted, and the liquid is *slowly* poured into a clean container, possibly with the aid of a glass stirring rod to minimize dripping down the sides of the flask (Fig. 2.54); the solid should remain in the original container. Decantation is preferable to gravity filtration when working with very volatile organic liquids, since filtration is likely to result in considerable evaporation and loss of material. When decanting from an Erlenmeyer flask, a *loosely packed* ball of glasswool can be put in the neck of the flask to help keep the solid in the flask.

One major disadvantage of decantation is that some liquid will remain in the flask; this problem may be minimized by using a Pasteur pipet to transfer the last few milli-liters and rinsing the residue with several milliliters of pure solvent that are also trans-ferred with a pipet.

Solid particles

FIGURE 2.54
Decantation.

2.20   Sublimation

Like the vapor pressure of a liquid, the vapor pressure of a solid increases with tem-perature. If the vapor pressure of a solid is greater than the ambient pressure at its melting point, then the solid undergoes a direct phase transition to the gas phase without first passing through the liquid state. This process is called **sublimation**.

The ease with which a molecule may escape from the solid to the vapor phase is determined by the strength of the intermolecular attractive forces between the molecules of the solid. Symmetrical structures have relatively uniform distributions of electron density, and they have small dipole moments compared to less symmetrical molecules. Since electrostatic interactions are the strongest intermolecular forces in the crystal lattice, molecules with smaller dipole moments will have higher vapor pressures. Sub-limation is thus generally a property of relatively nonpolar substances having fairly symmetrical structures. Van der Waals attractive forces are also important but less so than electrostatic ones. Van der Waals forces increase in magnitude with increasing molecular weight, and large molecules, even if symmetrical, tend not to sublime.

FIGURE 2.55   Miniscale sublimation apparatus: (a) using water as the coolant; (b) using low-temperature coolants.

FIGURE 2.56
Simple sublimation apparatus:
(a) test tube sublimator; (b) filter-flask sublimator.

In order to purify a compound by sublimation, it must have a relatively high vapor pressure, and the impurities must have vapor pressures significantly lower than the compound being purified. If the impurities in a compound have similar vapor pressures, recrystallization (Sec. 3.2) or column chromatography (Sec. 6.2) may be used to purify the compound. Since few organic solids exhibit vapor pressures high enough to sublime at atmospheric pressure, most sublimations are performed at reduced pressure; this is analogous to using vacuum distillation for high-boiling liquids.

Two common types of sublimation apparatus are shown in Figure 2.55. In each case there is a chamber that may be evacuated using a water aspirator, house vacuum, or vacuum pump and a cold-finger in the center of the vacuum chamber to provide a surface upon which the sublimed crystals may form. The cold-finger is cooled by water (Fig. 2.55a) or another cooling medium such as ice/water or Dry Ice/acetone (Fig. 2.55b). A simple sublimation apparatus may be assembled inexpensively from two test tubes, one having a side arm and rubber stoppers (Fig. 2.56a). In place of the test tube with a sidearm, a small vacuum filtration flask may also be used (Fig. 2.56b). The cold-finger may be cooled either with running water or with ice chips. If the cold-finger is cooled with running water, the water hoses must be securely attached to the inlet and outlet of the condenser by means of a piece of twisted copper wire or a hose clamp.

In order to purify an impure substance by sublimation, it is first placed at the bottom of the sublimation chamber. The sample is then heated under reduced pressure, using an oil or sand bath or a small flame, to a temperature *below the melting point* of the solid. The solid will be vaporized and transferred via the vapor phase to the surface of the cold-finger, where it condenses directly as a pure material to the solid state. After the sublimation is complete, the pressure must be released carefully to avoid dislodging the crystals from the cold-finger with a surge of air. For similar reasons, care must be exercised when removing the cold-finger from the sublimation apparatus. The pure crystals are scraped from the cold-finger with a spatula.

2.21   Extraction

A common technique for isolating and purifying the product of a chemical reaction involves liquid-liquid extraction, or simply **extraction**; its theory and applications are discussed in Chapter 5. This process involves transferring the solute from one solvent into another, based upon its greater solubility in the second. The two solvents must be immiscible and form two distinct layers, and in general one layer is aqueous and the other is an organic solvent such as diethyl ether, hexane, or dichloromethane. Depending upon the amounts of material, the physical separation of the two immiscible phases will be performed in **separatory funnels** or **conical vials**.

EXTRACTIONS USING SEPARATORY FUNNELS: MINISCALE TECHNIQUE

Separatory funnels are available in many different shapes, ranging from almost spherical to an elongated pear shape (Fig. 2.57); the more elongated the funnel, the longer the time required for the two liquid phases to separate after the funnel is shaken. Although separatory funnels may be equipped with either a glass or Teflon stopcock, the latter does not require lubrication and thus is the preferred type because the solutions being separated do not get contaminated with stopcock grease. Separatory funnels are most commonly used during work-up procedures after completion of a chemical reaction. For example, they are used for extracting the desired product from one immiscible liquid phase into another (Secs. 5.2 and 5.4) and for "washing" organic layers to remove undesired substances such as acids or bases from the desired organic compound.

There are a number of general guidelines for using separatory funnels that merit discussion:

1. **Filling Separatory Funnels.** The stopcock should be closed and a clean beaker placed under the funnel before any liquids are added to the funnel in case the stopcock leaks or is not completely closed. A separatory funnel should never be more than three-quarters full, especially when doing an extraction. The upper opening of the funnel is then stoppered either with a ground-glass, plastic, or rubber stopper; most separatory funnels are now fitted with a ground-glass or plastic stopper.
2. **Holding and Using Separatory Funnels.** If the contents of the funnel are to be shaken, it is held in a specific manner. If the user is right-handed, the stopper should be placed against the base of the index finger of the left hand and the funnel grasped with the first two fingers and the thumb. The thumb and the first two fingers of the right hand can then be curled around the stopcock (Fig. 2.58a). Holding the funnel in this manner permits the stopper and the stopcock to be held tightly in place during the shaking process. A left-handed person might find it easier to use the opposite hand for each position.
3. **Shaking Separatory Funnels.** A separatory funnel and its contents should be shaken as shown in Figure 2.58b to mix the immiscible liquids as intimately as possible. The shaking process increases the surface area of contact between the immiscible liquids so that the equilibrium distribution of the solute between the two layers will be attained quickly; however, overly vigorous or lengthy shaking may produce undesired **emulsions** (discussed below).

The funnel must be vented every few seconds to avoid the buildup of pressure within the funnel. Venting is accomplished by *inverting the funnel with the stopcock pointing upward and away from you and your neighbors* and slowly opening it to release any pressure (Fig. 2.58c). If the funnel is not carefully vented, liquid may be violently expelled, covering you and your laboratory partners with the contents. Venting is particularly important when using volatile, low-boiling solvents such as

FIGURE 2.57
Separatory funnels: (a) pear-shaped; (b) conical.

(a)   (b)

(a)

(b)

Point away from
yourself and others

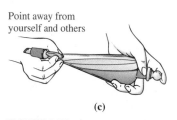

(c)

FIGURE 2.58
(a) Proper method for holding separatory funnel. (b) Shaking a separatory funnel. (c) Venting a separatory funnel.

diethyl ether or methylene chloride; it is also necessary whenever an acid is neutralized with either sodium carbonate or sodium bicarbonate, since CO_2 is produced. If the funnel is not vented frequently, the stopper may be accidentally blown out; under extreme circumstances the funnel might blow up. Any sudden release of pressure is likely to result in the contents of the flask being lost and spattered on you and your co-workers. The funnel may be vented simply by holding it as described previously and opening the stopcock by twisting the fingers curled around it without readjusting your grip on the funnel.

At the end of the period of shaking of (1–2 min are usually sufficient if the shaking is vigorous), the funnel is vented a final time. It is then supported on an iron ring (Fig. 2.59), and the layers are allowed to separate. When an iron ring is used to support a separatory funnel, the ring should be covered with a length of rubber tubing to prevent breakage. This may be accomplished by slicing the tubing along its side and slipping it over the ring. Copper wire may be used to fix the tubing permanently in place. The lower layer in the separatory funnel is then carefully dispensed into a flask through the stopcock while the interface between the two layers is watched. If small quantities of insoluble material collect at the boundary between the layers and make it difficult to see this interface, it is best to remove these solids with the *undesired* liquid layer; a small amount of the desired layer is inevitably lost by this procedure. An alternative procedure is to remove the solids by gravity or vacuum filtration before separating the layers.

4. **Layer Identification.** It is important to ascertain which of the two layers in a separatory funnel is the aqueous layer and which is the organic; this may be easily accomplished with a little care and thought. Since the layers will usually separate so that the denser solvent is on the bottom, knowledge of the densities of the liquids being separated provides an important clue for this identification. This generalization is not foolproof, however, because a high concentration of a solute in one layer may reverse the relative densities of the two liquids. You must not confuse the identity of the two layers in the funnel and then discard the layer containing your product. *Both layers should always be saved until there is no doubt about the identity of each and the desired product has been isolated.*

Since one of the layers is usually aqueous and the other is organic, there is a simple and foolproof method to identify the two layers. Withdraw a few drops of the upper layer with a pipet and add these drops to about 0.5 mL of water in a test tube. If the upper layer is aqueous, these drops will be miscible with the water in the test tube and will dissolve, but if the upper layer is organic, the droplets will not dissolve and will remain visible.

5. **Emulsions.** Occasionally the two immiscible liquids will not separate cleanly into two distinct layers after shaking because of an **emulsion** that results from a colloidal mixture of the two layers. If prior experience leads you to believe that an emulsion might form, you should avoid shaking the funnel too vigorously; instead, swirl it gently to mix the layers. Encountering an emulsion during an experiment can be extraordinarily frustrating, because there are no infallible convenient procedures for breaking up emulsions. An emulsion left unattended for an extended period of time sometimes separates. However, it is usually more expedient to attempt one or more of the following remedies:

   a. Add a few milliliters of a *saturated* solution of aqueous sodium chloride, commonly called brine, to the funnel and gently reshake the contents. This increases the ionic strength of the water layer, which helps force the organic material into the organic layer. This process can be repeated, but if it does not work the second time, other measures must be taken.

FIGURE 2.59
Separatory funnel positioned on iron ring with beaker located to catch liquid if funnel leaks.

b. Filter the heterogeneous mixture by vacuum filtration through a thin pad of a filter-aid (Sec. 2.17), and return the filtrate to the separatory funnel; if a filter-aid is not used, the pores of the filter paper may become clogged and the filtration will be slow. Sometimes emulsions are caused by small amounts of gummy organic materials whose removal will often remedy the problem.

c. Add a *small* quantity of water-soluble detergent to the mixture and reshake the mixture. This method is not as desirable as the first two techniques, particularly if the desired compounds are in the water layer, because the detergent adds an impurity that must be removed later.

d. Sometimes intractable emulsions that appear to be stabilized by small trapped air bubbles are encountered during the work-up of phase-transfer reactions. If the separatory funnel is thick-walled, apply a gentle vacuum with a water aspirator to speed the separation of the phases.

e. If all these procedures fail, it may be necessary to select a different extraction solvent.

EXTRACTIONS USING CONICAL VIALS AND CENTRIFUGE TUBES: MICROSCALE TECHNIQUE

The volumes of solvent are too small to use separatory funnels for extractions when reactions are conducted on a microscale level. In these instances, conical vials may be employed for volumes up to about 4 mL, and screw-cap centrifuge tubes may be used for volumes up to approximately 10 mL (Fig. 2.60). To avoid accidental spills place the conical vial or centrifuge tube in a small beaker when you are not handling it.

Before using either a conical vial or a centrifuge tube for an extraction, ensure that it does not leak when shaken. This may be done easily by placing about 1 mL of water in the vial or tube, and screwing the cap on securely. Shake the vial or tube vigorously and check for leaks.

(a)   (b)

FIGURE 2.60
Apparatus for microscale extraction: (a) conical vial; (b) screw-cap centrifuge tube.

Most extractions involve an aqueous and an immiscible organic phase, and it is important to mix the two layers thoroughly when performing an extraction. This may be achieved in several ways, the simplest of which is to shake the vial or tube vigorously. Carefully vent the vial or tube by loosening the cap to release any pressure. A vortex mixer works well if one is available in your laboratory. Adequate mixing can sometimes be achieved by stirring the mixture with a magnetic spinvane or a microspatula for 5–10 min. Another technique involves drawing the mixture into a Pasteur pipet and rapidly squirting it back into the vial or tube; be careful not to squirt so hard that the mixture spills out of the vial. Repeat this operation several times to ensure adequate extraction. Multiple extractions with fresh solvent should be performed to obtain maximum recovery of your product.

Depending upon the solvent used, the organic phase may be either the upper or lower layer, and a slightly different procedure is used for separating each. The more common situation is one in which the organic solvent is less dense than water, and it is this layer that is removed. For example, if you extract an aqueous solution with diethyl ether (ether), the organic layer is the upper one because ether is less dense than water. Alternatively, when a solvent that is heavier than water, such as dichloromethane, is used for the extraction, the lower layer is the organic layer, and it is the one removed.

Microscale Technique for Removing the Upper Organic Layer. The procedure for removing the upper organic layer using conical vials is outlined in Figure 2.61; centrifuge tubes may be substituted for the conical vials for larger volumes.

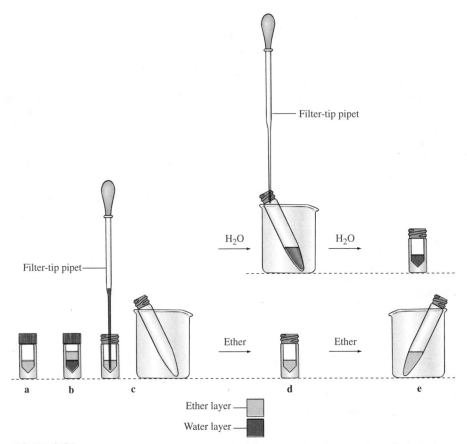

FIGURE 2.61 Extracting an aqueous solution using an organic solvent such as diethyl ether that is less dense than water: (a) the desired product is in the aqueous solution; (b) the aqueous phase is extracted with diethyl ether; (c) the lower (aqueous) phase is withdrawn; (d) the aqueous phase is transferred to a test tube; the ether layer remains in the conical vial; (e) the ether phase is transferred to a test tube, and the aqueous phase is returned to the extraction vial.

1. Place the aqueous phase containing the dissolved product in a 5-mL conical vial (Fig. 2.61a).
2. Add about 1 mL of diethyl ether, cap the vial securely, and mix the contents by vigorous shaking or one of the other techniques outlined previously. Unscrew the cap slightly to vent the pressure in the vial. Allow the phases to separate completely so you can see the two layers. If you are using centrifuge tubes, centrifugation may facilitate this separation. The diethyl ether phase is the upper layer (Fig. 2.61b).
3. Place an empty test tube or conical vial adjacent to the extraction vial. You will need a beaker or test tube rack to hold a test tube. Attach a rubber bulb to a Pasteur filter-tip pipet (Fig. 2.9a), squeeze the bulb, and insert the tip of the pipet into the vial until it touches the bottom. Slowly withdraw the lower aqueous layer into the pipet while keeping the tip of the pipet on the bottom of the conical vial (Fig. 2.61c). Be sure to avoid transferring any of the organic layer or any emulsion that may remain at the interface of the two phases. Although a regular Pasteur pipet may also be used for this transfer, a filter-tip pipet is easier to control when removing small volumes of liquid from the vial.

4. Transfer the aqueous phase you have drawn into the pipet into the empty test tube or conical vial leaving the ether layer behind in the extraction vial (Fig. 2.61d). It is best to position the second container as close as possible to the extraction vial to avoid messy transfers.

5. Use a clean filter-tip pipet to transfer the ether layer from the extraction vial to a test tube for storage. Although a regular Pasteur pipet may also be used for this transfer, a filter-tip pipet is easier to control when removing small volumes of volatile organic solvents such as diethyl ether from the vial. Return the aqueous layer to the extraction vial using a Pasteur pipet (Fig. 2.61e).

6. Repeat steps 2 to 5 one or two times to achieve complete extraction of the desired material from the aqueous layer. Dry the organic layer with a suitable drying agent such as anhydrous sodium sulfate or magnesium sulfate (Sec. 2.25).

Microscale Technique for Removing the Lower Organic Layer. The procedure for removing the lower organic layer using conical vials is outlined in Figure 2.62; centrifuge tubes may be substituted for the conical vials for larger volumes.

FIGURE 2.62   Extracting an aqueous solution using an organic solvent such as dichloromethane (CH_2Cl_2) that is denser than water: (a) the desired product is in the aqueous solution; (b) the aqueous phase is extracted with dichloromethane; (c) the lower (organic) phase is withdrawn; (d) the organic phase is transferred to a test tube; the aqueous phase remains in the conical vial.

1. Place the aqueous phase containing the dissolved product in a 5-mL conical vial (Fig. 2.62a).
2. Add about 1 mL of dichloromethane, cap the vial securely, and mix the contents by vigorous shaking or one of the other techniques outlined previously. Unscrew the cap slightly to vent the pressure in the vial. Allow the phases to separate completely so you can see the two layers. If you are using centrifuge tubes, centrifugation may facilitate this separation. The dichloromethane layer is the lower layer. Tapping the vial or gentle stirring with a microspatula may also help the layers separate (Fig. 2.62b).
3. Place an empty test tube or conical vial adjacent to the extraction vial. You will need a beaker or test tube rack to hold a test tube. Attach a rubber bulb to a Pasteur filter-tip pipet (Fig. 2.9a), squeeze the bulb, and insert the tip of the pipet into the vial until it touches the bottom. Slowly withdraw the lower organic layer into the pipet while keeping the tip of the pipet on the bottom of the conical vial (Fig. 2.62c). Be sure to avoid transferring any of the aqueous layer or any emulsion that may remain at the interface of the two phases. Although a regular Pasteur pipet may also be used for this transfer, a filter-tip pipet is easier to control when removing small volumes of volatile organic solvents such as dichloromethane from the vial.
4. Transfer the organic phase you have drawn into the pipet into the *dry* empty test tube or *dry* conical vial leaving the aqueous layer behind in the extraction vial (Fig.2.62d). It is best to position the second container as close as possible to the extraction vial to avoid messy transfers.
5. Repeat steps 2 to 4 one or two times to achieve complete extraction of the desired material from the aqueous layer. Dry the organic layer with a suitable drying agent such as anhydrous sodium sulfate or magnesium sulfate (Sec. 2.25).

Layer Identification. It is important to ascertain which of the two layers in a conical vial is the aqueous layer and which is the organic; this may be easily accomplished with a little care and thought. Since the layers will usually separate so that the denser solvent is on the bottom, knowledge of the densities of the liquids being separated provides an important clue for this identification. This generalization is not foolproof, however, because a high concentration of a solute in one layer may reverse the relative densities of the two liquids. You must not confuse the identity of the two layers in the conical vial or test tube and then discard the layer containing your product. *Both layers should always be saved until there is no doubt about the identity of each and the desired product has been isolated.*

Since one of the layers is usually aqueous and the other is organic, there is a simple and foolproof method to identify the two layers: Add a few drops of water to the layer you believe to be the aqueous one. Watch closely as you add the water drops to see whether they dissolve and increase the volume of the layer. If the drops form a new layer on either the top or bottom of the layer you are testing, then the test layer is organic. Similarly, you can test to see if a layer is organic by adding a few drops of the organic solvent you are using, for example diethyl ether or dichloromethane. If the volume of the layer increases, it is organic, but if a new layer forms, the layer is aqueous.

2.22 Heating Under Reflux

The term **heating under reflux** means that a reaction mixture is heated at its boiling point in a flask equipped with a **reflux condenser** to allow continuous return of the volatile materials to the flask; in this manner no solvent or reactant is removed or lost.

Since the reaction may be conducted at higher temperatures using this technique, less time is required for its completion.

MINISCALE APPARATUS

A typical apparatus used for heating a miniscale reaction under reflux is shown in Figure 2.63a (See Sec. 2.4). The solvent, if any, and reactants are placed in a boiling flask, together with a stirbar; this flask should be set up about 15 cm (6 in.) or more from the bench top to allow for fast removal of the heating source if necessary. The flask is then fitted with a reflux condenser with water running slowly through it; the water hoses should be secured to the inlet and outlet of the condenser with copper wire or a hose clamp. *Do not stopper the top of the condenser, since a closed system should never be heated.* The heating source is then placed under the flask, and the contents are slowly heated to the boiling point of the mixture; slow heating makes it possible to control any sudden exothermicity more readily. The volatile components of the mixture vaporize and reliquefy in the condenser, and the condensate returns to the boiling flask,

FIGURE 2.63   (a) Miniscale apparatus for heating a reaction mixture under reflux.
(b) Miniscale apparatus for a reaction being heated under reflux and equipped for adding a liquid reagent or solution.

where the process continues. A **reflux ring** appears in the condenser at the boundary of the two zones, and the condenser appears dry above this ring. The vapors should rise 3–6 cm (1–2 in.) above the bottom of the condenser. If they rise to the top of the condenser or are ''seen'' being emitted into the room, then too much heat is being supplied to the flask, the flow of water through the condenser is insufficient, or the condenser may be too small to permit adequate cooling of the vapors. The problem may usually be corrected by lowering the temperature of the heat source or adjusting the flow of water through the condenser. Upon completion of the reaction, the heat is removed, and the flask is allowed to cool to room temperature. This ensures that all of the material in the condenser has returned to the flask.

Many chemical reactions are executed by adding all the reagents, including solvent and catalyst, if any, to a reaction flask and heating the resulting mixture under reflux; however, sometimes one of the reactants must be added to the reaction during reflux. For example, if a reaction is highly exothermic, its rate may be controlled by adding one of the reagents slowly. Other reactions may require that one of the reagents be present in high dilution to minimize the formation of side products. Figure 2.63b shows a typical apparatus that allows a liquid reagent or solution to be added to the reaction flask using an addition funnel. The standard-taper separatory funnel available in many glassware kits can serve as an addition funnel, but there must be some provision for air to be admitted above the surface of the liquid to equalize the pressure inside; this can be accomplished by *not* placing the stopper on top of the funnel or by inserting a small piece of filter paper or copper wire between the stopper and the funnel. The flask can be heated to reflux during the addition, or the temperature in the flask can be maintained at room temperature or below with an ice-water bath and heated under reflux following completion of the addition.

MICROSCALE TECHNIQUE

The components of the apparatus for heating a reaction under reflux on the microscale are the same as for a preparative scale reaction. A common set up is shown in Figure 2.64. It is generally best to use a water-jacketed condenser to prevent all of the vapors

Glasswool

Drying tube (optional)

Clamps

Water out

Condenser

Thermometer

Water in

Conical reaction vial

Spinvane

Heating source

Magnetic stirrer

Laboratory jack

FIGURE 2.64
Microscale apparatus for heating a reaction mixture under reflux.

from escaping; a short air condenser is frequently inadequate for the more volatile organic solvents as diethyl ether and dichloromethane.

2.23 Gas Traps

Some organic reactions release noxious gases that should not escape into the laboratory atmosphere. Such reactions should be performed in the hood, if space is available. Alternatively, a **gas trap** can be used.

MINISCALE

If the gas being generated is water-soluble, the trap shown in Figure 2.65a may be suitable. In this example, the vapors are carried into a stemmed funnel whose rim is just above or resting on the surface of the water. A second modification evacuates the undesired water-soluble gas through a house vacuum or water aspirator protected by a trap (Fig. 2.65b). This style of gas trap uses a vacuum adapter attached at the top of a reflux condenser, for example. If the reaction being performed is sensitive to atmospheric moisture, the adapter can be filled with anhydrous calcium chloride that dries the air passing through it. Only a slight vacuum is needed to make this type of trap effective, so do *not* turn the house vacuum or water aspirator on full force.

MICROSCALE

The style of trap portrayed in Figure 2.65a can also be used with a microscale apparatus. A typical apparatus of this sort is shown in Figure 2.66a. A simplified analog is shown in Figure 2.66b, wherein the water-soluble gas is trapped in the water covering the surface of a cotton or glasswool packing. In some cases, simply putting moistened cotton or glasswool in a drying tube attached to the apparatus (Fig. 2.66c) will suffice to trap the gas being evolved.

FIGURE 2.65   Gas traps for miniscale apparatus: (a) trapping water-soluble gas in water; (b) removing gases through water aspirator.

FIGURE 2.66   Gas traps for microscale apparatus: (a) trapping water-soluble gas in water; (b) trapping water-soluble gas in moistened cotton or glasswool; (c) simplified modification for trapping water-soluble gas in moistened cotton or glasswool.

2.24   Drying Agents

Drying a reagent, solvent, or product will be encountered at some stage of nearly every reaction performed in the organic chemistry laboratory. The techniques of drying solids and liquids are described in this and the following sections.

DRYING AGENTS AND DESICCANTS

Most organic liquids are distilled at the end of the purification process, and any residual moisture that is present may react with the compound during the distillation; water may also co- or steam-distil with the liquid and contaminate the distillate. In order to remove these small traces of moisture before distillation, **drying agents**, sometimes called **desiccants**, are used. There are two general requirements for a drying agent: (1) neither it nor its hydrolysis product may react chemically with the organic liquid being dried, and (2) it must be *completely* and *easily* removed from the dry liquid. A drying agent should also be efficient so that the water is removed by the desiccant in a reasonably short period of time.

Some commonly used drying agents and their properties are listed in Table 2.1. These desiccants function in one of two ways: (1) the drying agent interacts *reversibly* with water by the process of adsorption or absorption (Eq. 2.1); or (2) it reacts irreversibly with water by serving as an acid or a base.

With drying agents that function by reversible hydration, a certain amount of water will remain in the organic liquid in equilibrium with the hydrated drying agent. The lesser the amount of water left at equilibrium, the greater the efficiency of the desiccant. A drying agent that forms a hydrate (Eq. 2.1) must be *completely* removed by gravity filtration (Sec. 2.17) or by decantation (Sec. 2.19) *before* the dried liquid is distilled, since most hydrates decompose with loss of water at temperatures above 30–40 °C.

$$\text{Drying agent (solid)} + x\,H_2O\text{ (liquid)} \rightleftharpoons \text{Drying agent} \cdot x\,H_2O\text{ (solid)} \qquad \textbf{(2.1)}$$
$$\quad\ \text{Anhydrous} \qquad\qquad\qquad\qquad\qquad\qquad\qquad\quad \text{Hydrate}$$

Table 2.1	Table of Common Drying Agents, Their Properties, and Uses	
Drying Agent	Acid-Base Properties	Comments[1]
$CaCl_2$	Neutral	High capacity and fast action with reasonable efficiency; good preliminary drying agent; readily separated from dried solution because $CaCl_2$ is available as large granules; cannot be used to dry either alcohols and amines (because of compound formation) or phenols, esters, and acids (because drying agent contains some $Ca(OH)_2$).
Na_2SO_4	Neutral	Inexpensive, high capacity; relatively slow action and low efficiency; good general preliminary drying agent; preferred physical form is that of small granules, which may be easily separated from the dry solution by decantation or filtration.
$MgSO_4$	Weakly acidic	Inexpensive, high capacity, rapid drying agent with moderate efficiency; excellent preliminary drying agent; requires filtration to remove drying agent from solution.
H_2SO_4	Acidic	Good for alkyl halides and aliphatic hydrocarbons; cannot be used with even such weak bases as alkenes and ethers; high efficiency.
P_2O_5	Acidic	See comments under H_2SO_4; also good for ethers, aryl halides, and aromatic hydrocarbons; generally high efficiency; preliminary drying of solution recommended; dried solution can be distilled from drying agent.
CaH_2	Basic	High efficiency with both polar and nonpolar solvents, although inexplicably it fails with acetonitrile; somewhat slow action; good for basic, neutral, or *weakly* acidic compounds; cannot be used for base-sensitive substances; preliminary drying of solution is recommended; dried solution can be distilled from drying agent. *Caution:* Hydrogen gas is evolved with this drying agent.
Na or K	Basic	Good efficiency but slow action; cannot be used on compounds sensitive to alkali metals or to base; care must be exercised in destroying excess drying agent; preliminary drying *required*; dried solution can be distilled from drying agent. *Caution:* Hydrogen gas is evolved with this drying agent.
BaO or CaO	Basic	Slow action but high efficiency; good for alcohols and amines; cannot be used with compounds sensitive to base; dried solution can be distilled from drying agent.
Molecular Sieve 3A or 4A[2]	Neutral	Rapid and highly efficient; preliminary drying recommended; dried solution can be distilled from drying agent if desired. *Molecular sieves* are aluminosilicates, whose crystal structure contains a network of pores of uniform diameter; the pore sizes of sieves 3A and 4A are such that only water and other small molecules such as ammonia can pass into the sieve; water is strongly adsorbed as water of hydration; hydrated sieves can be reactivated by heating at 300–320 °C under vacuum or at atmospheric pressure.

[1] Capacity, as used in this table, refers to the amount of water that can be removed by a given weight of drying agent; efficiency refers to the amount of water, if any, in equilibrium with the hydrated desiccant.

[2] The numbers refer to the nominal pore size, in Ångström units, of the sieve.

Drying agents that remove water by an irreversible chemical reaction are very efficient, but they are generally more expensive than other types of drying agents. Such drying agents are sometimes more difficult to handle and are normally used to remove *small* quantities of water from reagents or solvents prior to a chemical reaction. For example, phosphorus pentoxide, P_2O_5, removes water by reacting vigorously with it to form phosphoric acid (Eq. 2.2). Desiccants such as calcium hydride (CaH_2) and sodium (Na) metal also react vigorously with water. When CaH_2 or Na metal is used as a drying agent, hydrogen gas is evolved, and appropriate precautions must be taken to vent the hydrogen and prevent buildup of this highly flammable gas.

$$P_2O_5 \text{ (solid)} + 3\,H_2O \text{ (liquid)} \longrightarrow 2\,H_3PO_4 \qquad\qquad \textbf{(2.2)}$$

Of the drying agents listed in Table 2.1, *anhydrous* calcium chloride, sodium sulfate, and magnesium sulfate will generally suffice for the needs of this introductory laboratory course. Both sodium sulfate and magnesium sulfate have high capacities and absorb a large amount of water, but magnesium sulfate dries a solution more completely. Calcium chloride has a low capacity, but it is a more efficient drying agent than magnesium sulfate. Do *not* use an unnecessarily large quantity of drying agent when drying a liquid, since the desiccant may adsorb or absorb the desired organic product along with the water. Mechanical losses on filtration or decantation of the dried solution may also become significant. The amount of drying agent required depends upon the quantity of water present, the capacity of the drying agent, and the amount of liquid to be dried.

2.25 | Drying Organic Solutions

When an organic solvent is shaken with water or an aqueous solution during an extraction procedure, it will contain some dissolved water. The amount of water varies with the organic solvent. Of the solvents commonly used for extraction, diethyl ether dissolves the most water. If this water is not removed prior to evaporation of the organic solvent, significant quantities of water will remain in the product and may complicate further purification. It is therefore necessary to dry the solution using a suitable drying agent, usually *anhydrous* sodium sulfate or magnesium sulfate.

MINISCALE TECHNIQUE

Place the organic liquid to be dried in an Erlenmeyer flask of suitable size so that it will be no more than half-filled with liquid. Start by adding a small spatula-tip full of drying agent such as *anhydrous* sodium sulfate or magnesium sulfate and swirl the flask gently. Swirling increases the surface area for contact between the solid and liquid phases and generally facilitates drying. If the drying agent 'clumps' or if liquid still appears cloudy after the solid has settled to the bottom of the flask, add more drying agent and swirl again. Repeat this process until the liquid appears clear and some of the drying agent flows freely upon swirling the mixture. An amount of drying agent that covers the bottom of the flask should normally be sufficient. After drying is complete, remove the drying agent either by gravity filtration (Sec. 2.17) or by decantation (Sec. 2.19). Rinse the drying agent once or twice with a small volume of the organic solvent and transfer the rinse solvent to the organic solution.

MICROSCALE TECHNIQUE

Place the organic liquid to be dried in a conical vial or a test tube of suitable size so that it will be no more than half-filled with liquid. Start by adding a microspatula-tip full of drying agent such as *anhydrous* sodium sulfate or magnesium sulfate; granular anhydrous sodium sulfate is the most convenient for microscale manipulations. Stir the mixture with a microspatula. Stirring increases the surface area for contact between the solid and liquid phases and generally facilitates drying. If the drying agent 'clumps' or if the liquid still appears cloudy after the solid has settled to the bottom of the flask, add more drying agent and stir again. Repeat this process until the liquid appears clear and some of the drying agent flows freely upon stirring the mixture. After drying is complete, use a *dry* Pasteur filter-tip pipet (Fig. 2.9a) or a *dry* Pasteur pipet to transfer the organic solution to a *dry* conical vial or test tube. Rinse the drying agent once or twice with a small volume of the organic solvent and transfer the rinse solvent to the organic solution.

2.26 | Drying Solids

Solid organic compounds must be dried because the presence of water or organic solvents will affect their weight, melting point, quantitative elemental analysis, and spectra. Since proton sources must be excluded from some reactions, it is also necessary to remove all traces of moisture or protic solvents from a solid prior to performing such a reaction.

A solid that has been recrystallized from a volatile organic solvent can usually be dried satisfactorily by allowing it to air-dry at room temperature, provided it is not **hygroscopic** and thus absorbs moisture from the air. After the solid is collected on a Büchner or Hirsch funnel fixed on a filter flask, it is first pressed as dry as possible with a clean spatula or cork while air continues to be pulled through the funnel and solid by use of a water aspirator or house vacuum (Sec. 2.17). The solid is then spread on a piece of filter paper, which absorbs the excess solvent, or on a clean watchglass and allowed to stand overnight or longer.

Water is sometimes removed from a solid by dissolving the solid in a suitable organic solvent such as chloroform or toluene, removing any water by azeotropic distillation, and then recovering the solid by removal of the solvent. However, water is more commonly removed from organic solids using desiccators containing desiccants such as silica gel, phosphorus pentoxide, calcium chloride, or calcium sulfate. The desiccator may be used at atmospheric pressure or under a vacuum; however, if a vacuum is applied, the desiccator must be enclosed in a metal safety cage or wrapped with electrical or duct tape. Desiccators or tightly stoppered bottles containing one of these desiccants may also be used to store dry solids contained in small vials.

If the sample is hygroscopic or if it has been recrystallized from water or a high-boiling solvent, it must be dried in an oven operating at a temperature *below* the melting or decomposition point of the sample. The oven-drying process can be performed at atmospheric pressure or under vacuum. Air-sensitive solids must be dried either in an inert atmosphere, such as in nitrogen or helium, or under vacuum. Samples to be submitted for quantitative elemental analysis are normally dried to constant weight by heating them under vacuum.

2.27 Drying Tubes

(a)   **(b)**

FIGURE 2.67
Drying tubes: (a) straight; (b) bent.

Frequently it is necessary to protect a reaction mixture from moisture. There are advanced techniques for doing this that involve performing the reaction under an inert, dry atmosphere of nitrogen or argon gas, but a simple and moderately effective procedure simply utilizes a drying tube containing a drying agent. Two types of drying tubes are shown in Figure 2.67. The straight tube (Fig. 2.67a) and the bent tube are commonly found in glassware kits for miniscale and microscale experiments, respectively. A drying tube is prepared by placing a loose plug of glasswool at the bottom of the tube and filling the tube with a *granular* desiccant such as anhydrous calcium chloride or calcium sulfate. The drying agent commonly contains a blue indicator that turns pink when it is no longer effective. Although it is not necessary to cap the straight tube with glasswool, a loose plug of glasswool should be placed in the other end of the bent drying tube to prevent the drying agent from falling out. Neither the desiccant nor the glasswool should be packed too tightly.

2.28 Drying Apparatus

Reactions involving the use of water-sensitive reagents must be performed under anhydrous conditions. Thus prior to conducting such a reaction and prior to introducing any reagents, as much moisture as possible must be removed from the apparatus itself.

The glass parts of an apparatus such as round-bottom flasks, conical vials, and glass syringes may be conveniently dried by placing them for 15–20 min in a drying oven heated at about 150 °C. Remove the glass components from the oven using a pair of tongs, and if possible place them in a desiccator to keep them dry as they cool to room temperature. After the glassware is cool, remove it from the desiccator and assemble the apparatus as quickly as possible to avoid reintroduction of moisture on the glass surface. If a desiccator is not available, then allow the glassware to cool only to the point where you can handle it, and assemble the apparatus as quickly as possible to avoid reintroduction of moisture on the glass surface.

An alternative method that may be used is to 'flame-dry' the apparatus. If you are instructed to use this technique, be sure that no one is working with flammable solvents in the room. The apparatus, which must have one opening to the atmosphere, is gently warmed with a Bunsen burner or microburner (Sec. 2.9), a process called 'flaming.' A heat gun (Sec. 2.9) may also be used to heat the apparatus. The apparatus is heated first at the point most remote from the opening to the atmosphere, and the flame is gradually worked toward the opening. The moisture in the apparatus will be driven through the opening. If a condenser is part of the apparatus, no water should be in or flowing through it. While the apparatus is still warm, a filled drying tube (Sec. 2.27) is inserted into the opening, and the system is allowed to cool to room temperature. As cooling occurs, air from the atmosphere is drawn into the apparatus through the drying agent, which removes the moisture, and the dry apparatus is ready to be used.

2.29 Evaporating Solvents

For many experiments and laboratory operations, it is necessary to remove the excess solvents to recover the product. Although an Erlenmeyer flask or beaker containing the solution could simply be left unstoppered in the hood until the solvent has evaporated, this is rather slow and impractical. A variety of techniques may be used to speed the process. Several other methods for removing solvents in miniscale experiments are shown in Figure 2.68, and the related microscale techniques are illustrated in Figure 2.69.

MINISCALE METHODS

The solvents in solutions contained in Erlenmeyer flasks can be evaporated by adding several boiling stones or a wooden boiling stick to the flask and heating the flask with a suitable heat source such as a steam bath, a hot plate, or a sand or oil bath so that the solvent boils gently. Magnetic stirring may also be used to prevent bumping. The evaporation rate can be increased by directing a stream of air or nitrogen into the flask (Fig. 2.68a), or by applying a vacuum over the flask with an inverted funnel to help draw away solvent vapors (Fig. 2.68b). The solution may also be placed in a filter flask and a wooden boiling stick; after the flask is subjected to a gentle vacuum, it is swirled over the heat source to facilitate smooth evaporation and reduce the possibility of bumping (Fig. 2.68c).

Another common method for removing solvent is by simple distillation (Sec. 2.13). Discontinue heating when only a small amount of solvent remains, and *do not overheat*

(a)   (b)

(c)

FIGURE 2.68 Miniscale techniques for evaporating solvents: (a) heating under a stream of air or nitrogen gas, (b) heating with removal of vapors by vacuum, and (c) heating under vacuum.

FIGURE 2.69   Microscale techniques for evaporating solvents. (a) Heating under a stream of air or nitrogen gas, (b) heating with removal of vapors by vacuum, and (c) heating under vacuum.

the stillpot, as your product may decompose if heated too strongly. The stillpot should be cooled to room temperature, and the receiving flask containing the distillate should be replaced with an empty round-bottom flask. The apparatus should then be set up for vacuum distillation (Fig. 2.40b) and attached to a vacuum source, such as a water aspirator, for a few minutes to remove the last traces of solvent.

MICROSCALE METHODS

The solvents in solutions contained in a conical vial or test tube can be evaporated by heating the conical vial or test tube with a suitable heat source such as a sand bath or aluminum block. To prevent bumping, the solution should be agitated with magnetic stirring or with a boiling stone. The heat should also be adjusted so the solvent does not boil vigorously. The evaporation rate can be increased by directing a stream of air or nitrogen into the conical vial or test tube (Fig. 2.69a), or by applying a vacuum over the conical vial or test tube with an inverted funnel to help remove solvent vapors (Fig. 2.69b). The solution may also be placed in a test tube with a sidearm and a wooden boiling stick; after the test tube is subjected to a gentle vacuum, it is swirled over the heat source to facilitate smooth evaporation and reduce the possibility of bumping (Fig. 2.69c).

Another common method for removing solvent is by simple microscale distillation (Sec. 2.13). Discontinue heating when only a small amount of solvent remains, and *do not overheat the stillpot*, as your product may decompose if heated too strongly. The stillpot should be cooled to room temperature. Remove the Hickman stillhead and water condenser, and insert an air condenser into the vial. Attach a thermometer adapter fitted with a glass tube to the top of the condenser and attach to a vacuum source (Fig. 2.70).

FIGURE 2.70   Microscale apparatus for removing solvent under reduced pressure.

To vacuum

Glass tube

Thermometer adapter

Clamps

Air condenser

Thermometer

Conical reaction vial

Spinvane

Heating source

Magnetic stirrer

Laboratory jack

If you use a water aspirator, be sure to place a trap between the condenser and the aspirator pump (Fig. 2.40b). While stirring, apply a gentle vacuum to the apparatus so the remainder of the organic solvent evaporates slowly under reduced pressure.

Rotary Evaporators. In a research laboratory and in many undergraduate laboratories, solvents are removed under reduced pressure using an apparatus called a **rotary evaporator** (Fig. 2.71). This is a device that is especially designed for the rapid evaporation

Condenser

To vacuum

Water out

Variable speed motor

Keck clip

Water in

Keck clip

Evaporation flask

Receiver

Hot water bath

FIGURE 2.71   Rotary ev

of solvents without bumping. A variable speed motor is used to rotate the flask containing the solvent being evaporated. While the flask is being rotated, a vacuum is applied and the flask is heated. The rotation of the flask spreads a thin film of the solution on the inner surface of the flask to accelerate evaporation. This motion also agitates the contents of the flask to reduce the problem of bumping. The rate of evaporation may be controlled by adjusting the vacuum, the temperature of the water bath, and the rate of rotation of the flask.

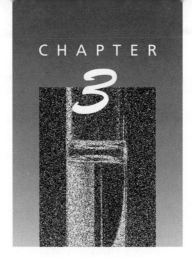

CHAPTER 3

Solids
Recrystallization and Melting Points

3.1 Introduction

The organic chemist usually works with substances that are in the liquid or solid state. The purpose of this and the succeeding two chapters is to present an introduction to the theory and practice of the most important methods for separation and purification of mixtures of organic compounds in these two physical states. As you read these chapters, you should also refer to the corresponding sections of Chapter 2 that describe the apparatus used for the various techniques. References to the appropriate sections have been provided for your convenience.

A pure compound is a homogeneous sample consisting only of molecules having the same structure. However, substances believed to be pure on the basis of certain criteria may actually contain small amounts of contaminants. Indeed, the presence of impurities in what were believed to be pure compounds has sometimes led to incorrect structural assignments and scientific conclusions. The possibility of making such errors was of particular concern prior to the advent of the powerful chromatographic (Chap. 6) and spectral (Chap. 8) techniques that have been developed since the 1950s. It is now relatively easy for the chemist to purify solids and liquids and to demonstrate their purity.

A compound formed in a chemical reaction or extracted from some natural source is rarely pure when initially isolated. For example, a chemical transformation intended to produce a single product almost invariably yields a reaction mixture containing a number of contaminants. These may include the products of side reactions proceeding concurrently with the main reaction, unchanged starting materials, inorganic materials, and solvents. Unfortunately, even chemicals purchased commercially are not always pure, owing to the expense of the needed purification process or to decomposition that may occur during storage.

Organic chemists devote considerable effort to the isolation of pure products, the ultimate goal being to obtain a substance that cannot be purified further by any known experimental techniques. This chapter focuses on the purification of solids by recrystallization and their characterization by the physical property of melting points.

3.2   Recrystallization

Recrystallization of solids is a valuable technique to master because it is one of the methods used most often for purification of solids. Other techniques for purifying solids include **sublimation** (Sec. 2.20), **extraction** (Chap. 5) and **chromatography** (Chap. 6). Nevertheless, even when one of these alternative methods of purification has been used, the solid material thus isolated may still be recrystallized to achieve the highest possible state of purity.

The process of recrystallization involves dissolution of the solid in an appropriate solvent at an elevated temperature and the subsequent re-formation of the crystals upon cooling, so that any impurities remain in solution. This technique, called **solution re-crystallization**, is discussed here. An alternative approach involves melting the solid in the absence of solvent and then allowing the crystals to re-form so that impurities are left in the melt. This method is seldom used in the organic laboratory because the crystals often form out of a viscous oil that contains the impurities and from which it is difficult to separate the desired pure solid. It is interesting to note, however, that this is the technique used to prepare the high-purity single crystals of silicon used in computer chips.

Almost all solids are *more* soluble in a *hot* than in a *cold* solvent, and solution crystallization takes advantage of this fact. Thus, if a solid is first dissolved in an amount of hot solvent insufficient to dissolve it when cold, crystals should form when the hot solution is allowed to cool. The extent of precipitation of the solid depends on the difference in its solubility in the particular solvent at temperatures between the extremes used. The upper extreme is determined by the boiling point of the solvent, whereas the lower limit is usually dictated by experimental convenience. For example, an ice-water bath is often used to cool the solution to 0 °C, whereas ice-salt and Dry Ice–acetone baths are commonly used to cool solutions to −20 °C and −78 °C, respectively (Sec. 2.10). The solid should be recovered with greater efficiency at these temperatures, provided the solvent itself does not freeze.

If the impurities present in the original solid mixture have dissolved and *remain* dissolved after the solution is cooled, isolation of the crystals that have formed should *ideally* provide pure material. Alternatively, the impurities may not dissolve at all in the hot solution and may be removed by filtration *before* the solution is cooled. The crystals that subsequently form should be purer than the original solid mixture. Solution recrystallization is seldom quite so simple in practice, but these two idealized generalizations do outline the basic principles of the technique.

Even after a solid has been recrystallized, it may still not be pure. Thus, it is important to determine the purity of the sample, and one of the easiest methods to do this is by determining the melting point of the solid. This technique is described in Section 3.3.

The technique of solution recrystallization involves the following steps:

1. **selection** of an appropriate solvent
2. **dissolution** of the solid to be purified in the solvent near or at its boiling point
3. **decoloration** with an activated form of carbon, if necessary, to remove colored impurities and **filtration** of the hot solution to remove insoluble impurities and the decolorizing carbon
4. **formation** of crystalline solid from the solution as it cools
5. **isolation** of the purified solid by filtration
6. **drying** the crystals

Each step of the sequence is discussed in the following subsections and representative experimental procedures are presented at the end of the discussion.

SELECTION OF SOLVENT

The choice of solvent is perhaps the most critical step in the process of recrystallization since the correct solvent must be selected to form a product of high purity and in good recovery or yield. Consequently a solvent should satisfy certain criteria for use in recrystallization: (**a**) The desired compound should be reasonably soluble in the *hot* solvent, about 1 g/20 mL (1 mg/20 μL) being satisfactory, and **insoluble** or *nearly* **insoluble** in the *cold* solvent. Note that the reference temperature for determination of the solubility in "cold" solvent is often taken to be room temperature. This combination of solute and solvent will allow dissolution to occur in an amount of solvent that is not unduly large and will also permit recovery of the purified product in high yield. A solvent having this type of solubility properties as a function of temperature would be said to have a favorable **temperature coefficient** for the desired solute. (**b**) Conversely, the impurities should either be insoluble in the solvent at all temperatures *or* must remain at least moderately soluble in the cold solvent. In other words, if the impurities are soluble, the temperature coefficient for them must be *unfavorable*; otherwise the desired product *and* the impurities would both crystallize simultaneously from solution. (**c**) The boiling point of the solvent should be low enough so that it can readily be removed from the crystals. (**d**) The boiling point of the solvent should generally be lower than the melting point of the solid being purified. (**e**) The solvent should not react chemically with the substance being purified.

The chemical literature is a valuable source of information about solvents suitable for recrystallizing known compounds. If the compound has *not* been prepared before, it is necessary to resort to trial-and-error techniques to find an appropriate solvent for recrystallization. The process of selection can be aided by consideration of some generalizations about solubility characteristics for classes of solutes. Polar compounds are normally soluble in polar solvents and insoluble in nonpolar solvents, for example, whereas nonpolar compounds are more soluble in nonpolar solvents. Such characteristics are summarized by the adage, "*like dissolves like*." Of course, although a highly polar compound is unlikely to be soluble in a hot, nonpolar solvent, it *may* be very soluble in a cold, very polar solvent. In this case, a solvent of intermediate polarity *may* be the choice for a satisfactory recrystallization.

The solvents commonly used in recrystallizations range widely in polarity, a property measured by the **dielectric constants** (ϵ) listed in Table 3.1. Those solvents with dielectric constants in the range of 2–3 are considered **nonpolar**, and those with constants above 10 as **polar**. Solvents in the 3–10 range are of intermediate polarity. Of the various solvents listed, *petroleum ether* deserves special mention because of its confusing common name. This solvent does *not* contain the *ether* functional group at all; rather it is a mixture of volatile aliphatic hydrocarbons obtained from the refining of petroleum. The composition and boiling point of the mixture depends on the particular distillation "cut" obtained. Thus, the boiling range of this type of solvent is usually given, as in the description, "petroleum ether, bp 60–80 °C (760 Torr)."

Occasionally a *mixture* of solvents is required for satisfactory recrystallization of a solute. The mixture is usually comprised of only two solvents; one of these dissolves the solute even when cold and the other one does not. The logic of this will become clear when reading about this technique under "Dissolution" (p. 91).

Table 3.1	**Common Solvents for Recrystallization[1]**				
Solvent	Boiling Point °C (760 Torr)	Freezing Point, °C[2]	Water Soluble	Dielectric Constant (ϵ)	Flammable
Petroleum ether[3]	Variable		No	1.9	Yes
Diethyl ether	35		Slightly	4.3	Yes
Dichloromethane[4]	41		No	9.1	No
Acetone[3]	56		Yes	20.7	Yes
Methanol[3]	65		Yes	32.6	Yes
Tetrahydrofuran	65		Yes	7.6	Yes
Ethyl acetate[3]	77		Yes	6.0	Yes
Ethanol (95%)[3]	78		Yes	24.6	Yes
Cyclohexane	81	6	No	1.9	Yes
Water[3]	100	0	N/A	78.5	No
1,4-Dioxane	101	11	Yes	2.2	Yes
Toluene[3]	111		No	2.4	Yes

[1] Benzene has been omitted from this list, owing to its toxicity. Cyclohexane can usually be substituted for it.
[2] Freezing points not listed are below 0 °C.
[3] Denotes solvents used most often.
[4] As a general rule, avoid the use of chlorocarbon solvents *if another equally good solvent is available*. Their toxicity is greater and their disposal is more difficult than for other types of solvents.

Chemical Structures of Common Recrystallization Solvents Found in Table 3.1

CH_2CH_3OH

Ethanol
(ethyl alcohol)

CH_3OH

Methanol
(methyl alcohol)

Cyclohexane

Toluene

$CH_3CH_2OCH_2CH_3$

Diethyl ether
(ether)

Tetrahydrofuran

1,4-Dioxane

CH_2Cl_2
Dichloromethane
(methylene chloride)

$CH_3COCH_2CH_3$
Ethyl acetate

CH_3CCH_3
Acetone

DISSOLUTION

This step may involve the handling of relatively large volumes of volatile solvents. Although most solvents used in the organic laboratory are of relatively low toxicity, it is prudent to avoid inhaling their vapors. For this reason, the following operations are best performed in a hood. If this is not possible, clamping an inverted funnel over the recrystallization flask and connecting the stem of the funnel to a source of a vacuum (Fig. 2.68b, p. 83) will help to minimize exposure. Another precaution is to choose the source of heat for the recrystallization carefully. As noted in Table 3.1, many solvents are highly flammable and should *never* be heated with an open flame; rather, water, steam, and oil baths or hot plates (Sec. 2.9) should be used in such cases.

The procedures for performing dissolutions in *single* solvents at the mini- and microscales differ slightly, as described below. Discussion of the strategies for using *mixed* solvents for recrystallizations is presented as well.

Miniscale Technique. The solid to be purified is weighed and placed in an Erlenmeyer flask of appropriate size. A beaker is *not* a suitable vessel for recrystallization because it has a relatively large surface area to which recrystallized product may stick, lowering the efficiency of recovery. Some crystals of the impure solid should always be retained, for they may be needed as **"seeds"** to induce crystallization. The flask should be equipped for magnetic stirring (Sec. 2.11) or contain either boiling stones or a wooden stick to prevent bumping of the solution while boiling.

A few milliliters of solvent are added to the flask, and the mixture is then heated to the boiling point (Sec. 2.9). More solvent is added to the mixture in *small* portions using a Pasteur pipet until the solid just dissolves. It is important to let boiling resume after each addition so that a minimum amount of solvent is used to effect dissolution; *using excessive amounts of solvent decreases the recovery of the solute.* If it necessary to perform a hot filtration (Sec. 2.17), it is prudent to add an additional 2–5% of solvent to prevent premature crystallization during this operation.

If adding solvent fails to dissolve any more solid, it is likely that insoluble impurities are present. These can be removed by hot filtration. Thus, to avoid using too much solvent and risking poor recovery of the purified solute, you should observe the dissolution process *carefully*. This is particularly important when only a relatively small quantity of solid remains, as this may be the material that is insoluble.

Microscale Technique. The protocol for dissolution at the microscale level mimics that at larger scales. The solid is weighed and placed in a test tube or Craig tube, with a few crystals of impure material being retained as "seeds" to induce crystallization. A boiling stone should be added to prevent bumping of the solution upon boiling. Alternatively, smooth boiling can be promoted by twirling a microspatula in the mixture.

A few drops of solvent, normally totaling no more than a milliliter are added, and the mixture is heated to the boiling point (Sec. 2.9). Additional solvent, as needed, is added dropwise using a Pasteur pipet or filter-tip pipet (Sec. 2.5) to the boiling mixture until the solid just dissolves. It is important to let boiling resume after each addition so that a minimum amount of solvent is used to effect dissolution; *using excessive amounts of solvent decreases the recovery of the solute.* If it necessary to perform a hot filtration (Sec. 2.17), it is prudent to add an additional 2–5% of solvent to prevent premature crystallization during this operation.

If adding solvent fails to dissolve any more solid, it is likely that insoluble impurities are present. These can be removed by hot filtration. Thus, to avoid using too much solvent and risking poor recovery of the purified solute, you should observe the dissolution process *carefully*. This is particularly important when only a relatively small quantity of solid remains, as this may be the material that is insoluble.

The same general approach used for single-solvent dissolution is followed at the mini- and microscale levels when mixed solvents are employed. However, there are two options for effecting dissolution once the solvents have been selected. In one, the solid to be purified is first dissolved in a *minimum* volume of the hot solvent in which it is soluble; the second solvent is then added to the *boiling* solution until it turns cloudy. The cloudiness signals initial formation of crystals, caused by the fact that addition of the second solvent results in a solvent mixture in which the solute is less soluble. Finally, more of the first solvent is added *dropwise* until the solution clears.

Two further aspects of this option should be noted. First, the solution must be cooled slightly below the lower boiling point before the second solvent is added *if* this solvent has a boiling point *lower* than the first; otherwise the addition of this solvent could cause sudden and vigorous boiling of the mixture and hot solvent might spew from the apparatus. Second, hot filtration should be performed if needed *before* addition of the second solvent; this will prevent crystallization during the filtration step. A potential disadvantage of this method for mixed solvent recrystallization is that unduly large volumes of the second solvent may be required if excessive amounts of the initial solvent have been used.

In the second option, the solute is added to the solvent in which it is insoluble, and the mixture is heated near the boiling point of the solvent; the second solvent is then added in small portions until the solid just dissolves. As with recrystallization from a single solvent, it is generally wise to add 2–5% of additional solvent to prevent premature crystallization of the solute during hot filtration, if this step is necessary. The use of this approach to mixed solvent recrystallization also has the disadvantage that using too much of the first solvent will require the addition of undesirably large volumes of the second solvent.

DECOLORATION AND HOT FILTRATION

After dissolution of the solid mixture, the solution may be colored. This signals the presence of impurities if the desired compound is known to be colorless. If the compound is colored, contaminants may alter the color of the solution; for example, impurities should be suspected if the substance is yellow but the solution is green. Of course, the decoloration step is unnecessary if the solution is colorless.

Colored impurities may often be removed by adding a small amount of decolorizing carbon (Sec. 2.8) to the hot, *but not boiling*, solution. Adding it to a boiling solution is likely to cause the liquid to froth over the top of the flask, resulting in the loss of product and possible injury. After the decolorizing carbon is added, the solution is heated to boiling for a few minutes while being continuously stirred or swirled to prevent bumping.

Completely removing decolorizing carbon by hot filtration (Sec. 2.17) may be difficult if the powdered rather than the pelletized form is used, because the finely divided solid particles may pass through filter paper or the cotton of a Pasteur filtering pipet (Sec. 2.5). If this occurs, the dark particles should be visible in the filtrate.

The following steps are recommended to assist in removing the decolorizing carbon during filtration. The hot decolorized solution is allowed to cool slightly below its boiling point, a small amount of a filter-aid such as Celite (Sec. 2.17) is added to *adsorb* the carbon, the mixture is briefly reheated to boiling, and then is subjected to hot filtration. It may be necessary to repeat this procedure if some of the decolorizing carbon still passes through the filter paper and remains in the filtrate.

This technique for decolorizing solutions works because colored impurities as well as the compound being purified are adsorbed on the surface of the carbon particles. For electronic reasons, the colored substances adsorb more strongly to the surface, and this factor, combined with the fact that the impurities are normally present in minor amounts, results in complete removal of the colored contaminants. Of course, because the desired product itself is adsorbed by the decolorizing carbon, less of it will be recovered if too much carbon is added.

Insoluble impurities, including dust and decolorizing carbon, if used, are removed by gravity filtration of the hot solution (Sec. 2.17); this step is not necessary if the hot solution is clear and homogeneous. Gravity filtration is normally preferred to vacuum filtration because the latter technique may cause cooling and concentration of the solution, owing to evaporation of the solvent, and this may result in premature crystallization. A short-stemmed or stemless glass funnel should be used to minimize crystallization in the funnel, and using fluted filter paper (Sec. 2.17) will minimize crystallization on the filter. To keep liquid from flowing over the top of the funnel, the top of the paper should not extend above the funnel by more than 1–2 millimeters.

CRYSTALLIZATION

The hot solution of solute is allowed to cool slowly to room temperature, and crystallization should occur. During cooling and crystallization, the solution should be protected from airborne contaminants by covering the opening with a piece of filter paper, an inverted beaker, or by loosely plugging it with a clean cork. Rapid cooling by immersing the flask in water or an ice-water bath is undesirable because the crystals formed tend to be very small, and their resulting large surface area may foster adsorption of impurities from solution. In this sense, the crystals are functioning like decolorizing carbon! Generally the solution should not be disturbed as it cools, since this also leads to production of small crystals. The formation of crystals larger than about 2 mm should be avoided because some of the solution may become occluded or trapped *within* the crystals. The drying of such crystals is more difficult, and impurities may be left in them. Should overly large crystals begin to form, brief, gentle agitation of the solution normally induces production of smaller crystals.

Failure of crystallization to occur after the solution has cooled somewhat usually means that either too much solvent has been used or that the solution is supersaturated. A supersaturated solution can usually be made to produce crystals by seeding. A crystal of the original solid is added to the solution to induce crystallization, which may then be quite rapid. If no solid is available and a volatile solvent is being used, it is sometimes possible to produce a seed crystal by immersing the tip of a glass stirring rod or metal spatula in the solution, withdrawing it, and allowing the solvent to evaporate. The crystals that form on the end of the rod or spatula are then reinserted into the solution to initiate crystallization. Alternatively, crystallization can often be induced by using a glass rod to rub the inside surface of the crystallization vessel *at* or *just above* the air/solution interface. This should be done carefully, as excessive force may scratch or break the vessel or result in a broken rod.

Occasionally the solute will separate from solution as an oil rather than a solid. This type of separation, which is sometimes called **oiling out**, is undesirable for purification of solutes because the oils usually contain significant amounts of impurities. Two general approaches are helpful in solving this problem: (1) Oils may persist on cooling with no evidence of crystallization. These may often be induced to crystallize by scratching the oil against the side of the flask with a glass stirring rod at the interface of the oil and the solution. If this fails, several small seed crystals of the original solid may be added to the oil, and the mixture allowed to stand for a period of time. Failure of these alternatives may necessitate separation of the oil from the solution and crystallization of it from another solvent. (2) Oils may form from the hot solution and then solidify to an amorphous mass at lower temperatures; in the meantime, crystals of the solute may precipitate from the mother liquor. Because the oil is not a pure liquid, the solid mass produced from it will be impure, as noted earlier. In a case such as this, the usual remedy is to reheat the entire mixture to effect dissolution, add a few milliliters of additional pure solvent, and allow the resulting solution to cool.

FILTRATION AND SOLVENT REMOVAL

The crystalline product is isolated by filtration. The technique for doing this varies depending on the scale on which the crystallization was performed.

Miniscale Technique. The solid product is isolated by vacuum filtration using a Büchner or Hirsch funnel and a clean, dry filter flask (Sec. 2.17). The crystals normally are washed with a small amount of *pure, cold* solvent, with the vacuum off; the vacuum is then reapplied to remove as much solvent as possible from the filter cake. Care must be taken in this step to ensure that the filter paper is not lifted off the bed of the filter while the vacuum is off; this could result in loss of product when the vacuum is reapplied to remove the washes.

Microscale Technique. The crystalline product, contained in a Craig tube fitted with a plug, is isolated by centrifugation (Sec. 2.17). The solid is scraped off the plug back into the Craig tube, and washed by stirring it briefly with a small amount of *pure, cold* solvent. The plug is then reinserted, and the solvent is once again removed by centrifugation.

Further cooling of the filtrate, sometimes called the **mother liquor**, in an ice-water or ice-salt bath (Sec. 2.10) may allow isolation of a second crop of crystals. The filtrate can also be concentrated by evaporating part of the solvent and cooling the residual solution. The crystals isolated as a second or even a third crop are likely to be less pure than those in the first. Consequently, the various crops should *not* be combined until their purity has been assessed by comparison of their melting points as described in Section 3.3 for example.

DRYING THE CRYSTALS

At the miniscale level, the final traces of solvent are removed by transferring the crystals from the filter paper of the Büchner or Hirsch funnel to a watchglass or vial. Alternatively, solids may also be transferred to fresh pieces of filter or weighing paper for drying. This is a less desirable option, however, because fibers of paper may contaminate the product when it is ultimately transferred to a container for submission to your instructor. For microscale operations, the solid may be left in the Craig tube used for the crystallization. It is good practice to protect the crystals from airborne contaminants by covering them with a piece of filter paper or using loosely inserted cotton or a cork to plug the opening of the vessel containing the solid.

Removing the last traces of solvent from the crystalline product may be accomplished by air- or oven-drying (Sec. 2.26). With the latter option, the temperature of the oven *must* be below the melting point of the product.

RECRYSTALLIZATION

Purpose: Exploring the techniques for recrystallizing solids.

> **Safety Alert**
>
> 1. Do not use a burner in these procedures unless instructed to do so. Most solvents used for recrystallization are flammable (Table 3.1).
> 2. When using a hot plate, do not set it at its highest value. A moderate setting will prevent overheating and the resultant bumping and splashing of materials from the flask. Do not employ hot plates for heating volatile or flammable solvents; rather use a steam bath.
> 3. Avoid inhalation of solvent vapors. If a hood is not available to you, clamp an inverted funnel just above the Erlenmeyer flask in which you will be heating solvents. Attach this funnel to a source of vacuum by means of rubber tubing (Fig. 2.68b).
> 4. When pouring or transferring solutions, either wear latex gloves or avoid getting these solutions on the skin. Organic compounds are much more rapidly absorbed through the skin when they are in solution, particularly in water-soluble solvents such as acetone and ethanol. For this reason do not rinse organic materials off your skin with solvents such as acetone; wash your hands thoroughly with soap and hot water instead.
> 5. *Never* add decolorizing carbon to a boiling solution; doing so may cause the solution to boil out of the flask. Add the carbon only when the temperature of the solvent is below the boiling temperature. This same precaution applies when using a filter-aid to assist in the removal of the carbon during the hot filtration step.

A · SOLVENT SELECTION

Procedure

Preparation   Answer the Pre-Lab Exercises on page PL. 1. Read or review Sections 2.9, and 3.2 and review the solvent properties listed in Table 3.1. Although different criteria are used for defining solubility, plan to use the following definitions in this experiment: (a) soluble—20 mg of solute will dissolve in 0.5 mL of solvent; (b) slightly soluble—some but not all of the 20 mg of solute will dissolve in 0.5 mL of solvent; (c) insoluble—none of the solute appears to dissolve in 0.5 mL of solvent. *Be certain to record all your observations regarding solubilities in your notebook!*

Apparatus   Test tubes (10-mm × 75-mm), hot-water (80–100 °C) or steam bath.

Protocol   For *known* compounds, place about 20 mg (a microspatula-tip full) of the finely crushed solid in a test tube and add about 0.5 mL of water using a calibrated Pasteur pipet. Stir the mixture with a glass rod or microspatula to determine whether the solid is soluble in water at room temperature. If the solid is not completely soluble at room temperature, warm the test tube in the hot-water or steam bath, and stir or swirl its contents to determine whether the solid is soluble in hot water.

If any of your solutes is soluble in the hot solvent but only slightly soluble or insoluble at room temperature, allow the hot solution to cool slowly to room temperature and compare the quantity, size, color, and form of the resulting crystals with the original solid material.

Repeat the solubility test for the solutes using 95% ethanol and then petroleum ether (bp 60–80 °C, 760 Torr). After completing these additional tests, record which of the three solvents you consider best suited for recrystallization of each of the solutes.

Compounds **1–4** contain a variety of functional groups that impart differing solubility properties to the molecules, and are possible substrates on which to practice the technique of determining solubilities. Other known compounds may be assigned by your instructor.

1	**2**	**3**	**4**
Benzoic Acid	Acetanilide	Naphthalene	Resorcinol

For *unknown* compounds, a *systematic* approach is important for determining their solubility, and the following protocol accomplishes this. From Table 3.1, select the solvents to be used in the tests. It should not be necessary to test all the solvents, but you should consider trying those that are denoted with an asterisk in the table. Your instructor may also suggest solvents to evaluate.

After selecting the solvents, obtain enough clean, dry test tubes so that there is one for each solvent to be tested. Place about 20 mg (a microspatula-tip full) of the finely crushed unknown in each test tube and add about 0.5 mL of a solvent to a tube containing the solid. Stir each mixture and determine the solubility of the unknown in each solvent at room temperature. Use the definitions of *soluble, slightly soluble*, or *insoluble* given earlier.

If the unknown is insoluble in a particular solvent, warm the test tube in the hot-water or steam bath. Stir or swirl the contents of the tube and note whether the unknown is soluble in hot solvent. If the solid is soluble in the hot solvent but only slightly soluble or insoluble at room temperature, allow the hot solution to cool to room temperature slowly. If crystals form in the cool solution, compare their quantity, size, color, and form with the original solid material and with those obtained from other solvents.

It is a good idea to test the solubility of a solute in a variety of solvents. Even though nice crystals may form in the first solvent you try, another one might prove better if it provides either better recovery or higher-quality crystals. To assist in determining the best solvent to use in recrystallizing an unknown, you should construct a table containing the solubility data you gather by the systematic approach described above.

If these solubility tests produce no clear choice for the solvent, mixed solvents might be considered. Review the discussion presented earlier in this section for the procedure for using a mixture of two solvents. Before trying any combinations of solvent pairs, take about 0.2 mL of each *pure* solvent being considered and mix them to ensure that they are miscible in one another. If they are not, that particular combination *cannot* be used.

B · RECRYSTALLIZING IMPURE SOLIDS

Miniscale Procedures

Preparation   Answer the Pre-Lab Exercises on page PL. 3. Read or review Sections 2.6, 2.7, 2.9, 2.11, 2.17, and 2.18. Plan to determine the melting points of both the purified and the crude solids as a way to show the benefits of recrystallization.

1. Benzoic Acid

Apparatus   Two 50-mL Erlenmeyer flasks, graduated cylinder, apparatus for magnetic stirring, vacuum filtration, and *flameless* heating.

Dissolution   Place 1.0 g of impure benzoic acid in an Erlenmeyer flask equipped for magnetic stirring or with boiling stones. Measure 25 mL of water into the graduated cylinder and add a 10-mL portion of it to the flask. Heat the mixture to a gentle boil and continue adding water in 0.5-mL portions until no more solid appears to dissolve in the boiling solution. Record the total volume of water used; no more than 10 mL should be required. *Caution:* Because the sample may be contaminated with insoluble material, pay close attention to whether additional solid is dissolving as you add more solvent; if it is not, *stop adding solvent.*

Decoloration   Pure benzoic acid is colorless, so a colored solution indicates that treatment with decolorizing carbon (Sec. 2.18) is necessary. *Caution:* Do *not* add decolorizing carbon to a *boiling* solution! Cool the solution slightly, add a microspatula-tip full of carbon, and reheat to boiling for a few minutes. To aid in removing the finely divided carbon by filtration, allow the solution to cool slightly, add about 0.2 g of a filter-aid (Sec. 2.17), and reheat.

Hot Filtration and Crystallization   If there are insoluble impurities or decolorizing carbon in the solution, perform a hot filtration (Sec. 2.17) using a 50-mL Erlenmeyer flask to receive the filtrate (Fig. 2.50). Rinse the empty flask with about 1 mL of *hot* water and filter this rinse into the original filtrate. If the filtrate remains colored, repeat the treatment with decolorizing carbon. Cover the opening of the flask with a piece of filter paper, an inverted beaker, or loose-fitting cork to exclude airborne impurities from the solution, and allow the filtrate to stand undisturbed until it has cooled to room temperature and no more crystals form.★ To complete the crystallization, place the flask in an ice-water bath for at least 15 min.

Isolation and Drying   Collect the crystals on a Büchner or Hirsch funnel by vacuum filtration (Fig. 2.52) and wash the filter cake with two small portions of *cold* water. Press the crystals as dry as possible on the funnel with a clean cork or spatula. Spread the benzoic acid on a watchglass, protecting it from airborne contaminants with a piece of filter paper, and air-dry it at room temperature or in an oven. Be certain that the temperature of the oven is below the melting point of the product!

Analysis   Determine the melting points of the crude and recrystallized benzoic acid, the weight of the latter material, and calculate your percent recovery using Equation 3.1.

$$\text{Percent recovery} = \frac{\text{weight of pure crystals recovered}}{\text{weight of original sample}} \times 100 \qquad \textbf{(3.1)}$$

2. Acetanilide

Apparatus   Two 50-mL Erlenmeyer flasks, graduated cylinder, apparatus for magnetic stirring, vacuum filtration, and *flameless* heating.

Dissolution   Place 1.0 g of impure acetanilide in an Erlenmeyer flask equipped for magnetic stirring or with boiling stones. Measure 20 mL of water into the graduated cylinder and add a 10-mL portion of it to the flask. Heat the mixture to a gentle boil.

A layer of oil should form when the stated amount of water is added. (If you have not done so already, review the discussion of *Crystallization* in this section with emphasis on how to crystallize compounds that form oils.) This layer consists of a solution of water in acetanilide. More water must be added to effect complete solution of the acetanilide in water. However, even if a homogeneous solution is produced at the boiling point of the mixture, an oil may separate from it as cooling begins. The formation of this second liquid phase is known to occur only under specific conditions: the acetanilide/water mixture must have a composition that is between 5.2% and 87% in acetanilide and be at a temperature above 80 °C. Because the solubility of acetanilide in water at temperatures near 100 °C exceeds 5.2%, a homogeneous solution formed by using the minimum quantity of water meets these criteria. Such a solution will yield an oil on cooling to about 83 °C; solid begins to form below this temperature.

Continue adding 3–5 mL of water in 0.5-mL portions to the boiling solution until the oil has completely dissolved. *Caution:* Because the sample may be contaminated with insoluble material, pay close attention to whether additional solid is dissolving as you add more solvent; if it is not, *stop adding solvent*. Once the acetanilide has just dissolved, add an additional 1 mL of water to prevent formation of oil during the crystallization step. If oil forms at this time, reheat the solution and add a little more water. Record the total volume of water used.

Continue the procedure by following the directions for *Decoloration*, *Hot Filtration and Crystallization*, and *Isolation and Drying* given for benzoic acid in Part 1.

Analysis   Determine the melting points of the crude and recrystallized acetanilide, the weight of the latter material, and calculate your percent recovery using Equation 3.1.

3. Naphthalene

Apparatus   Two 50-mL Erlenmeyer flasks, graduated cylinder, apparatus for magnetic stirring, vacuum filtration, and *flameless* heating.

Dissolution   Naphthalene may be conveniently recrystallized from methanol, 95% ethanol, or 2-propanol. Because these solvents are somewhat toxic and/or flammable, proper precautions should be taken. The sequence of steps up through the hot filtration should be performed in a hood if possible. Alternatively, if instructed to do so, position an inverted funnel connected to a vacuum source above the mouth of the flask being used for recrystallization (Fig. 2.68b).

Place 1.0 g of impure naphthalene in an Erlenmeyer flask equipped for magnetic stirring or with boiling stones and dissolve it in the minimum amount of boiling alcohol required. *Caution:* Because the sample may be contaminated with insoluble material, pay close attention to whether additional solid is dissolving as you add more solvent;

if it is not, *stop adding solvent*. Then add 0.5 mL of additional solvent to ensure that premature crystallization will not occur during subsequent transfers. Record the total volume of solvent used.

Continue the procedure by following the directions for *Decoloration*, *Hot Filtration and Crystallization*, and *Isolation and Drying* given for benzoic acid in Part 1; however, rather than water use the solvent in which you dissolved the naphthalene.

Analysis Determine the melting points of the crude and recrystallized naphthalene, the weight of the latter material, and calculate your percent recovery using Equation 3.1.

4. Unknown Compound

Apparatus Two 50-mL Erlenmeyer flasks, graduated cylinder, apparatus for magnetic stirring, vacuum filtration, and *flameless* heating.

Dissolution *Accurately* weigh about 1 g of the unknown compound and transfer it to an Erlenmeyer flask equipped for magnetic stirring or with boiling stones. Measure about 15 mL of the solvent you have selected on the basis of solubility tests into a graduated cylinder and add 10 mL of it to the flask. Bring the mixture to a gentle boil using *flameless* heating unless water is the solvent, add a 1-mL portion of the solvent, and again boil the solution. Continue adding 3-mL portions of solvent, one portion at a time, until the solid has completely dissolved. Bring the solution to boiling after adding each portion of solvent. *Caution:* Because the sample may be contaminated with insoluble material, pay close attention to whether additional solid is dissolving as you add more solvent; if it is not, *stop adding solvent*. Record the total volume of solvent that is added.

Continue the procedure by following the directions for *Decoloration*, *Hot Filtration and Crystallization*, and *Isolation and Drying* given for benzoic acid in Part 1.

Analysis Determine the melting points of the crude and recrystallized unknown, the weight of the latter material, and calculate your percent recovery using Equation 3.1.

5. Mixed-Solvent Crystallization

Apparatus Two 25-mL Erlenmeyer flasks, graduated cylinder or calibrated Pasteur pipet, apparatus for magnetic stirring, vacuum filtration, and *flameless* heating.

Dissolution Place 1.0 g of impure benzoic acid or acetanilide in an Erlenmeyer flask equipped for magnetic stirring or with boiling stones. Add 2 mL of 95% ethanol and heat the mixture to a gentle boil. If necessary, continue adding the solvent in 0.5-mL portions until no more solid appears to dissolve in the boiling solution. *Caution:* Because the sample may be contaminated with insoluble material, pay close attention to whether additional solid is dissolving as you add more solvent; if it is not, *stop adding solvent*. Record the total volume of 95% ethanol used.

Decoloration If the solution is colored, treat it with decolorizing carbon (Sec. 2.18) and a filter-aid according to the procedure given for benzoic acid in Part 1.

Hot Filtration and Crystallization If there are insoluble impurities or decolorizing carbon in the solution, perform a hot filtration (Sec. 2.17) using a 25-mL Erlenmeyer flask to receive the filtrate (Fig. 2.50). Rinse the empty flask with about 0.5 mL of *hot*

95% ethanol and filter this rinse into the original filtrate. If the filtrate remains colored, repeat the treatment with decolorizing carbon.

Reheat the decolorized solution to boiling and add water dropwise from a Pasteur pipet until the boiling solution remains cloudy or precipitate appears; this may require adding several milliliters of water. Then add a few drops of 95% ethanol to produce a clear solution at the boiling point. Remove the flask from the heating source, and follow the same directions as given for benzoic acid in Part 1 to complete both this stage of the procedure and *Isolation and Drying*.

Analysis   Determine the melting points of the crude and recrystallized product, the weight of the latter material, and calculate your percent recovery using Equation 3.1.

Microscale Procedures

Preparation   Answer the Pre-Lab Exercises on page PL. 5. Read or review Sections 2.5, 2.6, 2.7, 2.9, 2.17, and 3.3. Plan to determine the melting points of both the purified and the crude solids as a way to show the benefits of recrystallization.

1. Benzoic Acid

Apparatus   A 10-mm × 75-mm test tube, Pasteur filtering and filter-tip pipets, apparatus for Craig tube filtration and *flameless* heating.

Dissolution   Place 100 mg of benzoic acid in the test tube. Add 1 mL of water to the tube and heat the mixture to a gentle boil. To aid in dissolution, stir the mixture vigorously with a microspatula while heating; stirring also prevents bumping and possible loss of material from the test tube. If needed, add more solvent in 0.1–0.2 mL increments to dissolve any remaining solid. Bring the mixture to boiling and continue stirring after each addition. Once all the solid has dissolved, add an additional 0.1–0.2-mL portion of solvent to ensure that the solute remains in solution during transfer of the hot solution to a Craig tube. *Caution:* Because the sample may be contaminated with insoluble material, pay close attention to whether additional solid is dissolving as you add more solvent. If it is not, *stop adding solvent*. Record the approximate total volume of solvent used.

Decoloration   If the solution is colored and the recrystallized product is known or suspected to be colorless, treatment with decolorizing carbon (Sec. 2.18) is necessary. *Caution:* Do *not* add decolorizing carbon to a *boiling* solution! Cool the solution slightly, add a half of a microspatula-tip full of powdered or, preferably, a pellet or two of decolorizing carbon, and reheat to boiling for a few minutes.

Hot Filtration and Crystallization   Preheat a Pasteur filtering pipet by pulling hot solvent into the pipet. Then transfer the hot solution into this pipet with a Pasteur pipet or filter-tip pipet that has also been preheated with solvent, using the tared Craig tube as the receiver for the filtrate (Fig. 2.51). If decolorizing carbon or other insoluble matter appears in the Craig tube, pass the solution through the filtering pipet a second, and if necessary, a third time. Concentrate the clear solution in the Craig tube to the point of saturation by heating it to boiling. Rather than using a boiling stone to prevent superheating and possible bumping, continually stir the solution with a microspatula while heating. The saturation point will be signaled by the appearance of cloudiness in the solution and/or the formation of crystals on the microspatula at the air/liquid interface.

When this occurs, add solvent dropwise at the boiling point until the cloudiness is discharged. Then remove the tube from the heating source, cap the tube with a loose plug of cotton to exclude air-borne impurities, and allow the solution to cool to room temperature.★ If necessary, induce crystallization by gently scratching the surface of the tube at the air/liquid interface or by adding seed crystals, if these are available. To complete crystallization, cool the tube in an ice-water bath for at least 15 min.

Isolation and Drying   Affix a wire holder to the Craig tube and, using the wire as a hanger, invert the apparatus in a centrifuge tube (Fig. 2.53). Remove the solvent by centrifugation, carefully dissemble the Craig tube, and scrape any crystalline product clinging to its upper section into the lower part. Protect the product from airborne contaminants by plugging the opening of the tube with cotton or by covering it with a piece of filter paper held in place with a rubber band. Air-dry the crystals to constant weight either at room temperature or in an oven. Be certain that the temperature of the oven is below the melting point of the product!

Analysis   Determine the melting points of the crude and recrystallized benzoic acid, the weight of the latter material, and calculate your percent recovery using Equation 3.1.

2. Acetanilide

Apparatus   A 10-mm × 75-mm test tube, Pasteur filtering and filter-tip pipets, apparatus for Craig tube filtration and *flameless* heating.

Dissolution   Place 100 mg of acetanilide in the test tube. Add 1 mL of water to the tube and heat the mixture to a gentle boil. To aid in dissolution, stir the mixture vigorously with a microspatula while heating; stirring also prevents bumping and possible loss of material from the test tube. A layer of oil should form when the stated amount of water is added. An explanation for this phenomenon is found under the *Dissolution* stage of Part 2 of the *Miniscale Procedures*. Continue adding water in 0.1-mL portions to the boiling solution until the oil has completely dissolved. *Caution:* Because the sample may be contaminated with insoluble material, pay close attention to whether additional solid is dissolving as you add more solvent; if it is not, *stop adding solvent.*

Once the acetanilide has just dissolved, add an additional 0.1 mL of water to prevent formation of oil during the crystallization step. If oil forms at this time, reheat the solution and add a little more water. Record the approximate total volume of water used.

Continue the procedure by following the directions for *Decoloration*, *Hot Filtration and Crystallization*, and *Isolation and Drying* given for benzoic acid in Part 1.

Analysis   Determine the melting points of the crude and recrystallized acetanilide, the weight of the latter material, and calculate your percent recovery using Equation 3.1.

3. Naphthalene

Apparatus   A 10-mm × 75-mm test tube, Pasteur filtering and filter-tip pipets, apparatus for Craig tube filtration and *flameless* heating.

Dissolution   Naphthalene may be conveniently recrystallized from methanol, 95% ethanol, or 2-propanol. Because these solvents are somewhat toxic and/or flammable, proper precautions should be taken. If instructed to do so, position an inverted funnel connected to a vacuum source above the mouth of the test tube being used for recrystallization (Fig. 2.69b) to minimize release of vapors into the laboratory.

Place 100 mg of naphthalene in the test tube and dissolve it in a minimum amount of boiling solvent. *Caution*: Because the sample may be contaminated with insoluble material, pay close attention to whether additional solid is dissolving as you add more solvent; if it is not, *stop adding solvent*. Then add 0.1 mL of additional solvent to ensure that premature crystallization will not occur during subsequent transfers. Record the total volume of solvent used.

Continue the procedure by following the directions for *Decoloration*, *Hot Filtration and Crystallization*, and *Isolation and Drying* given for benzoic acid in Part 1; however, rather than water, use the solvent in which you dissolved the naphthalene.

Analysis   Determine the melting points of the crude and recrystallized naphthalene, the weight of the latter material, and calculate your percent recovery using Equation 3.1.

4. Unknown Compound

Apparatus   A 10-mm × 75-mm test tube, Pasteur filtering and filter-tip pipets, apparatus for Craig tube filtration and *flameless* heating.

Dissolution   If instructed to do so, position an inverted funnel connected to a vacuum source above the mouth of the test tube being used for recrystallization (Fig. 2.69b) to minimize release of vapors into the laboratory. Place 100 mg of the unknown in the test tube and dissolve it in a minimum amount of boiling solvent you selected on the basis of solubility tests. *Caution:* Because the sample may be contaminated with insoluble material, pay close attention to whether additional solid is dissolving as you add more solvent; if it is not, *stop adding solvent*. Then add 0.1 mL of additional solvent to ensure that premature crystallization will not occur during subsequent transfers. Record the total volume of solvent used.

Continue the procedure by following the directions for *Decoloration*, *Hot Filtration and Crystallization*, and *Isolation and Drying* given for benzoic acid in Part 1; however, if you are not using water, use the solvent in which you dissolved the unknown.

Analysis   Determine the melting points of the crude and recrystallized unknown, the weight of the latter material, and calculate your percent recovery using Equation 3.1.

5. Mixed-Solvent Crystallization

Apparatus   A 10-mm × 75-mm test tube, Pasteur filtering and filter-tip pipets, apparatus for Craig tube filtration and *flameless* heating.

Dissolution   Place 100 mg of benzoic acid or acetanilide in the test tube. Add 0.2 mL of 95% ethanol and heat the mixture to a gentle boil. If necessary, continue to add 95% ethanol dropwise by Pasteur pipet until a homogeneous solution is obtained. *Caution:* Because the sample may be contaminated with insoluble material, pay close attention to whether additional solid is dissolving as you add more solvent; if it is not, *stop adding solvent*. Once all the solid has dissolved, add an additional 0.1–0.2-mL portion of solvent to ensure that the solute remains in solution during transfer of the hot solution to a Craig tube and, if necessary, decolorize the solution according to the directions in Part 1 for benzoic acid.

Hot Filtration and Crystallization   Preheat a Pasteur filtering pipet by pulling hot solvent into the pipet. Then transfer the hot solution into this pipet with a Pasteur pipet or filter-tip pipet that has also been preheated with solvent, using the tared Craig tube as the receiver for the filtrate (Fig. 2.51). If decolorizing carbon or other insoluble matter appears in the Craig tube, pass the solution through the filtering pipet a second, and if necessary, a third time.

Reheat the decolorized solution to boiling and add water dropwise from a Pasteur pipet until the boiling solution remains cloudy or precipitate forms. Rather than using a boiling stone to prevent superheating and possible bumping, continually stir the solution with a microspatula while heating. Then add a drop or two of 95% ethanol to restore homogeneity. Remove the tube from the heating source, cap the tube with a loose plug of cotton to exclude airborne impurities, and allow the solution to cool to room temperature.★ Follow the same directions as given for benzoic acid in Part 1 to complete both this stage of the procedure and *Isolation and Drying*.

Analysis   Determine the melting points of the crude and recrystallized product, the weight of the latter material, and calculate your percent recovery using Equation 3.1.

Flush any *aqueous filtrates* or *solutions* down the drain. With the advice of your instructor, do the same with the *filtrates* derived from use of *alcohols, acetone,* or other *water-soluble solvents*. Use the appropriate containers for the *filtrates* containing *halogenated solvents* or *hydrocarbon solvents*. Put *filter papers* in the container for nontoxic waste, unless instructed to do otherwise.

Finishing Touches

EXERCISES

1. Define or describe each of the following terms as applied to recrystallization:

   a. solution recrystallization

   b. temperature coefficient of a solvent

   c. the relationship between dielectric constant and polarity of a solvent

   d. petroleum ether   j. washing crystals

   e. mixed solvents   k. mother liquors

   f. solvent selection   l. filtrate

   g. filter-aid   m. solute

   h. hot filtration   n. solvent

   i. seeding   o. occlusion

2. Briefly explain how a colored solution may be decolorized.

3. Briefly explain how insoluble particles can be removed from a hot solution.

4. List five criteria that should be used in selecting a solvent for a recrystallization.

5. When might Celite, a filter-aid, be used in recrystallization, and why is it used?

6. Briefly describe how a mixture of sand and benzoic acid, which is soluble in hot water, might be separated to provide pure benzoic acid.

7. The following solvent selection data were collected for two different impure solids:

Solid A

Solvent	Solubility at Room Temperature	Solubility When Heated	Crystals Formed When Cooled
Methanol	Insoluble	Insoluble	—
Ethyl acetate	Insoluble	Soluble	Very few
∨ Cyclohexane	Insoluble	Soluble	Many
Toluene	Insoluble	Soluble	Very few

Solid B

Solvent	Solubility at Room Temperature	Solubility When Heated	Crystals Formed When Cooled
Water	Soluble	—	—
Ethanol	Soluble	—	—
Dichloromethane	Insoluble	Insoluble	—
Petroleum ether	Insoluble	Insoluble	—
Toluene	Insoluble	Insoluble	—

Based on these results, what solvents or mixture of solvents might you consider using to recrystallize solids A and B? Explain.

8. The goal of the recrystallization procedure is to obtain purified material with a maximized recovery. For each of the items listed, explain why this goal would be adversely affected.

   a. In the solution step, an unnecessarily large volume of solvent is used.

   b. The crystals obtained after filtration are not washed with fresh cold solvent before drying.

   c. The crystals referred to in (b) are washed with fresh hot solvent.

   d. A large quantity of decolorizing carbon is used.

   e. Crystals are obtained by breaking up the solidified mass of an oil that originally separated from the hot solution.

   f. Crystallization is accelerated by immediately placing the flask of hot solution in an ice-water bath.

9. A second crop of crystals may be obtained by concentrating the vacuum filtrate and cooling. Why is this crop of crystals probably less pure than the first crop?

10. Explain why the rate of dissolution of a crystalline substance may depend on the size of its crystals.

11. The solubility of benzoic acid at 0 °C is 0.02 g per 100 mL of water, and that of acetanilide is 0.53 g per 100 mL of water. If you performed either of these recrystallizations, calculate, with reference to the total volume of water you used in

preparing the hot solution, the amount of material in your experiment that was unrecoverable by virtue of its solubility at 0 °C.

12. Assuming that either solvent is otherwise acceptable in a given instance, what advantages does ethanol have over 1-octanol as a crystallization solvent? hexane over pentane? water over methanol?

13. Look up the solubility of benzoic acid in hot water. According to the published solubility, what is the minimum amount of water in which 1 g of benzoic acid can be dissolved?

14. List the steps in the systematic procedure for miniscale recrystallization, briefly explaining the purpose of each step.

15. In performing a hot filtration at the miniscale level, what might happen if the filter funnel is not preheated before the solution is poured through it?

16. Describe the use of the following pieces of equipment during recrystallizations at the miniscale level.

    a. filter flask

    b. filter trap

    c. Büchner funnel

17. Why should the filter flask not be connected directly to a water aspirator pump when performing a vacuum filtration?

18. List the steps in the systematic procedure for microscale recrystallization, briefly explaining the purpose of each step.

19. Why is a Pasteur filter-tip pipet preferred to a regular Pasteur pipet for transferring hot volatile solutions?

20. In performing a hot filtration at the microscale level, what might happen if the Pasteur filtering and filter-tip pipets are not preheated before filtration?

21. Why is a Craig tube filtration ineffective for removing finely divided particles from a solution?

3.3 Physical Constants: Melting Points

PHYSICAL CONSTANTS

Physical constants of compounds are numerical values associated with measurable properties of these substances. These properties are *invariant* and are useful in the identification and characterization of substances encountered in the laboratory so long as accurate measurements are made under specified conditions such as temperature and pressure. Physical constants are useful only in the identification of *previously known* compounds, however, because it is not possible to predict accurately the values of such properties. Among the more frequently measured physical properties of organic compounds are **melting point (mp)**, **boiling point (bp)**, **index of refraction (n)**, **density (d)**, **specific rotation ([α])**, and **solubility**. Melting points, discussed below, boiling points, described in Section 4.1, and solubilities, outlined in Section 3.2, are the properties most commonly encountered. Index of refraction and density are mentioned in Chapter 23. Specific rotation is discussed in Chapters 7 and 22 but applies only to molecules that are **optically active**. Whether the substance is known or unknown, such values along with other properties like color, odor, and crystal form should be recorded in the laboratory notebook.

The values of one or two of the common physical properties *may* be identical for more than one compound, but it is most unlikely that values of several such properties will be the same for two different compounds. Consequently, a list of physical constants is a highly useful way to characterize a substance. Extensive compilations of the physical constants are available (Chap. 24). One of the most convenient is the *CRC Handbook of Chemistry and Physics*, which contains a tabulation of the physical constants and properties of a large number of inorganic and organic compounds. *The Handbook of Tables for Organic Compounds* is especially useful for organic compounds. Neither of these books is comprehensive; rather, they contain entries for only the more common organic and inorganic substances. So many compounds are known that multi-volume sets of books are required to list their physical properties (Chap. 24).

MELTING POINT OF A PURE SUBSTANCE

The melting point of a substance is defined as the temperature at which the liquid and solid phases exist in equilibrium with one another without change of temperature. Ideally, addition of heat to a mixture of the solid and liquid phases of a pure substance at the melting point will cause no rise in temperature until all the solid has melted. Conversely, removal of heat from the equilibrium mixture will produce no decrease in temperature until all the liquid solidifies. This means that the melting and freezing points of a pure substance are identical.

The melting point is expressed as the temperature *range* over which the solid starts to melt and then is completely converted to liquid. Consequently, rather than a melting *point*, what is actually measured is a **melting range**, although the two terms are used interchangeably. If a solid substance is pure, it should melt over a very narrow or sharp range, which should normally be no more than 1 °C if the melting point is determined carefully. The melting ranges reported for many ''pure'' compounds may be greater than 1 °C because the particular compound was not quite pure or the melting point was not measured properly. The process of melting may actually begin by ''softening,'' as evidenced by an apparent shrinking of the solid, but such softening is difficult to observe. Thus, for our purposes the start of melting is defined as the temperature at which the first droplet of liquid can be detected. Note that it is improper and inexact to report a single temperature, such as 118 °C for a melting point; rather, a range of 117–119 °C or 117.5–118.0 °C, for example, should be recorded.

EFFECT OF IMPURITIES ON MELTING POINTS; MIXTURES

Many solid substances prepared in the organic laboratory are initially impure, so the effect of impurities on melting-point ranges deserves further discussion. Although this topic is discussed in freshman chemistry textbooks, a brief review of its basic principles is given here.

The presence of an impurity generally *decreases* the melting point of a pure solid. This is shown graphically by the melting-point-composition diagram of Figure 3.1, in which points *a* and *b* represent the melting points of pure *A* and *B*, respectively. Point *E* is called the **eutectic point** and is determined by the equilibrium composition at which *A* and *B* melt in constant ratio. In Figure 3.1, this ratio is 60 mol % *A* and

FIGURE 3.1
Melting-point-composition diagram
for two hypothetical solids, *A* and *B*.

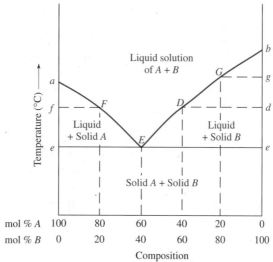

40 mol % *B*; an impure solid comprised of *A* and *B* in this ratio would be called a
eutectic mixture. The temperature at the eutectic point is designated by *e*.

Now consider the result of heating a solid mixture composed of 80 mol % *A* and
20 mol % *B*, a sample that might be considered as "impure *A*." As heat is applied to
the solid, its temperature will rise. When the temperature reaches *e*, *A* and *B* will both
begin to melt in the constant ratio defined by the composition at the eutectic point.
Once all of the "impurity" *B* has melted, only solid *A* will be left in equilibrium with
the melt. The remaining solid *A* will continue to melt as additional heat is supplied,
and the percentage of *A* in the melt will increase, changing the composition of the melt
from that of the eutectic mixture. This increases the vapor pressure of *A* in the solution
according to Raoult's law (Eq. 4.2) and raises the temperature at which solid *A* is in
equilibrium with the molten solution. The relationship between the equilibrium temperature and the composition of the molten solution is then represented by curve *EF* in
Figure 3.1. When the temperature reaches *f*, no solid *A* will remain and melting of the
sample will be complete. The impure sample *A* exhibits a melting "point" that extends
over the relatively broad temperature range *e–f*. Because melting both begins and ends
below the melting point of pure *A*, the melting point of *A* is said to be *depressed*.

The foregoing analysis is easily extended to the case in which substance *B* contains
A as an impurity. In Figure 3.1, this simply means that the composition of the solid
mixture is to the right of point *E*. The temperature during the melting process would
follow curve *ED* or *EG*, and the melting range would now be *e–d* or *e–g*.

A sample whose composition is exactly that of the eutectic mixture (point *E*, Fig.
3.1) will exhibit a sharp melting point at the eutectic temperature. This means a eutectic
mixture can be mistaken for a pure compound, because both have a sharp melting point.

From a practical standpoint, it may be very difficult to observe the initial melting
point of solid mixtures, particularly with the capillary tube melting-point technique used
in the Experimental Procedure that follows. This is because the presence of only a
minor amount of impurity means that only a tiny amount of liquid is formed in the
stage of melting that occurs at the eutectic temperature. In contrast, the temperature at
which the last of the solid melts (points *d* and *g*, Fig. 3.1) can be determined accurately.
Consequently, a mixture containing smaller amounts of impurities will generally have
both a higher final melting point and a narrower observed melting-point range than one
that is less pure.

The broadening of the melting-point range that results from introducing an impurity into a pure compound may be used to advantage for identifying a pure substance. The technique is commonly known as a **mixed melting-point** determination, and is illustrated by the following example. Assume that an unknown compound X melts at 134–135 °C, and you suspect it is either urea, H_2NCONH_2, or *trans*-cinnamic acid, $C_6H_5CH{=}CHCO_2H$, both of which melt in this range. If X is mixed intimately with urea and the melting point of this mixture is found to be lower than that of the pure compound and pure urea, then urea is acting as an impurity, and the compound cannot be urea. If the mixture melting point is identical to that of the pure compound and of urea, the compound is identified as urea. Obviously, this procedure is useful in identifying compounds only when authentic samples of the likely possibilities are available.

A convenient and rapid method for ascertaining the purity of a solid is measuring its melting point. A narrow melting-point range ordinarily signals that the sample is *pure*, although there is a *low* probability that the solid is a eutectic mixture. If recrystallizing a sample changes an originally broad melting range to a narrow one, the reasonable conclusion is that the recrystallization was successful in purifying the solid. Should the melting-point range remain broad after recrystallization, the sample may be contaminated with solvent and additional drying is required. It is also possible that the recrystallization was not completely successful in removing impurities, in which case the solid should be recrystallized using the same solvent. If this fails to narrow the melting range satisfactorily, recrystallization should be performed with a different solvent.

MICRO MELTING-POINT METHODS

The determination of highly accurate melting points of organic compounds can be time-consuming. Fortunately, micro methods are available that are convenient, require negligible amounts of sample, and give melting-point data that are satisfactory for most purposes. The technique using the capillary-tube melting-point procedure is the one used most commonly in the organic laboratory.

There are practical considerations in determining melting points, and some of them are briefly noted here. First, the observed melting-point range depends on several factors, including the quantity of sample, its state of subdivision, the rate of heating during the determination, and the purity and chemical characteristics of the sample. The first three factors can cause the observed melting-point range to differ from the true value because of the time lag for transfer of heat from the heating medium to the sample and for conduction of heat within the sample. For example, if the sample is too large, the distribution of heat within it may not be uniform, and inaccurate melting ranges will result. A similar problem of non-uniform heat distribution is associated with use of crystals of large size: It will be difficult to pack the sample tightly in the capillary melting tube, and the resulting airspace results in poor conduction of heat. If the rate of heating is too fast, the thermometer reading will lag behind the actual temperature of the heating medium and produce measurements that are low. The chemical characteristics of the sample may be important if the compound tends to decompose upon melting. When this occurs, discoloration of the sample is usually evident, and may be accompanied by gas evolution. The decomposition products constitute impurities in the sample, and the true melting point is lowered as a result. The reporting of melting points for compounds that melt with decomposition should reflect this, as in ''mp 195 °C (dec).''

In determining the melting point of a compound, much valuable time can be wasted waiting for melting to occur if the proper slow rate of heating is being used on a sample whose melting point is unknown to you. It is considerably more efficient to prepare two capillary tubes containing the compound being studied and determine the approximate melting point by rapidly heating one of them. Then the heating source is allowed to cool 10–15 °C below this approximate melting point, and an accurate melting point is obtained with the second sample.

The accuracy of any type of temperature measurement ultimately depends on the quality and calibration of the thermometer. A particular thermometer may provide accurate readings in some temperature ranges, but may be off by a degree or two in others. Melting points that have been determined using a calibrated thermometer may be reported in the form, ''mp 101–102 °C (corr.),'' where ''corr.'' is the abbreviation for ''corrected''; the corresponding abbreviation for values obtained with an uncalibrated thermometer is ''uncorr.'' for ''uncorrected.''

Calibration involves the use of standard substances for the measurement of the temperature at a series of known points within the range of the thermometer and the comparison of the observed readings with the true temperatures. The difference between the observed and the true temperature measurement provides a correction that must be applied to the observed reading. Calibration over a range of temperatures is necessary because the error is likely to vary at different temperatures.

EXPERIMENTAL PROCEDURES

MELTING POINTS

Purpose Determining melting points using the capillary-tube method.

Safety Alert

1. If a burner is used in this experiment, be sure that no flammable solvents are nearby. Keep the rubber tubing leading to the burner away from the flame. Turn off the burner when it is not being used.
2. Some kinds of melting-point apparatus, such as the Thiele tube, use mineral or silicone oils as the heat transfer medium. These oils may not be heated safely if they are contaminated with even a few drops of water. Heating these oils above 100 °C may produce splattering of hot oil as the water turns to steam. Fire can also result if splattered oil comes in contact with open flames. Examine your Thiele tube for evidence of water droplets in the oil. If any are observed, either change the oil or exchange tubes. Give the contaminated tube to your instructor.
3. Mineral oil is a mixture of high-boiling hydrocarbons and should not be heated above 200 °C because of the possibility of spontaneous ignition, particularly when a burner is used for heating. Some silicone oils may be heated to about 300 °C without danger (Sec. 2.9).
4. Be careful to avoid contact of chemicals with your skin. Clean up any spilled chemicals immediately with a brush or paper towel.
5. If you use a Thiele tube, handle it carefully when you are finished, because the tube cools slowly. To avoid burns, take care when removing it from its support.

Table 3.2	Standards for Thermometer Calibration
Compound	Melting Point (°C)
Ice water	0
3-Phenylpropanoic acid	48.6
Acetamide	82.3
Acetanilide	114
Benzamide	133
Salicylic acid	159
4-Chloroacetanilide	179
3,5-Dinitrobenzoic acid	205

A · CALIBRATION OF THERMOMETER

Procedure

Preparation   Answer the Pre-Lab Exercises on page PL. 7. Read or review Sections 2.7 and 2.9.

Apparatus   Capillary melting-point tubes, packing tube, and melting point apparatus.

Protocol   Carefully determine the capillary melting points of a series of standard substances. A list of suitable standards is provided in Table 3.2. The temperatures given in Table 3.2 correspond to the upper limit of the melting-point range for pure samples of these standards.

Plot the corrections in your notebook as deviations from zero versus the temperature over the range encompassed by the thermometer. This allows you to tell, for example, that at about 130 °C the thermometer gives readings that are 2 °C too low, or that at 190 °C the readings are about 1.5 °C too high. These values should then be applied to correct all temperature measurements taken. The corrections you obtain are valid only for the thermometer used in the calibration. If it is broken, the calibration must be repeated with a new one.

B · DETERMINING CAPILLARY-TUBE MELTING POINTS

Procedure

Preparation   Answer the Pre-Lab Exercises on page PL. 1. Read or review Sections 2.7 and 2.9. Plan to prepare two melting-point tubes if you are determining the melting points of unknown compounds.

Apparatus   Capillary melting-point tubes, packing tube, and melting point apparatus.

1. Pure Compound

Protocol   Select one or more compounds from a list of available compounds of known melting point and determine the melting-point ranges. Repeat as necessary until you obtain accurate results and are confident with the technique.

2. Unknown Compound

Protocol   Accurately determine the melting range of an unknown pure compound supplied by your instructor.

3. Mixed Melting Points

Sample Preparation   Prepare a sample for mixed melting-point determination by introducing 5–10% of a second substance as an impurity into a solid whose melting point was determined in Part 1. Intimately mix the two components by grinding them together with a small mortar and pestle. Alternatively, use a small, clean watchglass and glass stirring rod or metal spatula to mix the components. Be careful not to apply too much pressure to the glass rod, however, because it is more fragile than a pestle and may break. Do not perform this operation on a piece of filter paper because fibers from the paper may contaminate the sample.

Protocol   Accurately determine the melting range of the sample and compare it to that of the major component of the mixture, in order to study the effect of impurities on the melting range of a previously pure compound.

Return any *unused samples* to your instructor or dispose of them in the appropriate container for nonhazardous organic solids. Discard the used *capillary tubes* in a container for broken glass; do *not* leave them in the area of the melting-point apparatus or throw them in wastepaper baskets.

Finishing Touches

EXERCISES

1. Describe errors in procedure that may cause an observed capillary melting point of a pure compound

    a. to be *lower* than the correct melting point

    b. to be *higher* than the correct melting point

    c. to be *broad* in range (over several degrees)

2. Briefly define the following terms:

    a. vapor pressure as applied to melting   **d.** eutectic point

    b. melting point or melting-point range   **e.** eutectic mixture

    c. mixture or mixed melting point

3. Filter paper is usually a poor material on which to powder a solid sample before introducing it into a capillary melting-point tube because small particles of paper may end up in the tube along with the sample. Why is this undesirable, and how might the presence of paper in the sample make the melting-point determination difficult?

4. Criticize the following statements by indicating whether each is *true* or *false*, and if false, explain why.

    a. An impurity always lowers the melting point of an organic compound.

FIGURE 3.2

Melting-point-composition diagram for Exercise 7.

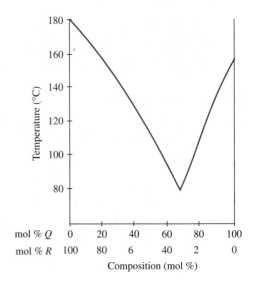

mol % Q: 0, 20, 40, 60, 80, 100
mol % R: 100, 80, 6, 40, 2, 0

Composition (mol %)

b. A sharp melting point for a crystalline organic substance always indicates a pure single compound.

c. If the addition of a sample of compound A to compound X does not lower the melting point of X, X must be identical to A.

d. If the addition of a sample of compound A lowers the melting point of compound X, X and A cannot be identical.

5. The melting points of pure benzoic acid and pure 2-naphthol are 122.5 °C and 123 °C, respectively. Given a pure sample that is known to be either pure benzoic acid or 2-naphthol, describe a procedure you might use to determine the identity of the sample.

6. A student used the Thiele micro melting-point technique to determine the melting point of an unknown and reported it to be 182 °C. Is this value believable? Explain why or why not.

7. The melting-point-composition diagram for two substances, Q and R, is provided in Figure 3.2, which should be used to answer the following questions.

a. What are the melting points of pure Q and R?

b. What are the melting point and the composition of the eutectic mixture?

c. Would a mixture of 20 mol % Q and 80 mol % R melt if heated to 120 °C? to 160 °C? to 75 °C?

d. A mixture of Q and R was observed to melt at 105–110 °C. What can be said about the composition of this mixture? Explain briefly.

8. For the following melting points, indicate what might be concluded regarding the purity of the sample:

a. 120–122 °C   **c.** 147 °C (dec)

b. 46–60 °C   **d.** 162.5–163.5 °C

9. Suppose that the observed melting-point range of a solid was originally 150–160 °C, but this became 145–145.5 °C after recrystallization. How might you account for this observation?

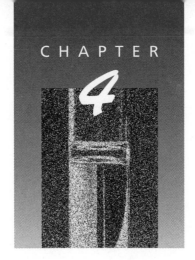

C H A P T E R

4

Liquids
Distillation and Boiling Points

The purification of solids by recrystallization and the use of melting points as a criterion of their purity are discussed in Chapter 3. The techniques used for the purification of liquids, namely, **simple**, **fractional**, **steam**, and **vacuum distillation**, are topics of this chapter. Boiling points are also discussed as a physical property that can be used as one means of determining the purity and identity of liquids.

4.1   Boiling Points of Pure Liquids

The molecules of a liquid are in constant motion, and those at the surface are able to escape into the vapor phase. The consequences of vaporization of a liquid contained in a closed, *evacuated* system are considered first, and then the situation in which the system is open to the atmosphere is discussed.

In a closed evacuated system, the number of molecules in the gas phase will initially increase until the rate at which they reenter the liquid becomes equal to the rate at which they escape from it. At this point, no further *net* change is observed, and the system is said to be in a state of **dynamic equilibrium**. The molecules in the gas phase are in rapid motion and continually collide with the walls of the vessel, which results in the exertion of pressure against the walls. The magnitude of this vapor pressure at a given temperature is called the **equilibrium vapor pressure** of the liquid at that temperature. Vapor pressure is temperature dependent, as shown in Figure 4.1. As the temperature of the system rises, the average kinetic energy of the liquid molecules increases, thus facilitating their escape into the gas phase. The rate of reentry of gaseous molecules into the liquid phase also increases, and equilibrium is reestablished at the higher temperature. Because there are now more molecules in the gas phase than there were at the lower temperature, the vapor pressure of the system is greater.

There is a very important safety rule based on the fact that the pressure in a closed system *increases* as the temperature of the system rises. Such a system should *not* be heated unless an apparatus designed to withstand the pressure is used; otherwise, an *explosion* will result. For purposes of the first laboratory course in organic chemistry, the rule is simply, ''*Never heat a closed system!*''

113

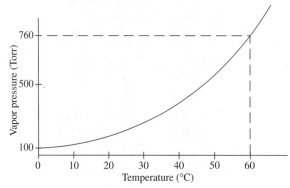

FIGURE 4.1   Graph of the dependence of vapor pressure on temperature for a liquid.

Now suppose that a liquid sample at a particular temperature is placed in an *open* container so that the molecules of the vapor over the liquid are mixed with air. The **total pressure** above the liquid is defined by **Dalton's law of partial pressures** (Eq. 4.1) and is equal to the sum of the partial pressures of the sample and of air. The partial pressure of the sample is equal to its **equilibrium vapor pressure** at the given temperature. When the temperature of the liquid is raised, the equilibrium vapor pressure of the sample will rise, and the number of gas molecules that have escaped from the liquid by the process of evaporation will increase in the space above the liquid. This will have the net effect of displacing some of the air. At the higher temperature, the partial pressure of the sample will be a *larger* percentage of the *total* pressure. This trend will continue as the temperature of the liquid is further increased, until the equilibrium vapor pressure of the sample equals the total pressure. At this point all of the air will have been displaced from the vessel containing the liquid. Entry of additional molecules from the liquid into the gas phase will only have the effect of further displacing those already in that phase; the partial pressure of the molecules of the sample will no longer increase.

$$P_{total} = P_{sample} + P_{air} \qquad \textbf{(4.1)}$$

When the temperature of the liquid is such that the equilibrium vapor pressure of the sample equals the total pressure, the rate of evaporation increases dramatically, and bubbles form in the liquid. This is the boiling process, and the temperature associated with it is the **boiling point** of the liquid. Since the boiling point is dependent upon the total pressure, that pressure must be specified when boiling points are reported, for example, "bp 132 °C (748 Torr)." The boiling point of water is 100 °C *only* when the external pressure is 760 Torr, or 1 atm. The **normal** boiling point of a liquid is measured at 760 Torr (1 atm) and is shown by the dashed lines in Figure 4.1 to be 60 °C. This figure also allows determination of the boiling point of the liquid at various total pressures. For example, at 500 Torr the boiling point will be about 50 °C; the decrease in temperature results from the decrease in the total pressure in the system.

The dependence of vapor pressure upon total pressure, as reflected by the boiling point of a liquid, can be used to advantage in the following way. Suppose that a liquid has a normal boiling point such that it decomposes appreciably when heated to this temperature. Reduction in the total pressure of the system reduces the boiling point of the sample to a temperature at which it no longer decomposes. This technique, called **vacuum distillation**, is discussed in Section 2.15.

A **pure** liquid generally boils at a constant temperature or over a narrow temperature range, provided the total pressure in the system remains constant. However, the boiling point of a liquid is affected by the presence of both nonvolatile and volatile impurities, and most **mixtures** of liquids boil over a fairly wide range. These effects are discussed in Section 4.2.

EXPERIMENTAL PROCEDURES

BOILING POINTS OF PURE LIQUIDS

Purpose: To determine the boiling points of pure liquids at atmospheric pressure.

1. Volatile organic liquids are flammable, so burners should be used carefully in this experiment. Consult with your instructor *before* lighting a burner.
2. Use paper towels to clean up spilled liquids; discard the towels as directed by your instructor. Avoid contact of organic liquids with your skin; if this happens, wash the affected area thoroughly with soap and water.

Miniscale Procedure

Preparation   Answer the Pre-Lab Exercises on page PL. 9. Read or review Sections 2.8 and 2.9.

Apparatus   Thermometer, 10- × 75-mm test tube, two clamps, and apparatus for *flameless* heating.

Setting Up   Obtain a liquid from your instructor and determine its boiling point following the technique described in Section 2.8 using the apparatus shown in Figure 2.20 and *flameless* heating.

Microscale Procedure

Preparation   Answer the Pre-Lab Exercises on page PL. 11. Read or review Sections 2.8 and 2.9.

Apparatus   Thiele melting-point apparatus, micro boiling-point apparatus, and a Bunsen or microburner.

Setting Up   Determine the boiling point of the liquid(s) assigned by your instructor. Follow the technique presented in Section 2.8 for using the micro boiling-point apparatus (Fig. 2.21). Use a Bunsen burner or a microburner for heating. In the event you do not know the boiling point of liquid, *first* determine an approximate boiling-point range by heating the Thiele tube (Fig. 2.17a) *fairly rapidly*. Repeat the measurement by heating the tube until the temperature is 20–30 °C below the approximate boiling point, and then heat the sample at a rate of 4–5 °C/min to obtain an accurate value. It may be desirable to repeat this procedure to obtain a more reliable boiling point.

Finishing Touches

Unless instructed otherwise, return the *organic liquids* to the appropriate bottle for either nonhalogenated liquids or halogenated liquids.

EXERCISES

1. Refer to Figure 4.1 and answer the following:
   a. What total pressure would be required in the system in order for the liquid to boil at 30 °C?
   b. At what temperature would the liquid boil when the total pressure in the system is 100 Torr?

2. Describe the relationship between escaping tendency of liquid molecules and vapor pressure.

3. Define the following terms:
   a. boiling point
   b. normal boiling point
   c. Dalton's law of partial pressure
   d. equilibrium vapor pressure

4. Explain why the boiling point at 760 Torr of a solution of water, bp 100 °C (760 Torr), and ethylene glycol ($HOCH_2 CH_2OH$), bp 196–198 °C (760 Torr), exceeds 100 °C. For purposes of your answer, consider ethylene glycol as a nonvolatile compound.

5. Why should there be no droplets of water in the oil of a heating bath?

6. Why is the micro boiling-point technique not applicable for boiling points in excess of 200 °C if *mineral* oil rather than *silicone* oil is the heating fluid in the Thiele tube?

4.2 Simple Distillation

Simple distillation allows separation of distillates from less-volatile substances that remain as pot residue at the completion of the distillation. In the ideal case, only a single component of the mixture to be distilled will be volatile, so the distillate will be a pure compound. More commonly, however, several volatile components comprise the mixture. Simple distillation allows isolation of the various components of the mixture in acceptable purity if the *difference* between the boiling points of each pure substance is greater than 40–50 °C. For example, a mixture of diethyl ether, bp 35 °C

(760 Torr), and toluene, bp 111 °C (760 Torr), could be separated by simple distillation, with the ether distilling first. Organic chemists frequently use this technique to separate a desired reaction product from the solvents used for the reaction or its work-up. The solvents are usually more volatile than the product and are readily removed from it by simple distillation.

To understand the principles of distillation, a review of the effect of impurities on the vapor pressure of a pure liquid is in order. The discussion starts with consideration of the consequences of having *nonvolatile* impurities present and then turns to the more common case of contamination of the liquid with other *volatile* substances.

Consider a homogeneous solution comprised of a nonvolatile impurity and a pure liquid; for the present purpose these are taken as sugar and water, respectively. The sugar reduces the vapor pressure of the water because this nonvolatile impurity lowers the concentration of the volatile constituent in the liquid phase. The consequence of this is shown graphically in Figure 4.2. In this figure, Curve 1 corresponds to the dependence of the temperature upon the vapor pressure of *pure* water and intersects the 760-Torr line at 100 °C. Curve 2 is for a *solution* having a particular concentration of sugar in water. Note that the vapor pressure at any temperature is reduced by a constant amount by the presence of the nonvolatile impurity, in accord with Raoult's law as discussed later. The temperature at which this curve intersects the 760-Torr line is higher because of the lower vapor pressure, and consequently the temperature of the boiling solution, 105 °C, is higher.

Despite the presence of the sugar in the solution, the **head temperature** (Sec. 2.13) in the distillation will be the same as for the pure water, namely, 100 °C (760 Torr), since the water condensing on the thermometer bulb is now uncontaminated by the impurity, sugar. The **pot temperature** will be *elevated*, however, owing to the decreased vapor pressure of the solution (Fig. 4.2). As water distils, the pot temperature will progressively rise because the concentration of the sugar in the stillpot increases, further lowering the vapor pressure of the water. Nevertheless, the head temperature will remain constant, just as though pure water were being distilled.

The quantitative relationship between vapor pressure and composition of homogeneous liquid mixtures is known as **Raoult's law** (Eq. 4.2). The factor P_X represents the partial pressure of component X, and is equal to the vapor pressure, P_X°, of pure X at a given temperature times the mole fraction N_X of X in the mixture. The **mole fraction** of X is defined as the fraction of *all* molecules present in the liquid mixture that are

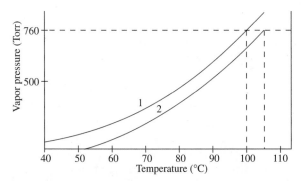

FIGURE 4.2 Diagram of the dependence of vapor pressure on temperature. Curve 1 represents pure water and curve 2 represents pure water to which sugar, a nonvolatile impurity, has been added.

molecules of X. It is obtained by dividing the number of moles of X in a mixture by the sum of the number of moles of all components (Eq. 4.3). Raoult's law is strictly applicable only to **ideal solutions**, which are defined as those in which the interactions between *like* molecules are the same as those between *unlike* molecules. Fortunately, many organic solutions approximate the behavior of ideal solutions, so the following mathematical treatment applies to them as well.

$$P_X = P_X^\circ N_X \tag{4.2}$$

$$N_X = \frac{nX}{nX + nY + nZ + \cdots} \tag{4.3}$$

Note that the partial vapor pressure of X above an ideal solution depends *only* on its mole fraction in solution and is completely independent of the vapor pressures of the other volatile components of the solution. If all components other than X are nonvolatile, the total vapor pressure of the mixture will be equal to the partial pressure of X, since the vapor pressure of nonvolatile compounds may be taken as zero. Thus the distillate from such a mixture will always be pure X. This is the case for the distillation of a solution of sugar and water, as discussed earlier.

When a mixture contains two or more volatile components, the *total* vapor pressure is equal to the *sum* of the partial vapor pressures of each such component. This is known as **Dalton's law** (Eq. 4.4), where P_X, P_Y, and P_Z refer to the vapor pressures of the volatile components. The process of distilling such a liquid mixture may be significantly different from that of simple distillation, because the vapors above the liquid phase will now contain some of each of the volatile components. Separation of the liquids in this case may require the use of fractional distillation, which is discussed in Section 4.3.

$$P_{\text{total}} = P_X + P_Y + P_Z + \cdots \tag{4.4}$$

EXPERIMENTAL PROCEDURES

SIMPLE DISTILLATION

Purpose: To demonstrate the technique for purification of a volatile organic liquid containing a nonvolatile impurity.

1. Cyclohexane is highly flammable, so be sure that burners are not being used in the laboratory. Use *flameless* heating (Sec. 2.9).
2. Examine your glassware for cracks and other weaknesses before assembling the distillation apparatus. Look for "star" cracks in round-bottom flasks, because these can cause a flask to break upon heating.
3. Proper assembly of glassware is important in order to avoid possible breakage and spillage or the release of distillate vapors into the

room. Be certain that all connections in the apparatus are tight before beginning the distillation. Have your instructor examine your
set-up after it is assembled.

4. The apparatus used in these experiments *must* be open to the atmosphere at the receiving end of the condenser. *Never heat a
closed system* because the pressure buildup may cause the apparatus to explode.

5. Be certain that the water hoses are *securely* fastened to your condensers so that they will not pop off and cause a flood. If heating
mantles or oil baths are used for heating in this experiment, water
hoses that come loose may cause water to spray onto electrical connections or into the heating sources, either of which is potentially
dangerous.

6. Avoid excessive inhalation of organic vapors at all times.

Miniscale Procedure

Preparation   Answer the Pre-Lab Exercises on page PL. 13. Read or review Sections
2.2, 2.4, 2.9, 2.11, and 2.13.

Apparatus   A 25-mL round-bottom flask and apparatus for simple distillation, magnetic stirring, and *flameless* heating.

Setting Up   Place 10 mL of cyclohexane containing a nonvolatile dye in the round-
bottom flask. Add a stirbar to the flask to ensure smooth boiling, and assemble the
simple distillation apparatus shown in Figure 2.35a. Be sure to position the thermometer
in the stillhead so the *top* of the mercury thermometer bulb is level with the *bottom* of
the sidearm of the distillation head. Have your instructor check your apparatus *before*
you start heating the distilling flask.

Distillation   Begin heating the stillpot, and as soon as the liquid begins to boil *and
the condensing vapors have reached the thermometer bulb*, regulate the heat supply so
that distillation continues steadily at a rate of *2–4 drops per second*; if a drop of liquid
cannot be seen suspended from the end of the thermometer, the rate of distillation is
too *fast*. As soon as the distillation rate is adjusted and the head temperature is constant,
note and record the temperature. Continue the distillation and periodically record the
head temperature. Discontinue heating when only 2–3 mL of impure cyclohexane remains in the distillation flask. Record the volume of distilled cyclohexane that you
obtain.

Optional Procedure

You may be required to perform this distillation using the short-path apparatus discussed
in Section 2.13 and illustrated in Figure 2.35b. After assembling the equipment, make
certain that the *top* of the mercury in the thermometer bulb is level with the *bottom* of
the sidearm of the distillation head. The preferred way to collect the distillate in this
distillation is to attach a dry round-bottom flask to the vacuum adapter and put a drying
tube containing calcium chloride on the sidearm of the adapter to protect the distillate
from moisture. The receiver should be cooled in an ice-water bath to prevent loss of
product by evaporation and to ensure complete condensing of the distillate.

Microscale Procedure

Preparation: Answer the Pre-Lab Exercises on page PL. 15. Read or review Sections 2.3, 2.4, 2.9, 2.11, and 2.13.

Apparatus. A Pasteur pipet and apparatus for simple distillation, magnetic stirring, and *flameless* heating.

Setting Up Transfer 2 mL of impure cyclohexane containing a nonvolatile dye to a 5-mL conical vial. Equip the vial with the Hickman stillhead and a condenser as shown in Figure 2.36. Place the apparatus in the heating source and have your instructor check your apparatus before you start heating the vial.

Distillation Increase the temperature of the heating source until vapors of distillate begin to rise into the Hickman stillhead and condense into the flared portion of the head. Control the rate of heating so that the vapor line rises no more than halfway up the upper portion of the head; otherwise, distillate may be lost to the atmosphere. Be certain to record the bath temperature at which the cyclohexane distils. You may also measure the approximate temperature of the vapors by inserting a thermometer through the condenser and stillhead to a point in the top third of the conical vial just above the boiling liquid. This will give you an approximate boiling point of the cyclohexane. Terminate heating when about 0.3 mL of the original solution remains. Disconnect the vial from the Hickman stillhead and, using a Pasteur pipet, transfer the distillate to a properly labeled screw-cap vial.

Finishing Touches Unless directed otherwise, return the *distilled* and *undistilled cyclohexane* to a bottle marked "Recovered Cyclohexane."

EXERCISES

1. Define the following terms:
   a. simple distillation
   b. head temperature
   c. pot temperature
   d. Raoult's law
   e. ideal solution
   f. Dalton's law

2. Sketch and completely label the apparatus required for a simple distillation.

3. Explain the role of the stirbar or spinvane that is normally added to a liquid that is to be heated to boiling.

4. In a miniscale distillation, the top of the mercury bulb of the thermometer placed at the head of a distillation apparatus should be adja ︶ the exit opening to the condenser. Explain the effect on the observed tem︐ re reading if the bulb is placed (a) below the opening to the condenser ( ) ᷇ the opening.

5. At 100 °C, the vapor pressures for water, methan︰ ᷇hanol are 760, 2625, and 1694 Torr, respectively. Which compound has the highe᷇ normal boiling point and which the lowest?

It is easy to separate a volatile compound from a nonvolatile one by simple distillation (Sec. 4.2). The same technique may also be used to separate volatile compounds from one another if their boiling points differ by at least 40–50 °C. If this is not the case, the technique of fractional distillation is normally used to obtain each volatile component of a mixture in pure form. The theoretical basis of this technique is the subject of the following discussion.

THEORY

For simplicity we shall consider only *ideal solutions* (Sec. 4.2) consisting of two volatile components, designated X and Y. Solutions containing more than two such components are often encountered, but their behavior on distillation may be understood by extension of the principles developed here for a binary system.

The vapor pressure of a compound is a measure of the ease with which its molecules escape the surface of a liquid. When the liquid is comprised of two volatile components, in this case X and Y, the number of molecules of X and of Y in a given volume of the vapor above the mixture will be proportional to their respective partial vapor pressures. This relationship is expressed mathematically by Equation 4.5, where N_X'/N_Y' is the ratio of the mole fractions of X and Y in the *vapor* phase. The mole fraction of each component may be calculated from the equations $N_X' = P_X/(P_X + P_Y)$ and $N_Y' = P_Y/(P_X + P_Y)$. The partial vapor pressures, P_X and P_Y, are determined by the composition of the liquid solution according to Raoult's law (Eq. 4.2). Since the solution boils when the sum of the partial vapor pressures of X and Y is equal to the external pressure, as expressed by Dalton's law (Eq. 4.4), the boiling temperature of the solution is determined by its composition.

$$\frac{N_X'}{N_Y'} = \frac{P_X'}{P_Y'} = \frac{P_X^\circ N_X}{P_Y^\circ N_Y} \qquad (4.5)$$

The relationship between temperature and the composition of the liquid and vapor phases of ideal binary, two-component, solutions is illustrated in Figure 4.3 for mixtures of benzene, bp 80 °C (760 Torr), and toluene, bp 111 °C (760 Torr). The lower curve is called the **liquid line** and gives the boiling points of all mixtures of these two compounds. The upper curve, the **vapor line**, is calculated using Raoult's law and defines the composition of the vapor phase in equilibrium with the boiling liquid phase at the same temperature. For example, a mixture whose composition is 58 mol % benzene and 42 mol % toluene will boil at 90 °C (760 Torr), as shown by point A in Figure 4.3. The composition of the vapor in equilibrium with the solution when it *first* starts to boil can be determined by drawing a horizontal line from the *liquid line* to the *vapor line*; in this case, the vapor has the composition 78 mol % benzene and 22 mol % toluene, as shown by point *B* in Figure 4.3. This is a key point, for it means that at any given temperature the *vapor phase is richer in the more volatile component than is the boiling liquid with which the vapor is in equilibrium*. This phenomenon provides the basis of fractional distillation.

When the liquid mixture containing 58 mol % benzene and 42 mol % toluene is heated to its boiling point, which is 90 °C (760 Torr), the vapor formed initially contains 78 mol % benzene and 22 mol % toluene. If this first vapor is condensed, the condensate

FIGURE 4.3 Temperature–composition diagram for a binary mixture of benzene and toluene.

would also have this composition and thus would be much richer in benzene than the original liquid mixture from which it was distilled. After this vapor is removed from the original mixture, the liquid remaining in the stillpot will contain a smaller mol % of benzene and a greater mol % of toluene because more benzene than toluene was removed by vaporization. The boiling point of the liquid remaining in the distilling flask will rise as a result. As the distillation continues, the boiling point of the mixture will steadily rise until it approaches or reaches the boiling point of pure toluene. The composition of the distillate will change as well and will ultimately consist of ''pure'' toluene at the end of the distillation.

Now let us return to the first few drops of distillate that could be obtained by condensing the vapor initially formed from the original mixture. This condensate, as noted earlier, has a composition identical to that of the vapor from which it is produced. Were this liquid to be collected and then redistilled, its boiling point would be the temperature at point C, namely 85 °C; this boiling temperature is easily determined by drawing a vertical line from the vapor line at point B to the liquid line at point C, which corresponds to the composition of the distillate initially obtained. The first distillate obtained at this temperature would have the composition D, 90 mol % benzene and 10 mol % toluene; this composition is determined from the intersection with the vapor line of the horizontal line from point C on the liquid line.

In theory, this process could be repeated again and again to give a very small amount of pure benzene. Similarly, collecting the *last* small fraction of each distillation and redistilling it in the same stepwise manner would yield a very small amount of pure toluene. If larger amounts of the initial and final distillates were collected, reasonable quantities of materials could be obtained, but a large number of individual simple distillations would be required. This process would be extremely tedious and time-consuming. Fortunately, the repeated distillation can be accomplished almost automatically in a single step by using a **fractional distillation column**, the theory and use of which are described next.

FRACTIONAL DISTILLATION COLUMNS AND THEIR OPERATION

There are many types of fractional distillation columns, but all can be discussed in terms of a few fundamental characteristics. The column provides a vertical path through which the vapor must pass from the stillpot to the condenser before being collected in the receiver (Fig. 2.37). This path is significantly longer than in a simple distillation apparatus. As the vapor from the stillpot rises up the column, some of it condenses *in the column* and returns to the distilling flask. *If the lower part of the distilling column is maintained at a higher temperature than the upper part of the column*, the condensate will be partially revaporized as it flows down the column. The uncondensed vapor, together with that produced by revaporization of the condensate in the column, rises higher and higher in the column and undergoes a repeated series of condensations and revaporizations. This repetitive process is equivalent to performing a number of simple distillations *within* the column, with the vapor phase produced in each step becoming increasingly richer in the *more* volatile component; the condensate that flows down the column correspondingly becomes richer in the less volatile component.

Each step along the path *A-B-C-D-E-F* of Figure 4.3 represents a *single* ideal distillation. One type of fractional distillation column, the bubble-plate column, was designed to effect one such step for each **plate** it contained. This led to the description of the efficiency of any fractional distillation column in terms of its equivalency to such a column in **theoretical plates**. Another index of the separating efficiency of a fractional distillation column is the **HETP**, which stands for ***h**eight **e**quivalent to a **t**heoretical **p**late* and is the vertical length of a column that is necessary to obtain a separation efficiency of one theoretical plate. For example, a column 60 cm long with an efficiency of 30 plates has an HETP value of 2 cm. Such a column would usually be better for research purposes than a 60-plate column that is 300 cm long (HETP = 5 cm) because of the small liquid capacity and **hold-up** of the shorter column. Hold-up refers to the condensate that remains in a column during and after distillation. When small amounts of material are to be distilled, a column must be chosen that has an efficiency, HETP, adequate for the desired separation and also a moderate to low hold-up.

As stated earlier, equilibrium between liquid and vapor phases must be established in a fractional distillation column so that the more volatile component is selectively carried to the top of the column and into the condenser, where the vapor condenses into the distillate. After all of the more volatile component is distilled, the less volatile one remains in the column and the stillpot; the heat supplied to the stillpot is then further increased in order to distil the second component. The most important requirements for performing a successful fractional distillation are (a) intimate and extensive contact between the liquid and the vapor phases in the column, (b) maintenance of the proper temperature gradient along the column, (c) sufficient length of the column, and (d) sufficient difference in the boiling points of the components of the liquid mixture. Each of these factors is considered here.

a. The desired contact between the liquid and vapor phases can be achieved by filling the column with an inert material having a large surface area. Examples of suitable packing materials include glass, ceramic, or metal pieces. Figure 2.38a shows a Hempel column packed with Raschig rings, which are pieces of glass tubing approximately 6 mm long. This type of column will have from two to four theoretical plates per 30 cm of length, if the distillation is carried out sufficiently slowly to maintain equilibrium conditions. Another type of fractional distillation column is the Vigreux column (Fig. 2.38b), which is useful for small-scale distillations of liquid where low hold-up is of paramount importance. A 30-cm Vigreux column will only have about 1 or 2 theoretical plates and consequently will be less efficient than the corresponding Hempel column. The Vigreux column has the advantage of a hold-up of less than 1 mL as compared to 2–3 mL for a Hempel column filled with Raschig rings.

b. Temperature gradient refers to the difference in temperature between the top and bottom of the column. The maintenance of the proper temperature gradient within the column is particularly important for an effective fractional distillation. Ideally, the temperature at the bottom of the column should be approximately equal to the boiling temperature of the solution in the stillpot, and it should decrease continually in the column until it reaches the boiling point of the more volatile component at the head of the column. The significance of the temperature gradient is seen in Figure 4.3, where the boiling temperature of the distillate decreases with each succeeding step, for example, A (90 °C) to C (85 °C) to E (82 °C).

The necessary temperature gradient from stillpot to stillhead in most distillations will be established *automatically* by the condensing vapors *if* the rate of distillation is properly adjusted. Frequently, this gradient can be maintained only by insulating the column with a material such as glasswool around the outside of the column. Insulation helps reduce heat losses from the column to the atmosphere. Even when the column is insulated, an insufficient amount of vapor may be produced to heat the column if the stillpot is heated too slowly, so that little or no condensate reaches the head. This rate must then be increased, but it must be kept below the point where the column is flooded. A **flooded column** is characterized by a column or ''plug'' of liquid that may be observed within the distillation column, often at the joint between it and the stillpot.

Factors directly affecting the temperature gradient in the column are the rate of heating of the stillpot and the rate at which vapor is removed at the stillhead. If the heating is too vigorous or the vapor is removed too rapidly, the entire column will heat up almost uniformly, and there will be no fractionation and thus no separation of the volatile components. On the other hand, if the stillpot is heated too strongly and if the vapor is removed too slowly at the top, the column will flood with returning condensate. Proper operation of a fractional distillation column thus requires *careful* control of the heat supplied to the stillpot and of the rate at which the distillate is removed at the stillhead. This rate should be *no more than* 1 drop every 2–3 sec.

The ratio of the amount of condensate returning to the stillpot and the amount of vapor removed as distillate per unit time is defined as the **reflux ratio**. A ratio of 10:1, for example, means that 10 drops of condensate return to the stillpot for each drop of distillate that is obtained. In general, the higher the reflux ratio the more efficient is a fractional distillation.

c. Correct column length is difficult to determine in advance of performing a fractional distillation. The trial-and-error technique must normally be used, and if a particular column does not efficiently separate a certain mixture, a longer column or a different type of column or column packing must be selected.

d. The difference in boiling points between the two pure components of a mixture should be no less than 20–30 °C in order for a fractional distillation to be successful when a Hempel column packed with Raschig rings or a similar type of packing is used. As mentioned previously, modifications in column length and type may result in the successful separation of mixtures having smaller boiling point differences.

In summary, the most important variables that can be controlled experimentally in a fractional distillation are correct selection of the column and column packing, adequate insulation of the column, and careful control of the rate of heating so as to provide the proper reflux ratio and a favorable temperature gradient within the column. Under such conditions, two different temperature **plateaus** will be observed in the fractional distillation of a typical binary mixture (Fig. 4.4). The head temperature should first rise to the normal boiling point of the more volatile component and remain there until that component is completely removed (Fig. 4.4, Fraction 1). The head temperature may then drop somewhat, indicating that the more volatile component has been removed. As additional heat is provided to the stillpot, the less volatile component will begin to distil, and the head temperature will rise to the boiling point of the second component (Fig. 4.4, Fraction 2). If the separation is efficient, the volume of this fraction, which contains a mixture of the two components, will be small. The head temperature should then remain constant until all of the second component has distilled (Fig. 4.4, Fraction 3).

FIGURE 4.4
Progress curve for a typical fractional distillation.

EXPERIMENTAL PROCEDURES

FRACTIONAL DISTILLATION OF A BINARY MIXTURE

Purpose: To demonstrate the technique for the separation of two volatile organic liquids.

1. Cyclohexane and toluene are highly flammable, so be sure that burners are not being used in the laboratory. Use *flameless* heating (Sec. 2.9).
2. Examine your glassware for cracks and other weaknesses before assembling the distillation apparatus. Look with particular care for "star" cracks in round-bottom flasks, because these can cause a flask to break upon heating.
3. Proper assembly of glassware is important in order to avoid possible breakage and spillage and to avoid the release of distillate vapors into the room. Be certain that all connections in the apparatus are tight *before* beginning the distillation. Have your instructor examine your set-up after it is assembled.
4. The apparatus used in these experiments *must* be open to the atmosphere at the receiving end of the condenser. *Never heat a closed system* because the pressure buildup may cause the apparatus to explode.
5. Be certain that the water hoses are securely fastened to your condensers so that they will not pop off and cause a flood. If heating mantles or oil baths are to be used for heating in this experiment, water hoses that come loose may cause water to spray onto electrical connections or into the heating sources, either of which is potentially dangerous to you and to those who work around you.
6. Avoid excessive inhalation of organic vapors at all times.

Safety Alert

Miniscale Procedure

Preparation   Answer the Pre-Lab Exercises on page PL. 13. Read or review Sections 2.2, 2.4, 2.9, 2.11, and 2.14.

Apparatus   A 50-mL round-bottom flask, apparatus for fractional distillation, magnetic stirring, and *flameless* heating.

Setting Up   Place 10 mL of cyclohexane and 20 mL of toluene in the round-bottom flask, and add a stirbar to ensure smooth boiling. Equip this flask for fractional distillation as shown in Figure 2.37. Pack a Hempel or similar distillation column, using the type of packing specified by your instructor. When packing the column, *be careful not to break off the glass indentations at the base of the column*. Do not pack the column too tightly, because heating a fractional distillation apparatus equipped with a column that is too tightly packed is analogous to heating a closed system. Insulate the fractionating column by wrapping it with glasswool. The position of the thermometer in the stillhead is particularly important; the *top* of the mercury thermometer bulb should be level with the *bottom* of the sidearm of the distillation head. Clean and dry three 25-mL containers, which may be bottles or Erlenmeyer flasks, for use as receiving flasks and label them *A*, *B*, and *C*. Place receiver *A* so that the tip of the vacuum adapter extends inside the neck of the container to minimize evaporation of the distillate. Have your instructor check your assembled apparatus before heating the distillation flask.

Distillation   Begin heating the stillpot using the heating method specified by your instructor. As the mixture is heated, the head temperature will rise to 81 °C (760 Torr), which is the normal boiling point of cyclohexane, and distillation will begin. Regulate the heat so that distillation continues steadily at a rate *no faster* than *1 drop of distillate every 1 to 2 sec*; if a drop of liquid cannot be seen suspended from the end of the thermometer, the rate of distillation is *too fast*.

The head temperature will remain at 81 °C for a period of time, but eventually will either rise or may drop slightly. Receiver *A* should be left in place until this increase or decrease is observed. As soon as the temperature deviates from 81 °C by more than ±3 °C, change to receiver *B* and increase the amount of heat supplied to the distillation flask. The temperature will again start to rise, and more liquid will distil. Leave receiver *B* in place until the temperature reaches 110 °C (760 Torr), which is the normal boiling point of toluene, and change to receiver *C*. Continue the distillation until 1–2 mL of liquid remains in the pot, and then discontinue heating.

Analysis   Record the volumes of the distillate collected in each receiver by means of a graduated cylinder. Allow the liquid in the column to drain into the distillation flask, then record the volume of this pot residue. If instructed to do so, submit or save 0.2-mL samples of each fraction *A*, *B*, and *C* for GLC analysis (Sec. 6.4).

Finishing Touches   Unless directed to do otherwise, pour the *pot residue* into the container for nonhalogenated organic liquids and return the *distillation fractions* to a bottle marked "Recovered Cyclohexane and Toluene."

EXERCISES

1. Define the following terms:

   a. fractional distillation   f. mole fraction

   b. head temperature   g. height equivalent to a theoretical plate (HETP)

   c. pot temperature   h. temperature gradient

   d. Raoult's law   i. Dalton's law

   e. ideal solution   j. reflux ratio

2. Sketch and completely label the apparatus required for fractional distillation.

3. Explain why a packed fractional distillation column is more efficient than an unpacked column for separating two closely boiling liquids.

4. If heat is supplied to the stillpot too rapidly, the ability to separate two liquids by fractional distillation may be drastically reduced. In terms of the theory of distillation presented in the discussion, explain why this is so.

5. Explain why the column of a fractional distillation apparatus should be aligned as near to the vertical as possible.

6. Explain the role of the stirbar normally added to a liquid that is to be heated to boiling.

7. The top of the mercury bulb of the thermometer placed at the head of a distillation apparatus should be adjacent to the exit opening to the condenser. Explain the effect on the observed temperature reading if the bulb is placed (a) below the opening to the condenser or (b) above the opening.

8. In the fractional distillation of your mixture of cyclohexane and toluene, what can be learned about the efficiency of the separation on the basis of the relative volumes of Fractions *A*, *B*, and *C*?

9. a. A mixture of 80 mol % *n*-propylcyclohexane and 20 mol % *n*-propylbenzene is distilled through a simple distillation apparatus; assume that no fractionation occurs during the distillation. The boiling temperature is found to be 157 °C (760 Torr) as the first small amount of distillate is collected. The standard vapor pressures of *n*-propylcyclohexane and *n*-propylbenzene are known to be 769 Torr and 725 Torr, respectively, at 157.3 °C. Calculate the percentage of each of the two components in the first few drops of distillate.

   b. A mixture of 80 mol % benzene and 20 mol % toluene is distilled under exactly the same conditions as in Part **a**. Using Figure 4.3, determine the distillation temperature and the percentage composition of the first few drops of distillate.

   c. The normal boiling points of *n*-propylcyclohexane and *n*-propylbenzene are 156 °C and 159 °C, respectively. Compare the distillation results in Parts **a** and **b**. Which of the two mixtures would require the more efficient fractional distillation column for separation of the components? Why?

10. Examine the boiling-point-composition diagram for mixtures of toluene and benzene given in Figure 4.3.

    a. Assume you are given a mixture of these two liquids of composition 80 mol % toluene and 20 mol % benzene and that it is necessary to effect a fractional distillation that will afford at least some benzene of greater than 99% purity. What would be the *minimum* number of theoretical plates required in the fractional distillation column chosen to accomplish this separation?

b. Assume that you are given a 20-cm Vigreux column having an HETP of 10 cm in order to distil a mixture of 58 mol % benzene and 42 mol % toluene. What would be the composition of the first small amount of distillate that you obtained?

11. At 50 °C, the vapor pressures for methanol and ethanol are 406 and 222 Torr, respectively. Given a mixture at 50 °C that contains 0.1 mol of methanol and 0.2 mol of ethanol, compute the partial pressures of each liquid and the total pressure.

12. Figure 4.5 shows a temperature–composition diagram for a mixture of carbon tetrachloride and toluene. Answer the following questions, assuming that these substances form an ideal solution:

a. What are the boiling points of pure carbon tetrachloride and toluene?

b. Answer the following questions for a mixture that contains 50 mol % carbon tetrachloride and 50 mol % toluene:

  i. At what temperature will it boil?

  ii. What is the composition, in mol %, of the first small amount of vapor that forms?

  iii. If the vapor in **ii** is condensed and the resulting liquid is heated to its boiling point, what is the composition, in mol %, of the new vapor?

c. A mixture of unknown composition is heated until it just starts to boil, and the composition of the resulting vapor is found to be 50 mol % toluene and 50 mol % carbon tetrachloride. What is the percent composition of the original mixture, and at what temperature does the mixture start to boil?

d. Calculate the amount of carbon tetrachloride and toluene in the liquid and vapor phases in a boiling mixture of the two at 85 °C and at 100 °C. Express your answer both in mole percent and in mole fraction.

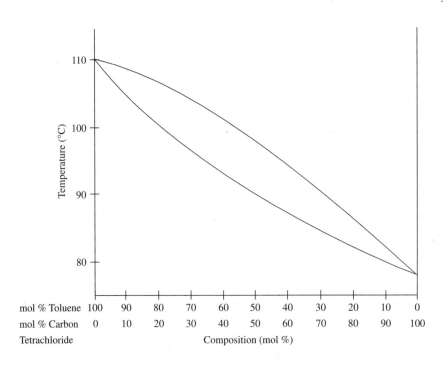

FIGURE 4.5 Temperature–composition diagram for Exercise 12.

The separation and purification of *volatile* organic compounds that are **immiscible** or nearly immiscible with water can often be accomplished by steam distillation. The technique normally involves the codistillation of a mixture of organic liquids and water, although some organic solids can also be separated and purified by this means. Of the various distillation methods, steam distillation is utilized least frequently, owing to the rather stringent limitations on the types of substances for which it can be used. These limitations as well as the virtues of this technique are revealed by considering the principles underlying steam distillation.

THEORY AND DISCUSSION

The partial pressure P_i of each component i of a mixture of *immiscible*, volatile substances at a given temperature is equal to the vapor pressure P_i° of the pure compound at the same temperature (Eq. 4.6) and does not depend on the mole fraction of the compound in the mixture; in other words, each component of the mixture vaporizes independently of the others. This behavior contrasts sharply with that exhibited by solutions of miscible liquids, for which the partial pressure of each constituent of the mixture depends on its mole fraction in the solution (Raoult's law, Eq. 4.2).

$$P_i = P_i^\circ \tag{4.6}$$

Now recall that, according to Dalton's law (Eq. 4.4), the total pressure, P_T, of a mixture of gases is equal to the sum of the partial pressures of the constituent gases, meaning that the total vapor pressure of a mixture of immiscible, volatile compounds is given by Equation 4.7. This expression shows that the total vapor pressure of the mixture at any temperature is always higher than the vapor pressure of even the most volatile component at that temperature, owing to the contributions of the vapor pressures of the other constituents in the mixture. The boiling temperature of a mixture of immiscible compounds must then be *lower* than that of the lowest-boiling component.

$$P_T = P_a^\circ + P_b^\circ + \cdots + P_i^\circ \tag{4.7}$$

Application of the principles just outlined is seen in an analysis of the steam distillation of an immiscible mixture of water, bp 100 °C (760 Torr), and bromobenzene, bp 156 °C (760 Torr). Figure 4.6 is a plot of the vapor pressure versus temperature for each pure substance and for a mixture of these compounds. Analysis of this graph shows that the mixture should boil at about 95 °C (760 Torr), the temperature at which the total vapor pressure equals standard atmospheric pressure. As theory predicts, this temperature is below the boiling point of water, which is the lowest-boiling component in this example.

The ability to distil a compound at the relatively low temperature of 100 °C or less by means of a steam distillation is often of great use, particularly in the purification of substances that are heat-sensitive and would therefore decompose at higher temperatures. It is useful also in the separation of compounds from reaction mixtures that contain large amounts of nonvolatile residues such as inorganic salts.

The composition of the condensate from a steam distillation depends upon the molecular weights of the compounds being distilled and upon their respective vapor

FIGURE 4.6   Vapor pressure–temperature graph for pure bromobenzene, pure water, and a mixture of bromobenzene and water.

pressures at the temperature at which the mixture steam-distils. To illustrate this, consider a mixture of two immiscible components, A and B. If the vapors of A and B behave as ideal gases, the ideal gas law can be applied and Equations 4.8a and 4.8b are obtained. In these two expressions, $P°$ is the vapor pressure

$$P_A^\circ V_A = (g_A/M_A)(RT) \tag{4.8a}$$

$$P_B^\circ V_B = (g_B/M_B)(RT) \tag{4.8b}$$

of the pure liquid, V is the volume in which the gas is contained, g is the weight in grams of the component in the gas phase, M is its molecular weight, R is the gas constant, and T is the absolute temperature in kelvins (K). Dividing Equation 4.8a by Equation 4.8b gives Equation 4.9.

$$\frac{P_A^\circ V_A}{P_B^\circ V_B} = \frac{g_A/M_A(RT)}{g_B/M_B(RT)} \tag{4.9}$$

Because the RT factors in the numerator and denominator are identical and because the volume in which the gases are contained is the same for both ($V_A = V_B$), these terms in Equation 4.9 cancel to yield Equation 4.10.

$$\frac{\text{grams of } A}{\text{grams of } B} = \frac{(P_A^\circ)(\text{molecular weight of } A)}{(P_B^\circ)(\text{molecular weight of } B)} \tag{4.10}$$

Now let the immiscible mixture of A and B consist of bromobenzene and water, whose molecular weights are 157 g/mol and 18 g/mol, respectively, and whose vapor pressures at 95 °C, as determined from Figure 4.6, are 120 Torr and 640 Torr, respectively. The composition of the distillate at this temperature can be calculated from Equation 4.10 as shown in Equation 4.11. This calculation

$$\frac{g_{\text{bromobenzene}}}{g_{\text{water}}} = \frac{(120)(157)}{(640)(18)} = \frac{1.64}{1} \qquad (4.11)$$

indicates that on the basis of weight *more* bromobenzene than water is contained in the steam distillate, even though the vapor pressure of the bromobenzene is *much* lower at the temperature of the distillation.

Organic compounds generally have molecular weights much higher than that of water, so it is possible to steam-distil compounds having vapor pressures of only about 5 Torr at 100 °C with a fair efficiency on a weight-to-weight basis. Thus, *solids* that have vapor pressures of at least this magnitude can even be purified by steam distillation. Examples are camphor, used in perfumes, and naphthalene, present in some brands of mothballs. The rather high vapor pressures of these solids is evidenced by the fact that their odors can be detected at room temperature.

In summary, steam distillation provides a method for separating and purifying moderately volatile liquid and solid organic compounds that are insoluble or nearly insoluble in water from nonvolatile compounds. Although relatively mild conditions are used in steam distillation, it cannot be used for substances that decompose on prolonged contact with steam or hot water, that react with water, or that have vapor pressures of 5 Torr or less at 100 °C, all of which are significant limitations to the method.

4.5 Steam Distillation: Isolation of Citral from Lemon Grass Oil

Citral is a natural oil that is actually a mixture of two **diastereomers**, the **geometric isomers** geranial (**1**) and neral (**2**), which are extremely difficult to separate. Citral possesses a lemonlike odor and taste. Although this odor and taste is pleasant to humans, it is less attractive to other species. For example, ants secrete citral to ward off potential predators. Thus, citral serves as a **defense pheromone**. There are several commercial applications of citral. For example, it may be added to perfumes whenever a lemonlike essence is desired, and it is used as an intermediate for the synthesis of vitamin A (**3**).

1
Geranial

2
Neral

3
Vitamin A

The commercial importance of citral has stimulated an extensive search for its presence in natural products. One source is the oil from the skins of lemons and oranges, although it is only a minor component of this oil. However, citral is the major constituent of the oil obtained from lemon grass, and in fact 75–85% of crude oil derived from pressing lemon grass is the desired natural product.

Citral is a chemically labile substance, and its isolation therefore presents a challenge. This task is simplified, however, by the fact that citral, bp 229 °C (760 Torr), is relatively volatile and has a low solubility in water. These properties make it a suitable candidate for steam distillation, a technique that allows distillation of citral from crude lemon grass oil at a temperature below 100 °C, which is far below its normal boiling point. Neutral conditions are maintained in steam distillation as is the partial exclusion of oxygen, so the possibility of oxidation and/or polymerization of citral is minimal.

The steam distillation can be performed by two different methods, as described in Section 2.16. One involves using an external steam source, which may be either a laboratory steam line or a steam generator, and passing the steam through a mixture of lemon grass oil and water. This method has some advantages but is experimentally more difficult than an alternate method, which involves heating a mixture of lemon grass oil and water and collecting the distillate. Although this latter procedure is simpler from the experimental standpoint, there are some limitations to its application. For example, in the steam distillation of only slightly volatile substances, a large initial volume of water will be required, or water must be added as the distillation proceeds, perhaps by means of an addition funnel. However, the technique for internal generation of steam is well-suited for this experiment.

EXPERIMENTAL PROCEDURES

STEAM DISTILLATION OF CITRAL FROM LEMON GRASS OIL

Purpose: To isolate the natural product citral using the technique of steam distillation.

Safety Alert

1. Steam distillation involves the use of glassware that becomes very hot. Exercise care when handling hot glassware.
2. Be certain that the steam distillate is cooled below 30 °C before extracting it with diethyl ether; otherwise excessive pressure may develop in the separatory funnel and may blow out the stopper.
3. Diethyl ether is *extremely* flammable. Be certain that there are no flames in the vicinity during its use and its removal from the citral.

Miniscale Procedure

Preparation   Answer the Pre-Lab Exercises on page PL. 17. Read or review Sections 2.2, 2.4, 2.9, 2.13, 2.16, 2.19, 2.21, 2.29, and 4.4.

Apparatus   A 125-mL separatory funnel, a 125-mL round-bottom flask, and apparatus for steam distillation using an internal steam source, simple distillation, and *flameless* heating.

Setting Up   Determine the weight of 2.5 mL of lemon grass oil, and then place it and 60 mL of water in the round-bottom flask. Equip this flask for steam distillation (Fig. 2.42); use the heating source suggested by your instructor. Have your instructor check your assembled apparatus before continuing the experiment.

Distillation   Heat the flask to boiling and adjust the heat source so that the distillation proceeds as rapidly as possible; avoid applying excess heat to the distillation flask, especially when the volume of water in the flask has been reduced below about 25 mL. Continue the distillation until oil droplets no longer appear in the distillate, which should occur after about 30 mL of distillate has been collected. Allow the distillate to cool to room temperature or below, using an ice-water bath if necessary.★

Isolation   Transfer the cooled distillate to a separatory funnel and extract it sequentially with two 15-mL portions of diethyl ether. The funnel should be vented *frequently* to avoid the buildup of pressure. Transfer the organic layer from the separatory funnel to a 50-mL Erlenmeyer flask, and add about 0.5–1.0 g of anhydrous calcium chloride. Allow the ethereal solution to remain in contact with the drying agent until the organic layer is dry, as evidenced by observing that it is completely clear. If the experiment is stopped at this point, loosely stopper the flask and store it in a hood; *never leave flasks containing diethyl ether in your locker drawer.*★

   Decant the dried organic solution into a *tared* 125-mL round-bottom flask, and equip the flask for simple distillation. Evaporate the solvent using simple distillation. Alternatively, rotary evaporation or one of the other techniques in Section 2.29 as directed by your instructor may be used to concentrate the solution. After the ether is completely removed, the pot residue is crude citral.

Analysis   Determine the weight of citral isolated and calculate the percentage recovery of citral based on the weight of the original sample of lemon grass oil. If instructed to do so, save or submit a sample of citral for GLC analysis (Sec. 6.4). You may be asked to perform one or more of the following tests on citral.

1. Test for a carbon-carbon double bond. Perform the tests for unsaturation described in Sections 23.8A and 23.8B
2. Test for an aldehyde. Perform the chromic acid test outlined in Section 23.7D.
3. Analyze the citral using gas chromatography to assess the purity of the product.

Flush the *aqueous solution* remaining in the distillation flask down the drain. Do the same with the *aqueous steam distillate* once you have completed the extraction. Pour the *solution from the test for unsaturation* into the container for halogen-containing liquids. Neutralize the *solution for the chromic acid test* and then pour it into the container for hazardous heavy metals.

Finishing Touches

EXERCISES

1. What type of product is expected from the reaction of citral with Br_2/CH_2Cl_2? with chromic acid?

2. Provide structures for the semicarbazone and the 2,4-dinitrophenylhydrazone of citral.

3. Why does the citral float on the surface of the aqueous distillate rather than sinking to the bottom?

4. Explain why the substitution of 1-propanol, bp 97 °C (760 Torr), for water in a steam distillation would not work.

5. Suppose that you are to steam-distil a sample of a natural product whose vapor pressure at 100 °C is known to be half that of citral. What consequence would this have on the amount of distillate required per mole of the natural product present?

6. Both geranial (**1**) and vitamin A (**3**) are members of the class of natural products called **terpenes**. This group of compounds have the common characteristic of being biosynthesized by linkage of the appropriate number of five-carbon units having the skeletal structure shown below. Determine the number of such units present in each of these terpenes and indicate the bonds linking the various individual units.

CHAPTER

5

Extraction

The desired compound from a reaction is frequently part of a mixture, and its isolation in pure form can be a significant experimental challenge. Two of the more common methods for separating and purifying organic liquids and solids are **recrystallization** and **distillation**; these procedures are discussed in Chapters 3 and 4, respectively. Many other techniques are also available for these purposes, two very important examples being **extraction** and **chromatography**. As you will see here and in Chapter 6, both of these methods involve partitioning of compounds between two *immiscible* phases. This process is termed **phase distribution** and can result in separation of compounds if they distribute differently between phases.

Distribution of solutes between phases is the result of **partitioning** or **adsorption** phenomena. Partitioning involves the difference in solubilities of a substance in two immiscible solvents—in other words, *selective dissolution*. Adsorption, on the other hand, is based on the selective attraction of a substance in a liquid or gaseous mixture to the surface of a solid phase. The various chromatographic techniques depend on both of these processes, whereas the extraction process relies only on partitioning.

The process of extraction involves *selectively* removing one or more components of a solid, liquid, or gaseous mixture into a separate phase. The substance being extracted will partition between the two immiscible phases that are in contact. As would be expected, the ratio of its distribution between the phases will depend on the relative solubility of the solute in each phase.

5.2 | Theory of Extraction

Liquid-liquid extraction is one of the most common methods for removing an organic compound from a mixture. This process is used by chemists not only in the isolation of natural products but also in the isolation and purification of products in virtually *every* organic synthesis.

The technique involves distributing a solute A between two immiscible liquids, S_x, the **extracting phase**, and S_o, the **original phase**. The immiscible liquids normally encountered in the organic laboratory are water and some organic solvent, such as diethyl ether, $(C_2H_5)_2O$, or dichloromethane, CH_2Cl_2. At a given temperature the amount of A, in g/mL, in each phase is expressed *quantitatively* in terms of a constant, K, commonly called the **partition coefficient** (Eq. 5.1). Strictly speaking, the volumes of *solution* should be used in the definition of $[A]$ but if the solutions are dilute, only slight errors result if volumes of *solvent* are used. Furthermore, a close *approximation* of the partition coefficient K may be obtained by simply dividing the solubility of A in the extracting solvent S_x by the solubility of A in the original solvent S_o.

$$K = \frac{[A] \text{ in } S_x}{[A] \text{ in } S_o} \tag{5.1}$$

The process of liquid-liquid extraction can be considered a competition between two immiscible liquids for solute A, with solute A distributing or partitioning between these two liquids when it is intimately contacted with them. The mathematical expression of Equation 5.1 shows that *at equilibrium* the ratio of concentrations of A in the two phases will always be constant.

Equation 5.1 leads to several important predictions. For example, if $K > 1$, solute A will be mainly in the *extracting* solvent, S_x, so long as the *volume*, V_x, of this solvent is at least equal to the volume, V_o, of the original solvent S_o; the amount of solute remaining in S_o will depend on the value of K. By recasting Equation 5.1 into Equation 5.2, it is easily seen that if the volumes of S_x and S_o are equal, the value of K is simply the ratio of the grams of A in each solvent. Because the product of the ratios A_x/A_o and V_o/V_x must be constant for a given solvent pair, increasing the volume of extracting solvent S_x will result in a net *increase* in the amount of solute A in S_x.

$$K = \frac{\text{grams of } A \text{ in } S_x}{\text{grams of } A \text{ in } S_o} \times \frac{\text{mL of } S_o}{\text{mL of } S_x} = \frac{A_x}{A_o} \times \frac{V_o}{V_x} \tag{5.2}$$

There are practical, economic, and environmental reasons, however, that limit the quantities of organic solvents that can realistically be used in extractions. The question thus arises as to whether it is better to perform a single extraction with all of the solvent, or to perform several extractions with smaller volumes. For example, if an organic compound is dissolved in 10 mL of water, and only 30 mL of diethyl ether is used, will *three* extractions with 10-mL portions of ether provide better recovery of solute than a *single* one with 30 mL?

Applying Equation 5.2 would provide the answer, but the process is tedious for multiple extractions. This equation can be generalized, however, to accommodate multiple extractions in terms of the fraction, F_A, of solute A remaining in the original solvent S_o after n extractions, using a constant volume of an immiscible solvent S_x for each extraction. Thus, the amount of solute remaining in S_o after one extraction with S_x is obtained by rearranging Equation 5.3 to Equation 5.4. The fraction, F_A, of A still in the

$$K = \left(\frac{A_o - A_1}{V_x}\right)\left(\frac{V_o}{A_1}\right) \tag{5.3}$$

where A_o = amount (grams) of solute in S_o before extraction

$A_1 = (A_o - A_x)$ = amount (grams) of solute in S_o after extraction

V_o and V_x = volume (mL) of original and extracting solvents, respectively

$$A_1 = \frac{(A_o V_o)}{K V_x + V_o} \tag{5.4}$$

$$F_A = \frac{C_f}{C_i} = \frac{V_o}{K V_x + A_o} \tag{5.5}$$

where $C_i = A_o/S_o$ and $C_f = A_1/S_o$, respectively

original solvent is obtained through Equation 5.5, in which C_i and C_f are the *initial* and *final* concentrations, respectively. For a second extraction, Equations 5.6–5.8 result. Equation 5.8 can be generalized to Equation 5.9 when n extractions are performed.

$$K = \left(\frac{A_1 - A_2}{V_x} \right) \left(\frac{V_o}{A_2} \right) \tag{5.6}$$

$$K = A_o \left(\frac{V_o}{K V_x + V_o} \right)^2 \tag{5.7}$$

$$F_{A'} = \frac{C_f}{C_i} = \left(\frac{V_o}{K V_x + A_o} \right)^2 \tag{5.8}$$

$$F_A = \frac{C_f}{C_i} = \left(\frac{V_o}{K V_x + A_o} \right)^n \tag{5.9}$$

Performing calculations according to Equation 5.9 for the case of $K = 5$, we obtain $F_A = 1/16$ for a single extraction with 30 mL of diethyl ether, and 1/216 for three extractions with 10-mL portions. This means that 6.3% of A remains in the aqueous phase when *one* extraction is performed, whereas the value drops to only 0.5% in the case of *three* successive extractions with the *same* total volume of solvent. The amount of A that could be isolated is increased by some 6% with multiple extractions. Setting K at 2 gives corresponding values for F_A of 1/7 and 1/27, respectively, which translates to a 10% increase in the quantity of A that is removed from the aqueous layer with multiple extractions.

From these calculations, it is clear that multiple extractions become increasingly important as the value of K decreases. The improved recovery of solute from multiple extractions, however, must be balanced with the practical consideration that the relatively small increase in recovery may not justify the additional time required to do multiple extractions unless the product is of great value. The partition coefficients K are generally large for the extractions required in the procedures provided in this textbook, so only one or two extractions are usually required.

The selection of the appropriate extracting solvent is obviously a key to the success of this technique of isolating and purifying compounds, and important guidelines for making the choice are summarized here.

1. The extracting solvent *must not react* in a chemically irreversible way with the components of the mixture.
2. The extracting solvent *must be immiscible*, or nearly so, with the original solution.
3. The extracting solvent *must selectively remove* the desired component of the solution being extracted. That is, the partition coefficient K of the component being removed must be *high* while the partition coefficients of all other components should be *low*.
4. The extracting solvent *should be readily* separable from the solute. Use of a volatile solvent is advantageous because the solvent often is removed by simple distillation.

5.3   Base Extractions

The discussion of Section 5.2 focuses on partitioning of *one* substance between two *immiscible* solvents. Now consider what happens if a mixture of *two or more* compounds is present in a given volume of solvent S_o and an extraction using a solvent S_x is performed. If the partition coefficient K of one of the components, let us call it A, is *significantly* greater than 1.0 and if those of the others are *significantly* less than 1.0, the majority of A will be in S_x whereas most of the other compounds will remain in S_o. Physical separation of the two solvents will give at least partial separation, and thus purification, of the solute A from the other components of the mixture.

Solutes differing significantly in polarity should have very different coefficients K for partitioning between nonpolar and polar solvents. An example of this would be the distribution of two organic compounds, the first neutral and nonpolar, and the second ionic and polar, between a nonpolar solvent and a polar solvent. If the compounds are originally dissolved in the nonpolar solvent, and this solution is shaken with the polar solvent, the neutral compound would preferentially partition into the *nonpolar* phase, whereas the polar constituent would preferentially partition into the *polar* phase. Separating the two phases effects a separation of the two solutes.

Carboxylic acids and phenols (**1** and **3**, Eqs. 5.10 and 5.11, respectively) are two classes of organic compounds containing **functional groups** that are **polar** and **hydrophilic** (water-loving). Unless they contain fewer than about six carbon atoms, such compounds are generally either *insoluble* or only *slightly soluble* in water because of the **hydrophobic** (water-avoiding) properties of the carbon-containing portion, R or Ar, of the molecule. They are *soluble* in common organic solvents like dichloromethane or diethyl ether that have at least modest polarity (Table 3.1, p. 90). Consequently, carboxylic acids or phenols dissolved in diethyl ether, for example, will largely remain in that phase when the solution is extracted with water.

$$R-C\overset{O}{\underset{O}{\big\langle}}O-H \quad + \quad B^-\,Na^+\,(aq) \;\rightleftharpoons\; R-C\overset{O^-}{\underset{O}{\big\langle}}\;Na^+ \; + \; B-H \qquad (5.10)$$

1		**2**
A carboxylic acid	A base	A carboxylate
Water-*insoluble*	(as sodium salt)	(as sodium salt)
$K_{water/org} < 1$		Water-*soluble*
		$K_{water/org} > 1$

$$Ar-O\overset{\frown}{\underset{H}{\big)}} \quad B^-\,Na^+\,(aq) \;\rightleftharpoons\; Ar-O^-\,Na^+ \; + \; B-H \qquad (5.11)$$

3	**4**
A phenol (Ar = aryl)	A phenoxide
Water-*insoluble*	(as sodium salt)
$K_{water/org} < 1$	Water-*soluble*
	$K_{water/org} > 1$

(A Base (as sodium salt))

Now consider what happens if the organic solution is extracted with a *basic* aqueous solution. If the base, B^-Na^+, is strong enough, the organic acid **1** or **3** will be converted into the corresponding **conjugate bases 2** or **4** (Eqs. 5.10 and 5.11). Because it is a salt, the conjugate base is highly polar, and $K_{water/org} > 1$. Thus the conjugate bases of organic acids may be selectively extracted from an organic phase into an aqueous phase. If the basic extract is then neutralized by an acid such as hydrochloric acid, the conjugate base **2** or **4**, will be protonated to regenerate the organic acid **1** or **3** (Eqs. 5.12 and 5.13). Because the acid is water-insoluble, it will appear as either a precipitate or a second layer, if it is a liquid. The desired organic acid may then be recovered by filtration or separation of the layers.

$$\text{R—C} \begin{matrix} \text{O}^- \\ \\ \text{O} \end{matrix} \text{Na}^+ + \text{H—Cl (aq)} \rightleftharpoons \text{R—C} \begin{matrix} \text{O—H} \\ \\ \text{O} \end{matrix} + \text{Na}^+\,\text{Cl}^- \qquad \textbf{(5.12)}$$

2   **1**

Water-*soluble*   Water-*insoluble*

$$\text{Ar—O}^-\;\text{Na}^+ + \text{H—Cl} \rightleftharpoons \text{Ar—O} \begin{matrix} \\ \text{H} \end{matrix} + \text{Na}^+\,\text{Cl}^- \qquad \textbf{(5.13)}$$

4   **3**

Water-*soluble*   Water-*insoluble*

Thus, two water-insoluble organic compounds, **HA**, which is acidic, and **N**, which is neutral, that are dissolved in an organic solvent may be separated by selectively extracting the acidic compound into an basic aqueous phase. After the aqueous and organic phases are separated, **HA** is recovered from the aqueous phase upon neutralization, and **N** is obtained by removing the solvent (Fig. 5.1).

The choice of the base is determined by the acidity of the organic acid **HA**. This is defined in aqueous solution by K_a in Equation 5.14, and is often expressed as pK_a (Eq. 5.15).

$$K_{a(\text{HA})} = \frac{[\text{A}^-][\text{H}_3\text{O}^+]}{[\text{HA}]} \qquad \textbf{(5.14)}$$

$$pK_a = -log_{10}\,K_a \qquad \textbf{(5.15)}$$

The next step in determining what base is needed is to consider the equilibrium shown in Equation 5.16, in which **HA** is the organic acid, **B⁻** is the base being used, and **HB** is the **conjugate acid** of this base. The **equilibrium constant**, K_{eq}, for this process is defined by Equation 5.17. Given the expressions for K_a for **HA** and **HB** (Eq. 5.14 and 5.18, respectively), Equation 5.17 transforms to Equation 5.19. From this equation we see that log_{10} of the equilibrium constant, K_e, equals $pK_{a(\textbf{HB})} - pK_{a(\textbf{HA})}$. Thus, we

H A   Extract with
N   aqueous base   ⎯→ *Aqueous phase* ⎯→ **A⁻** ⎯ *Neutralization with acid* ⎯→ **H A**

                             ⎯→ *Organic phase* ⎯→ **N** ⎯ *Removal of solvent* ⎯→ **N**

FIGURE 5.1   Separating an acidic compound and a neutral compound.

can predict how effective a particular base **B⁻** will be in converting an organic acid **HA** to its anion simply by knowing the relative acidities of **HA** and **HB**! The equilibrium for Equation 5.14 will lie to the right, that is $log_{10} K_e$ will be > 1 whenever the acidity of **HA** is *greater* than that of **HB**.

$$\textbf{HA} + \textbf{B}^- \underset{}{\overset{K_{eq}}{\rightleftharpoons}} \textbf{A}^- + \textbf{HB} \tag{5.16}$$

$$K_{eq} = \frac{[\text{A}^-][\text{HB}]}{[\text{HA}][\text{B}^-]} \tag{5.17}$$

$$K_{a(\text{HB})} = \frac{[\text{B}^-][\text{H}_3\text{O}^+]}{[\text{HB}]} \tag{5.18}$$

$$K_{eq} = \frac{K_{a(\text{HA})}}{K_{a(\text{HB})}} \text{ and } log_{10} K_{eq} = log_{10} K_{a(\text{HA})} - log_{10} K_{a(\text{HB})} \tag{5.19}$$

Let us consider two specific examples of mixtures of acidic and neutral compounds. Assume that you have a solution of benzoic acid (**5**) and naphthalene (**7**), a neutral compound, in diethyl ether and that you wish to separate the two. Both compounds and the solvent are insoluble in water at room temperature. The strategy is to convert **5** to its water-soluble conjugate base **6** by using an aqueous base to extract the organic solution. The pK_a of benzoic acid is 4.19, so any base whose *conjugate acid* has a pK_a *greater* than this will yield a value of $log_{10} K_{eq} > 1$ (Eq. 5.19). Thus aqueous hydroxide, whose conjugate acid is water (pK_a 15.7) would give $log_{10} K_{eq} = 10.5$, which means that this base would be extremely effective for the extraction of **5**. Indeed, aqueous hydroxide is the most commonly used solution for extracting organic acids into an aqueous phase.

| **5** | **6** | **7** |
| Benzoic acid | Benzoate | Naphthalene |

Other aqueous bases could also be used. For example, carbonate (**8**) and bicarbonate (**9**) would work because their conjugate acids, **9** and **10** (Eq. 5.20), have pK_as

| **8** | **9** | **10** |
| Carbonate | Bicarbonate | Carbonic acid |

of 10.25 and 6.37, respectively. These bases are not as strong as hydroxide, so on the purely *thermodynamic* basis described by Equation 5.19, they would provide a lower

value of $\log_{10} K_{eq}$. However, both **8** and **9** may be converted to carbonic acid (**10**) upon di- and monoprotonation, respectively (Eq. 5.20). Carbonic acid is *kinetically* unstable and decomposes to carbon dioxide and water (Eq. 5.21), and this disrupts the equilibrium of Equation 5.16, driving it to the *right* as **HB**, in the form of **10**, undergoes decomposition. The net result would be efficient deprotonation of **5** by these bases.

$$H_2CO_3 \longrightarrow CO_2 + H_2O \qquad \text{(5.21)}$$

The decomposition of **10** presents a practical problem when **8** and **9** deprotonate acids for purposes of extraction, however. If a separatory funnel is being used for the extraction, the gas pressure that results from the formation of carbon dioxide can blow out the stopcock and stopper. Alternatively, if a screw-cap vessel is being used, product may be ejected when the cap is loosened to relieve the gas pressure. Thus as a general rule, it is preferable to use hydroxide for basic extractions.

Turning to a second example, assume you have a solution of 2-naphthol (**11**), which is a phenol, and naphthalene (**7**) in diethyl ether. The pK_a of **11** is 9.51, so in this case either aqueous hydroxide or carbonate would give $K_{eq} > 1$; bicarbonate (**8**) is too weak a base to effect the deprotonation, however. Because aqueous hydroxide would provide $K_{eq} \gg 1$, it is the base of choice for this extraction.

| **11** | **12** |
| 2-Naphthol | 2-Naphthoxide |

In the experimental procedures that follow, two types of separations are described. One involves separating a mixture of benzoic acid (**5**) and naphthalene (**7**), using aqueous hydroxide for the extraction. The flow chart for this separation corresponds to that of Figure 5.1. The second procedure involves separating a mixture of **5**, **7**, and 2-naphthol (**11**). In this case, **5** is first *selectively* removed from the solution by extraction with bicarbonate, which *does not* deprotonate **11**. A second basic extraction using hydroxide removes **11** from the organic solution. The flow chart for this sequence is depicted in Figure 5.2. Both procedures show the power of using the pH-dependence of water solubility of organic compounds as a basis for a separation technique.

FIGURE 5.2 Separating two acidic compounds and a neutral compound.

EXPERIMENTAL PROCEDURES

BASE EXTRACTION

Purpose: To separate multi-component mixtures as a function of pH of solution.

1. Avoid letting any of the chemicals used in this experiment come in contact with your skin. If they do, *immediately* wash the affected areas with soap and water.
2. Diethyl ether is *extremely* volatile and flammable. Be certain there are *no flames* in your vicinity when using it for extraction and when removing it from the "Neutral Compound."
3. Wear latex gloves for the extraction, and use caution in handling wet glassware because it can easily slip out of your grasp.
4. When extracting a solution vent the separatory funnel or vial *often* to avoid a build-up of pressure.

Miniscale Procedures

Preparation   Answer the Pre-Lab Exercises on pages PL. 19 and/or 21. Read or review Sections 2.7, 2.10, 2.11, 2.13, 2.17, 2.21, 2.24, 2.25, 2.29, 3.2, and 3.3.

A. One-Base Extraction

Apparatus   Separatory funnel, ice-water bath, apparatus for magnetic stirring, vacuum filtration, simple distillation, and *flameless* heating.

Dissolution   Obtain 2 g of a mixture comprised of benzoic acid and naphthalene. Dissolve the mixture by swirling it with 30 mL of diethyl ether in an Erlenmeyer flask. If any solids remain, add more diethyl ether to effect complete dissolution. Transfer the solution to the separatory funnel.

Extraction   Extract the solution with a 15-mL portion of 2.5 M (10%) aqueous sodium hydroxide. *Be sure to hold both the stopcock and the stopper of the funnel tightly and frequently vent the funnel by opening its stopcock* (Fig. 2.58). Identify the aqueous layer by the method described in Section 2.21, transfer it to an Erlenmeyer flask labeled "Basic Extract," and transfer the organic solution into a clean Erlenmeyer flask containing three or four spatula-tips full of anhydrous sodium sulfate and labeled "Neutral Compound." Let this solution stand for about 15 min, occasionally swirling it to hasten the drying process. If the solution remains cloudy, add additional portions of sodium sulfate to complete the drying process.★

Precipitating and Drying   Cool the *basic* extract in an ice-water bath. Carefully acidify this solution with 3 M hydrochloric acid, so that the solution is distinctly acidic to pHydrion™ paper.★ Upon acidification, a precipitate should form.

Collect the precipitate by vacuum filtration (Fig. 2.52) using a Büchner or Hirsch funnel. Wash the solid on the filter paper with a small portion of *cold* water. Transfer it to a labeled watchglass, cover it with a piece of filter or weighing paper, and allow the product to air-dry until the next laboratory period. Alternatively, dry the solid in an oven having a temperature of 90–100 °C for about 1 h. After drying, transfer the benzoic acid to a dry, *tared* vial.

Separate the "Neutral Compound" from the drying agent by decantation (Fig. 2.54) or gravity filtration through a cotton plug (Fig. 5.3) into a 100-mL round-bottom flask. Remove the solvent by simple distillation (Fig. 2.53). Alternatively, use rotary evaporation or one of the other techniques described in Section 2.29. Allow the residue to cool to room temperature to solidify, scrape the contents of the flask onto a piece of weighing paper to air-dry, and then transfer it to a dry, *tared* vial.

FIGURE 5.3
Miniscale gravity filtration through a cotton plug.

Analysis and Recrystallization Determine the weight and melting point of each crude product. If you know the relative amounts of **5** and **7** in the original mixture, calculate the percent recovery of each compound. Recrystallize them according to the general procedures provided in Section 3.2 and determine the melting points of the purified materials.

B. Two-Base Extraction

Apparatus Separatory funnel, ice-water bath, apparatus for vacuum filtration, simple distillation, and *flameless* heating.

Dissolution Obtain 2 g of a mixture comprised of benzoic acid, 2-naphthol, and naphthalene. Dissolve the mixture by swirling it with 30 mL of diethyl ether in a 125-mL Erlenmeyer flask. If any solids remain, add more diethyl ether to effect complete dissolution.

Extraction Extract the solution with a 20-mL portion of 1.25 *M* (10%) aqueous sodium bicarbonate (Fig. 2.58). *Caution:* Gaseous carbon dioxide is generated and care must be taken in this step! To prevent accidental loss of material, do the following: After adding the aqueous bicarbonate to the funnel, swirl the *unstoppered* funnel until all foaming has subsided. Then stopper the funnel and, *holding both the stopcock and the stopper of the separatory funnel tightly*, invert the funnel, *immediately* venting it by opening the *stopcock*. Finally, shake the funnel with *frequent* venting until gas is no longer evolved. Separate the layers, transferring the aqueous layer to a 125-mL Erlenmeyer flask labeled "Bicarbonate Extract."★ To ensure properly identifying the aqueous layer, consult Section 2.21. Return the organic solution to the funnel.

Now extract the organic solution with a 20-mL portion of *cold* 2.5 *M* (10%) aqueous sodium hydroxide, venting the funnel frequently during the process. Transfer the aqueous layer to a 125-mL Erlenmeyer flask labeled "Hydroxide Extract," and the organic solution to a 50-mL Erlenmeyer flask labeled "Neutral Compound" and containing one or two spatula-tips full of anhydrous sodium sulfate. Loosely cork the flask containing the organic solution, and let it stand for at least 15 min with occasional swirling to promote drying. If the solution remains cloudy, add additional portions of sodium sulfate to complete the drying process.★

Precipitating and Drying For the two *basic* extracts, follow the directions given in Part A for the *basic* extract. Also apply the protocol provided in Part A for the "Neutral Compound" to the corresponding solution obtained in this procedure.

Analysis and Recrystallization Determine the weight and melting point of each crude product. If you know the relative amounts of **5**, **7**, and **11** in the original mixture, calculate the percent recovery of each compound. Recrystallize them according to the general procedures provided in Section 3.2 and determine the melting points of the purified materials.

Microscale Procedures

Preparation   Answer the Pre-Lab Exercises on pages PL. 19 and/or 21. Read or review Sections 2.7, 2.9, 2.10, 2.17, 2.21, 2.25, 2.29, 3.2, and 3.3.

A. One-Base Extraction

Apparatus   A 5-mL conical vial, Pasteur filter-tip pipets, ice-water bath, 3-mL Craig tube, apparatus for Craig tube filtration and *flameless* heating.

Dissolution   Obtain 0.2 g of a mixture comprised of benzoic acid and naphthalene. Dissolve the mixture by swirling it with 2 mL of diethyl ether in the conical vial. If any solids remain, use a filter-tip pipet to add enough diethyl ether to effect complete dissolution.

Extraction   Add 0.5 mL of *cold* 2.5 *M* (10%) aqueous sodium hydroxide to the vial and cap it. Shake the vial vigorously for a minute or two, occasionally unscrewing the cap slowly to relieve any pressure. Allow the layers to separate and, using a Pasteur pipet, transfer the *aqueous* layer to a Craig tube labeled ''Basic Extract.'' To ensure properly identifying the aqueous layer, consult Section 2.21.

Precipitating and Drying   Add a microspatula-tip full of sodium sulfate to the organic solution in the conical vial and dry the solution according to the directions provided in Section 2.25. Once the organic solution is dry, use a filter-tip pipet to transfer it to a test tube. Evaporate the diethyl ether using *flameless* heating; stir the solution with a microspatula or use a boiling stone to prevent bumping during evaporation. Perform this step at a hood or use a funnel attached to a vacuum source for removing vapors (Fig. 2.69b). Scrape the isolated naphthalene onto a piece of weighing paper to air-dry and then transfer it to a dry, *tared* vial.

Cool the *basic* extract in an ice-water bath and, with stirring, add concentrated hydrochloric acid dropwise with a Pasteur pipet until the solution is distinctly acidic to pHydrion paper.★ Allow the acidified solution to remain in the bath to complete the crystallization. Then isolate the benzoic acid by Craig tube filtration (Fig. 2.53). Rinse the solid back into the Craig tube, with 0.5 mL of cold water, stir the mixture briefly, and repeat the filtration. Transfer the benzoic acid to a watchglass, cover it with a piece of filter paper, and allow the product to air-dry until the next laboratory period. After drying, put the isolated solid in a dry, tared vial.

Analysis and Recrystallization   Determine the weight and melting points of each crude product. If you know the relative amounts of **5** and **7** in the original mixture, calculate the percent recovery of each compound. Recrystallize the products according to the general microscale procedures provided in Section 3.2 and determine the melting points of the purified materials.

B. Two-Base Extraction

Apparatus   A 5-mL conical vial, Pasteur filter-tip pipets, ice-water bath, 3- and 5-mL Craig tubes, apparatus for Craig tube filtration and *flameless* heating.

Dissolution   Obtain 0.3 g of a mixture comprised of benzoic acid, 2-naphthol, and naphthalene. Dissolve the mixture by swirling it with 2 mL of diethyl ether in the conical vial. If any solids remain, use a filter-tip pipet to add enough diethyl ether to effect complete dissolution.

Extraction   Extract the solution with a 1-mL portion of 1.25 M (10%) aqueous sodium bicarbonate. *Caution:* Gaseous carbon dioxide is generated and care must be taken in this step! To prevent accidental loss of material, do the following: After adding the first portion of aqueous bicarbonate to the vial, stir the two-phase mixture vigorously with a microspatula or stirring rod until all foaming has subsided. Then cap the vial, shake it gently for 2–3 sec, and loosen the cap *slowly* to relieve any pressure. Retighten the cap and shake the vial vigorously for about 1 min, occasionally venting the vial by loosening its cap. Allow the layers to separate and, using a Pasteur pipet, transfer the *aqueous* layer to a 5-mL Craig tube labeled "Bicarbonate Extract." To ensure properly identifying the aqueous layer, consult Section 2.21.★

Now extract the organic solution with a 0.5-mL portion of *cold* 2.5 M (10%) aqueous sodium hydroxide, following the same protocol as that given in Part A, placing the basic extract in a 3-mL Craig tube.

Precipitating and Drying   Use the same procedure as that given in Part A for *separately* acidifying the two *basic* extracts and isolating and drying the two solids that result.

Analysis and Recrystallization   Determine the weight and melting points of each crude product. If you know the relative amounts of **5**, **7**, and **11** in the original mixture, calculate the percent recovery of each compound. Recrystallize the products according to the general microscale procedures provided in Section 3.2 and determine the melting points of the purified materials.

Place the used *filter papers* in a container for nontoxic solid waste. Flush the *acidic* and *basic* filtrates down the drain. Pour any *diethyl ether* that has been isolated into a container for nonhalogenated organic liquids.

Finishing Touches

EXERCISES

General Questions

1. Using Equation 5.9, show that three extractions with 5-mL portions of a solvent give better recovery than a single extraction with 15 mL of solvent when $K = 0.5$.

2. Show mathematically how Equation 5.17 can be converted to Equation 5.19.

3. Define the following terms:

   a. immiscible liquid phases   **e.** conjugate base of an acid, HA.

   b. distribution or partition coefficient   **f.** conjugate acid of a base, B⁻

   c. adsorption   **g.** liquid-liquid extraction

   d. absorption   **h.** hydrophilic

4. Explain why swirling or shaking a solution and its drying agent hastens the drying process.

5. Assume that the partition coefficient, K, for partitioning of compound A between diethyl ether and water is 3; that is, A preferentially partitions into ether.

   a. Given 400 mL of an aqueous solution containing 12 g of compound A, how many grams of A could be removed from the solution by a *single* extraction with 200 mL of diethyl ether?

   b. How many total grams of A can be *removed* from the aqueous solution with three successive extractions of 67 mL each?

6. a. The pK_a of benzoic acid (**5**) is 4.19. Show mathematically that this acid is 50% ionized at pH 4.19.

   b. Use the result of Part **a** to explain why precipitation of **5** is incomplete if the pH of an aqueous solution of benzoate ion (**6**) is lowered only to pH 7 by adding acid.

7. Consider the base, monosodium phosphate, Na^+ $(HO)_2P(=O)O^-$. (a) Write the structure of the conjugate acid of this base. (b) Given that the pK_a of this conjugate acid is 2.12, explain why an aqueous solution of monosodium phosphate would be ineffective for extracting benzoic acid (**5**) from a diethyl ether solution.

8. Amines, which are organic *bases*, are protonated by aqueous acid to produce a water-soluble ammonium salt (Eq. 5.22). With this in mind, provide a flow chart analogous to those in Figures 5.1 and 5.2 for separating a diethyl ether solution containing benzoic acid (**5**), 4-nitroaniline (**13**), an amine, and *N*-(4-nitrophenyl)benzamide (**14**), a neutral compound. Both **13** and **14** are solids when pure.

$$RNH_2 \text{ or } ArNH_2 + H_3O^+ \longrightarrow RNH_3^+ \text{ or } ArNH_3^+ + H_3O \qquad \textbf{(5.22)}$$

| **13** | **14** |
| 4-Nitroaniline | *N*-(4-nitrophenyl)-benzamide |

One-Base Extraction

9. What practical consideration makes aqueous hydroxide rather than aqueous bicarbonate (**9**) the preferred base for extracting benzoic acid (**5**) from diethyl ether?

10. Naphthalene (**7**) has a relatively high vapor pressure for a solid (because of its volatility, naphthalene is the active ingredient in some brands of mothballs). In view of this, what might happen if you placed naphthalene under vacuum or in an oven for several hours to dry it?

Two-Base Extraction

11. Why does the sequence for extracting the diethyl ether solution of benzoic acid (**5**), naphthalene (**7**), and 2-naphthol (**11**) start with aqueous bicarbonate and follow with aqueous hydroxide rather than the reverse order?

12. What would be the consequence of acidifying the basic extract containing **12** only to pH 10?

13. Why is caution advised when acidifying the bicarbonate solution of sodium benzoate (**6**)?

5.4 Extraction of a Natural Product: Trimyristin

Solvent extraction is a valuable method for obtaining a desired substance from its natural source. A familiar example is the hot-water extraction from coffee beans and tea leaves of caffeine (**15**) and the various oils that constitute the flavors of freshly brewed coffee and tea. As contrasted to the **liquid-liquid** extractions described in Section 5.2, this process is an example of **solid-liquid** extraction. The theory underlying it is the same, however. Because most organic compounds we wish to isolate are *insoluble* in water, a variety of organic solvents such as diethyl ether, dichloromethane, ethanol, and acetone are used for extracting **natural products**.

Extraction of natural materials often produces complex mixtures of products, as described in the ''Historical Highlight'' for this chapter, so additional procedures are required to separate and purify individual compounds. These may involve techniques in which the compounds are separated on the basis of their acidic or basic properties (Sec. 5.3) or by chromatographic methods (Chap. 6). Thus, the isolation of pure natural products normally involves a series of complex and tedious operations. Fortunately there are some exceptions to this general rule. The isolation of trimyristin (**16**) from nutmeg is one of them.

Trimyristin, which occurs in many vegetable **fats** and **oils**, is a **triester** of glycerol (**17**) and the **fatty acid** myristic acid (**18**). Most naturally occurring fats and oils are esters of **17** and straight-chain carboxylic acids, the most common of which contain 14 to 20 carbon atoms. In fact, the only difference between a fat and an oil is whether it is a solid or liquid at room temperature! The lack of any double bonds in the carbon chain of myristic acid makes **16** a member of the family of **saturated fats**, which allegedly increase the risk of heart disease if present in the diet in excessive amounts.

Nutmeg, a hard, aromatic seed of an East Indian tree (*Myristica fragrans*) that is the source of trimyristin in this experiment, has been a valued spice ever since its discovery in the Spice Islands of Indonesia by Portuguese sea captains over four centuries ago. The chemical makeup of nutmeg is somewhat unusual in that extraction of the ground seeds with diethyl ether yields **16** in high purity without contamination by other structurally related esters of glycerol and fatty acids. Thus, the procedure described here is *not* typical of that usually required for isolating natural products, but its simplicity makes it suitable for demonstrating the general technique of extracting biological materials from natural sources.

15
Caffeine

16
Trimyristin

17
Glycerol

18
Myristic acid

ISOLATION OF TRIMYRISTIN FROM NUTMEG

Purpose: To apply the technique of extraction for isolating a natural product.

Safety Alert	

1. Diethyl ether and acetone are both highly volatile and flammable solvents. *Use no flames or electrical dryers in the laboratory during this experiment.*
2. Monitor the heating of diethyl ether under reflux so that the ring of condensate *remains in the lower third of the condenser*. This will ensure that no vapors escape from the condenser into the room. Very little heat will be required regardless of the heating source used to maintain reflux.

Miniscale Procedure

Preparation   Answer the Pre-Lab Exercises on page PL. 23. Read or review Sections 2.7, 2.9, 2.10, 2.11, 2.13, 2.17, 2.21, 2.22, 2.29, 3.2, and 3.3.

Apparatus   A 25- and a 50-mL round-bottom flask, ice-water bath, apparatus for heating under reflux, magnetic stirring, simple distillation, vacuum filtration, and *flame-less* heating.

Setting Up   Place about 4 g of ground nutmeg in the 50-mL round-bottom flask containing a stirbar and add 10 mL of diethyl ether. If only seeds of nutmeg are available, grind the seeds in a mortar and pestle or a blender. Attach a water-cooled condenser to the flask, and set up the apparatus for heating under reflux (Fig. 2.63a).

Extraction and Isolation   Heat the mixture under gentle reflux for approximately 0.5 h. Allow the contents of the flask to cool to room temperature.★ Filter the mixture through a fluted filter paper into the 25-mL round-bottom flask (Fig. 2.48), rinsing any residue remaining in the 50-mL flask onto the filter paper with an additional 2–4 mL of diethyl ether.

    Equip the flask containing the filtrate for simple distillation (Fig. 2.35a), and remove the diethyl ether by distillation. Alternatively, use rotary evaporation or one of the other techniques described in Section 2.29. Dissolve the resulting yellow oil in 3–4 mL of acetone while warming the mixture (*no flames!*). Immediately pour the hot solution into a 25-mL Erlenmeyer flask. Allow the solution to stand at room temperature for about 0.5 h. If crystallization has not occurred during this time, gently scratch the flask at the air/liquid interface.★ Complete crystallization by cooling the flask in an ice-water bath for an additional 15 min. Collect the trimyristin on a Büchner or Hirsch funnel by vacuum filtration (Fig. 2.52). Transfer it to a watchglass or piece of clean filter or weighing paper for air-drying.

Analysis   Weigh the trimyristin and determine its melting point. The reported melting point is 55–56 °C. Based on the original weight of nutmeg used, calculate the percent recovery.

Finishing Touches

Once any residual diethyl ether has evaporated from the filter cake, place the used filter paper and its contents in a container for nonhazardous solids. Any diethyl ether that has been isolated should be poured into a container for nonhalogenated organic liquids. Either flush the acetone-containing filtrate down the drain or pour it into a container for nonhalogenated organic liquids.

Microscale Procedure

Preparation   Answer the Pre-Lab Exercises on page PL. 25. Read or review Sections 2.5, 2.7, 2.9, 2.10, 2.13, 2.17, 2.21, 2.22, 2.29, 3.2, and 3.3.

Apparatus   A 5-mL conical vial or 10-mL round-bottom flask, ice-water bath, Pasteur filter-tip and filtering pipets, tared 5-mL Craig tube, apparatus for heating under reflux, magnetic stirring, and *flameless* heating.

Setting Up   Place 1 g of ground nutmeg in the vial or flask, equipped for magnetic stirring. Add 2–3 mL of diethyl ether, attach the reflux condenser, and set up the apparatus for heating under reflux (Fig. 2.64).

Extraction and Isolation   Heat the heterogeneous mixture for 0.5 h under *gentle* reflux with stirring, being certain not to lose diethyl ether through the top of the condenser. Allow the mixture to cool to room temperature★ and then gravity-filter the ethereal solution into a Craig tube, using Pasteur filter-tip and filtering pipets (Fig. 2.49). Use a filter-tip pipet to rinse the reaction vessel with 0.5–1 mL of technical diethyl ether, and use the pipet to transfer the rinse to the Craig tube.★

Add a small boiling stone or a boiling stick to the Craig tube and, working at the hood, gently evaporate the diethyl ether using *flameless* heating. If hood space is not available, perform the evaporation under an inverted funnel attached to a water aspirator (Fig. 2.69b). After the diethyl ether has evaporated, add 1 mL of acetone to the residual yellow oil and warm the mixture briefly (*no flames!*) to effect solution. Allow the solution to cool to room temperature and remain at that temperature for about 0.5 h.★ If crystallization has not occurred during this time, gently scratch the tube at the air/liquid interface.

Complete crystallization by cooling the Craig tube in an ice-water bath for an additional 15 min, and then isolate the product by Craig tube filtration (Fig. 2.53). Carefully dissemble the Craig tube, scrape any crystalline material clinging to the glass or Teflon plug into the tube, and air-dry the product.

Analysis   Weigh the trimyristin and determine its melting point. The reported melting point is 55–56 °C. Based on the original weight of nutmeg used, calculate the percent recovery.

Finishing Touches

Either flush the *acetone-containing filtrate* down the drain or pour it into a container for nonhalogenated organic liquids.

EXERCISES

1. Why is diethyl ether rather than acetone chosen as the extraction solvent?

2. Why is finely divided rather than whole nutmeg specified in this extraction?

3. Considering the solubility properties of trimyristin (**16**), would water be a good choice as an extracting solvent for obtaining **16** from nutmeg?

4. Judging from your examination of trimyristin (**16**), is this substance responsible for the odor of nutmeg?

5. A certain plant material is known to contain mainly trimyristin and tripalmitin, in approximately equal amounts. Tripalmitin, the structure of which you should look up in a reference source, has mp 66–67 °C. Extraction of these two compounds from the plant material with diethyl ether gave an oil after removal of the solvent, and this oil was difficult to crystallize. Explain this result. (*Hint:* see Sec. 3.3.)

6. In the miniscale procedure,

   a. why is gravity rather than vacuum filtration used to separate the ethereal extract from the residual nutmeg?

   b. why is fluted rather than plain filter paper used for the filtration procedure?

7. In the microscale procedure,

   a. why is it important that a gentle rather than vigorous reflux rate be used?

   b. why is it preferred to transfer the ethereal extract with a Pasteur filter-tip pipet rather than a regular Pasteur pipet?

▬▬▬ H I S T O R I C A L   H I G H L I G H T ▬▬▬

Natural Products

The naturally occurring organic compounds produced by living organisms have fascinated human beings for centuries. Indeed, traditional folk medicines from many cultures are effective because individual organic compounds that occur in the native plants exhibit biological activities that may be used to treat various maladies. Compounds of nature have been widely used in modern medicine as analgesics to relieve pain and as antibacterial, antiviral, anticancer, and cholesterol-lowering agents to treat or cure numerous ailments and diseases. On the other hand, some plants produce compounds that are toxic and may even cause death. For example, Socrates drank a fatal potion prepared from an extract of hemlock. Other plants have attracted attention because of the flavors they possessed or the odors they released. Organic compounds found in these plants are often incorporated as additives in the food and perfume industries. Some natural products such as those found in flowering plants are highly colored, and these may be used to prepare the dyes that color paints and fabrics.

However, natural products have not only drawn attention because of their practical applications in our daily lives; they are also important for the various scientific challenges they present. Such challenges include understanding the basis of their sensory and biological properties and developing means for their chemical synthesis. Indeed, natural products provided chemists with many interesting experimental problems during the period when chemistry was emerging from alchemy into a more exact science. Today the isolation, identification, and synthesis of natural products, especially those having interesting and useful biological activity, remain important areas of research for modern organic chemists.

Historically, most natural products were extracted from plants rather than from animals but more recently, microorganisms are assuming increased importance as sources of natural products, particularly since the advent of molecular biology. The isolation of a natural product in pure form normally represents a significant challenge, mainly because even the simplest plants and microorganisms contain a multitude

HISTORICAL HIGHLIGHT
Natural Products (continued)

of organic compounds, often in only minute amounts. For example, a kilogram of plant material may only yield milligram quantities, or even less, of the desired natural product. The general approach used to isolate and purify a natural product is labor-intensive and typically starts with grinding a plant or other organism into fine particles. The resulting material is then extracted with a solvent or a mixture of solvents. Volatile natural products in the extract can be detected and isolated by gas chromatography. However, many natural products are relatively nonvolatile, and removing the solvent used for the extraction yields an oil or gum that requires further separation by chromatographic methods to obtain the various components in pure form. Often a series of chromatographic separations are required before the individual compounds are pure.

The next step is to determine the structure of the isolated products. Traditional approaches to structural elucidation involved performing standard qualitative tests for the various functional groups to identify those present in the molecule (Chap. 23). The compound was then subjected to a series of simple chemical degradations to obtain simpler compounds. If a known compound was formed as a result of these transformations, it was then possible to work backward and deduce the structure of the original natural substance. More recently, instrumental techniques including infrared and nuclear magnetic resonance spectroscopy (Chap. 8) and mass spectrometry have greatly facilitated the determination of structures. However, the most reliable means of determining the structure of an unknown substance is by X-ray diffraction. In this technique, a crystal of the sample is irradiated with X-rays to provide a three-dimensional map of the positions of each atom within the molecule. Applying this method requires that the unknown itself, or a suitable derivative of it, be a crystalline solid, and this is not always possible.

The final goal of organic chemists working in this exciting field is often the laboratory synthesis of the natural product. While the total synthesis of a natural product may represent mainly an intellectual challenge, more often the undertaking is a unique opportunity to develop and demonstrate the utility of new synthetic techniques in organic chemistry. In cases where the natural product has medicinal applications, the development of an efficient synthetic route may be of great importance, especially when the supply of the material from the natural source is severely limited or its isolation is difficult. Through synthesis it is often possible to prepare derivatives of the natural product that possess improved biological properties.

There are a number of classes of natural products, and a detailed discussion of these compounds may be found in your lecture textbook. Representative compounds from the different classes range from the simple, as in the case of formic acid, HCO_2H, the irritating ingredient in ant venom, to the extraordinarily complex, as engendered in DNA, for example. The examples of reserpine (1), Taxol® (2), and calicheamicin γ^I_1 (3) are illustrative of the diversity and complexity of the molecules of nature that have inspired research efforts of contemporary organic chemists and tested their creativity.

1
Reserpine

2
Taxol®

HISTORICAL HIGHLIGHT
Natural Products *(continued)*

3
Calicheamicin γ_1^I

Reserpine (**1**) is an alkaloid that was isolated from the Indian snakeroot *Rauwolfia serpentina* Benth in 1952. This alkaloid has a potent effect upon the central nervous system, where it depletes the postganglionic adrenergic neurons of the neurotransmitter, norepinephrine. This novel mode of action was responsible for its clinical use as an agent for treating high blood pressure. Recently, however, antihypertensive drugs with improved pharmacological profiles have replaced reserpine as a drug. It is normally more economical to isolate complex natural products from their source than it is to prepare them by total synthesis. Thus, it is noteworthy that the reserpine that was used for medical purposes was prepared by *total synthesis* using slight modifications of the original route that was developed by Professor Robert B. Woodward of Harvard University in 1956.

Taxol (**2**) was isolated from the bark of the Pacific yew tree, *Taxus brevifolia*, in 1962 as part of an extensive program directed toward the discovery of novel anticancer agents that was conducted jointly by the Department of Agriculture and the National Cancer Institute. Because extracts isolated from the bark of this tree were observed to exhibit potent cytotoxic activity, there was an intense effort to identify the active principle, which was isolated in only minute quantities. Finally in 1971, the structure of Taxol, a complex diterpene, was elucidated by X-ray crystallography.

The clinical use of Taxol was initially hampered by its low solubility, but its scarcity rendered its use even more problematic. For example, it was estimated that the bark of a 100-year old yew tree would yield only enough Taxol for a *single dose* for a cancer patient. Stripping the bark from a tree is a death sentence for it, and environmentalists became extremely concerned that the quest for sufficient amounts of Taxol to treat needy cancer patients would destroy the ancient forests of the Pacific northwest, which are home to the endangered spotted owl. Fortunately for humans and spotted owls alike, a related diterpene was found in the needles and twigs, which are more readily renewable than the tree itself, of the European yew tree, *Taxus baccata*. This compound may be converted into Taxol using organic reactions related to those you learn in the lecture portion of this course. Consequently, Taxol is now available in sufficient quantities for treating ovarian cancer and was approved for such use by the U. S. Food and Drug Administration in 1992. It also holds promise as a chemotherapeutic agent for the treatment of other types of cancer including lung and breast.

The biological importance of Taxol coupled with its unusual and complex structure served as the stimulus for synthetic investigations by a large number of research groups, and it has been recently prepared by several syntheses that start with commercially available materials unrelated to Taxol itself. These efforts uncovered a significant amount of novel and useful chemistry that may be applied to other important problems in organic chemistry. However, they did not provide a *practical* synthesis of Taxol; this goal remains a challenge to the ingenuity of future organic chemists.

Calicheamicin γ_1^I (**3**), is perhaps the most prominent member of a recently discovered class of anticancer agents that contain an enediyne moiety, which consists of a double and two triple bonds. Its discovery is an interesting story that typifies the ongoing search for medicinally active compounds in Nature. Employees of many pharmaceutical companies make it a practice to collect soil samples during their travels around the world. When these scientists return, these samples are placed in different growth cultures to see whether the microorganisms living in the soil produce compounds with interesting biological activity. One such sample was collected by a scientist from some chalky rocks, which are commonly called caliche, along a highway in Central Texas. Within these rocks lived bacteria (*Micromonospora echinospora ssp* calichenis) that produced calicheamicin γ_1^I and other related compounds upon cultivation in laboratory cultures.

The complexity of calicheamicin γ_1^I is awesome. The core of the bicyclic skeleton contains a con-strained enediyne array, whose reactivity and DNA-cleaving ability is unleashed by the cellular cleavage of the trisulfide. Although the exact mechanism of action of **3** *in vivo* is unknown, it is a highly efficient agent for inducing the selective cleavage of double-stranded DNA, an event that leads to cell death. While calicheamicin γ_1^I itself is too toxic for use as an anticancer agent in humans, related compounds possessing enediyne arrays are currently attractive drug candidates and are involved in clinical trials at several pharmaceutical companies.

Relationship of Historical Highlight to Experiments

In one of the experiments in this chapter you isolate the naturally occurring fat trimyristin from nutmeg, a valuable spice. In other experiments in this textbook you have the opportunity to isolate citral, the odoriferous component of lemon grass oil (Sec. 4.5) and to separate different pigments found in green leaves (Sec. 6.3). You may also examine the two enantiomers of carvone, one of which is found in spearmint and the other in caraway seed oil (Sec. 7.4). The techniques you will use in these experiments serve as an introduction to the methods that are used in research laboratories to isolate natural products having potential value to society.

CHAPTER 6

Chromatography

The common laboratory techniques of extraction, recrystallization, and distillation as methods for purifying organic compounds are discussed in Chapters 3 through 5. In many instances, however, the mixtures of products obtained from chemical reactions do not lend themselves to easy separation by any of these techniques because the appropriate physical properties of the individual components are too similar. In this chapter we learn some of the basic techniques that are used to separate mixtures by taking advantage of the differential distribution of the individual compounds in two immiscible phases, a principle that also forms the basis of extraction (Chap. 5).

6.1 Introduction

The word **chromatography** was first used to describe the colored bands observed when a solution containing plant pigments is passed through a glass column containing an adsorbent packing material. From that origin, the term now encompasses a variety of separation techniques that are widely used for analytical and preparative purposes. All methods of chromatography operate on the principle that the components of a mixture will distribute unequally between two immiscible phases. The **mobile phase** is generally a liquid or a gas that flows continuously over the fixed **stationary phase**, which may be a solid or a liquid. The individual components of the mixture have different affinities for the mobile and stationary phases so a dynamic equilibrium is established in which each component is selectively, but temporarily, removed from the mobile phase by binding to the stationary phase. When the equilibrium concentration of that substance in the mobile phase decreases, it is released from the stationary phase and the process continues. Since each component partitions between the two phases with a different equilibrium constant or **partition coefficient**, the components separate into separate regions termed **migratory bands** (Fig. 6.1). *The component that interacts with or binds more strongly to the stationary phase moves more slowly in the direction of the flow of the mobile phase.* The attractive forces that are involved in this selective adsorption are the same forces that cause attractive interactions between any two molecules: electrostatic and dipole–dipole interactions, hydrogen bonding, complexation, and van der Waals forces.

Direction of
solvent
flow

Increasing time

FIGURE 6.1
Separation of components by column chromatography.

154

The chromatographic methods used by modern chemists to identify and/or purify components of a mixture are characterized by the nature of the mobile and stationary phases. For example, the techniques of **column**, **high-pressure** (or **high-performance**) **liquid** (HPLC), and **thin-layer chromatography** (TLC) each involve *liquid–solid* phase interactions. **Gas–liquid partition chromatography** (GLC), also known as **gas chromatography** (GC), involves distributions between a mobile *gas* phase and a stationary *liquid* phase coated on a solid support. These important techniques can be used as tools to analyze and identify the components in a mixture as well as to separate the mixture into its pure components for preparative purposes. Although there are other chromatographic techniques, such as ion exchange and paper chromatography, a review of those methods is beyond the scope of this discussion.

6.2   Column Chromatography

Column chromatography is a form of **solid–liquid adsorption** chromatography and is a powerful technique in preparative synthetic organic chemistry. In industry and research, it is frequently necessary to separate the components of a reaction mixture so that one of these may be used in a subsequent chemical reaction. Typically, a multicomponent mixture is dissolved in a small amount of an appropriate solvent and applied to the top of a packed column containing a finely divided, active solid **adsorbent** that serves as the stationary phase (Fig. 6.2). Next an **eluant** or **eluting solvent**, which is the mobile phase, is passed through the column. The individual components of the mixture, which were initially adsorbed on the stationary phase at the top of the column, begin to move downward with the eluting solvent. These components travel at different rates depending on their relative affinities for the packing material; a more weakly adsorbed compound is eluted faster from the column than is a more strongly adsorbed compound. As the individual components exit from the bottom of the column in bands, they are collected in separate containers. The solvent is then removed from each fraction by evaporation to provide the pure components, which are characterized and identified by determining their physical constants (Chaps. 3 and 4) and spectral properties (Chap. 8).

When all of the components of a mixture are colored, their bands are easily observed. Many organic compounds are colorless, however, and other methods are required for detecting the bands as they elute from the column. For those organic compounds that absorb ultraviolet or visible light, **electronic detectors** that measure differences in the absorption of light as the solvent exits the column are used to locate the bands of the individual components. Detectors that measure differences in the **refractive index** of the **eluate** are also used to identify the different bands; such detectors do not rely on absorption of light by the organic components.

If a detector is not available, the progress of the chromatographic separation can be conveniently followed using thin-layer chromatography (TLC) to analyze the eluate at regular intervals (Sec. 6.3). Another method, albeit more laborious, involves collecting small, equal fractions of the eluate from the column in a series of flasks. The solvent in each flask is then evaporated, and the presence or absence of a solute in the residue then provides a means of locating the bands of each component. If there is adequate separation of the different bands, a given flask will often contain only one constituent of the original mixture. That component will normally appear in a number of consecutive fractions, however.

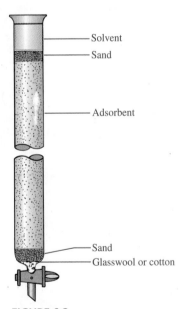

Solvent
Sand

Adsorbent

Sand
Glasswool or cotton

FIGURE 6.2
Chromatography column.

ADSORBENTS

Selecting the proper adsorbent as the stationary phase is one important decision that must be made before performing a chromatographic separation. The polar adsorbents most commonly used to separate organic compounds are alumina (Al_2O_3) and silica gel (silicic acid, $SiO_2 \cdot x\ H_2O$). Alumina, the more polar of the two, is a highly active and strongly adsorbing polar substance that is commercially available in three forms: neutral, acidic, and basic. Acidic and basic alumina are sometimes used to separate basic and acidic compounds, respectively, but neutral alumina is frequently the packing of choice and is used in the experimental portion of this section. Commercial silica gel is slightly acidic.

Since the attraction of solute to adsorbent is a surface phenomenon, the most effective solid adsorbents have *uniform particle size* and *high specific area*. The higher the specific area, the faster the equilibrium of the solute between the mobile and solid phases is established and the narrower the bands. High specific areas on the order of several hundred m^2/g are common for good grades of alumina and silica gel.

The strength of the adsorption of an organic compound to the solid support depends not only on the polarity and nature of the adsorbent but also on the nature of the functional groups present in the molecule. In **normal-phase chromatography** the packing material for the stationary phase is a polar adsorbent such as alumina and silica gel, and organic solvents are used as the mobile phase. Under these conditions, compounds containing carboxyl groups and other polar functional groups are more tightly held than those containing halogen, as the **elutropic series** shown below indicates. More polar solvents must then be used as the mobile phase to elute highly polar molecules from the column.

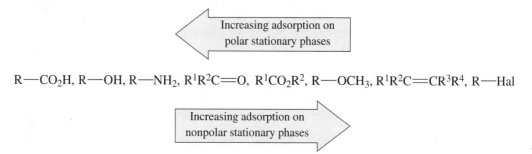

Increasing adsorption on polar stationary phases

$$R—CO_2H,\ R—OH,\ R—NH_2,\ R^1R^2C{=}O,\ R^1CO_2R^2,\ R—OCH_3,\ R^1R^2C{=}CR^3R^4,\ R—Hal$$

Increasing adsorption on nonpolar stationary phases

In **reverse-phase chromatography**, the packing material for the stationary phase consists of glass beads coated with a nonpolar, hydrocarbon film, and mixtures of water and organic solvents are generally used as the eluting solvents. Under these conditions, nonpolar organic molecules are more strongly attracted to the nonpolar stationary phase, whereas polar solutes are more strongly attracted to the mobile phase. The order of elution is then the *reverse* of that shown earlier, with the more polar components of a mixture eluting more rapidly than the less polar ones. Reverse-phase chromatography may sometimes be used to separate mixtures that are inseparable by normal-phase chromatography.

ELUANT

Selecting the eluant or the mobile liquid phase is the other important decision that must be made in planning a chromatographic separation. The best solvent or combination of solvents for a given separation is determined by trial and error. Often the most efficient method to determine the optimal solvent system(s) for column chromatography is to perform a series of trial runs using TLC (Sec. 6.3). These experiments may be performed quickly, and the amount of material required for the trials is small.

Certain criteria guide the selection of the solvent. For example, an effective eluting solvent must readily dissolve the solute but not compete with it for binding sites on the stationary phase. If the mixture to be separated is not soluble in the solvent, the individual components may remain permanently adsorbed on the stationary phase. Another criterion is that a solvent should not be too polar because it may bind strongly to the adsorbent and force the solute to remain in the mobile phase. In such circumstances, the components will move rapidly on the column, offering little opportunity to establish the equilibria required for separation. Consequently, the eluting solvent must be significantly *less* polar than various components of the mixture to obtain an effective separation. As a rule, the relative ability of different solvents to move a given substance down a column of alumina or silica gel, is termed **eluting power** and is generally found to follow the order shown below. In reverse-phase chromatography the situation is just the opposite, with less-polar solvents having the greater eluting power.

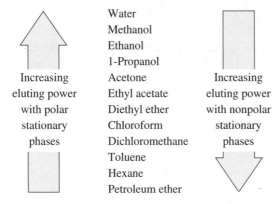

Increasing eluting power with polar stationary phases

Water
Methanol
Ethanol
1-Propanol
Acetone
Ethyl acetate
Diethyl ether
Chloroform
Dichloromethane
Toluene
Hexane
Petroleum ether

Increasing eluting power with nonpolar stationary phases

EXPERIMENTAL TECHNIQUE

With this general discussion as background, we now discuss some of the experimental aspects of performing column chromatography. The optimal adsorbent and eluting solvent(s) typically are first determined using TLC (Sec. 6.3), and then the column is packed with the adsorbent. The amount of adsorbent used to prepare the column varies according to the differences in partition coefficients and polarities of the individual components in the chromatographic system. For simple separations, it is possible to use as little as 10 g of adsorbent for 1 g of the mixture, but when the components of the mixture have similar polarities, this ratio must be increased to as much as 100–200:1; a ratio of about 25:1 is a convenient starting point. As a general rule of thumb, the height-to-diameter ratio for the packed column should be about 8:1.

The column is made of glass and is fitted with a stopcock or a segment of polyethylene tubing with a screw clamp to control the flow of solvent through the column. Using a long glass rod or a piece of wire, a plug of cotton or glasswool is first inserted into the small end of the column to prevent the adsorbent from washing out the bottom. A layer of white sand, approximately 1 cm deep, is then added to provide an even bed for the adsorbent (Fig. 6.2).

Proper packing of the column is vital to the success of column chromatography because this step ultimately determines the efficiency of separation. Two general protocols are followed for this operation. The first of these, the **dry pack** method, involves pouring the dry adsorbent slowly into a vertical glass column half-filled with the solvent that will serve as the eluant. The other technique is the **wet pack** method, in which a slurry of the adsorbent in the eluting solvent is added to the column; this is the preferred procedure when silica gel is the adsorbent. With both methods, the column is constantly tapped as the solid settles through the liquid to ensure even and firm packing of the adsorbent and to remove any trapped air bubbles. Some solvent may be drained from the column during this operation, but the liquid level in the column should *never* be allowed to fall below the top of the adsorbent. If this occurs, the air bubbles that form in the column will allow *channeling*, which results in poor separations because the components are eluted as ragged rather than sharp bands. *Uniform packing of the adsorbent is essential* so the solvent will move down the column with a horizontal front.

To complete packing of the column, a layer of sand is normally placed on top of adsorbent, as shown in Figure 6.2. The purpose of the sand is twofold: (1) It allows the sample to flow evenly onto the surface of the adsorbent, and (2) it prevents disruption of the packing material as eluting solvent is added.

Mixtures of solids are typically dissolved in a *minimal* volume of a solvent before being transferred to the top of the column; a liquid mixture may be transferred directly to the column. It is important to distribute the sample evenly on the surface of the adsorbent and to use as little solvent as possible in loading the column. This procedure ensures that with the proper choice of eluting solvent, the bands that form during the development of the chromatogram will be narrow, thereby providing the best possible separation. If too much solvent is used to dissolve the sample, the initial band will be broad, and poor resolution of the mixture may result.

Once the sample has been loaded onto the column, there are several different techniques that may be used to elute its components. In a **simple elution** experiment, a *single* solvent is passed through the column during the entire course of the separation. This procedure works well for the separation of mixtures containing only two or three compounds having similar polarities. However, the more common chromatographic procedure is **stepwise** or **fractional elution**. In this technique, a series of increasingly more polar solvents is used to elute the mixture from the column. A nonpolar solvent such as petroleum ether, which is a mixture of low-boiling hydrocarbons, or hexane is first used to move the least polar component of the mixture down the column while the others remain at or near the top of the column. After elution of the first band, the polarity of the eluant is increased using either pure solvents or mixed solvent combinations so that the bands are individually eluted from the column. Systematic and gradual increases in solvent polarity are sometimes essential so that individual bands remaining on the column separate and do not coelute. As the polarity of the solvent system increases, those components of the mixture that are more tightly adsorbed on the column will begin to move. As a rule of thumb, a volume of solvent approximately equal to three times the column volume is passed through the column prior to switching to a solvent of higher polarity.

The separation is monitored using one of a variety of methods. Unfortunately, most organic compounds are not highly colored, and sophisticated devices for their detection are rarely available in the undergraduate laboratory. The most effective technique for following the separation is to collect fractions of equal volume in tared flasks, to concentrate the solvent, and to reweigh the flasks. The fractions containing the different bands may then be easily identified by the relative amounts of solute in each flask. One may also use TLC (Sec. 6.3) to monitor the separation.

In the experiment that follows, column chromatography will be used to separate fluorene (**1**) from an oxidation product, fluorenone (**2**). One of these compounds is white, the other is yellow. Consequently, the progress of the chromatography may be followed by evaporation of the solvent at periodic intervals as well as by observing the slower-moving yellow band.

1
Fluorene
White

2
Fluorenone
Yellow

EXPERIMENTAL PROCEDURES

COLUMN CHROMATOGRAPHY

Purpose: To separate fluorene and fluorenone by column chromatography.

> **Safety Alert**
>
> Petroleum ether is a highly volatile and flammable mixture of low-boiling hydrocarbons. During the preparation and development of the chromatographic column, be certain that there are *no flames* in the vicinity.

Procedure

Preparation   Answer the Pre-Lab Exercises on page PL. 27. Read or review Sections 2.13 and 2.29.

Apparatus   A *dry* 50-mL glass buret or chromatography column about 1 cm in diameter and 25 cm long fitted with either a Teflon stopcock or a short piece of polyethylene tubing and a screw clamp, and apparatus for simple distillation.

Setting Up   Clamp the glass buret or chromatography column in a vertical position with its *ungreased* stopcock, preferably Teflon, *closed*. Using a piece of glass tubing, insert a small plug of cotton or glasswool loosely into the bottom of the column. Cover the cotton or glasswool plug with enough clean sand to form a layer about 1 cm thick, and add approximately 25 mL of petroleum ether, bp 30–60 °C (760 Torr). Place a funnel on top of the column, and *slowly* add 5 g of dry neutral alumina to the column while constantly tapping the buret. A rubber "tapping mallet" may be made by placing a pencil in a one-hole rubber stopper. When this process has been completed, wash the inner walls of the column with additional petroleum ether to remove any alumina that may adhere to the sides. Cover the alumina with a 1-cm layer of clean sand, and open the stopcock to allow the solvent to drain into an Erlenmeyer flask until the solvent level reaches just to the *top of the alumina*. The column is now ready for the addition of the sample mixture (Fig. 6.3).

FIGURE 6.3
Set-up for column chromatography.

Separation   Obtain a sample of an approximately 1:1 mixture of fluorene (**1**) and fluorenone (**2**) and accurately determine its melting-point range. In a small test tube, dissolve about 0.1 g of this mixture in 0.5 mL of dichloromethane. Using a Pasteur or filter-tip pipet, carefully transfer this solution directly to the top of the column. Open the stopcock until the liquid level is at the top of the alumina. *Do not allow the solvent to drain below the level of the alumina*, as air bubbles and channels might develop in the solid support. Add approximately 1–2 mL of fresh petroleum ether to the top of the column, and again allow the liquid to drain to the top of alumina.

Fill the buret with approximately 20 mL of fresh petroleum ether, open the stopcock, and collect the eluant in an Erlenmeyer flask. Follow the progress of the chromatography by collecting a drop or two of eluant on a watchglass with every 5 mL that elutes from the column. When the solvent evaporates, any white solid present will be visible on the watchglass. Using this visualization technique, you can determine when all of the white solid has been eluted. Your instructor might also direct you to follow the chromatography by TLC using 15% dichloromethane in petroleum ether as the developing solvent (Sec. 6.3). Most of the white solid should elute in a volume of 15–20 mL of petroleum ether, and slow movement of a yellow band down the column should occur. Wash any of the white solid from the tip of the column into your collection flask with fresh petroleum ether.

When all of the white solid has eluted from the column, change the collection flask to another clean Erlenmeyer flask. Elute the column with about 5 mL of petroleum ether, and then change the eluant to dichloromethane, a more polar solvent. Watch the progress of the yellow band as it now proceeds rapidly down the column. When this yellow band just reaches the bottom of the column, change to a third clean Erlenmeyer flask. The intermediate fraction should not contain significant amounts of solid; verify this by evaporating a few drops on a watchglass. Continue eluting the column with dichloromethane until the elutant is colorless; approximately 10 mL will be required.

Isolation   The first and third fractions should contain pure samples of fluorene and fluorenone. Most of the solvent in these fractions may be removed by simple distillation (Sec. 2.13) or according to one of the other techniques outlined in Section 2.29. Attach the flask to a water aspirator or house vacuum to remove the last traces of solvent under reduced pressure.

Analysis   When the crystals of each of the purified compounds are completely dry, determine their melting points and weights. Use your experimentally determined melting points to identify the order of elution of fluorene and fluorenone from the column and record your observations and conclusions in your notebook. Based upon the weight of the mixture used, calculate the percent recovery of fluorene and fluorenone.

Finishing Touches   Place recovered *petroleum ether* in the container for nonhalogenated organic liquids, but pour the recovered *dichloromethane* into a container for halogenated organic liquids. Spread out the *alumina adsorbent*, which is wet with organic solvent, in a hood to dry and then place it in the nonhazardous solid waste container.

EXERCISES

1. Why is it preferable to use a Teflon or an ungreased stopcock rather than a greased stopcock on a column used for column chromatography?

2. Why should care be exercised in the preparation of the column to prevent air bubbles from being trapped in the adsorbent?

3. Why is a layer of sand placed above the cotton plug prior to the addition of the column packing material?

4. Does fluorene or fluorenone move faster down the column when petroleum ether is used as the eluant? Why?

5. When separating a mixture by standard column chromatography, why is it better to change from a less-polar solvent to a more-polar solvent rather than the reverse?

6. Normally, when increasing the polarity of the eluant, the increase is made gradually; however, in this procedure, *pure* dichloromethane rather than intermediate mixtures of petroleum ether and dichloromethane was added to the column. Why is this variation from the usual technique acceptable in this case?

7. The observed melting point of the 1:1 mixture of fluorene and fluorenone should be relatively sharp, although lower than the melting point of either of the pure compounds. On the other hand, a 3:1 mixture has a broad melting-point range of about 60–90 °C. Explain these observations. (*Hint:* See Sec. 3.3.)

8. Define the following terms.

   a. eluate   **b.** eluant   **c.** adsorbent

6.3   Thin-Layer Chromatography

Thin-layer chromatography (TLC) is another form of **solid–liquid adsorption** chromatography and involves the same fundamental principles as column chromatography (Sec. 6.2). The properties of the adsorbents and the relative eluting abilities of the solvents given for column chromatography also apply to TLC. However, unlike column chromatography where the mobile phase moves down the column, the mobile phase in TLC *ascends* the thin layer of adsorbent. Another significant difference is that a TLC experiment may be performed more rapidly and requires much less sample, sometimes as little as 10^{-9} g. Thin-layer chromatography is an excellent analytical tool and is frequently used to determine the optimal combinations of solvent and adsorbent for preparative column chromatographic separations as well as to monitor the progress of a separation by column chromatography. Note that TLC is limited to relatively *non-volatile* compounds whose boiling points are above 150 °C (760 Torr).

EXPERIMENTAL TECHNIQUE

For TLC, the adsorbent, usually alumina or silica gel, is mixed with a small quantity of a **binder**, such as starch or calcium sulfate, and spread as a layer approximately 250 μ thick on either a glass or plastic plate. The binder is necessary for proper adhesion of the thin layer of adsorbent to the plate. Thin-layer chromatography plates should be dried in an oven for an hour or more at 110 °C prior to use to remove any adsorbed moisture. The presence of water molecules on the surface of the adsorbent decreases its activity and effectiveness in separating the components of the mixture.

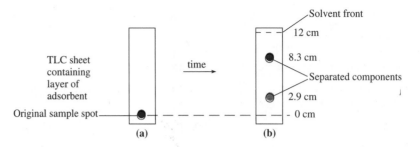

FIGURE 6.4   Thin-layer chromatography. (a) Original plate. (b) Developed chromatogram.

In a typical TLC analysis, a *small* drop of a solution of a mixture is carefully applied to a rectangular chromatographic plate near the edge of a narrow end as shown in Figure 6.4a. The plate is placed with spotted end down in a closed jar, called a **developing chamber**, containing a *saturated atmosphere* of a suitable eluting solvent, which may be either a pure solvent or a mixture of two or more solvents; the level of the eluant should be *just below* that of the spot (Fig. 6.4a). As the solvent moves up the plate, the components of the mixture are carried along at different rates to produce a series of spots on the plate as depicted in Figure 6.4b. When the solvent front has advanced nearly to the top of the TLC plate, the development of the chromatogram is complete, and the plate is withdrawn from the developing chamber.

The components of the mixture may be detected in a variety of ways. If the compounds being separated are colored, visual detection of the spots is easy. Many organic compounds are colorless, however, and a variety of methods have been developed to detect their presence on the plate:

1. Compounds that **fluoresce** may be located by placing the plate under an ultraviolet light. Since the spots disappear when the light is removed, it is necessary to circle the spots with a pencil in order to have a permanent record of the chromatogram. There are also commercially available plates that contain a fluorescent material as part of their coating; compounds that do not fluoresce but do absorb ultraviolet light then appear as dark spots under ultraviolet light.
2. The chromatographic plate may be sprayed with a variety of reagents such as sulfuric acid, potassium permanganate, phosphomolybdic acid, and ninhydrin; these reagents will react with the individual components to produce colored or dark spots.
3. The chromatographic plate may be exposed to iodine vapor by placing it in a closed chamber containing several crystals of iodine. As the iodine forms complexes with the various organic compounds, the spots become brown. Since the process is reversible and the spots fade, it is necessary to circle the spots with a pencil in order to have a permanent record of the chromatogram.

Once the separation of the components of the mixture is complete and the individual spots have been detected, the **retention factor** (R_f) of each compound may be calculated as shown below for the chromatogram pictured in Figure 6.4b:

$$R_f = \frac{\text{distance traveled by substance}}{\text{distance traveled by solvent}}$$

$$R_f(\text{compound 1}) = \frac{2.9 \text{ cm}}{12 \text{ cm}} = 0.24$$

$$R_f(\text{compound 2}) = \frac{8.3 \text{ cm}}{12 \text{ cm}} = 0.69$$

The R_f value for a compound is a physical constant for a given set of chromatographic conditions, so the adsorbent and the eluting solvent should be recorded along with the experimentally determined R_f values.

There are many important applications of TLC in modern organic chemistry. For example, TLC is commonly used to identify components of an unknown mixture by running chromatograms of the unknown sample side by side with known standards. Multiple aliquots of samples collected from chromatographic columns (Sec. 6.2) may be analyzed by TLC to follow the chromatographic separation. Alternatively, it is possible to follow the progress of a reaction by TLC by monitoring the disappearance of starting material or the appearance of product. Samples are simply withdrawn from a reaction mixture and subjected to TLC analysis.

APPLICATIONS

Two experiments are presented here to demonstrate the TLC technique. The first involves the separation of the pigments present in spinach leaves. A variety of other sources including crushed tomato pulp or carrot scrapings as well as leaves from grasses, shrubs, and trees may be substituted; however, waxy leaves are not acceptable.

In the second experiment, the **geometric isomers** *syn*-azobenzene (**3**) and *anti*-azobenzene (**4**) are separated. Commercially available azobenzene consists predominantly of the more stable *anti*- form, but this isomer may be isomerized to the less stable *syn*- isomer by irradiation with ultraviolet or sunlight. Since the colors of the two isomers differ, they may be detected visually. The course of the reaction and the effectiveness of the irradiation is followed by placing spots of irradiated and nonirradiated samples side-by-side on the TLC strip.

3
Syn-azobenzene

4
Anti-azobenzene

EXPERIMENTAL PROCEDURES

SEPARATION OF SPINACH PIGMENTS BY TLC

Purpose: To identify solvent mixtures that will separate the colored components in spinach leaves using thin layer chromatography.

Petroleum ether, ethanol, and acetone are highly volatile and flammable solvents. Be certain there are *no flames* in the vicinity during this experiment.

Safety
Alert

(a)

——Adsorbent

——Solvent front
——Solvent
——Spot

(b)

FIGURE 6.5 TLC chamber.
(a) Folded filter paper to be placed
in developing chamber for solvent
equilibration. (b) Developing
chamber with chromatogram.

Procedure

Preparation   Answer the Pre-Lab Exercises on page PL. 29. Read or review Section
2.21.

Apparatus   A wide-mouth bottle with a screw-top cap, small mortar and pestle, and
capillary pipets.

Setting Up   Prepare a developing chamber by placing a folded filter paper lengthwise
in a wide-mouth bottle (Fig. 6.5). As directed by your instructor, prepare three or four
mixtures of eluants that contain different ratios of varying pairs of the following solvents: petroleum ether, bp 60–80 °C (760 Torr), chloroform, acetone, and ethanol. For
example, prepare 10 mL of a 70:30 mixture of petroleum ether and acetone to use as
one eluant. Add an amount of the eluant to the developing chamber so that it forms a
1-cm layer on the bottom of the container. Screw the cap onto the bottle *tightly*, and
shake the container *well* to saturate the atmosphere of the chamber with vapors of the
solvent.

Preparing the Sample   Using a small mortar and pestle, thoroughly grind a spinach
leaf in a mixture of 4 mL of petroleum ether and 2 mL of ethanol. Transfer the liquid
extract to a test tube using a Pasteur pipet and *swirl* the extract gently with an equal
volume of water. Do not shake the test tube because emulsions are readily formed.
Remove and discard the aqueous layer; if you do not know which layer is the aqueous
one, perform the necessary test (Sec. 2.21). Wash the organic layer with water two
more times to remove the ethanol and any water-soluble materials that are present in
the leaves. Transfer the petroleum ether extract to the Erlenmeyer flask and add several
spatula-tips full of anhydrous sodium sulfate. After 5 to 10 min, decant the solution
from the drying agent. If the solution is not deeply colored, concentrate it using a gentle
stream of air or nitrogen to remove some of the solvent.

Preparing and Developing a Plate   Obtain a 2-cm × 10-cm strip of silica gel chromatogram sheet *without* a fluorescent indicator. Handle the strip *only* by the sides to
avoid contaminating the plate with oils from your hands. Place a pencil dot in the middle
of the sheet about 1 cm from one end. Using a capillary pipet, apply a spot of pigment
solution over the pencil dot by *lightly* and *briefly* applying the tip of the pipet to the
surface of the plate; you may blow gently on the plate as the sample is applied. Do not
allow the spot to diffuse to a diameter of more than 1–2 mm during application of the
sample. Perform the spotting process an additional four or five times, allowing the
solvent of each drop to evaporate before adding the next. When the spot has thoroughly
dried, place the strip in the developing chamber containing the first eluant; be careful
not to splash solvent onto the plate. The spot *must be above* the solvent level. Allow
the solvent front to move to within 2–3 mm of the top of the strip and then remove the
strip. Mark the position of the solvent front with a pencil, and allow the plate to air-
dry. Repeat this process with other eluants as directed by your instructor.

Analysis   With a good separation, one may observe as many as eight colored spots.
These are, in order of decreasing R_f values, the carotenes (two spots, orange), chlorophyll *a* (blue-green), the xanthophylls (four spots, yellow), and chlorophyll *b* (green).
Calculate the R_f values of all spots on your developed plate. In your notebook, include
a picture of the developed plate drawn to scale as a permanent record. Identify the
solvent mixture that gave the best separation of the pigments.

Put the unused eluants containing only *petroleum ether*, *acetone*, and *ethanol* in the container for nonhalogenated organic liquids and any unused eluants containing *chloroform* in the container for halogenated organic liquids. Discard the dry chromatographic plates in the nonhazardous solid waste container.

Finishing Touches

EXPERIMENTAL PROCEDURE

SEPARATION OF *SYN-* AND *ANTI-*AZOBENZENES BY TLC

Purpose: To identify solvent mixtures that will separate *syn-* and *anti-*azobenzenes using thin layer chromatography.

Safety Alert

1. Petroleum ether, ethanol, and acetone are highly volatile and flammable solvents. Be certain there are *no flames* in the vicinity during this experiment.
2. Since azobenzene is a suspected carcinogen, avoid contacting it with your skin or ingesting it.

Procedure

Preparation   Answer the Pre-Lab Exercises on page PL. 31.

Apparatus   A wide-mouth bottle with a screw-top cap, and a capillary pipet.

Setting Up   Prepare a developing chamber by placing a folded filter paper lengthwise in a wide-mouth bottle (Fig. 6.5). As directed by your instructor, prepare three or four mixtures of eluants that contain different ratios of varying pairs of the following solvents: petroleum ether, bp 60–80 °C (760 Torr), chloroform, acetone, and ethanol. For example, prepare 10 mL of a 90:10 mixture of petroleum ether and chloroform to use as one eluant. Add an amount of the eluant to the developing chamber so that it forms a 1-cm layer on the bottom of the container. Screw the cap onto the bottle *tightly*, and *shake* the container *well* to saturate the atmosphere of the chamber with vapors of the solvent.

Preparing and Developing a Plate   Obtain three or four 3-cm × 10-cm strips of silica gel chromatogram sheets *without* a fluorescent indicator. Handle the strip *only* by the sides in order to avoid contaminating the plate with oils from your hands. Place one pencil dot about 1 cm from the left side and about 1 cm from one end of one sheet and another about 1 cm from the right side the same distance from the bottom as the first. Using a capillary pipet, carefully apply a *small* spot of a 10% solution of commercial azobenzene in toluene, which you should obtain from your instructor, over one of the pencil dots. Do not allow the spot to diffuse to a diameter of more than 1–2 mm during application of the sample. Repeat this process for each strip. Allow the spots to dry and then expose the plates to sunlight for one to two hours (or a sun lamp for about 20 min).

When the irradiation is complete, apply another spot of the *original* solution on the plate over the second pencil dot in the same manner as just described and allow each strip to dry. Place a strip in the developing chamber, being careful not to splash solvent onto the plate. Both spots *must be above* the solvent level. Allow the solvent to move to within approximately 2–3 mm of the top of the strip and then remove the strip. Repeat this process for each additional strip using a different eluting solvent as directed by your instructor. Mark the position of the solvent front with a pencil, and allow the plate to air-dry.

Analysis   Note the number of spots arising from each of the two original spots. Pay particular attention to the relative intensities of the two spots nearest the starting point in each of the samples; these are *syn*-azobenzene. Calculate the R_f values of each of the spots on your developed plate. In your notebook include a picture of the developed plate drawn to scale as a permanent record. Identify the solvent mixture that gave the best separation of *syn*- and *anti*-azobenzene.

Finishing Touches   Put the *unused eluants* containing mixtures of only *petroleum ether*, *acetone*, and *ethanol* in the container for nonhalogenated organic liquids and any unused mixtures containing *chloroform* in the container for halogenated organic liquids. Put the dry chromatographic plates in the hazardous solid waste container, since they contain small amounts of azobenzene.

EXERCISES

1. In a TLC experiment, why should the spot not be immersed in the solvent in the developing chamber?

2. Explain why the solvent must not be allowed to evaporate from the plate during development.

3. Explain why the diameter of the spot should be as small as possible.

4. Which of the two diastereomers of azobenzene would you expect to be more thermodynamically stable? Why?

5. From the results of the TLC experiment with the azobenzenes, describe the role of sunlight.

6.4   Gas-Liquid Chromatography

Gas-liquid chromatography (GLC), which is also called **gas chromatography** (GC), is a technique that may be used to separate mixtures of volatile compounds whose boiling points may differ by less than 0.5 °C. It can also be applied as an analytical tool to identify the components of a mixture or in preparative applications when quantities of the pure components are desired.

BASIC PRINCIPLES OF GAS-LIQUID CHROMATOGRAPHY

Gas-liquid chromatography operates on the principle of partitioning the components of a mixture between a mobile gaseous phase and a stationary liquid phase. In practice, a sample is injected into a heated chamber where it is immediately vaporized and carried through a column by a flowing inert gas such as helium or nitrogen, which is called the **carrier gas**. This gaseous mixture is the **mobile phase**. The column is packed with a finely divided solid support that has been coated with a viscous, high-boiling liquid, which serves as the **stationary phase**. As the mobile phase moves through the column, its components are continuously partitioned between the two phases. Those components that show a higher affinity for the mobile phase move through the column more quickly, whereas those with a stronger attraction to the stationary phase migrate more slowly, and separation occurs. As with fractional distillation (Sec. 4.3), a given column may contain a large number of theoretical plates.

The **retention time** of a component is the elapsed time required for the compound to pass from the point of injection to the detector, and represents a physical property that may be used for purposes of identification. The retention time of a component is *independent* of the presence or absence of other components in the sample mixture. Four experimental factors influence a compound's retention time: (1) the *nature* of the stationary phase, (2) the *length* of the column, (3) the *temperature* of the column, and (4) the *rate of flow* of the inert carrier gas. Thus, for a particular column, temperature, and flowrate, the retention time will be the same for a specific compound.

Although a large number of **stationary liquid phases** are available, only a few are widely used (Table 6.1). Each liquid phase has a maximum temperature limit above which it cannot be used. This temperature depends upon the stability and volatility of the liquid phase; at higher temperatures the liquid phase will vaporize and ''bleed'' from the column with the mobile phase.

The differences in the *partition coefficients* (Secs. 5.2 and 5.3) of the individual components of a mixture in GLC depend primarily upon the differences in solubility of each of the components in the liquid phase. Two important factors determining the solubility of a gas in a liquid are its vapor pressure at the **ambient temperature**, which is the temperature of the column, and the magnitude of its interactions with the liquid phase. Regarding the first factor, the solubility of a gas in a liquid decreases as its vapor pressure increases. This means the more volatile components of a gaseous mixture tend to pass through the column more rapidly and elute before the less volatile ones. The impact of the second factor is understandable from the principle that *like dissolves like*. Polar samples are most effectively separated by using a polar liquid phase, whereas nonpolar compounds are best separated using a nonpolar liquid phase.

Table 6.1	Common Stationary Phases for Gas-Liquid Chromatography		
Liquid Phase	Type	Property	Maximum Temperature Limit, °C
Carbowax 20M	Hydrocarbon wax	Polar	250
OV-17	Methylphenyl Silicone	Intermediate polarity	300
QF-1	Fluorosilicone	Intermediate polarity	250
SE-30	Silicone gum rubber	Nonpolar	375

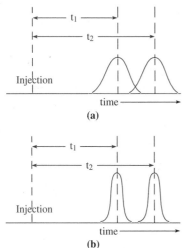

FIGURE 6.6

Effect of band-width on resolution.
(a) Wide bands. (b) Narrow bands.

The stationary liquid phase is normally coated as a thin film on an inert solid support. The support is composed of small, uniformly meshed granules, so that a large surface area of the liquid phase is available for contact with the vapor phase to ensure efficient separation. Some common types of solid supports include Chromosorb P and Chromosorb W, which are composed of diatomaceous earth. The surface areas of these supports varies from $1-6$ m^2/g. Columns are now commercially available with a wide variety of liquid phases on different solid supports. An alternative method of supporting the liquid phase is used in capillary columns. In these columns, the liquid is coated directly onto the inner walls of the tubing. These columns are highly efficient but relatively expensive.

In general, the efficiency, or **resolution**, of a column increases with increasing length and decreasing diameter. Increasing the path length increases the difference in retention times between bands, whereas decreasing the diameter of the column gives rise to narrower bands. With a small band separation, as measured from the band centers, wide bands are more likely to overlap (Fig. 6.6a) than narrow bands (Fig. 6.6b).

The two other experimental factors that may be varied to alter the degree of separation of the bands are the *temperature* at which the column is maintained and the *flowrate* of the carrier gas. Increasing the temperature results in shorter retention times, because the solubility of gases in liquids decreases with increasing temperature. The partition coefficients are thus affected, and the bands move through the column at a faster rate. Higher flow rates also cause retention times to decrease. In spite of the decreased resolution and band separation obtained at higher temperatures and flowrates, these conditions are sometimes necessary for substances otherwise having very long retention times.

INSTRUMENTATION

All commercially available gas-liquid chromatographs (GLCs) have a number of basic features in common. These are illustrated schematically in Figure 6.7. Parts 1–5 are associated with supplying the dry **carrier gas**, usually helium or nitrogen, and allowing an operator to control its flow. The mixture to be separated is injected using a gas-tight

FIGURE 6.7   Schematic diagram of apparatus for gas-liquid chromatography. (1) Carrier gas supply. (2) Pressure-reducing valve. (3) Desiccant. (4) Fine-control valve. (5) Flowmeter. (6) Heated injection port. (7) Column. (8) Oven. (9) Detector. (10) Electronic recorder. (11) Exit port.

syringe through a septum into the **injection port** (6), an individually heated chamber in which the sample is immediately vaporized. The sample then enters the flowing stream of carrier gas and is swept into and through the **column** (7), which is located in an **oven** (8) and consists of coiled aluminum, stainless steel, or glass tubing containing an appropriate stationary phase. In the column, the individual components separate into bands that ultimately pass through a **detector** (9) producing a signal whose voltage is proportional to the amount of material other than carrier gas present in the mobile phase. One type of detector that is commonly used is the **thermal conductivity detector** (TCD), which operates on the basis of differences in the thermal conductivity of the mobile phase as a function of its composition. A **flame ionization detector** (FID) is much more sensitive and operates by detecting the number of ions produced by passing the mobile phase through a hydrogen flame. The **recorder** (10) plots the changes in voltage measured by the detector as a function of time to give the **gas chromatogram**. The vapors then pass from the detector into either the atmosphere or a collection device at the **exit port** (11).

EXPERIMENTAL TECHNIQUES

Qualitative Analysis. The *retention time* of a pure compound is constant under a specified set of experimental conditions, including the column, temperature, and flowrate. Consequently this property may be used as a *first* step to identify an unknown compound or the individual components in a mixture. In a typical experiment, an unknown compound or mixture is injected into the injection port of a GLC, and the retention time(s) of the component(s) is (are) measured. A series of known samples are then injected under the same conditions. Comparison of the retention times of the standard samples with those of the unknown allows a preliminary identification of the component(s) of the unknown. A convenient way of confirming that the retention times of a standard and the unknown are the same involves injecting a sample prepared by combining equal amounts of the two. If a *single* peak is observed in the chromatogram, the *retention times* of the standard and the unknown are identical. However, observation of the same retention time for a known and an unknown substance is a *necessary but not sufficient* condition to establish identity, because it *is* possible for two different compounds to have the same retention time. Independent confirmation of the identity of the unknown by spectral (Chap. 8) or other means is imperative.

An example of the use of GLC as a qualitative, analytical tool is illustrated in Figure 6.8. These sets of peaks represent a gas chromatographic separation of the distillation fractions of a mixture of cyclohexane and toluene similar to those obtained in the distillation experiment described in Section 4.3. The notations *A*, *B*, and *C* refer to the three fractions taken in that experiment. The individual peaks in the mixture may be identified by comparing their retention times with those of pure cyclohexane and pure toluene; the peak with the shorter retention time in the mixture is cyclohexane, whereas the peak with the longer retention time is toluene.

Quantitative Analysis. The voltage output of the detector is related to the mole fraction of the material being detected in the vapor, so there is a correlation between the *relative areas* under the peaks in the chromatogram and the *relative amounts* of each of the components in the mixture. The quantitative evaluation of the chromatogram thus requires reliable methods for determining these peak areas.

An **electronic integrator**, which measures the intensity of detector output as a function of time, is the most accurate method for determining peak areas. However,

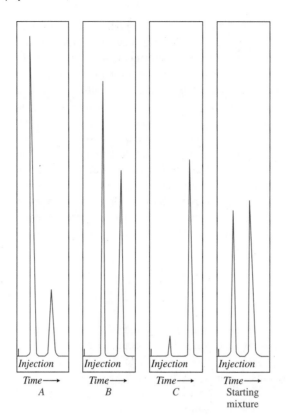

FIGURE 6.8
GLC analysis of the distillation fractions from the distillation experiment in Section 4.3.

these devices are expensive, so they are usually found only in research laboratories. Since the thickness and density of chart paper is reasonably uniform, another means of determining the relative areas involves carefully cutting the peaks out with scissors; the peak areas are then assumed to be proportional to their weight, as measured on an analytical balance. The original chromatogram should be saved as a permanent record, so the peaks should be cut from a photocopy of the chromatogram.

If the peaks are symmetrical, as are those shown in Figure 6.8, the areas may be approximated by assuming them to be equilateral triangles. The area of a symmetrical peak is then determined by multiplying the width of the peak at its half-height times its height. The percentage of each component in the mixture may be computed as the area of the peak corresponding to that component, expressed as a percentage of the sum of the areas of all peaks in the chromatogram. A sample calculation of this type is shown in Figure 6.9.

Although the peak areas are related to the mole fraction of the component in the mobile phase, they are not *quantitatively* related, since detector response varies with the class of compound. Not all compounds have the same thermal conductivity (TCD), nor do they ionize in a hydrogen flame to form the same types or number of ions (FID). Thus, it is necessary to *correct* the measured areas in the chromatogram using the appropriate **response factor** to obtain an accurate quantitative analysis of the mixture. Although response factors for different compounds may be determined experimentally, approximate values are published in monographs on gas chromatography. The response factors for thermal conductivity and flame ionization detectors for compounds that you may encounter in this experiment are given in Table 6.2. Notice that the correction

FIGURE 6.9
Determination of percentage composition
of a mixture by GLC.

Areas: $A = 17 \times 8 = 136$ mm^2

$B = 34 \times 17 = 578$

$C = 25 \times 12 = 300$
_____

Total area = 1014 mm^2

$$\%A = \frac{136}{1014} \times 100 = 13.4\%$$

$$\%B = \frac{578}{1014} \times 100 = 57.0\%$$

$$\%C = \frac{300}{1014} \times 100 = 29.6\%$$
_____

100%

factors vary more widely for flame ionization detectors than for thermal conductivity detectors. In the experimental section, you have the opportunity to calculate the response factor for an unknown compound.

To analyze a mixture of substances quantitatively when **weight factors** are known, the peak area for each component is simply *multiplied* by the weight factor for that particular compound. The resulting *corrected* areas are used to calculate the percentage composition of the mixture according to the procedure outlined in Figure 6.9. Note that using these factors provides the composition on a *weight percentage* basis.

Calculation of the composition of a mixture on a mole percentage basis requires the use of **mole factors**, M_f. These are obtained by dividing the weight factors by the

Table 6.2	Weight (W_f) and Mol (M_f) Correction Factors for Some Representative Substances[1]			
	Thermal Conductivity		**Flame Ionization**	
Substance	W_f	M_f	W_f	M_f
Benzene	1.00	1.00	1.00	1.00
Toluene	1.02	0.86	1.01	0.86
Ethylbenzene	1.05	0.77	1.02	0.75
Isopropylbenzene	1.09	0.71	1.03	0.67
Ethyl acetate	1.01	0.89	1.69	1.50
n-Butyl acetate	1.10	0.74	1.48	0.99
Heptane	0.90	0.70	1.10	0.86
o-Xylene	1.08	0.79	1.02	0.75
m-Xylene	1.04	0.76	1.02	0.75
p-Xylene	1.04	0.76	1.02	0.75
Ethanol	0.82	1.39	1.77	3.00
Water	0.71	3.08	—	—

[1] McNair, H.M.; Bonelli, E.J. *Basic Gas Chromatography*, 5th ed., Varian Aerograph, Walnut Creek, CA, 1969.

molecular weights of each component of the standard solution and normalizing the resulting numbers. A sample calculation utilizing mole correction factors is provided below in the analysis of a mixture of ethanol, heptane, benzene, and ethyl acetate with a GLC equipped with a thermal conductivity detector. The last column shows the percentage composition calculated directly from the measured peak areas, without correction for detector response. The dramatic differences in the calculated composition with and without this correction, as noted in the last two columns, underscore the importance of this correction for quantitative analysis.

Compound	Area (A) (mm^2)	M_f	$A \times M_f$	Mol % ($A \times M_f/194 \times 100$)	Uncorrected % ($A/207.1 \times 100$)
Ethanol	44.0	1.39	61.2	31.5	21.2
Heptane	78.0	0.70	54.6	28.1	37.7
Benzene	23.2	1.00	23.2	11.9	11.2
Ethyl acetate	61.9	0.89	55.4	28.5	29.9
Total	207.1		194.4	100	100

Use of Syringes. Liquid samples are injected into the heated injection port of a gas chromatograph using a gas-tight syringe with a capacity of 1 to 10 μL. The sample is either injected neat or dissolved in a volatile liquid such as diethyl ether or pentane. The sample should not contain nonvolatile substances that may eventually clog the injection port or contaminate the stationary phase of the column.

Gas-tight syringes are precision-made and expensive. They should be handled with care and used according to the following procedure. To fill the syringe draw slightly more than the desired volume of the sample into the barrel by withdrawing the plunger as needed, point the syringe needle-up, push out the excess liquid, and then wipe the tip of the needle with a tissue. To inject the sample, insert the needle *straight* into the septum as far as it will go, *push the plunger all the way in with one quick motion*, and remove the syringe from the septum while holding the plunger in place. If the sample is injected slowly, it will produce a wider band on the column and the peaks will be broadened. Be careful to *avoid bending the needle or the plunger* during the injection. It is important to *clean the syringe immediately after use*. Rinse it with a volatile solvent such as acetone and dry it by drawing a stream of air through it.

EXPERIMENTAL PROCEDURES

A · QUALITATIVE AND QUANTITATIVE ANALYSES OF A MIXTURE OF COMPOUNDS BY GLC

Purpose: To identify individual components of a mixture and to quantify the relative amounts of each by gas-liquid chromatography.

The solvents and other liquids that are used in sample mixtures for GLC analysis are flammable and volatile. *No flames* should be used in the vicinity of these liquids.

Procedure

Preparation Answer the Pre-Lab questions on page PL. 33. Read or review Section 2.5.

Apparatus Gas chromatograph and a gas-tight syringe.

Setting Up Since there is considerable variation in the operating procedures of different commercial gas chromatographs, consult your instructor for specific directions to use the instrument in your laboratory. Obtain a mixture of liquid compounds from your instructor. A suitable mixture would be one containing varying amounts of some or all of the following compounds: ethanol, ethyl acetate, toluene, *n*-butyl acetate, ethylbenzene, and isopropylbenzene.

Injection Following the directions of your instructor, use the syringe to inject a 1–5-μL sample of your mixture into the gas-liquid chromatograph to obtain a gas chromatogram. In the same way, obtain a chromatogram of pure samples of ethanol, ethyl acetate, toluene, *n*-butyl acetate, ethylbenzene, and isopropylbenzene under the same instrumental conditions used for the mixture.

Qualitative Analysis Identify each of the components in your mixture as follows: (1) Compare the retention time of each of the components of the mixture with the retention times of the authentic samples of pure compounds. (2) Verify your assignments for one or more components of the mixture by preparing a series of new sample mixtures, each of which will contain one volume of the pure known compound with two volumes of the original mixture. Prepare a gas chromatogram of this new mixture. The peak of the mixture that has been amplified in the new chromatogram corresponds to the pure compound that has been added.

Quantitative Analysis Use the quantitative method recommended by your instructor to determine the relative areas under the peaks in your chromatogram of the mixture. Correct these values by multiplying the measured area by the appropriate weight or molar correction factors for the type of detector in the GLC. These factors may be found in Table 6.2 or will be provided by your instructor. Determine the relative amounts, in weight and mole percentages, of each of the components of the mixture.

Pour any unused *volatile organic compounds* into the appropriate container for nonhalogenated liquids or halogenated hydrocarbons; put any mixture that contains a halogenated hydrocarbon in the container for halogenated hydrocarbons.

Finishing Touches

B · DETERMINING GLC RESPONSE FACTORS FOR KNOWN COMPOUNDS

Purpose: To determine the weight and mole correction factors for compounds using flame ionization or thermal conductivity detectors.

The solvents and other liquids that are used in sample mixtures for GLC analysis are flammable and volatile. *No flames* should be used in the vicinity of these liquids.

Procedure

Preparation   Answer the Pre-Lab questions on page PL. 35. Read or review Section 2.5.

Apparatus   Gas chromatograph, 1-mL syringe, and a gas-tight syringe.

Setting Up   Since there is considerable variation in the operating procedures of different commercial gas chromatographs, consult your instructor for specific directions to use the instrument in your laboratory. Look up the densities of the compound whose weight and mole response factors are to be determined and the standard substance whose response factors are known. For example, the standard sample may be any of the compounds in Table 6.2. Prepare a test solution by combining 0.5 mL of the compound whose response factors are to be determined and 0.5 mL of the standard in a small vial or test tube and swirl the solution to mix the two components *completely*. Assuming the volumes of the two liquids are additive, calculate the weight fraction and mole fraction composition of the mixture.

Injection   Following the directions of your instructor, use a gas-tight syringe to inject a 1–5-μL sample of the test mixture into the GLC to obtain a chromatogram. Repeat this process for each of the compounds for which you are to obtain weight and mole response factors.

Analysis   Use the quantitative method recommended by your instructor to determine the areas under the two peaks in your gas chromatogram of the test mixture. Divide the weight fraction and mole fraction of each of the two components by the area under the peak for that component. Normalize the value for the weight and mole factors for the unknown by the corresponding weight and mole factors for the standard and record these values in your notebook.

Finishing Touches   Pour any unused *volatile organic compounds* into the appropriate container for nonhalogenated liquids or halogenated hydrocarbons; put any mixture that contains a halogenated hydrocarbon in the container for halogenated hydrocarbons.

EXERCISES

1. Define the following terms.

   a. stationary phase **d.** retention time

   b. mobile phase **e.** solid support

   c. carrier gas **f.** thermal conductivity

2. Refer to the GLC traces given in Figure 6.8. These are analyses of the various fractions collected during the fractional distillation of the mixture of cyclohexane and toluene. The weight and mole correction factors (flame ionization detector) for cyclohexane are 0.84 and 0.78, respectively, and those for toluene are given in Table 6.2. Determine accurately both the weight percent and the mole percent compositions of the distillation fractions A, B, and C.

3. Benzene (4 g, 0.05 mol) is subjected to reaction with 1-chloropropane (3.9 g, 0.05 mol) and $AlCl_3$. The product (4.8 g) is subjected to analysis on a GLC equipped with a thermal conductivity detector. The chromatogram shows two product peaks identified as n-propylbenzene (area = 60 mm^2; W_f = 1.06) and isopropylbenzene (area = 108 mm^2; W_f = 1.09). Calculate the percent yield of each of the two isomeric products obtained in this reaction. Note that since each of the products has the same molecular weight of 120, the use of weight factors gives both weight and mole percent composition.

Benzene   1-Chloropropane

n-Propylbenzene   Isopropylbenzene

CHAPTER 7

Stereoisomers

Knowledge of the three-dimensional characteristics of organic molecules is an important part of organic chemistry. The experiments in this chapter are designed to provide an introduction to **stereoisomers**. Such isomers have molecular skeletons that are identical, but they differ in the three-dimensional orientation of their atoms in space. The two broad subclasses of stereoisomers that are of importance in organic chemistry are **conformational** isomers and **configurational** isomers.

Conformational isomers, as illustrated by two **Newman projections 1** and **2** for 1,2-dibromoethane, are stereoisomers that are interconverted by rotation about single bonds. The interconversion of such isomers is usually a low-energy process and occurs rapidly at room temperature and below. Consequently, isolation of conformational isomers is *rarely* possible.

Br   Br
H   H   H   Br
H   H   H   H
Br   H
1   **2**

Configurational isomers, in contrast, are stereoisomers whose interconversion normally requires breaking and remaking of a chemical bond, usually a high-energy process. An example of this type of stereoisomerism is seen with carvone, a natural product for which the two representations **3** and **4** can be written. These are related as **nonsuperimposable mirror images**, so **3** and **4** are defined as **enantiomers**. Molecules having a nonsuperimposable mirror image are said to be **chiral**, a term that is derived from the Greek word, *cheir*, meaning ''hand.'' Because hands are perhaps the best-known example of nonsuperimposable mirror images, use of the word ''chiral'' to characterize molecules that have a "handedness" is certainly appropriate. A companion term, **achiral**, describes all molecules having **superimposable mirror images**.

3
Carvone

4
Carvone

Enantiomers have identical physical and chemical properties in an achiral environment. Their differential effect on plane-polarized light, however, is an important exception to this general rule: Enantiomers rotate the plane of such light an equal number of degrees but in *opposite* directions. For this reason they are sometimes called **optical isomers** and are said to possess **optical activity**. Under most circumstances, the chiral compounds that you will prepare or use in the laboratory will be 50:50 mixtures of the two enantiomers, a composition referred to as the **racemate** or the **racemic mixture**. An equimolar mixture of **3** and **4** would thus be called a racemate and would produce *no* net rotation of plane-polarized light.

Frequently the source of chirality in a molecule is the presence of a **stereocenter**, which in organic compounds is usually a carbon atom bearing four different substituents. This is the case for the carbon atom marked with an asterisk in **3** and **4**. The substituents are a hydrogen atom, an isopropenyl group, and two methylene groups. The methylene groups are distinct from each other because one of them is bound to a carbon atom of a carbon-carbon double bond, and the other to a carbon atom doubly bound to oxygen.

A chiral molecule containing two or more stereocenters may exist as **diastereomers**, which are configurational isomers that are *not* enantiomers. Consider 1,2-cyclohexanediol, **5**, for example, which contains the stereocenters marked by asterisks. Each center consists of a carbon atom bound to a hydroxyl group, a hydrogen atom, a methylene group, and the carbon atom of the other stereocenter. Three configurational isomers, **6–8**, can then be written for **5**. The *cis* isomer **6** is clearly different from the *trans* isomers **7** and **8**, which are enantiomers of one another and diastereomers of **6**. Moreover, even though it contains two stereocenters, **6** has a *superimposable mirror image*. We could therefore variously describe this isomer as being achiral, optically inactive, or **meso**. The last term describes those configurational isomers that have two or more stereocenters *and* a superimposable mirror image. Although the identity of the physical and chemical properties of the enantiomers **7** and **8** might make their separation from one another difficult, the isolation of **6** free from contamination by **7** and **8** would be comparatively easy because *diastereomers have different physical and chemical properties.*

5
1,2-
Cyclohexanediol

6
cis-1,2-
Cyclohexanediol

7
trans-1,2-
Cyclohexanediol

8
trans-1,2-
Cyclohexanediol

Even isomeric pairs of *achiral* molecules may exist as separable stereoisomers. Examples include *cis-* and *trans-*, or (*Z*)- and (*E*)-, alkenes such as **9** and **10**, and cyclic compounds such as **11** and **12**. The molecules are nonsuperimposable, yet are *not* mirror images; that is, they are *not* enantiomers. Thus they are diastereomers and have different chemical and physical properties.

9

Dimethyl maleate

10

Dimethyl fumarate

11

cis-1,4-Dimethylcyclohexane

12

trans-1,4-Dimethylcyclohexane

In contrast to the case with diastereomers, separation or **resolution** of enantiomers is generally difficult because of the identity of their chemical properties. Moreover, their differing effect on plane-polarized light is not helpful for separating enantiomers. Nonetheless, various experimental techniques have been developed to accomplish resolutions and one of them is presented in Section 7.6.

The procedures in this chapter are chosen to study: (a) separating diastereomers by chromatography (Sec. 7.2); (b) converting one diastereomer into another (Sec. 7.3); (c) evaluating some chemical and physical properties of enantiomers (Sec. 7.4); and (d) resolving a racemate (Sec. 7.6). A description of the technique of polarimetry is given in Section 7.5 to support this last study.

7.2   Separation of Diastereomeric 1,2-Cyclohexanediols

Cyclic 1,2-diols can be produced stereoselectively from cycloalkenes. For example, a racemic mixture of the enantiomeric *trans*-1,2-cyclohexanediols **7** and **8** is produced by reaction of a peracid with cyclohexene and hydrolysis of the intermediate epoxide **13** (Eq. 7.1). Oxidation of this alkene by permanganate, on the other hand, gives *cis*-1,2-cyclohexanediol (**6**), as shown in Equation 7.2. The intermediate in this process is presumably the cyclic manganese-containing species **14**, which is not isolated.

$$\text{(7.1)}$$

13

Cyclohexene   Cyclohexene oxide

$$\text{(7.2)}$$

Permanganate ion   **14**

In this experiment a commercial mixture containing the *meso*-isomer **6** and the *racemate* **7/8** will be separated into its components; the mixture is produced by the catalytic hydrogenation of catechol (**15**) (Eq. 7.3). Because the enantiomeric *trans*-diols **7** and **8** have identical physical properties, except for their effect on plane-polarized light, they will not be resolved from one another in the procedure of this section. The *meso*-isomer **6**, on the other hand, being diastereomeric to **7** and **8**, has physical and chemical properties that are sufficiently different that it can be separated from the other two stereoisomers.

$$\text{(7.3)}$$

15
Catechol

The separation of diastereomers from one another may often be done solely on the basis of differing solubilities, but this is not the case with the 1,2-cyclohexanediols, which have similar solubilities. The desired separation is possible by chromatographic means, however, and the technique of TLC (Sec. 6.3) is used here, although column (Sec. 6.2) or gas chromatography (Sec. 6.4) would also work.

A challenge to the experimentalist is determining which spot on the TLC plate corresponds to the racemic mixture of *trans* isomers and which to the *cis* diastereomer. In other words, it is one thing to separate different compounds and quite another to assign specific structures to them. This common problem can be addressed in a variety of ways. For example, you might extract the portions of the TLC plate containing each of the isomers and determine the melting point or IR or NMR spectral characteristics (Chap. 8) of the solid left after removal of solvent. Spectral characterization is difficult to do in this case because only extremely small amounts of material would be obtained from the TLC plate. In addition, the melting points of the diastereomers are similar, being 100–101 °C for **6** and 103–104 °C for **7** and **8**; thus, traces of impurities in the *trans* isomer could easily lower its melting range (Sec. 3.2) to that of the *cis*, making unambiguous assignment of structure impossible.

The method used to make the needed structural assignment in this experiment is to compare the "unknown" compounds with an authentic specimen of "known" material. Specifically, by spotting the TLC plate with authentic samples of either *cis-* or *trans*-1,2-cyclohexanediol as well as the commercial mixture, you are able to make a direct comparison of R_f values and determine which component of the mixture is which.

EXPERIMENTAL PROCEDURE

SEPARATION OF DIASTEREOMERIC 1,2-CYCLOHEXANEDIOLS

Purpose: To demonstrate that diastereomers may be readily separated by chromatography.

Safety Alert

1. Acetone and petroleum ether are *highly* flammable; *do not have any flames in the vicinity* when they are being used.
2. Do not allow iodine to come in contact with your skin and do not inhale its vapor, as it is corrosive and *toxic*.

Procedure

Preparation   Answer the Pre-Lab Exercises on page PL. 37. Read or review Section 6.3.

Apparatus   TLC chamber.

Setting Up   Obtain or prepare approximately 5% solutions in acetone of a commercial mixture of *cis-* and *trans*-1,2-cyclohexanediol and of pure *trans*-1,2-cyclohexanediol. Place small spots of each solution side by side on a 3-cm × 10-cm silica gel TLC plate.

Elution and Analysis   Develop the chromatogram (Fig. 6.5b) using a mixture of 75% petroleum ether, bp 60–80 °C (760 Torr), and 25% 2-propanol by volume. Remove the plate from the solvent mixture when the solvent front approaches the top of the plate, mark the position reached by the solvent front, air-dry the plate for a few minutes, and place it in a closed container with a few iodine crystals to make the spots visible. Remove the plate and circle the location of any spots. Record the R_f-value of each spot.

Finishing Touches

Transfer any excess of the *eluting solvents* to a container for nonhalogenated organic liquids. Dispose of the used *TLC plates* in a container for nonhazardous materials.

EXERCISES

1. By comparing the R_f-values of the separated spots from the mixture of the *cis* and *trans* diols with the R_f-value of the *trans* isomer, decide which isomer is adsorbed

more strongly by the silica gel. Give a reason for the difference in adsorptivity of the two isomers.

2. It may happen that the separation of the isomers by TLC is not complete—one of the spots may be found to be smeared out to some extent. Considering this, which one of the isomers could more easily be obtained pure by column chromatography?

3. What would be the consequence of using pure petroleum ether as the eluting solvent?

4. What would be the consequence of using pure 2-propanol as the eluting solvent?

5. In Section 7.1 it was stated that interconversion of configurational isomers requires the cleavage and re-formation of a chemical bond. Indicate any such bonds in *cis*-1,2-cyclohexanediol (**6**) that upon being broken and then reformed would afford the enantiomers **7** and **8**. Be sure to describe what must occur stereochemically during the process of bond cleavage and remaking to effect the interconversion of the **6** to the other two diastereomers.

6. (a) Consider the bond dissociation energies of the various bonds in **6** and then predict which bond would be easiest to break for the conversion of **6** to **7** and **8**.

   (b) Estimate the $\Delta H^{\ddagger}$ for this process.

7.3 Isomerization of Dimethyl Maleate to Dimethyl Fumarate

Dimethyl maleate (**9**) and dimethyl fumarate (**10**) are examples of *cis* and *trans* isomers, called **geometric isomers**. Their names do not reflect this isomeric relationship, however, because the parent diacids from which they can be prepared were discovered and named before this stereochemical fact was known. Being diastereomers, these diesters would be expected to have different physical and chemical properties, and indeed they do. The difference in their physical states at room temperature is the most dramatic: The maleate **9** is a *liquid* with a freezing point of about $-19\ °C$, whereas the fumarate **10** is a *solid* with a melting point of $103-104\ °C$.

9
Dimethyl maleate

10
Dimethyl fumarate

As discussed in Section 7.1, the interconversion of configurational isomers involves the cleavage and subsequent remaking of a chemical bond. The diastereomers **9** and **10** could be equilibrated by breaking the π part of the carbon-carbon double bond, rotating about the single bond that remains, and remaking the π bond. The required bond cleavage can be promoted (a) thermally, (b) photochemically, or (c) chemically. The latter possibility involves the reversible addition of a reagent to a carbon atom involved in the carbon-carbon double bond, resulting in breaking of the π bond.

The goal of this experiment is to formulate a mechanism by which dimethyl maleate (**9**) is isomerized to dimethyl fumarate (**10**) by the action of heat, light, or molecular bromine. Three different procedures are followed to study the reaction: exposure to light of two samples of **9**, one of which contains a catalytic amount of bromine, and treating a third sample with bromine but keeping the mixture in the dark at room temperature. By thoughtful analysis of your results, it is possible to discover which of the three modes of breaking the π-bond, which were designated (a)–(c) in the previous paragraph, is responsible for the isomerization. The analysis also allows you to determine whether polar intermediates or radical species are involved in the isomerization.

There will be residual bromine at the end of the procedure, and it is necessary to destroy it prior to isolation of the crystalline **10**. A convenient way to do this is by adding cyclohexene to the reaction mixture, which reacts with bromine to produce a dibromide (Eq. 7.4).

$$\text{Cyclohexene} + Br_2 \longrightarrow \text{Dibromocyclohexane} \tag{7.4}$$

EXPERIMENTAL PROCEDURES

ISOMERIZATION OF DIMETHYL MALEATE TO DIMETHYL FUMARATE

Purpose: To study the mechanism of a geometric isomerization.

> **Safety Alert**
>
> 1. Bromine is a hazardous chemical. Do not breathe its vapors or allow it to come into contact with the skin because it may cause serious chemical burns. Perform all operations involving the transfer of solutions of bromine at a hood and wear latex gloves when handling bromine or solutions of it. If you get bromine on your skin, *immediately* wash the affected area with soap and warm water and soak the skin in 0.6 *M* sodium thiosulfate solution, for up to 3 h if the burn is particularly serious.
> 2. Bromine reacts with acetone to form α-bromoacetone, $BrCH_2COCH_3$, a powerful *lachrymator*. Do not rinse glassware that might contain residual bromine with acetone! Rather, follow the procedure described in FINISHING TOUCHES.

Miniscale Procedure

Preparation   Answer the Pre-Lab Exercises on page PL. 39. Read or review Sections 2.9, 2.17, and 3.2.

Apparatus   Three 13-mm × 100-mm test tubes, incandescent light source, apparatus for vacuum filtration and *flameless* heating.

Setting Up   Measure 0.5 mL of dimethyl maleate into each of the test tubes. Working at the hood and wearing latex gloves, add enough of a 0.6-*M* solution of bromine in dichloromethane dropwise with the filter-tip pipet to *two* of the tubes to give an orange solution. Add an equal volume of dichloromethane to the third test tube. Loosely cork all the test tubes.

Isomerization and Isolation   Place one of the tubes containing bromine in the dark, and expose the other two tubes to strong light. Place the tubes about 20 cm from the light source to prevent boiling the solvent and popping the cork off the tubes. If decoloration of a solution occurs, add an additional portion of the bromine solution. After 30 min, record in which test tube(s) crystals have appeared and add 1–2 drops of cyclohexene to the tubes containing bromine to destroy it. Add 2–3 mL of 95% ethanol to any tubes containing solid, and heat the mixture to effect dissolution. Then allow the solution to cool slowly to room temperature. Isolate the precipitate by vacuum filtration using a Hirsch funnel (Fig. 2.52). Press the crystals as dry as possible on the filter disk with the aid of a clean cork or spatula, transfer it to a watchglass or piece of weighing paper, and air-dry it.

Analysis   Weigh the dimethyl fumarate (**10**) and calculate its yield, based on the amount of dimethyl maleate used in the solution(s) that produced **10**. Determine its melting point and compare the result with the reported melting point. If the observed melting point is low, recrystallize the product from ethanol.

Microscale Procedure

Preparation   Answer the Pre-Lab Exercises on page PL. 39. Read or review Sections 2.9, 2.17, and 3.2.

Apparatus   Two Craig tubes, 10-mm × 75-mm test tube, incandescent light source, apparatus for Craig tube filtration and *flameless* heating.

Setting Up   Using a Pasteur pipet, dispense two drops of dimethyl maleate into each of the Craig tubes and the test tube. Working at the hood, add enough of a 0.1-*M* solution of bromine in dichloromethane dropwise to the *Craig tubes* to produce an orange-colored solution. Add sufficient dichloromethane to the *test tube* so that the volume of solution in the tube approximately equals that in the Craig tubes. Loosely cork each of the tubes.

Isomerization and Isolation   Place one of the Craig tubes in a dark location, and expose the other two solutions to strong light. Place the tubes about 20 cm from the light source to prevent boiling the solvent and popping the cork off the tubes. Should decoloration of a solution occur, add an additional portion of the bromine solution. After 20 min, record in which test tube(s) crystals have appeared and add 1 drop of cyclohexene to the tubes containing bromine to destroy it. Add 0.5–1 mL of 95% ethanol to any tubes containing solid, and heat the mixture to effect dissolution. Then allow the solution to cool slowly to room temperature.

Filter any solutions containing solids by Craig tube filtration (Fig. 2.53). Transfer the product to a watchglass or piece of weighing paper, and air-dry it.

Analysis   Weigh the dimethyl fumarate (**10**). Determine its melting point and compare the result with the reported melting point. If the observed melting point is low, recrystallize the product from ethanol using a Craig tube.

Finishing Touches

Decolorize any *solutions or containers in which the color of bromine* is visible by dropwise addition of cyclohexene, and then discard the resulting solutions in a container for halogenated organic liquids. Flush the *ethanolic filtrate* from the recrystallization process down the drain with water.

EXERCISES

1. What observation allows elimination of thermal cleavage of the π-bond as the initiation step for isomerization of **9** to **10**?

2. What observation allows exclusion of direct photoexcitation of **9** as the means of initiating its conversion to **10**?

3. Write a stepwise mechanism for the isomerization of **9** to **10**. Be certain to account specifically for the roles, if any, that light and bromine play in promoting this process.

4. Explain why decoloration of a solution of bromine and **9** is slow.

5. State which one of the two diesters **9** and **10** is predicted to be more stable thermodynamically and support your prediction.

6. What type of configurational isomers are **9** and **10**, enantiomers or diastereomers? How is your answer consistent with the fact that these two compounds have different physical properties?

7. The double bonds of **9** and **10** contain both a σ- and a π-component. Explain why rupture of the π-bond rather than the σ-bond is the more probable event in the isomerization process.

7.4   Properties of the Enantiomers of Carvone

Enantiomers have *identical* physical properties except with respect to their effect on plane-polarized light. Their chemical properties are also identical, with one *important* exception: One enantiomer undergoes a chemical change at a rate *different* from that of its mirror image *if* the reaction occurs in a stereochemically asymmetric environment, such as the active site of an enzyme or a biological receptor. This is because the active sites of enzymes and receptors are themselves chiral and, as a consequence, complex preferentially with one member of a pair of enantiomers. The complexed enantiomer will then undergo the enzyme-promoted reaction faster than will its mirror image.

A dramatic example of such differential complexation is the olfactory response to the two enantiomers of carvone. (*R*)-(−)-Carvone (**3**) smells like spearmint and is the principal component of spearmint oil, which also contains minor amounts of limonene (**16**) and α- and β-phellandrene (**17** and **18**). On the other hand (*S*)-(+)-carvone (**4**), along with limonene, is found in caraway seed and dill seed oils, and has been shown to be the compound largely responsible for the characteristic odor of these oils. This remarkable difference in how we sense these two enantiomers is because the odor receptors in our noses are chiral environments linked to the nervous system. The type of receptor-site that complexes with a particular enantiomer will determine what odor is detected by the brain.

CH₃   CH₃   CH₃   CH₂   CH₃

3   **4**   **16**   **17**   **18**

(R)-(−)-Carvone   (S)-(+)-Carvone   Limonene   α-Phellandrene   β-Phellandrene

It is unusual for both enantiomers of a compound to occur naturally, and extremely rare for each of them to be available from readily accessible sources such as spearmint and caraway oils. This fortunate situation enables the separate isolation of **3** and **4** from their respective sources by vacuum distillation (Sec. 2.15). Because of their ready isolation, both of these enantiomers are commercially available.

The investigative experiments that you perform on the enantiomeric carvones will depend upon the time and facilities that are available. If samples of both enantiomers are available, you can verify their different odors. Bear in mind, however, that 8–10% of the population cannot detect the difference, and you may be in this group. Should a gas-liquid chromatograph (Sec. 6.4) be available, analyses may be performed on samples of spearmint, caraway oils, and dill seed as well as solutions containing one or both of the enantiomers **3** and **4**. Typical chromatograms are given in Figures 7.1–7.3.

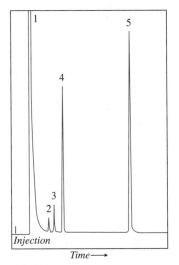

FIGURE 7.1
GLC analysis of spearmint oil (numbers in parentheses are retention times in minutes and peak areas, respectively). Peak 1: solvent; peak 2: unknown; peak 3: phellandrenes (2.16, 17819); peak 4: limonene (2.46, 84889); peak 5: (R)-(−)-carvone (5.13, 248316).

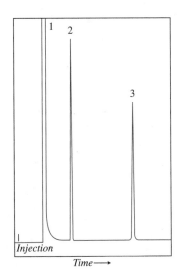

FIGURE 7.2
GLC analysis of caraway seed oil (numbers in parentheses are retention times in minutes and peak areas, respectively). Peak 1: solvent; peak 2: limonene (2.46, 125719); peak 3: (S)-(+)-carvone (5.13, 385950).

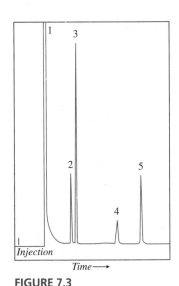

FIGURE 7.3
GLC analysis of dill seed oil (numbers in parentheses are retention times in minutes and peak areas, respectively). Peak 1: solvent; peak 2: unknown (2.27, 49418); peak 3: limonene (2.44, 139972); peak 4: unknown (4.15, 17091); peak 5: (S)-(+)-carvone (5.07, 78223).

Note that under the conditions of these analyses limonene (**16**) and the phellandrenes **17** and **18** elute *before* **3** and **4** because their boiling points, 170–177 °C (760 Torr), are lower than that of the enantiomers (bp 232–233 °C, 760 Torr).

The presence of a carbon-carbon double bond in carvone can be shown by performing a test for unsaturation using bromine in dichloromethane (Eq. 7.5a) and aqueous potassium permanganate (Eq. 7.5b). Confirmation of the carbonyl function in **3** and/or **4** and of the structural assignment is obtained by making the 2,4-dinitrophenylhydrazone (2,4-DNP), of carvone (Eq. 7.6) and comparing its melting point to that reported (193–194 °C) for authentic material. By doing a mixture melting point (Sec. 3.3) of the 2,4-DNPs prepared from both **3** and **4**, you can determine whether one enantiomeric derivative depresses the melting point of the other.

(7.5a)

(7.5b)

(7.6)

2,4-Dinitrophenylhydrazine

Carvone 2,4-dinitrophenylhydrazone

EXPERIMENTAL PROCEDURES

PROPERTIES OF THE ENANTIOMERIC CARVONES

Purpose: To explore physical and chemical properties of enantiomers.

Safety Alert

1. Bromine is a hazardous chemical. Do not breathe its vapors or allow it to come into contact with the skin because it may cause serious

chemical burns. Perform all operations involving the transfer of solutions of bromine at a hood and wear latex gloves when handling bromine or solutions of it. If you get bromine on your skin, *immediately* wash the affected area with soap and warm water and soak the skin in 0.6 *M* sodium thiosulfate solution, for up to 3 h if the burn is particularly serious.

2. Bromine reacts with acetone to form α-bromoacetone, $BrCH_2COCH_3$, a powerful *lachrymator*. Do not rinse glassware that might contain residual bromine with acetone! Rather, follow the procedure described in FINISHING TOUCHES.

3. Wear latex gloves to prevent solutions of potassium permanganate from contacting your skin. These solutions cause unsightly stains on your skin and last several days. If contact occurs, wash the affected area thoroughly with warm water.

Procedures

Preparation   Answer the Pre-Lab Exercises on page PL. 41. Read or review Sections 6.3, 23.7, and 23.8.

1. Odor.   Compare the odor of the carvone obtained from spearmint oil with that of the carvone from caraway seed oil.

2. Tests for Unsaturation.   Obtain samples of **3** and/or **4**, and test for the presence of a carbon-carbon double bond using procedures A and/or B in Section 23.8. Record your observations.

3. Gas-Liquid Chromatography.   Obtain gas chromatograms of spearmint and caraway oils and of the enantiomeric carvones. Co-inject a sample of an authentic sample of **3** or **4** with one of the natural oils to determine which peak in the oil corresponds to carvone. Use the data to determine (a) the retention time of the carvones and the relative percentage of carvone **3** or **4** compared to the total amount of the three compounds **16–18** that may be present in the essential oils. A capillary column with dimensions of 0.25-mm $\times$ 25-m with 100% dimethylpolysiloxane as the stationary phase, a flowrate of 1.2 mL/min, and a column temperature of 110 °C is suitable for the analysis.

4. Carvone 2,4-Dinitrophenylhydrazone.   Using a 0.5-g sample of **3** or **4**, or of the oils from which they are obtained, prepare this derivative according to procedure A in Section 23.7 and determine its melting point. If recrystallization is necessary, use a mixture of ethanol and ethyl acetate as the solvent. Obtain a sample of the 2,4-DNP prepared from the enantiomer opposite from yours. Make a 1:1 mixture of the enantiomeric 2,4-DNPs by intimately mixing the two, and determine the melting point of the mixture. Compare the result with the melting point you originally obtained.

EXERCISES

1. Why would a positive test for unsaturation on spearmint, caraway seed, or dill seed oil *not* prove that the enantiomeric carvones contain unsaturation?

2. Based on the results of the GLC analysis of the enantiomeric carvones, would you expect the *boiling point* of a *mixture* of the enantiomers to be the same as that of a *single* enantiomer? Explain.

3. Assume that the peak areas given in Figures 7.1 and 7.2 for limonene (**16**), the phellandrenes **17** and **18**, and the carvones **3** and **4** reflect the relative *molar* amounts of these compounds. Compute the percentage of the carvone present in spearmint and caraway seed oils.

4. Figure 7.3 is a gas chromatogram of dill seed oil.

   a. Using the same assumption as that in Exercise 3, compute the percentage of the carvone present in this oil.

   b. Explain why the retention time of the carvone in this oil does *not* allow assignment of which enantiomer is present.

5. When a student attempted to prepare the 2,4-DNP of (*S*)-(+)-carvone directly from dill seed oil, a red-orange solid was obtained that melted at 110–115 °C even after recrystallization. This suggests that a mixture of DNPs has been produced, and that they are not easily separated by recrystallization. How does the GLC analysis of this oil (Fig. 7.3) support the proposition that there could be a significant amount of a carbonyl-containing contaminant of carvone in dill seed oil? What might you do experimentally to support your proposal?

6. What melting point should you expect from a mixture of the 2,4-DNPs of (*R*)-(−)-carvone (**3**) and (*S*)-(+)-carvone (**4**) if both derivatives are pure?

7. Identify the chiral center in limonene (**16**) and draw the (*R*)-enantiomer of this molecule.

8. Show that the phellandrenes **17** and **18** are chiral.

9. In Section 7.1 it was stated that interconversion of configurational isomers involves the cleavage and remaking of a chemical bond. Indicate the four different bonds in **3** that could be broken and then remade to produce **4**. For each bond be sure to describe what must occur stereochemically during the process of bond cleavage/remaking that effects the interconversion of one enantiomer to the other.

7.5 Polarimetry

Chiral molecules are optically active, as was noted in Section 7.1; that is, they rotate the plane of polarized light as the light passes through the sample. An explanation of the physical basis for this phenomenon is beyond the scope of this discussion, but it is important to understand that achiral molecules do *not* exhibit this same property.

Of primary interest to organic chemists is the fact that each member of a pair of enantiomers rotates plane-polarized light by exactly the same amount but in *opposite* directions. This means that a solution containing equal amounts of two enantiomers will cause *no* net rotation of the light because the rotation in one direction by molecules of one enantiomer will be exactly offset with an opposite rotation by those of the other. Thus, a solution containing enantiomers can produce a rotation only if the amount of one enantiomer exceeds that of its mirror image.

The value α of an **observed rotation** depends on several factors: (a) the nature of the chiral compound; (b) its *concentration*, if in solution, or *density*, if a neat liquid; (c) the *pathlength* of the cell containing the sample; (d) the *temperature* at which the measurement is taken; (e) the nature of the *solvent*, if one is used; and (f) the *wavelength* of plane-polarized light used in the measurement. Factors (b) and (c) determine the average number of chiral molecules in the path of the light, and, in combination with the wavelength, have the greatest effect on the magnitude of the rotation of a sample.

The sign of α is defined by convention. When the observer views through the sample *toward* the source of the plane-polarized light, a clockwise rotation of the plane of the light is taken as positive or **dextrorotatory**, a counterclockwise rotation as negative or **levorotatory**.

Optical rotation represents a **physical constant** of a chiral compound *if* the variables listed above are considered. By specifying the temperature and wavelength at which the measurement is taken, and dividing the observed rotation by the factors that define the average number of molecules in the light path, a constant called the **specific rotation**, [α], is obtained. This is expressed mathematically by Equation 7.7.

$$[\alpha]_\lambda^T = \frac{\alpha}{l \times c} \text{ or } \frac{\alpha}{l \times d} \qquad (7.7)$$

where

$[\alpha]$ = specific rotation (degrees)

λ = wavelength (nanometers, nm)

T = temperature (°C)

α = observed rotation (degrees)

l = path length of cell (decimeters, dm)

c = concentration (g/mL of solution)

d = density (g/mL, neat sample)

A specific rotation might be reported as $[\alpha]_{490}^{25}$ −19.6° (c = 1.8, CHCl$_3$). Alternatively, the specific rotation might be written as $[\alpha]_D^{23}$ +39.2° (c = 12.1, CH$_3$OH). The D refers to the fact that a sodium lamp emitting radiation at 589 nm, the D line of sodium, was used as the light source.

The basic components of all **polarimeters**, the instruments used for measuring optical rotations, are shown in Figure 7.4. Ordinary light first passes through a polarizer, such as a Nicol® prism or a sheet of Polaroid® film, and the plane-polarized light that emerges passes through the sample tube. If the tube contains an optically active substance, the plane of the light will be rotated either clockwise or counterclockwise. The

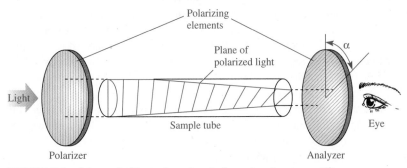

FIGURE 7.4 Schematic illustration of optical rotation (α = ~−45°).

degree of rotation can be determined by turning the analyzer, another Nicol prism or piece of Polaroid film, until the intensity of the light reaching the detector matches that observed in the absence of any of the optically active compound in the sample tube. The human eye or a photoelectric cell constitutes the detector. Finally, using Equation 7.7 allows the conversion of the observed rotation α to the specific rotation $[\alpha]$.

7.6   Resolution of Racemic 1-Phenylethanamine

Enantiomers cannot normally (see Historical Highlight at the end of this chapter) be physically separated because they have identical properties, except toward plane-polarized light (Sec. 7.1). However, *diastereomers* do have different physical properties, such as solubility, melting and boiling points, and adsorptive characteristics in chromatography. Recognition of this characteristic of diastereomers has led to a general strategy for the separation of enantiomers that involves three steps: (1) conversion of the enantiomers to diastereomers by way of a chemical reaction; (2) separation of these diastereomers by any of a number of standard techniques, such as recrystallization and chromatography; and (3) regeneration of one or both of the pure enantiomers from the corresponding diastereomer. This is illustrated schematically in Equation 7.8, where **R** and **S** represent the enantiomers to be separated and **R'** stands for a *single* enantiomer of a compound, termed a **resolving agent**, with which both **R** and **S** can react. To convince yourself that **RR'** and **SR'** are actually *diastereomers*, you need only convince yourself that the *enantiomer* of **RR'**, for instance, would be written as **SS'**.

$$\mathbf{R + S} \xrightarrow{\ \ R'\ \ } \mathbf{RR' + SR'} \xrightarrow[\text{technique}]{\text{Separation}} \begin{array}{l} \mathbf{RR'} \xrightarrow{\text{Regeneration}} \mathbf{R + R'} \\ \mathbf{SR'} \xrightarrow{\text{Regeneration}} \mathbf{S + R'} \end{array} \qquad (7.8)$$

| Racemic mixture | Diastereomeric mixture | Separated diastereomers | Pure Enantiomers |

Nature is the source of a number of different resolving agents that are used to prepare separable diastereomeric pairs from enantiomers. In this experiment, we use (+)-tartaric acid (**19**), which is produced from grapes in the production of wine, to resolve racemic 1-phenylethanamine (**20**), as shown in the resolution scheme outlined in Figure 7.5. Conversion of the enantiomers of **20** into separable diastereomers involves an acid-base reaction with **19**. The diastereomeric salts that result have different solubilities in methanol and are separable by fractional crystallization. The individual enantiomers of **20** are then regenerated by neutralization of the corresponding salts.

In principle, it would be possible to obtain both enantiomers of a chiral compound in high optical purity by this technique, but this generally would require an undesirably large number of crystallizations. Fortunately, one or two crystallizations usually suffice to allow isolation, in reasonable optical purity, of the enantiomer derived from the less soluble salt.

Measuring the optical rotation (Sec. 7.5) of the 1-phenylethanamine recovered in this experiment allows determination of whether it is the (+)- or (−)-enantiomer and the extent to which optical purification is achieved. If you are to do this part of the experiment, consult with your instructor for special experimental directions.

$$(+)C_6H_5-\underset{\underset{\displaystyle CH_3}{|}}{C}H-NH_2 \ + \ (^-)C_6H_5-\underset{\underset{\displaystyle CH_3}{|}}{C}H-NH_2$$

20

1–Phenylethanamine
Racemic mixture of enantiomers

$$(+)HO_2C-\underset{\underset{\displaystyle OH}{|}}{C}H-\underset{\underset{\displaystyle OH}{|}}{C}H-CO_2H$$

19

$$\left[(+)HO_2C\underset{\underset{\displaystyle OH}{|}}{C}H-\underset{\underset{\displaystyle OH}{|}}{C}H-CO_2^- \quad (+)C_6H_5-\underset{\underset{\displaystyle CH_3}{|}}{C}H-\overset{+}{N}H_3\right]$$

\+

$$\left[(+)HO_2C\underset{\underset{\displaystyle OH}{|}}{C}H-\underset{\underset{\displaystyle OH}{|}}{C}H-CO_2^- \quad (^-)C_6H_5-\underset{\underset{\displaystyle CH_3}{|}}{C}H-\overset{+}{N}H_3\right]$$

Diastereomers

Separate by fractional crystallization from methanol

$$\left[(+)HO_2C\underset{\underset{\displaystyle OH}{|}}{C}H-\underset{\underset{\displaystyle OH}{|}}{C}H-CO_2^- \quad (+)C_6H_5-\underset{\underset{\displaystyle CH_3}{|}}{C}H-\overset{+}{N}H_3\right]$$

$$\left[(+)HO_2C\underset{\underset{\displaystyle OH}{|}}{C}H-\underset{\underset{\displaystyle OH}{|}}{C}H-CO_2^- \quad (^-)C_6H_5-\underset{\underset{\displaystyle CH_3}{|}}{C}H-\overset{+}{N}H_3\right]$$

↓ NaOH   ↓ NaOH

$$(+)NaO_2C\underset{\underset{\displaystyle OH}{|}}{C}H-\underset{\underset{\displaystyle OH}{|}}{C}H-CO_2Na \ + \ (+)C_6H_5-\underset{\underset{\displaystyle CH_3}{|}}{C}H-NH_2$$

$$(+)NaO_2C\underset{\underset{\displaystyle OH}{|}}{C}H-\underset{\underset{\displaystyle OH}{|}}{C}H-CO_2Na \ + \ (^-)C_6H_5-\underset{\underset{\displaystyle CH_3}{|}}{C}H-NH_2$$

↓ (1) Extract with diethyl ether   ↓ (1) Extract with diethyl ether
  (2) Distil   (2) Distil

$$(+)C_6H_5-\underset{\underset{\displaystyle CH_3}{|}}{C}H-NH_2$$

$$(^-)C_6H_5-\underset{\underset{\displaystyle CH_3}{|}}{C}H-NH_2$$

FIGURE 7.5   Resolution of (±)-1-phenylethanamine.

EXPERIMENTAL PROCEDURES

RESOLUTION OF RACEMIC 1-PHENYLETHANAMINE

Purpose: To demonstrate a method for resolving enantiomers.

Safety Alert

1. Methanol is flammable; when heating it use *flameless* heating.
2. Avoid contact of 14 *M* NaOH with the skin as it is corrosive. Wear latex gloves when handling this solution. If skin contact occurs, wash the area with copious amounts of water.
3. Diethyl ether is extremely volatile and flammable. When using it for extractions or distilling it, be certain that there are *no flames* in the vicinity.

Miniscale Procedure

Preparation   Answer the Pre-Lab Exercises on page PL. 000. Read or review Sections 2.6, 2.9, 2.11, 2.13, 2.15, 2.17, 2.25, and 2.29.

Apparatus   A 500-mL Erlenmeyer flask, apparatus for magnetic stirring, vacuum filtration, and *flameless* heating.

Setting Up   Dissolve an *accurately* weighed sample of approximately 5 g of racemic 1-phenylethanamine in 35 mL of methanol and determine its specific rotation [α] using a polarimeter and Equation 7.5. The amounts of 1-phenylethanamine and methanol specified here give satisfactory results when a polarimeter sample tube with a light path of 2 dm or more and a volume of 35 mL or less is used. If a tube of shorter light path and/or larger volume is used, either increase the amount of racemic 1-phenylethanamine or combine your resolved product with that of other students to give a solution having sufficient concentration to allow accurate measurement of the optical rotation. Consult with your instructor about the type of polarimeter sample tube that is available.

Diastereomer Formation and Isolation   Place 15.6 g of (+)-tartaric acid and 210 mL of methanol in the Erlenmeyer flask and heat the mixture to boiling using *flameless* heating. To the stirred hot solution, *cautiously* add the solution recovered from the polarimeter and 7.5 g of racemic 1-phenylethanamine. Allow the solution to cool slowly to room temperature and to stand *undisturbed* for 24 h or until the next laboratory period.★ The amine hydrogen tartrate should separate in the form of white prismatic crystals. If the salt separates in the form of needlelike crystals, the mixture should be reheated *until all* the crystals have dissolved and then allowed to cool slowly. If any prismatic crystals of the salt are available, use them to seed the solution.

Collect the crystals of the amine hydrogen tartrate by vacuum filtration using a Büchner funnel (Fig. 2.52); wash them with a small volume of *cold* methanol. Do *not* discard the filtrate (see FINISHING TOUCHES). Dissolve the crystals in four times their weight of water, and add 8 mL of 14 *M* sodium hydroxide solution. Extract the resulting mixture with four 35-mL portions of diethyl ether. Do *not* discard the aqueous solution (see FINISHING TOUCHES). Wash the combined ethereal extracts with 25 mL of saturated aqueous sodium chloride, and then dry the ethereal solution over several spatula-tips full of anhydrous sodium sulfate for 10–15 min; swirl the flask if the solution remains cloudy after this period of time, add additional portions of sodium sulfate. Decant or filter the dried solution from the desiccant.★

Remove the diethyl ether from the solution by simple distillation.★ Alternatively, use rotary evaporation or one of the other techniques discussed in Section 2.29. Then distil the residual 1-phenylethanamine under vacuum with a short-path apparatus (Fig. 2.35b). It will be necessary to use electrical heating for the distillation because the amine has a boiling point of 94–95 °C (28 Torr).

Analysis   Determine the yield of the resolved 1-phenylethanamine (**20**). Combine your product with that of another student to give a total of 3 g of **20**. Weigh the combined product accurately, and then transfer it *quantitatively* into about 30 mL of absolute ethanol. Measure the volume of the methanolic *solution* accurately. It is the *weight*, in grams, of the 1-phenylethanamine divided by the *volume* of the *solution*, in mL, that provides the concentration *c* for Equation 7.7. Transfer the solution to the polarimeter sample tube, measure the observed rotation, and determine the specific rotation of the sample by using Equation 7.7. Optically pure (R)-(+)-1-phenylethanamine has $[\alpha]_D^{20}$ +30° (c = 10, ethanol).

Finishing Touches

Transfer the recovered *diethyl ether* to a container for nonhalogenated organic liquids, and put the *methanolic filtrate* from isolation of the less-soluble amine tartrate either in this same container or in a special container from which the more-soluble amine tartrate could be recovered; consult your instructor for specific directions. Place the *methanolic solution of 1-phenylethanamine* either in a container so labeled or in the one labeled for nonhalogenated organic liquids. Transfer the *sodium hydroxide solution* into a container so labeled; the tartaric acid salt it contains can be recovered.

EXERCISES

1. When a label is prepared for the product that has been isolated, it should include the magnitude of the specific rotation as well as its sign. Suppose the observed rotation was found to be 180°. How could you determine whether the rotation was (+) or (−)?

2. Given the value for the specific rotation of optically pure 1-phenylethanamine, calculate the optical purity of the sample of resolved amine for which you obtained a rotation.

3. How might the optical purity of the product be increased?

4. Describe clearly the point in the experimental procedure at which the major part of the other enantiomer was removed from your product.

5. How could the (+)-tartaric acid be recovered so that it can be reused to resolve additional racemic amine?

6. The absolute configuration of one of the stereocenters in (+)-tartaric acid is R. Explain why the absolute configuration at the other stereocenter must also be R and provide a perspective drawing of the (R, R)-enantiomer of tartaric acid.

7. The absolute configuration of (+)-1-phenylethanamine is R. Make a perspective drawing of this configuration.

8. Suppose you had prepared a racemic organic *acid* and it was necessary to resolve it. How could this be done?

■ HISTORICAL HIGHLIGHT ■

Discovery of Stereoisomers

Louis Pasteur (1822–1895) is better known for his contributions to microbiology than to chemistry, although he began his professional career as a chemist. In 1848, when Pasteur was only 24 years old and had just received his doctorate from the Sorbonne in Paris, he undertook the study of a salt of **racemic acid**, a substance deposited on wine casks during the fermentation of grape juice. (The word *racemic* is

---

HISTORICAL HIGHLIGHT
Discovery of Stereoisomers *(continued)*

derived from the Latin word *racemus*, meaning a bunch of grapes.) Another chemist, E. Mitscherlich, had earlier reported that the sodium ammonium salt of racemic acid was identical in all ways to a salt of **tartaric acid** that is also found on wine casks, *except* that the salt of tartaric acid was optically active, whereas the salt of racemic acid was not.

Pasteur was puzzled by the fact that although the salts of tartaric acid and racemic acid were said to be identical in many ways, including chemical composition and even crystalline form, they had different effects on plane-polarized light. Specifically, the salt of racemic acid had *no* effect, whereas the salt of tartaric acid was dextrorotatory. When Pasteur examined the crystals of the salt of racemic acid produced according to Mitscherlich's description, he observed something that Mitscherlich had not. There were, in fact, *two* kinds of crystals present that were related to one another as a left hand is to a right hand. Pasteur carefully separated the ''left-handed'' crystals from the ''right-handed'' crystals using tweezers under a microscope. When he had a sufficient quantity of each type of crystal, he did something that arose from either a hunch or a flash of genius. He separately dissolved some of each kind of crystal in water and placed the solutions in turn in a polarimeter. He found that *both* solutions were optically active, the solution of the ''left-handed'' crystals rotating the polarized light counterclockwise, and the solution of the ''right-handed'' crystals rotating the polarized light clockwise. The author Vallery-Radot, in *The Life of Pasteur* (1902), reported that the young scientist was so excited by his discovery that he, ''not unlike Archimedes, rushed out of the laboratory exclaiming 'I have it!' ''

When Pasteur carefully measured the amounts of each kind of crystal for making the solutions, he found that equal amounts produced exactly the same degree of rotation, *but in opposite directions*. Moreover, the magnitude of rotation by the ''right-handed'' crystals was the same as that given by a similar solution of the salt of tartaric acid. Thus, Pasteur demonstrated that his ''right-handed'' racemic acid salt was actually identical with the dextrorotatory tartaric acid salt, and his ''left-handed'' racemic acid salt was a previously unknown mirror-image

form of tartaric acid salt, as shown below. Finally, Pasteur made a mixture of equal amounts of the two kinds of crystals and found, as he expected, that a solution of this mixture was optically inactive.

Mirror
(+)-Tartrate   plane   (−)-Tartrate

By separating the two kinds of crystals of the salt of racemic acid, Pasteur accomplished the first and most famous example of what chemists now call **resolution** of a racemic mixture. The name of the specific acid, *racemic*, that Pasteur studied is now used generally to describe any equimolar mixture of **enantiomers** such as the ''left-handed'' and ''right-handed'' salts of Pasteur.

One is likely to think that the separation of the crystals of different shapes and the demonstration that their solutions rotate the plane of polarized light in opposite directions was a clever but rather trivial experiment. In a limited sense it was, but from a broader perspective the experiments had tremendous importance. This was the first time it had been shown that organic compounds exist in enantiomeric forms at the molecular level. Before Pasteur's experiments, the effect of quartz crystals on polarized light could be explained on the basis of the shape of the crystals themselves, because when the quartz crystals were melted the optical activity disappeared. Pasteur's work showed that the difference in the crystalline form of the salts of racemic acid reflected a difference in the three-dimensional shape of the molecules themselves. Thus, despite the fact that dissolving the salts of racemic acid in water destroyed their crystal structures just as that of quartz is destroyed by melting it, the resulting *solutions* of the two kinds of crystals of Pasteur's salt still exhibited optical activity. Pasteur later converted the separated salts of racemic acid into the corresponding acid forms and showed that there were indeed two isomeric forms, ''left-

HISTORICAL HIGHLIGHT
Discovery of Stereoisomers *(continued)*

handed'' or (−)-tartaric acid and "right-handed'' or (+)-tartaric acid.

Pasteur's observation of the different shapes of the crystals, his separation of them, and his deductions about the meaning of their opposite effects on polarized light were indeed acts of genius, but accidents played a large part in his discovery. There were two remarkable elements of chance that entered into the findings. The first was that the sodium ammonium salt of racemic acid, which was the one Pasteur examined, has been shown to be one of the only salts of this acid that crystallizes in mirror-image forms that are visually different and can be separated mechanically. Second, the crystallization in these two forms occurs only at temperatures below 26 °C (79 °F); above 26 °C the crystals that form are identical and have no optical activity. To obtain his crystals, Pasteur had placed the flask containing the solution of racemic acid on a cool window ledge in his laboratory in Paris and left it there until the next day for the crystallization to occur. Except for the combination of a fortuitous choice of the proper salt of racemic acid and the cool Parisian climate, Pasteur would not have made his important observations.

By his discovery, he recognized that a direct relationship exists between molecular geometry and optical activity. This led him to propose that molecules that rotate plane-polarized light in equal but opposite directions are related as an object and its mirror image. However, it remained for two other chemists, Van't Hoff and LeBel, to explain 25 years later exactly how the atoms could be assembled into such molecular structures. In the interim, Pasteur turned his attention with great success to the biological problems mentioned earlier, but it was his pioneering work on the resolution of racemic acid that led the way for other chemists to explain the relationship of chirality or "handedness'' in molecular structure to biological activity, and this is the real significance of Pasteur's work.

The importance of the chirality of molecules is tragically illustrated by the case of thalidomide, shown below. In the 1950s and 1960s, this drug was prescribed to pregnant women as the racemate before it was known that only the (+)-enantiomer of the molecule is safe and effective against morning sickness. Unfortunately, the (−)-enantiomer is a potent **teratogen**, an agent that causes fetal malformities. Numerous serious birth defects occurred before the problem was recognized and thalidomide was banned from the marketplace.

Thalidomide

Ibuprofen

Ibuprofen is an analgesic (pain-suppressing) agent that is a pharmacologically active component in over-the-counter products such as Nuprin® and Advil®. Like thalidomide, the molecule is chiral and only the (S)-enantiomer is effective for relief of pain. In this case, however, the (R)-isomer is slowly converted to the (S)-isomer in the body, so it is also a therapeutically useful species. The commercially available drug is most commonly sold as the racemate, but the (S)-isomer may now be purchased.

That the different enantiomers may have vastly different pharmacological properties has led to great interest in developing methods for the synthesis of single enantiomers. An informative article describing the growing importance of chiral drugs in the development of new medicines is provided in *Chemical and Engineering News*, September 28, 1992, pages 46–78.

Relationship of the Historical Highlight to the Experiments

Enantiomers of chiral compounds can sometimes be distinguished not only by their equal, but opposite, effects on plane-polarized light, as first demonstrated by Pasteur, but also by their odors, if they are volatile enough. The enantiomers of carvone (Sec. 7.4) fit

HISTORICAL HIGHLIGHT
Discovery of Stereoisomers *(continued)*

this description, and their differing odors can be distinguished by most persons because the nasal receptors responsible for the detection of smells are chiral. The manner in which chiral odor receptors differentiate between enantiomers can be illustrated in the following way. Assume that your receptors are right-handed, symbolized as (+). Combining these with the (+)- and (−)-enantiomers of carvone gives complexes that can be represented as (+)/(−) for spearmint oil and (+)/(+) for caraway seed oil and dill seed oil, respectively. These complexes are diastereomeric, and they trigger unique nervous responses that we sense as different aromas. Thus, your nose is able to distinguish between diastereomers just as a chromatographic plate or column is able to differentiate between the diastereomers of 1,2-cyclohexanediol (Sec. 7.2).

Pasteur's observation of the fortuitous resolution of the tartrates by crystallization of the racemate is an exception to the general methods required for separating enantiomers. The more usual case involves converting the racemic mixture into a mixture of diastereomers, which may then be separated on the basis of differing physical properties, such as solubility. The resolution of 1-phenylethanamine (Sec. 7.6) is representative of this approach.

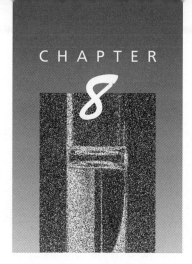

CHAPTER 8

Spectral Methods

Introduction

As a beginning organic chemist you will first master the skills and techniques for executing organic reactions and isolating the products in pure form. The structures of the starting materials and products that you encounter will usually be known so you will not normally be faced with the necessity of determining the structures of your products. However, you may be presented with a compound whose structure is known by your instructor but not by you. You would then be required to determine the structure of the molecule. This is a common challenge faced by organic chemists who must identify compounds that are isolated from new chemical reactions or from natural sources.

The classic method for determining the exact structures of organic compounds was a difficult and time-consuming undertaking. A variety of solubility and classification tests were first performed to identify the functional groups present (Chap. 23). Typically the structure of the unknown substance was then elucidated by converting it via a series of chemical reactions into a compound whose structure was known. Fortunately, powerful spectroscopic methods are now available that have greatly simplified the task of structural elucidation. Perhaps the techniques of greatest importance to the practicing organic chemist are **infrared** (IR) and **nuclear magnetic resonance** (NMR) spectroscopy, although **ultraviolet** (UV) spectroscopy, **mass spectrometry** (MS), and **X-ray diffractometry** are also widely used for structure determination. With the notable exception of X-ray diffractometry, unequivocal definition of a structure is seldom possible by performing only a single type of spectroscopic analysis; rather, a combination of different analyses is generally required. In this textbook we emphasize IR and NMR spectroscopy because the necessary instrumentation is more commonly available to students in the introductory organic laboratory course.

The spectral methods described in this chapter depend on the absorption of radiation to produce an excitation that transforms the original molecule to a state of higher energy. The particular type of excitation that occurs is determined by the amount of energy associated with the radiation. The relationship between the energy of radiation (in kcal/mol) and its frequency ν, or wavelength, λ, is expressed in Equation 8.1, where N is Avogadro's number (6.023×10^{23}), h is Planck's constant ($1.5825 \times$

10^{-37} kcal sec), c is the velocity of light (2.998×10^{14} μm/sec), and ν and λ are in Hertz (1.0 Hz $= 1.0$ cycle/sec) and micrometers (μm $= 10^{-6}$ m), respectively. Because of the reciprocal relationship between ν and λ, the *higher* the frequency or the *lower* the wavelength, the *higher* is the energy associated with the radiation.

$$E = Nh\nu = Nhc/\lambda \tag{8.1}$$

In IR spectroscopy, the absorption of radiation corresponds to transitions among different molecular **vibrational-rotational levels** within the same electronic state, normally the ground state, of the molecule; these excitations require energies of 1–36 kcal/mol. The data from IR spectroscopy are most useful for determining the presence or absence of **functional groups** in a molecule. For instance, examining the appropriate regions of an IR spectrum will show whether or not carbon-carbon multiple bonds, aromatic rings, carbonyl groups, or hydroxyl groups are present. This technique does not give quantitative information regarding the elemental composition of a compound, nor does it allow assignment of an exact structure to an unknown compound unless the IR spectrum of the unknown is shown to be *identical* to that of a known compound.

The phenomenon associated with NMR spectroscopy involves transitions among **nuclear spin states** in a magnetic field; these excitations are of very low energy, less than 1 cal/mol. NMR spectroscopy provides information regarding the number and arrangement of various atoms in a molecule, although not all types of nuclei found in organic molecules can be detected by this method. Application of this technique to the analysis of hydrogen atoms in molecules, which is called ^{1}H NMR spectroscopy, allows determination of the number of **nearest neighbors** to a hydrogen atom as well as the presence of certain functional groups, such as carbon-carbon multiple bonds and carbonyl groups. A companion method involving carbon-13 atoms, and abbreviated as ^{13}C NMR spectroscopy, permits analysis of the different types of carbon atoms present in a molecule and may be useful for detecting certain functional groups whose presence is not indicated in the ^{1}H NMR spectrum.

8.2   Infrared (IR) Spectroscopy

INTRODUCTION

The infrared portion of the electromagnetic spectrum spans from about 0.8 μm (4×10^{10} Hz) to 40 μm (7.5×10^8 Hz). Within this range, the region from 2.5 μm to 20 μm is of most interest to organic chemists because it is here that most of the common functional groups absorb IR radiation. Although wavelength, in μm, was once used to express the location of an IR absorption, current practice is to do so with another unit, **wavenumber** or **reciprocal centimeter**, written as $\tilde{\nu}$ or cm^{-1}. The relationship between wavelength and wavenumber is expressed by Equation 8.2. Thus, the normal IR range of 4000 to 500 cm^{-1} corresponds to 2.5–20 μm. As with frequency ν, the energy associated with radiation *increases* as $\tilde{\nu}$ increases. It is imprecise to call wavenumbers "frequencies," despite the mathematical relationship between the two, although we do so for convenience.

$$\text{wavenumber (cm}^{-1}) = 10{,}000/\lambda \; (\mu\text{m}) \tag{8.2}$$

PRINCIPLES

The interaction of IR radiation with an organic molecule may be qualitatively understood by imagining that the covalent bonds between atoms are analogous to groups of

Symmetric stretch   Asymmetric stretch   Symmetric in-plane bend   Asymmetric in-plane bend

FIGURE 8.1

Some vibrational modes of a group of covalently bonded atoms.

molecular springs that are constantly undergoing **stretching**, **twisting**, and **bending** (Fig. 8.1). The characteristic frequencies, and thus the energies, of these vibrations depend on the masses of the atoms involved and the type of chemical bond joining the atoms. Indeed, a good approximation of the *stretching* frequency of bonds can be obtained by applying **Hooke's law** for simple **harmonic oscillators** (Eq. 8.3). The value of the **force constant** k, which essentially reflects the strength of the bond between the atoms A and B, is about 5×10^5 dyne cm^{-1} for single bonds and twice and three times this figure for double and triple bonds, respectively.

$$\tilde{\nu} = \frac{1}{2\pi c} \sqrt{\frac{k}{m^*}} \qquad (8.3)$$

where $\tilde{\nu}$ = frequency of absorption in cm^{-1}

   c = speed of light

   k = force constant of bond

   m^* = reduced mass of atoms joined by bond = $\dfrac{m_A m_B}{m_A + m_B}$

The validity of Equation 8.3 for predicting stretching frequencies is illustrated by its use to give $\tilde{\nu}_{C-H}$ as $3040\ cm^{-1}$; the experimental value for this vibration is in the range of $2960–2850\ cm^{-1}$. Other consequences of Equation 8.3 are noteworthy. For instance, the wavenumber for stretching of a C–D bond is about $2150\ cm^{-1}$, a lower value than that for a C–H bond; this difference reflects the effect of atomic mass on the absorption frequency as accounted for by m^* in Equation 8.3. Similarly, the appearance of the stretching vibration of a carbon-carbon double bond at about $1620\ cm^{-1}$ and that of a carbon-carbon triple bond at approximately $2200\ cm^{-1}$ is because k (Eq. 8.3) approximately doubles in going from a double to a triple bond; this increases the energy required to stretch the bond between the atoms.

Because different energies are required to cause the molecular vibrations associated with various types of bonds, each of these functional groups has a characteristic vibrational frequency. When the energy of the radiation corresponds to that associated with a particular stretching or bending vibration, energy from the incident light will be absorbed. The efficiency of this absorption depends on several factors that are beyond the scope of the present discussion. Suffice it to say that the greater the change in the **dipole moment** associated with a particular vibration, the more efficient is the transfer of energy to the molecule and the more intense is the observed absorption band.

Most modern IR instruments are designed to measure the amount of light transmitted *through* the sample, the **transmittance**, rather than the **absorbance** of light *by*

FIGURE 8.2

IR spectrum of 3-methyl-2-cyclohexenone (neat).

the sample. Consequently, the **infrared spectra** portrayed in this textbook are plots of transmittance, expressed in *percent* as defined in Equation 8.4, versus the wavenumber

$$\%T = I/I_o \times 100 \qquad\qquad (8.4)$$

I = intensity of radiation transmitted through sample at a particular wavelength

I_o = intensity of radiation incident on sample at same wavelength

of the **incident radiation** (I_o), as seen in Figure 8.2. The lower is $\%T$, the greater is the amount of incident radiation being absorbed by the sample. For example, in the IR spectrum shown in Figure 8.2 the strongest absorption occurs at about 1675 cm^{-1}. The energy required to produce the observed molecular excitations *decreases* in going from left to right in the spectrum.

It should be noted that the absorption of infrared radiation by a molecule increases its total energy, which results in an increase in **amplitude** of the particular molecular vibration that was responsible for absorption of the radiation. However, no irreversible change occurs in the molecule, as the excess energy is quickly dissipated in the form of heat. Obtaining an IR spectrum thus does *not* permanently change the structure of a molecule.

PRACTICAL CONSIDERATIONS

Transmittance. As a general rule the IR spectrum of a liquid or solid sample should show detectable peaks for relatively weak absorptions, but there should be no absorptions so strong that they go off-scale. Experimentally, a good operating principle is to modify the sample being analyzed so the *most* intense peak in its spectrum is close to 0 $\%T$; this maximizes the possibility that even weak peaks will be observable. This practice is exemplified in Figure 8.2, where the strongest absorption has a $\%T$ of nearly zero. The two experimental variables determining the absolute intensities of the peaks in an IR spectrum are the concentration of the sample and the length of the cell containing it. These factors define the number of molecules that are in the path of the radiation as it passes through the sample; the more molecules there are, the greater the absolute intensity of a given absorption. With modern IR spectrometers, the absolute intensities of the peaks on the printed spectrum may also be amplified electronically.

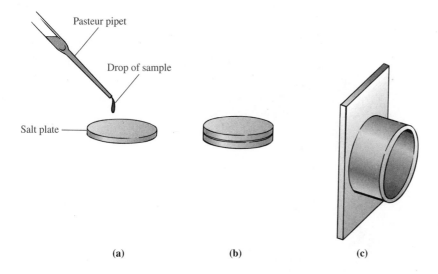

(a) **(b)** **(c)**

FIGURE 8.3
Simple demountable IR cell.
(a) Salt plate for sample.
(b) Assembled pair of salt plates.
(c) Cell holder.

Cells. Two general types of cells are used to contain samples for analysis by IR spectroscopy: Cells in which the pathlength or thickness of the sample is fixed, and those in which the pathlength is variable. The latter type of cell can be very expensive, but a simple version of it is two transparent salt plates, or windows, between which the sample is sandwiched (Figs. 8.3 and 8.4). The cell is prepared by putting a drop of sample on one plate and placing the second plate on top of the film that results. The assembled plates are then put into a holder that fits into the spectrometer. If a sample is too thick, some of it may be removed or the plates may be squeezed together, *carefully* to avoid breaking them, to produce a thinner film. However, if the sample is too thin, the plates can be separated and more material can be added. The plates can then be rejoined so as to leave the film of sample as thick as possible.

A **fixed-thickness cell** is constructed of two transparent salt plates separated by a plastic or metal gasket, which defines the thickness of the sample contained in the cell. One such cell is shown in Figure 8.5. A syringe is normally used to load the cell, which should be laid at an angle on top of a pen or pencil to keep an air bubble from forming

(a) **(b)**

FIGURE 8.4
A demountable IR cell. (a) Details showing correct assembly of cell. (b) Completely assembled cell.

FIGURE 8.5

A fixed-thickness IR cell. (a) Cell with plugs removed for cleaning or filling. (b) Completely assembled cell.

within the cell as sample is introduced. If a bubble is observed, more sample must be added until the bubble disappears.

Proper care and handling of the windows used in IR cells are *very* important. The plates are generally a clear fused salt such as sodium chloride or potassium bromide, and contact with moisture must be minimized to keep the windows from becoming cloudy; this would decrease the amount of light transmitted through the cell. Consequently, breathing on the windows or touching their faces must be avoided when handling assembled IR cells or the salt plates themselves. Moreover, any substances, including samples, that come in contact with the cells must be *dry*. This means that any materials used to clean an IR cell or salt plates must also be *dry*. This excludes 95% ethanol as a solvent for cleaning the windows, although *absolute* ethanol *can* be used. Other solvents suitable for cleaning cells are dichloromethane and chloroform. Permanent fixed-thickness cells are dried by passing a slow stream of *dry* nitrogen or air through them, whereas demountable cells and salt plates may also be dried by *gently* wiping or blotting them with soft laboratory tissue to avoid scratching the faces of the cells or plates. The clean, dry cells or plates are stored in a desiccator for protection from atmospheric moisture.

Liquid Samples. The IR spectrum of a liquid sample is most conveniently obtained on the pure liquid, called a **neat sample**. Since concentration is not a variable in such samples, peak intensities may be modified only by changing the pathlength. This is done in one of two ways. One approach is to use a fixed-thickness cell with a pathlength of 0.020–0.030 mm. This short pathlength generally provides a spectrum in which strong absorptions, such as those associated with carbonyl groups, give peaks that remain on the chart paper, rather than going to 0 %T. If the spectrum is not suitable, however, different fixed thicknesses may be used. The other option is to use salt plates and to adjust the thickness of the film between them by gently pressing the plates together. If the sample is volatile, having a boiling point of less than about 50 °C, salt plates would *not* be appropriate because some or all of the sample will evaporate as the spectrum is run, and inaccurate relative peak intensities will result. Alternatively, but less conveniently, a liquid sample may be analyzed as a solution or absorbed into a microporous film, as described in the next section.

Solid Samples. There are several options for preparing a solid sample for IR spectroscopy. These include the preparation of **solutions**, **KBr pellets**, and **microporous polyethylene films**. Each of these techniques is described in turn.

a. Solution. A common technique for obtaining an IR spectrum of a solid is to analyze it in solution. The preferred solvents are chloroform and carbon disulfide, in that order. The solubility of the sample in the selected solvent must be in the range of 5–10 weight percent because solutions of such concentration generally provide suitable spectra in a fixed-thickness cell having a pathlength of 0.1 mm.

Two fixed-thickness cells having identical pathlengths, so called **matched cells**, are used if a *double beam* IR instrument is being used. One cell contains the solution, and is referred to as the **sample cell**, whereas the other contains pure solvent and is called a **reference cell**. The purpose of the reference cell is to allow cancellation of the absorption bands due to the solvent itself so that they do not appear in the spectrum.

If a **Fourier transform (FT) IR** instrument is available, a reference cell is not required because the IR spectrum of the pure solvent is previously stored in the memory of the spectrometer. The software that controls the instrument then *substracts* those absorptions that are due to the solvent from the spectrum. However, as is evident from the IR spectrum in Figure 8.6a, this subtraction may be imperfect, most notably at about 750 and 1220 cm^{-1} where the solvent chloroform strongly absorbs.

b. Potassium Bromide (KBr) Pellet. Potassium bromide may be fused to produce a nearly transparent salt plate. This is done by subjecting the KBr to high pressures, in the range of 5–10 tons/in^2, to generate the heat required for the salt to become plastic and flow; releasing the pressure then provides a clear plate. A KBr pellet containing the compound is thus prepared by intimately mixing about 1 mg of sample with 100 mg of *dry*, powdered KBr and subjecting the resulting mixture to high pressure in a die. Mixing is done with a mortar and pestle made of agate rather than ceramic materials, because ceramics may be abraded and contaminate the pellet with nontransparent particles. Once a pellet having satisfactory transparency has been produced, the IR spectrum is obtained. If the intensity of the bands in the spectrum is unsatisfactory, another pellet must be prepared in which the concentration of the solid sample has been modified accordingly.

Potassium bromide is *extremely* **hygroscopic**, so it should be stored in a tightly capped bottle in a desiccator except when needed for preparation of a pellet. Even with the greatest precautions, it is difficult to prepare a pellet that is completely free of water. Possible contamination of a KBr pellet with water may cause problems in interpreting IR absorptions in the range 3600–3200 cm^{-1} because the O–H stretching vibration of water occurs in this region. Consequently, moisture in the pellet may partially or completely obscure the absorptions due to the O–H and N–H stretching vibrations of alcohols and amines, which occur in this same region (Fig. 8.6b). Further, the appearance of an absorption in the aforementioned range may lead to an erroneous conclusion about the possible presence of O–H and N–H groups in the sample.

c. Microporous Polyethylene Films. Disposable IR cards have recently been developed to which samples can be applied for infrared analysis (Fig. 8.7). These convenient IR cards have a 19-mm circular aperture containing a thin microporous film of chemically resistant polyethylene. Cards having microporous polytetrafluoroethylene are also

FIGURE 8.6
IR spectra of acetanilide.

(a) in chloroform.

(b) as KBr pellet.

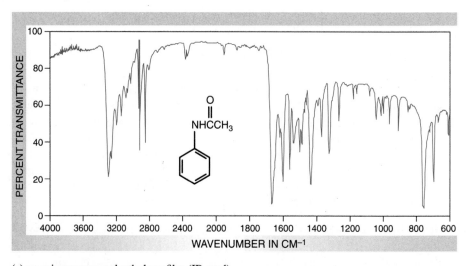

(c) on microporous polyethylene film (IR card).

available for special applications. The cards with polyethylene films may be used for infrared analysis from 4000–400 cm^{-1} except for the region of aliphatic C–H stretching that occurs between 3000–2800 cm^{-1}. In this region of an FT-IR spectrum, there are two sharp spikes at about 2840 cm^{-1} and 2910 cm^{-1} that result from the imperfect subtraction of the stored spectrum of the polyethylene film, as shown in Figure 8.6c.

To prepare a sample for IR analysis, approximately 50 μL of a solution of the compound in a volatile organic solvent is simply applied to the film using a pipet or syringe. Because of the unique nature of the microporous films, the solution is absorbed into the film. The solvent evaporates within minutes at room temperature leaving the solute evenly dispersed in the polyethylene matrix. Non-volatile liquid samples may also be applied to the film. The cards are then mounted directly into the slot on the spectrometer that is normally used for holding reusable cells.

FIGURE 8.7
Application of sample to disposable IR card.

Spectral Calibration. If you are not using an FT-IR spectrometer, it is good practice to calibrate the experimentally obtained IR spectra against a standard that has absorption bands at known positions. Calibration is necessary because problems such as misalignment of the chart paper in the instrument or mechanical slippage of gears may result in absorptions appearing at wavenumbers that are not correct. Polystyrene (Fig. 8.8) is a commonly used standard and is typically available as a thin, transparent film in a cardboard holder that fits into the spectrometer. After the spectrum of the sample is obtained, the standard is inserted in place of the sample cell, the reference cell, if used, is removed, and at least one band of the standard is recorded on the spectrum of the sample. In the case of polystyrene, the absorption at 1601 cm^{-1} is typically recorded, although other bands may be used.

Once the reference peak of the standard is measured, the reported positions of the absorptions of the sample are adjusted accordingly. For example, if the 1601 cm^{-1} peak of polystyrene appears at 1595 cm^{-1} on the spectrum, the wavenumbers for the bands in the sample are corrected by *adding* 6 cm^{-1} to the values recorded on the chart paper. Because the source of the error may vary over the range, it is not strictly correct to use the same correction factor throughout the spectrum, but for most purposes this is a satisfactory protocol. For more precise work, calibration peaks should be recorded at several different locations in the IR spectrum.

FIGURE 8.8
Infrared spectrum of polystyrene film.

USE AND INTERPRETATION OF IR SPECTRA

An IR spectrum of a compound is used to *identify* the *functional groups* present in a molecule, *evaluate* the *purity* of a known compound, and *determine* the *structure* of an unknown substance. These valuable applications are discussed in the following paragraphs.

Analysis of the IR spectrum of an unknown compound provides useful information regarding the presence or absence of different functional groups. As a general rule, absorptions in the approximate range of $4000–1250$ cm^{-1} which is referred to as the **functional group region**, involve vibrational excitations that are associated with specific functional groups. Thus, the stretching mode of the carbonyl group of a ketone appears in the region of $1760–1675$ cm^{-1} and that of a carbon-carbon double bond occurs in the vicinity of $1680–1610$ cm^{-1}. A compilation of various functional groups and the region of the IR spectrum in which they absorb is provided in Table 8.1. Table 8.2 is a more detailed version of Table 8.1. Note that the *absence* of an absorption may have as much significance as the *presence* of one. For example, if there is no strong band in the carbonyl region of the spectrum, this functionality is probably absent; this eliminates a number of functional groups such as aldehyde, ketone, and ester from consideration.

Table 8.1	Abbreviated Table of Infrared Absorption Ranges of Functional Groups		
Bond	**Type of Compound**	**Frequency Range, cm^{-1}**	**Intensity**
C–H	Alkanes	2850–2970	Strong
		1340–1470	Strong
C–H	Alkenes $\left(\begin{smallmatrix} \diagdown \\ / \end{smallmatrix} C=C \begin{smallmatrix} ^H \\ \diagdown \end{smallmatrix}\right)$	3010–3095	Medium
		675–995	Strong
C–H	Alkynes (–C≡C–H)	3300	Strong
C–H	Aromatic rings	3010–3100	Medium
		690–900	Strong
O–H	Monomeric alcohols, phenols	3590–3650	Variable
	Hydrogen-bonded alcohols, phenols	3200–3600	Variable, sometimes broad
	Monomeric carboxylic acids	3500–3650	Medium
	Hydrogen-bonded carboxylic acids	2500–3200	Broad
N–H	Amines, amides	3300–3500	Medium
C=C	Alkenes	1610–1680	Variable
C=C	Aromatic rings	1500–1600	Variable
C≡C	Alkynes	2100–2260	Variable
C–N	Amines, amides	1180–1360	Strong
C≡N	Nitriles	2210–2280	Strong
C–O	Alcohols, ethers, carboxylic acids, esters	1050–1300	Strong
C=O	Aldehydes, ketones, carboxylic acids, esters	1675–1760	Strong
NO$_2$	Nitro compounds	1500–1570	Strong
		1300–1370	Strong

Table 8.2	Detailed Table of Characteristic Infrared Absorption Frequencies

The hydrogen stretch region (3600–2500 cm^{-1}). Absorption in this region is associated with the stretching vibration of hydrogen atoms bonded to carbon, oxygen, and nitrogen. Care should be exercised in interpreting very weak bands because these may be overtones of strong bands occurring at frequencies from 1800–1250 cm^{-1}, which is one-half the value of the weak absorption. Overtones of bands near 1650 cm^{-1} are particularly common.

$\bar{\nu}$ (cm^{-1})	Functional Group	Comments
(1) 3600–3400	O–H stretching Intensity: variable	3600 cm^{-1} (sharp) unassociated O–H, 3400 cm^{-1} (broad) associated O–H; both bands frequently present in alcohol spectra; with strongly associated O–H (CO$_2$H or enolized β-dicarbonyl compound), band is very broad (about 500 cm^{-1} with its center at 2900–3000 cm^{-1}).
(2) 3400–3200	N–H stretching Intensity: medium	3400 cm^{-1} (sharp) unassociated N–H, 3200 cm^{-1} (broad) associated N–H; an NH$_2$ group usually appears as a doublet (separation about 50 cm^{-1}); the N–H of a secondary amine is often very weak.
(3) 3300	C–H stretching of an alkyne Intensity: strong	The *complete* absence of absorption in the region from 3300–3000 cm^{-1} indicates the absence of hydrogen atoms bonded to C=C or C≡C and *usually* signals the lack of
(4) 3080–3010	C–H stretching of an alkene Intensity: strong to medium	unsaturation in the molecule. Because this absorption may be very weak in large molecules, care should be exercised in this interpretation. In addition to the absorption at about 3050 cm^{-1}, aromatic compounds
(5) 3050	C–H stretching of an aromatic compound Intensity: variable; usually medium to weak	frequently show *sharp* bands of medium intensity at about 1500 *and* 1600 cm^{-1}.
(6) 3000–2600	OH strongly hydrogen-bonded Intensity: medium	A very broad band in this region superimposed on the C–H stretching frequencies is characteristic of carboxylic acids.
(7) 2980–2900	C–H stretching of an aliphatic compound Intensity: strong	As in previous C–H entries (3–5), *complete* absence of absorption in this region indicates absence of hydrogen atoms bound to tetravalent carbon atoms. The tertiary C–H absorption is weak.
(8) 2850–2760	C–H stretching of an aldehyde Intensity: weak	Either one or two bands *may* be found in this region for a single aldehyde function in the molecule.

(continued)

Table 8.2	*(continued)*

The triple-bond region (2300–2000 cm^{-1}). Absorption in this region is associated with the stretching vibration of triple bonds.

$\bar{\nu}$ (cm^{-1})	Functional Group	Comments
(1) 2260–2215	C≡N Intensity: strong	Nitriles conjugated with double bonds absorb at *lower* end of frequency range; nonconjugated nitriles appear at *upper* end of range.
(2) 2150–2100	C≡C Intensity: strong in *terminal* alkynes, variable in others.	This band is absent if the alkyne is symmetrical, and will be very weak or absent if the alkyne is nearly symmetrical.

The double-bond region (1900–1550 cm^{-1}). Absorption in this region is *usually* associated with the stretching vibration of carbon-carbon, carbon-oxygen, and carbon-nitrogen double bonds.

$\bar{\nu}$ (cm^{-1})	Functional Group	Comments
(1) 1815–1770	C=O stretching of an acid chloride Intensity: strong	Carbonyls conjugated with double bonds absorb at *lower* end of range; nonconjugated carbonyls appear at *upper* end of range.
(2) 1870–1800 and 1790–1740	C=O stretching of an acid anhydride Intensity: strong	*Both bands* are present; *each band* is altered by ring size and conjugation to approximately the same extent noted for ketones.
(3) 1750–1735	C=O stretching of an ester or lactone Intensity: very strong	This band is subject to all the structural effects discussed for ketones; thus, a conjugated ester absorbs at about 1710 cm^{-1} and a γ-lactone absorbs at about 1780 cm^{-1}.
(4) 1725–1705	C=O stretching of an aldehyde or ketone Intensity: very strong	This value refers to carbonyl absorption frequency of acyclic, nonconjugated aldehyde or ketone having no electronegative groups, for example halogens, near the carbonyl group; because this frequency is altered in a predictable way by structural alterations, the following generalizations are valid: (a) *Effect of conjugation:* Conjugation of carbonyl group with an aryl ring or carbon-carbon double or triple bond *lowers* the frequency by about 30 cm^{-1}. If the carbonyl group is part of cross-conjugated system (unsaturation on each side of the carbonyl group), the frequency is lowered by about 50 cm^{-1}. (b) *Effect of ring size:* Carbonyl groups in six-membered and larger rings exhibit

(continued)

$\bar{\nu}$ (cm^{-1})	Functional Group	Comments
		approximately the same absorption as acyclic ketones; carbonyl groups in rings smaller than six absorb at *higher* frequencies: for example, a cyclopentanone absorbs at about 1745 cm^{-1} and a cyclobutanone at about 1780 cm^{-1}. The effects of conjugation and ring size are additive: A 2-cyclo-pentenone absorbs at about 1710 cm^{-1}, for example.
		(c) *Effect of electronegative atoms:* An electronegative atom, especially oxygen or halogen, bound to the α-carbon atom of an aldehyde or ketone usually raises the position of the carbonyl absorption frequency by about 20 cm^{-1}.
(5) 1700	C=O stretching of an acid Intensity: strong	Conjugation *lowers* this absorption frequency by about 20 cm^{-1}.
(6) 1690–1650	C=O stretching of an amide or lactam Intensity: strong	Conjugation *lowers* the frequency of this band by about 20 cm^{-1}. The frequency of the band is *raised* about 35 cm^{-1} in γ-lactams and 70 cm^{-1} in β-lactams.
(7) 1660–1600	C=C stretching of an alkene Intensity: variable	Conjugated alkenes appear at *lower* end of range, and absorptions are medium to strong; nonconjugated alkenes appear at *upper* end of range, and absorptions are usually weak. The absorption frequencies of these bands are raised by ring strain but to a lesser extent than noted with carbonyl functions.
(8) 1680–1640	C=N stretching Intensity: variable	This band is usually weak and difficult to assign.

The hydrogen bending region (1600–1250 cm^{-1}). Absorption in this region is commonly due to bending vibrations of hydrogen atoms attached to carbon and to nitrogen. These bands generally do not provide much useful structural information. In the listing, the bands that are most useful for structural assignment are marked with an asterisk.

$\bar{\nu}$ (cm^{-1})	Functional Group	Comments
(1) 1600	$-NH_2$ bending Intensity: strong to medium	In conjunction with bands in the 3300 cm^{-1} region, this band is often used to characterize primary amines and amides.
(2) 1540	$-NH-$ bending Intensity: generally weak	In conjunction with bands in the 3300 cm^{-1} region, this band is used to characterize secondary amines and amides. This band, like the N−H stretching band in the 3300 cm^{-1} region, may be very weak in secondary amines.

(continued)

Table 8.2 *(continued)*		
$\bar{\nu}$ (cm^{-1})	Functional Group	Comments
(3) *1520 and 1350	NH_2 coupled stretching bands Intensity: strong	This pair of bands is usually very intense.
(4) 1465	$-CH_2-$ bending Intensity: variable	Intensity of this band varies according to the number of methylene groups present; the more such groups, the more intense the absorption.
(5) 1410	$-CH_2-$ bending of carbonyl-containing component Intensity: variable	This absorption is characteristic of methylene groups adjacent to carbonyl functions; its intensity depends on the number of such groups present in the molecule.
(6) *1450 and 1375	$-CH_3$ Intensity: strong	The band of lower frequency (1375 cm^{-1}) is usually used to characterize a methyl group. If two methyl groups are bound to one carbon atom, a characteristic doublet (1385 and 1365 cm^{-1}) is observed.
(7) 1325	$-\overset{\mid}{C}H$ bending Intensity: weak	This band is weak and often unreliable.

The fingerprint region (1250–600 cm^{-1}). The fingerprint region of the spectrum is generally rich in detail, with many bands appearing. This region is particularly diagnostic for determining whether an unknown substance is identical to a known substance, the IR spectrum of which is available. It is not practical to make assignments to all these bands because many of them represent combinations of vibrational modes and therefore are very sensitive to the overall molecular structure; moreover, many single-bond stretching vibrations and a variety of bending vibrations also appear in this region. Suggested structural assignments in this region must be regarded as tentative and are generally taken as corroborative evidence in conjunction with assignments of bands at higher frequencies.

$\tilde{\nu}$ (cm^{-1})	Functional Group	Comments
(1) 1200	⬡–O– Intensity: strong	It is not certain whether these strong bands arise from C–O bending or C–O stretching vibrations. One or more strong bands are found in this region in the spectra of alcohols, ethers, and esters. The relationship indicated between structure and band location is only approximate, and any structural assignment based on this relationship must be regarded as tentative. Esters often exhibit one or two strong bands between 1170 and 1270 cm^{-1}.
(2) 1150	$-\overset{\mid}{\underset{\mid}{C}}-O$ Intensity: strong	
(3) 1100	$-\overset{\mid}{C}H-O$ Intensity: strong	
(4) 1050	$-CH_2-O-$ Intensity: strong	

(continued)

$\tilde{\nu}$ (cm^{-1})	Functional Group	Comments
(5) 985 and 910	H$_2$C=CH$_2$ (terminal vinyl, H on both carbons) C—H bending Intensity: strong	This pair of strong bands characterizes a terminal vinyl group.
(6) 965	H—C=C—H (trans) C—H bending Intensity: strong	This strong band is present in the spectra of *trans*-1,2-disubstituted ethylenes.
(7) 890	C=CH$_2$ C—H bending Intensity: strong	This strong band characterizes a 1,1-disubstituted ethylene group and may be *raised* by 20–80 cm^{-1} if the methylene group is bound to an electronegative group or atom.
(8) 810–840	H—C=C (trisubstituted) Intensity: strong	Very unreliable; this band is not always present and frequently seems to be outside this range because substituents are varied.
(9) 700	C=C (cis, H on both carbons) Intensity: variable	This band, attributable to a *cis*-1,2-disubstituted ethylene, is unreliable because it is frequently obscured by solvent absorption or other bands.
(10) 750 and 690	monosubstituted benzene C—H bending Intensity: strong	These bands are of limited value because they are frequently obscured by solvent absorption or other bands. They are most useful when independent evidence leads to a structural assignment that is complete except for positioning of aromatic substituents.
(11) 750	ortho-disubstituted benzene C—H bending Intensity: very strong	
(12) 780 and 700	meta-disubstituted benzene and 1, 2, 3, Intensity: very strong	

(continued)

Table 8.2	(Continued)	
$\tilde{\nu}$ (cm^{-1})	Functional Group	Comments
(13) 825	and 1, 2, 4, Intensity: very strong	
(14) 1400–1000	C–F Intensity: strong	The position of these bands is quite sensitive to structure. As a result, they are not particularly useful because the presence of halogen is more easily detected by chemical methods. The bands are usually strong.
(15) 800–600	C–Cl Intensity: strong	
(16) 700–500	C–Br Intensity: strong	
(17) 600–400	C–I Intensity: strong	

Assessing purity involves comparing the spectrum of the sample with that of the pure material; the observation of extra bands in the sample signals that contaminants are present. On the whole, the technique is *not* sensitive to low levels of impurities, however, since levels of contamination in the range of 1–5% may easily go undetected.

Determining the structure of an unknown depends on the generally accepted premise that *if the IR spectra of two samples are superimposable, then the samples are identical*. The criterion of **superimposability** is a stiff requirement in the strictest sense because it means that the intensity, shape, and location of *every* absorption in the two spectra *must* be the same. This is difficult to achieve unless the *same* cells and instrument are used to obtain the IR spectra of *both* samples. Nevertheless, finding that the IR spectrum of an unknown material is very similar to the spectrum of a known material, even if the spectra are not obtained with the same instrument and cells, makes it probable that the structures of two samples are similar, if not identical. Alternative methods, such as NMR spectroscopy, preparing derivatives (Secs. 23.6–18), and mixed melting points (Sec. 3.3) are then used to determine whether or not the samples are indeed identical.

Absorptions that occur in the region from 1250–500 cm^{-1} usually arise from a complex *vibrational-rotational* excitation of the *entire* molecule. This portion of the spectrum typically is *unique* for a particular compound and is aptly described as the **fingerprint region**. Although the IR spectra of similar molecules may be comparable in the functional group region from 4000–1250 cm^{-1}, it is highly unlikely that this similarity will extend throughout the fingerprint region, as is illustrated in Figure 8.9.

FIGURE 8.9   Infrared spectrum of (a) 2,3-dimethylbutane and (b) 2-methylbutane (neat).

It is important to learn to identify functional groups on the basis of the IR spectra of organic compounds. A useful way to do this is to examine the spectra of known compounds and correlate the key absorption bands with the functional groups responsible for them with the aid of Tables 8.1 and 8.2. Analysis of the IR spectra of the starting materials and products associated with the various experimental procedures in this textbook is an excellent way to sharpen your skills at interpreting such spectra.

EXERCISES

1. Derive the mathematical relationship between ν and $\tilde{\nu}$ by using Equations 8.1 and 8.2 and point out why it is imprecise to use the terms "frequency" and "wavenumber" interchangeably.

2. Compute the reduced mass, m^*, of a C–H and a C–D bond and then use Equation 8.3 to determine the wavenumber at which the C–D stretching vibration occurs, assuming the corresponding vibration for the C–H group is 3000 cm^{-1}.

3. Assume that the force constant k for a carbon-carbon double bond is twice that of a carbon-carbon single bond. Use Equation 8.3 to determine the wavenumber at which the C–C stretching vibration occurs, assuming the corresponding vibration for the C=C group is 1640 cm^{-1}. Repeat this calculation for the case of a carbon-carbon triple bond, for which k is the appropriate multiple of that for the carbon-carbon double bond.

4. Water has its O–H stretching mode at about 3600 cm^{-1}. Use Equation 8.3 to calculate at approximately what wavenumber ''heavy'' water, D_2O, would have the corresponding O–D stretch. Show your work.

5. **a.** If the strengths of a carbon-carbon triple bond and a carbon-nitrogen triple bond are assumed to be the same, as reflected in the force constant k (Eq. 8.3), predict which of the two functionalities would appear at the higher wavenumber in the IR spectrum and support your prediction.

   b. Is this in agreement with experimental fact? (Table 8.2.)

6. Define the following terms.

a. percent transmittance	**e.** functional group
b. neat liquid	**f.** stretching vibrational mode
c. hygroscopic	**g.** bending vibrational mode
d. fixed-thickness cell	**h.** superimposability (as applied to IR spectra)

7. Explain why the windows of IR cells and KBr pellets should not be exposed to moisture.

8. Explain why the windows of IR cells should not be subjected to excessive pressure.

9. Why and when is an IR spectrum calibrated?

10. Name two solvents that are appropriate and two that are inappropriate for cleaning the faces of IR cells.

11. What are the two main drawbacks to obtaining IR spectra of solids as KBr pellets?

12. When obtaining the IR spectra of liquids or solids using IR cards with microporous polyethylene films, there are two sharp spikes at 2840 cm^{-1} and 2910 cm^{-1}. What causes these spikes?

13. What portions of an IR spectrum are referred to as the functional group region and the fingerprint region?

14. What difficulties would be associated with obtaining a solution IR spectrum on a double beam spectrometer without using a reference cell containing the pure solvent?

15. What problems might be encountered in attempting to make a KBr pellet of an organic solid that has crystallized as a hydrate?

16. Figures 8.10–8.16 are the IR spectra of the compounds shown on the individual spectra. Assign as many of the major absorptions as possible in the spectrum of each compound, using Tables 8.1 and 8.2.

FIGURE 8.10
IR spectrum of phenylethyne (phenylacetylene).

FIGURE 8.11
Infrared spectrum of *n*-butyl ethanoate (*n*-butyl acetate) (neat).

FIGURE 8.12
Infrared spectrum of *N*, *N*-dimethylacetamide (neat).

FIGURE 8.13
Infrared spectrum of benzoyl chloride (neat).

FIGURE 8.14
Infrared spectrum of diethyl ether (neat).

$CH_3CH_2OCH_2CH_3$

FIGURE 8.15
Infrared spectrum of benzonitrile (neat).

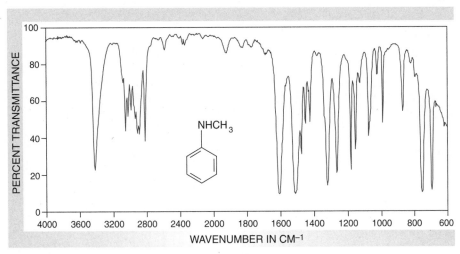

FIGURE 8.16   Infrared spectrum of *N*-methylaniline (neat).

REFERENCES

1. Silverstein, R. M.; Bassler, G. C.; Morrill, T. C. *Spectrometric Identification of Organic Compounds*, 5th ed., John Wiley & Sons, New York, **1992**, Chap. 3.
2. Colthup, N. B.; Daly, L. H.; Wiberley, S. E. *Introduction to Infrared and Raman Spectroscopy*, 3rd ed., Academic Press, New York, **1990**.
3. Pouchert, C. J., ed. *Aldrich Library of FTIR Infrared Spectra*, vol. I–II, Aldrich Chemical Co., Milwaukee, WI, **1985**.
4. *Sigma Library of FT–IR Spectra*, Keller, R. J., ed. Sigma Chemical Company, St. Louis, Mo. **1986**.
5. *Sadtler Standard Infrared Prism Spectra*, vol. 1–123, Sadtler Research Laboratories, Philadelphia, PA. Compilation of 91,000 spectra, as of 1996.
6. *Sadtler Standard Infrared Grating Spectra*, vol. 1–123, Sadtler Research Laboratories, Philadelphia, PA. Compilation of 91,000 spectra, as of 1996.

8.3   Nuclear Magnetic Resonance (NMR) Spectroscopy

Nuclear magnetic resonance (NMR) spectroscopy is probably the single most powerful spectroscopic method available to the modern organic chemist for analyzing compounds. This technique depends on the property of **nuclear spin** that is exhibited by certain nuclei when they are placed in a magnetic field. Some such nuclei commonly found in organic compounds are ^1H, ^2H, ^{19}F, ^{13}C, ^{15}N, and ^{31}P. Nuclei such as ^{12}C, ^{16}O, and ^{32}S do *not* have nuclear spin and thus cannot be studied by NMR spectroscopy.

GENERAL PRINCIPLES

The theory underlying the technique of NMR spectroscopy is presented in any comprehensive lecture textbook of organic chemistry and is not discussed in detail here. It is important to recall, however, that the observation of absorptions in the NMR spectrum

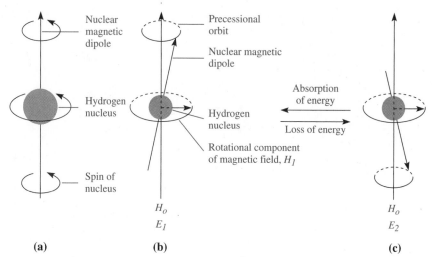

FIGURE 8.17   Spin properties of the hydrogen nucleus. (a) Rotation of hydrogen nucleus and its magnetic moment. (b) Magnetic moment of nucleus aligned with applied external magnetic field. (c) Magnetic moment of nucleus aligned against applied magnetic field.

of a particular nucleus is due to *realignment of nuclear* spins in an **applied magnetic field**, H_o, as shown in Parts (b) and (c) of Figure 8.17. The energy associated with the realignment of spin or spin flip as it is often called, is expressed in Equation 8.5 and depends on the strength of the applied field H_o and the **magnetogyric ratio** γ that is characteristic of the particular nucleus being examined. In a modern NMR spectrometer operating at 90–500 MHz, the value of H_o is in the range of 21,000 to 117,000 gauss (2.1–11.7 Tesla); this means that the energy required for the transition is less than 0.1 cal/mol for a nucleus such as ^{1}H.

Another way of illustrating the relationship between the strength of the magnetic field H_o, and the energy difference between spin states is given in Figure 8.18, where it is seen that for a nucleus that has two spin states, labeled $\pm 1/2$, ΔE increases as H_o increases. Energies of this sort are associated with radiation in the **radio-frequency (rf) range** of the electromagnetic spectrum; consequently, a radio-frequency oscillator is incorporated into the NMR spectrometer to provide the energy for the nuclear spin excitation. When the relationship between the magnetic field and the frequency of the oscillator is that defined by Equation 8.6, the **resonance condition** is achieved, and the transition from one nuclear spin state to the other can occur.

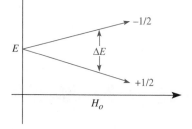

FIGURE 8.18
Energy difference between spin states as a function of strength of magnetic field, H_o.

$$\Delta E = h\gamma H_o/2\pi \tag{8.5}$$

$$h\nu = h\gamma H_o/2\pi \quad \text{or} \quad \nu = \gamma H_o/2\pi \tag{8.6}$$

where ΔE = energy difference between two spin states

h = Planck's constant

γ = magnetogyric ratio (a constant characteristic of a particular type of nucleus)

H_o = strength of applied external magnetic field

ν = frequency of oscillator

To achieve the resonance condition of Equation 8.6, H_o could be held constant and ν varied, or ν could be kept constant and H_o changed. In **continuous wave** (CW) spectrometers, the earliest form of commercial NMR instruments, the latter option is more common, and the spectrum is obtained by slowly sweeping through the range of field strengths required to produce resonance at a particular oscillator frequency. Modern instruments, called **Fourier transform** (FT) spectrometers, operate with a pulse technique in which all resonance frequencies are produced simultaneously while H_o is held constant; this technique allows collection of spectral data in much less time than with a CW instrument.

PROTON MAGNETIC RESONANCE (^{1}H NMR) SPECTROSCOPY

The NMR spectral technique involving the hydrogen nucleus, ^{1}H, which has a nuclear spin, I_z, of 1/2 is called **proton magnetic resonance** spectroscopy and is abbreviated as ^{1}H NMR. Before discussing this type of NMR spectroscopy further, it is important to note that in the terminology of NMR spectroscopy a covalently bound hydrogen atom is often imprecisely called a proton.

Considering Equations 8.5 and 8.6 reveals that the *same* energy would be required to produce the resonance condition for *all* hydrogen nuclei in a molecule *if* all of these nuclei were magnetically identical. In other words, if the magnetic environment as defined by H_o were identical for all of the hydrogen atoms, the same frequency ν would be needed for the spin flip. The consequence of this would be the appearance of a *single* absorption band in the ^{1}H NMR spectrum, a very uninformative result. Fortunately the nuclei are *not* all magnetically equivalent because the three-dimensional electronic structure of a molecule produces variations in the magnetic environments of the hydrogen atoms in it (see "**Chemical Shift**" later in this section). Thus, achieving the resonance condition requires *varying* amounts of energy and this results in a spectrum containing valuable information for determining molecular structure.

Consider the ^{1}H NMR specturm of 1-nitropropane (**1**), shown in Figure 8.19. The significant features illustrated in this spectrum are **chemical shift**, **spin-spin splitting**,

FIGURE 8.19 ^{1}H NMR spectrum of 1-nitropropane (300 MHz, CDCl$_3$).

and **peak integration**, each of which is discussed in detail. First, however, a brief overview of the general format for presenting NMR data is necessary.

$$
\begin{array}{ccc}
H_a & H_b & H_c \\
| & | & | \\
H_a\!-\!C\!-\!\!-\!C\!-\!\!-\!C\!-\!NO_2 \\
| & | & | \\
H_a & H_b & H_c
\end{array}
$$

1

In Figure 8.19, the strength of the applied magnetic field H_o is plotted along the horizontal axis and *increases* from left to right in the spectrum. This direction is called **upfield**; the opposite one is termed **downfield**. As an example, the group of peaks labeled B could be described as being upfield of the resonances labeled C. The *energy* needed to achieve resonance also *increases* in going from left to right. The intensity of the absorption band is plotted on the vertical axis and *increases* from the bottom, or baseline, of the plot.

A scale is printed horizontally across the bottom of the spectrum to define the location of peaks in the spectrum relative to those of a reference compound. The **delta-** or **δ-scale**, which is in units of **parts per million**, ppm, is used for the ^{1}H NMR spectra in this textbook. The reference compound most commonly used is **tetramethylsilane**, $(CH_3)_4Si$, which is abbreviated as TMS. This compound is an inert, volatile liquid that is added directly to solutions of the sample and thus serves as an internal standard. The essential aspects of ^{1}H NMR spectra are presented under the headings ''**Chemical Shift**,'' ''**Spin-Spin Splitting**,'' and ''**Integration**,'' respectively.

Chemical Shift. Three groups of peaks, centered at about 1.0, 2.0, and 4.4 on the δ-scale, are seen in the spectrum of Figure 8.19. These values are the **chemical shifts**, in ppm, of the three chemically distinct types of hydrogen nuclei in 1-nitropropane (**1**). Predicting whether the various protons in a molecule are chemically distinct, and thus will *probably* appear at different chemical shifts, is an important part of understanding ^{1}H NMR spectra. Fortunately, determining how many chemically different protons are present in a molecule is straightforward.

The first step in the analysis is to group the protons according to their **connective equivalency**, placing in the same set all protons having identical molecular connectivities. The next step is to assess whether *connectively* equivalent nuclei are **chemically equivalent** and thus likely to have identical chemical shifts. This procedure is illustrated by further consideration of **1**. Examining **1** reveals that it contains three connectively unique sets of protons, labeled H_a, H_b, and H_c, respectively. *Hydrogen nuclei that are connectively distinct are chemically nonequivalent* and are thus defined as being **heterotopic**. As a general rule, such nuclei have *different* chemical shifts from one another, although some of the peaks may overlap.

The second step in the analysis involves evaluating connectively *identical* nuclei for their possible chemical nonequivalency. This is done by a **substitution test** in which it is supposed that a hydrogen nucleus is replaced with a **probe nucleus** that can be distinguished from other nuclei in the same set. The stereoisomeric relationship between such substituted species is then assessed.

The substitution procedure is presented in Figure 8.20 for protons of types H_a and H_c in **1**. As shown, the hypothetical replacement of H_{a1}, H_{a2}, and H_{a3} with deuterium (**D**) as the probe nucleus provides three new representations that are *identical* to one another; the identity is established by rotating about the bond joining the methyl group

(a)

FIGURE 8.20
Topicity analysis of 1-nitropropane by substitution.

(b)

adjacent to the methylene group and seeing that the resulting molecules are *superimposable* upon one another. Such hydrogens are defined as being **homotopic** and are *chemically equivalent*; they will all have the *same* chemical shift in the ^{1}H NMR spectrum of **1**.

Applying the same substitution process to H_{c1} and H_{c2} gives *enantiomeric* representations, and the two hydrogen atoms are thus described as being **enantiotopic**. Enantiotopic nuclei are also *chemically equivalent* and will have the same chemical shift, provided they are in an *achiral* environment, such as that provided by the solvents commonly used in NMR studies. It is left as an exercise to demonstrate that the hydrogen atoms of type H_b are also enantiotopic.

Another type of topicity is possible when a molecule contains a **stereogenic center** (Sec. 7.1), or is subject to **restricted rotation**, a phenomenon usually associated with the presence of a π-bond or a ring. The first case is illustrated by considering 2-chlorobutane (**2**) as seen in Figure 8.21; although only a single enantiomer is shown, the ensuing analysis gives the same result for the other one as well. The molecule has a chiral center at C-2, the effect of which makes the connectively equivalent hydrogen atoms at C-3 chemically *nonequivalent*. The nonequivalency is revealed by performing the substitution test on these protons, as illustrated in Figure 8.21. This generates **diastereomers**, thereby defining the two protons at C-3 as **diastereotopic**. Because diastereotopic nuclei *may* have chemical shifts that are different from one another, there are *five* rather than four sets of chemically nonequivalent protons in **2**: the two different

FIGURE 8.21
Topicity analysis of 2-chlorobutane by substitution.

methyl groups, each of which contains three *homotopic* protons, the methine hydrogen at C-2, and the two *diastereotopic* nuclei at C-3.

Chemical nonequivalency arising from restricted rotation is seen in 4-methyl-1-pentene (**3**) (Fig. 8.22). By applying the substitution test (Fig. 8.22), the two vinylic hydrogen atoms at C-1 are found to be *diastereotopic* and thus chemically nonequivalent. This means that **3** contains *three* distinct sets of vinylic protons, each of which appears at a different chemical shift, as seen in Figure 8.23. Applying the substitution test in this instance is simply a formal way of proving that the geminal vinylic protons of an *unsymmetrically* substituted alkene, $H_2C=CR^1R^2$, are not equivalent to one another. For example, one of these hydrogen atoms will be *cis* to R^1, and the other one *trans* to this substituent; their nonequivalency can be determined rapidly and simply from this stereochemical relationship.

The occurrence of diastereotopic hydrogens in a molecule may complicate the appearance of a ^{1}H NMR spectrum. However, the simple procedure of looking for chiral centers and for the features that may enforce restricted rotation in the molecule minimizes erroneous analyses caused by the increased number of resonances associated with diastereotopicity.

In 1-nitropropane (**1**) there are three chemically nonequivalent sets of protons. As a result, three different resonances are predicted, as is observed (Fig. 8.19). The relative locations of the three groups of resonances in the ^{1}H NMR spectrum of **1** varies according to the magnetic environment of each hydrogen nucleus in the molecule. This primarily depends on two factors: (1) the externally applied magnetic field, and (2) circulation of the electrons within the molecule, which provides an **electronic shield** about the atoms.

The field H_o is of *uniform* strength throughout the molecule, so it *cannot* be responsible for the magnetic nonequivalence of the different nuclei. On the other hand,

FIGURE 8.22 Topicity analysis of 4-methyl-1-pentene by substitution.

FIGURE 8.23 ^{1}H NMR spectrum of 4-methyl-1-pentene (300 MHz, CDCl$_3$).

the shielding effect produced by circulation of electrons is *not* uniform because the electron density varies within the molecule. This is a critical point because the **internal electric field** of the molecule induces an **internal magnetic field**, H_i, about its nuclei. The induced magnetic field acts to *oppose* the applied magnetic field, so the **effective magnetic field**, H_e, at the nuclei is *less* than H_o (Eq. 8.7); this phenomenon is the source of **diamagnetic shielding** in the molecule.

The strength of H_i varies in the same way as that of the electric field and thus makes H_e different for the various types of protons in the sample. For example, the greater is H_i the less is H_e, so a larger external field must be applied to achieve resonance. In other words, the more magnetic shielding experienced by a nucleus, the higher is the external field H_o required to produce the resonance condition at a particular oscillator frequency.

$$H_e = H_o - H_i \tag{8.7}$$

The consequences of the magnitude of H_i on the location of resonances in ^{1}H NMR spectra are seen in Figure 8.19. The nitro group, NO$_2$, serves as an electron sink, and this causes the electric field to be lower for those nuclei near it. Correspondingly, the value of H_i is less for these nuclei than for others more distant from the nitro function. The protons labeled H_c in **1** thus experience the *least* diamagnetic shielding and resonate at the *lowest* value of H_o. The methyl protons H_a are the *most* **shielded** because they are furthest from the nitro group and appear at the highest field (Fig. 8.19). Put another way, the nearer a proton is to the electronegative nitro group, the more **deshielded** it is relative to other hydrogen atoms in the molecule, and the further **downfield** it appears. On this basis, the various groups of peaks in the spectrum are assigned as shown below.

The locations of the three groups of peaks in 1-nitropropane are expressed in units of ppm on the δ-*scale*, as was noted earlier. This scale, calculated according to Equation 8.8, is defined such that measurement of chemical shifts is *independent* of the frequency of the rf oscillator of the NMR instrument. The numerator is the shift, in Hz, of a resonance of the sample relative to that of the reference compound, which is usually TMS for ^{1}H NMR spectra. The sign of the shift by convention is taken to be *positive* if the resonance appears *downfield* of TMS and *negative* if it is *upfield*. The shift is divided by the corresponding oscillator frequency, and for convenience in reporting chemical shifts, the result is multiplied by 10^6 to provide units of ppm. Alternatively, the relative positions of peaks in the spectrum could simply be expressed in Hz relative to TMS. This is unsatisfactory because the chemical shift would then depend on the frequency of oscillator being used in the spectrometer. Consequently, reporting shifts in Hz would require specifying the frequency at which the data were obtained.

$$\delta = \frac{\text{chemical shift (in Hz)} \times 10^6}{\text{oscillator frequency (in Hz)}} \tag{8.8}$$

To illustrate the difference between measuring chemical shifts in terms of δ versus Hz, consider the case in which a resonance is observed *100 Hz* downfield of TMS when a spectrometer operating at 100 MHz is used. On the δ-scale, this corresponds to a shift of *1.0 ppm*. Had the spectrum been measured on a 300-MHz instrument, the resonance would appear *300 Hz*, rather than 100 *Hz*, downfield of TMS; however, the shift would remain at *1.0 ppm* on the δ-scale.

Almost all types of protons appear *downfield* of TMS and thus have chemical shifts that are *positive* values because of the sign convention noted above. However, some organic compounds contain hydrogen nuclei that are *more* shielded than those in TMS and therefore appear *upfield* of it. The chemical shifts are *negative* in such cases.

The consistent effect that particular functional groups have on chemical shifts has led to tabulations correlating these shifts with the structural features responsible for them. Table 8.3 contains one such compilation, and Table 8.4 is an expanded listing of the chemical shifts typical of a variety of specific types of hydrogen atoms. Because the shifts are given in units of ppm, they apply to all NMR spectrometers regardless of the frequency at which the instrument is operating.

The *magnitude* of chemical shifts is primarily related to two factors, namely, the *electronegativity* of any functional groups that are near the proton being observed and the nature of *induced magnetic fields*, H_i, in molecules having π-bonds. Both effects are associated with circulation of electrons in the molecule and the interaction of the resulting induced magnetic field with the applied field H_o (Eq. 8.7).

The influence of electronegativity is evident from comparison of the chemical shifts of the methyl halides, $CH_3 - X$, in which the δ-values are 4.3, 3.0, 2.7, and 2.2 for X = F, Cl, Br, and I, respectively (Table 8.4). That of π-electrons is seen in the relative chemical shifts of the aliphatic protons of alkanes versus the vinylic protons of alkenes, as in ethane (δ 0.9) and ethylene (δ 5.25). The hydrogen nuclei of ethylene are *deshielded* partly because an sp^2-hybridized carbon atom is more electronegative than one that is sp^3-hybridized. Moreover, the vinylic protons are further deshielded because they lie in a region of the induced field H_i where the lines of force *add* to H_o, as shown in Figure 8.24a (see page 228). A similar analysis of the induced field accounts for the fact that an alkynic proton, as in acetylene (δ 1.80), resonates at *higher* field than does a vinylic hydrogen, even though an sp-hybridized carbon atom is the *most* electronegative of all carbon atoms found in organic compounds. In alkynes, the acetylenic proton is in a portion of the magnetic field induced by the circulating π-electrons where the

Table 8.3	**Chemical Shifts of Hydrogen Atoms Attached to Various Functional Groups**

Functional Group; Hydrogen Type Shown as H	δ, ppm	Functional Group; Hydrogen Type Shown as H	δ, ppm
TMS, $(CH_3)_4Si$	0	**Alcohols, ethers**	
Cyclopropane	0–1.0	$HO-\overset{\mid}{\underset{\mid}{C}}-H$	3.4–4
Alkanes			
RCH_3	0.9	$RO-\overset{\mid}{\underset{\mid}{C}}-H$	3.3–4
R_2CH_2	1.3		
R_3CH	1.5	Acetals $\}-O\diagdown\diagup$	
Alkenes		$\}-O\diagup^C\diagdown H$	5.3
$-\overset{\mid}{C}=\overset{\mid}{C}-H$ (vinyl)	4.6–5.9	**Esters**	
$-\overset{\mid}{C}=\overset{\mid}{C}-CH_3$ (allyl)	1.7	$R-\overset{O}{\overset{\parallel}{C}}-O-\overset{\mid}{\underset{\mid}{C}}-H$	3.7–4.1
Alkynes			
$-C\equiv C-H$	2–3	$RO-\overset{O}{\overset{\parallel}{C}}-\overset{\mid}{\underset{\mid}{C}}-H$	2–2.6
$-C\equiv C-CH_3$	1.8		
Aromatic		**Carboxylic acids**	
$Ar-H$	6–8.5	$HO-\overset{O}{\overset{\parallel}{C}}-\overset{\mid}{\underset{\mid}{C}}-H$	2–2.6
$Ar-\overset{\mid}{\underset{\mid}{C}}-H$ (benzyl)	2.2–3	$R-\overset{O}{\overset{\parallel}{C}}-O-H$	10.5–12
Fluorides, $F-\overset{\mid}{\underset{\mid}{C}}-H$	4–4.45	**Ketones**	
Chlorides		$R-\overset{O}{\overset{\parallel}{C}}-\overset{\mid}{\underset{\mid}{C}}-H$	2–2.7
$Cl-\overset{\mid}{\underset{\mid}{C}}-H$	3–4	**Aldehydes**	
$Cl-\overset{\overset{Cl}{\mid}}{\underset{\mid}{C}}-H$	5.8	$R-\overset{O}{\overset{\parallel}{C}}-H$	9–10
Bromides, $Br-\overset{\mid}{\underset{\mid}{C}}-H$	2.5–4	**Amides**	
Iodides, $I-\overset{\mid}{\underset{\mid}{C}}-H$	2–4	$R-\overset{O}{\overset{\parallel}{C}}-\overset{\mid}{N}-H$	5–8
		Alcohols, $R-O-H$	4.5–9
		Phenols, $Ar-O-H$	4–12
Nitroalkanes, $O_2N-\overset{\mid}{\underset{\mid}{C}}-H$	4.2–4.6	Amines, $R-NH_2$	1–5

lines of force *oppose* the applied field (Fig. 8.24b), thereby increasing the shielding of this type of nucleus. The protons of aromatic compounds, as in benzene (δ 7.27), resonate at lower fields than the vinylic protons of alkenes because the induced magnetic field responsible for deshielding is greater owing to the *cyclic* nature of the circulation of π-electrons (Fig. 8.24c); this is called the **ring-current effect**.

Table 8.4	Compilation of ^{1}H NMR Absorptions for Various Molecules

Listed below are the ^{1}H NMR chemical shifts observed for the protons of a number of organic compounds. The shifts are classified according to whether they are methyl, methylene, or methine types of hydrogen atoms. The atom shown in **bold** is responsible for the absorptions listed below.

METHYL ABSORPTIONS

Compound	δ, ppm	Compound	δ, ppm
CH_3NO_2	4.3	CH_3CHO	2.2
CH_3F	4.3	CH_3I	2.2
$(CH_3)_2SO_4$	3.9	$(CH_3)_3N$	2.1
$C_6H_5COOCH_3$	3.9	$CH_3CON(CH_3)_2$	2.1
$C_6H_5-O-CH_3$	3.7	$(CH_3)_2S$	2.1
CH_3COOCH_3	3.6	$CH_2=C(CN)CH_3$	2.0
CH_3OH	3.4	CH_3COOCH_3	2.0
$(CH_3)_2O$	3.2	CH_3CN	2.0
CH_3Cl	3.0	CH_3CH_2I	1.9
$C_6H_5N(CH_3)_2$	2.9	$CH_2=CH-C(CH_3)=CH_2$	1.8
$(CH_3)_2NCHO$	2.8	$(CH_3)_2C=CH_2$	1.7
CH_3Br	2.7	CH_3CH_2Br	1.7
CH_3COCl	2.7	$C_6H_5C(CH_3)_3$	1.3
CH_3SCN	2.6	$C_6H_5CH(CH_3)_2$	1.2
$C_6H_5COCH_3$	2.6	$(CH_3)_3COH$	1.2
$(CH_3)_2SO$	2.5	$C_6H_5CH_2CH_3$	1.2
$C_6H_5CH=CHCOCH_3$	2.3	CH_3CH_2OH	1.2
$C_6H_5CH_3$	2.3	$(CH_3CH_2)_2O$	1.2
$(CH_3CO)_2O$	2.2	$CH_3(CH_2)_3Cl, Br, I$	1.0
$C_6H_5OCOCH_3$	2.2	$CH_3(CH_2)_4CH_3$	0.9
$C_6H_5CH_2N(CH_3)_2$	2.2	$(CH_3)_3CH$	0.9

METHYLENE ABSORPTIONS

Compound	δ, ppm	Compound	δ, ppm
$EtOCOC(CH_3)=CH_2$	5.5	$CH_3CH_2CH_2Cl$	3.5
CH_2Cl_2	5.3	$(CH_3CH_2)_4N^+I^-$	3.4
CH_2Br_2	4.9	CH_3CH_2Br	3.4
$(CH_3)_2C=CH_2$	4.6	$C_6H_5CH_2N(CH_3)_2$	3.3
$CH_3COO(CH_3)C=CH_2$	4.6	$CH_3CH_2SO_2F$	3.3
$C_6H_5CH_2Cl$	4.5	CH_3CH_2I	3.1
$(CH_3O)_2CH_2$	4.5	$C_6H_5CH_2CH_3$	2.6
$C_6H_5CH_2OH$	4.4	CH_3CH_2SH	2.4
$CF_3COCH_2C_3H_7$	4.3	$(CH_3CH_2)_3N$	2.4
$Et_2C(COOCH_2CH_3)_2$	4.1	$(CH_3CH_2)_2CO$	2.4
$HC\equiv C-CH_2Cl$	4.1	$BrCH_2CH_2CH_2Br$	2.4
$CH_3COOCH_2CH_3$	4.0	Cyclopentanone (α-CH_2)	2.0
$CH_2=CHCH_2Br$	3.8	Cyclohexene (α-CH_2)	2.0
$HC\equiv CCH_2Br$	3.8	Cyclopentane	1.5
$BrCH_2COOCH_3$	3.7	Cyclohexane	1.4
CH_3CH_2NCS	3.6	$CH_3(CH_2)_4CH_3$	1.4
CH_3CH_2OH	3.6	Cyclopropane	0.2

(continued)

| Table 8.4 | *(continued)* | | |

METHINE ABSORPTIONS

Compound	δ, ppm	Compound	δ, ppm
C_6H_5CHO	10.0	C_6H_5Cl	7.2
$4\text{-}ClC_6H_4CHO$	9.9	$CHCl_3$	7.2
$4\text{-}CH_3OC_6H_4CHO$	9.8	$CHBr_3$	6.8
CH_3CHO	9.7	*p*-Benzoquinone	6.8
Pyridine (α-**H**)	8.5	$C_6H_5NH_2$	6.6
$1,4\text{-}C_6H_4(NO_2)_2$	8.4	Furan (β-**H**)	6.3
$C_6H_5CH{=}CHCOCH_3$	7.9	$CH_3CH{=}CHCOCH_3$	5.8
C_6H_5CHO	7.6	Cyclohexene (vinylic **H**)	5.6
Furan (α-**H**)	7.4	$(CH_3)_2C{=}CHCH_3$	5.2
Naphthalene (β-**H**)	7.4	$(CH_3)_2CHNO_2$	4.4
$1,4\text{-}C_6H_4I_2$	7.4	Cyclopentyl bromide (**H** at C-1)	4.4
$1,4\text{-}C_6H_4Br_2$	7.3	$(CH_3)_2CHBr$	4.2
$1,4\text{-}C_6H_4Cl_2$	7.2	$(CH_3)_2CHCl$	4.1
C_6H_6	7.3	$C_6H_5C{\equiv}C{-}H$	2.9
C_6H_5Br	7.3	$(CH_3)_3C{-}H$	1.6

Spin-Spin Splitting. Information beyond that regarding the magnetic environment of a nucleus is available from analysis of the NMR spectrum. For example, the **spin-spin splitting** pattern shown by a particular type of hydrogen atom provides information about the *number* of its **nearest neighbors**. Nearest neighbors are defined as elements that have nuclear spin and, in most cases, are *no more* than *three* bonds, or *two* atoms, away from the atom of interest. For our purposes, hydrogen is the primary element of interest for analyzing splitting patterns because compounds containing other magnetically active nuclei are only rarely encountered in the experiments in this textbook. The presence of ^{13}C in organic compounds is *not* a factor in the analysis because this isotope, which also has a nuclear spin of 1/2, is present at the level of only about 1%. Thus spin-spin splitting associated with it is generally *not* observed.

Observation of coupling between atoms having nuclear spins requires that the nuclei must *not* be magnetically equivalent. By definition, **magnetic equivalency** means that *every* nucleus in a particular set of chemical shift-equivalent nuclei is coupled equally to *every* other nucleus of the spin system. Chemical shift-nonequivalent nuclei are always magnetically nonequivalent and couple with one another. As discussed earlier, this type of nonequivalence is generally seen for protons that are chemically distinct by virtue of being *heterotopic* or *diastereotopic*. That protons may be chemically equivalent but magnetically nonequivalent is illustrated by a discussion later in this section.

Analysis of 1-nitropropane (**1**) in this context reveals that the type *a* hydrogen atoms have *two* nearest neighbors in the form of the two protons of type *b* that are magnetically active and chemically nonequivalent. Nuclei of type *c* also have two such nearest neighbors, whereas those of type *b* nuclei have *five*.

The **splitting** or **coupling pattern** of each chemically distinct type of nucleus may be predicted in the following way. In the general case, an atom A having nuclear spin I_z, where $I_z = 1/2, 1, 3/2$, and so on, that couples with another atom B splits the

FIGURE 8.24

Effect on chemical shifts of induced field from circulation of π-electrons. (a) Deshielding of vinylic protons. (b) Shielding of acetylenic protons. (c) Deshielding of aromatic protons.

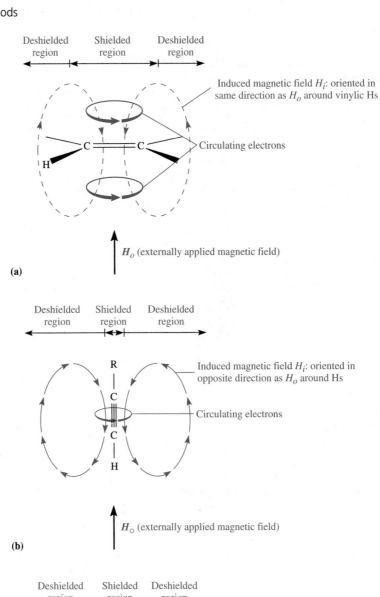

Deshielded region   Shielded region   Deshielded region

Induced magnetic field H_i: oriented in same direction as H_o around vinylic Hs

Circulating electrons

H_o (externally applied magnetic field)

(a)

Deshielded region   Shielded region   Deshielded region

Induced magnetic field H_i: oriented in opposite direction as H_o around Hs

Circulating electrons

H_O (externally applied magnetic field)

(b)

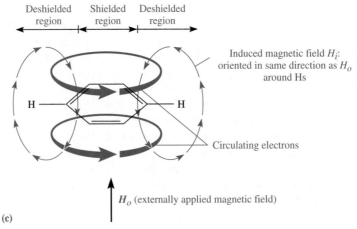

Deshielded region   Shielded region   Deshielded region

Induced magnetic field H_i: oriented in same direction as H_o around Hs

Circulating electrons

H_o (externally applied magnetic field)

(c)

resonance of B into the number of peaks given by Equation 8.9. When A is ^{1}H, ^{13}C, ^{19}F, or ^{31}P, this expression simplifies to the $n + 1$ rule expressed in Equation 8.10, because $I_z = 1/2$ for all these nuclei. The expected splitting patterns for 1-nitropropane (**1**), according to Equation 8.10, are *three* peaks each for H_a and H_c and six peaks for H_b. This is precisely the splitting pattern appearing in Figure 8.19.

$$N = 2nI_z + 1 \qquad (8.9)$$

where N = number of peaks observed
         for absorbing atom B

    n = number of magnetically equivalent
         nearest neighbor atoms A

    I_z = nuclear spin of A

    $N = n + 1 \qquad (8.10)$

Strictly speaking, the $n + 1$ rule for predicting splitting patterns of coupled hydrogen nuclei applies only under the following circumstances: (1) The **coupling constants**, J, of all nearest neighbor hydrogen nuclei coupling with the nucleus of interest must have the same value, and (2) the ratio of the difference in chemical shift $\Delta\nu$, in Hz, of the coupled nuclei to their coupling constant J must be greater than about 10 (Eq. 8.11). When these criteria are met, a **first-order analysis** of the multiplicity of each resonance is possible. For first-order ^{1}H NMR spectra, Equation 8.10 may be used to determine the number, n, of nearest neighbors to a hydrogen nucleus. For example, if a *sextet* is observed in the ^{1}H NMR spectrum, there are *five* nearest neighbors to the proton(s) responsible for that sextet because N of Equation 8.10 is six.

$$n + 1 \text{ rule valid if: } \Delta\nu(\text{Hz})/J(\text{Hz}) \geq \sim 10 \qquad (8.11)$$

The relative intensities of the peaks in a *first-order* multiplet that obeys Equation 8.10 may be *approximated* by applying Pascal's triangle (Fig. 8.25). For example, if a hydrogen atom has two nearest neighbors, then the triplet that results is predicted to have intensities in the ratio of about 1:2:1. If there are five nearest neighbors, the corresponding ratio for the sextet is 1:6:15:20:15:6:1. Prediction of relative peak intensities by this method becomes more accurate as the ratio $\Delta\nu/J$ increases.

Because the ^{1}H NMR spectra in this book were all taken at 300 MHz, many may be interpreted as first-order spectra, and the peak intensities of the multiplets are predictable by applying Pascal's triangle. For example, reconsidering the spectrum of Figure 8.19, we find that $J_{ab} = J_{bc} \cong 7$ Hz. The difference in chemical shifts for the sets of coupled nuclei are $\Delta\nu_{AB} = 315$ Hz, and $\Delta\nu_{BC} = 700$ Hz, respectively. Application of Equation 8.11 gives values of $315/7 = 45$ and $700/7 = 100$, respectively, for the ratios of interest. The fact that both ratios exceed the criterion of Equation 8.11 means that the splitting patterns may be predicted from the $n + 1$ rule, as evidenced in Figure 8.19. Moreover, the relative intensities of each of the peaks in the multiplets in Figure 8.19 have the approximate values as predicted by applying Pascal's triangle.

As the term $\Delta\nu/J$ decreases, the observed spectra become **second-order** and are more complicated. The splitting patterns may no longer be predictable by the $n + 1$ rule, and the intensities of the peaks in the multiplets cannot be approximated by

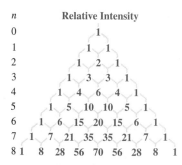

FIGURE 8.25 Pascal's triangle showing relative intensities of first-order multiplets; coupling constants of all nearest neighbors, n, are assumed to be equal.

FIGURE 8.26   ¹H NMR spectrum of 1-butanol in deuterochloroform (300 MHz).

applying Pascal's triangle. This trend may be illustrated by examining the ¹H NMR spectra for 1-butanol (**4**) in Figures 8.26 and 8.27 and pentane (**5**) in Figure 8.28.

$$
\begin{array}{c}
H_e \quad H_d \quad H_c \quad H_b \\
| \quad\quad | \quad\quad | \quad\quad | \\
H_e\!-\!\mathrm{C}\!-\!\mathrm{C}\!-\!\mathrm{C}\!-\!\mathrm{C}\!-\!\mathrm{OH}_a \\
| \quad\quad | \quad\quad | \quad\quad | \\
H_e \quad H_d \quad H_c \quad H_b
\end{array}
\qquad\qquad
\begin{array}{c}
H_a \quad H_b \quad H_c \quad H_b \quad H_a \\
| \quad\quad | \quad\quad | \quad\quad | \quad\quad | \\
H_a\!-\!\mathrm{C}\!-\!\mathrm{C}\!-\!\mathrm{C}\!-\!\mathrm{C}\!-\!\mathrm{C}\!-\!H_a \\
| \quad\quad | \quad\quad | \quad\quad | \quad\quad | \\
H_a \quad H_b \quad H_c \quad H_b \quad H_a
\end{array}
$$

4   **5**

1-Butanol   Pentane

In 1-butanol (**4**), nuclei of type H_e resonate at δ 0.94 ppm with those of type H_d appearing at about δ 1.40 ppm; J_{de} is about 7 Hz (Fig. 8.26). Using a 300 MHz spectrometer, the ratio $\Delta\nu/J_{de}$ is greater than 10, so the criterion of Equation 8.11 is satisfied, and the nuclei of type H_e appear as a triplet. The nuclei of type H_c appear at about

FIGURE 8.27   ¹H NMR spectrum of 1-butanol in dimethyl sulfoxide-d₆ (300 MHz).

FIGURE 8.28 ^{1}H NMR spectrum of pentane (300 MHz, CDCl$_3$).

δ 1.56 ppm and J_{cd} is also about 7 Hz. Because the ratio $\Delta\nu/J_{cd}$ is only about 7, the criterion of Equation 8.11 is *not* satisfied, and the appearance of the multiplets centered at δ 1.40 and 1.56 ppm is no longer predicted by simple first-order rules. The number of peaks in the multiplets remains that predicted by the $n + 1$ rule, but the peak intensities are no longer approximated by applying Pascal's triangle.

Assuming that $J_{bc} = J_{ab}$, application of the $n + 1$ rule to nuclei of type H$_b$ and H$_a$ leads to a predicted splitting pattern of a quartet for H$_b$ and a triplet for H$_a$. However, somewhat broadened and distorted quartets and triplets at δ 3.65 and 1.63 ppm are actually observed as seen in Figure 8.26. The broadening of these peaks is not due to any second-order effects. Rather the spectrum was obtained under conditions in which the hydroxylic proton H$_a$ is undergoing rapid exchange with other molecules of the alcohol or with adventitious water present in the sample. As a result of this rapid exchange, the spin information carried by the hydroxylic proton is partially lost, and the peaks arising from protons that are spin-coupled with the hydroxyl proton are broadened. Spectra obtained under conditions that suppress this exchange, such as using CD$_3$SOCD$_3$, DMSO-d$_6$, rather than CDCl$_3$ as the solvent, allow the coupling to be observed, and the triplet at δ 4.33 and the quartet at δ 3.38 are sharp as illustrated in Figure 8.27. Comparison of the spectra of **4** in Figures 8.26 and 8.27 also reveals that the chemical shifts for protons in a molecule *can* vary with the solvent, so it is always important to indicate the solvent used to prepare the sample when reporting NMR spectral data.

In the spectrum of pentane (**5**) (Fig. 8.28), nuclei of type H$_a$ resonate at about δ 0.90 ppm, whereas all of the nuclei of types H$_b$ and H$_c$ resonate at about δ 1.30 ppm. Because the ratio $\Delta\nu/J_{ab}$ is greater than 10, the condition of Equation 8.11 is satisfied, and the peak for the H$_a$ is a triplet according to the $n + 1$ rule. However, the ratio $\Delta\nu/J_{bc}$ is much less than 10, and the splitting pattern for the nuclei of types H$_b$ and H$_c$ is complex and cannot be interpreted by first-order analysis and the $n + 1$ rule.

Thus far, we have predicted splitting patterns by grouping hydrogens according to their chemical shift equivalence and then determining the number of nearest neighbors n. There are cases, however, in which nuclei may have equivalent chemical shifts but are *not* magnetically equivalent; this may complicate the appearance and interpretation of the spectrum. This phenomenon is often seen in the ^{1}H NMR spectra of aromatic compounds.

FIGURE 8.29 ^{1}H NMR spectrum of 4-bromonitrobenzene (300 MHz, CDCl$_3$).

For example, analyzing 4-bromonitrobenzene (**6**) by the substitution test leads to the conclusion that there are two sets of homotopic hydrogen atoms, H$_a$H$_{a'}$ and H$_b$H$_{b'}$. The two protons in each set have equivalent chemical shifts. However, the coupling constants between all members of the spin set are not equivalent. Namely, $J_{H_aH_b} \equiv J_{H_{a'}H_{b'}}$ and $J_{H_{ab'}} \equiv J_{H_{a'b}}$, but $J_{H_aH_b} \neq J_{H_{a'}H_b}$ and $J_{H_aH_b} \neq J_{H_aH_{b'}}$. This makes H$_a$ and H$_{a'}$ **magnetically nonequivalent**, as are H$_b$ and H$_{b'}$. It should be mentioned that the pairs of nuclei H$_a$H$_{b'}$ and H$_{a'}$ H$_b$ are not nearest neighbors, yet coupling may occur between them. Nuclear spin interactions that occur over more than three bonds are termed **long-range coupling**, a phenomenon that is most frequently observed in systems having **conjugated π-bonds**.

Br

H$_a$ H$_{a'}$

H$_b$ H$_{b'}$

NO$_2$

6

4-Bromonitrobenzene

The effect of magnetic nonequivalency is evident in the ^{1}H NMR spectrum of **6** (Fig. 8.29). Careful examination of the resonances for the aromatic protons reveals additional splitting for each of the doublets that is the result of magnetic nonequivalency within the set. Nevertheless, the aromatic multiplets still have the overall appearance of the doublets expected from the simple first-order analysis using the $n + 1$ rule.

The *magnitude* of coupling constants, J_{HH}, is *independent* of the oscillator frequency, so units of Hz are used. The measured values depend on two main factors: the number of bonds intervening between the coupled nuclei, and the bond angle or dihedral angle between the nuclei. The angular dependence of **geminal coupling**, which is coupling between magnetically different protons on the *same* atom, is shown graphically in Figure 8.30, in which θ is the bond angle between the coupled nuclei. The situation for **vicinal coupling**, in which the coupled nuclei are on *adjacent* atoms and have a

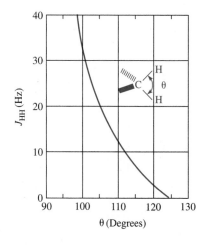

FIGURE 8.30
The *geminal* Karplus correlation. J_{HH} for CH_2 groups as a function of H–C–H angle, θ.

dihedral angle, ϕ, is presented in Figure 8.31. Some typical ranges for coupling constants are provided in Figure 8.32.

Coupling constants provide valuable information about the structure of organic compounds. If two different protons are on adjacent carbon atoms, the coupling constant between the two has only one value, that is, $J_{ab} \equiv J_{ba}$, because the protons are mutually coupled with each other. Thus, if the coupling constants of two multiplets suspected to be the result of two particular nuclei coupling with one another are *not* the same, the two nuclei are *not* coupled. This fact helps in making structural assignments.

It is difficult to measure coupling constants using the 300 MHz spectra reproduced in this textbook because the spectra have been photographically reduced, making measurement of the distance between peaks subject to large error. However, using modern high field NMR spectrometers, the values of coupling constants for simple splitting

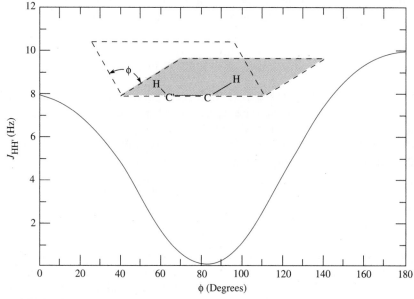

FIGURE 8.31   The *vicinal* Karplus correlation. J_{HH} for H–C–C–H groups as a function of dihedral angle, ϕ.

FIGURE 8.32
Examples of the values of coupling constants J_{HH} for selected functional groups.

patterns are easily determined from digital printouts of chemical shifts of the various peaks. To enable you to gain experience in interpreting ^1H NMR data, the ^1H NMR spectra of all compounds are found on the CD-ROM that accompanies this text. By expanding the spectra, you will be able to measure the coupling constants accurately.

Data provided in the form of $\Delta\delta$ must first be converted to Hz before computation of the coupling constant, J. For example, if the peaks of a quartet are found to differ by 0.04 ppm in a spectrum obtained on a 300-MHz instrument, the coupling constant would be 12 Hz by application of Equation 8.12, where $\Delta\delta$ is 0.04 and *instrument frequency* is 300×10^6 Hz.

$$J(\text{Hz}) = (\Delta\delta \times 10^{-6}) \times \text{instrument frequency} \qquad (8.12)$$

Integration. The areas of the peaks in a ^1H NMR spectrum are important measures of the *relative* numbers of the different types of hydrogen nuclei in a molecule because these peak areas are a linear function of the number of nuclei producing the peaks. The areas are determined by an electronic integration that is typically plotted as a stepped trace that appears on the spectrum, as seen in Figure 8.19.

In practice, the ^1H NMR spectrum is first recorded and the integration is then plotted on the same sheet of paper. The vertical distance that the integration rises over a peak or group of peaks is a direct measure of the area under the peak(s) and thus of the relative number of protons producing the resonances. The integrator is usually reset to zero after each set of resonances has been measured, as in Figure 8.19, but this need not be done, as seen in Figure 8.26. Resetting the integrator allows the height of each step in the integration curve to be higher, so the experimental error in measuring the step is less. In this textbook, the numerical values of the integrations rather than the plotted integrals will normally be provided below sets of peaks. You may verify these values by examining the ^1H NMR spectra on the CD-ROM that accompanies this text.

The height of each step is then measured and the *relative* number of hydrogen atoms producing the steps is computed by dividing the integration heights by the smallest height measured; this gives a ratio of the various types of protons. For example, in the spectrum of 1-nitropropane (**1**) the heights of the steps in Figure 8.19 are represented by the distances *ab*, *ac*, and *ad*. The relative heights are determined to obtain the ratio of 1.55 : 1.0 : 1.0 for groups *A*, *B*, and *C* in the spectrum. Because the number of hydrogen atoms present in a molecule must be integral, this ratio is converted to whole numbers by multiplying *each* value by an *integer* that gives the **relative ratio** of the types of protons in whole numbers. *If* the molecular formula of the molecule is known, the integer is chosen to give a *sum* of the values in the ratio corresponding to the *absolute* number of hydrogen nuclei present. There are seven hydrogen atoms in 1-nitropropane, so *two* is used as the multiplier to provide the **absolute ratio** 3.10 : 2.0 : 2.0, the sum of which is 7.1. Because the accuracy of electronic integration is generally no better than 5–10%, this is within experimental error of the total number of protons in the molecule.

ANALYSIS OF ^{1}H NMR SPECTRA

General Procedure. The proper analysis of the ^{1}H NMR spectrum of a compound may provide a wealth of information regarding its structure. Determining the *chemical shift* of each peak or group of peaks allows speculation about the *type* of functional group to which the hydrogen nucleus producing the resonance is attached. The *spin-spin splitting pattern* and the *n* + 1 rule give information concerning the *number of nearest neighbor hydrogen atoms* for the proton(s) producing a particular absorption. *Integration* allows evaluation of the peak areas of each set of peaks and defines the *relative numbers of each type* of proton present in the molecule; if the molecular formula of the compound is known, the *absolute* number of each type may also be computed.

Appropriate steps for complete analysis of a ^{1}H NMR spectrum are as follows:

1. Determine the relative numbers of the different types of hydrogen nuclei by measuring the integrated step height for each peak or group of peaks. Convert the resulting *relative* ratio to an *absolute* ratio if the molecular formula is known. The *absolute* values of the integrations are provided for most of the spectra in this textbook by the numbers in parentheses located below the respective sets of peaks in the spectra.
2. Measure the *chemical shifts* of the different sets of resonances and use the data to try to deduce the functional groups that might be present in the molecule. This process is greatly aided if the molecular formula and/or other spectral data, such as IR and ^{13}C NMR spectra, are available.
3. Analyze the *spin-spin splitting* patterns of each group of resonances to determine the number of *nearest neighbors* associated with each type of hydrogen nucleus.

Applying these steps *may* allow assignment of the structure of an unknown compound. However, without additional information about a substance, such as its molecular formula or the chemical reaction(s) by which it was formed, it may only be possible to make a partial structural determination. Other chemical or spectral data may be required to make a complete assignment. In any case, structural assignments based *exclusively* on interpretation of spectra *must* be confirmed by comparing the spectra of the unknown with those of an authentic specimen, if the compound is known, or by chemically converting the unknown to a known compound.

FIGURE 8.33   ¹H NMR spectrum for unknown compound (300 MHz, CDCl₃).

Analysis of an Unknown. An example of the use of this procedure for making a structural assignment follows. Other examples may be found in the exercises at the end of this section.

Problem. Provide the structure of the compound having the molecular formula C_4H_9Br and whose ¹H NMR spectrum is given in Figure 8.33.

Solution. For purposes of this analysis, the groups of peaks are referred to as representing nuclei of types *A–C* in going from left to right, or upfield, in the spectrum.

Integration. The *relative* ratio of the integrals for H_a, H_b, and H_c, is $2.1 : 1.0 : 5.7$, respectively. Given the molecular formula, the *absolute* ratio must be $2 : 1 : 6$, for a total of nine hydrogen atoms. Thus, there are two hydrogen atoms of type *A*, *one* of type *B*, and *six* of type *C*.

Chemical Shifts. The δ-values at the *center* of the individual multiplets are approximately 3.3 (H_a), 2.0 (H_b), and 1.0 (H_c). The low-field doublet for H_a presumably reflects the presence of the electronegative bromine atom nearby. Since there are *two* such nuclei, the partial structure, $–CH_2Br$, may be written. The appearance of H_c at about δ 1.0 is consistent with these nuclei being associated with a methyl group. Because there are six such hydrogen atoms, *two* methyl groups must be present. Another partial structure, $(CH_3)_2C–$, may then be written. Because these two partial structures account for *all* of the carbon atoms required by the molecular formula and all but *one* of the hydrogen atoms, the missing atom must be a *methine* hydrogen atom located on the carbon atom bearing the two methyl groups and the bromomethyl substituent. Although the chemical shift of this methine hydrogen atom is outside the range for such nuclei, as given in Table 8.3, reference to Table 8.4 shows that the methyl resonances are shifted *downfield* when there is a halogen β to the carbon atom bearing the nuclei of interest; this is evidenced in the δ-value of 1.7 for the methyl group of ethyl bromide, CH_3CH_2Br. The same effect applies to methine hydrogen atoms. Enough information is now available to enable assignment of the structure of the unknown as 1-bromo-2-methylpropane (isobutyl bromide), $(CH_3)_2CHCH_2Br$.

In this case, the determination of the structure may be made simply on the basis of molecular formula, chemical shifts, and integration; analysis of the spin-spin splitting patterns, as in Part (c), is required only to confirm the conclusion regarding the structure of the unknown and not for initially generating the structure. More typically, the information contained in the splitting patterns is crucial to a first formulation of the structure of an unknown.

Spin-Spin Splitting Patterns. The fact that the nuclei of types A and B appear as a doublet means that they have *one* nearest neighbor, which is consistent with the fact that both H_a and H_c in 1-bromo-2-methylpropane should be coupled with the methine hydrogen atom H_b but *no* others. Conversely, H_b must be coupled with both the other types of protons, which total eight. Application of the $n + 1$ rule dictates nine peaks in the splitting pattern of the methine hydrogen atom. The two ''missing'' members of this multiplet are not apparent because they are the weak, outermost peaks, and the amplification at which the spectrum was plotted was insufficient for them to be seen. If the multiplet for H_b is viewed at significantly higher amplification, all nine peaks may be observed.

There is an important lesson to be learned from the fact that fewer peaks are observed than would be expected from the $n + 1$ rule: When multiplets of six or more are predicted, one or more of the outermost absorptions of the multiplet is frequently too weak (see Pascal's triangle, Figure 8.25) to appear under the normal operating conditions for the spectrometer. This fact must be taken into account when using spin-spin splitting patterns to interpret the spectrum of an unknown compound.

Determination of Structure. Although the present problem was solved by using only two of the three basic types of information available from an integrated ^{1}H NMR spectrum, the usual case would require careful interpretation of all three. The strategy is to generate partial structures that are consistent with the available data and then to attach these pieces in various ways that are consistent with the observed splitting patterns and the valences of the bonded atoms. In a majority of cases, particularly with molecules having a total of only twenty atoms or so and one or more functional groups, the actual structure of the compound may be correctly assigned. Once proposed, a possible structure must be checked carefully to ascertain that it is consistent with all of the available data. If not, the structure is incorrect, and alternate possibilities must be explored and tested against the data until one is found that is consistent with *all* of the spectral information.

EXERCISES

1. Define, explain, or give an example of the following:

   a. tetramethylsilane (TMS) as an internal standard

   b. delta (δ) scale

   c. upfield shift

   d. $n + 1$ rule

   e. chemical shift

   f. integration curve

   g. *relative* ratio of integrated peaks

   h. *absolute* ratio of integrated peaks

   i. homotopic hydrogen atoms

   j. enantiotopic hydrogen atoms

   k. diastereotopic hydrogen atoms

   l. coupling constant

   m. chemically equivalent protons

   n. diamagnetic shielding

   o. heterotopic hydrogen atoms

2. Demonstrate that the H_b hydrogen atoms in 1-nitropropane (**1**) are enantiotopic.

3. Label the sets of connectively equivalent sets of protons in the molecules (**a**)–(**i**) according to whether they are homotopic, enantiotopic, or diastereotopic. Using Tables 8.3 and 8.4, predict the approximate chemical shifts, in δ, and the splitting patterns expected for each chemically distinct type of hydrogen atom that you identify. Assume in this analysis that the coupling constants of all nearest neighbors are identical.

a. CH_3CHCH_3
    |
    Cl

b. $C_6H_5CHCH_3$
    |
    Br

c. $(CH_3)_2CHCH(CH_3)_2$

d. $(CH_3)_3CCCH_2CH_3$ (with O double-bonded to the central C)

e.

f. $ClCH_2CH_2CH_2Cl$

g.

h. $(CH_3)_2CHC\!-\!OH$ (with O double-bonded to C)

i.

4. By applying the criterion of Equation 8.10, predict whether the multiplicities of the following sets of coupled nuclei, H_a and H_b, could be predicted by a first-order analysis. Note the units used to report the chemical shifts of the nuclei and the frequency of the rf oscillator for the instrument used to obtain the shifts.

   a. δ_{H_a} 1.3 ppm, δ_{H_b} 3.9 ppm, J_{ab} = 8.5 Hz, 200-MHz instrument.

   b. δ_{H_a} 0.8 ppm, δ_{H_b} 1.0 ppm, J_{ab} = 2.5 Hz, 200-MHz instrument.

   c. δ_{H_a} 0.8 ppm, δ_{H_b} 1.0 ppm, J_{ab} = 8.5 Hz, 200-MHz instrument.

   d. δ_{H_a} 0.8 ppm, δ_{H_b} 1.0 ppm, J_{ab} = 8.5 Hz, 500-MHz instrument.

   e. δ_{H_a} 0.8 ppm, δ_{H_b} 1.0 ppm, J_{ab} = 2.5 Hz, 200-MHz instrument.

5. The ^1H NMR chemical shifts, splitting patterns, and *relative* numbers of hydrogen atoms for three compounds are provided in **a**–**c**. Deduce one or more structures consistent with these data. You may find it helpful to sketch these spectra on a sheet of paper.

   a. C_4H_9Br: δ 1.04 (6 H) doublet, δ 1.95 (1 H) multiplet, δ 3.33 (2 H) doublet.

   b. $C_3H_6Cl_2$: δ 2.2 (2 H) quintet, δ 3.75 (1 H) triplet.

   c. $C_5H_{11}Br$: δ 0.9 (3 H) doublet, δ 1.8 (1.5 H) complex multiplet, δ 3.4 (1 H) triplet.

6. Compute the *relative* and *absolute* ratios for the different types of hydrogens in each pair of molecules shown below.

a. $CH_3-CH_2-C\overset{\displaystyle O}{\underset{\displaystyle H}{<}}$ $\overset{CH_3}{\underset{CH_3}{>}}$ b.

7. Figures 8.34–8.40 are of the compounds shown on the spectra. Interpret these spectra as completely as possible in terms of the observed chemical shifts, integrations, and splitting patterns.

FIGURE 8.34 ^{1}H NMR spectrum of carvone (300 MHz, CDCl$_3$).

FIGURE 8.35 ^{1}H NMR spectrum of ethyl phenyl ether (300 MHz, CDCl$_3$).

FIGURE 8.36
¹H NMR spectrum of 2-butanone
(300 MHz, CDCl₃).

FIGURE 8.37
¹H NMR spectrum of butanoic acid
(butyric acid) (300 MHz, CDCl₃).

FIGURE 8.38
¹H NMR spectrum of *N*-
ethylaniline (300 MHz, CDCl₃).

FIGURE 8.39   ¹H NMR spectrum of ethyl acetate (300 MHz, CDCl₃).

FIGURE 8.40   ¹H NMR spectrum of butanal (300 MHz, CDCl₃).

8. Provide structures of the compounds **7–9**, whose ¹H NMR spectra are given in Figures 8.41–8.43, respectively.

a. Compound **7**, $C_{10}H_{14}$

b. Compound **8**, C_3H_8O

c. Compound **9**, C_8H_9Cl

FIGURE 8.41
¹H NMR spectrm (300 MHz, CDCl₃) for compound **7**.

FIGURE 8.42
¹H NMR spectrum (300 MHz, CDCl₃) for compound **8**.

FIGURE 8.43
¹H NMR spectrum (300 MHz, CDCl₃) for compound **9**.

REFERENCES

1. Silverstein, R. M.; Bassler, G. C.; Morrill, T. C. *Spectrometric Identification of Organic Compounds*, 5th ed., John Wiley & Sons, New York, **1992**, Chap. 4.

2. Yoder, C. H.; Schaeffer, Jr., C. D. *Introduction to Multinuclear NMR*, Benjamin/Cummings Publishing Co., Menlo Park, CA, **1987**.

3. Abraham, R. J.; Fisher, J.; Loftus, P. *Introduction to NMR Spectroscopy*, 2nd ed., John Wiley & Sons, New York, **1988**.

4. King, R. W.; Williams, K. R. "The Fourier Transform in Chemistry-NMR," *J. Chem. Ed.* **1989**, *66*, A213–A219, A243–A248; King, R. W.; Williams, K. R. "The Fourier Transform in Chemistry-NMR," *J. Chem. Ed.* **1990**, *67*, A100–A105, A125–A137; Williams, K. R.; King, R. W. "The Fourier Transform in Chemistry-NMR," *J. Chem. Ed.* **1990**, *67*, A93–A99.

5. *Sadtler Nuclear Magnetic Resonance Spectra*, vol. 1–113, Sadtler Research Laboratories, Philadelphia, PA. Compilation of 64,000 ^{1}H NMR spectra, as of 1996.

6. *Aldrich Library of NMR Spectra*, Pouchert, C. J., ed. vol. I–II, Aldrich Chemical Co., Milwaukee, WI, **1983**.

7. *Aldrich Library of ^{13}C and ^{1}H NMR Spectra*, Pouchert, C. J.; Behnke, J., eds., Aldrich Chemical Co., Milwaukee, WI, **1993**.

CARBON-13 NUCLEAR MAGNETIC RESONANCE (^{13}C NMR) SPECTROSCOPY

Carbon-12, ^{12}C, has no **nuclear spin**, but its carbon-13 isotope, ^{13}C, has a nuclear spin of 1/2, as does hydrogen, ^{1}H. The **natural abundance** of ^{13}C is about 1.1%, so unless substances are prepared from precursors having artificially high levels of this isotope, only 1.1% of the carbon atoms in a compound will undergo the *spin flip* or *resonance* that is characteristic of the NMR experiment. In other words, only about one in a hundred carbon atoms of a sample having ^{13}C present at natural abundance will be the proper isotope for producing an absorption in the NMR spectrum.

This low level of the NMR-active isotope of carbon makes it more difficult to obtain a suitable **carbon nuclear magnetic resonance** (^{13}C NMR) spectrum. For example, whereas it is usually possible to measure a ^{1}H NMR spectrum in a few minutes, it may take tens of minutes or even hours to accumulate enough data to produce a ^{13}C NMR spectrum in which the signal to noise ratio is high enough for the resonances due to the carbon atoms to be seen. Nonetheless, modern spectrometers and the sophisticated computers associated with them allow acquisition of the data necessary for a ^{13}C NMR spectrum on samples of 1–5 mg, which is about an order of magnitude greater than the amount needed for a ^{1}H NMR spectrum.

The principles of ^{13}C NMR spectroscopy are the same as those for ^{1}H NMR spectroscopy. When placed in an external magnetic field, ^{13}C nuclei adopt one of two spin states, whose energy difference is determined by the strength of the field (Fig. 8.18). Exposing such nuclei to electromagnetic radiation that has the appropriate energy produces the resonance condition (Eq. 8.6). This means that the spin flip associated with transition from one energy level to another (Fig. 8.18) may occur.

The ^{13}C nuclei exist in different electronic environments in the molecule, as do hydrogen nuclei, and this gives rise to *chemical shifts* that are characteristic of the magnetic environments of the various types of carbon atoms. As in ^{1}H NMR spectroscopy, tetramethylsilane, $(CH_3)_4Si$, is used as the reference compound for measuring chemical shifts in ^{13}C NMR spectra. Most ^{13}C resonances are *downfield* from that of TMS and are given *positive* values by convention. These chemical shifts are computed according to Equation 8.8 and are reported as parts per million (ppm) on a δ scale (Fig. 8.44), in analogy to the procedure used with ^{1}H NMR shifts.

FIGURE 8.44 ^{13}C NMR spectrum of 2-butanone (75 MHz, CDCl$_3$).

Modern instruments are capable of measuring both ^{1}H and ^{13}C NMR spectra. When used for the latter purpose, the spectrometer is usually operated in a mode that permits decoupling of all of the proton(s) that are *nearest neighbors* to the ^{13}C atom, a process termed **broadband proton decoupling**. Applying this technique provides ^{13}C NMR spectra that do *not* show the effects of the spin-spin splitting observed in ^{1}H NMR spectra. Moreover, possible coupling between adjacent ^{13}C atoms is of no concern: Because of the low natural abundance of this isotope in organic compounds, the probability of having two such atoms bound to one another is only *one* in *ten thousand*. Proton-decoupled ^{13}C NMR spectra are thus extremely simple, as they consist of a single sharp resonance line for each magnetically distinct carbon atom.

This is seen in the spectrum of 2-butanone (Fig. 8.44), in which the four different types of carbon atoms produce four separate resonances. Note that three peaks in the spectrum are associated with the solvent, deuterochloroform, CDCl$_3$. The single carbon atom of this solvent produces more than one peak because of *deuterium*-carbon splitting, which is *not* eliminated by the broadband decoupling technique. Deuterium has a nuclear spin, I_z, of 1, so applying Equation 8.9 predicts a *triplet* for the carbon atom, as is observed experimentally.

In contrast to decoupled spectra, *coupled* ^{13}C NMR spectra may be extremely complex and difficult to interpret because the magnitudes of ^{1}H–^{13}C coupling constants are large, on the order of 120–230 Hz, so overlapping of peaks may become a problem. However, special instrumental techniques provide the type of information that would normally be derived by analysis of splitting patterns, namely the number of hydrogen atoms bound to the carbon atom of interest. For example, **off-resonance decoupling** simplifies the spectrum as is illustrated in Figure 8.45, where the multiplets for each carbon atom correspond to the number of peaks predicted by the same $n + 1$ rule (Eq. 8.10) that is used to predict spin-spin splitting patterns in ^{1}H NMR spectra. The difference is that n corresponds to the number of hydrogen nuclei that are *directly bound* to the carbon atom rather than the number of *nearest neighbors* as defined for predicting the splitting patterns for proton spectra. Even *off-resonance decoupled* spectra may be complicated, so only the *chemical shifts* derived from *broadband decoupled* spectra are given for the starting materials and products in this textbook.

The chemical shifts of carbons atoms are *much more sensitive* to their molecular environments than are those of protons. Consequently, the range of chemical shifts for

FIGURE 8.45 Proton off-resonance decoupled ^{13}C NMR spectrum of 2-butanone (75 MHz, CDCl$_3$).

^{13}C resonances is much greater than for ^{1}H. Specifically, whereas most types of hydrogens resonate within 10 ppm downfield of TMS, the chemical shifts of carbon atoms occur over a range of some 220 ppm downfield of TMS. This means that it is unlikely that two different types of carbon atoms will resonate at exactly the same chemical shift. Consequently, the number of peaks in a proton-decoupled ^{13}C NMR spectrum may *tentatively* be interpreted as being equal to the number of connectively different types of carbon atoms present in the molecule. This principle is illustrated both by the observation of four peaks in the ^{13}C NMR spectrum of 2-butanone (Fig. 8.45) and of six peaks in that of methyl benzoate (Fig. 8.46). In the latter case, the carbon atoms *ortho* and *meta* to the ester function are magnetically equivalent, so there are only four resonances for the carbon atoms of the aromatic ring.

As with ^{1}H NMR spectroscopy, tables of chemical shifts for carbon atoms in different environments have been developed; one example is provided by Table 8.5. Examination of the data in the table shows that some of the structural features in a molecule that produce downfield chemical shifts in ^{1}H NMR spectra do the same in ^{13}C NMR spectra. For example, an electronegative substituent such as a carbonyl group or

FIGURE 8.46 ^{13}C NMR spectrum of methyl benzoate (75 MHz, CDCl$_3$).

Table 8.5	^{13}C Chemical Shifts in Carbon-13 Magnetic Resonance (^{13}C NMR) Spectroscopy
Absorbing Carbon Atom (shown as **C** or **Ar**)	Approximate δ, ppm
RCH_2CH_3	13–16
RCH_2CH_3	16–25
R_3CH	25–38
$CH_3\overset{\displaystyle O}{\overset{\|}{C}}-R$	30–32
$CH_3\overset{\displaystyle O}{\overset{\|}{C}}-OR$	20–22
RCH_2-Cl	40–45
RCH_2-Br	28–35
RCH_2-NH_2	37–45
RCH_2-OH	50–65
$RC{\equiv}CH$	67–70
$RC{\equiv}CH$	74–85
$RCH{=}CH_2$	115–120
$RCH{=}CH_2$	125–140
$RC{\equiv}N$	118–125
ArH	125–150
$R\overset{\displaystyle O}{\overset{\|}{C}}-OR'$	170–175
$R\overset{\displaystyle O}{\overset{\|}{C}}-OH$	175–185
$R\overset{\displaystyle O}{\overset{\|}{C}}-H$	190–200
$R\overset{\displaystyle O}{\overset{\|}{C}}-CH_3$	205–210

heteroatom on a carbon atom causes a downfield shift in the resonance, relative to a saturated hydrocarbon, just as it would with hydrogen nuclei. This effect results from the *deshielding* effect of the electronegative moiety. A second factor in defining chemical shifts in ^{13}C NMR spectra is the hybridization of the carbon atom, as is seen by comparing the range of chemical shifts for the sp^3- (10–65 ppm), sp^2- (115–210 ppm), and sp-hybridized (65–85 ppm) carbon atoms of the general structures shown in the table.

Finally, the nature of a substituent G that is zero (α-effect), one (β-effect), or two (γ-effect) atoms away from the carbon atom of interest, as shown in **10**, may also affect the chemical shift. Such substituent effects are remarkably additive in nature, so it becomes possible to predict with reasonable accuracy the expected position of the chemical shifts for carbon atoms in a molecule whose spectrum has not been reported. Table 8.6 is a compilation of some of these additivity effects for acyclic alkanes. These may be used in conjunction with the data in Table 8.7 to predict chemical shifts for a compound. Similar tables of additivity effects for other classes of organic compounds are published in the references at the end of this section.

$$\gamma \quad \alpha$$
$$\beta \quad G$$
10

Table 8.6	Incremental Substituent Effects (ppm) on Replacement of H by Y in Alkanes. Y is Terminal or Internal* (+ downfield, − upfield)

Terminal   Internal

| Y | α | | β | | γ |
	Terminal	Internal	Terminal	Internal	
CH_3	+ 9	+ 6	+10	+ 8	−2
$CH{=}CH_2$	+20		+ 6		−0.5
$C{\equiv}CH$	+ 4.5		+ 5.5		−3.5
COOH	+21	+16	+ 3	+ 2	−2
COOR	+20	+17	+ 3	+ 2	−2
COR	+30	+24	+ 1	+ 1	−2
CHO	+31		0		−2
Phenyl	+23	+17	+ 9	+ 7	−2
OH	+48	+41	+10	+ 8	−5
OR	+58	+51	+ 8	+ 5	−4
OCOR	+51	+45	+ 6	+ 5	−3
NH_2	+29	+24	+11	+10	−5
NHR	+37	+31	+ 8	+ 6	−4
NR_2	+42		+ 6		−3
CN	+ 4	+ 1	+ 3	+ 3	−3
F	+68	+63	+ 9	+ 6	−4
Cl	+31	+32	+11	+10	−4
Br	+20	+25	+11	+10	−3
I	− 6	+ 4	+11	+12	−1

* Add these increments to the shift values of the appropriate carbon atom in Table 8.7.
Source: Reference 1.

The following example illustrates the method for making these predictions. Consider 1-chlorobutane (**11**), $CH_3CH_2CH_2CH_2Cl$. The computed chemical shifts, δ, are determined from the observed chemical shifts (Table 8.7) of 13.4 and 25.2 ppm for C-1 (C-4) and C-2 (C-3), respectively, for butane as follows:

C-1: 13.4 + 31 = 44.4 ppm   C-2: 25.2 + 11 = 36.2 ppm

C-3: 25.2 − 4 = 21.2 ppm   C-4: 13.4 + 0 = 13.4 ppm

The observed chemical shifts for **11** are provided below the structure. The agreement is remarkable and shows the impressive predictive power of this approach.

$$CH_3{-}CH_2{-}CH_2{-}CH_2{-}Cl$$
13.4   20.4   35.0   44.6
11
1-Chlorobutane

In the spectrum of 2-butanone presented earlier (Fig. 8.44), the four different resonances of the sample appear as singlets, owing to broadband decoupling of the

Table 8.7		^{13}C Shifts for Some Linear and Branched-Chain Alkanes (ppm from TMS)			
Compound	C-1	C-2	C-3	C-4	C-5
Methane	−2.3				
Ethane	5.7				
Propane	15.8	16.3	15.8		
Butane	13.4	25.2	25.2		
Pentane	13.9	22.8	34.7	22.8	13.9
Hexane	14.1	23.1	32.2	32.2	23.1
Heptane	14.1	23.2	32.6	29.7	32.6
Octane	14.2	23.2	32.6	29.9	29.9
Nonane	14.2	23.3	32.6	30.0	30.3
Decane	14.2	23.2	32.6	31.1	30.5
2-Methylpropane	24.5	25.4			
2-Methylbutane	22.2	31.1	32.0	11.7	
3-Methylpentane	11.5	29.5	36.9	18.8 (3−CH$_3$)	

Source: Reference 1.

protons. With the exception of the carbonyl carbon atom itself, which appears far down-field at δ 208.3, the carbon atom responsible for each resonance may be assigned either (a) by recognizing that the carbonyl group causes the carbon atoms to which it is directly attached (α-effect) to be shifted downfield farther than those atoms farther removed from it or (b) by using the data of Tables 8.6 and 8.7 to compute the expected chemical shifts. These computations are provided below. Again, comparing with the experimentally determined shifts listed in Figure 8.44 shows the value of the method of substituent additivity shown on the spectrum.

C-1: −2.3 + 30 = 27.7 ppm   C-2: cannot be computed from data in Tables 8.6 and 8.7

C-3:   5.7 + 31 = 36.7 ppm   C-4: 5.7 + 1 = 6.7 ppm

It is apparent in the ^{13}C NMR spectrum of 2-butanone that the relative intensities of the various peaks are *not* identical, even though a single carbon atom is responsible for each resonance. The physical basis for this is beyond the scope of this discussion, but there is *not* a 1:1 correlation between the number of carbon atoms producing a particular resonance and the intensity of that resonance. However there is a *rough* correlation between intensities as reflected in peak heights, and whether or not any hydrogen atoms are on the carbon atom resonating at a particular chemical shift: As a general rule, if there are *no* hydrogen atoms attached to the carbon atom, the intensity of the resonance associated with that atom will be relatively low. This relationship is illustrated in Figure 8.44, although it should be noted that the signal at δ 208.3, which is the resonance for the carbonyl carbon atom, has been electronically amplified by a factor of two relative to all other resonances in the spectrum. In contrast, note that the intensity of the methylene carbon atom is greater than that of either of the carbon atoms of the methyl groups in the molecule, a fact showing that there is *no general* relationship between the number of hydrogen nuclei on a carbon atom and the intensity of the absorption for that atom.

Considering the ^{13}C NMR spectrum (Fig. 8.46) of methyl benzoate (**12**) shows how assigning peaks may be done on the basis of information in Table 8.6 and use of intensities as a measure of whether or not a carbon atom bears any hydrogens. First of all, there are a total of *six* magnetically distinct carbon atoms in the molecule because of its symmetry. According to the data in Table 8.6, the resonance at δ 166.8 must be

Table 8.8	Incremental Shifts of the Aromatic Carbon Atoms of Monosubstituted Benzenes (ppm from Benzene at 128.5 ppm, + downfield, − upfield). Carbon Atom of Substituents (ppm from TMS)				
Substituent	C-1 (Attachment)	C-2	C-3	C-4	C of Substituent (ppm from TMS)
H	0.0	0.0	0.0	0.0	
CH_3	9.3	+ 0.7	−0.1	− 2.9	21.3
CH_2CH_3	+15.6	− 0.5	0.0	− 2.6	29.2 (CH_2), 15.8 (CH_3)
$CH(CH_3)_2$	+20.1	− 2.0	0.0	− 2.5	34.4 (CH), 24.1 (CH_3)
$C(CH_3)_3$	+22.2	− 3.4	−0.4	− 3.1	34.5 (C), 31.4 (CH_3)
$CH=CH_2$	+ 9.1	− 2.4	+0.2	− 0.5	137.1 (CH), 113.3 (CH_2)
$CH\equiv CH$	− 5.8	+ 6.9	+0.1	+ 0.4	84.0 (C), 77.8 (CH)
C_6H_5	+12.1	− 1.8	−0.1	− 1.6	
CH_2OH	+13.3	− 0.8	−0.6	− 0.4	64.5
OH	+26.6	−12.7	+1.6	− 7.3	
OCH_3	+31.4	−14.4	+1.0	− 7.7	54.1
OC_6H_5	+29.0	− 9.4	+1.6	− 5.3	
$\overset{\overset{O}{\|\|}}{OCCH_3}$	+22.4	− 7.1	−0.4	− 3.2	23.9 (CH_3), 169.7 (C=O)
$\overset{\overset{O}{\|\|}}{CH}$	+ 8.2	+ 1.2	+0.6	+ 5.8	192.0
$\overset{\overset{O}{\|\|}}{CCH_3}$	+ 7.8	− 0.4	−0.4	+ 2.8	24.6 (CH_3), 195.7 (C=O)
$\overset{\overset{O}{\|\|}}{CC_6H_5}$	+ 9.1	+ 1.5	−0.2	+ 3.8	196.4 (C=O)
$\overset{\overset{O}{\|\|}}{COH}$	+ 2.9	+ 1.3	+0.4	+ 4.3	168.0
$\overset{\overset{O}{\|\|}}{COCH_3}$	+ 2.0	+ 1.2	−0.1	+ 4.8	51.0 (CH_3), 166.8 (C=O) 168.5
$\overset{\overset{O}{\|\|}}{CCl}$	+ 4.6	+ 2.9	+0.6	+ 7.0	
$C\equiv N$	−16.0	+ 3.6	+0.6	+ 4.3	119.5
NH_2	+19.2	−12.4	+1.3	− 9.5	
$N(CH_3)_2$	+22.4	−15.7	+0.8	−11.8	40.3
$\overset{\overset{O}{\|\|}}{NHCCH_3}$	+11.1	− 9.9	+0.2	− 5.6	
NO_2	+19.6	− 5.3	+0.9	+ 6.0	
F	+35.1	−14.3	+0.9	− 4.5	
Cl	+ 6.4	+ 0.2	+1.0	− 2.0	
Br	− 5.4	+ 3.4	+2.2	− 1.0	
I	−32.2	+ 9.9	+2.6	− 7.3	
SO_2NH_2	+15.3	− 2.9	+0.4	+ 3.3	

Source: Reference 1.

due to the carbonyl carbon atom, C-1′, and, as expected, is of relatively low intensity. Similarly, the table supports assigning the absorption at δ 51.8 to the methoxy carbon atom, C-2′.

The remaining four resonances are in the aromatic region of the ^{13}C NMR spectrum and may be assigned as follows. Based on its low intensity, the peak at δ 130.5 is probably due to C-1 of the aromatic ring. The carbon atoms *ortho* and *para* to an ester group should be *downfield* of those for the *meta* carbon atoms because the ester moiety deshields the *ortho* and *para* positions by delocalizing π-electrons, as illustrated by resonance structures **12a–12c** and by inductively withdrawing electrons from the ring via σ-bonds. This means that the resonance at δ 128.5 is for the two meta carbon atoms. A *tentative* assignment of the remaining two resonances at δ 129.7 and 132.9 in the aromatic region is made on the basis that there are twice as many *ortho* as *para* carbon atoms. Thus the more intense resonance at δ 129.7 is for C-2 and that at δ 132.9 for C-4. These assignments are consistent with those calculated using the substituent additivity effects provided in Table 8.8.

12   **12a**   **12b**   **12c**

Methyl benzoate

EXERCISES

1. Based on the substituent effects shown in Table 8.8, calculate the chemical shifts expected for the various carbon atoms of the aromatic ring of methyl benzoate (**12**).

2. The ^{13}C NMR spectrum of 3-methyl-2-butanone, $CH_3 C (=\!\!\!=O) CH (CH_3)_2$, exhibits peaks at δ 18.2, 27.2, 41.6, and 211.2. Provide assignments of the various resonances to the carbon atoms responsible for them, using Tables 8.5 through 8.7.

3. Compute the ^{13}C NMR chemical shifts expected for 2-methyl-1-propanol, $(CH_3)_2CHCH_2OH$.

REFERENCES

1. Silverstein, R. M.; Bassler, G. C.; Morrill, T. C. *Spectrometric Identification of Organic Compounds*, 5th ed., John Wiley & Sons, New York, **1992**, Chap. 5.
2. Yoder, C. H.; Schaeffer, Jr., C. D. *Introduction to Multinuclear NMR*, The Benjamin/Cummings Publishing Co., Menlo Park, **1987**.
3. Abraham, R. J.; Fisher, J.; Loftus, P. *Introduction to NMR Spectroscopy*, 2nd ed., John Wiley & Sons, New York, **1988**.
4. Whitesell, J. K.; Minton, M. A. *Stereochemical Analysis of Alicyclic Compounds by C-13 NMR Spectroscopy*, Chapman & Hall, London, **1987**.
5. Breitmaier, E.; Voelter, W. *Carbon-13 Nuclear Magnetic Resonance*, 3rd ed., VCH Publishers, New York, **1987**.
6. Wehrli, F. W.; Marchand, A. P.; Wehrli, S. *Interpretation of Carbon-13 NMR Spectra*, 2nd ed., John Wiley & Sons, New York, **1988**.
7. ^{13}*C NMR Spectra*, vol. 1–210, Sadtler Research Laboratories, Philadelphia, PA. Compilation of 42,000 ^{13}C NMR spectra, as of 1996.

PREPARING SAMPLES FOR NMR SPECTROSCOPY

Organic chemists normally obtain NMR spectra of liquid samples contained in special glass tubes of high-precision bore. Although spectra may be obtained of pure liquids having low viscosities, the substrate of interest, regardless of its normal physical state, is generally dissolved in an appropriate solvent. The spectra of viscous samples are unsatisfactory because broad, not sharp, absorptions may be observed, so there is a loss of resolution between the peaks.

All modern NMR spectrometers operate in the pulsed Fourier transform (FT) mode; for technical reasons this means that deuterated solvents must be used. Deuterochloroform, $CDCl_3$, is the most common such solvent although a number of other, albeit generally more expensive, solvents such as acetone-d_6, $(CD_3)_2CO$, dimethyl sulfoxide-d_6, CD_3SOCD_3, and benzene-d_6, C_6D_6, are also available. Deuterium oxide, D_2O, is often used if the sample is water-soluble.

Proton-containing solvents are generally not appropriate because the intense resonance due to the hydrogen atoms of the solvent may obscure absorptions of the sample itself. Although deuterium, ^2H, also possesses the property of nuclear spin, its resonance appears in a region of the NMR spectrum different from that of ^1H. Replacing all of the hydrogen atoms of solvent with deuterium removes such resonances from the ^1H NMR spectrum, but for practical reasons weak absorptions due to residual protium are still usually observed. This is because common organic solvents containing 100 atom % D are too expensive for routine spectral work. Thus, $CDCl_3$ of this purity costs about \$2.00/g, whereas material having 99.6 atom % D is only 22¢/g.

Minor resonances due to solvent must therefore be taken into account when interpreting ^1H NMR spectra. Moreover, in solvents in which there is more than one deuterium atom on a carbon atom, the residual *protium* on that atom will have a splitting pattern that reflects the presence of the deuterium atoms, which have a nuclear spin, I_z, of 1. The resonance of the residual hydrogen present in deuteroacetone thus appears as a *quintet* centered at δ 2.17. This pattern arises from the coupling of a single residual hydrogen with the two deuteriums that are present on the same carbon atom in acetone-d_6 having 99 atom % or greater deuterium content.

Of course, the *carbon* atom(s) of an NMR solvent appear in the ^{13}C NMR spectrum, whether or not the solvent is deuterated. As with residual protium, splitting of the carbon resonance(s) of the solvent by attached deuterium atoms is observed. The appearance of *three* peaks for the carbon atom of $CDCl_3$ illustrates this (Figs. 8.44–8.46). Care should therefore be exercised when interpreting ^{13}C NMR spectra to avoid mistaking peaks due to solvent for those of the sample itself. If overlap of sample and solvent peaks is suspected, it may be necessary to select a different solvent for a ^{13}C NMR spectrum.

The concentration of the solutions appropriate for obtaining NMR spectra is in the range of 5–15% by weight; about 1 mL of solution is needed to fill an NMR tube to the proper level, although as little as 0.6 mL can be used. The tube must be scrupulously clean and dry, and the solution must be free of undissolved solids arising from the sample itself or even dust. Furthermore, as a consequence of contact of the solute or the solvent with metals such as the iron in "tin" cans, trace amounts of ferromagnetic impurities may contaminate the solution; the result is a spectrum that has only broad, weak absorptions.

A solution that contains any solid material must be filtered *prior* to measuring its NMR spectrum. This is done most easily by inserting a small plug of glasswool into a Pasteur pipet and then filtering the solution through the plug into the NMR tube.

Following their use, NMR tubes should be cleaned and dried thoroughly. They are best stored in a closed container or in an inverted position so as to minimize the possibility of particles of dust entering them.

HISTORICAL HIGHLIGHT

Medical Diagnostics via Nuclear Magnetic Resonance Spectroscopy

Sometimes discoveries made as part of doing "pure" or basic scientific research rather than "applied" or goal-oriented studies ultimately evolve into valuable practical applications. Such is the case with nuclear magnetic resonance (NMR) spectroscopy.

In the early 1940s physicists observed that certain nuclei can be made to undergo transitions between two or more energy states, so-called "spin-flips," when placed in an applied magnetic field, the so-called "NMR effect." Discovery of this fundamental property for nuclei prompted considerable experimental and theoretical research of this effect. By the late 1950s, chemists were hard at work making the phenomenon of nuclear spin-flip the basis for the extremely powerful analytical tool that nuclear magnetic resonance spectroscopy is today. In fact, many present-day chemists believe that it is the single most powerful technique for elucidating the structures of unknowns containing nuclei susceptible to the NMR effect.

Throughout the 1960s, the focus of NMR spectroscopy was on analysis of solutions of samples contained in small glass tubes. The samples were then inserted into a probe within the magnet of a spectrometer so that the NMR phenomenon could be observed. By the 1970s, the theory of NMR, the availability of powerful computers that could process the data, and the quality of the magnets had all progressed to the point that an exciting new application of the technique was envisioned. This new dimension of NMR involved the direct observation of NMR-active nuclei in *living* organisms.

All organisms are comprised of a myriad of organic compounds, and the human body is no exception. Among other elements, our bodies contain carbon, nitrogen, hydrogen, and phosphorus in molecules such as nucleic acids, proteins, fats, and carbohydrates, and certain isotopes of each of these elements have nuclear spin. Furthermore, water is a universal component of fluids throughout the body, and its protons are susceptible to the NMR effect. Our bodies thus serve as "containers" for a plethora of compounds containing nuclei that respond to an externally applied magnetic field. This sets the stage for the medical diagnostic procedure termed "Magnetic Resonance Imaging," or MRI, a name that is a surrogate for nuclear magnetic resonance spectroscopy but avoids the negative connotation the average citizen associates with the word "nuclear."

From a practical standpoint, the only NMR-active isotopes in the body present in concentrations high enough for detection in a typical MRI scan are ^{31}P and ^{1}H. As you know, the usual *in vitro* ^{1}H NMR experiment provides chemists with information regarding **chemical shifts, relative numbers** of protons in different magnetic environments, and the **number of nearest neighbors** to a particular type of proton. In contrast, the ^{1}H MRI experiment yields information on **hydrogen density** and **spin-lattice (T_1)** and **spin-spin relaxation times (T_2)**, which define the rate and mode by which a nucleus relaxes to a lower-energy spin state.

The value of MRI for medical diagnostic purposes is seen in a variety of applications. For example, tumors may be detected because the relaxation times T_1 and T_2 for water molecules contained in certain types of tumors is different from that for water in healthy cells. In another application, anomalies in the brain are detectable because white and gray matter respond differently to the MRI experiment; a typical image is shown on page 253. Moreover, certain metabolic irregularities may be diagnosed by monitoring the conversion of adenosine triphosphate (ATP) to adenosine monophosphate (AMP) and of organic phosphate to inorganic phosphate. Yet another application is producing images of the heart, showing that the technique be used for an organ that is constantly in motion and undergoing changes in its shape.

HISTORICAL HIGHLIGHT
Medical Diagnostics via Nuclear Magnetic Resonance Spectroscopy *(continued)*

Many, but not all, of the images of internal structures that are provided by the MRI technique are also available by using x-ray technology. However, the latter method, albeit currently less expensive, is less applicable to soft tissue structures such as the brain and internal organs. X-rays are also a high-energy form of radiation that can damage tissue; MRI does not suffer this disadvantage. Thus, MRI has become an exceedingly valuable tool in the arsenal of diagnostic methods available to medical researchers and practitioners, and there are now over 10,000 MRI devices installed worldwide

The quality of imaging that is possible is illustrated in the two figures below. They are both horizontal ''slices'' of the head of a living person and clearly show the soft tissues of the brain. One of the slices portrays tissue associated with the optic system as well. The images result from portraying a particular NMR property as a gray scale color.

There are a number of websites that provide further information on applications of MRI to biomolecules. URLs for two such sites are http://galaxy.einet.net/editors/doug_morris/mri.html and http://www~wilson.ucsd.edu/education/spectroscopy/nmr.html.

Relationship of Historical Highlight to Chapter

The theory and interpretation of ^1H and ^{13}C NMR spectra are presented in this chapter. The importance of this technique for confirming the structures of both known and unknown organic compounds is established, and NMR spectra of reactants and products are provided throughout the textbook and on the CD-ROM that accompanies the text. This Historical Highlight describes the evolution of a phenomenon that was once primarily of interest to physicists to one that is an increasingly important tool for medical applications.

(a)

(b)

(a) MRI of the human brain, axial view, showing optic nerves. (b) MRI of the human brain, axial view. (Courtesy of Professor M. R. Willcott, University of Texas, Galveston)

CHAPTER

9

Alkanes

Alkanes are saturated hydrocarbons having the general molecular formula C_nH_{2n+2} and are typically unreactive. This characteristic lack of reactivity, even toward strong acids and bases, limits the use of alkanes as practical starting materials for most laboratory syntheses. However, some chemical reactions that occur by **free-radical chain** mechanistic pathways do allow the introduction of functional groups onto an alkane. For example, alkanes are converted into alkyl chlorides or bromides, R–X (X = Cl or Br, respectively) by a free-radical process in which a mixture of the alkane and halogen is heated at 200–400 °C or is irradiated with ultraviolet light (Eq. 9.1). Under these conditions, the σ-bond of molecular chlorine or bromine undergoes **homolytic cleavage** (Eq. 9.2) to generate chlorine and bromine atoms, which are **free radicals**. Generating chlorine and bromine atoms is essential to initiating the reaction between an alkane and molecular chlorine or bromine to form alkyl halides. These may then be transformed into a variety of other functional groups.

$$R\!-\!H + X_2 \xrightarrow{\text{heat or } h\nu} R\!-\!X + H\!-\!X \qquad (9.1)$$
$$X = Cl \text{ or } Br \qquad \text{An alkyl halide}$$
$$\text{(mixture)}$$

$$X\!-\!X \xrightarrow{\text{heat or } h\nu} X^{\bullet} + X^{\bullet} \qquad (9.2)$$

The procedures described in Sections 9.2 and 9.3 illustrate methods for transforming alkanes to alkyl chlorides and bromides by **free-radical substitution** reactions. These experiments give you an opportunity to analyze the mixtures of products that are obtained and to gain insights about the relative reactivities of different types of hydrogen atoms toward chlorine and bromine radicals.

9.2 Chlorination Using Sulfuryl Chloride

Chlorine atoms may be generated from molecular chlorine under mild conditions using a *catalytic* amount of an **initiator**, In-In. Thus, homolysis of a molecule of initiator occurs upon heating or irradiation to give a pair of free radicals, In˙ (Eq. 9.3). These free radicals, In˙, may then react with molecular chlorine to produce In–Cl and a chlorine atom (Eq. 9.4), to initiate the **free-radical chain reaction**. For safety and convenience, sulfuryl chloride, SO_2Cl_2, rather than molecular chlorine is used in this experiment as the source of chlorine radicals.

$$In{-}In \xrightarrow{\text{heat or h}\nu} In^{\bullet} + In^{\bullet} \tag{9.3}$$

$$In^{\bullet} + Cl{-}Cl \longrightarrow In\text{-}Cl + Cl^{\bullet} \tag{9.4}$$

The first step in initiating the free-radical chain reaction involves the homolysis of 2,2′-azobis[cyclohexanenitrile] (**1**), abbreviated as ABCN, to form nitrogen and the free radical **2** (Eq. 9.5); this reaction proceeds at a kinetically acceptable rate at 80–100 °C. The radical **2** then attacks sulfuryl chloride to generate chlorine atoms and SO_2 according to Equations 9.6 and 9.7. The series of reactions depicted in Equations 9.5–9.7 comprise the **initiation** steps of the reaction.

Initiation

$$\tag{9.5}$$

1
ABCN

2

$$\tag{9.6}$$

Sulfuryl chloride

$$\tag{9.7}$$

The next stage of the reaction involves the **propagation** steps. These include abstraction of a hydrogen atom from the hydrocarbon by a chlorine atom to produce a new free radical, R˙ (Eq. 9.8), followed by attack of this radical upon sulfuryl chloride to yield the alkyl chloride and the radical ˙SO_2Cl (Eq. 9.9). The latter fragments into SO_2 and another chlorine atom as the final propagation step (Eq. 9.10), thereby producing a radical chain reaction.

Propagation

$$\text{Cl}^{\bullet} + \text{H}\!-\!\text{R} \longrightarrow \text{R}^{\bullet} + \text{H}\!-\!\text{Cl} \tag{9.8}$$

$$\text{R}^{\bullet} + \text{Cl}\!-\!\overset{\displaystyle O}{\underset{\displaystyle O}{\overset{\|}{\underset{\|}{S}}}}\!-\!\text{Cl} \longrightarrow \text{R}\!-\!\text{Cl} + {}^{\bullet}\overset{\displaystyle O}{\underset{\displaystyle O}{\overset{\|}{\underset{\|}{S}}}}\!-\!\text{Cl} \tag{9.9}$$

$${}^{\bullet}\overset{\displaystyle O}{\underset{\displaystyle O}{\overset{\|}{\underset{\|}{S}}}}\!-\!\text{Cl} \longrightarrow \text{SO}_2 + \text{Cl}^{\bullet} \tag{9.10}$$

One might assume that, once initiated, this reaction continues until either sulfuryl chloride or the alkane, whichever is the **limiting reagent** (Sec. 1.6), is completely consumed. In practice, the chain reaction is interrupted by a series of side reactions known as **termination** reactions (Eqs. 9.11–9.13), and the initiation process must be continued throughout the course of the reaction.

Termination

$$\text{Cl}^{\bullet} + {}^{\bullet}\text{Cl} \longrightarrow \text{Cl}\!-\!\text{Cl} \tag{9.11}$$

$$\text{R}^{\bullet} + {}^{\bullet}\text{Cl} \longrightarrow \text{R}\!-\!\text{Cl} \tag{9.12}$$

$$\text{R}^{\bullet} + {}^{\bullet}\text{R} \longrightarrow \text{R}\!-\!\text{R} \tag{9.13}$$

In summary, the three distinct stages that are involved in a **free-radical chain mechanism** are

1. *Initiation:* Radicals are formed in *low* concentration from neutral molecules resulting in an *increase* of the concentration of free radicals within the system.
2. *Propagation:* Radicals produced in the first step react with molecules to yield new molecules and new radicals; there is *no net change* in radical concentration.
3. *Termination:* Various radicals combine to give molecules resulting in a net *decrease* in radical concentration and a decrease in the rate of the reaction.

Free-radical halogenation of saturated compounds generally yields products of monosubstitution as well as of polysubstitution. For example, chlorination of 1-chlorobutane (**3**), the substrate used in the experimental procedure in this section, gives the *di*chloro compounds **4–7**, which are products of monosubstitution, as well as *tri*chloro compounds (Eq. 9.14). As a general rule, polysubstitution is minimized by using an excess of the substrate undergoing halogenation relative to the amount of halogenating agent. In our procedure, the molar ratio of substrate to chlorinating agent is more than two. This strategy has a logical basis: Statistically speaking, a chlorine atom is more likely to collide with a molecule of **3** than with one of the product dichlorobutanes **4–7**.

$$CH_3CH_2CH_2CH_2Cl \xrightarrow[\text{SO}_2\text{Cl}_2/\text{heat}]{\text{ABCN}} ClCH_2(CH_2)_2CH_2Cl + CH_3\overset{\overset{\displaystyle Cl}{|}}{C}HCH_2CH_2Cl$$

3   **4**   **5**

1-Chlorobutane   1,4-Dichlorobutane   1,3-Dichlorobutane

$$\text{(9.14)}$$

$$+ CH_3CH_2\overset{\overset{\displaystyle Cl}{|}}{C}HCH_2Cl + CH_3(CH_2)_2CHCl_2$$

6   **7**

1,2-Dichlorobutane   1,1-Dichlorobutane

The mechanistic step determining which product will ultimately be formed in a free-radical halogenation reaction is the one in which a hydrogen atom is abstracted (Eq. 9.8). In the case of 1-chlorobutane, the ratio of the different isomers **4–7** is defined by a **statistical factor** and an **energy factor**. The statistical factor is determined by the number of hydrogen atoms whose replacement will give a specific *constitutional isomer*, whereas the energy factor is related to the strength of the type of C—H bond being broken and is called the **relative reactivity**. In hydrocarbons such as butane, $CH_3(CH_2)_2CH_3$, the relative reactivity of **secondary** (2°) and **primary** (1°) hydrogen atoms (Fig. 9.5, Sec. 9.3) toward chlorine atoms is 3.3 : 1.0 reflecting the fact that a secondary C–H bond is *weaker* than a primary one. However, these values do *not* apply to the corresponding types of hydrogens in **3** because of the presence of the chloro substituent. The electron-withdrawing effect of this substituent is known to *increase* the strength of the C–H bonds nearby it, thereby *lowering* their relative reactivity. By analyzing the data from this experiment, you should be able to confirm this fact and to derive corrected values of the relative reactivities.

Using sulfuryl chloride for chlorination results in formation of equimolar amounts of sulfur dioxide and hydrochloric acid (Eq. 9.15), both of which are lost as gases from the reaction mixture. The resulting loss of weight provides a convenient means for monitoring the progress of the chlorination in that the reaction is complete when the theoretical loss of weight has occurred. This can be calculated on the basis that sulfuryl chloride is the *limiting reagent* in the reaction.

$$R\text{—}H + SO_2Cl_2 \xrightarrow{\text{Initiator}} R\text{—}Cl + SO_2 + HCl$$

$$\text{(9.15)}$$

EXPERIMENTAL PROCEDURES

FREE-RADICAL CHAIN CHLORINATION OF 1-CHLOROBUTANE

Purpose: To perform a free-radical chain reaction and determine relative reactivities of C–H bonds toward chlorine atoms.

Safety Alert

1. Sulfuryl chloride is a potent **lachrymator** and reacts violently with water to generate sulfuric acid and hydrochloric acid. *Exercise great care* when handling this chemical, *making sure that your apparatus*

is dry and that you do *not* either allow it to contact your skin or inhale its vapors. Measure out this chemical in a hood, and wear latex gloves when transferring it. Vessels containing sulfuryl chloride should be kept stoppered whenever possible!

2. Be certain that all connections in your apparatus are tight prior to heating the reaction mixture and that your gas trap is functioning to prevent venting of noxious gases into the laboratory.

3. When the reaction mixture is washed with aqueous sodium carbonate, carbon dioxide is generated in the separatory funnel. To relieve any gas pressure that develops, be sure to vent the funnel frequently when shaking it.

Miniscale Procedure

Preparation   Answer the Pre-Lab Exercises on page PL. 45. Review Sections 2.11, 2.14, 2.21, 2.22, 2.23, 2.25, 2.28, and 6.4.

Apparatus   A *dry* 25-mL round-bottom flask, gas trap, separatory funnel, apparatus for magnetic stirring, heating under reflux, and *flameless* heating.

Setting Up   Assemble the apparatus for heating under reflux, and attach a gas trap such as one shown in Figure 2.65, to the top of the condenser. Equip the round-bottom flask with a stirbar and add 0.1 g of 2,2′-azobis(cyclohexanenitrile) (ABCN). Working *at the hood*, sequentially measure 5 mL of 1-chlorobutane, and 2 mL of sulfuryl chloride into the flask. Stopper and weigh the flask, and then attach it to the reflux condenser, working quickly to minimize the introduction of the vapors of sulfuryl chloride into the atmosphere.

Reaction   With stirring, heat the solution under gentle reflux for 20 min and then allow the reaction mixture to cool below the reflux temperature. Add a second 0.1-g portion of ABCN and heat the mixture under reflux for another 10 min. Weigh the flask and consult with your instructor if the resulting mixture has not lost at least 90% of the expected weight after this second period of reflux.

Work-Up and Isolation   Cool the reaction mixture in an ice-water bath and, working at the hood, *cautiously* pour it with stirring into 15 mL of ice-cold saturated aqueous sodium chloride (brine) contained in an Erlenmeyer flask. Transfer the resulting two-phase solution to a separatory funnel and separate the layers.★ Wash the organic layer in the funnel with 10 mL of 0.5 *M* sodium carbonate solution, venting the funnel *frequently*. Using pHydrion paper, determine whether the aqueous layer (Sec. 2.21) is basic. If it is not, repeat washing the organic layer with 10-mL portions of sodium carbonate solution until the aqueous washes are *basic*. After washing the reaction mixture with 15 mL of brine, transfer the organic layer to an Erlenmeyer flask containing several spatula-tips full of anhydrous sodium sulfate.★ Swirl the flask occasionally during 10–15 min to hasten drying, which is indicated by the liquid becoming clear. If the liquid remains cloudy, add additional portions of sodium sulfate with swirling. Carefully decant the organic layer into a dry, *tared* container.

Analysis   Weigh the reaction mixture and use the result to calculate your percentage yield of material recovered. This calculation requires that you first calculate the theo-

FIGURE 9.1
GLC analysis of mixture from chlorination of 1-chlorobutane (numbers in parentheses are boiling points at 760 Torr and peak areas, respectively). Peak 1: solvent; peak 2: 1-chlorobutane (77–78 °C, 2719); peak 3: 1,1-dichlorobutane (114–115 °C, 487); peak 4: 1,2-dichlorobutane (121–123 °C, 1771); peak 5: 1,3-dichlorobutane (131–133 °C, 3367); peak 6: 1,4-dichlorobutane (161–163 °C, 1137). Column and conditions: 0.5-mm × 30-m, 0.25 μ film of SE-54; 80 °C, 40 mL/min.

retical weight of product dichlorides *and* unreacted 1-chlorobutane expected from the reaction.

Analyze the reaction mixture by gas chromatography or submit samples to your instructor for such analysis. The peak corresponding to 1-chlorobutane should be ''off-scale'' so that the peaks for the isomeric dichlorobutanes are sufficiently large for an accurate analysis of their areas. Assign the four product peaks to the isomers responsible for them and calculate the relative amounts of each isomer present in the mixture. A typical GLC trace of the reaction mixture is depicted in Figure 9.1.

Obtain IR and ^{1}H NMR spectra of your starting material and an NMR spectrum of your product mixture; compare them with those of authentic samples (Figs. 9.2–9.4).

Microscale Procedure

Preparation Answer the Pre-Lab Exercises on page PL. 47. Review Sections 2.4, 2.11, 2.14, 2.21, 2.22, 2.23, 2.25, 2.28, and 6.4.

Apparatus *Dry* 3-mL screw-cap conical vial, gas trap, 5-mL conical vial or screw-cap centrifuge tube, apparatus for magnetic stirring, heating under reflux, and *flameless* heating.

Assembly Assemble the apparatus for heating under reflux, and attach a gas trap, such as one shown in Figuare 2.66, to the top of the condenser. Place a spinvane in the vial and add 10 mg (0.010 g) of 1, 1'-azobis(cyclohexanecarbonitrile) (ABCN). Working *at the hood*, sequentially add 0.5 mL of 1-chlorobutane and 0.2 mL (200 μL) of sulfuryl chloride to the vial. Cap and weigh the vial, and then attach it to the reflux condenser, working quickly to minimize the introduction of the vapors of sulfuryl chloride into the atmosphere.

Reaction Heat the solution under gentle reflux for about 20 min, and then allow the reaction mixture to cool below the reflux temperature. Add a second 0.010-g portion of ABCN and continue heating under reflux for an additional 10 min. Weigh the flask, and consult with your instructor if the resulting mixture has not lost at least 90% of the expected weight after this second period of reflux.

Work-Up and Isolation   Remove the vial from the apparatus and cap it.★ Working at the hood, *cautiously* and with stirring, pour the reaction mixture into 1 mL of ice-cold saturated aqueous sodium chloride (brine) contained in the 5-mL conical vial or centrifuge tube. Allow the layers to separate and stir or shake the mixture briefly. Remove the lower layer with a Pasteur pipet and determine whether it is the organic or aqueous layer (Sec. 2.21).★ Do *not* discard *either* layer until you know which one is the organic layer!

Carefully add 1 mL of aqueous 0.5 *M* sodium carbonate to the organic layer (*Caution:* foaming), shake the mixture with venting as necessary, and remove the lower layer. Using pHydrion paper, determine whether the aqueous layer is basic. If it is not, repeat washing the organic layer with 1-mL portions of sodium carbonate solution until the aqueous washes are *basic*. After shaking the organic layer with 1 mL of brine, carefully transfer the organic layer to a dry vial, and add one or two microspatula-tips full of anhydrous sodium sulfate to it. Stir the resulting mixture occasionally during 10–15 min to hasten drying, which is indicated by the liquid becoming clear. If the liquid remains cloudy, add additional portions of sodium sulfate with stirring. Carefully transfer the dried mixture to a dry, *tared* sample vial using a Pasteur filter-tip pipet.

Analysis   Weigh the reaction mixture and use the result to calculate your percentage yield of material recovered. This calculation requires that you first calculate the theoretical weight of product dichlorides *and* unreacted 1-chlorobutane expected from the reaction.

Analyze the reaction mixture by gas chromatography or submit samples to your instructor for such analysis. The peak corresponding to 1-chlorobutane should be ''off-scale'' so that the peaks for the isomeric dichlorobutanes are sufficiently large for an accurate analysis of their areas. Assign the four product peaks to the isomers responsible for them and calculate the relative amounts of each isomer present in the mixture. A typical GLC trace of the reaction mixture is depicted in Figure 9.1.

Obtain IR and NMR spectra of your starting material and an NMR spectrum of your product mixture; compare them with those of authentic samples (Figs. 9.2–9.4).

Finishing Touches

Combine *all of the aqueous solutions*. If the resulting solution is acidic, neutralize it carefully with solid sodium carbonate and flush it down the drain with a large excess of water. Rinse the *glasswool* from the gas trap with aqueous 0.5 *M* sodium carbonate and discard it in the trash. Put the *sodium sulfate* in a container for solids contaminated with alkyl halides.

EXERCISES

1. Define and provide a specific example of each of the following terms.

    a. propagation step   **d.** limiting reagent

    b. termination step   **e.** secondary hydrogen

    c. bond homolysis

2. Why is the amount of sulfuryl chloride used *less* than the amount theoretically required to convert all the starting material to monosubstituted products?

3. Why is only a catalytic amount of ABCN (**1**) needed in this reaction?

4. Write at least one product that might be derived from ABCN (**1**) and explain why the potential contamination of the reaction mixture by such a product is of little concern?

5. Why is it prudent to use caution when pouring the crude reaction mixture into brine? Why should the reaction mixture and brine be *ice-cold* for this step?

6. How might a greater than theoretical loss of weight occur in a chlorination reaction performed according to the procedure of this section?

7. Using a relative reactivity for 1° to 2° hydrogen atoms of 1:3.3,

    a. calculate the percentage of each chlorobutane expected from monochlorination of 1-chlorobutane and

    b. determine the ratio of 1° to 2° dichlorobutanes to be expected.

8. a. Using your own data or those provided in Figure 9.1, calculate the ratio of 1° to 2° dichlorobutanes obtained from monochlorination of 1-chlorobutane.

    b. Why does this ratio *not* agree with that found in Exercise 7?

9. Based on your calculation of the percentage of the various dichlorobutanes formed in the reaction, determine the values for the relative reactivity for the various types of hydrogen atoms in 1-chlorobutane. Assign the relative reactivity of the methyl hydrogens as 1.0.

10. Specify the factors determining the proportion of monochlorinated products derived from 1-chlorobutane.

11. Indicate from which of the three compounds shown below it should be easiest to prepare a *single* pure monochlorinated product and explain why.

$$\text{CH}_3 \qquad\qquad \text{CH}_3-\overset{\displaystyle \text{CH}_3}{\underset{\displaystyle \text{CH}_3}{\text{C}}}-\text{CD}_3$$

Refer to Tables 8.1–8.8 as needed for answering the following questions on spectroscopy.

12. Consider the NMR data for 1-chlorobutane (Fig. 9.3).

    a. In the ^{1}H NMR spectrum, assign the various resonances to the hydrogen nuclei responsible for them.

    b. For the ^{13}C NMR data, assign the various resonances to the carbon nuclei responsible for them.

13. Figure 9.4 is the ^{1}H NMR spectrum of a mixture of 1,x-dichlorobutanes obtained by free-radical chlorination of 1-chlorobutane.

    a. Assign the triplets centered near δ 1.0 and 1.1 and the doublet centered near δ 1.6 to the hydrogen nuclei of the three different isomers from which they arise. (*Hint:* Write the structures of the isomeric 1,x-dichlorobutanes and predict the multiplicity and approximate chemical shift of each group of hydrogens.)

    b. The integrated areas of the selected multiplets in the spectrum are as follows: δ 1.0, 2 units, δ 1.1, 9 units, δ 1.6, 32 units, δ 3.6, 26.8 units, δ 3.7, 26.8 units, δ 4.0, 3.3 units, δ 5.8 (a triplet *not* visible in the spectrum), 0.7 units. Based on your assignments in part (**a**) and on the integration data, calculate the approximate percentage of each isomer present in the mixture. Explain your determination as completely as possible.

14. Compare the ^{1}H NMR spectrum of the mixture of 1,*x*-dichlorobutanes (Fig. 9.4) with that of 1-chlorobutane (Fig. 9.3). What features in Figure 9.4 are consistent with the incorporation of a second chlorine atom in this experiment and with the proposal that a *mixture* of dichlorobutanes, rather than a single isomer, is formed?

Spectra   *Starting Material and Products*

FIGURE 9.2
IR spectrum of 1-chlorobutane (neat).

FIGURE 9.3
NMR data for 1-chlorobutane (CDCl$_3$).

(a) ^{1}H NMR spectrum (300 MHz);

(b) ^{13}C NMR data: δ 13.4, 20.3, 35.0, 44.6.

FIGURE 9.4
^{1}H NMR spectrum of mixture of 1,x-dichlorobutanes (CDCl$_3$, 300 MHz).

9.3 Bromination: Selectivity of Hydrogen Atom Abstraction

Hydrocarbons undergo free-radical chain bromination (Eq. 9.16) through a mechanism analogous to that for free-radical chlorination (Sec. 9.2). The **initiation** step involves the thermal or photochemical **homolysis** of molecular bromine to form bromine atoms, Br· (Eq. 9.17). **Propagation** steps then produce the **substitution** product, R–Br, and regenerate Br· to continue the chain reaction (Eqs. 9.18 and 9.19). The **termination** steps are analogous to those of Equations 9.11–9.13 (Sec. 9.2).

$$\text{R—H} + \text{Br}_2 \xrightarrow{h\nu} \text{R—Br} + \text{H—Br} \qquad (9.16)$$

Colorless Reddish   Colorless Colorless
             orange

$$\text{Br—Br} \xrightarrow{\text{heat or } h\nu} \text{Br}^{\bullet} + \text{Br}^{\bullet} \qquad (9.17)$$

$$\text{Br}^{\bullet} + \text{H—R} \longrightarrow \text{R}^{\bullet} + \text{HBr} \qquad (9.18)$$

$$\text{Br}^{\bullet} + \text{Br—Br} \longrightarrow \text{R—Br} + \text{Br}^{\bullet} \qquad (9.19)$$

As with chlorination, the mechanistic step determining which product will be formed in a free-radical bromination reaction is the one in which a hydrogen atom is abstracted by a bromine atom (Eq. 9.18). The likelihood that a particular type of hydrogen atom will be successfully abstracted is defined by a **statistical factor** and an **energy factor**. The statistical factor is determined by the number of hydrogen atoms whose replacement will give a specific *constitutional isomer*, whereas the energy factor is related to the strength of the type of C–H bond being broken and is called the **relative reactivity**. In the experiment of this section, you will determine the relative rates of the free-radical chain reaction of molecular bromine with several hydrocarbons containing different types of hydrogen atoms. You should be able to deduce the order of relative reactivity of *seven* different types of hydrogens toward Br· by carefully considering the results of your observations.

FIGURE 9.5   Classification of hydrogen atoms in hydrocarbons.

As summarized by the structures in Figure 9.5, hydrogen atoms in hydrocarbons are classified according to the nature of the carbon atom to which they are attached. Specifically, hydrogens bound to sp^3-hybridized carbons are categorized as **aliphatic**; all hydrogen atoms in the hydrocarbon family of alkanes belong to this group. Aliphatic hydrogens may be further classified as being **primary (1°)**, **secondary (2°)**, or **tertiary (3°)** depending upon the number of other carbon atoms attached to the reference carbon atom; primary carbons are bonded to one other carbon, secondary carbons to two, and tertiary carbons to three. An aliphatic hydrogen atom attached to an sp^3-hybridized carbon atom that in turn is bound to a vinylic or aromatic carbon is called **allylic** or **benzylic**, respectively. These various classifications can be combined, so that one refers to primary, secondary, or tertiary aliphatic, allylic, or benzylic hydrogens. Hydrogens attached to sp^2-hybridized carbons are classified as either **vinylic** or **aromatic**, depending upon whether the carbon atom is part of a normal double bond or an aromatic ring. Hydrogens are called **acetylenic** when bonded to an sp-hybridized carbon atom.

The hydrocarbons **8–13**, which contain various types of hydrogen atoms, are the substrates for our experiment, and measuring their relative reactivities toward Br· in free-radical chain bromination is rather straightforward for two reasons. First, molecular bromine is colored, so the progress of bromination is easily monitored by observing the disappearance of color as the bromine is consumed (Eq. 9.16). Second, because the relative rates of hydrogen abstraction by Br· are highly sensitive to the strength of the C–H bond being broken, the substrates have rates of free-radical chain bromination that differ enough to allow ready determination of the qualitative order of their reactivity.

| **8** | **9** | **10** | **11** |
| Toluene | Ethylbenzene | Isopropylbenzene | *tert*-Butylbenzene |

CH₃

12
Cyclohexane

13
Methylcyclohexane

Experimentally, an *excess* of a hydrocarbon is allowed to react with molecular bromine, and the time required for disappearance of the color of the bromine is measured. The reactions are carried out in dichloromethane solution using two different conditions: (1) room temperature with no illumination; and (2) room temperature with illumination. The period of time required for the decoloration depends upon the reaction conditions and the relative reactivities of the various hydrocarbons; some of the substrates may not react at all under the specified conditions. The ratios of the different reaction times represent the relative rates of the free-radical chain bromination reaction.

EXPERIMENTAL PROCEDURE

RELATIVE RATES OF FREE-RADICAL CHAIN BROMINATION

Purpose: To determine the relative reactivities of different types of hydrogens toward bromine atoms.

1. *Bromine is a hazardous chemical.* Do *not* breathe its vapors or allow it to come into contact with the skin because it may cause serious chemical burns. Perform all operations involving the transfer of the pure liquid or its solutions at a hood, and wear latex gloves. If you get bromine on your skin, wash the area immediately with soap and warm water and soak the affected area in 0.6 *M* sodium thiosulfate solution for up to 3 h if the burn is particularly serious. 2. The 1 *M* bromine in dichloromethane solution should be kept in the hood. Dispense 0.5-mL portions of it, using a pipet pump, calibrated Pasteur filter-tip pipet, or a buret fitted with a Teflon stopcock. 3. Bromine reacts with acetone to produce the powerful *lachrymator*, α-bromoacetone, BrCH₂COCH₃. Do *not* rinse glassware containing residual bromine with acetone! 4. Avoid inhalation of the vapors of any of the materials being used in this experiment.	**Safety Alert**

Procedure

Preparation   Answer the Pre-Lab Exercises on page PL. 49. Review Sections 2.5 and 2.10.

Construct a table in your notebook with the following four main headings: (1) "Hydrocarbon"; (2) "Types of Hydrogen," entries under which will include the terms, 1° aliphatic, 2° benzylic, and so on; (3) "Conditions," with the subheadings, RT, RT/$h\nu$; and (4) "Elapsed Time," entries under which will be the time required for reaction as measured by decoloration.

Plan carefully for executing this experiment because proper labeling of test tubes and managing of time once the reactions have been started are critical. Two separate experimental trials are to be performed, with one trial starting 5–10 min after the other.

Apparatus   Fourteen 13-mm × 100-mm test tubes and corks, six 10-mm × 75-mm test tubes, six Pasteur pipets, filter-tip pipet, ice-water bath, light source.

Setting Up   Organize *twelve* 13-mm x 100-mm test tubes into pairs and label each tube in the pair with the name or number of the hydrocarbon **8–13** that you will add to it. Measure 2.5 mL of dichloromethane into each of these tubes and then add 0.5 mL or 10 drops of the hydrocarbon for which it is labeled. Label the *two* remaining 13-mm × 100-mm test tubes "Control" and add 3 mL of dichloromethane to them. You now should have *seven* pairs of labeled tubes; place them in a beaker. Working at the hood, dispense 0.5 mL of a 1 *M* solution of bromine in dichloromethane into each of the *six* small test tubes and into one of the larger tubes labeled "Control."

Bromination   In rapid succession, add the solution of bromine to each of the hydro-carbon-containing tubes. Do this with agitation to ensure good mixing and record the time of the additions. After all the additions are made, loosely cork each tube to prevent loss of solvent and bromine. Monitor all the tubes, including that labeled "Control," and record the elapsed time required for the discharge, if any, of the color of each solution.

After about 5 min, again dispense 0.5 mL of 1 *M* bromine solution into the six small test tubes used originally for this reagent *and* also into the remaining tube labeled "Control." Working as before, add the solution of bromine to each of the hydrocarbon-containing tubes. Place the six tubes containing the resulting solutions and the tube labeled "Control" in a beaker located 14–16 cm above an unfrosted 100- or 150-watt light bulb so that all the solutions are exposed to the same amount of light. Monitor these seven tubes and record the elapsed time required for the discharge, if any, of color.

Terminate your observations of both sets of samples after about 1 h.

Analysis   Evaluate the intensity of color remaining in any of the tubes relative to that of the "Control" samples and record the result in your notebook. Obtain IR and ^{1}H NMR spectra of your starting materials and compare them with those of authentic samples (Figs. 9.6–9.17).

Finishing Touches

By the dropwise addition of cyclohexene, decolorize any *solutions* in which the color of bromine is visible; then discard the resulting solutions together with *all other solutions* in a container for halogenated organic liquids.

EXERCISES

1. Arrange the six hydrocarbons **8–13** in *increasing* order of reactivity in free-radical chain bromination.

2. **a.** On the basis of the observed order of reactivity of the hydrocarbons, deduce the order of reactivity of the seven different types of hydrogens found in these compounds, that is, (1) primary aliphatic, (2) secondary aliphatic, (3) tertiary aliphatic, (4) primary benzylic, (5) secondary benzylic, (6) tertiary benzylic, and (7) aromatic.

   b. Outline the logic you used to arrive at your sequence in part **a**.

3. Draw the structure of the major monobrominated product expected from each of the hydrocarbons used in this experiment.

4. Comment on the need for light in order for bromination to occur with some of the hydrocarbons.

5. Why is a "Control" sample needed in this experiment?

6. Perform the calculations necessary to demonstrate that bromine is indeed the limiting reagent (Sec. 1.6). The densities and molecular weights needed to complete the calculations can be found in various handbooks of chemistry.

7. Consider the IR spectra of compounds **8–11** (Figs. 9.6, 9.8, 9.10, and 9.12), all of which contain a phenyl ring. What absorption(s) in the spectra denote the presence of this ring?

8. Consider the NMR data for compounds **8–13** (Figs. 9.7, 9.9, 9.11, 9.13, 9.15, and 9.19).

   a. In the ^1H NMR spectra, assign the various resonances to the hydrogen nuclei responsible for them.

   b. In the ^{13}C NMR spectra of these compounds assign the various resonances to the carbon nuclei responsible for them.

Starting Materials

Spectra

FIGURE 9.6
IR spectrum of toluene (neat).

FIGURE 9.7
NMR data for toluene (CDCl₃).

(a) ¹H NMR spectrum (300 MHz).

(b) ¹³C NMR data: δ 21.4, 125.3, 128.2, 129.0, 137.8.

FIGURE 9.8
IR spectrum of ethylbenzene (neat).

FIGURE 9.9
NMR data for ethylbenzene (CDCl₃).

(a) ¹H NMR spectrum (300 MHz).

(b) ¹³C NMR data: δ 15.6, 29.1, 125.7, 127.9, 128.4, 144.2.

FIGURE 9.10
IR spectrum of isopropylbenzene (neat).

(a) ^{1}H NMR spectrum (300 MHz).

FIGURE 9.11
NMR data for isopropylbenzene (CDCl$_3$).

(b) ^{13}C NMR data: δ 24.1, 34.2, 125.8, 126.4, 128.4, 148.8.

FIGURE 9.12
IR spectrum of *tert*-butylbenzene (neat).

FIGURE 9.13
NMR data for *tert*-butylbenzene
(CDCl$_3$).

(a) ^{1}H NMR spectrum (300 MHz).

(b) ^{13}C NMR data: δ 31.3, 34.5, 125.1, 125.3, 128.0, 150.8.

FIGURE 9.14
IR spectrum of cyclohexane (neat).

FIGURE 9.15
NMR data for cyclohexane
(CDCl$_3$).

(a) ^{1}H NMR spectrum (300 MHz).

(b) ^{13}C NMR datum: δ 27.3.

FIGURE 9.16
IR spectrum of methylcyclohexane (neat).

(a) ^{1}H NMR spectrum (300 MHz).

FIGURE 9.17
NMR data for methylcyclohexane (CDCl$_3$).

(b) ^{13}C NMR data: δ 22.9, 26.6, 26.7, 33.0, 35.6.

CHAPTER 10

Alkenes

Introduction

A **functional group** is an atom or group of atoms that governs the chemical and physical properties of a family of compounds. The introduction and manipulation of these functional groups are major objectives in modern organic chemistry. In this chapter, the chemistry of **alkenes 1**, which are organic compounds possessing a polarizable carbon-carbon π-bond as the functional group, is explored. Methods for introducing a carbon-carbon double bond into a molecule from alkyl halides and alcohols are developed first, and some of the reactions characteristic of this functional group are then examined.

Elimination reactions are among the most common ways to produce a carbon-carbon π-bond. For example, the elements of hydrogen halide, H–X, may be eliminated from an **alkyl halide 2**. The functional group of an alkyl halide is a carbon-halogen, **C–X**, single bond, and the process by which the carbon-halogen bond is converted into a carbon-carbon π-bond is an example of a **functional group transformation**. A carbon-carbon π-bond may also be formed upon removing the elements of water from an **alcohol 3**, in which a **C–OH** single bond is the functional group; this reaction is called **dehydration**. Although other aspects of the chemistry of alkyl halides and alcohols will be presented in Sections 14.1–14.5 and 16.2, a brief introduction to these families is essential to understand how they may be used as starting materials for the synthesis of alkenes.

$$
\underset{\substack{1 \\ \text{An alkene}}}{\overset{\displaystyle R^2}{\underset{\displaystyle R^1}{\Large C}}\!\!=\!\!CHR^3}
\qquad
\underset{\substack{2 \\ X = Cl,\,Br,\,I \\ \text{An alkyl halide}}}{R^1\!-\!\overset{\displaystyle R^2}{\underset{\displaystyle R^3}{\Large CH}}\!-\!\overset{\delta^+}{CH}\!-\!\overset{\delta^-}{X}}
\qquad
\underset{\substack{3 \\ \text{An alcohol}}}{R^1\!-\!\overset{\displaystyle R^2}{\underset{\displaystyle R^3}{\Large CH}}\!-\!\overset{\delta^+}{CH}\!-\!\overset{\delta^-}{OH}}
$$

10.2 Dehydrohalogenation of Alkyl Halides

The electronegative halogen atom of an alkyl halide **2** polarizes the carbon-halogen bond so the carbon atom bears a partial positive charge, δ^+, and the halogen atom a partial negative charge, δ^-. This polarization may be transmitted through the σ-bond network, a phenomenon referred to as an **inductive effect,** to enhance the acidity of hydrogen atoms on the β-carbon atom. Removing a proton from this atom, and simultaneous departure of the halide ion, which is a **leaving group**, from the α-carbon atom forms the carbon-carbon π-bond of the alkene **4** in a **concerted reaction** (Eq. 10.1). An important characteristic of a leaving group is that it should be a *weak* base; in contrast, the base B:$^-$ should be *strong*. The **transition state 4** for the reaction is shown in Equation 10.2, in which the curved arrows symbolize the flow of electrons. As shown in **4**, the preferred **dihedral angle** between the β-hydrogen being removed and the leaving group X is 180°, an angular relationship called **anti-periplanar**. This transformation of alkyl halides to alkenes is called **dehydrohalogenation**.

$$\overset{\beta}{R}\overset{\alpha}{CH}CH_2{-}X + B{:}^- \longrightarrow RCH{=}CH_2 + B{-}H + X{:}^- \tag{10.1}$$

$$\underset{H}{\underset{|}{}}$$

An alkyl halide   An alkene

4

Transition state

The concerted reaction depicted in Equation 10.2 is classified as an **E2** process, where **E** stands for **elimination** and **2** refers to the **molecularity** of the rate-determining step of the reaction. For E2 processes, the rate of the reaction depends upon the concentrations of the organic substrate and the base (Eq. 10.3), so *both* reactants are involved in the transition state of the rate-determining step. This **bimolecularity** is illustrated in **4** (Eq. 10.2).

$$\text{Rate} = k_2[\text{alkyl halide}][B{:}^-] \tag{10.3}$$

Because it bears a partial positive charge, the α-carbon atom of an alkyl halide is **electrophilic** and thus also subject to attack by **nucleophiles**, which, as **Lewis bases**, are electron-rich and frequently anionic species. This process produces **substitution** rather than elimination products (Eq. 10.4), so that a possible competition between the two types of reactions must be considered. Examining transition states **4** and **5** for the two reactions (Eqs. 10.2 and 10.4) suggests that steric factors may play a critical role in the competition. In support of this hypothesis, it is observed experimentally that elimination is favored as the degree of substitution on the α-carbon atom increases. The resulting steric hindrance inhibits direct attack of the base on the α-carbon atom, instead fostering removal of the β-proton and formation of elimination products. Thus, when the α-carbon atom is *tertiary*, elimination to form alkenes is generally the only observable reaction.

$$
\text{B:}^- \quad \overset{\text{RCH}_2}{\underset{\text{H}}{\overset{|}{\underset{|}{\text{H}''}}}}\overset{\delta^+}{\text{C}}\overset{\delta^-}{\text{—X}} \longrightarrow \left[\underset{\overset{|}{\text{H}}\,\text{H}}{\overset{\text{CH}_2\text{R}}{\underset{}{\overset{|}{\text{B}}}}}\overset{\delta^-}{\underset{}{\text{---C---}}}\overset{\delta^-}{\text{X}} \right]^{\ddagger} \longrightarrow \text{B—C}\underset{\overset{|}{\text{H}}}{\overset{\text{CH}_2\text{R}}{\underset{}{\overset{}{''''\text{H}}}}} + \text{X:}^- \qquad (10.4)
$$

5

Transition state

Dehydrohalogenation may give a mixture of products if the halogen is unsymmetrically located on the carbon skeleton. For example, 2-bromo-2-methylbutane (**6**), the substrate for this experiment, yields both 2-methyl-2-butene (**7**) and 2-methyl-1-butene (**8**) on reaction with base (Eq. 10.5). Such elimination reactions are normally

$$
\underset{\overset{|}{\text{H}}}{\text{CH}_2}\text{—}\underset{\overset{|}{\text{Br}}}{\overset{\overset{\text{CH}_3}{|}}{\text{C}}}\text{—}\underset{\overset{|}{\text{H}}}{\text{CH}}\text{—CH}_3 \xrightarrow{\text{base}} \underset{\overset{|}{\text{H}}}{\text{CH}_2}\text{—}\overset{\overset{\text{CH}_3}{|}}{\text{C}}\text{=CH—CH}_3 + \text{CH}_2\text{=}\overset{\overset{\text{CH}_3}{|}}{\text{C}}\text{—}\underset{\overset{|}{\text{H}}}{\text{CH}}\text{—CH}_3 \qquad (10.5)
$$

6   **7**   **8**

2-Bromo-2-methylbutane   2-Methyl-2-butene   2-Methyl-1-butene

irreversible under the experimental conditions used, so the alkenes **7** and **8** are *not* produced by elimination to give either **7** or **8**, *followed* by an equilibration of these isomers. Consequently, the ratio of **7** and **8** is defined by the relative rates of their formation. These rates, in turn, are determined by the relative free energies of the two corresponding transition states, **9** and **10**, respectively, leading to them rather than by the relative free energies of the alkenes themselves.

$$
\left[\underset{\underset{\text{B:}\,\delta^-}{\overset{|}{\text{H}}}}{\overset{\overset{\overset{\text{Br}\,\delta^-}{|}}{\text{H}_3\text{C}\diagdown\alpha\diagdown\text{CH}_3}}{\text{H}_3\text{C}\diagup\underset{\beta}{\diagup}\text{H}}} \right]^{\ddagger} \qquad \left[\underset{\underset{\text{B:}\,\delta^-}{\overset{|}{\text{H}}}}{\overset{\overset{\overset{\text{Br}\,\delta^-}{|}}{\text{H}_3\text{C}\diagdown\alpha\diagdown\text{CH}_2\text{CH}_3}}{\text{H}\diagup\underset{\beta}{\diagup}\text{H}}} \right]^{\ddagger}
$$

9   **10**

In the absence of complicating factors, the predominant product in an E2 elimination is the more highly substituted alkene, which is **7** in the present example. An increase in the number of alkyl substituents on the double bond almost always increases the stability—lowers the free energy—of an alkene, so elimination preferentially produces the more stable alkene. The relative free energies of the alkenes **7** and **8** reflect the relative energies of the corresponding transition states **9** and **10** because those factors that stabilize the product alkenes also play a role in stabilizing the respective transition states in which *partial* double-bond character is developing between the two carbon atoms. Thus, the free energy of activation for forming the more stable alkene **7** is less than that for the less stable alkene **8**.

The relative free energies of transition states of competing elimination reactions may be influenced by **steric factors**, which increase the energies of some transition states relative to others. For instance, the less stable alkene may become the major product if steric factors raise the energy of the transition state leading to the more highly substituted alkene more than that for forming the less highly substituted. This may occur when sterically demanding substituents on the carbon atom from which the proton is being removed offer hindrance to the approaching base, for example. Removal of a proton from a less substituted carbon to give a less substituted alkene then becomes more favorable. As more sterically demanding, bulky bases are used, abstraction of the more accessible proton will also be favored, thereby leading to the formation of even greater amounts of the less substituted alkene.

Transition states **9** and **10** serve as an illustrative example of these principles. In **9**, which is the transition state leading to the formation of **7**, there can be an unfavorable steric interaction between the methyl group on the β-carbon atom and the base. There is no comparable steric interaction in **10**, which produces the thermodynamically less stable **8**. Consequently, as the steric bulk of the base is increased, **8** will be formed in increased amounts.

In the experiments that follow, general techniques for performing base-promoted elimination reactions of 2-bromo-2-methylbutane (**6**) are demonstrated. Using this tertiary alkyl bromide eliminates problems that might be associated with forming by-products from substitution reactions and allows a study of effects on product ratios when bases having different steric bulks are used. The dehydrohalogenation occurs cleanly using base/solvent combinations such as potassium hydroxide in 1-propanol or potassium *tert*-butoxide in *tert*-butyl alcohol.

EXPERIMENTAL PROCEDURES

BASE-PROMOTED ELIMINATION OF AN ALKYL HALIDE

Purpose: To demonstrate the formation of alkenes by dehydrohalogenation and assess product distributions using different bases.

Safety Alert

1. Most of the chemicals used in this experiment are highly flammable, so do not handle these liquids near open flames.
2. The solutions used in this experiment are *highly* caustic. *Do not allow them to come in contact with your skin.* If this should happen, flood the affected area with water and then thoroughly rinse the area with a solution of dilute acetic acid. Wear latex gloves when preparing and transferring all solutions.

A · ELIMINATION WITH ALCOHOLIC POTASSIUM HYDROXIDE

Miniscale Procedure

Preparation Answer the Pre-Lab Exercises on page PL. 51. Review Sections 2.2, 2.4, 2.9, 2.10, 2.11, 2.14, 2.22, and 6.4.

Apparatus   A 10-mL and a 50-mL round-bottom flask, drying tube, apparatus for fractional distillation, magnetic stirring, and *flameless* heating.

Setting Up   Place 25 mL of a 4 *M* solution of potassium hydroxide in 1-propanol in the 50-mL flask containing a stirbar. Attach a drying tube to the flask. Add 2.5 mL of 2-bromo-2-methylbutane and then equip the flask for fractional distillation. Be sure to lubricate the ground-glass joint connecting the Hempel column to the flask with a thin layer of stopcock grease. Lubricating this joint is *particularly* important in this experiment because the strong base being used may cause the joint to freeze. To increase the cooling efficiency of the Hempel column, fill it with Raschig rings, coarsely broken glass tubing, coarse steel wool, or another packing material as directed by your instructor. Using a short piece of tubing, fit the vacuum adapter holding a 10-mL receiving flask with a drying tube and immerse the receiving flask in an ice-water bath. Attach water hoses to the *Hempel column* and circulate water through the jacket of the column during the period of reflux for this reaction.

Elimination and Isolation   Heat the reaction mixture under gentle reflux with stirring for a period of 1–1.5 h.★ Allow the mixture to cool below its boiling point, and drain the water from the jacket of the Hempel column. Connect water hoses to the *condenser* so that the apparatus is now set for fractional distillation. Leave in the receiving flask any low-boiling distillate that has condensed and continue to cool this flask in an ice-water bath. Distil the product mixture, collecting all distillate boiling *below* 45 °C, and transfer the product to a tared sample bottle with a tight-fitting stopper or cap.

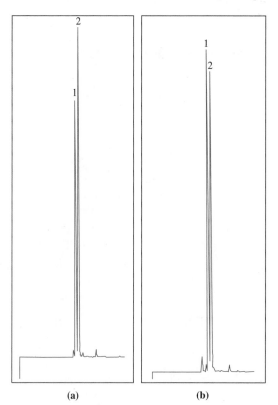

FIGURE 10.1

Typical GLC traces of the products of elimination of 2-bromo-2-methylbutane. Assignments and peak areas (in parentheses): peak 1: 2-methyl-1-butene; peak 2: 2-methyl-2-butene.

(a) Elimination with KOH; peak 1 (82210 units), peak 2 (120706 units).

(b) Elimination with KOC(CH₃)₃; peak 1 (94260 units), peak 2 (85140 units); Column and conditions: 50-mm × 30-m, 0.25 *μ* film of SE-54; 35 °C, 1.5 mL/min.

(a)   (b)

Analysis   Weigh the distillate and calculate the yield of products. Then put the container in an ice-water bath until all tests on the distillate have been completed. Test the distillate for unsaturation using the bromine in dichloromethane (Sec. 23.8A) and the Baeyer tests (Sec. 23.8B). Analyze your distillate by GLC or submit a sample of it for analysis. After obtaining the results, calculate the relative percentages of the two isomeric alkenes formed; assume that the response factors are identical for the two alkenes. A typical GLC analysis of the products from this elimination is shown in Figure 10.1a. Obtain IR and NMR spectra of your starting material and product. Compare your spectra with those of authentic samples of 2-bromo-2-methylbutane (Figs. 10.2 and 10.3) and of the two possible alkenes (Figs. 10.4–10.7). The ^{1}H NMR spectrum of the product mixture obtained from a representative experiment is given in Figure 10.8.

Microscale Procedure

Preparation   Answer the Pre-Lab Exercises on page PL. 53. Review Sections 2.3, 2.4, 2.9, 2.11, 2.13, 2.22, 2.27, and 6.4.

Apparatus   A 5-mL conical vial, apparatus for simple distillation, magnetic stirring, and *flameless* heating.

Setting Up   Place a spinvane in the conical vial. Using a calibrated pipet, add 2.5 mL of a 4 *M* solution of potassium hydroxide in 1-propanol. Avoid getting any of this solution on the ground glass joint of the vial. With a clean syringe or calibrated pipet, measure 1 mL of 2-bromo-2-methylbutane into the vial. Thoroughly mix the contents by stirring for a few seconds. Equip the vial for microscale distillation and circulate cold water through the condenser. Be sure to lubricate the ground-glass joint connecting the Hickman distillation head to the vial with a thin layer of stopcock grease. Lubricating this joint is *particularly* important because the strong base being used may cause the joint to freeze.

Elimination and Isolation   Heat the reaction vial with a bath that has been preheated to about 75 °C (bath temperature). The vial should be positioned so that the levels of the top of the bath and of the liquid in the reaction vial are equal. Stir the mixture at 75 °C for 1 h and then slowly raise temperature of the bath to about 90 °C over a period of about 20 min, collecting all of the distillate until about 2.5 mL of liquid remains in the vial. It should not be necessary to heat the bath higher than 90–95 °C. Using a Pasteur pipet, transfer the distillate to a dry sample vial or a test tube with a tight-fitting cap or stopper.

Analysis   Weigh the distillate and calculate the yield of products. Then put the container in an ice-water bath until all tests on the distillate have been completed. Test the distillate for unsaturation using the bromine in dichloromethane (Sec. 23.8A) and the Baeyer tests (Sec. 23.8B). Analyze your distillate by GLC or submit a sample of it for analysis. After obtaining the results, calculate the relative percentages of the two isomeric alkenes formed; assume that the response factors are identical for the two alkenes. A typical GLC trace of the products from this elimination is shown in Figure 10.1a. Obtain IR and NMR spectra of your starting material and product. Compare your spectra with those of authentic samples of 2-bromo-2-methylbutane (Figs. 10.2 and 10.3) and of the two possible alkenes (Figs. 10.4–10.7). The ^{1}H NMR spectrum of the product mixture obtained from a representative experiment is given in Figure 10.8.

Spectra   *Starting Material and Products*

FIGURE 10.2
IR spectrum of 2-bromo-
2-methylbutane (neat).

FIGURE 10.3
NMR data for 2-bromo-
2-methylbutane (CDCl₃).

(a) ¹H NMR spectrum (300 MHz).

(b) ¹³C data: δ 10.7, 33.8, 40.3, 69.2.

FIGURE 10.4
IR spectrum of 2-methyl-1-butene
(neat).

(a) ^{1}H NMR spectrum (300 MHz).

(b) ^{13}C data: δ 12.5, 22.3, 31.0, 108.8, 147.5.

FIGURE 10.5
NMR data for 2-methyl-1-butene (CDCl$_3$).

FIGURE 10.6
IR spectrum of 2-methyl-2-methylbutene (neat).

(a) ^{1}H NMR spectrum (300 MHz).

(b) ^{13}C data: δ 13.4, 17.3, 25.6, 118.8, 132.0.

FIGURE 10.7
NMR data for 2-methyl-2-methylbutene (CDCl$_3$).

FIGURE 10.8
¹H NMR spectrum (CDCl₃, 300 MHz) of the product mixture from the elimination of 2-bromo-2-methylbutane with potassium hydroxide.

FIGURE 10.9
¹H NMR spectrum (CDCl₃, 300 MHz) of the product mixture from the elimination of 2-bromo-2-methylbutane with potassium *tert*-butoxide.

B · ELIMINATION WITH POTASSIUM *tert*-BUTOXIDE

Miniscale Procedure

Preparation   Answer the Pre-Lab Exercises on page PL. 51. Review Sections 2.2, 2.4, 2.9, 2.10, 2.11, 2.14, 2.22, 2.27, and 6.4.

Apparatus   A *dry* 50-mL round-bottom flask, drying tube, apparatus for fractional distillation, magnetic stirring, and *flameless* heating.

Setting Up   Equip the flask with a stirbar. Working quickly, add 25 mL of a 1 *N* solution of potassium *tert*-butoxide in *tert*-butyl alcohol. This reagent is *very* moisture-sensitive so exposure to the atmosphere *must* be minimized. Then add 2.5 mL of 2-bromo-2-methylbutane to the solution and equip the flask for fractional distillation. Complete setting up according to the directions in the **Miniscale Procedure** of Part A.

Elimination and Isolation   Follow the procedure given in the **Miniscale Procedure** of Part A for completing the reaction.

Analysis   Weigh the distillate and calculate the yield of products. Then put the container in an ice-water bath until all tests on the distillate have been completed. Test the distillate for unsaturation using the bromine in dichloromethane (Sec. 23.8A) and the Baeyer tests (Sec. 23.8B). Analyze your distillate by GLC or submit a sample of it for analysis. After obtaining the results, calculate the relative percentages of the two isomeric alkenes formed; assume that the response factors are identical for the two alkenes. A typical GLC trace of the products from this elimination is shown in Figure 10.1b. Obtain IR and NMR spectra of your starting material and product. Compare your spectra with those of authentic samples of 2-bromo-2-methylbutane (Figs. 10.2 and 10.3) and of the two possible alkenes (Figs. 10.4–10.7). The ^{1}H NMR spectrum of the product mixture obtained from a representative experiment is given in Figure 10.9.

Finishing Touches

Neutralize the *residue remaining in the stillpot* with 10% aqueous hydrochloric acid; then flush it down the drain. Place the *dichloromethane solution* from the bromine test for unsaturation in a container for halogenated organic liquids; put the *manganese dioxide* from the Baeyer test for unsaturation in a container for heavy metals.

EXERCISES

1. Demonstrate that the amount of 4 *M* potassium hydroxide solution used for the elimination provides an excess of base relative to the alkyl bromide.

2. How might the ratio of alkenes obtained in the *tert*-butoxide-promoted elimination change if the *tert*-butyl alcohol contains water?

3. Why does the excess of base used in these eliminations favor the E2 over the E1 mechanism for elimination?

4. What is the solid material that precipitates as the eliminations proceed?

5. If all the elimination reactions in the experimental section had proceeded by the E1 mechanism, would the results have been different from those actually obtained? Why?

6. From the results of the experiments that were performed and/or from the GLC data in Figure 10.1, what conclusions can be drawn concerning the effect of relative base size upon product distribution?

7. If the steric demands of the leaving group in the 2-methyl-2-butyl system were greater than those of a methyl group, why would 2-methyl-1-butene be expected to be formed in greater amounts than if the leaving group were smaller than methyl, regardless of which base is used? Use Newman projections to support your explanation.

8. In the miniscale procedures, what purpose is served by packing the Hempel column with Raschig rings or other packing materials?

Refer to Tables 8.1–8.8 as needed for answering the following questions on spectroscopy.

9. Consider the NMR data for 2-bromo-2-methylbutane (Fig. 10.3).

    a. In the ^{1}H NMR spectrum, assign the various resonances to the hydrogen nuclei responsible for them.

    b. For the ^{13}C NMR data, assign the various resonances to the carbon nuclei responsible for them.

10. Consider the spectral data for 2-methyl-1-butene (Figs. 10.4 and 10.5).

    a. In the functional group region of the IR spectrum, specify the absorptions associated with the carbon-carbon double bond and the hydrogens attached to the C-1 carbon atom.

    b. In the ^{1}H NMR spectrum, assign the various resonances to the hydrogen nuclei responsible for them.

    c. For the ^{13}C NMR data, assign the various resonances to the carbon nuclei responsible for them.

11. Consider the spectral data for 2-methyl-2-butene (Figs. 10.6 and 10.7).

    a. In the functional group region of the IR spectrum, specify the absorptions associated with the carbon-carbon double bond and the hydrogen attached to the C-3 carbon atom.

    b. In the ^{1}H NMR spectrum, assign the various resonances to the hydrogen nuclei responsible for them.

    c. For the ^{13}C NMR data, assign the various resonances to the carbon nuclei responsible for them.

12. Discuss the differences observed in the IR and NMR spectra of 2-bromo-2-methylbutane and of 2-methyl-1-butene and 2-methyl-2-butene that are consistent with dehydrohalogenation occurring in this experiment.

13. Referring to Figures 10.5 and 10.7–10.9, calculate the percentage compositions of the mixtures of isomeric methylbutenes obtained from the reaction of 2-bromo-2-methylbutane with potassium hydroxide and potassium *tert*-butoxide. In the spectra of the mixtures, the integration of the resonances in the region of δ 5.0 ppm has been electronically amplified in the upper stepped line so that the relative areas of the two multiplets in that region can be more accurately measured.

10.3   Dehydration of Alcohols

? Proton
OH donated

The nonbonding lone pairs of electrons on the oxygen atom of the hydroxyl group, which is the functional group of alcohols, serve as a **Lewis base** and may complex with **Brønsted** and **Lewis acids**, which are defined as proton-donating and electron pair-accepting acids, respectively. For example, the protonation of an alcohol **11** is rapid and reversible and produces an **oxonium ion 12** in which the positively charged oxygen atom further polarizes the carbon-oxygen single bond. In contrast to hydroxide ion, a strong base and a poor leaving group, water is a weak base and an excellent leaving group, and **12** may then undergo facile elimination and substitution reactions upon reaction with a base or **nucleophile** Nu:$^-$ (Eq. 10.6).

$$R^2\diagdown C=C \diagup R^3 \diagdown R \qquad \text{Alkene.}$$

Equation structures (10.6), (10.7), (10.8) with labels:

11 An alcohol → **12** An oxonium ion → **13** Elimination + H_2O / **14** Substitution + H_2O (Eq. **10.6**)

12 → **12‡** Transition state → **15** + H_2O → **13** + $R—\overset{+}{O}H_2$ (Eq. **10.7**)

12 → alkene + H_2O + $R—OH_2$ (Eq. **10.8**)

The loss of water that effects **dehydration** of the oxonium ion **12** to produce an alkene **13** may occur by mechanistic pathways like those shown in Equations 10.7 and 10.8. With *secondary* and *tertiary* alcohols, **12** undergoes **endothermic** heterolysis via the transition state **12‡** to produce the **carbocation 15** and water. Rapid transfer of a proton from an adjacent carbon atom to a weak base such as water then gives **13** (Eq. 10.7). The fragmentation of **12** to give **15** is kinetically a **first-order** process and is the **rate-determining step** (rds) of the overall reaction. Consequently, the reaction is classified as **E1**, which stands for *elimination unimolecular*. Because the formation of the carbocation **15** is an endothermic process, the transition state **12‡** is product-like and is characterized by extensive heterolysis of the carbon-oxygen bond with simultaneous development of partial positive charge on the carbon atom. Thus, the ease of dehydration of alcohols by the E1 mechanism is in the order, $3° > 2° >> 1°$, paralleling the relative stabilities of carbocations. In the case of *primary* alcohols, ionization of the oxonium ion **12** would produce a highly unstable 1° carbocation (**12**, $R^3 = R^4 = H$), so the alternate E2 (*elimination bimolecular*) mechanism of Equation 10.8 applies.

As illustrated in Equations 10.6–10.8, each of the steps along the reaction pathway is reversible, so the desired alkene may also undergo acid-catalyzed hydration to form an alcohol. In practice, reversal of the dehydration may be avoided by removing the alkene, whose boiling point is always *lower* than the parent alcohol, from the reaction mixture by distillation. This technique shifts the equilibrium to the right and maximizes the yield of alkene. Shifting an equilibrium of a reversible reaction to the right by removing one of the products as it is formed follows from the **LeChatelier principle**.

According to Equation 10.6, elimination and substitution reactions can be competing pathways (see also Sec. 14.3), but the relative importance of each may often be controlled through proper choice of reaction conditions and reagents. In the case of the acid-catalyzed dehydration of secondary and tertiary alcohols, these competing reactions involve combining the intermediate carbocation with the **conjugate base** of the acid catalyst or with some other nucleophile such as solvent. For example, if hydrochloric acid were chosen as the catalyst for dehydration, the chloride ion could trap the carbocation to give an alkyl chloride (Eq. 10.9). This side reaction thus not only produces an undesired by-product but also consumes the acid catalyst. Consequently, the desired dehydration is terminated because formation of the alkyl chloride is irreversible under the conditions normally used for dehydration.

$$R-\underset{\underset{R}{|}}{\overset{\overset{R}{|}}{C}}-OH \xrightarrow[-H_2O]{HCl} R-\underset{\underset{R}{|}}{\overset{\overset{R}{|}}{C}}{}^+ \underset{Cl:^-}{\overset{}{}} \longrightarrow R-\underset{\underset{R}{|}}{\overset{\overset{R}{|}}{C}}-Cl \qquad (10.9)$$

The side reaction depicted in Equation 10.9 is avoided when the formation of the substitution product is reversible. For this reason, the catalysts recommended to effect the dehydration of alcohols in the laboratory are sulfuric or phosphoric acid, since the intermediate alkyl bisulfate (Nu:$^-$ = HSO_3O^-) or dihydrogen phosphate (Nu:$^-$ = $H_2PO_3O^-$) ions readily reionize to the intermediate carbocation under the reaction conditions (Eq. 10.10). Subsequent loss of a proton then gives the desired alkene.

$$R^1-\underset{\underset{H}{|}}{\overset{\overset{R^2}{|}}{C}}-\overset{\overset{R^3}{|}}{CH}-OH \underset{-H_2O}{\overset{H_3O^+}{\rightleftharpoons}} R^1-\underset{\underset{H}{|}}{\overset{\overset{R^2}{|}}{C}}-\overset{\overset{R^3}{|}}{\underset{+}{CH}} \overset{-H^+}{\rightleftharpoons} \underset{R^1}{\overset{R^2}{}}C{=}CHR^3 \qquad (10.10)$$

Elimination product

$$\Big\downarrow Nu:^-$$

$$R^1-\underset{\underset{H}{|}}{\overset{\overset{R^2}{|}}{C}}-\underset{\underset{Nu}{|}}{\overset{\overset{R^3}{|}}{CH}}$$

Substitution product

Two or more isomeric alkenes may be formed in E1 dehydration reactions. For example, consider the unsymmetrically substituted carbocation **17**, formed from alcohol **16** (Eq. 10.11). Loss of H_A produces **18**, whereas loss of H_B yields **19**. The relative amounts of the two alkenes depends upon the relative free energies of the transition states for the loss of the H_A and H_B. In these transition states, the C–H bond is partially broken and the C–C double bond is partially formed, as depicted in **20** and **21**. Since the major difference between these two transition states is the substitution on the incipient carbon-carbon double bond, their relative energies parallel the relative stabilities

(handwritten note top right)
Ⓚ ↓ lower energy
transition state
is more stable.

of the product alkenes. Consequently <u>the lower energy transition state is the one leading</u> <u>to the more highly substituted, and more stable alkene,</u> which is **19** in this generalized case. Furthermore, as a general rule the (*E*)-isomer of a pair of geometric isomers predominates over the (*Z*)-isomer for steric reasons.

$$R^1-CH=CH-\underset{\underset{H_B}{|}}{\overset{\overset{R^2}{|}}{C}}-R^3$$

18

$$R^1-\underset{\underset{H_A}{|}}{\overset{\overset{OH}{|}}{CH}}-\underset{\underset{H_B}{|}}{\overset{\overset{R^2}{|}}{C}}-R^3 \underset{-H_2O}{\overset{H_3O^+}{\rightleftharpoons}} R^1-\underset{\underset{H_A}{|}}{CH}-\overset{+}{CH}-\underset{\underset{H_B}{|}}{\overset{\overset{R^2}{|}}{C}}-R^3$$

$\boxed{-H_A^+}$ ↗

$\boxed{-H_B^+}$ ↘

16   **17**   (10.11)

$$R^1-\underset{\underset{H_A}{|}}{CH}-CH=\overset{\overset{R^2}{|}}{C}-R^3$$

19

$$\left[R^1-\underset{\delta^+ H_A}{\overset{\delta^+}{CH{=\!=}CH}}-\underset{\underset{H_B}{|}}{\overset{\overset{R^2}{|}}{C}}-R^3\right]^{\ddagger} \quad \left[R^1-\underset{\underset{H_A}{|}}{CH}-\overset{\delta^+}{CH{=\!=}}\underset{\underset{\delta^+ H_B}{\vdots}}{\overset{\overset{R^2}{|}}{C}}-R^3\right]^{\ddagger}$$

20   **21**

The distribution of alkenes from the acid-catalyzed dehydration of alcohols is not always predictable from the orientational factors just described. Rather, the intermediate carbocation may rearrange through the migration of either a **hydride**, H:⁻, or an **alkyl group**, R:⁻, from a carbon atom adjacent to the cationic center; the rearrangement is more likely if the new carbocation is more stable than the original ion. <u>The subsequent</u> <u>loss of a proton from the rearranged carbocation according to the orientation effects</u> <u>already discussed then leads to the additional alkenes.</u>

The acid-catalyzed dehydration of 4-methyl-2-pentanol (**22**), which is one of the experiments in this section, is shown in Equation 10.12 and nicely illustrates the preceding principles. Dehydration of **22** produces a complex mixture of isomeric alkenes, including 4-methyl-1-pentene (**23**), *trans*-4-methyl-2-pentene (**24**), *cis*-4-methyl-2-

$$(CH_3)_2CHCH_2\underset{\underset{OH}{|}}{CH}-CH_3 \xrightarrow[\Delta]{H_3O^+} (CH_3)_2CHCH_2CH=CH_2 + \underset{H}{\overset{(CH_3)_2CH}{>}}C=C\underset{CH_3}{\overset{H}{<}}$$

(10.12)

22   **23**   **24**
4-Methyl-2-pentanol   4-Methyl-1-pentene   *trans*-4-Methyl-2-pentene

$$+ \underset{H}{\overset{(CH_3)_2CH}{>}}C=C\underset{H}{\overset{CH_3}{<}} + \underset{H_3C}{\overset{H_3C}{>}}C=C\underset{H}{\overset{CH_2CH_3}{<}} + \underset{H}{\overset{H}{>}}C=C\underset{CH_2CH_2CH_3}{\overset{CH_3}{<}}$$

25   **26**   **27**
cis-4-Methyl-2-pentene   2-Methyl-2-pentene   2-Methyl-1-pentene

pentene (**25**), 2-methyl-2-pentene (**26**), and 2-methyl-1-pentene (**27**). A pathway to **23–25** is by way of deprotonation of the intermediate carbocation **28**. However, forming **26** and **27** requires intervention of carbocations **29** and **30**.

$$(CH_3)_2CH—CH_2—\overset{+}{C}HCH_3 \qquad (CH_3)_2CH—\overset{+}{C}H—CH_2CH_3 \qquad (CH_3)_2\overset{+}{C}—CH_2—CH_2CH_3$$
$$\textbf{28} \qquad\qquad\qquad \textbf{29} \qquad\qquad\qquad \textbf{30}$$

In contrast to **22**, cyclohexanol (**31**) undergoes acid-catalyzed dehydration *without* rearrangement to yield a single product, cyclohexene (Eq. 10.13). This dehydration reaction may be accompanied by some acid-catalyzed polymerization of the cyclohexene.

$$\text{(10.13)}$$

31
Cyclohexanol   Cyclohexene

EXPERIMENTAL PROCEDURES

DEHYDRATION OF ALCOHOLS

Purpose: To demonstrate the acid-catalyzed dehydration of alcohols to alkenes.

Safety Alert

1. The majority of materials, particularly the product alkenes, that will be handled during this experiment are highly flammable so use *flameless* heating and be certain that there are no open flames in your vicinity.

2. Several experimental operations require pouring, transferring, and weighing chemicals and reagents that cause burns on contact with your skin. Wear latex gloves when handling the strongly acidic catalysts and during the work-up and washing steps in which a separatory funnel is used. Should acidic solutions accidentally come in contact with your skin, immediately flood the affected area with water, then wash it with 5% sodium bicarbonate solution.

A · DEHYDRATION OF 4-METHYL-2-PENTANOL

Miniscale Procedures

Preparation   Answer the Pre-Lab Exercises on page PL. 55. Review Sections 2.4, 2.9, 2.10, 2.11, 2.13, 2.14, 2.22, 2.25, 2.27, and 6.4.

Apparatus   A 25-mL and two 10-mL round-bottom flasks, drying tube, ice-water bath, apparatus for simple and fractional distillation, magnetic stirring, and *flameless* heating.

Setting Up   Place a stirbar and 4.0 mL of 4-methyl-2-pentanol in the 25-mL round bottom flask. Add 2.5 mL of 9 *M* sulfuric acid and thoroughly mix the liquids by gently swirling the flask. Equip the flask for fractional distillation and use an ice-water bath to cool the receiving flask.

Elimination   Heat the reaction mixture and collect all distillates while maintaining the head temperature below 90 °C. If the reaction mixture is not heated too strongly, the head temperature will stay below 80 °C for most of the reaction. When about 2.5 mL of liquid remains in the reaction flask, discontinue heating. Transfer the distillate to an Erlenmeyer flask and add several spatula-tips full of anhydrous potassium carbonate to neutralize any acid and to dry the distillate.★ Occasionally swirl the mixture during a period of 10–15 min to hasten drying; add additional portions of anhydrous potassium carbonate if the liquid remains cloudy.

Isolation   Transfer all of the dried organic mixture into a dry 10-mL round-bottom flask by decantation or by using a Pasteur pipet. Add a stirbar and equip the flask for simple distillation (Fig. 2.35a). Use a tared 10-mL flask as the receiver. Protect the receiver from atmospheric moisture by equipping the vacuum adapter with a drying tube and cool it in an ice-water bath.

Isolate the alkenes by simple distillation. The expected products of the reaction are 4-methyl-1-pentene, *cis*-4-methyl-2-pentene, *trans*-4-methyl-2-pentene, 2-methyl-1-pentene, and 2-methyl-2-pentene. Collect the fraction boiling between 53 and 69 °C (760 Torr) in a pre-weighed, dry 10-mL receiving flask.

Analysis   Weigh the distillate and determine the yield of products. Test the distillate for unsaturation using the bromine in dichloromethane (Sec. 23.8A) and the Baeyer tests (Sec. 23.8B). Analyze your distillate by GLC or submit a sample of it for analysis. After obtaining the results, calculate the relative percentages of the isomeric alkenes formed; assume that the response factors are identical for the isomers. A typical GLC trace of the alkenes from this elimination is shown in Figure 10.10. Obtain IR and NMR spectra of your starting material and compare your spectra with those of an authentic sample (Figs. 10.11 and 10.12).

FIGURE 10.10

GLC analysis of the product mixture from dehydration of 4-methyl-2-pentanol. Assignments and peak areas (in parentheses): peak 1: diethyl ether (solvent for analysis); peak 2: 4-methyl-1-pentene (3030); peak 3: *cis*- and *trans*-4-methyl-2-pentene (51693); peak 4: 2-methyl-1-pentene (2282); peak 5: 2-methyl-2-pentene (9733). Column and conditions: 0.25-mm × 25-m AT-1 Helifex 100% dimethylsiloxane; 37 °C, 35 mL/min.

B · DEHYDRATION OF CYCLOHEXANOL

Preparation   Answer the Pre-Lab Exercises on page PL. 55. Review Sections 2.9, 2.10, 2.11, 2.13, 2.14, 2.22, 2.27, and 6.4.

Apparatus   A 25-mL and two 10-mL round-bottom flasks, drying tube, ice-water bath, apparatus for simple and fractional distillation, magnetic stirring, and *flameless* heating.

Setting Up   Place a stirbar and 5.0 mL of cyclohexanol in the 25-mL round-bottom flask. Add 2.5 mL of 9 *M* sulfuric acid and thoroughly mix the liquids by gently swirling the flask. Equip the flask for fractional distillation and use an ice-water bath to cool the receiving flask.

Elimination and Isolation   Follow the same protocol as provided in Part A with the *single exception* that the product should be collected over the range of 80–85 °C (760 Torr). Note that water and cyclohexene form a **minimum-boiling azeotrope**, which is a mixture of the two compounds that distils *below* the boiling point of either pure compound. This makes it vital that the crude product be dried *thoroughly* prior to the simple distillation.

Analysis   Weigh the distillate, and assuming that it is pure cyclohexene, determine the yield of product. Test the distillate for unsaturation using the bromine in dichloromethane (Sec. 23.8A) and the Baeyer tests (Sec. 23.8B). Analyze your distillate by GLC or submit a sample of it for analysis. Obtain IR and ^{1}H NMR spectra of your starting material and product and compare them with those of authentic samples (Figs. 10.23–10.26).

Microscale Procedure

Preparation   Answer the Pre-Lab Exercises on page PL. 55. Review Sections 2.4, 2.9, 2.13, 2.22, 2.25, and 6.4.

Apparatus   A 3-mL conical vial, apparatus for magnetic stirring, simple distillation, and *flameless* heating.

Setting Up   Place a spinvane in the conical vial. Using a calibrated Pasteur pipet, add 1 mL of cyclohexanol and 0.5 mL of 9 *M* sulfuric acid to the vial. Thoroughly mix the contents by briefly stirring or swirling the liquid for a few seconds. Equip the vial for microscale distillation and circulate cold water through the condenser.

Elimination and Isolation   Heat the reaction vial in a bath that has been preheated to about 120 °C (bath temperature). The vial should be positioned so that the levels of the top of the bath and of the liquid in the reaction vial are equal. Slowly raise the temperature of the bath as necessary to maintain distillation of the product and continue until about 0.5-0.7 mL of liquid remains in the flask. It should not be necessary to heat the bath higher than about 130–135 °C. Dismantle the apparatus, and remove the distillate from the Hickman stillhead using a Pasteur pipet. Transfer the distillate to a dry sample vial or test tube, and add several microspatula-tips full of anhydrous potassium carbonate to neutralize any acid and to dry the distillate. Occasionally swirl the mixture for 5–10 min to hasten drying and add additional portions of anhydrous potassium carbonate if the liquid remains cloudy.

Analysis Test the distillate for unsaturation using the bromine in dichloromethane (Sec. 23.8A) and the Baeyer tests (Sec. 23.8B). Analyze your distillate by GLC or submit a sample of it for analysis. Obtain IR and ^{1}H NMR spectra of your starting material and product and compare them with those of authentic samples (Figs. 10.23–10.26).

Dilute the *residue remaining in the stillpot* with water, carefully neutralize it with sodium carbonate, and flush it down the drain with large quantities of water. Dry the *potassium carbonate* on a tray in the hood and flush it down the drain or place it in a container for nonhazardous solids. Pour the *dichloromethane solution* from the bromine test for unsaturation in a container for halogenated organic liquids, and put the *manganese dioxide* from the Baeyer test for unsaturation in a container for heavy metals.

Finishing Touches

EXERCISES

Refer to Tables 8.1–8.8 as needed for answering the questions on spectroscopy.

General Questions

1. Why is the boiling point of the parent alcohol higher than that of the product alkene?

2. Define the LeChatelier principle.

3. Why are the distillates obtained in the initial step of the dehydration reaction dried over anhydrous potassium carbonate?

4. Why is the head temperature kept below 90 °C in the initial step of the dehydration reaction?

5. Why is it necessary to separate a dried organic solution from the drying agent prior to distilling the solution?

6. In principle the equilibrium in the dehydration of an alcohol could be shifted to the right by removal of water. Why is this tactic not a good option for the dehydration of 4-methyl-2-pentanol and cyclohexanol?

7. The loss of a proton from suitably substituted carbocations can provide both the *trans*- and *cis*-isomers of the resulting alkene, but the *trans*-isomer normally predominates. For example, deprotonation at C-3 of the 2-pentyl carbocation produces mainly *trans*-pentene. By analyzing the relative energies of the conformational isomers of the carbocation that lead to the two isomeric 2-pentenes, explain why the *trans*-isomer is formed preferentially. This analysis is aided by the use of Newman projections based on the partial structure shown.

Questions for Dehydration of 4-Methyl-2-pentanol

8. Give a detailed mechanism explaining formation of **23–25** from the dehydration of 4-methyl-2-pentanol (**22**). Use curved arrows to symbolize the flow of electrons.

9. A mechanism for formation of **26** and **27** can be developed from rearrangement of the carbocation **28** by way of hydride shifts.

   a. Provide a mechanism by which **26** may be formed from **28** through a hydride shift followed by deprotonation. Use curved arrows to symbolize the flow of electrons.

   b. Show how your mechanism in part **a** can be elaborated to provide for formation of **27**.

   c. An alternative pathway for formation of the alkene **26** involves addition of a proton to **24** or **25** and deprotonation of the resulting carbocation. Similarly, **27** could be produced by the addition of a proton to **26** followed by deprotonation of the intermediate carbocation. Write out mechanisms for these transformations using curved arrows to symbolize the flow of electrons.

   d. How might you experimentally distinguish the two types of mechanisms for the formation of **26** and **27**?

10. Near the end of the dehydration of 4-methyl-2-pentanol, a white solid may precipitate from the reaction mixture. What is the solid likely to be?

11. Give the structure, including stereochemistry, of the product of adding bromine (refer to Sec. 10.6) to *trans*-4-methyl-2-pentene (**24**) and to *cis*-4-methyl-2-pentene (**25**). Relative to one another, are these products identical, enantiomeric, or diastereomeric (Sec. 7.1)? Give your reasoning.

12. Consider the spectral data for 4-methyl-2-pentanol (Figs. 10.11 and 10.12).

   a. In the functional group region of the IR spectrum, specify any absorptions associated with the alcohol function.

   b. In the ^{1}H NMR spectrum, assign the various resonances to the hydrogen nuclei responsible for them.

   c. For the ^{13}C NMR data, assign the various resonances to the carbon nuclei responsible for them.

13. Consider the spectral data for 4-methyl-1-pentene (Figs. 8.23, 10.13, and 10.14).

   a. In the IR spectrum, specify the absorptions associated with the carbon-carbon double bond and the vinylic hydrogen atoms.

   b. In the ^{1}H NMR spectrum, assign the various resonances to the hydrogen nuclei responsible for them.

   c. For the ^{13}C NMR data, assign the various resonances to the carbon nuclei responsible for them.

14. Consider the spectral data for *trans*-4-methyl-2-pentene (Figs. 10.15 and 10.16).

   a. In the IR spectrum, specify the absorptions, if any, associated with the carbon-carbon double bond and with the *trans* relationship of the vinylic hydrogen atoms.

   b. In the ^{1}H NMR spectrum, assign the various resonances to the hydrogen nuclei responsible for them.

   c. For the ^{13}C NMR data, assign the various resonances to the carbon nuclei responsible for them.

15. Consider the spectral data for *cis*-4-methyl-2-pentene (Figs. 10.17 and 10.18).

   a. In the IR spectrum, specify the absorptions, if any, associated with the carbon-carbon double bond and with the *cis* relationship of the vinylic hydrogen atoms.

b. In the ^{1}H NMR spectrum, assign the various resonances to the hydrogen nuclei responsible for them.

c. For the ^{13}C NMR data, assign the various resonances to the carbon nuclei responsible for them.

16. Consider the spectral data for 2-methyl-1-pentene (Figs. 10.19 and 10.20).

a. In the IR spectrum, specify the absorptions associated with the carbon-carbon double bond and with the hydrogens attached to the C-1 carbon atom.

b. In the ^{1}H NMR spectrum, assign the various resonances to the hydrogen nuclei responsible for them.

c. For the ^{13}C NMR data, assign the various resonances to the carbon nuclei responsible for them.

17. Consider the spectral data for 2-methyl-2-pentene (Figs. 10.21 and 10.22).

a. In the IR spectrum, specify the absorptions associated with the carbon-carbon double bond and with the vinylic hydrogen atom.

b. In the ^{1}H NMR spectrum, assign the various resonances to the hydrogen nuclei responsible for them.

c. For the ^{13}C NMR data, assign the various resonances to the carbon nuclei responsible for them.

18. Discuss the differences observed in the IR and NMR spectra of 4-methyl-2-pentanol and the various methylpentenes that are consistent with dehydration occurring in this experiment.

Questions for Dehydration of Cyclohexanol

19. Why is it particularly important that the crude cyclohexene be dry prior to its distillation?

20. Give the structure, including stereochemistry, of the product of addition of bromine (refer to Sec. 10.6) to cyclohexene. Should it be possible, at least in principle, to resolve this dibromide into separate enantiomers? Explain your answer.

21. Provide a detailed mechanism for the acid-catalyzed dehydration of cyclohexanol. Use curved arrows to symbolize the flow of electrons.

22. Cyclohexyl carbocation undergoes attack by nucleophiles. With this in mind, propose a mechanism whereby cyclohexene might form the dimer **31** in the presence of H_3O^+. Note that this type of reaction leads to polymerization of cyclohexene.

31

23. Consider the spectral data for cyclohexanol (Figs. 10.23 and 10.24).

a. In the functional group region of the IR spectrum, specify any absorptions associated with the alcohol function.

b. In the ^{1}H NMR spectrum, assign the various resonances to the hydrogen nuclei responsible for them.

c. For the ^{13}C NMR data, assign the various resonances to the carbon nuclei responsible for them.

24. Consider the spectral data for cyclohexene (Figs. 10.25 and 10.26).

   a. In the IR spectrum, specify the absorptions associated with the carbon-carbon double bond and with the vinylic hydrogen atoms.

   b. In the ^{1}H NMR spectrum, assign the various resonances to the hydrogen nuclei responsible for them.

   c. For the ^{13}C NMR data, assign the various resonances to the carbon nuclei responsible for them.

25. Discuss the differences observed in the IR and NMR spectra of cyclohexanol and cyclohexene that are consistent with dehydration occurring in this experiment.

Spectra   *Starting Materials and Products*

FIGURE 10.11
IR spectrum of 4-methyl-2-pentanol (neat).

FIGURE 10.12
NMR data for 4-methyl-2-pentanol (CDCl$_3$).

(a) ^{1}H NMR spectrum (300 MHz).

(b) ^{13}C data: δ 22.5, 23.2, 23.9, 24.9, 48.8, 65.8.

FIGURE 10.13
IR spectrum of 4-methyl-1-pentene (neat).

$(CH_3)_2\ CHCH_2CH=CH_2$ ^{13}C data: δ 22.3, 28.1, 43.7, 115.5, 137.8.

FIGURE 10.14
NMR data for 4-methyl-1-pentene (CDCl$_3$).

FIGURE 10.15
IR spectrum of *trans*-4-methyl-2-pentene (neat).

(a) ^1H NMR spectrum (300 MHz).

FIGURE 10.16
NMR data for *trans*-4-methyl-2-pentene (CDCl$_3$).

(b) ^{13}C data: δ 17.6, 22.7, 31.5, 121.6, 139.4.

FIGURE 10.17
IR spectrum of *cis*-4-methyl-
2-pentene (neat).

FIGURE 10.18
NMR data for *cis*-4-methyl-
2-pentene (CDCl₃).

(a) ¹H NMR spectrum (300 MHz).

(b) ¹³C data: δ 12.7, 23.1, 26.4, 121.4, 138.6.

FIGURE 10.19
IR spectrum of 2-methyl-1-pentene
(neat).

(a) ¹H NMR spectrum (300 MHz).

(b) ¹³C data: δ 13.8, 20.8, 22.3, 40.0, 109.7, 146.0.

FIGURE 10.20
NMR data for 2-methyl-1-pentene (CDCl₃).

FIGURE 10.21
IR spectrum of 2-methyl-2-pentene (neat).

(a) ¹H NMR spectrum (300 MHz).

(b) ¹³C data: δ 14.5, 17.5, 21.6, 25.7, 126.9, 130.6.

FIGURE 10.22
NMR data for 2-methyl-2-pentene (CDCl₃).

FIGURE 10.23
IR spectrum of cyclohexanol (neat).

FIGURE 10.24
NMR data for cyclohexanol
(CDCl₃).

(a) ¹H NMR spectrum (300 MHz).

(b) ¹³C data: δ 24.5, 25.9, 35.5, 70.1.

FIGURE 10.25
IR spectrum of cyclohexene (neat).

(a) ^{1}H NMR spectrum (300 MHz).

FIGURE 10.26
NMR data for cyclohexene
(CDCl$_3$).

(b) ^{13}C data: δ 22.9, 25.2, 127.3.

10.4 Addition Reactions of Alkenes: Overview

Alkenes are useful starting materials for organic syntheses because they undergo a large variety of reactions involving their functional group, the carbon-carbon π-bond. One of the typical reactions of alkenes is the **addition** of a reagent X–Y across the π-bond of the alkene (Eq. 10.14). Such additions occur because of two principal factors. First, the strength of the carbon-carbon π-bond is 60–65 kcal/mol, a value substantially lower than the 80–100 kcal/mol that is typical for the σ-bond strengths of other atoms bound to carbon. Adding a reagent across the double bond is therefore usually *exothermic* because one π-bond is being replaced by two σ-bonds. Second, the π-electrons of the double bond are more loosely held than σ-electrons, so they are more polarizable. Consequently, the π-electron "cloud" is readily distorted through electrostatic interaction with an electron-deficient reagent, and this enhances the reactivity of the alkene toward attack.

$$
\begin{array}{c}
R^1 \\
\diagdown \\
C{=}C \\
\diagup \diagdown \\
R^2 R^4
\end{array}
R^3
\;+\; X{-}Y \;\longrightarrow\;
R^2{-}\underset{\underset{X}{|}}{\overset{\overset{R^1}{|}}{C}}{-}\underset{\underset{R^4}{|}}{\overset{\overset{Y}{|}}{C}}{-}R^3
\qquad (10.14)
$$

The most common mode of addition of X–Y across the π-bond of alkenes involves an *ionic* **stepwise mechanism** in which an **electrophile**, E^+, first attacks the π-bond to produce a carbocation. In such cases, X–Y may be rewritten as E–Nu, wherein E is the *less* electronegative of the two atoms. The **nucleophile** Nu$^-$ then attacks the carbocation to produce the final product. The overall process is termed **electrophilic addition**.

Considering this type of addition in terms of sequential reactions of Lewis acids and Lewis bases is helpful for understanding the mechanism. In the first step, the electrons of the polarizable π-bond, a Lewis base, attacks the electrophile E^+, an electron-deficient species and a Lewis acid, to give either an acyclic planar carbocation

32 or a cyclic cation **33** (Eq. 10.15). If carbocation **32** is formed as an intermediate in the electrophilic addition to an *unsymmetrically substituted* alkene, the electrophile E^+ generally adds to the terminus of the double bond that produces the *more stable carbocation* **32** according to **Markovnikov's rule** (see the Historical Highlight at the end of this chapter). Subsequent reaction of this carbocation, a Lewis acid, with the nucleophile, Nu^-, which is electron-rich and a Lewis base, then gives the addition product. When the positively charged carbon of **32** is substituted with three different groups, the nucleophile may attack from *both* faces of the planar ion by paths **a** and **b** to give a mixture of **34** and **35**, which are stereoisomers (Sec. 7.1). Acyclic carbocations like **32** may also rearrange via 1,2-hydride or 1,2-alkyl migration to form a more stable carbocation (Sec. 10.3); subsequent reaction of this isomeric carbocation with the nucleophilic partner produces adducts isomeric with **34** and **35**. Some electrophilic reagents add to a carbon-carbon double bond to form a cyclic cation **33**; such cations tend *not* to rearrange. The nucleophile $Nu:^-$ may then attack **33** by paths **c** and **d** to give mixtures of isomers **36** and **37**.

$$(10.15)$$

The general principles set forth in Equation 10.15 may be illustrated by considering several examples. The addition of the unsymmetrical reagent hydrogen bromide, H–Br, to an alkene is shown in Equation 10.16, and the acid-catalyzed addition of water, or **hydration**, of an alkene is depicted in Equation 10.17. In both reactions, the π-bond is protonated to form the more stable intermediate carbocation according to Markovnikov's rule. This carbocation may then rearrange to a more stable carbocation (Sec. 10.3), or, as shown in the present case, may undergo direct nucleophilic attack from *both* faces of the planar carbocation to give the observed alkyl bromide or alcohol. These two examples of electrophilic additions to alkenes are **regioselective** processes because the two unsymmetrical reactants combine *predominantly* in one orientational sense to give one **regioisomer** preferentially.

$$(10.16)$$

As mixture of stereoisomers

$$R^1—\underset{\underset{R^2}{|}}{C}=\underset{\underset{R^3}{|}}{CH} + H^+ \xrightarrow[H_2O]{} \left[R^1—\underset{\underset{H}{|}}{\overset{+}{C}}—\underset{\underset{R^3}{|}}{\overset{R^2}{|}}CH\right] \longrightarrow R^1—\underset{\underset{HO^+}{|}}{\overset{R^2}{|}}C—\underset{\underset{H}{|}}{\overset{R^3}{|}}CH \longrightarrow R^1—\underset{\underset{HO}{|}}{\overset{R^2}{|}}C—\underset{\underset{H}{|}}{\overset{R^3}{|}}CH + B—H \quad (10.17)$$

As mixture of stereoisomers

Since the ionic addition of hydrogen bromide and the acid-catalyzed addition of water to an alkene proceed via planar carbocations, the formation of products derived from rearranged carbocations is relatively common. In contrast, the bromination of alkenes, which is illustrated by the addition of bromine to cyclopentene (Eq. 10.18), occurs via a **cyclic bromonium ion 38** related to **33**; skeletal rearrangements of such cations are *not* generally observed. Moreover, the intervention of this cyclic ion dictates that the stereochemistry of the addition is completely in the *anti* sense as shown. The reddish color of bromine is discharged upon addition to an alkene, making this reaction a useful qualitative test for unsaturation (Sec. 23.8A).

Cyclopentene	Bromine	A cyclic bromonium ion	*trans*-1,2- Dibromocyclopentene
Colorless	Red-Brown		Colorless

The addition of reagents X–Y to carbon-carbon π-bonds may also proceed via a **concerted mechanism** in which each new σ-bond is formed simultaneously on the *same* face of the π-bond. The stereochemistry of such reactions is *syn*. For example, the reaction of potassium permanganate, which is purple, with an alkene such as cyclohexene proceeds via initial *syn*-addition of permanganate ion across the π-bond to give **39**, which is colorless. Subsequent decomposition of **39** gives a *cis*-1,2-diol and manganese dioxide, the brown precipitate that is observed as the other product of the reaction (Eq. 10.19). This decoloration of potassium permanganate by alkenes forms the basis of the Baeyer qualitative test for the presence of carbon-carbon π-bonds (Sec. 23.8B).

Cyclohexene	Purple		Colorless	*cis*-Cyclohexane- 1,2-diol	Brown

10.5 Addition of Hydrobromic Acid to Alkenes

In this experiment, the **ionic addition** of hydrobromic acid, H–Br, to 1-hexene to yield 2-bromohexane (**41**) (Eq. 10.20) according to Markovnikov's rule is examined. This type of reaction is normally rather difficult to perform in the undergraduate laboratory for several reasons. First, common alkenes are immiscible with the concentrated aqueous hydrobromic acid, and the reaction is sluggish if the layers are not mixed efficiently. Second, H–Br is a *strong* acid that protonates water extensively to give the hydronium ion, H_3O^+, which is a *weaker* acid than undissociated hydrobromic acid and is unable to protonate the alkene rapidly under mild reaction conditions. Moreover, the presence of water in the reaction mixture introduces the possibility of competing acid-catalyzed addition of water to the alkene (Eq. 10.17, Sec. 10.4). These problems are reduced by using anhydrous hydrogen bromide, but the highly corrosive nature of this gas makes it difficult to handle and use.

$$CH_3(CH_2)_3CH\!\!=\!\!CH_2 + H\!-\!Br \longrightarrow CH_3(CH_2)_3CH\!-\!CH_2 \qquad (10.20)$$
$$\underset{\substack{\textbf{40}\\ \text{1-Hexene}}}{} \qquad\qquad\qquad \underset{\substack{\overset{|}{Br}\;\;\overset{|}{H}\\ \textbf{41}\\ \text{2-Bromohexane}}}{}$$

A convenient solution to these experimental difficulties entails the addition of a catalytic amount of a quaternary ammonium salt such as methyltrioctylammonium chloride, $CH_3(n\text{-}C_8H_{17})_3N^+Cl^-$, also known as Aliquat 336®, to the heterogeneous mixture of the aqueous acid and the alkene. The tetraalkyl ammonium salt partitions between the aqueous and organic phases because of the amphoteric nature of the catalyst: It is **lipophilic** or **nonpolar-loving** due to the alkyl groups and **hydrophilic** or **polar-loving** because of the ionic ammonium function, respectively. By forming a complex with the H–Br, for example **42**, the quaternary ammonium salt extracts H–Br from the aqueous phase, transporting it into the organic phase and the presence of the alkene, as depicted in Scheme 10.1. The quaternary ammonium salt repartitions into the aqueous phase to complete the catalytic cycle. The transfer of the H–Br into the organic phase essentially dehydrates the acid, making it more reactive toward the alkene so that the addition becomes possible.

SCHEME 10.1

$$H\!-\!Br + R_4\overset{+}{N}X^- \rightleftharpoons R_4\overset{+}{N}X^- \text{----} H\!-\!Br$$
$$\textbf{42}$$

Aqueous Phase

Phase Interface

Organic Phase

$$CH_3(CH_2)_3\underset{\overset{|}{Br}}{C}H\!-\!\underset{\overset{|}{H}}{C}H_2 + R_4\overset{+}{N}X^- \xleftarrow{\text{1-Hexene}} R_4\overset{+}{N}X^- \text{----} H\!-\!Br$$
$$\textbf{42}$$

Compounds that promote the transport of reagents between immiscible layers by means of **ion pairs** like **42** are called **phase-transfer catalysts**. Their presence can have dramatic effects on the rates of bimolecular reactions between reagents contained in immiscible phases: Rate accelerations of 10^4 to 10^9 are common. One factor that determines the overall rate of a reaction involving phase-transfer catalysis is the efficiency of the partitioning of the reagents and reactants between the two phases. This is a function of the total surface area at the interface of the phases. To increase this area, the reaction mixture must be *vigorously* agitated by stirring or shaking to promote emulsification and the formation of tiny droplets of the immiscible layers.

As depicted in Scheme 10.2, the mechanism of the reaction presumably involves protonation of 1-hexene to afford the secondary carbocation **43**. Attack of bromide ion on this ion then leads to 2-bromohexane (**41**). The carbocation may also rearrange by way of a hydride shift (Sec. 10.3) to provide a different secondary carbocation, **44**, which would provide 3-bromohexane (**45**) upon reaction with bromide ion. Alternatively, it may deprotonate to 2-hexene (**46**), addition of H–Br to which could afford both **41** and **45**. In this experiment you will have the opportunity to assess whether or not any 3-bromohexane is produced in the reaction of 1-hexene with H–Br.

SCHEME 10.2

EXPERIMENTAL PROCEDURES

ADDITION OF HYDROGEN BROMIDE TO 1-HEXENE

Purpose: To demonstrate polar addition of H–Br to an alkene and Markovnikov's rule.

Safety Alert

1. Concentrated hydrobromic acid (47–49%) is a *highly corrosive* and *toxic* material. Working at a hood, measure out the amount of acid required and avoid inhaling its vapors. Use latex gloves when transferring the acid between containers. If the acid comes in contact with your skin, flood the affected area immediately and thoroughly with water and rinse it with 5% sodium bicarbonate solution.

2. Quaternary ammonium salts are *toxic* substances and can be absorbed through the skin. Should they accidentally come in contact with your skin, wash the affected area immediately with large amounts of water.

3. *If* a flame must be used in the distillation step, be certain that all joints in the apparatus are well lubricated and tightly mated because 1-hexene and the solvents used, particularly petroleum ether, are highly flammable. *Flameless* heating is preferred.

Miniscale Procedure

Preparation   Answer the Pre-Lab Exercises on page PL. 57. Review Sections 2.4, 2.8, 2.9, 2.11, 2.13, 2.17, 2.19, 2.21, 2.22, and 6.4.

Apparatus   A 25- and a 50-mL round-bottom flask, separatory funnel, apparatus for heating under reflux, simple distillation, magnetic stirring, and *flameless* heating.

Setting Up   Combine 3.0 mL of 1-hexene, 14 mL of 48% aqueous hydrobromic acid, and 1.0 g of methyltrioctylammonium chloride in the 50-mL round-bottom flask. Equip the flask with a water-cooled condenser and set up the apparatus heating under reflux; the drying tube shown in the figure is unnecessary for this set-up.

Addition and Work-Up   With *rapid* stirring, heat the heterogeneous reaction mixture under gentle reflux for 2 h and then allow the mixture to cool to room temperature.★

Carefully transfer the two-phase mixture to a separatory funnel, rinse the reaction flask with 15 mL of petroleum ether (bp 60–80 °C, 760 Torr), and add the rinse to the separatory funnel. Shake the funnel thoroughly and allow the layers to separate. Verify that the lower one is the aqueous phase and remove it. Sequentially wash the organic phase with two 15-mL portions of 10% sodium bicarbonate solution. Vent the funnel *frequently* because gas is evolved and excessive pressure must *not* develop in the funnel. Transfer the organic layer to an Erlenmeyer flask and dry it with swirling over several spatula-tips full of anhydrous sodium sulfate for at least 0.5 h.★ Add additional portions of anhydrous sodium sulfate if the liquid remains cloudy.

Isolation   Decant or gravity-filter the dried solution into a 25-mL round-bottom flask, equip the flask for simple distillation, and distil the product. Carefully control the rate of heating throughout the course of this distillation because severe foaming can occur. Do not attempt to remove the solvent too rapidly, as this may result in excessive loss of product due to foaming.

After the first fraction containing solvent and unreacted 1-hexene is removed over a range of room temperature–80 °C (760 Torr),★ change the receiving flask and collect the bromohexane as a single fraction in a tared container. Because the volume of product is likely to be only 1–2 mL, an accurate boiling point may be difficult to obtain, and all material that distils above 110–115 °C (760 Torr) should be collected in order to obtain a reasonable yield.

Analysis   Weigh the distillate and determine the yield of crude 2-bromohexane. Verify that the principal product is 2-bromohexane rather than 1- or 3-bromohexane by IR and/or NMR spectroscopic methods or by GLC, assuming authentic samples of the

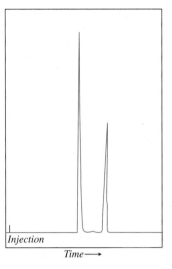

FIGURE 10.27
GLC analyses of the bromohexanes.

(a) Mixture of diethyl ether
(peak 1), 2-bromohexane (peak 2)
and 1-bromohexane (peak 3).

(b) Distillate from addition of
HBr to 1-hexene. Column and
conditions: 0.25-mm × 25-m AT-1
Helifex 100% dimethylsiloxane;
110 °C, 40 mL/min.

isomers are available. A GLC trace of a mixture of 1- and 2-bromohexanes is given in Figure 10.27a, along with a chromatogram of the product of the reaction (Fig. 10.27b); note that 2- and 3-bromohexane are *not* separated under the GLC conditons used. Obtain IR and ^{1}H NMR spectra of your starting material and product and compare them with those of authentic samples (Figs. 10.28–10.33). Alternatively, demonstrate whether or not Markovnikov addition has occurred by subjecting the product to the silver nitrate or sodium iodide/acetone tests for classifying alkyl halides (Secs. 23.9A and 23.9B, respectively).

Microscale Procedure

Preparation   Answer the Pre-Lab Exercises on page PL. 57. Review Sections 2.4, 2.8, 2.9, 2.11, 2.13, 2.17, 2.19, 2.21, 2.22, and 6.4.

Apparatus   A 5-mL conical vial, screw-cap centrifuge tube, apparatus for heating under reflux, simple distillation, magnetic stirring, and *flameless* heating.

Setting Up and Addition   Equip the conical vial with a spinvane and add to it 500 μL (0.5 mL) of 1-hexene, 2 mL of 48% aqueous hydrobromic acid, and 150 mg of methyltrioctylammonium chloride. Equip the vial with a water-cooled condenser and assemble the apparatus for heating under reflux; the drying tube shown in the figure is unnecessary for this set-up. With *rapid* stirring, heat the heterogeneous mixture under reflux for about 2 h, and then allow the reaction mixture to cool to room temperature.★

Work-Up   Add 1 mL of petroleum ether (bp 60–80 °C, 760 Torr) to the reaction mixture, and shake the vial, venting as necessary to relieve pressure. Allow the layers to separate. Remove the aqueous layer with a Pasteur pipet. Do *not* discard this liquid until you are certain that it is indeed the aqueous layer!

Add 0.5 mL of 10% sodium bicarbonate solution, stir the mixture (*Caution:* gas evolution!), and remove the aqueous layer after allowing separation. Using pHydrion paper, check to determine whether this layer is basic. If it is not, repeat washing the organic layer with aqueous sodium carbonate until the aqueous washes are basic.

Finally, dry the reaction mixture over several microspatula-tips full of anhydrous sodium sulfate for about 15 min; swirl the vial occasionally to hasten drying. Add additional portions of anhydrous sodium sulfate if the liquid remains cloudy.

Distillation   Using a filter-tip pipet, transfer the dried organic layer into a clean 5-mL conical vial equipped with a spinvane. Attach the vial to a Hickman stillhead equipped with a water-cooled condenser (Fig. 2.36) and perform a microscale distillation. The first fraction should contain solvent and unchanged 1-hexene and should distil at a bath temperature below 110 °C. Remove this fraction from the still, and collect a second fraction, using a bath temperature of 165 °C. This fraction should mainly be comprised of the desired product. Transfer this distillate to a tared screw-cap vial.

Analysis   Weigh the distillate and determine the yield of crude 2-bromohexane. Verify that the principal product is 2-bromohexane rather than 1- or 3-bromohexane by IR and/or NMR spectroscopic methods or by GLC, assuming authentic samples of the isomers are available. A GLC trace of a mixture of 1- and 2-bromohexanes is given in Figure 10.27a, along with a chromatogram of the product of the reaction (Fig. 10.27b); note that 2- and 3-bromehexane are *not* separated under the GLC conditions used. Obtain IR and ¹H NMR spectra of your starting material and product and compare them with those of authentic samples (Figs. 10.28–10.33). Alternatively, demonstrate whether or not Markovnikov addition has occurred by subjecting the product to the silver nitrate or sodium iodide/acetone tests for classifying alkyl halides (Secs. 23.9A and 23.9B, respectively).

Finishing Touches

Carefully (foaming may occur!) combine the *aqueous layers and washes*, neutralize if necessary with sodium carbonate, and flush them down the drain with excess water. Since the *first distillation fraction* may contain some halogenated product, pour it in a container for halogenated organic solvents. Evaporate the petroleum ether from the *sodium sulfate* used as drying agent in the hood, and place the solid in a container for nonhazardous solids. Dilute the *silver nitrate test solution* with water and flush it down the drain with excess water unless instructed to do otherwise.

EXERCISES

1. It was stated that aqueous HBr exists predominantly as $H_3O^+ Br^-$. Confirm the validity of this statement by calculating the equilibrium constant for the process shown below. The values in parenthesis are the pK_as of the indicated acids. (*Hint:* See Section 5.3.)

$$H\!-\!Br + H_2O \rightleftharpoons H_3O^+ + Br^-$$
$$(-8) \qquad\qquad\qquad (-1.7)$$

2. The yield of 2-bromohexane is much higher when the reaction mixture is stirred than when it is not. Why is this?

3. Account mechanistically for the fact that polar addition of H–Br to 1-hexene preferentially forms 2- rather than 1-bromohexane.

4. When methyltrioctylammonium chloride is used as the phase transfer catalyst, formation of 2-*chloro*hexane as a by-product is possible. Why would contamination of the desired 2-bromohexane by the chloro compound be a minor concern in this experimental procedure?

5. Why is methyltrioctylammonium chloride only partially soluble in 1-hexene? In aqueous hydrobromic acid?

6. Outline a procedure that would allow monitoring of the course of this addition reaction as a function of time. In other words, how might the reaction mixture be analyzed periodically so that you could determine when the reaction was complete?

7. Devise an experiment that would demonstrate that a phase-transfer catalyst accelerates the rate of reaction between 1-hexene and aqueous hydrobromic acid. Why would it be difficult to perform the ionic addition of hydrogen bromide to 1-pentene using the procedure outlined in this experiment?

8. Refer to Figure 10.27a and determine whether the stationary phase used for the GLC analysis shown separates the bromohexanes on the basis of their relative boiling points or of some other property. Explain your answer.

9. What is the mechanistic basis (a) of the silver nitrate and (b) of the sodium iodide in acetone tests for differentiating 1° from 2° alkyl bromides from each other?

Refer to Tables 8.1–8.8 as needed for answering the following questions on spectroscopy.

10. Consider the spectral data for 1-hexene (Figs. 10.28 and 10.29).

   a. In the IR spectrum, specify the absorptions associated with the carbon-carbon double bond and the vinylic hydrogen atoms.

   b. In the ^{1}H NMR spectrum, assign the various resonances to the hydrogen nuclei responsible for them.

   c. For the ^{13}C NMR data, assign the various resonances to the carbon nuclei responsible for them.

11. Consider the NMR data for 1-bromohexane (Fig. 10.31).

   a. In the ^{1}H NMR spectrum, assign the various resonances to the hydrogen nuclei responsible for them.

   b. For the ^{13}C NMR data, assign the various resonances to the carbon nuclei responsible for them.

12. Consider the NMR data for 2-bromohexane (Fig. 10.33).

   a. In the ^{1}H NMR spectrum, assign the various resonances to the hydrogen nuclei responsible for them.

   b. For the ^{13}C NMR data, assign the various resonances to the carbon nuclei responsible for them.

13. Discuss the differences observed in the IR and NMR spectra of 1-hexene and 2-bromohexane that are consistent with addition of HBr occurring in this experiment.

14. Assume you obtained the ^{1}H NMR spectrum of a mixture of 1- and 2-bromohexane and found that the integrations for the multiplets centered at δ 3.4 and 4.1 were 20 units and 40 units, respectively. What is the relative ratio of these two compounds in the sample?

15. The ^{1}H NMR spectrum of the crude bromohexane isolated by distillation in one experiment reveals a multiplet centered at δ 4.00, which is slightly upfield of the multiplet centered at δ 4.13 for 2-bromohexane.

a. Determine whether this multiplet may signal the presence of 3-bromohexane by looking up the ^{1}H NMR spectrum of this isomer in either the Aldrich or Sadtler catalogs of reference spectra.

b. The integrated area of the multiplet centered at δ 4.00 is 45 units, whereas that of the multiplet centered at δ 4.13 is 200 units. Assuming that the upfield multiplet represents 3-bromohexane, what is its percentage in the mixture of bromohexanes?

Spectra   *Starting Materials and Products*

FIGURE 10.28
IR spectrum of 1-hexene (neat).

FIGURE 10.29
NMR data for 1-hexene (CDCl$_3$).

(a) ^{1}H NMR spectrum (300 MHz).

(b) ^{13}C data: δ 14.0, 22.5, 31.6, 33.8, 114.3, 139.2.

FIGURE 10.30
IR spectrum of 1-bromohexane
(neat).

(a) ^{1}H NMR spectrum (300 MHz).

FIGURE 10.31
NMR data for 1-bromohexane
(CDCl$_3$).

(b) ^{13}C data: δ 14.0, 22.7, 28.1, 31.2, 33.1, 33.4.

FIGURE 10.32
IR spectrum of 2-bromohexane
(neat).

FIGURE 10.33
NMR data for 2-bromohexane
(CDCl₃).

(a) ^{1}H NMR spectrum (300 MHz).

$CH_3(CH_2)_3\overset{\underset{\displaystyle |}{Br}}{C}HCH_3$

4.22 ppm   1.9   1.7   1.5   1.3   1.1 ppm

δ, ppm

(b) ^{13}C data: δ 13.9, 22.1, 26.5, 29.9, 41.0, 51.0.

10.6 Bromination of Alkenes

The reaction of bromine with (*E*)-stilbene (**47**) to give *meso*-stilbene dibromide (**48**) as the major product (Eq. 10.21) is another example of an electrophilic addition reaction of alkenes. The addition of bromine to many alkenes is a **stereospecific** reaction that proceeds by *anti* addition to the double bond. However, the addition of bromine to **47** is *not* stereospecific because small amounts of *dl*-stilbene dibromide (**49**) are also formed in this reaction. The formation of *meso*-stilbene dibromide presumably occurs via the nucleophilic attack of bromide on an intermediate cyclic bromonium ion **46**. The possible interconversion of **50** and the acyclic carbocation **51** (Eq. 10.22) is one possible way to account for the presence of *dl*-stilbene dibromide in the product.

The dibromides **48** and **49** are *diastereomers* and thus have different physical and chemical properties (Sec. 7.1). Indeed, their melting points differ by 100 °C. This makes it is easy to identify the major product of the bromination.

$$\text{(structures)} \qquad (10.21)$$

47
(*E*)-Stilbene
Colorless

48
meso-Stilbene dibromide
Colorless

49
dl-Stilbene dibromide
Colorless

(+ enantiomer)

(10.22)

50   **51**

EXPERIMENTAL PROCEDURES

BROMINATION OF (*E*)-STILBENE

Purpose: To demonstrate the *anti* stereochemistry of addition of bromine to an alkene.

> **Safety Alert**
>
> 1. *Bromine is a hazardous chemical.* Do not breathe its vapors or allow it to come into contact with the skin because it may cause serious chemical burns. Perform all operations involving the transfer of the pure liquid or its solutions at a hood, and wear latex gloves. If you get bromine on your skin, wash the area immediately with soap and warm water and soak the affected area in 0.6 *M* sodium thiosulfate solution, for up to 3 h if the burn is particularly serious.
> 2. Dispense the 1 *M* bromine in dichloromethane solution from burets or similar devices fitted with Teflon stopcocks and located in hoods.
> 3. Bromine reacts with acetone to produce the powerful **lachrymator**, α-bromoacetone, $BrCH_2COCH_3$. Do *not* rinse glassware containing residual bromine with acetone!
> 4. All parts of this experiment should be conducted in a hood if possible.

Miniscale Procedure

Preparation   Answer the Pre-Lab Exercises on page PL. 59. Review Sections 2.7, 2.11, and 2.17.

Apparatus   A 25-mL round-bottom flask, apparatus for magnetic stirring and vacuum filtration.

Setting Up   Equip the flask with a stirbar and add to it 0.9 g of (*E*)-stilbene and 10 mL of dichloromethane. Stir or swirl the mixture to effect dissolution.

Bromination and Isolation   Measure 5 mL of freshly prepared 1 *M* bromine in dichloromethane directly from the dispenser into the round-bottom flask. Swirl the flask gently during the addition to mix the contents. After the addition of the bromine solution is complete, stopper the reaction flask loosely and stir the mixture for 15 min. Isolate the product by vacuum filtration. Wash the product with one or two 1-mL portions of *cold* dichloromethane until it is white. Transfer the product to a watchglass or a piece of filter or weighing paper and allow it to air-dry. Recrystallize the product from xylene.

Analysis   Weigh the product and determine the yield. Measure the melting point to confirm the stereochemistry of the bromination. *Caution:* Do *not* use mineral oil as the heating fluid for this determination! Obtain IR and ^{1}H NMR spectra of your starting material and product and compare them with those of authentic samples (Figs. 10.34–10.37).

Microscale Procedure

Preparation   Answer the Pre-Lab Exercises on page PL. 59. Review Sections 2.7, 2.11, and 2.17.

Apparatus   A 5-mL conical vial, apparatus for magnetic stirring, vacuum filtration, and Craig tube filtration.

Setting Up   Equip the vial with a spinvane and add to it 180 mg of (*E*)-stilbene and 2 mL of dichloromethane. Stir or swirl the mixture to effect dissolution.

Bromination and Isolation   Measure 1 mL of freshly prepared 1 *M* bromine in dichloromethane directly from the dispenser into the vial. Swirl the vial gently during the addition to mix the contents. After the addition of the bromine solution to the vial is complete, loosely screw a cap onto the vial and stir the mixture for 15 min. Isolate the product by vacuum filtration using a Hirsch funnel. Wash the product with several drops of *cold* dichloromethane until it is white. Transfer the product to a watchglass or a piece of filter or weighing paper and allow it to air-dry. Using a Craig tube, recrystallize the product from xylene.

Analysis   Weigh the product and determine the yield. Measure the melting point to confirm the stereochemistry of the bromination. *Caution:* Do *not* use mineral oil as the heating fluid for this determination! Obtain IR and ^{1}H NMR spectra of your starting material and product and compare them with those of authentic samples (Figs. 10.34–10.37).

Finishing Touches   Decolorize any *solutions* in which the color of bromine is visible by the dropwise addition of cyclohexene; then discard the resulting solutions together with *all other solutions* in a container for halogenated organic liquids.

EXERCISES

1. Why does solid not separate immediately after you begin adding the solution of bromine to that containing (*E*)-stilbene?

2. Why should mineral oil *not* be used as the heating fluid for determining the melting point of *meso*-stilbene dibromide?

3. Write the structure of the product obtained when cyclohexene is used to decolorize bromine-containing solutions.

4. Using suitable stereochemical structures, write the mechanism for the addition of bromine to (*E*)-stilbene to give *meso*-stilbene dibromide via the intermediate cyclic bromonium ion **50**. Use curved arrows to symbolize the flow of electrons.

5. Draw suitable three-dimensional structures of the two products obtained by attack of bromide ion at both of the carbon atoms in the cyclic bromonium ion **50** and show that they are identical.

6. Based upon the mechanism you provided in Exercise 4, predict the major product of the addition of bromine to (*Z*)-stilbene.

7. Demonstrate that attack of bromide ion, Br^-, on the carbocation **51** can provide *both dl-* and *meso*-stilbene dibromide.

8. The reaction of bromine with cyclopentene according to Equation 10.18 (p. 299) is stereospecific and proceeds by *anti* addition. On the other hand the addition of bromine to (*E*)-stilbene gives both *meso*-stilbene dibromide (**48**) and *dl*-stilbene dibromide (**49**). Rationalize this difference by comparing the relative stabilities of the carbocations resulting from ring-opening of the cyclic bromonium ions **38** and **50**.

9. Write the expected products for the addition of bromine to the following alkenes.

a. $CH_3CH{=}CH_2$

b.

c.

Refer to Tables 8.1–8.8 as needed for answering the following questions on spectroscopy.

10. Consider the spectral data for (*E*)-stilbene (Figs. 10.34 and 10.35).

 a. In the IR spectrum, identify the absorption(s) consistent with the *trans* relationship of the vinylic hydrogen atoms and explain the absence of a significant absorption in the range of $1600-1700 \, cm^{-1}$ normally characteristic of alkenes.

 b. In the ^1H NMR spectrum, assign the various resonances to the hydrogen nuclei responsible for them.

 c. For the ^{13}C NMR data, assign the various resonances to the carbon nuclei responsible for them.

11. Consider the ^1H NMR spectrum of *meso*-stilbene dibromide (Fig. 10.37a).

 a. In the ^1H NMR spectrum, assign the various resonances to the hydrogen nuclei responsble for them.

 b. For the ^{13}C NMR data, assign the various resonances to the carbon nuclei responsible for them.

12. Discuss the differences observed in the IR and NMR spectra of (*E*)-stilbene and *meso*-stilbene dibromide that are consistent with addition of Br_2 occurring in this experiment.

Spectra   *Starting Materials and Product*

FIGURE 10.34
IR spectrum of (*E*)-stilbene
(IR card).

FIGURE 10.35
NMR data for (*E*)-stilbene (CDCl₃).

(a) ¹H NMR spectrum (300 MHz).

(b) ¹³C data: δ 126.3, 127.8, 128.9, 129.0, 137.6.

FIGURE 10.36
IR spectrum of *meso*-stilbene
dibromide (IR card).

(a) ^{1}H NMR spectrum (300 MHz).

FIGURE 10.37
NMR data for *meso*-stilbene dibromide (CDCl$_3$).

(b) ^{13}C data: δ 56.8, 129.1, 129.6, 129.7, 141.7

10.7 Acid-catalyzed Hydration of Alkenes

The acid-catalyzed addition of the elements of water across a carbon-carbon π-bond to give an alcohol is referred to as **hydration** of an alkene (Eq. 10.17). Mechanistically, this process is simply the reverse of the acid-catalyzed **dehydration** of alcohols (Sec. 10.3). The position of the equilibrium for these two competing processes depends upon the reaction conditions. Hydration of a double bond requires the presence of excess water to drive the reaction to completion, whereas the dehydration of an alcohol requires removing water to complete the reaction. In the experiment that follows, the acid-catalyzed hydration of norbornene (**52**) to give exclusively *exo*-norborneol (**53**) (Eq. 10.23) is examined.

$$\text{52} \xrightarrow{\text{H}_3\text{O}^+} \text{53} \qquad \qquad \text{(10.23)}$$

52
Norbornene

53
exo-Norborneol

The mechanism of this reaction and the structure of the intermediate carbocation have been the subject of intense controversy for over 30 years. When the double bond of norbornene is protonated, a carbocation is produced (Eq. 10.24). The question of whether this cation is **54a** and is in rapid equilibrium with the isomeric cation **54b**, or whether **54a** and **54b** are simply two contributing resonance structures to the resonance hybrid **55**, has been the focus of the debate. Note that the contributing structures **54a** and **54b** *differ in the location of a σ-bond*, not in the location of π-bonds as is more commonly encountered in **resonance structures**. The delocalized structure **55** is thus referred to as a **nonclassical carbocation**, and most chemists now accept this formulation as the more likely representation of this intermediate. It is evident from empirical results that the sterically more accessible side of the carbocation is the one away from, or *exo* to the bridging carbon atom as shown. Nucleophilic attack of water solely from this side leads to *exo*-norborneol (**53**) rather than *endo*-norborneol (**56**).

$$\text{(10.24)}$$

52   **54a**   **54b**

56   **55**   **53**
endo-Norborneol   *exo*-Norborneol

EXPERIMENTAL PROCEDURES

HYDRATION OF NORBORNENE

Purpose: To demonstrate the acid-catalyzed addition of water to an alkene.

Safety Alert

1. Norbornene is a volatile solid. Perform all weighings in a hood if possible.
2. If the acidic solution used in the first part of the experiment comes in contact with your skin, immediately flood the affected area with water and thoroughly rinse it with 5% sodium bicarbonate solution.
3. If the basic solution used in this experiment comes in contact with your skin, immediately flood the affected area with water and thoroughly rinse the area with a dilute solution (1%) of acetic acid.

Miniscale Procedure

Preparation   Answer the Pre-Lab Exercises on page PL. 61. Review Sections 2.7, 2.20, 2.21, and 2.29.

Apparatus   A 25-mL Erlenmeyer flask, separatory funnel, ice-water bath, apparatus for simple distillation, magnetic stirring, sublimation, and *flameless* heating.

Setting Up   Slowly add 2 mL of concentrated sulfuric acid to 1 mL of water in a 25-mL Erlenmeyer flask containing a stirbar and cool the solution to room temperature.

Hydration   Working at the hood, add 1.0 g of norbornene in small pieces to the stirred solution of aqueous sulfuric acid. Continue stirring the mixture until all of the norbornene dissolves, cooling the flask *briefly* in an ice-water bath if the mixture becomes noticeably warm; *do not cool the contents of the flask below room temperature.* Stir this solution 15–30 min. Prepare a solution of 1.5 g of potassium hydroxide in 7.5 mL of water.

Work-Up and Isolation   Cool both solutions in an ice-water bath and slowly add the base to the acidic reaction mixture to partially neutralize the acid. Be sure that the resulting mixture is at room temperature or below, then transfer it to a small separatory funnel, and add 15 mL of diethyl ether. If any solid is present in the bottom of the separatory funnel, add 1–2 mL of water to dissolve it. After shaking the funnel, *periodically venting it to release pressure*, separate the aqueous layer (*Save!*) and pour the ethereal solution into an Erlenmeyer flask. Return the aqueous layer to the separatory funnel, and extract again with a fresh 10-mL portion of diethyl ether. Separate the aqueous layer, and combine the two ethereal extracts in the separatory funnel.

Wash the combined extracts with 5 mL of water. Separate the aqueous layer, and wash the ethereal solution sequentially with 5-mL portions of saturated sodium bicarbonate solution and saturated sodium chloride solution; *periodically vent the separatory funnel* to release pressure that may develop. Separate the layers, place the ethereal solution in a dry 50-mL Erlenmeyer flask, and dry it over several spatula-tips full of anhydrous sodium sulfate with occasional swirling for 10–15 min.★ Add additional portions of anhydrous sodium sulfate if the liquid remains cloudy.

Transfer the dried ethereal solution to a 50-mL round-bottom flask equipped for simple distillation and concentrate the solution by distillation. Alternatively, rotary evaporation or other techniques (Sec. 2.29) may be used to concentrate the solution. Transfer the residue to a 25- or 50-mL filter flask.

Sublimation   Equip the filter flask with a filter adapter or rubber stopper containing a centrifuge tube or a 13-mm × 120-mm test tube that extends to within about 2 cm of the bottom of the flask (Fig. 2.56b). The adapter or stopper should be lightly greased to prevent water from entering the sublimation chamber. Evacuate the flask using a water aspirator equipped with a trap and *then* half-fill the centrifuge tube or test tube with tightly packed chipped ice. *Gently* heat the flask using a steam bath or the sweeping motion of a heat gun or small flame. Add ice as necessary to the cold finger on which the sublimed alcohol collects. When the alcohol has completely sublimed, cool the flask, break the vacuum to the aspirator, and disconnect the flask. Carefully decant the water out of the cold finger *before* carefully removing it from the flask. Maintain the cold finger in a horizontal or slightly inverted position. Scrape the soft crystalline product from the test tube.

Analysis   Weigh the purified solid and determine the yield. Measure the melting point by placing the sample in a capillary tube that has been sealed about 2 cm from the open end. Obtain IR and ^{1}H NMR spectra of your starting material and product and compare them with those of authentic samples (Figs. 10.38–10.41).

Finishing Touches

Slowly combine the *aqueous layers and washes*. If necessary, neutralize them with sodium carbonate, and flush them down the drain with excess water. Pour any *diethyl ether* obtained during the concentration step into a container for flammable organic solvents. After the diethyl ether has evaporated from the *sodium sulfate* in the hood, place the solid in a container for nonhazardous solids.

EXERCISES

1. What purpose is served by washing the ethereal solution of product with bicarbonate solution? Saturated sodium chloride solution?

2. How would the yield of *exo*-norborneol be affected if it was left under vacuum for an excessive period of time to remove all traces of solvent?

3. Why is it necessary to seal the capillary tube before taking the melting point of *exo*-norborneol?

4. *exo*-Norborneol undergoes reaction with concentrated H–Br to form a single diastereomer. Write the structure of this bromide and propose a stepwise mechanism for its formation, using curved arrows to symbolize the flow of electrons. Why is only one diastereomer produced?

5. *exo*-Norborneol contains three chiral or stereogenic centers, but the product obtained from the hydration of norbornene in this experiment is optically inactive. Identify these three centers with asterisks and explain why the product is optically inactive.

6. For each of the alkenes **a**–**e**, write the structure of the alcohol that should be the major product obtained upon acid-catalyzed hydration.

Refer to Tables 8.1–8.8 as needed for answering the following questions on spectroscopy.

7. Consider the spectral data for norbornene (Figs. 10.38 and 10.39).

   a. In the functional group region of the IR spectrum, identify any absorptions associated with the carbon-carbon double bond.

   b. In the ^1H NMR spectrum, assign the various resonances to the hydrogen nuclei responsible for them.

   c. For the ^{13}C NMR data, assign the various resonances to the carbon nuclei responsible for them. Explain why there are only four resonances in the spectrum.

Starting Material and Product

FIGURE 10.38
IR spectrum of norbornene
(IR card).

(a) ^{1}H NMR spectrum (300 MHz)

FIGURE 10.39
NMR data for norbornene (CDCl$_3$).

(b) ^{13}C data: δ 24.8, 42.0, 48.8, 135.4.

8. Consider the spectral data for *exo*-norborneol (Figs. 10.40 and 10.41).

   a. In the functional group region of the IR spectrum, specify any absorptions associated with the alcohol function.

   b. In the ^{1}H NMR spectrum, assign the various resonances to the hydrogen nuclei responsible for them.

   c. For the ^{13}C NMR data, assign the various resonances to the carbon atom responsible for them.

9. Discuss the differences observed in the IR and NMR spectra of norbornene and *exo*-norborneol that are consistent with the addition of water occurring in this experiment.

FIGURE 10.40

IR spectrum of *exo*-norborneol (IR card).

FIGURE 10.41

NMR data for *exo*-norborneol (CDCl₃).

(a) ^{1}H NMR spectrum (300 MHz).

(b) ^{13}C data; δ 24.7, 28.4, 34.5, 35.5, 42.2, 44.2, 74.6.

10.8 Hydroboration-Oxidation of Alkenes

The **regiochemistry** of acid-catalyzed hydration of alkenes to give alcohols is that predicted by Markovnikov's rule (Sec. 10.4) because the more stable intermediate carbocation is preferentially formed. Sometimes it is desirable to add the elements of water across a carbon-carbon π-bond in the opposite regiochemical sense to provide the **anti-Markovnikov** product (see the Historical Highlight at the end of this chapter). In order to accomplish this goal, a process termed **hydroboration-oxidation** is available that involves the reaction of an alkene sequentially with diborane, B_2H_6, and basic hydrogen peroxide (Eq. 10.25).

$$\underset{R^1}{\overset{R^2}{\diagup}}C{=}CHR^3 \xrightarrow[\text{2. } H_2O_2/HO^-]{\text{1. } B_2H_6} R^1{-}\underset{\underset{H}{|}}{\overset{\overset{R^2}{|}}{C}}{-}\underset{\underset{OH}{|}}{\overset{\overset{R^3}{|}}{C}}H \qquad (10.25)$$

Diborane is a dimer of borane, BH_3. The bonding in diborane is unusual because the hydrogen atoms bridge the two boron atoms with the two monomeric BH_3 subunits being bound with *two-electron, three-center bonds*. Because the boron atom in borane possesses an empty *p*-orbital, borane is a Lewis acid, and it forms stable complexes upon reaction with tetrahydrofuran (THF) and other ethers, which function as Lewis bases as illustrated by the formation of a borane-THF complex (Eq. 10.26).

$$\begin{array}{ccc} \text{Diborane} & \text{Tetrahydrofuran} & \text{A borane-THF complex} \end{array}$$

(10.26)

Alkenes may serve as Lewis bases and react with borane by the net addition of the boron-hydrogen bond across the carbon-carbon π-bond to give a monoalkylborane as the initial product. This process, known as **hydroboration**, is illustrated for 1-butene in Scheme 10.3. Depending upon steric factors and the degree of substitution on the double bond, the intermediate monoalkylborane, **57**, may add successively to additional molecules of the same alkene to form a dialkylborane **58** or a trialkylborane **59**.

SCHEME 10.3

$$CH_3CH_2CH{=}CH_2 \longrightarrow CH_3CH_2CH{-}CH_2 \xrightarrow{\text{1-butene}} [CH_3(CH_2)_2CH_2]_2BH \xrightarrow{\text{1-butene}} [CH_3(CH_2)_2CH_2]_3B$$

$$\begin{array}{cccc} \text{1-Butene} & \underset{\text{A monoalkylborane}}{\textbf{57}} & \underset{\text{A dialkylborane}}{\textbf{58}} & \underset{\text{A trialkylborane}}{\textbf{59}} \end{array}$$

As observed in the electrophilic additions to alkenes depicted in Equations 10.16 and 10.17, the addition of borane to alkenes is *regioselective* (Sec. 10.4) and occurs so the *electrophilic boron atom* preferentially attaches to the *less substituted carbon atom* of the carbon-carbon π-bond. Understanding this selectivity rests on the mechanism of hydroboration. As borane approaches the double bond, a Lewis acid–Lewis base interaction occurs in which the π-electrons of the double bond begin to interact with the empty *p*-orbital on the Lewis acidic boron atom, resulting in the development of a partial positive charge on the more highly substituted carbon atom. The boron atom then assumes a partial negative charge that facilitates the transfer of the hydride ion, $H:^-$, from the boron to the carbon atom bearing the partial positive charge; this occurs through the four-centered transition state **60**, in which all four atoms simultaneously undergo changes in bonding.

Thus, the regiochemistry of hydroboration is predicted by the same general rule that applies to all electrophilic additions to alkenes: *The reaction of an electrophile with a carbon-carbon double bond occurs preferentially via the transition state in which a partial positive charge develops on that carbon atom better able to accommodate it.* Geometric constraints inherent in the cyclic transition state **60** require that the addition of borane to the alkene proceed so that both the boron and the hydrogen add from the same face of the double bond, a process called *syn* **addition**.

$$\left[\begin{array}{c} \delta^- \\ H\text{-}\text{-}BH_2 \\ \\ \delta^+ \\ C_2H_5 \end{array} \right]^{\ddagger}$$

60

Treating a trialkylborane with alkaline hydrogen peroxide promotes oxidative cleavage to form three moles of an alcohol and boric acid, as exemplified in the conversion of **59** into 1-butanol (**64**), shown in Scheme 10.4. The first step of the reaction involves nucleophilic attack of the hydroperoxide anion on the boron atom of **59** to give **61**. In the next step, migration of the alkyl group from boron to oxygen proceeds with displacement of hydroxide ion and cleavage of the weak oxygen-oxygen bond to give a monoalkoxyborane **62**. An important consequence of the *intramolecularity* of this rearrangement is that it proceeds with *retention* of configuration at the carbon atom. The sequence is repeated until all of the boron-carbon bonds have been converted into oxygen-carbon bonds, whereupon the intermediate trialkoxyborane **63** undergoes alkaline hydrolysis to give 1-butanol and boric acid. Thus, hydroboration–oxidation of an alkene gives an alcohol that is the product of overall *anti-Markovnikov* addition of the elements of H–OH to the carbon-carbon double bond. Note, however, that the key step determining the regiochemistry of the alcohol produced is the *Markovnikov* addition of the hydrogen-boron bond across the π-bond.

SCHEME 10.4

$$[CH_3(CH_2)_2CH_2]_3B \; + \; \overset{\cdot\cdot}{\underset{\cdot\cdot}{O}}{-}OH \longrightarrow [CH_3(CH_2)_2CH_2]_2\bar{B} \overset{CH_2(CH_2)_2CH_3}{\underset{O-OH}{\Big\langle}} \xrightarrow{\;-HO^-\;}$$

59   61

$$[CH_3(CH_2)_2CH_2]_2B{-}O{-}CH_2(CH_2)_2CH_3 \overset{\bar{\overset{\cdot\cdot}{O}}-OH}{\longrightarrow} CH_3(CH_2)_2CH_2B \overset{O(CH_2)_3CH_3}{\underset{O(CH_2)_3CH_3}{\Big\langle}}$$

62   A dialkoxyborane
A monoalkoxyborane

$$\overset{\bar{\overset{\cdot\cdot}{O}}-OH}{\longrightarrow} [CH_3(CH_2)_3O]_3B \xrightarrow[H_2O]{HO^-} 3 \; CH_3(CH_2)_2CH_2{-}OH \; + \; B(OH)_3$$

63   64
A trialkoxyborane   1-Butanol   Boric acid

In this experiment, the regiochemistry and stereochemistry of the sequence of hydroboration-oxidation are examined using (+)-α-pinene (**65**) as the alkene and a solution of borane-tetrahydrofuran complex as the hydroborating reagent (Eq. 10.27). Using this complex avoids the need of handling diborane, B_2H_6, a potentially hazardous reagent. Since the double bond in **65** is trisubstituted and sterically hindered, only *two* rather than three molecules of this alkene react with each molecule of borane, forming the intermediate dialkylborane **61**. Subsequent oxidation of **61** with alkaline hydrogen peroxide then gives (−)-isopinocampheol (**67**) as the sole product.

(10.27)

65   66   67
(+)-α-Pinene   (−)-Isopinocampheol

The hydroboration of (+)-α-pinene provides experimental proof that the addition of borane to alkenes occurs in a highly *regioselective* fashion with exclusive *syn* ad-

Table 10.1

Structure	Melting Point, °C	Melting Point, °C, of α-Naphthylurethane
67	51–53 (55–56)	87.5–88
68	45–47 (48)	88
69	27	148
70	67	91
71	78–79	not formed
72	58–59	not formed

dition. Another stereochemical feature of this reaction is that the borane reagent approaches the π-bond of **65** from only one face. Because only one of several possible stereoisomeric products is formed, the hydroboration of (+)-α-pinene is an example of a **stereoselective** reaction. The identity of the product of this reaction may be verified by determining its melting point and the melting point of a suitable derivative. Comparing these data with those (Table 10.1) for the alternative products **68–70**, which could also arise from adding borane in a Markovnikov sense, and the possible products **71** and **72** derived by anti-Markovnikov hydroboration, provides compelling evidence that the structure of the product is indeed **67**.

68	**69**	**70**	**71**	**72**
(+)-Neoisopinocampheol	(−)-Neopinocampheol	(+)-Pinocampheol	(+)-*cis*-2-Pinanol	(+)-*trans*-2-Pinanol

EXPERIMENTAL PROCEDURES

HYDROBORATION–OXIDATION OF (+)-α-PINENE

Purpose: To demonstrate hydration of an alkene by means of hydroboration–oxidation and to assess the regiochemistry and stereochemistry of the process.

Safety Alert

1. Tetrahydrofuran is a flammable and potentially toxic solvent. Do not use this solvent near open flames.
2. The 30% hydrogen peroxide used is a *strong oxidant* and may blister the skin on contact. If you accidentally spill some on your skin, wash the affected area with copious amounts of water.
3. The complex of borane and tetrahydrofuran reacts *vigorously* with water. Use care when disposing of any excess reagent and when quenching the reaction.

Rubber septum

FIGURE 10.42
Miniscale hydroboration apparatus.

Miniscale Procedure

Preparation   Answer the Pre-Lab Exercises on page PL. 63. Review Sections 2.5, 2.10, 2.11, 2.13, 2.21, 2.27, 2.28, and 2.29.

Apparatus   A 50-mL round-bottom flask, Claisen adapter, separatory funnel, two 5-mL glass syringes with needles, drying tube, water bath, apparatus for simple distillation, magnetic stirring, and *flameless* heating.

Setting Up   Equip the round-bottom flask with the Claisen adapter, and prepare a drying tube filled with calcium chloride for attachment to the *sidearm* of the adapter. Fit a rubber septum onto the joint of the adapter that is directly in line with the lower male joint (Fig. 10.42). Dry the glass parts of the syringes in an oven at 110 °C for at least 15 min. Using a small flame or a heat gun, gently heat the flask to remove traces of water from the apparatus; do *not* overheat the joint with the septum. After the flask has been heated for a few minutes, discontinue the heating, add a stirbar, attach the drying tube, and allow the flask to *cool to room temperature*. Remove the glass syringes from the oven and allow them to *cool to room temperature*, preferably in a desiccator. Prepare a solution of 1.6 mL of $(+)$-α-pinene in 2 mL of dry tetrahydrofuran.

Hydroboration   Draw 5 mL of a 1.0 M solution of borane-tetrahydrofuran complex in tetrahydrofuran into the *dry* 5-mL glass syringe. Pierce the rubber septum with the needle of the syringe, being careful not to press down on the plunger, and add the solution directly to the round-bottom flask. Place the apparatus in a water bath made from cold tap water and located on a magnetic stirrer. Use another dry 5-mL glass syringe to slowly transfer *dropwise* the solution of $(+)$-α-pinene to the stirred solution of borane-tetrahydrofuran complex; this addition should require about 10–15 min. Remove the water bath and stir the reaction mixture at room temperature for about 1.5 h. Remove the septum from the reaction flask, and place the flask in an ice-water bath on a magnetic stirrer. Slowly add 0.5 mL of water with stirring to the reaction mixture to decompose any unreacted borane-tetrahydrofuran complex.

Oxidation   Prepare 3 mL of an alkaline solution of hydrogen peroxide by mixing equal parts of 30 weight % aqueous hydrogen peroxide and 3 *M* aqueous NaOH. Using a Pasteur pipet, *slowly* add this solution to the stirred reaction mixture. Perform the addition *cautiously*, as vigorous gas evolution may occur. Once the addition is complete, stir the resulting mixture for about 5–10 min.

Work-Up and Isolation   Transfer the reaction mixture to a small separatory funnel. Add 5 mL of water and separate the layers. Extract the aqueous layer with two separate 10-mL portions of diethyl ether. Wash the combined organic layers sequentially with two 10-mL portions of water and two 10-mL portions of saturated brine, and dry them over several spatula-tips full of anhydrous sodium sulfate.★ Add additional portions of anhydrous sodium sulfate if the liquid remains cloudy. Filter or decant the dried solution into a dry 50-mL round-bottom flask and remove the volatile solvents by simple distillation. Alternatively, rotary evaporation or other techniques (Sec. 2.29) may be used to remove the solvents. The final traces of solvents may be removed by attaching the flask to a vacuum source, and gently swirling the contents as the vacuum is applied.★ Transfer the solid residue to a tared container.

Analysis   Weigh your product and determine the yield. Obtain IR and NMR spectra of your starting material and product and compare them with those of authentic samples

(Figs. 10.44–10.47). You may also verify the identity of the product by comparing its properties and the melting point of its α-naphthylurethane derivative, prepared according to the procedure in Section 23.11D, with the data summarized in Table 10.1.

Microscale Procedure

Preparation   Answer the Pre-Lab Exercises on page PL. 63. Review Sections 2.5, 2.10, 2.11, 2.13, 2.27, 2.28, and 2.29.

Apparatus   A 3- and 5-mL conical vial, two 1-mL glass syringes with needles, screw-cap centrifuge tube Claisen adapter, drying tube, water bath, apparatus for magnetic stirring, simple distillation, and *flameless* heating.

Setting Up   Dry the 5-mL conical vial, the Claisen adapter, and the glass parts of the two syringes in an oven at 110 °C for at least 15 min. Remove the vial and adapter from the oven, and after they have cooled, preferably in a desiccator, place a spinvane in the vial, attach the adapter, and fit its sidearm with a drying tube. Seal the other opening of the adapter with a rubber septum (Fig. 10.43). Place the apparatus in a water bath made from cold tap water and located on a magnetic stirrer. Remove the glass syringes from the oven and allow them to *cool to room temperature*, preferably in a desiccator. Prepare a solution of 0.3 mL of (+)-α-pinene in 0.5 mL of dry tetrahydrofuran.

FIGURE 10.43
Microscale hydroboration apparatus.

Hydroboration   Draw 1 mL of a 1.0 *M* solution of borane-tetrahydrofuran complex in tetrahydrofuran into a *dry* 1-mL glass syringe. Pierce the rubber septum with the needle of the syringe, being careful not to press down on the plunger, and add the solution directly to the conical vial. Use the other dry 1-mL glass syringe to slowly transfer *dropwise* the solution of (+)-α-pinene to the stirred solution of borane-tetrahydrofuran complex; this addition should require about 5–10 min. The rate of addition may be controlled by slowly twisting the plunger down the body of the syringe. Remove the water bath and stir the reaction mixture at room temperature for about 1.5 h. Remove the septum from the reaction vial and place the vial in an ice-water bath on a magnetic stirrer. Slowly add 2–3 drops of water with stirring to the reaction mixture to decompose any unreacted borane-tetrahydrofuran complex.

Oxidation   Prepare 0.6 mL of an alkaline solution of hydrogen peroxide by mixing equal parts of 30 weight % aqueous hydrogen peroxide and 3 *M* aqueous NaOH. Using a Pasteur pipet, *slowly* add this solution to the stirred reaction mixture. Perform this addition *cautiously*, as vigorous gas evolution may occur. Once the addition is complete, stir the resulting mixture for about 5 min and then add 1 mL of water followed by 1 mL of diethyl ether.

Work-Up and Isolation   Stir the mixture in the vial for about 5 min in order to mix the liquid and solid materials. Remove the lower aqueous layer using a Pasteur pipet and transfer it to a 3-mL conical vial; transfer the organic layer to a screw-cap centrifuge tube. Extract the aqueous layer three times with separate 1 mL-portions of diethyl ether, each time transferring the organic layer to the centrifuge tube. Vent the tube as necessary to relieve pressure. Wash the combined organic layers sequentially with two 2-mL portions each of water and saturated brine. Transfer the organic layer to a dry test tube using a Pasteur pipet, and add several microspatula-tips full of anhydrous sodium sulfate. Allow the solution to stand for 10–15 min, with occasional swirling to hasten drying.★ Add additional portions of anhydrous sodium sulfate if the liquid remains cloudy.

Using a Pasteur pipet, transfer the organic solution to a clean 5-mL conical vial and add a spinvane to the vial. Remove the volatile solvents by simple distillation. Alternatively, rotary evaporation or other techniques (Sec. 2.29) may be used to remove the solvent. Following removal of solvents, transfer the solid residue to a tared container.

Analysis Weigh your product and determine the yield. Obtain IR and NMR spectra of your starting material and product and compare them with those of authentic samples (Figs. 10.44–10.47). You may also verify the identity of the product by comparing its properties and the melting point of its α-naphthylurethane derivative, prepared according to the procedure in Section 23.11, with the data summarized in Table 10.1.

Finishing Touches

Dilute the *aqueous layer* with water, neutralize it with acetic acid, and then flush it down the drain with excess water. Pour the *distillate* containing THF and ether in the container for nonhalogenated organic solvents.

EXERCISES

1. Why is it necessary to maintain the pH of the reaction mixture above 7 during the oxidation of an alkyl borane with hydrogen peroxide?

2. Why was it necessary to use dry apparatus and solvents for the preparation and reaction of the complex of borane with THF?

3. Predict the structure of the major product that would be formed upon hydroboration-oxidation of each of the alkenes **a–e**. Include the correct stereochemistry wherever appropriate.

a.
b.
c.
d.
e.

4. For each of the alkenes in Exercise 3, write the structure of the alcohol that should be the major product upon *acid-catalyzed* hydration.

5. An important advantage of hydroboration-oxidation for hydration of alkenes is that rearrangements of the carbon skeleton do not occur, whereas they do in acid-catalyzed hydration. For example, acid-catalyzed hydration of (+)-α-pinene (**65**) proceeds with skeletal rearrangements to produce a number of isomeric alcohols.

   a. Draw the structure(s) of the product(s) you would expect to obtain upon acid-catalyzed hydration of (+)-α-pinene if no skeletal rearrangements occurred.

   b. Draw the structures of the isomeric alcohols having the formula $C_{10}H_{18}O$ that you might expect to obtain upon treatment of (+)-α-pinene with aqueous acid.

6. The hydroboration of **65** is a stereoselective reaction in which borane approaches from only one face of the carbon-carbon bond to give the dialkylborane **66**. Provide a rationale for this observed stereoselectivity.

7. Neither of the alcohols **71** or **72** forms an α-naphthylurethane derivative, whereas the isomers **67–70** each do. Explain the difference in reactivities.

Refer to Tables 8.1–8.8 as needed for answering the following questions on spectroscopy.

8. Consider the spectral data for (+)-α-pinene (Figs. 10.44 and 10.45).

   a. In the functional group region of the IR spectrum, identify any absorptions associated with the carbon-carbon double bond.

   b. In the ^{1}H NMR specturm, assign the various resonances to the hydrogen nuclei responsible for them.

   c. For the ^{13}C NMR data, assign the resonances at δ 116.2 and 144.1 to the carbon nuclei responsible for them.

9. Consider the spectral data for (−)-isopinocampheol (Figs. 10.46 and 10.47).

   a. In the functional group region of the IR spectrum, specify any absorptions associated with the alcohol function.

   b. In the ^{1}H NMR spectrum, assign the various resonances to the hydrogen nuclei responsible for them.

   c. For the ^{13}C NMR data, assign the resonance at δ 71.7 to the carbon atom responsible for it.

10. Discuss the differences observed in the IR and NMR spectra of (+)-α-pinene and (−)-isopinocampheol that are consistent with overall addition of water to the double bond occurring in this experiment.

Starting Materials and Products

Spectra

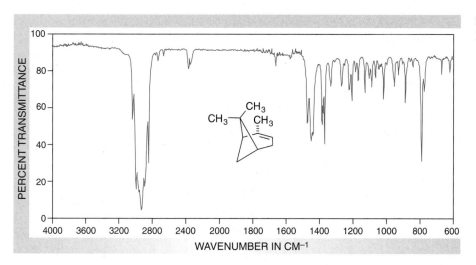

FIGURE 10.44
IR spectrum of (+)-α-pinene (neat).

FIGURE 10.45
NMR data for (+)-α-pinene
(CDCl₃).

(a) ^{1}H NMR spectrum (300 MHz).

(b) ^{13}C data: δ 20.8, 22.8, 26.8, 31.4, 31.5, 38.0, 41.5, 42.2, 116.2, 144.1.

FIGURE 10.46
IR spectrum of (−)-
isopinocampheol (IR card).

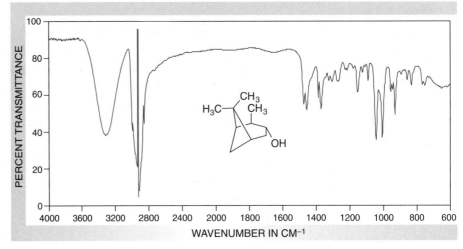

FIGURE 10.47
NMR data for (−)-isopinocampheol
(CDCl₃).

(a) ^{1}H NMR spectrum (300 MHz).

(b) ^{13}C data: δ 20.7, 23.7, 27.7, 34.4, 38.1, 39.1, 41.8, 47.8, 47.9, 71.7.

■ H I S T O R I C A L   H I G H L I G H T ■

Additions Across Carbon-Carbon π-Bonds

A characteristic reaction of molecules containing carbon-carbon multiple bonds is the addition of a generalized reagent "X–Y" across the π-bond. If both this reagent and the π-bond to which it is adding are unsymmetrical, two orientations of addition are possible, as shown below for the possible modes of addition of H–Br to 1-methylcyclohexene. The mode of addition providing the 1-bromo-1-methyl isomer is known as "Markovnikov addition" in recognition of the contributions of V. Markovnikov, who was the first chemist to recognize that when a reagent "H–Y" adds to an unsymmetrically substituted alkene, the proton becomes attached to the carbon atom already bearing the larger number of hydrogen atoms.

1-Methylcyclohexene   1-Bromo-1-methylcyclohexane

1-Bromo-2-methylcyclohexane

Markovnikov, a Russian chemist, was on the faculty of Kazan State University in the second half of the 19th century, and it was during this period that he performed the research that led to his "rule." Although not a well-known academic institution in the West, this university fathered a long line of chemists whose names are familiar to us, among which are Saitsev, Arbuzov, and Reformatsky, all of whom have a rule or reaction bearing their names. The university also has the distinction of having expelled Lenin in 1887 for taking part in an antigovernment students' rally!

The theoretical basis for Markovnikov's rule was unknown at the time of its formulation, and indeed there were important apparent exceptions to it. Thus in 1929, Frank Mayo, a graduate student of M. S.

Kharasch at the University of Chicago, found that adding H–Br to allyl bromide provided *either* 1,2-dibromopropane, the "Markovnikov adduct," *or* 1,3-dibromopropane, the "anti-Markovnikov adduct," as the principal product, depending on the particular reaction conditions used. Thus, bubbling nitrogen through a commercial sample of allyl bromide, followed by the addition of anhydrous H–Br afforded 1,2-dibromopropane almost exclusively, whereas substituting oxygen for nitrogen provided mainly 1,3-dibromopropane. Indeed, treating various samples of allyl bromide under seemingly identical reaction conditions provided widely differing ratios of the two adducts: Sealing mixtures of allyl bromide and anhydrous H–Br under vacuum provided ratios of 1,2-dibromopropane to 1,3-dibromopropane ranging from 4:1 to 1:5. Such variations were confusing, to say the least, and Kharasch was led to on one occasion to comment, "You know, Mayo, maybe it's [the variation in percentages] the phase of the moon!".

$$CH_2 \!=\! CHCH_2Br + H\!-\!Br \rightarrow CH_2\!-\!CHCH_2Br$$
$$\qquad\qquad\qquad\qquad\qquad\quad | \qquad |$$
$$\qquad\qquad\qquad\qquad\qquad\; H \quad\; Br$$

Allyl bromide   1,2-Dibromopropane

$$CH_2 \!=\! CHCH_2Br + H\!-\!Br \rightarrow CH_2\!-\!CHCH_2Br$$
$$\qquad\qquad\qquad\qquad\qquad\quad | \qquad |$$
$$\qquad\qquad\qquad\qquad\qquad\; Br \quad\; H$$

1,3-Dibromopropane

Mayo continued his efforts to define the variables dictating the orientation of addition of H–Br to allyl bromide, performing nearly 500 different trials, some of which lasted for over a year. He systematically varied a number of parameters such as temperature and the presence of the water, air, hydrogen, and light to determine their effects on the ratio of the two isomers. His persistence ultimately paid off, as he was able to show that the anti-Markovnikov mode of addition that provided 1,3-dibromopropane was due to contamination of the allyl bromide with small amounts of peroxides, although a mechanistic

HISTORICAL HIGHLIGHT
Additions Across Carbon-Carbon π-Bonds *(continued)*

understanding of the ''peroxide effect'' was lacking at the time he and Kharasch published their results (Kharasch, M. S.; Mayo, F. R. *J. Am Chem. Soc.* **1933**, *55*, 2468). It is not surprising that they failed to propose free radicals as being responsible for the peroxide effect: It was only in 1929 that a report from another research group showed that radicals played a role in organic reactions, so free radical chemistry was in its infancy at the time of Mayo's experiments. Indeed, in the opinion of many chemists of that era, a unified basis for the whole of organic chemistry would be founded on ''quasi-ionic'' or ''heterolytic'' reaction mechanisms.

Based on modern concepts of reaction mechanisms, the peroxide effect arises from decomposition of a peroxide to peroxy radicals, which then react with hydrogen bromide to produce bromine atoms, as shown below. This initiates free-radical chain addition rather than polar addition of H–Br to the

RO—OR ⟶ 2 RO·
A peroxide   Peroxy radicals

RO· + H—Br ⟶ RO—H + Br·

π-bond. Consequently, the product-determining step in free-radical chain addition of H–Br to an alkene is attack of Br· on the π-bond, whereas that for polar addition of this reagent is electrophilic attack of a proton. This results in opposite regiochemistries of the overall addition since in both cases the more stable tertiary species is produced.

RADICAL ADDITION OF HBr TO AN ALKENE

A tertiary radical

⟶ 1-Bromo-2-methylcyclohexane

POLAR ADDITION OF HBr TO AN ALKENE

A tertiary carbocation

⟶ 1-Bromo-1-methylcyclohexane

Although it was the experimental findings of Mayo that revealed the ''peroxide effect'' on the regiochemistry of addition across carbon-carbon double bonds, the insights and guidance of his mentor were critical to the success of the project. For example, Prof. Kharasch had considerable experience with chemicals used by the rubber industry, which had been undergoing rapid development as the automobile became increasingly popular and affordable. Among these chemicals were anti-oxidants, a class of molecules that increased the durability of tires. Kharasch had Mayo study the effect of several of these substances on the addition of H–Br to allyl bromide, and Mayo found that they suppressed the formation of 1,3-dibromopropane, the anti-Markovnikov product. We now know that this is because anti-oxidants function by trapping free radicals and thereby stopping free-radical chain reactions. By continuing their investigations of the role of peroxides and anti-oxidants on additions to alkenes, Prof. Kharasch and his students made many contributions to the understanding of free-radical chain reactions. Their accomplishments in this area of research are considered by many scientists to have had the most influence of any group on the development of the organic chemistry of radicals than any other group.

Interestingly, Kharasch, whose research ultimately led to the understanding of anti-Markovnikov addition to a carbon-carbon double bond, was born in the Ukraine and thus shares a Russian heritage with Markovnikov himself. These two eminent chemists laid the foundation for the understanding of addition reactions to carbon-carbon π-bonds.

HISTORICAL HIGHLIGHT
Additions Across Carbon-Carbon π-Bonds *(continued)*

Relationship of Historical Highlight to Experiments

Several of the procedures in this chapter involve addition reactions characteristic of alkenes. Two of them, the addition of hydrogen bromide to 1-hexene (Sec. 10.5) and of borane to α-pinene (Sec. 10.8),

represent examples in which both the reagent and the alkene to which it is adding are unsymmetrical. Consequently, identifying the products from these reactions provides a means of testing whether a Markovnikov or an anti-Markovnikov mode of addition has occurred.

CHAPTER 11

Alkynes

Unsaturated organic compounds that contain a carbon-carbon triple bond as the **functional group** are called **alkynes**. Acetylene (ethyne), $H\!-\!C\!\equiv\!C\!-\!H$, is the simplest alkyne and is widely used in industry as a fuel and as a chemical feedstock for the preparation of other organic compounds such as acetic acid, vinyl chloride ($CH_2\!=\!CHCl$), a monomer used in the manufacture of polyvinyl chloride, and chloroprene ($CH_2\!=\!CCl\!-\!CH\!=\!CH_2$), which polymerizes to give neoprene.

The carbon-carbon triple bond in an alkyne is formed by the overlap of two orthogonal pairs of p-orbitals on adjacent sp-hybridized carbon atoms. Because the functional group in alkynes is related to that of the carbon-carbon double bond in alkenes (Chap. 10), the chemistry of these two classes of compounds is similar. For example, in much the same manner as one **elimination reaction** may be used to form the double bond in alkenes, *two* sequential elimination reactions yield the triple bond in alkynes (Eq. 11.1).

$$R\!-\!\underset{\underset{H}{|}}{\overset{\overset{X}{|}}{C}}\!-\!\underset{\underset{H}{|}}{\overset{\overset{X}{|}}{C}}\!-\!R \xrightarrow[\Delta]{\text{strong base}} R\!-\!C\!\equiv\!C\!-\!R \qquad (11.1)$$

A 1,2-dihaloalkane   An alkyne

The π-electrons in both alkenes and alkynes provide a Lewis-base site for interaction with electrophilic reagents, which are Lewis acids, so one of the typical reactions of alkenes *and* alkynes is **electrophilic addition** as illustrated in Equations 11.2 and 11.3. Such reactions of alkynes can be stopped after the addition of one equivalent of a reagent, but the use of excess reagent leads to the formation of saturated products. Some aspects of the chemistry of alkynes will be explored in the experiments in this chapter.

$$R-C\equiv C-R + 2\,H-X \longrightarrow R-\overset{\overset{\displaystyle X}{|}}{\underset{\underset{\displaystyle X}{|}}{C}}-\overset{\overset{\displaystyle H}{|}}{\underset{\underset{\displaystyle H}{|}}{C}}-R \qquad (11.2)$$

An alkyne   A hydrogen halide   A 1,1-dihaloalkane

$$R-C\equiv C-R + 2\,X_2 \longrightarrow R-\overset{\overset{\displaystyle X}{|}}{\underset{\underset{\displaystyle X}{|}}{C}}-\overset{\overset{\displaystyle X}{|}}{\underset{\underset{\displaystyle X}{|}}{C}}-R \qquad (11.3)$$

An alkyne   A halogen   A 1,1,2,2-tetrahaloalkane

11.2 Dehydrohalogenation of 1,2-Dihaloalkanes

The dehydrohalogenation of alkyl halides (Sec. 10.2) is a general method to produce a double bond (Eq. 11.4). By analogy, the sequential elimination of two molecules of a hydrogen halide will lead to the formation of a triple bond as shown by the general reaction in Equations 11.5 and 11.6. The first stage of the reaction involves the base-induced elimination of a hydrogen halide to give a haloalkene, sometimes referred to as a vinyl halide. This step proceeds by a **concerted E2 mechanism** with the base abstracting the proton on the carbon atom β to the leaving halide ion. This reaction involves a transition state in which the hydrogen and the departing halogen are in the **anti-periplanar** orientation shown in Equation 11.5. The second step of the sequence (Eq. 11.6) requires a strong base and more forcing conditions such as heating. Although the relationship between the hydrogen and the leaving halide ion in the reaction shown in Equation 11.6 is anti-periplanar, elimination may also occur when the orientation of these two groups on the vinyl halide is **syn-periplanar.** However, under these circumstances even higher temperatures are required to effect dehydrohalogenation.

$$\underset{\substack{| \\ \displaystyle X}}{\overset{\displaystyle H}{\underset{\displaystyle RCH-CHR}{|}}} \xrightarrow{\text{base}} RCH=CHR \qquad (11.4)$$

A haloalkane   An alkene

$$\underset{\substack{\displaystyle B:^- \\ \text{A 1,2-dihaloalkane}}}{\overset{R}{\underset{\displaystyle H}{\overset{X}{C}}}\overset{X}{\underset{\displaystyle H}{\overset{X}{C}}}R} \longrightarrow \underset{\substack{\displaystyle R \\ \text{A haloalkene}}}{\overset{X}{C}=\overset{R}{\underset{H}{C}}} + B-H + X:^- \qquad (11.5)$$

$$\underset{\substack{\displaystyle B:^- \\ \text{A haloalkene}}}{\overset{X}{\underset{R}{C}}=\overset{R}{\underset{H}{C}}} \xrightarrow{\Delta} R-C\equiv C-R + B-H + X:^- \qquad (11.6)$$

An alkyne

In the experiment that follows, the dehydrobromination of *meso*-stilbene dibromide (**1**), which can be prepared by the addition of bromine to (*E*)-stilbene (Sec. 10.6), to give diphenylacetylene (**2**) will be performed according to Equation 11.7. Thus, the combination of the bromination of an alkene followed by the double dehydrobromination of the intermediate *vicinal* dibromide represents a useful means of converting an alkene into an alkyne. Because the temperature required to effect elimination of hydrogen bromide from the intermediate vinyl bromide is nearly 200 °C, it is necessary to use a high-boiling solvent such as triethylene glycol so the reaction may be carried out without special apparatus.

$$ \text{1} + 2\ \text{J.OH} \xrightarrow[\text{190 °C}]{\text{HO(CH}_2\text{CH}_2\text{O)}_2\text{CH}_2\text{CH}_2\text{OH}} \text{2} \qquad (11.7) $$

1
meso-Stilbene dibromide

2
Diphenylacetylene

EXPERIMENTAL PROCEDURES

DEHYDROBROMINATION OF *MESO*-STILBENE DIBROMIDE

Purpose: To demonstrate the preparation of an alkyne by a double dehydrohalogenation.

> **Safety Alert**
>
> **1.** The potassium hydroxide solution used in this experiment is *highly* caustic. *Do not allow it to come in contact with your skin.* If this should happen, flood the affected area with water and then thoroughly rinse the area with a solution of dilute acetic acid. Wear latex gloves when preparing and transferring all solutions.

Miniscale Procedure

Preparation   Answer the Pre-Lab Exercises on page PL. 65. Review Sections 2.9, 2.10, and 2.17.

Apparatus   A sand bath, hot plate, 25-mL Erlenmeyer flask, apparatus for vacuum filtration.

Setting Up   Preheat the sand bath to about 190–200 °C. Place 800 mg of *meso*-stilbene dibromide and 5 pellets (about 400 mg) of commercial potassium hydroxide in the Erlenmeyer flask. Add 4 mL of triethylene glycol and a carborundum boiling stone to the flask.

Dehydrobromination and Isolation   Place the flask in the sand bath and heat the mixture. After potassium bromide begins to separate from solution, heat the mixture for an additional 5 min. Remove the flask from the sand bath and allow it to cool to room temperature. Add 10 mL of water and place the flask in an ice-water bath for 5 min. Collect the diphenylacetylene that precipitates by vacuum filtration. Wash the solid with about 2–3 mL of cold water. Recrystallize the product from 95% ethanol. If the solution is allowed to cool slowly undisturbed, you should obtain large spar-like, colorless crystals.

Analysis   Determine the melting point, weight, and percent yield of the recrystallized product. Test the product for unsaturation using the bromine in dichloromethane (Sec. 23.8A) and the Baeyer tests (Sec. 23.8B). Obtain IR and ^{1}H NMR spectra of your starting material and product and compare them with those of authentic samples (Figs. 10.36, 10.37, 11.1, and 11.2).

Microscale Procedure

Preparation   Answer the Pre-Lab Exercises on page PL. 65. Review Sections 2.9, 2.10, and 2.17.

Apparatus   A sand bath, hot plate, 13-mm × 120-mm Pyrex test tube, apparatus for vacuum filtration.

Setting Up   Preheat the sand bath to about 190–200 °C. Place 150 mg of *meso*-stilbene dibromide and 1 pellet (about 80 mg) of commercial potassium hydroxide in the Pyrex test tube. Add 1 mL of triethylene glycol and a carborundum boiling stone to the test tube.

Dehydrobromination and Isolation   Place the test tube in the sand bath and heat the mixture. After potassium bromide begins to separate from solution, heat the mixture for an additional 5 min. Remove the test tube from the sand bath and allow it to cool to room temperature. Add 2 mL of water and place the test tube in an ice-water bath for 5 min. Collect the diphenylacetylene that precipitates by vacuum filtration. Wash the solid with about 1 mL of cold water. Recrystallize the product from 95% ethanol.

Analysis   Determine the melting point, weight, and percent yield of the recrystallized product. Test the product for unsaturation using the bromine in dichloromethane (Sec. 23.8A) and the Baeyer tests (Sec. 23.8B). Obtain IR and ^{1}H NMR spectra of your starting material and product and compare them with those of authentic samples (Figs. 10.36, 10.37, 11.1, and 11.2).

Finishing Touches

Combine the *filtrate* from the reaction mixture with the *mother liquor* from the recrystallization, dilute with water and neutralize with 10% aqueous hydrochloric acid; then flush the solution down the drain. Place the *dichloromethane solution* from the bromine test for unsaturation in a container for halogenated organic liquids; put the *manganese dioxide* from the Baeyer test for unsaturation in a container for heavy metals.

EXERCISES

1. Write three-dimensional structures of *meso*-stilbene dibromide and one enantiomer of *dl*-stilbene dibromide.

2. Give a stepwise mechanism showing the base-induced formation of diphenylacetylene from *meso*-stilbene dibromide. Use curved arrows to symbolize the flow of electrons.

3. The E2 reaction of most compounds is known to proceed preferentially by removal of a proton *anti-periplanar* to the leaving group. Based upon this generalization, predict the geometry of the 1-bromo-1,2-diphenylethylene that is produced as the intermediate in the double dehydrobromination of *meso*-stilbene dibromide.

4. If *meso*-stilbene dibromide is treated with KOH in ethanol, it is possible to isolate the 1-bromo-1,2-diphenylethylene that is formed from the first dehydrobromination. The E2 elimination of the second molecule of hydrogen bromide from this intermediate alkene to give diphenylacetylene has a higher activation energy than the first elimination and thus requires a higher reaction temperature. Explain.

5. Provide an explanation for why ethanol or ethylene glycol are not suitable solvents for the second dehydrobromination.

Refer to Tables 8.1–8.8 as needed for answering the following questions on spectroscopy.

6. Consider the NMR spectral data for *meso*-stilbene dibromide (Fig. 10.37).

   a. In the ^{1}H NMR spectrum, assign the various resonances to the hydrogen nuclei responsible for them.

   b. For the ^{13}C NMR data, assign the various resonances to the carbon nuclei responsible for them.

7. Consider the spectral data for diphenylacetylene (Figs. 11.1 and 11.2).

   a. In the functional group region of the IR spectrum, identify the absorptions associated with the aromatic rings. Why is there no absorption for the carbon-carbon triple bond?

   b. In the ^{1}H NMR spectrum, assign the various resonances to the hydrogen nuclei responsible for them.

   c. For the ^{13}C NMR data, assign the various resonances to the carbon nuclei responsible for them.

8. Discuss the differences observed in the NMR spectra of *meso*-stilbene dibromide and diphenylacetylene that are consistent with the double dehydrobromination in this experiment.

Starting Material and Product

   The IR and NMR spectra of meso-stilbene dibromide are provided in Figures 10.36 and 10.37, respectively.

FIGURE 11.1
IR spectrum of diphenylacetylene (IR card).

(a) ¹H NMR spectrum (300 MHz).

FIGURE 11.2
NMR data for diphenylacetylene (CDCl₃).

(b) ¹³C NMR data: δ 89.6, 123.3, 128.1, 128.2, 131.6.

11.3. Addition Reactions of Alkynes

One of the general classes of reactions that characterizes the chemistry of alkynes as well as alkenes (Sec. 10.4) is the **electrophilic addition** of reagents generalized as **E–Nu** (Eq. 11.8). This reaction typically proceeds in two stages. Under carefully controlled reaction conditions in which only one equivalent of electrophilic reagent **E–Nu** is used, it is sometimes possible to stop the addition reaction at the alkene stage. However, if more than one equivalent of the reagent is present, further addition occurs, and substituted alkanes are formed. Typical reagents that add to alkynes include the halogens chlorine and bromine and acids such as hydrogen chloride and hydrogen bromide.

$$R^1-C\equiv C-R^2 \xrightarrow[\text{1 mole}]{\text{E-Nu}} \underset{E}{\overset{R^1}{>}}C=C\underset{R^2}{\overset{Nu}{<}} \xrightarrow[\text{1 mole}]{\text{E-Nu}} R^1-\underset{\underset{E}{|}}{\overset{\overset{E}{|}}{C}}-\underset{\underset{Nu}{|}}{\overset{\overset{Nu}{|}}{C}}-R^2 \qquad (11.8)$$

An alkyne   A substituted alkene   A substituted alkane

The *regiochemistry* of the electrophilic addition of unsymmetrical reagents to terminal alkynes follows *Markovnikov's rule* in the same fashion as observed for alkenes (Sec. 10.4). Thus, the Lewis acid component, E^+, of the reagent E–Nu adds to the electron-rich triple bond to form the more stable, or more substituted, of the two possible intermediate vinyl carbocations; this carbocation then undergoes nucleophilic attack by the Lewis base component, Nu^-, of the reagent (Eq. 11.9). The addition of the second mole of E–Nu to the resulting alkene occurs with the same regiochemistry as the first (Eqs. 11.8 and 11.9).

$$R-C\equiv C-H \xrightarrow[\text{1 mole}]{\text{E-Nu}} \underset{Nu}{\overset{R}{>}}C=C\underset{H}{\overset{E}{<}} \xrightarrow[\text{1 mole}]{\text{E-Nu}} R-\underset{\underset{Nu}{|}}{\overset{\overset{Nu}{|}}{C}}-\underset{\underset{E}{|}}{\overset{\overset{E}{|}}{C}}-H \qquad (11.9)$$

An terminal alkyne   A substituted alkene   A substituted alkane

Alkynes undergo electrophilic addition reactions with hydrogen halides and bromine, but these reactions have limited synthetic utility. However, one reaction of alkynes that is commonly used in organic chemistry is the **hydration** of the carbon-carbon triple bond to give a **ketone**; this transformation is catalyzed by mercuric ion in the presence of sulfuric acid (Eq. 11.10).

$$R-C\equiv C-H + H_2O \xrightarrow[\text{Hg}^{2+}]{\text{H}_2\text{SO}_4} R-\overset{\overset{O}{\|}}{C}-CH_3 \qquad (11.10)$$

An alkyne   A ketone

The role that the mercuric ion, a polarizable Lewis acid, plays in this reaction is not completely understood, but a reasonable hypothesis is that it first coordinates with the triple bond to generate a vinyl cation that may be stabilized by conversion to a cyclic mercurinium ion (Eq. 11.11). Attack by water on the cyclic ion then occurs at the more substituted carbon, and following several proton transfers, mercuric ion is lost to give a substituted vinyl alcohol, which is unstable and tautomerizes to give the **carbonyl group**, C=O, of the product ketone (Eq. 11.12).

$$R-C\equiv C-H \xrightarrow{\text{Hg}^{2+}} R-\overset{+}{C}=C\underset{H}{\overset{\ddot{\text{Hg}}^+}{<}} \rightleftharpoons R-C=C\underset{H}{\overset{\text{Hg}^{2+}}{<}} \qquad (11.11)$$

An alkyne   A vinyl cation   A cyclic mercurinium ion

(11.12)

A vinyl alcohol (enol)   A ketone
(*unstable; not isolated*)

In the experiment that follows, the hydration of a terminal alkyne is illustrated by the conversion of 2-methyl-3-butyn-2-ol (**3**) to 3-hydroxy-3-methyl-2-butanone (**4**), as shown in Equation 11.13. The presence of a hydroxyl group in **3** has little effect on the chemical properties of the carbon-carbon triple bond. Rather, the main effect of the polar hydroxyl group is on the physical properties of the molecule, with the melting and boiling points of **3** being considerably higher than those for acetylenic hydrocarbons having the same molecular weight.

(11.13)

3   **4**
2-Methyl-3-butyn-2-ol   3-Hydroxy-3-methyl-2-butanone

5
(not formed)

This experiment provides an opportunity to verify experimentally that the hydration of a terminal alkyne occurs in accordance with Markovnikov's rule (see Historical Highlight in Chap. 10). For example, that the structure of the product of the hydration of **3** is **4** rather than **5** may be confirmed by comparing the IR, ^{1}H and ^{13}C NMR spectra of the product of the reaction with known standard samples of **4** and **5**, the spectra of which are clearly distinct. When *all* of the spectral properties of a compound are superimposable with those of a standard sample, the two are identical.

Another common method of assigning a structure to the product of a reaction involves preparing a solid derivative and comparing its melting point with the melting point of the same derivative of a series of known compounds. If the melting points of the derivatives of the unknown and a known sample are identical, then a mixed melting point (Sec. 3.3) is taken to confirm the identity of the two samples. In the present case, a solid derivative of the product of the reaction can be made for comparison with the same derivatives of authentic samples of **4** and **5**. For example, forming the semicarbazone **6** from **4** proceeds smoothly, but converting the aldehyde **5** to its semicarbazone occurs with concomitant dehydration to give **7**, which is the semicarbazone of the unsaturated aldehyde $(CH_3)_2C{=}CHCHO$.

$$\underset{\textbf{6}}{CH_3-\overset{\overset{\displaystyle HO}{|}}{\underset{\underset{\displaystyle CH_3}{|}}{C}}-\overset{\overset{\displaystyle NNHCONH_2}{\|}}{C}-CH_3}$$

$$\underset{\textbf{7}}{\overset{CH_3}{\underset{CH_3}{>}}C{=}CH-\overset{NNHCONH_2}{\underset{H}{C}}}$$

EXPERIMENTAL PROCEDURE

PREPARATION OF 3-METHYL-3-HYDROXY-2-BUTANONE

Purpose: To demonstrate the hydration of a terminal alkyne and to establish the regiochemistry of the addition.

> **Safety Alert**
>
> 1. Concentrated sulfuric acid may cause severe chemical burns so do not allow it to contact your skin. Wear latex gloves when transferring this reagent. Wash any affected area immediately with water and then with 5% sodium bicarbonate solution.
> 2. Be sure that you add the concentrated *acid* slowly *to water* and *not* the reverse.

Miniscale Procedure

Preparation   Answer the Pre-Lab Exercises on page PL. 67. Review Sections 2.9, 2.10, 2.11, 2.13, 2.16, 2.21, 2.22, and 2.29.

Apparatus   A 10-mL, a 50-mL, and a 100-mL round-bottom flask, separatory funnel, ice-water bath, 13-mm × 120-mm test tube, apparatus for heating under reflux, simple distillation, magnetic stirring, steam distillation with internal steam generation, vacuum filtration, and *flameless* heating.

Setting Up   *Carefully* add 3 mL of *concentrated* sulfuric acid *to* 20 mL of water contained in the 100-mL round-bottom flask equipped with a stirbar. Dissolve 0.2 g of reagent grade mercuric oxide in the resulting warm solution and cool the flask to about 50 °C. Attach a reflux condenser to the flask, and add 3.6 mL of 2-methyl-3-butyn-2-ol in one portion through the top of the condenser. The cloudy white precipitate that forms is presumably the mercury complex of the alkyne.

Reaction   Stir the reaction flask to mix the contents thoroughly, whereupon a mildly exothermic reaction may ensue. Allow the reaction to proceed by itself for about 2 min. Then heat the mixture under reflux for 30 min, before allowing it to cool to room temperature.★

Isolation   While the reaction mixture is cooling, prepare an apparatus for steam distillation *with internal steam generation* . Add about 35–40 mL of water to the reaction flask. Heat the flask and steam-distil the product until about 30–35 mL of distillate has been collected.★

Transfer the distillate to a small separatory funnel, add 2–3 g of potassium carbonate and then carefully saturate the solution with sodium chloride; avoid adding more sodium chloride than will dissolve readily. If a second layer forms at this point, do not separate it but continue with the extraction as described. Extract the mixture sequentially with two 10-mL portions of dichloromethane and combine the extracts. Dry the organic extracts over several spatula-tips full of anhydrous sodium sulfate. Swirl the flask occasionally for a period of 10–15 min. to facilitate drying; add further portions of anhydrous sodium sulfate if the solution does not become clear.

Decant the solution into a 50-mL round-bottom flask and remove most of the dichloromethane by simple shortpath distillation; cool the receiving flask in an ice-water bath. Alternatively rotary evaporation or other techniques (Sec. 2.29) may be used to remove most of the dichloromethane. Using a Pasteur pipet to minimize material loss, transfer the crude product into a 10-mL round-bottom flask. Remove the last traces of dichloromethane from the crude product by connecting the flask to a vacuum source. Then continue the distillation using the shortpath apparatus fitted with a clean, tared receiving flask. Since the volume of product is likely to be only 1–2 mL, an accurate boiling point may be difficult to obtain, and all material that distils above 110–115 °C should be collected in order to obtain a reasonable yield.

Derivatization   Add 0.5 mL of **4** to a solution prepared from 0.5 g of semicarbazide hydrochloride and 0.8 g of sodium acetate dissolved in 2.5 mL of water contained in a test tube. Shake the mixture vigorously. Collect the crude solid semicarbazone that forms by vacuum filtration; use a glass rod to scratch at the air-liquid interface if necessary to initiate precipitation. Dissolve the solid product in a minimum volume of hot 2-propanol, cool the mixture to room temperature and then in an ice-water bath, and collect the crystals by vacuum filtration. Air-dry the crystals.

Analysis   Determine the weight and percent yield of the distilled product. Determine the melting point of the semicarbazone of your product. Use this information to confirm the identity of the product of hydration of 2-methyl-3-butyn-2-ol. The melting point of **6** is 162–163 °C, whereas **7** melts at 222–223 °C. Obtain IR and ^{1}H NMR spectra of your starting material and product and compare them with those of authentic samples (Figs. 11.3–11.6).

Since the *residue in the stillpot* from the steam distillation contains soluble mercury salts, place it in a special container for solutions of mercuric salts; do *not* pour it down the drain. If directed to do so by your instructor, you may convert the *soluble mercuric salts* into insoluble mercuric sulfide by reaction with sodium sulfide or hydrosulfide; by vacuum filtration collect the *mercuric sulfide* for recycling. Flush the *other aqueous solutions* down the drain with water. Pour the *dichloromethane* recovered as the forerun in the distillation into a container for halogenated organic solvents. Dilute the *filtrate* from the preparation of the semicarbazone derivative and the *mother liquor* from the recrystallization with water, and flush the solutions down the drain.

Finishing Touches

EXERCISES

1. Give the structures for the products that you would expect to obtain on hydration of 1-octyne and of 2-octyne.

2. Consider the compounds hexane, 1-hexene, and 1-hexyne; see Chapters 9 and 10 for a review of reactions of alkanes and alkenes. What similarities and what differences would you expect upon treatment of these compounds with:

   a. bromine in dichloromethane at room temperature in the dark

   b. an aqueous solution of potassium permanganate

   c. an aqueous solution of sulfuric acid?

   Give the structures of the products, if any, that would be obtained from each of these reactions.

3. The reaction of 1-butyne with hydrogen bromide gives 2,2-dibromobutane, as shown in Equation 11.14. Write a stepwise mechanism for this process, using curved arrows to symbolize flow of electrons. Also provide a rationale for the observed regioselectivity for addition of *each* mole of hydrogen bromide.

$$CH_3CH_2C\equiv C-H \; + \; 2\,H-Br \longrightarrow CH_3CH_2CBr_2CH_3 \qquad (11.14)$$

Refer to Tables 8.1–8.8 as needed for answering the following questions on spectroscopy.

4. Consider the spectral data for 2-methyl-3-butyn-2-ol (Figs. 11.3 and 11.4).

   a. In the functional group region of the IR spectrum, identify the absorptions associated with the carbon-carbon triple bond and the hydroxyl group.

   b. In the ^{1}H NMR spectrum, assign the various resonances to the hydrogen nuclei responsible for them.

   c. For the ^{13}C NMR data, assign the various resonances to the carbon nuclei responsible for them.

5. Consider the spectral data for 3-hydroxy-3-methyl-2-butanone (Figs. 11.5 and 11.6).

   a. In the functional group region of the IR spectrum, identify the absorptions associated with the carbonyl and hydroxyl groups.

   b. In the ^{1}H NMR spectrum, assign the various resonances to the hydrogen nuclei responsible for them.

   c. For the ^{13}C NMR data, assign the various resonances to the carbon nuclei responsible for them.

6. ^{1}H NMR spectroscopy could be used to determine whether **4** or **5** is produced by the hydration of **3**. Sketch the ^{1}H NMR spectrum you would expect from **5**, and discuss the differences that you would expect in the ^{1}H NMR spectra of **4** and **5**.

7. IR spectroscopy may also be used to determine whether **4** or **5** is produced by the hydration of **3**. Describe the important difference(s) you would expect in the IR spectra of the two compounds.

Starting Material and Product

Spectra

FIGURE 11.3
IR spectrum of 2-methyl-3-butyn-2-ol (neat).

(a) ^{1}H NMR spectrum (300 MHz).

FIGURE 11.4
NMR data for 2-methyl-3-butyn-2-ol (CDCl$_3$).

(b) ^{13}C NMR data: δ 31.3, 64.8, 70.4, 89.1.

FIGURE 11.5
IR spectrum of 3-hydroxy-3-methyl-2-butanone (neat).

FIGURE 11.6
NMR data for 3-hydroxy-3-methyl-2-butanone (CDCl$_3$).

(a) ^{1}H NMR spectrum (300 MHz).

(b) ^{13}C NMR data: δ 23.8, 26.4, 76.7, 213.2.

11.4   **Terminal Alkynes as Acids**

The acidity of the acetylenic hydrogen of terminal alkynes (R—C≡C—H), pK$_a$ ≈ 25, provides the basis of a simple and specific qualitative test for such compounds. Treating a terminal alkyne with a solution containing silver ammonia complex, which is prepared by dissolving silver nitrate in ammonia, provides a solid silver salt, as illustrated by the general reaction shown in Equation 11.15. This test differentiates terminal alkynes from alkenes, internal alkynes, dienes, and allenes.

$$\text{R—C≡C—H} + [\text{Ag(NH}_3)_2]^- \longrightarrow \underline{\text{R—C≡C—Ag}} + \text{NH}_3 + \overset{+}{\text{NH}}_4 \quad \text{(11.15)}$$

A terminal alkyne   A silver acetylide

This unique reaction of terminal alkynes may also be used to separate them from non-terminal alkynes, which do *not* form insoluble silver salts and remain in solution. The silver acetylide salts can be reconverted to the terminal alkynes on treatment with hydrochloric acid, as shown in Equation 11.16. *Dry silver salts of this type are quite sensitive to shock and tend to decompose explosively.* Therefore they should *never* be allowed to dry out before being decomposed by hydrochloric acid.

$$\underline{\text{R—C≡C—Ag}} + \text{HCl} \longrightarrow \text{R—C≡C—H} + \underline{\text{AgCl}} \quad \text{(11.16)}$$

E X P E R I M E N T A L   P R O C E D U R E

FORMATION OF A SILVER ACETYLIDE AND ITS DECOMPOSITION

Purpose: To demonstrate the formation and decomposition of the silver acetylide of a terminal alkyne.

1. Avoid spilling solutions of silver salts on your skin. If you do come into contact with the solutions, wash the affected area immediately with copious amounts of water.
2. Take special care to destroy *all* of the silver salt acetylide with hydrochloric acid *before* discarding it.

Preparation   Answer the Pre-Lab Exercises on page PL. 69. Review Section 2.17.

Apparatus   Two 13-mm × 120-mm test tubes, apparatus for vacuum filtration.

Setting Up   In a 10-mL Erlenmeyer flask, prepare a solution of silver ammonia complex from 2.5 mL of 0.1 *M* silver nitrate solution by adding ammonium hydroxide solution *dropwise*. Brown silver oxide forms first; add *just enough* ammonium hydroxide to dissolve the silver oxide. Dilute the solution by adding 1.5 mL of water.

Formation and Decomposition   Add 3 mL of the diluted silver ammonia complex solution to about 0.1 mL of 2-methyl-3-butyn-2-ol in a test tube, and note the formation of the silver acetylide salt. Collect the silver salt by vacuum filtration of the aqueous solution; *be careful not to let the silver salt dry because the dry salt is explosive.* Transfer the *wet* silver salt to a clean test tube, and add a small amount of dilute hydrochloric acid. Observe what changes occur, especially in the color and form of the precipitate.

Dilute the remaining *silver ammonia complex test solution* with water and flush it down the drain with water. Destroy all *solid silver acetylide salt* by treatment with hydrochloric acid. Collect the *silver chloride* and put it in a container for heavy metals, and then flush the *filtrate* down the drain.

Finishing Touches

EXERCISES

1. The pK_a of a terminal acetylene is 25 and the pK_a of NH_4^+ is about 9.

   a. Use these values to predict the K_{eq} for Equation 11.17 to prove that the acid-base equilibrium lies to the left as shown.

   $$R—C{\equiv}C—H + NH_3 \rightleftharpoons R—C{\equiv}\overset{-}{C} + \overset{+}{N}H_4 \qquad (11.17)$$

   b. Provide an explanation for the observation that the equilibrium for the reaction of a terminal alkyne with silver ammonia complex (Eq. 11.15) favors the formation of silver acetylide.

2. How might you separate, and obtain in pure form, 1-octyne and 2-octyne from a mixture containing both?

CHAPTER 12

Dienes: The Diels-Alder Reaction

In synthetic organic chemistry, reactions that produce carbocyclic rings from acyclic precursors are of great importance because they lead to the easy formation of complex molecular structures. Rings may be produced from the cyclization of an acyclic starting material by forming a *single* carbon-carbon bond. However, one of the most useful methods for constructing six-membered rings from acyclic starting materials involves the reaction of a 1,3-diene with an alkene to give a derivative of cyclohexene with the formation of *two* new carbon-carbon bonds, as illustrated by Equation 12.1. In this reaction, the alkene partner is generally referred to as a **dienophile**. This **cycloaddition** reaction is called the **Diels-Alder reaction** in honor of Otto Diels and Kurt Alder, the

$$\text{A 1,3-diene} \quad \text{A dienophile} \qquad \text{A cycloadduct} \tag{12.1}$$

two German chemists who recognized its importance and shared the 1950 Nobel Prize in chemistry for their extensive development of this reaction. A more detailed account of their elegant work is included in the Historical Highlight in this chapter. The Diels-Alder reaction also belongs to a class of reactions termed **1,4-additions** because the two new carbon-carbon σ-bonds are formed between the 1- and 4-carbon atoms of the diene and the two π-bonded carbon atoms of the dienophile.

The scope of the Diels-Alder reaction is broad, and many combinations of dienes and dienophiles furnish cycloadducts in good yields. For example, the presence of electron-withdrawing substituents such as cyano, $C\equiv N$, and carbonyl, $C=O$, groups on the dienophile increases the rate and the yield of the reaction. The Diels-Alder

344

reaction is remarkably free of complicating side reactions, and the yields of the desired product are often high. Probably the single most important side reaction that may be encountered is dimerization or polymerization of the diene and/or dienophile.

12.2 Mechanistic and Stereochemical Aspects

The mechanism of the Diels-Alder reaction is now fairly well understood. The accumulated evidence favors a picture in which reorganization of the π-electrons occurs in the transition state for the reaction so the two new σ-bonds are formed in a more or less **synchronous** or **concerted** fashion; little charge or free-radical character is developed at any of the terminal carbon atoms (Eq. 12.2). The reaction is one of a number of **pericyclic reactions** that are controlled by **orbital symmetry**. Detailed discussions of orbital symmetry and its ramifications may be found in all modern organic textbooks.

$$(12.2)$$

| Reactants | Six membered transition state | Product |

A consequence of the concerted nature of the Diels-Alder reaction is that the diene must be able to adopt a *planar conformation* in which the dihedral angle between the two double bonds is 0°. Such a conformation is designated as *s-cis*, meaning **cisoid** about the single bond. Dienes that have difficulty achieving this conformation are expected to undergo Diels-Alder reactions slowly or not at all. A concerted reaction of a dienophile with a diene in its *s-trans* conformation would lead to a six-membered ring containing a highly strained *trans* double bond, an improbable structure.

s-cis conformation   *s-trans* conformation

A consequence of the highly ordered relationship between the diene and dienophile in the transition state depicted in Equation 12.2 is that Diels-Alder reactions are highly **stereoselective** with the formation of the new carbon-carbon σ-bonds occurring on the *same* face of *both* the diene and the dienophile. Such addition reactions are termed *syn-additions*. Thus, the cycloadduct resulting from reaction of *trans,trans*-2,4-hexadiene and a dienophile is *exclusively cis*-3,6-dimethylcyclohexene (Eq. 12.3), whereas the product obtained from *cis, trans*-2,4-hexadiene and a dienophile is *trans*-3,6-dimethylcyclohexene. Both of these reactions proceed by *syn*-1,4-addition to the diene, with the stereochemical relationship between the two methyl groups on the diene being maintained. Similarly, the reaction between 1,3-butadiene and dimethyl maleate gives solely the *cis*-diester **1** (Eq. 12.4). This example illustrates that the Diels-Alder reaction proceeds by exclusive *syn*-1,2-addition to the double bond of the dienophile; no isomerization to the more stable *trans*-diester is observed. The results of the reactions depicted in Equations 12.3 and 12.4 are in accord with the hypothesis that the Diels-Alder reaction is concerted, with both new carbon-carbon bonds forming simultaneously.

(12.3)

trans-, *trans*-2,4-Hexadiene   An alkene   A *cis*-3,6-dimethylcyclohexene

(12.4)

s-cis-1,3-butadiene   Dimethyl maleate   Dimethyl 4-cyclohexene-*cis*-1,2-dicarboxylate

There is a second type of stereoselectivity that is characteristic of the Diels-Alder reaction. The addition of a dienophile such as maleic anhydride to a *cyclic* diene like 1,3-cyclopentadiene could in principle provide two products, the *endo*-adduct **2** and the *exo*-adduct **3** (Eq. 12.5). However, only the *endo*-cycloadduct **2** is observed experimentally, and its preferential formation follows what is now commonly termed the

(12.5)

Maleic   1,3-   Bicyclo[2.2.1]hept-5-ene-   Bicyclo[2.2.1]hept-5-ene-
Anhydride   Cyclopentadiene   *endo*-2,3-dicarboxylic   *exo*-2,3-dicarboxylic
              anhydride   anhydride
              2   **3**

Alder rule. The basis for this result is believed to be stabilization of the transition state **4** by **through-space**, or **secondary**, **orbital interactions** between the *p*-orbitals on the internal carbons of the diene and the carbonyl carbon atoms of the dienophile, as shown by the dashed lines in **4**. Analogous stabilization is not possible in transition state **5**. Transition state **4** is characterized as the one having the *maximum accumulation of double bonds*, an alternate way of stating that stabilization results from development of the through-space overlap of the *p*-orbitals. It should be noted that not all Diels-Alder reactions are as stereoselective as the one between 1,3-cyclopentadiene and maleic anhydride; mixtures of *endo*- and *exo*-products are sometimes obtained.

$$2 \longleftarrow \quad \mathbf{4} \qquad \mathbf{5} \quad \longrightarrow 3$$

12.3 **Applications of Diels-Alder Reactions**

Two experimental procedures nicely illustrate various features of the Diels-Alder reaction. In both cases the highly reactive dienophile maleic anhydride is used. In Part A, the Diels-Alder reaction of 1,3-butadiene with maleic anhydride to give 4-cyclohexene-*cis*-1,2-dicarboxylic anhydride (**6**) is performed (Eq. 12.6). Since 1,3-butadiene is a gas at room temperature (bp -4.4 °C, 760 Torr), Diels-Alder reactions involving this diene are normally performed in sealed steel pressure vessels into which the diene is introduced under pressure. However, in this experiment 1,3-butadiene is conveniently generated *in situ* by thermally decomposing 3-sulfolene (Eq. 12.7); it then reacts immediately with a dienophile that is present.

$$(12.6)$$

1,3-Butadiene   Maleic anhydride

6
4-Cyclohexene-*cis*-1,2-
dicarboxylic anhydride

$$(12.7)$$

3-Sulfolene   1,3-Butadiene

In Part B, the cycloaddition of cyclopentadiene with maleic anhydride is examined (Eq. 12.5). This experiment provides an opportunity to examine the stereoselectivity of the Diels-Alder reaction. Monomeric 1,3-cyclopentadiene cannot be purchased because it readily dimerizes by a Diels-Alder reaction at room temperature to give dicyclopentadiene (Eq. 12.8), which is commercially available. Fortunately, the equilibrium between the monomer and the dimer can be established at the boiling point of the dimer (170 °C, 760 Torr) by a process commonly called **cracking**, and the lower-boiling 1,3-cyclopentadiene can then be isolated by fractional distillation. The diene *must* be kept cold in order to prevent its redimerization prior to use in the Diels-Alder reaction.

$$(12.8)$$

1,3-Cyclopentadiene   Dicyclopentadiene

The Diels-Alder adducts **2** and **6** retain the anhydride function present in the dienophile, maleic anhydride, from which they are produced. As a rule, this functionality is unstable in the presence of water even at pH 7 and hydrolyzes to a dicarboxylic acid according to the general sequence outlined in Scheme 12.1. This process is initiated by nucleophilic attack of water on the carbonyl function, followed by cleavage of a C—O bond of the anhydride. Because anhydrides are so reactive toward water, it is important that the apparatus and reagents you use be dry to maximize the yield of cycloadduct. You may be instructed to prepare the dicarboxylic acids **7** and **8** by intentional hydrolysis of the corresponding anhydrides. Protocols for doing so are included in Part C of the experimental procedures.

SCHEME 12.1

4-Cyclohexene-*cis*-
1,2-dicarboxylic acid
7

Bicyclo[2.2.1]hept-5-en-
endo-2,3-dicarboxylic acid
8

EXPERIMENTAL PROCEDURES

DIELS-ALDER REACTION

Purpose: To demonstrate the formation of six-membered rings by a cycloaddition reaction.

> **Safety Alert**
> 1. Be certain that all joints in the apparatus are tight and well lubricated before heating the reaction mixture so that sulfur dioxide, a toxic and foul-smelling gas, does not escape into the laboratory.
> 2. The organic solvents are highly flammable, so use *flameless* heating.

A • REACTION OF 1,3-BUTADIENE AND MALEIC ANHYDRIDE

Miniscale Procedure

Preparation   Answer the Pre-Lab Exercises on page PL. 71. Review Sections 2.9, 2.11, 2.17, 2.22, and 2.23.

Apparatus   A 25-mL round-bottom flask, gas trap, apparatus for heating under reflux, magnetic stirring, vacuum filtration, and *flameless* heating.

Setting Up   Place 2.5 g of 3-sulfolene, 1.5 g of finely pulverized maleic anhydride, and 1 mL of dry xylene in the flask and equip it with a stirbar. Assemble the apparatus for heating under reflux. Attach the condenser to the gas trap specified by your instructor.

Cycloaddition and Isolation   Gently warm the flask with stirring to effect *complete* dissolution of all solids; failure to do so will lower the yield of the reaction *significantly*. Then heat the mixture under gentle reflux for about 0.5 h.★ Allow the reaction mixture to cool below the boiling point and add 10 mL of xylene. Reheat the stirred mixture to dissolve all of the solids, and then pour the hot solution into an Erlenmeyer flask. If any product crystallizes during this operation, heat the mixture gently until all of the solid redissolves.★ Carefully add petroleum ether (bp 60–80 °C, 760 Torr) to the hot solution until cloudiness develops, and then set the solution aside to cool to room temperature.★ Collect the crystals by vacuum filtration, transfer them to a piece of filter or weighing paper, and air-dry them thoroughly.

Analysis   Weigh your product to determine the yield and measure its melting point. Test the product for unsaturation using the bromine in dichloromethane (Sec. 23.8A) and the Baeyer tests (Sec. 23.8B). Obtain IR and ^{1}H NMR spectra of your starting materials and product and compare them with those of authentic samples (Figs. 12.1–12.6).

Microscale Procedure

Preparation   Answer the Pre-Lab Exercises on page PL. 71. Review Sections 2.9, 2.11, 2.17, 2.22, and 2.23.

Apparatus   A 3-mL conical vial, gas trap, apparatus for heating under reflux, magnetic stirring, Craig tube filtration, and *flameless* heating.

Setting Up   Equip the conical vial with a spinvane and add 250 mg of 3-sulfolene, 150 mg of finely pulverized maleic anhydride, and 0.5 mL of dry xylene. Assemble the apparatus for heating under reflux using the gas trap specified by your instructor.

Cycloaddition and Isolation   Gently warm the vial with stirring to effect *complete* dissolution of all solids; failure to do so will lower the yield of the reaction *significantly*. Then heat the mixture under gentle reflux for about 0.5 h. Allow the reaction mixture to cool below the boiling point and add 1 mL of xylene.★ Reheat the mixture to effect dissolution of all solids and then rapidly transfer the contents of the vial to a Craig tube. If any product crystallizes during this operation, heat the mixture gently until all of the solid redissolves. Carefully add petroleum ether (bp 60–80 °C, 760 Torr) to the hot solution until cloudiness develops, and then set the solution aside to cool to room temperature.★ Complete the crystallization by cooling the mixture in an ice-water bath, equip the Craig tube for filtration, and isolate the product by centrifugation.

Analysis   Weigh your product to determine the yield and measure its melting point. Test the product for unsaturation using the bromine in dichloromethane (Sec. 23.8A) and the Baeyer tests (Sec. 23.8B). Obtain IR and ^{1}H NMR spectra of your starting materials and product and compare them with those of authentic samples (Figs. 12.1–12.6).

Finishing Touches

Pour the *mixture of xylene and petroleum ether* recovered from the reaction into a container for nonhalogenated organic solvents. Pour the *dichloromethane solution* from the bromine test for unsaturation in a container for halogenated organic liquids, and put the *manganese dioxide* from the Baeyer test for unsaturation in a container for heavy metals.

B • REACTION OF 1,3-CYCLOPENTADIENE AND MALEIC ANHYDRIDE

Safety Alert

1. 1,3-Cyclopentadiene is a mildly toxic, foul-smelling, and volatile substance. Prepare and use it in a hood if possible; avoid inhaling its vapors. Keep the diene cold at all times to minimize vaporization and dimerization.
2. Be certain that all joints in the apparatus are tight before heating dicyclopentadiene to produce the monomer.
3. The 1,3-cyclopentadiene and organic solvents used are highly flammable, so use *flameless* heating.

Miniscale Procedure

Preparation   Answer the Pre-Lab Exercises on page PL. 73. Review Sections 2.9, 2.10, 2.11, 2.14, and 2.17.

Apparatus   A 10- and a 25-mL round-bottom flask, 25-mL Erlenmeyer flask, drying tube, ice-water bath, apparatus for fractional distillation, magnetic stirring, vacuum filtration, and *flameless* heating.

Setting Up   Place a stirbar and 7 mL of dicyclopentadiene in a 25-mL round-bottom flask, and attach the flask to a fractional distillation apparatus that is equipped with an *unpacked* fractionating column and the 10-mL round-bottom flask as the receiver. Cool the receiver in an ice-water bath and fit the vacuum adapter with a drying tube.

Preparing Reactants   Heat the dimer until the solution is briskly refluxing (*Caution*: occasional foaming) and the monomer begins to distil in the range 40–42 °C. Distil the monomer as *rapidly* as possible, but do *not* permit the head temperature to exceed 43–45 °C. About 2.5 mL of 1,3-cyclopentadiene should be obtained from the distillation, which will require about 0.5 h. If the distillate is cloudy because of condensation of moisture in the cold receiver, add about 0.5 g of *anhydrous* calcium chloride.

While the distillation is in progress, dissolve 1.5 g of maleic anhydride in 5 mL of ethyl acetate contained in a 25-mL Erlenmeyer flask; gently heat the mixture to hasten dissolution. Add 5 mL of petroleum ether (bp 60–80 °C, 760 Torr) and cool the solution thoroughly in an ice-water bath. Be sure the solution remains homogeneous upon cooling; if not, add up to one additional milliliter of ethyl acetate to effect dissolution.

Cycloaddition and Isolation   Add 1.5 mL of dry 1,3-cyclopentadiene in one person to the cooled solution of maleic anhydride, and swirl the resulting solution gently until the exothermic reaction subsides and the cycloadduct separates as a white solid.★ Heat the mixture until the solid has redissolved.

To prevent the desired product from oiling out of solution rather than crystallizing, the reaction mixture *must* cool slowly. To accomplish this, place the Erlenmeyer flask in a water bath warmed to about 60 °C, and let the bath and flask cool to room temperature.★ Seeding the solution to effect crystallization may also be required and can be performed according to the procedure described in Section 3.2. Once the solution has cooled to room temperature and no more crystallization is apparent, cool the flask in a ice-water bath for a few minutes to complete the process. Then isolate the product by vacuum filtration, and air-dry it.

Analysis   Weigh your product to determine the yield and measure its melting point. Test the product for unsaturation using the bromine in dichloromethane (Sec. 23.8A) and the Baeyer tests (Sec. 23.8B). Obtain IR and ^{1}H NMR spectra of your starting materials and product and compare them with those of authentic samples (Figs. 12.1, 12.2, and 12.7–12.10).

Finishing Touches

Pour the *pot residue of dicyclopentadiene*, any unused *1,3-cyclopentadiene*, and the *mother liquors* from the filtration into a container for recovered dicyclopentadiene or a container for nonhalogenated organic solvents. Allow any volatiles on the *calcium chloride* that was used to dry the 1,3-cyclopentadiene to evaporate in the hood, and then place the *residual solid* in a container for nonhazardous solids. Pour the *dichloromethane solution* from the bromine test for unsaturation in a container for halogenated organic liquids, and put the *manganese dioxide* from the Baeyer test for unsaturation in a container for heavy metals.

Microscale Procedure

Preparation   Answer the Pre-Lab Exercises on page PL. 75. Review Sections 2.9, 2.10, and 2.17.

Apparatus   Water bath and apparatus for Craig tube filtration.

Setting Up   Place 100 mg of maleic anhydride and 0.4 mL of ethyl acetate in the Craig tube and mix the contents of the tube carefully to effect dissolution; warming may be helpful. Then add 0.4 mL of petroleum ether (bp 60–90 °C, 760 Torr), and mix the contents of the tube by gentle swirling.

Cycloaddition and Isolation   In one portion, add 0.1 mL of dry 1,3-cyclopentadiene, which may be obtained from your instructor, to the Craig tube and mix the contents by swirling the tube until a homogeneous solution results. The exothermicity of the reaction may keep the desired product in solution, but if not, heat the tube gently to dissolve any solids that form.

To prevent the desired product from oiling out of solution rather than crystallizing, the reaction mixture *must* cool slowly. To accomplish this, place the Craig tube in water bath warmed to about 60 °C, and let the bath and tube cool to room temperature.★ Seeding the solution to effect crystallization may also be required and can be performed according to the procedure described in Section 3.2. Once the solution has cooled to room temperature and no more crystallization is apparent, cool the Craig tube in an ice-water bath for a few minutes to complete the process. Then equip the Craig tube for filtration and isolate the product by centrifugation.

Analysis   Weigh your product to determine the yield and measure its melting point. The reported melting point is 164–165 °C. Test the product for unsaturation using the bromine in dichloromethane (Sec. 23.8A) and the Baeyer tests (Sec. 23.8B). Obtain IR and ^{1}H NMR spectra of your starting materials and products and compare them with those of authentic samples (Figs. 12.1, 12.2, and 12.7–12.10).

Finishing Touches   **P**our the *mother liquors* from the filtration into a container for nonhalogenated organic solvents. Pour the *dichloromethane solution* from the bromine test for unsaturation in a container for halogenated organic liquids, and put the *manganese dioxide* from the Baeyer test for unsaturation in a container for heavy metals.

C · HYDROLYSIS OF ANHYDRIDES

1. 4-Cyclohexene-*cis*-1,2-dicarboxylic Acid

Miniscale Procedures

Preparation   Answer the Pre-Lab Exercises on page PL. 77. Review Sections 2.9, 2.10, and 2.17.

Apparatus   A 25-mL Erlenmeyer flask, ice-water bath, apparatus for vacuum filtration and *flameless* heating.

Setting Up   Place 1.0 g of anhydride **6** and 5 mL of distilled water in the Erlenmeyer flask and add a stirbar.

Hydrolysis and Isolation   Heat the mixture until it boils and all the oil that forms initially dissolves. Allow the solution to cool to room temperature and then induce crystallization by scratching the flask at the air-liquid interface.★ After crystallization begins, cool the flask in an ice-water bath to complete the process, collect the solid by vacuum filtration, and air-dry it. Recrystallize the diacid from water.

Analysis   Weigh the product to determine the yield and measure its melting point. Test the product for unsaturation using the bromine in dichloromethane (Sec. 23.8A) and the Baeyer tests (Sec. 23.8B) to determine whether hydrolysis has affected the carbon-carbon π-bond. Also test a saturated aqueous solution of the diacid with pHydrion paper and record the result. Obtain IR and ^{1}H NMR spectra of your product and compare them with those of an authentic sample (Figs. 12.11 and 12.12).

Optional Procedure   Hydrolysis of the anhydride **6** to the diacid **8** without isolating **6** is possible as follows. Modify the procedure in Part A such that after the initial 0.5 h period of reflux the reaction mixture is allowed to cool below the boiling point. Then add about 5 mL of water to the flask through the top of the condenser and heat the resulting mixture under gentle reflux with stirring for at least 0.5 h. Allow the mixture to cool to room temperature.★ If no crystals form, acidify the solution with about 0.3 mL (6 drops) of concentrated sulfuric acid, and cool the resulting mixture in an ice-water bath. Isolate and analyze the diacid according to the directions given above.

Microscale Procedure

Preparation   Answer the Pre-Lab Exercises on page PL. 77. Review Sections 2.9, 2.10, and 2.17.

Apparatus   A 10-mm × 75-mm test tube, ice-water bath, apparatus for Craig tube filtration and *flameless* heating.

Setting Up, Hydrolysis, and Isolation   Place 100 mg of the anhydride **6** in the test tube and add 0.5 mL of water and a boiling stone. Heat the mixture until it boils and all the oil that forms initially dissolves. Using a preheated Pasteur pipet, transfer the hot solution to a Craig tube. Allow the solution to cool to room temperature. If necessary, induce crystallization by gently scratching the tube at the air-liquid interface.★ Complete crystallization by cooling the mixture in an ice-water bath.

   Equip the Craig tube for filtration, isolate the crystalline product by centrifugation, and transfer it to a piece of filter or weighing paper for air-drying. Recrystallize the diacid from water.

Analysis   Weigh the product to determine the yield and measure its melting point. Test the product for unsaturation using the bromine in dichloromethane (Sec. 23.8A) and the Baeyer tests (Sec. 23.8B) to determine whether hydrolysis has affected the carbon-carbon π-bond. Also test a saturated aqueous solution of the diacid with pHydrion paper and record the result. Obtain IR and ^{1}H NMR spectra of your product and compare them with those of an authentic sample (Figs. 12.11 and 12.12).

2. Bicyclo[2.2.1]hept-5-en-*endo*-2,3-dicarboxylic Acid

Miniscale Procedure

Preparation   Answer the Pre-Lab Exercises on page PL. 77. Review Sections 2.9, 2.10, and 2.17.

Apparatus   A 25-mL Erlenmeyer flask, ice-water bath, apparatus for vacuum filtration and *flameless* heating.

Setting Up   Place 1.0 g of anhydride **2** and 5 mL of distilled water in the Erlenmeyer flask and add a stirbar.

Hydrolysis and Isolation   Follow the miniscale procedure of Part 1 to effect hydrolysis and isolation.

Analysis   Perform the same analysis as that described in the miniscale procedure of Part 1. The reported melting point of bicyclo[2.2.1]hept-5-en-*endo*-2,3-dicarboxylic acid is 180–182 °C. Obtain IR and ^{1}H NMR spectra of your product and compare them with those of an authentic sample (Figs. 12.13 and 12.14).

Microscale Procedure

Preparation   Answer the Pre-Lab Exercises on page PL. 77. Review Sections 2.9, 2.10, and 2.17.

Apparatus   A 10-mm × 75-mm test tube, ice-water bath, apparatus for Craig tube filtration and *flameless* heating.

Setting Up, Hydrolysis, and Isolation   Follow the microscale procedure of Part 1 to effect hydrolysis of **Z** and isolation of the diacid.

Analysis   Perform the same analysis as that described in the microscale procedure of Part 1. The reported melting point of bicyclo[2.2.1]hept-2-en-*endo*-5,6-dicarboxylic acid is 180–182 °C. Obtain IR and ^{1}H NMR spectra of your product and compare them with those of an authentic sample (Figs. 12.13 and 12.14).

Finishing Touches

Using sodium carbonate, neutralize the *aqueous mother liquors* from hydrolysis of the anhydride, dilute the solution, and flush it down the drain. Pour the *dichloromethane solution* from the bromine test for unsaturation in a container for halogenated organic liquids, and put the *manganese dioxide* from the Baeyer test for unsaturation in a container for heavy metals.

EXERCISES

Refer to Tables 8.1–8.8 as needed for answering the questions on spectroscopy.

General Questions

1. Explain why a *cyclic* 1,3-diene like 1,3-cyclopentadiene dimerizes much more readily than does an *acyclic* one like *trans*-1,3-pentadiene?

2. Write structures for the products expected in the following possible Diels-Alder reactions. If no reaction is anticipated, write ''N.R.''

3. When producing Diels-Alder adducts using maleic anhydride as the dienophile, why is it important that the reagents and apparatus be dry?

4. Consider the spectral data for maleic anhydride (Figs. 12.1 and 12.2).

   a. In the functional group region of IR spectrum, specify the absorptions associated with the carbon-carbon and carbon-oxygen double bonds.

   b. The ^{1}H NMR spectrum shows the chemical shift of the vinylic protons to be δ 7.10. Explain why this resonance is downfield of the normal range observed for vinyl hydrogen atoms.

   c. For the ^{13}C NMR data, assign the various resonances to the carbon nuclei responsible for them.

Questions for Part A

5. Why should 3-sulfolene and maleic anhydride be *completely* dissolved in xylene before heating the mixture to effect reaction?

6. Write the structure, including stereochemistry, of the expected addition product of bromine to the Diels-Alder adduct obtained by this procedure.

7. Consider the spectral data for 3-sulfolene (Figs. 12.3 and 12.4).

   a. In the IR spectrum, specify the absorption associated with the carbon-carbon double bond.

   b. In the ^{1}H NMR spectrum, assign the various resonances to the hydrogen nuclei responsible for them.

   c. For the ^{13}C NMR data, assign the various resonances to the carbon nuclei responsible for them.

8. Consider the spectral data for 4-cyclohexene-*cis*-1,2-dicarboxylic anhydride (Figs. 12.5 and 12.6).

   a. In the functional group region of IR spectrum, specify the absorptions associated with the carbon-carbon double bond and the carbonyl groups of the anhydride function.

   b. In the ^{1}H NMR spectrum, assign the various resonances to the hydrogen nuclei responsible for them.

   c. For the ^{13}C NMR data, assign the various resonances to the carbon nuclei responsible for them.

9. a. Given that sulfones typically show two strong absorptions near 1300 and 1150 cm^{-1} discuss the differences observed in the IR and NMR spectra of 3-sulfolene and of 4-cyclohexene-*cis*-1,2-dicarboxylic anhydride that are consistent with the loss of sulfur dioxide in this experiment.

   b. Discuss the differences observed in the IR and NMR spectra of maleic anhydride and of 4-cyclohexene-*cis*-1,2-dicarboxylic anhydride that are consistent with the involvement of the carbon-carbon π-bond of maleic anhydride in this experiment.

Questions for Part B

10. Why should the head temperature be maintained below about 45 °C when dicyclopentadiene is being cracked?

11. Why is it good technique to induce crystallization of the diacid derived from the Diels-Alder adduct *before* cooling the solution to 0 °C?

12. Why does 1,3-cyclopentadiene react more rapidly with maleic anhydride than with another molecule of itself?

13. The "*cracking*" of dicyclopentadiene to two moles of 1,3-cyclopentadiene (Eq. 12.8) is an example of a retro-Diels-Alder reaction. Predict the products to be anticipated from an analogous reaction with the compounds (**a**)–(**d**).

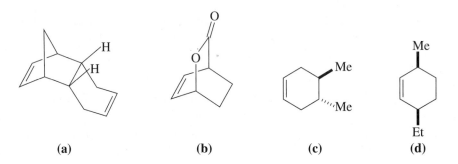

(**a**)	(**b**)	(**c**)	(**d**)

14. Consider the spectral data for 1,3-cyclopentadiene (Figs. 12.7 and 12.8).

   a. In the functional group region of the IR spectrum, specify the absorption associated with the carbon-carbon π-bond.

   b. In the ^{1}H NMR spectrum, assign the various resonances to the hydrogen nuclei responsible for them. Explain why the resonance associated with the two C-5 hydrogen atoms is not a triplet.

   c. For the ^{13}C NMR data, assign the various resonances to the carbon nuclei responsible for them.

15. Consider the spectral data for bicyclo[2.2.1]hept-5-en-*endo*-2,3-dicarboxylic anhydride (Figs. 12.9 and 12.10).

   a. In the functional group region of the IR spectrum, specify the absorption associated with the carbon-carbon and carbon-oxygen double bonds.

   b. In the ^{1}H NMR spectrum, assign the various resonances to the hydrogen nuclei responsible for them.

   c. For the ^{13}C NMR data, assign the various resonances to the carbon nuclei responsible for them.

16. **a.** Discuss the differences observed in the IR and NMR spectra of 1,3-cyclopentadiene and of bicyclo[2.2.1]hept-5-en-*endo*-2,3-dicarboxylic anhydride that are consistent with the loss of the 1,3-diene function in this experiment.

   b. Discuss the differences observed in the IR and NMR spectra of maleic anhydride and of bicyclo[2.2.1]hept-2-en-*endo*-5,6-dicarboxylic anhydride that are consistent with the involvement of the carbon-carbon π-bond of maleic anhydride in this experiment?

Questions for Part C

17. Consider the spectral data for 4-cyclohexene-*cis*-1,2-dicarboxylic acid (Figs. 12.11 and 12.12).

   a. In the functional group region of the IR spectrum, specify the absorptions associated with the carbon-carbon double bond and the carboxylic acid functions.

  b. In the ^{1}H NMR spectrum, assign the various resonances to the hydrogen nuclei responsible for them.

  c. For the ^{13}C NMR data, assign the various resonances to the carbon nuclei responsible for them. Does the number of resonances observed in the ^{13}C NMR spectrum allow you to prove that the *cis-* rather than the *trans*-dicarboxylic acid was produced by the hydrolysis of anhydride **6**? Explain.

18. Discuss the differences observed in the IR and NMR spectra of 4-cyclohexene-*cis*-1,2-dicarboxylic anhydride and of 4-cyclohexene-*cis*-1,2-dicarboxylic acid that are consistent with the conversion of an anhydride to a diacid function in this experiment.

19. Consider the spectral data for bicyclo[2.2.1]hept-5-en-*endo*-2,3-dicarboxylic acid (Figs. 12.13 and 12.14).

  a. In the functional group region of the IR spectrum, specify the absorptions associated with the carbon-carbon double bond and the carboxylic acid functions.

  b. In the ^{1}H NMR spectrum, assign the various resonances to the hydrogen nuclei responsible for them.

  c. For the ^{13}C NMR data, assign the various resonances to the carbon nuclei responsible for them. Does the number of resonances observed in the ^{13}C NMR spectrum allow you to prove that the *cis-* rather than the *trans*-dicarboxylic acid was produced by the hydrolysis of anhydride **2**? Explain.

20. Discuss the differences observed in the IR and NMR spectra of bicyclo[2.2.1]hept-5-en-*endo*-2,3-dicarboxylic anhydride and of bicyclo[2.2.1]hept-2-en-*endo*-5,6-dicarboxylic acid that are consistent with the conversion of an anhydride to a diacid function in this experiment.

Spectra   *Starting Materials and Products*

FIGURE 12.1
IR spectrum of maleic anhydride (IR card).

FIGURE 12.2
NMR data for maleic anhydride (DMSO-d$_6$).

(a) ^{1}H NMR datum (300 MHz).

CD$_3$SOCD$_2$H

δ, ppm

(b) ^{13}C NMR data. δ 137.0, 165.0.

FIGURE 12.3
IR spectrum of 3-sulfolene (IR card).

SO$_2$

PERCENT TRANSMITTANCE

WAVENUMBER IN CM^{-1}

FIGURE 12.4
NMR data for 3-sulfolene (CDCl$_3$).

(a) ^{1}H NMR spectrum (300 MHz).

SO$_2$

δ, ppm

(b) ^{13}C NMR data. δ 55.7, 124.7.

FIGURE 12.5
IR spectrum of 4-cyclohexene-*cis*-dicarboxylic anhydride (IR card).

(a) ^{1}H NMR spectrum (300 MHz).

5.90 ppm

3.48 ppm

δ, ppm

(b) ^{13}C NMR data. δ 26.1, 39.3, 125.5, 174.7.

FIGURE 12.6
NMR data for 4-cyclohexene-*cis*-dicarboxylic anhydride (CDCl$_3$).

FIGURE 12.7
IR spectrum of 1,3-cyclopentadiene (neat).

FIGURE 12.8

NMR data for 1,3-cyclopentadiene (CDCl$_3$).

(a) ^{1}H NMR spectrum (300 MHz).

(b) ^{13}C NMR data. δ 42.2, 133.0, 134.4.

FIGURE 12.9

IR spectrum of bicyclo[2.2.1]hept-5-en-*endo*-2,3-dicarboxylic anhydride (IR card).

FIGURE 12.10

NMR data for bicyclo[2.2.1]hept-5-en-*endo*-2,3-dicarboxylic anhydride (CDCl$_3$).

(a) ^{1}H NMR spectrum (300 MHz).

(b) ^{13}C NMR data. δ 46.8, 48.7, 49.1, 135.6, 174.4

FIGURE 12.11
IR spectrum of 4-cyclohexene-*cis*-1,2-dicarboxylic acid (IR card).

(a) ¹H NMR spectrum (300 MHz).

FIGURE 12.12
NMR data for 4-cyclohexene-*cis*-1,2-dicarboxylic acid (DMSO-d₆).

(b) ¹³C NMR data. δ 25.9, 39.2, 125.2, 174.9.

FIGURE 12.13
IR spectrum of bicyclo[2.2.1]hept-5-en-*endo*-2,3-dicarboxylic acid (IR card).

FIGURE 12.14
NMR data for bicyclo[2.2.1]hept-5-en-*endo*-2,3-dicarboxylic acid (DMSO-d$_6$).

(a) ^{1}H NMR spectrum (300 MHz).

(b) ^{13}C NMR data. δ 46.8, 48.7, 49.1, 135.6, 174.4.

■ ■ ■ ■ ■   H I S T O R I C A L   H I G H L I G H T   ■ ■ ■ ■ ■

Discovery of the Diels-Alder Reaction

The Diels-Alder reaction was discovered by the two German chemists whose names it bears. Otto Diels was a professor at the University of Kiel, and Kurt Alder was a postdoctoral student working with him. Diels, who was born in Hamburg in 1866, studied at the University of Berlin, where he received a Ph.D.; he later worked as an assistant to the world-famous chemist Emil Fischer (see Historical Highlight in Chap. 22). Alder, born in Upper Silesia in 1902, studied chemistry at the University of Berlin and then at the University of Kiel, where he became associated with Diels and received his Ph.D. in 1926.

One of the subjects that intrigued Diels and Alder was a novel addition of one or two molecules of 1,3-cyclopentadiene to one of *p*-benzoquinone, a reaction that had been investigated in other laboratories for over twenty years. For example, H. Albrecht, another German chemist, was apparently the first to observe this reaction, and he proposed in 1906 that the structures of the products were **1** and **2**. Between 1906 and 1928, this reaction was studied in several laboratories by a number of renowned chemists, and several other structures were proposed for the products. Diels and Alder conducted extensive investi-

gations of this and similar reactions, and in a classic paper published in 1928, they questioned the assignment of the structures **1** and **2**. In a series of reports in succeeding years, they showed conclusively that the structures proposed by Albrecht and others were incorrect and that the products from the reaction of 1,3-cyclopentadiene with *p*-benzoquinone were in fact **3** and **4**.

HISTORICAL HIGHLIGHT
Discovery of the Diels-Alder Reaction *(continued)*

3

4

As a result of their experiments, Diels and Alder suggested that this reaction and numerous other related processes belonged to common reaction type in which a 1,3-diene undergoes cycloaddition to a π-bond of a dienophile to form a product containing a new six-membered ring. This key feature is illustrated by the highlighted bonds in the structures of **3** and **4**. At the time of its discovery, this was an entirely new and important type of organic reaction for the construction of such rings.

The Diels-Alder reaction is considered to be one of the few truly new organic reactions of the 20th century, and it ranks in synthetic importance with the Wittig and Grignard reactions (see Historical Highlights for Chaps. 18 and 19, respectively). In recognition of their research and development of this process, Diels and Alder were awarded the Nobel Prize in 1951. In his Nobel Prize lecture, Diels commented that he thought it plausible that nature might use this type of reaction to fabricate some of the complicated polycyclic compounds that are so important to living organisms. The Diels-Alder reaction often does not require temperatures that lie outside of the range at which life exists, and it frequently proceeds in almost quantitative yield without using any special catalysts, reagents, or solvents. Despite Diels' suggestion that the Diels-Alder reaction would be a common transformation in nature, the biosyntheses of only a few natural products appear to involve such a process.

On the other hand, this reaction has been extensively used as a key step in the laboratory syntheses of a number of important naturally occurring substances—including alkaloids, terpenes, dyes, vitamins, and hormones.

The Diels-Alder reaction has also been a key reaction in industrial-scale syntheses of substances that are *not* naturally occurring. Prominent among these are certain bicyclic polychlorinated compounds, such as Aldrin, Dieldrin, and Chlordane, that are potent insecticides. The synthesis of each of these materials relies on perchloro-1,3-cyclopentadiene as the diene for the cycloaddition. Unfortunately, the beneficial uses of members of this class of compounds is often out-weighed by their negative environmental impact. Like DDT, whose use is now banned in the U.S. although not throughout the world, they tend to persist for months, if not years, after being applied. The resulting long-term, low-level exposure of many generations of the target pest to the insecticide may allow the species to develop biological resistance to it. Moreover, organisms that feed on the insects may accumulate the compound in their own tissues, and suffer a variety of harmful effects as a result. Consequently, alternative chemical insect controls are constantly being sought, an important criterion for such alternatives being that they are readily biodegradable.

Chlordane

Aldrin

HISTORICAL HIGHLIGHT
Discovery of the Diels-Alder Reaction *(continued)*

Dieldrin

DDT

An idea of the growth in the importance of the Diels-Alder reaction is gained by looking at the Cumulative Indexes of *Chemical Abstracts*. Only two entries were listed under the name "Diels-Alder reaction" in the ten-year period, 1927–1936, whereas such entries had increased to 180 by the decade span-ning 1947–1956, during which time Diels and Alder shared the Nobel Prize in Chemistry. By the five-year period of 1982–1986, a total of about 1300 entries appeared under this heading, and this figure nearly doubled to some 2400 in the most recent cumulative index, which covers 1987–1991. The number of citations presently exceeds 500 per year.

A recent and more extensive description of the history of the Diels-Alder reaction may be found in: Berson, J. *Tetrahedron* **1992**, *48*, 3–17.

Relationship of Historical Highlight to Experiments

The procedures in this chapter illustrate the basic principles of the Diels-Alder reaction, many of which were first unveiled by the extensive research efforts of Diels and Alder. In all cases, the reactions form products containing new, unsaturated six-membered rings, which is the fundamental structural outcome of the Diels-Alder process.

CHAPTER *13*

Kinetic and Thermodynamic Control of a Reaction

Organic chemists frequently need to predict which of several possible competing reactions will predominate when a specified set of reactants is used. The problem is illustrated by considering the general situation where reagents **W** and **X** can combine to produce either **Y** (Eq. 13.1) or **Z** (Eq. 13.2). Predicting which product will be formed preferentially under a particular set of conditions requires developing an **energy diagram** for the two reactions. Strictly speaking, such a diagram should be based on the **free energy changes**, ΔG, that occur during the conversion of reactants to products. To generate the needed diagram would require knowledge of both the **enthalpic**, ΔH, and **entropic**, ΔS, changes involved. If the reactions being considered are similar, as we assume is the case for the processes of Equations 13.1 and 13.2, entropic changes are not too important, so using values for ΔH gives a close approximation to the changes occurring in ΔG. This is the basis on which our potential energy diagram will be constructed.

$$W + X \rightleftharpoons Y \tag{13.1}$$

$$W + X \rightleftharpoons Z \tag{13.2}$$

First assume for the purposes of this discussion that both of the reactions are *exothermic and* that **Y** is *thermodynamically less stable* than **Z**. These assumptions mean that (1) the energy levels of both products are below that of the starting materials and (2) the heat of reaction for formation of **Z** is greater than that for **Y**, as symbolized by ΔH_Z and ΔH_Y in Figure 13.1. Now let the activation barriers for conversion of the reagents to the two transition states, $Y^{\ddagger}$ and $Z^{\ddagger}$, for forming **Y** and **Z**, be measured by $\Delta H_Y^{\ddagger}$ and $\Delta H_Z^{\ddagger}$ respectively. The latter barrier has been set *lower* on the diagram than that for **Y**. This ordering of the relative energies of the two transition states is arbitrary but is consistent with the *empirical* observation that when comparing two *similar* organic reactions, the *more exothermic process often has the lower enthalpy of activation,*

FIGURE 13.1
Typical reaction profile for competing reactions.

$\Delta H^{\ddagger}$. In other words, the relative energies of the transition states reflect the relative energies of the products that are produced from the transition states. This generalization is useful because it allows us to complete the potential energy diagram and to predict that because less energy is required for reactants **W** and **X** to reach the transition state leading to **Z**, this product will be formed faster than **Y**.

For all practical purposes, most organic reactions are either irreversible or are performed under conditions such that equilibrium between products and starting materials is *not* attained. The ratio of products obtained from reactions run under these circumstances is said to be subject to **kinetic control**. This means that the relative amounts of the various products are determined by the relative rates of their formation rather than by their relative thermodynamic stabilities. The major product of such a reaction is often called the **kinetically controlled product** even though other compounds may be formed. In the example of Figure 13.1, **Z** would represent the kinetically controlled product; note that $\Delta H_{\mathbf{Z}}^{\ddagger} < \Delta H_{\mathbf{Y}}^{\ddagger}$.

Organic reactions that are performed under conditions in which *equilibrium* is established between reactants and products produce product ratios that are subject to **thermodynamic control**. In such cases, the predominant product is called the **thermodynamically controlled product** and would be the more stable **Z** under the circumstances portrayed in Figure 13.1. Note that the ratio **Y** : **Z** would probably *not* be the same under both sets of reaction conditions because it is unlikely that the difference in enthalpies of activation, $\Delta\Delta H_{\mathbf{YZ}}^{\ddagger}$, between the two transition states would be equal to the enthalpy difference, $\Delta\Delta H_{\mathbf{YZ}}$, between the two ground states of the products.

The situation shown in Figure 13.1 is the one most frequently encountered in organic chemistry; namely, that the major product of kinetic control and thermodynamic control is the same. There are exceptions to this, however, and the reaction profiles of Figure 13.2 depict just such a case. To generate these profiles, it is again assumed that the reactions of Equations 13.1 and 13.2 are exothermic and that **Z** is more stable than **Y**. What is changed is that the activation barrier, $\Delta H_{\mathbf{Y}}^{\ddagger}$, for forming **Y** is now *less* than that for forming **Z**, $\Delta H_{\mathbf{Z}}^{\ddagger}$. The lower barrier means that **Y** would now be the product of kinetic control, although **Z** would remain the product of thermodynamic control.

Note that in developing Figures 13.1 and 13.2, conclusions about the preferred products of kinetic control and thermodynamic control do not depend on the exothermicity or endothermicity of the overall reactions. *All that matters are the relative*

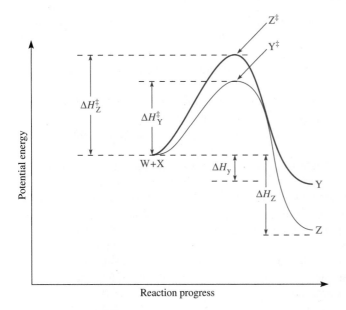

energies of the transition states leading to products, if the reaction is under kinetic control, or the relative energies of the products themselves, if the conditions are those of thermodynamic control. Whether different major products are produced under reversible as compared to irreversible reaction conditions is determined by measuring the ratios of products formed under the two different circumstances. Each product can also be isolated and then subjected to the reaction conditions under which they are produced. If equilibration is occurring, analysis of the product mixture will demonstrate this. Both types of procedures are described in this experiment.

13.2 Formation of Semicarbazones under Kinetic and Thermodynamic Control

The example chosen to illustrate the principle of kinetic and thermodynamic control of products involves the competing reactions of semicarbazide (**1**) with two carbonyl compounds, cyclohexanone (**2**) and 2-furaldehyde (**4**) as shown in Equations 13.3 and 13.4, respectively. In this case, one compound, semicarbazide, reacts with two different

$$H_2NCNHNH_2 + O = \bigcirc \rightleftharpoons H_2NCNHN = \bigcirc + H_2O \quad (13.3)$$

1	**2**	**3**
Semicarbazide	Cyclohexanone	Cyclohexanone semicarbazone

$$H_2NCNHNH_2 + \bigcirc \rightleftharpoons H_2NCNHN \bigcirc + H_2O \quad (13.4)$$

1	**4**	**5**
	2-Furaldehyde	2-Furaldehyde semicarbazone

compounds, cyclohexanone and 2-furaldehyde, to give two different **semicarbazones**, **3** and **5**, respectively. Both are crystalline solids that have distinctive melting points by which they may be easily identified. This makes it simple to determine experimentally which compound is the product of kinetic control and which is the product of thermodynamic control.

The rates and equilibrium constants of reactions of carbonyl compounds with nucleophiles such as semicarbazide are affected significantly by variations in the pH of the reaction medium. This may be understood by examining the results of protonating the reactants.

First, let us consider the effect that protonation of the carbonyl function has on the rate of reaction of a carbonyl compound with nucleophiles. The addition of a proton to the oxygen atom of a carbonyl group produces **6** (Eq. 13.5). This makes the carbonyl carbon atom much more *electrophilic* than the unprotonated compound because the partial positive charge on the carbon atom is increased, as shown by consideration of the resonance structures in Equation 13.5. Since the rate-determining step for the formation of semicarbazones is the nucleophilic attack of semicarbazide (**1**) on the carbonyl carbon atom of the substrate, the enhanced electrophilicity of **6** *increases* the rate of formation of the semicarbazone.

$$R_2C{=}\ddot{O} + H^+ \rightleftharpoons \left[R_2C{=}\overset{+}{O}H \longleftrightarrow R_2\overset{+}{C}{-}\ddot{O}H\right] \qquad \textbf{(13.5)}$$

Carbonyl
compound
6

Now let us assess what happens if the nucleophile is protonated. Adding of a proton to **1** produces its conjugate acid **7** (Eq. 13.6), which is *not* nucleophilic. Consequently the rate of the desired reaction (Eq. 13.7) is *decreased*, an outcome just the *reverse* of that described for protonation of the carbonyl compound. If the medium in which the reaction was performed is sufficiently acidic, essentially all of the **1** will be converted to **7**, thereby lowering the concentration of the nucleophilic **1**. The formation of the semicarbazone would then become extremely slow or might not occur at all. Thus both the rate and equilibrium constant for the reaction of Equation 13.7 are affected significantly by the pH at which the process is conducted.

$$\underset{\textbf{1}}{H_2N\overset{\overset{\displaystyle O}{\|}}{C}NHN H_2} + H^+ \rightleftharpoons \underset{\textbf{7}}{H_2N\overset{\overset{\displaystyle O}{\|}}{C}NH\overset{+}{N}H_3} \qquad \textbf{(13.6)}$$

$$\underset{\textbf{1}}{H_2N\overset{\overset{\displaystyle O}{\|}}{C}NHN H_2} \quad \underset{\textbf{6}}{R_2C{=}\overset{+}{O}H} \rightleftharpoons \left[H_2N\overset{\overset{\displaystyle O}{\|}}{C}NHN\underset{H}{\overset{H}{\underset{|}{|}}}\overset{OH}{\underset{+}{C}R_2}\right] \rightleftharpoons \left[H_2N\overset{\overset{\displaystyle O}{\|}}{C}NHN\overset{H}{\underset{|}{|}}\overset{+}{\overset{OH_2}{C}R_2}\right]$$

$$\qquad \qquad \qquad \qquad \qquad \qquad \qquad \qquad \qquad \qquad \qquad \qquad \qquad \textbf{(13.7)}$$

$$H_3O^+ + H_2N\overset{\overset{\displaystyle O}{\|}}{C}NH\ddot{N}{=}CR_2 \underset{H_2\ddot{O}:}{\rightleftharpoons} H_2N\overset{\overset{\displaystyle O}{\|}}{C}NH\underset{+}{N}{=}CR_2 + H_2O$$

The preceding discussion makes clear that with respect to Equation 13.7, the concentration of the nucleophile **1** is reduced at low pH whereas that of the activated electrophile **6** is reduced at high pH. It follows that there is an optimum pH at which the mathematical product of the concentrations of the nucleophilic reagent and conjugate acid of the carbonyl compound, [**1**] × [**6**], is maximized.

$$C_6H_5-NH\ddot{N}H_2 \qquad\qquad HO\ddot{N}H_2$$

$$\textbf{8} \qquad\qquad\qquad \textbf{9}$$

For reactions of aldehydes and ketones with reagents such as semicarbazide (**1**), phenylhydrazine (**8**), and hydroxylamine (**9**), the desired pH is produced and maintained by **buffer solutions**. A buffer solution resists changes in its pH. In general, this requires the presence of a pair of substances in the solution, one that is a weak acid and neutralizes hydroxide ions and another that is a weak base and neutralizes protons. For example, the $H_2PO_4^-/HPO_4^{2-}$ **buffer system** is produced by the addition of dibasic potassium phosphate, K_2HPO_4, a weak base, to semicarbazide hydrochloride. The weakly acidic $H_2PO_4^-$ component of the buffer system is produced as shown in Equation 13.8. The function of the two ions in neutralizing hydroxide and hydrogen ions is illustrated in Equations 13.9 and 13.10.

Each different buffer system functions to maintain the pH within a rather narrow range characteristic of the weak acid and weak base of that particular system. In the case of the phosphate system, the range is pH 6.1–6.2. By comparison, a carbonate/bicarbonate buffer system (CO_3^{2-}/HCO_3^-) maintains a pH of 7.1–7.2.

$$H_2NCNHNH_3^+\ Cl^- + HPO_4^{2-} \rightleftharpoons H_2NCNHNH_2 + H_2PO_4^- + Cl^- \qquad \text{(13.8)}$$
$$\textbf{1}$$

$$H_2PO_4^- + HO^- \rightleftharpoons HPO_4^{2-} + H_2O \qquad\qquad \text{(13.9)}$$

$$HPO_4^{2-} + H_3O^+ \rightleftharpoons H_2PO_4^- + H_2O \qquad\qquad \text{(13.10)}$$

The maximum rates of the reactions of most aldehydes and ketones with semicarbazide occur in the pH range of 4.5–5.0. For the purpose of making derivatives of carbonyl compounds (Sec. 23.7), semicarbazide is best used in an acetate buffer ($CH_3CO_2H/CH_3CO_2^-$) solution, which maintains a pH in the maximum rate range, 4.5–5.0. However, to demonstrate the principle of kinetic and thermodynamic control of reactions, buffers that maintain higher pHs, and thus produce lower rates, are more desirable. Parts A–C of the experimental procedure involve a phosphate buffer system, whereas the bicarbonate system is used in Part D, a strategy that allows a comparison of how the difference in rates in the two buffer systems affects the product ratio. Analysis of the products from the various parts of these experiments provides strong clues as to which of the semicarbazones is the product of kinetic control and which is the product of thermodynamic control.

In Section 13.1, it was noted that implicit in the theory of kinetic and thermodynamic control is the fact that the *kinetic product* (**Z** in Fig. 13.1 and **Y** in Fig. 13.2), which is produced more rapidly, is also reconverted to starting material more rapidly. The *thermodynamic product*, on the other hand, being more stable, is not so readily reconverted to starting material. Experimental procedure E provides tests of the relative stabilities of the two semicarbazone products toward the reverse reaction. The results of these experiments should provide additional evidence of the identity of one of the semicarbazones as the product of kinetic control and the other as the product of thermodynamic control.

EXPERIMENTAL PROCEDURES

KINETIC AND THERMODYNAMIC CONTROL OF A REACTION

Purpose: To study the phenomenon of kinetic and thermodynamic control of competing reactions.

> **Safety Alert**
>
> Because the melting point of one of the compounds is near 200 °C, do *not* use a liquid-filled apparatus such as a Thiele tube *unless* the heating medium is silicone oil. A metal block melting-point apparatus is satisfactory for determining the melting points of the semicarbazones in this experiment.

Preparation   Answer the Pre-Lab Exercises on page PL. 79. Review Sections 2.9, 2.10, and 2.17.

You may be assigned only some parts of these procedures. If so, plan to share results with other students who perform the other parts. This should enable you to draw your own conclusions and report them accordingly.

Apparatus   Two 13-mm × 100-mm test tubes, two 25-mL, seven 5-mL, and two 125-mL Erlenmeyer flasks, 1-mL graduated pipet, ice-water bath, apparatus for vacuum filtration and *flameless* heating.

A · PREPARATION OF CYCLOHEXANONE SEMICARBAZONE

Dissolve 0.5 g of semicarbazone hydrochloride and 1.0 g of dibasic potassium phosphate (K_2HPO_4) in 6 mL of water contained in a 25-mL Erlenmeyer flask. Using a 1-mL graduated pipet, deliver 0.5 mL of cyclohexanone into a test tube containing 2.5 mL of 95% ethanol. Pour the ethanolic solution into the aqueous semicarbazide solution, and swirl or stir the mixture immediately. Allow 5–10 min for crystallization of the semicarbazone to reach completion; then collect the crystals by vacuum filtration and wash them with a little cold water. *Thoroughly* air-dry the crystals★ and determine their weight and melting point. Obtain IR and NMR spectra of your starting material and product. Compare your spectra with those of authentic samples (Figs. 13.3–13.6).

B · PREPARATION OF 2-FURALDEHYDE SEMICARBAZONE

For best results, the 2-furaldehyde should be freshly distilled immediately before use. Prepare the semicarbazone of 2-furaldehyde by following the procedure of Part A exactly, except use 0.4 mL of 2-furaldehyde instead of 0.5 mL of cyclohexanone.★ Recrystallization of the product is unnecessary. Obtain IR and NMR spectra of your starting material and product. Compare your spectra with those of authentic samples (Figs. 13.7–13.10).

C · REACTIONS OF SEMICARBAZIDE WITH CYCLOHEXANONE AND 2-FURALDEHYDE IN PHOSPHATE BUFFER SOLUTION

Dissolve 3.0 g of semicarbazide hydrochloride and 6.0 g of dibasic potassium phosphate in 75 mL of water. Because this is an aqueous solution, it is referred to as *solution W*. Prepare a solution of 3.0 mL of cyclohexanone and 2.5 mL of 2-furaldehyde in 15 mL of 95% ethanol. Because this is an ethanolic solution, it is referred to as *solution E*.★

1. Cool a 25-mL portion of solution W and a 5-mL portion of solution E separately in an ice-water bath to 0–2 °C. Add solution E to solution W and swirl the mixture; crystals should form almost immediately. Place the mixture in an ice-water bath for 3–5 min; then collect the crystals by vacuum filtration and wash them on the filter with 2–3 mL of cold water. *Thoroughly* air-dry the crystals★ and determine their weight and melting point.

2. Add a 5-mL portion of solution E to a 25-mL portion of solution W at room temperature; crystals should be observed in 1–2 min. Allow the mixture to stand at room temperature for 5 min, cool it in an ice-water bath for about 5 min, and then collect the crystals by vacuum filtration and wash them on the filter with 2–3 mL of cold water. *Thoroughly* air-dry the crystals★ and determine their weight and melting point.

3. Warm a 25-mL portion of solution W and a 5-mL portion of solution E *separately* to 80–85 °C; add solution E to solution W and swirl the mixture. Continue to heat the solution for 10–15 min, cool it to room temperature, and then place it in an ice-water bath for about 5–10 min. Collect the crystals by vacuum filtration, and wash them on the filter with 2–3 mL of cold water. *Thoroughly* air-dry the crystals★ and determine their weight and melting point.

D · REACTIONS OF SEMICARBAZIDE WITH CYCLOHEXANONE AND 2-FURALDEHYDE IN BICARBONATE BUFFER SOLUTION

Dissolve 2.0 g of semicarbazide hydrochloride and 4.0 g of sodium bicarbonate in 50 mL of water. Prepare a solution of 2.0 mL of cyclohexanone and 1.6 mL of 2-furaldehyde in 10 mL of 95% ethanol. Divide each of these solutions into two equal portions.

1. Mix half of the aqueous solution and half of the ethanolic solution and allow the mixture to stand at room temperature for 5 min. Collect the crystals by vacuum filtration and wash them on the filter with 2–3 mL of cold water. *Thoroughly* air-dry the crystals★ and determine their weight and melting point.

2. Warm the other portions of the aqueous and ethanolic solutions *separately* to 80–85 °C; then combine them and continue heating the mixture for 10–15 min. Cool the solution to room temperature, and place it in an ice-water bath for 5–10 min. Collect the crystals by vacuum filtration and wash them on the filter with a 2–3 mL of cold water. *Thoroughly* air-dry the crystals★ and determine their weight and melting point.

E • TESTS OF REVERSIBILITY OF SEMICARBAZONE FORMATION

1. Place 0.3 g of cyclohexanone semicarbazone, prepared in Part A, 0.3 mL of 2-furaldehyde, 2 mL of 95% ethanol, and 10 mL of water in a 25-mL Erlenmeyer flask. Warm the mixture until a homogeneous solution is obtained (about 1 or 2 min should suffice), and continue warming an additional 3 min. Cool the mixture to room temperature and then place it in an ice-water bath. Collect the crystals on a filter, and wash them with 2–3 mL of cold water. *Thoroughly* air-dry the crystals★ and determine their weight and melting point.

2. Repeat the preceding experiment, but use 0.3 g of 2-furaldehyde semicarbazone (prepared in Part B) and 0.3 mL of cyclohexanone in place of the cyclohexanone semicarbazone and 2-furaldehyde.

On the basis of your results from experiments C, D, and E, deduce which semicarbazone is the product of kinetic control and which is the product of thermodynamic control. To do this, first use the observed melting points of the crystals produced in parts C1, C2, and C3 to deduce whether the product in each part is the semicarbazone of cyclohexanone, of 2-furaldehyde, or of a mixture of the two. Note that in C1 the crystals of product separate almost immediately, in C2 after 1 or 2 min, and in C3 only after 10–15 min at a higher temperature; thus the reaction time is shortest in C1, intermediate in C2, and longest in C3. Then compare the product of D1 with that of C2 and the product of D2 with that of C3.

When analyzing the results of the experiments in part E, remember that the thermodynamic product, being the more stable, is not easily transformed into the less stable kinetic product. In contrast, the kinetic product is more readily converted into the more stable thermodynamic product. These circumstances arise because the reverse reaction from the less stable kinetic product has a lower activation energy than the reverse reaction of the thermodynamic product.

Your completed laboratory report should include the diagram specified in Exercise 2 and answers to Exercises 3 and 4 as well, unless you were instructed to omit some parts of the experiment.

Finishing Touches

Neutralize the various *filtrates* and then flush them down the drain.

EXERCISES

1. Why is it particularly important in this experiment that the semicarbazones be dry before their melting points are determined?

2. On the basis of the results from experiments C, D, and E, draw a diagram similar to Figure 13.2 and clearly label the products corresponding to those obtained in experiments A and B.

3. On the basis of the results from the experiments of Part D, explain the effect of the higher pH on the reactions between semicarbazide and the two carbonyl compounds.

4. What results would be expected if sodium acetate buffer, which provides a pH of approximately 5, were used in experiments analogous to those of Part C? Explain.

5. Show how you could calculate the exact pH of a buffer solution prepared by adding equimolar amounts of glacial acetic acid and sodium acetate to distilled water; the pK_a of acetic acid is 4.75.

6. What different result, if any, is expected if the heating period at 80–85 °C is extended to one hour in experiment D2?

7. There are two NH_2 groups in semicarbazide (**1**), yet only one of them reacts with the carbonyl group to form a semicarbazone. Explain.

Refer to Tables 8.1–8.8 as needed for answering the following questions on spectroscopy.

8. Consider the spectral data for cyclohexanone (Figs. 13.3 and 13.4).

   a. In the functional group region of the IR spectrum, specify the absorption associated with the carbon-oxygen double bond.

   b. In the ^1H NMR spectrum, assign the various resonances to the hydrogen nuclei responsible for them and explain why the multiplet integrating for four protons is downfield of that integrating for six.

   c. For the ^{13}C NMR data, assign the various resonances to the carbon nuclei responsible for them.

9. Consider the NMR data for cyclohexanone semicarbazone (Fig. 13.6).

   a. In the ^1H NMR spectrum, assign the various resonances to the hydrogen nuclei responsible for them.

   b. For the ^{13}C data, assign the various resonances to the carbon nuclei responsible for them.

10. Consider the spectral data for 2-furaldehyde (Figs. 13.7 and 13.8).

   a. In the functional group region of the IR spectrum, specify the absorptions associated with the carbon-oxygen and carbon-carbon double bonds and with the aldehydic C–H bond.

   b. In the ^1H NMR spectrum, assign the various resonances to the hydrogen nuclei responsible for them.

   c. For the ^{13}C NMR data, assign the various resonances to the carbon nuclei responsible for them.

11. a. The IR spectra portrayed in Figures 13.5 and 13.9 contain a sharp absorption at approximately 3500 cm^{-1}. Specify the functional group responsible.

   b. Both of these spectra have complex absorptions in the range of 3100–3400 cm^{-1}. What functional group is responsible?

   c. In Figure 13.9, what functionality is responsible for the absorption near 1700 cm^{-1}?

12. Consider the NMR data for 2-furaldehyde semicarbazone in Figure 13.10.

   a. In the ^1H NMR spectrum, assign the various resonances to the hydrogen nuclei responsible for them.

   b. For the ^{13}C data, assign the various resonances to the carbon nuclei responsible for them.

13. Discuss the differences observed in the IR and NMR spectra of cyclohexanone and of its semicarbazone that are consistent with the replacement of the carbonyl function by a carbon-nitrogen double bond in this experiment.

14. Discuss the differences observed in the IR and NMR spectra of 2-furaldehyde and of its semicarbazone that are consistent with the replacement of the carbonyl function by a carbon-nitrogen double bond in this experiment.

Spectra

Starting Materials and Products

FIGURE 13.3

IR spectrum of cyclohexanone (neat).

FIGURE 13.4

NMR data for cyclohexanone (CDCl₃).

(a) ¹H NMR spectrum (300 MHz).

(b) ¹³C NMR data: δ 25.1, 27.2, 41.9, 211.2.

FIGURE 13.5
IR spectrum of cyclohexanone semicarbazone (IR card).

(a) ^{1}H NMR spectrum (300 MHz).

FIGURE 13.6
NMR data for cyclohexanone semicarbazone (CDCl$_3$).

(b) ^{13}C NMR data: δ 25.6, 25.8, 26.5, 27.0, 35.4, 153.4, 158.8.

FIGURE 13.7
IR spectrum of 2-furaldehyde (neat).

FIGURE 13.8
NMR data for 2-furaldehyde (CDCl₃).

(a) ^{1}H NMR spectrum (300 MHz).

(b) ^{13}C NMR data: δ 112.9, 121.9, 148.6, 153.3, 178.1.

FIGURE 13.9
IR spectrum of 2-furaldehyde semicarbazone (IR card).

FIGURE 13.10
NMR data for 2-furaldehyde semicarbazone (DMSO-d₆)

(a) ^{1}H NMR spectrum (300 MHz).

(b) ^{13}C NMR data: δ 110.6, 111.8, 129.8, 143.8, 149.9, 156.4.

Nucleophilic Aliphatic Substitution

Preparation of Alkyl Halides

Nucleophilic aliphatic substitution reactions constitute an important class of reactions in organic chemistry. Compounds containing a wide diversity of different functional groups can frequently be prepared from a single starting material simply by varying the nature of the nucleophilic partner. In this chapter we review the general features of this reaction and conduct experiments to study the conversion of alcohols into alkyl halides, one representative process of this type.

14.1 General Concepts

The substitution of one group for another at a saturated, sp^3-hybridized, carbon atom is a reaction commonly used as a means to interconvert different functional groups. This process is named **nucleophilic aliphatic substitution** and is exemplified in Equation 14.1, where **Nu:** is a symbol for a generalized **nu**cleophile and represents a molecule or ion that has Lewis basic character, and **L:** represents a leaving group. The reaction may be considered to be a Lewis acid-base reaction because the carbon atom bearing the electronegative leaving group has Lewis acid character. This is associated with the partial positive charge on carbon resulting from the polarization of the C–L bond.

$$\text{Nu:}^- + \text{R}-\overset{|}{\underset{|}{\text{C}}}\overset{\delta^+\ \delta^-}{-\text{L}} \longrightarrow \text{Nu}-\overset{|}{\underset{|}{\text{C}}}-\text{R} + \text{L:}^- \qquad \textbf{(14.1)}$$

Nucleophiles have in common the property of bearing *at least one nonbonding pair of electrons* and being either *neutral or negatively charged*. During the course of the substitution reaction, the nonbonding pair of electrons of the nucleophile is donated to an electrophilic, Lewis acidic, carbon atom with concomitant formation of a new covalent bond. Examples of some typical nucleophiles include $Cl:^-$, $Br:^-$, $I:^-$, $HO:^-$, $N\equiv C:^-$, $H_2O:$, $:N_3^-$, $:NR_3$, and $RS:^-$.

The leaving group, L:, which like Nu: may be neutral or negatively charged, must be able to accept the pair of bonding electrons from the carbon atom as the C–L bond breaks. The ease with which various groups leave in nucleophilic substitution reactions has been determined experimentally by studying the rates of reaction of a specific nucleophile with substrates having different leaving groups. These investigations show that the leaving ability of a particular group L: correlates with the strength of its conjugate acid: The better leaving groups are those that are **conjugate bases** of the stronger acids. For example, a leaving group such as Cl^- is the conjugate base of hydrochloric acid, a strong acid, and so Cl^- is a good leaving group. On the other hand, HO^- is a poor leaving group because it is the conjugate base of the weak acid water; however, H_2O is a good leaving group because it is the conjugate base of the strong acid hydronium, H_3O^+.

14.2 Classification of Nucleophilic Substitution Reactions

Nucleophilic substitution is a general reaction for aliphatic compounds in which the leaving group is attached to an sp^3-hybridized carbon (Eq. 14.1); however, the mechanism for a given transformation depends upon the structure of the alkyl group bearing the leaving group. The two different mechanistic pathways that apply to such substitutions are designated by the symbols S_N1 (**S** for substitution, **N** for nucleophilic, and **1** for unimolecular) and S_N2 (**2** for bimolecular). These two mechanisms are depicted in general terms in Equations 14.2 and 14.3, respectively.

S_N1 MECHANISM

$$R_3C-L \underset{\text{slow}}{\overset{\text{rds}}{\rightleftharpoons}} L:^- + R_3C^+ \xrightarrow[\text{fast}]{Nu:^-} R_3C-Nu \qquad (14.2)$$

S_N2 MECHANISM

$$\underset{R^2}{\overset{R^1}{Nu:^- \underset{H}{\overset{}{C}}-L}} \xrightarrow{\text{rds}} \left[\underset{H \quad R^2}{\overset{R^1}{Nu:^{\delta-}--C---L^{\delta-}}} \right]^{\ddagger} \longrightarrow \underset{R^2}{\overset{R^1}{Nu-C_{''''H}}} + L:^- \qquad (14.3)$$

Transition state

When nucleophilic substitution occurs by an S_N1 mechanism, the reaction proceeds in two successive steps, as illustrated in Equation 14.2. The first step involves the **heterolytic cleavage**, or **ionization**, of the bond between the carbon atom and the leaving group. This step is assisted by the polar interactions between solvent molecules and the incipient cationic and anionic centers. Since the leaving group acquires the pair of bonding electrons, the organic fragment becomes a **carbocation**, with the carbon atom formerly bound to L now bearing a positive charge. The intermediate carbocation may then undergo the usual reactions of carbocations: (1) rearrangement to a more stable carbocation (Sec. 10.3); (2) loss of a proton to give an alkene by net **elimination** of the elements of **H–L** (Sec. 10.3); or (3) combination with a nucleophile to form the substitution product, as indicated in Equation 14.2. Normally, the concentration of the nucleophile, Nu:$^-$, is high compared to that of the L:$^-$ that has been produced, so the reaction of the carbocation with L:$^-$ to give the starting material is relatively unimportant. If the nucleophile is the solvent, the reaction is known as a **solvolysis**.

The first step of an S_N1 reaction is slower than the second because it involves breaking the C–L bond to generate an unstable carbocation, an endothermic process. The second step is a fast, exothermic process involving bond formation. Thus, the first step of an S_N1 reaction is the **rate-determining step** (**rds**) of the reaction, and the rate of the reaction depends only upon the concentration of the substrate, R–L. This is expressed mathematically in Equation 14.4, where k_1 is the first-order rate constant.

$$\text{Rate} = k_1[\text{R–L}] \qquad \textbf{(14.4)}$$

When substitution occurs by an S_N2 mechanism, the nucleophile attacks the substrate directly, with the angle of approach being 180° to the C–L bond. This is called "backside attack", and the reaction proceeds with inversion of stereochemistry. The C–L bond is being broken concurrently with the formation of the C–Nu bond, so both the substrate, R–L, *and* the nucleophile are involved in the transition state of the rate-determining step. Reactions in which two reactants are involved in the transition state of the rate-determining step are termed bimolecular, and the rate of such processes depends on the concentration of the substrate and the nucleophile, as shown in Equation 14.5, where k_2 is the second-order rate constant.

$$\text{Rate} = k_2[\text{R–L}][\text{Nu:}] \qquad \textbf{(14.5)}$$

It is useful to summarize some of the important factors that dictate whether a particular substrate undergoes substitution preferentially by an S_N1 or S_N2 mechanism.

1. As more alkyl groups are attached to the carbon atom **C–L** undergoing substitution, it becomes sterically more difficult for the nucleophile to attack from the backside, thereby decreasing the ease with which the S_N2 process can occur.
2. With increasing substitution of alkyl groups on the carbon atom **C–L**, the incipient carbocation in the S_N1 reaction becomes more stable, thereby increasing its ease of formation along the S_N1 pathway.

These two effects reinforce one another and yield the following general trends:

$$CH_3\text{—L} \quad RCH_2\text{—L} \quad R_2CH\text{—L} \quad R_3C\text{—L}$$

Primary (1°, RCH_2–L) substrates undergo nucleophilic substitution predominantly by an S_N2 mechanism, whereas tertiary (3°, R_3C–L) substrates tend to react via an S_N1 mechanism. Secondary substrates (2°, R_2CH–L) may react by both mechanisms, and the specific pathway is dictated by factors such as solvent, reaction conditions, and nature of the nucleophile. Such generalizations must always be applied cautiously, and the mechanism by which a particular reaction occurs *must* be confirmed *experimentally*. For example, if the structures of the starting substrate and the product(s) are determined, it may be possible to deduce the mechanism by which the process occurred. Since carbocations are involved in the S_N1 mechanism, the formation of a product arising from a rearranged carbocation points to this mechanism; however, the lack of rearrangement does not necessarily exclude an S_N1 process.

14.3 Competition Between Substitution and Elimination

In all reactions in which nucleophilic aliphatic substitution occurs, elimination reactions may compete to produce alkenes. Unimolecular elimination reactions, **E1**, compete with S_N1 substitutions, and bimolecular elimination processes, **E2** (E stands for elimination and 2 for bimolecular), compete with S_N2 transformations. These competitions are shown in Equations 14.6 and 14.7. The nature of E1 reactions is discussed in detail in Section 10.3 and that of E2 processes in Section 10.2.

S_N2 VERSUS E2

Elimination: $\diagdown C{=}C\diagup + L{:}^{-} + HNu$

Substitution: $-\underset{Nu}{\overset{}{C}}-\underset{H}{\overset{}{C}}- + L{:}^{-}$

Substitution (S_N2)   Elimination (E2)

(14.6)

S_N1 VERSUS E1

$\xrightarrow[-L{:}^{-}]{\text{slow}}$

Elimination: $\diagdown C{=}C\diagup + HNu$

Substitution: $-\underset{Nu}{\overset{}{C}}-\underset{H}{\overset{}{C}}-$

Substitution (S_N1)   Elimination (E1)

(14.7)

The course of the reaction is influenced to some extent by the nature of the nucleophile or Lewis base that is present. *Substitution is favored* with weakly basic and highly polarizable nucleophiles such as I^-, Br^-, Cl^-, H_2O, and $CH_3CO_2^-$, whereas *elimination is favored* when strongly basic and only slightly polarizable nucleophiles such as $RO:^-$, $H_2N:^-$, $H:^-$, and $HO:^-$ are used. **Polarizability** is a measure of the ease with which the electron cloud of the Lewis base is distorted by a nearby center that bears a partial or full positive charge. Furthermore, bulky nucleophiles are more prone toward inducing elimination because the hydrogen atom is more sterically accessible than the carbon atom bearing the leaving group.

14.4 Preparation of 1-Bromobutane: An S_N2 Reaction

A common technique for converting a primary alcohol to an alkyl halide involves treating the alcohol with a hydrogen halide H–X (H–X = H–Cl, H–Br, or H–I) as shown in Equation 14.8. This reaction is *reversible*, and displacing the equilibrium to the right normally involves using a large excess of the acid, a strategy in accord with the **LeChatlier principle**.

$$H—X + R—OH \underset{\Delta}{\rightleftharpoons} R—X + H_2O \qquad (14.8)$$

where X = Cl, Br, I

1-Bromobutane may be prepared by heating 1-butanol with hydrobromic acid, H–Br, in the presence of sulfuric acid (Eq. 14.9). The mechanism for this reaction has been shown to occur in two steps (Eq. 14.10). The alcohol is protonated in the first

$$CH_3CH_2CH_2CH_2—OH + HBr \underset{\Delta}{\overset{H_2SO_4}{\rightleftharpoons}} CH_3CH_2CH_2CH_2—Br + H_2O \qquad (14.9)$$

1-Butanol   1-Bromobutane
(*n*-butyl alcohol)   (*n*-butyl bromide)

step via a Lewis acid-base reaction to give the oxonium ion **1**. This oxonium ion then undergoes displacement by the bromide ion to form 1-bromobutane and water. This process is an S_N2 reaction in which *water is the leaving group and bromide ion is the nucleophile*. The sulfuric acid serves two important purposes: (1) It is a dehydrating agent that reduces the activity of water and shifts the position of equilibrium to the right, and (2) it provides an added source of hydrogen ions to increase the concentration of oxonium ion **1**. The use of *concentrated* hydrobromic acid also helps to establish a favorable equilibrium.

$$n\text{-}C_3H_7CH_2—\overset{..}{O}H + H^+ \rightleftharpoons n\text{-}C_3H_7CH_2\overset{+}{\underset{H}{\overset{H}{—O}}} \overset{Br:^-}{\rightleftharpoons} n\text{-}C_3H_7CH_2—Br + H_2O \qquad (14.10)$$

1

No reaction occurs between 1-butanol and NaBr in the absence of strong acid, because leaving groups in nucleophilic substitution reactions must be *weakly* basic (Sec.

14.1), as is water in Equation 14.10. If the reaction of 1-butanol and NaBr were to occur (Eq. 14.11), the leaving group would necessarily be the *strongly* basic hydroxide ion; thus the forward reaction depicted in Equation 14.11 does not occur. On the other hand, the reverse reaction between 1-bromobutane and hydroxide proceeds readily, since the leaving group in this reaction is the weakly basic bromide ion.

$$CH_3CH_2CH_2CH_2—OH + NaBr \rightleftharpoons CH_3CH_2CH_2CH_2—Br + NaOH \quad \textbf{(14.11)}$$

The mixture of hydrobromic acid and sulfuric acid may be prepared by either adding concentrated sulfuric acid to concentrated hydrobromic acid or by generating the hydrobromic acid *in situ* by adding concentrated sulfuric acid to aqueous sodium bromide (Eq. 14.12). Both of these methods work well and give good yields of the alkyl bromide from low-molecular-weight alcohols. The method of generating H–Br *in situ* is not effective with higher-molecular-weight alcohols because of their low solubility in concentrated salt solutions, so concentrated (48%) hydrobromic acid is used instead.

$$NaBr + H_2SO_4 \rightleftharpoons HBr + NaHSO_4 \quad \textbf{(14.12)}$$

Although the presence of concentrated sulfuric acid facilitates the formation of the alkyl bromide, several side reactions involving sulfuric acid and the alcohol can occur. For example, esterification of the alcohol by sulfuric acid can form an alkyl hydrogen sulfate **2** (Eq. 14.13). This reaction is reversible, and the position of the equilibrium is

$$RO—H + H_2SO_4 \rightleftharpoons RO—SO_3H + H_2O \quad \textbf{(14.13)}$$
$$\textbf{2}$$

An alkyl hydrogen
sulfate

shifted to the left, regenerating the alcohol from **2** as the alkyl bromide is produced. The formation of the alkyl hydrogen sulfate **2** itself does not directly decrease the yield of alkyl bromide, but **2** undergoes other reactions to give undesired by-products. For example, **2** suffers elimination on heating to give a mixture of alkenes (Eq. 14.14). It may also react with another molecule of alcohol to give a dialkyl ether by an S_N2 reaction in which the nucleophile is ROH (Eq. 14.15). Both of these side reactions consume alcohol and decrease the yield of alkyl bromide. These side reactions may be minimized for *primary* alcohols by controlling the temperature of the reaction and the concentration of sulfuric acid.

$$RO—SO_3H \xrightarrow{\text{heat}} \text{alkenes} + H_2SO_4 \quad \textbf{(14.14)}$$

$$RO—SO_3H + ROH \xrightarrow{\text{heat}} R—O—R + H_2SO_4 \quad \textbf{(14.15)}$$

It is necessary to use different procedures to prepare secondary alkyl bromides from *secondary* alcohols because such alcohols are easily dehydrated by concentrated sulfuric acid to give alkenes by way of Equations 14.13 and 14.14. In fact, the acid-catalyzed dehydration of secondary and tertiary alcohols is a common method for synthesizing alkenes (Sec. 10.3). This problem may be circumvented by using concentrated hydrobromic acid; however, secondary alkyl bromides are generally better prepared by the reactions of secondary alcohols with phosphorus tribromide, PBr_3 (Eq. 14.16).

$$3\ R\text{—}OH + PBr_3 \longrightarrow 3\ R\text{—}Br + H_3PO_3 \qquad \textbf{(14.16)}$$

Phosphorus
tribromide

EXPERIMENTAL PROCEDURES

PREPARATION OF 1-BROMOBUTANE

Purpose: To demonstrate the conversion of a primary alcohol to a primary bromoalkane using hydrobromic acid.

Safety Alert

1. Examine your glassware for cracks and chips. This experiment involves heating *concentrated* acids, and defective glassware could break under these conditions and spill hot corrosive chemicals on you and those working around you.
2. *Concentrated* sulfuric acid and water mix with the evolution of substantial quantities of heat. *Always add the acid to the water* in order to disperse the heat through warming of the water. Add the acid slowly and with swirling to ensure continuous and thorough mixing.
3. If possible, wear latex gloves in this experiment. Be very careful when handling concentrated sulfuric acid. When it is poured from the reagent bottle, some may run down the outside of the bottle. If any concentrated sulfuric acid comes in contact with your skin, immediately wash it off with copious amounts of cold water and then with dilute sodium bicarbonate solution.

Miniscale Procedure

Preparation   Answer the Pre-Lab exercises on page PL. 81. Review Sections 2.9, 2.10, 2.11, 2.13, 2.21, and 2.22.

Apparatus   A 25-mL and a 100-mL round-bottom flask, separatory funnel, ice-water bath, apparatus for heating under reflux, simple distillation, magnetic stirring, and *flameless* heating.

Setting Up   Place 11.1 g of sodium bromide in the 100-mL round-bottom flask, and add about 10 mL of water and 10 mL of 1-butanol. Mix the contents of the flask thoroughly by swirling; then cool the flask in an ice-water bath. *Slowly* add 10 mL of *concentrated* sulfuric acid to the cold mixture with swirling and continuous cooling. Remove the flask from the ice-water bath, add a stirbar, and set up the apparatus for heating under reflux.

Reaction, Work-Up, and Isolation   Warm the flask gently until most of the salts have dissolved, and then heat the mixture under gentle reflux with stirring. Continue heating under reflux for 45 min.★ Equip the flask for simple distillation. Distil the mixture rapidly, and collect the distillate, which contains water and 1-bromobutane, in an ice-cooled receiver. Continue distilling until the distillate is clear; the head temperature should be around 115 °C at this point because sulfuric acid and hydrobromic acid are now codistilling with water.★

Transfer the distillate to a separatory funnel and add about 10 mL of water. Shake the funnel gently with venting. Separate the layers and determine which of these is the organic layer (Sec. 2.21). Wash the organic layer sequentially with two 5-mL portions of cold 2 *M* aqueous sodium hydroxide solution, 10 mL of water and 10 mL of saturated aqueous sodium chloride. Transfer the cloudy 1-bromobutane layer to an Erlenmeyer flask, and dry it over several spatula-tips full of anhydrous sodium sulfate.★ Swirl the flask occasionally for a period of 10–15 min until the crude 1-bromobutane is clear; add further small portions of anhydrous sodium sulfate if the solution appears cloudy.

Using a Pasteur pipet, carefully transfer all of the crude 1-bromobutane, without the drying agent, to a clean, dry 25-mL round-bottom flask. Add a stirbar, equip the flask for shortpath distillation, and carefully distil the product into a tared receiver. Because of the relatively small quantity of material, an accurate boiling point may be difficult to measure, and you should collect the fraction having a boiling point greater than about 90 °C (760 Torr).

Analysis   Determine the weight of the 1-bromobutane isolated and compute the yield. Perform the alcoholic silver nitrate and sodium iodide/acetone classification tests on your product (Sec. 23.9). Obtain IR and ^{1}H NMR spectra of your starting material and product and compare them with those of authentic samples (Figs. 8.26 and 14.1–14.4).

Microscale Procedure

Preparation   Answer the Pre-Lab exercises on page PL. 81. Review Sections 2.9, 2.10, 2.11, 2.13, 2.21, and 2.22.

Apparatus   A 3-mL and a 5-mL conical vial, ice-water bath, apparatus for heating under reflux, simple distillation, magnetic stirring, and *flameless* heating.

Setting Up   Place 1.1 g of sodium bromide in the 5-mL conical vial, and add 1.0 mL of water, 1.0 mL of 1-butanol and a spinvane. Mix the contents of the vial thoroughly by stirring and then cool the vial in an ice-water bath. *Slowly* add 1.0 mL of *concentrated* sulfuric acid to the cold mixture with stirring and continuous cooling. Remove the vial from the bath and set up the apparatus for heating under reflux.

Reaction, Work-Up, and Isolation   Warm the vial gently until most of the salts have dissolved, and then heat the mixture under gentle reflux. Continue heating under reflux for 45 min.★ Equip the vial for simple distillation. Distil the mixture rapidly, and collect the distillate, which contains water and 1-bromobutane, in the Hickman stillhead. Continue the distillation, increasing the temperature of the heating source as necessary until there is about 1.5–2.0 mL remaining in the vial and the distillate is clear.★

Using a Pasteur pipet, transfer the distillate to the 3-mL conical vial and add 1.0 mL of water. Cap the vial and shake it gently with venting to mix the contents. Separate the layers, and determine which of these is the organic layer (Sec. 2.21). Wash the organic layer sequentially with two 0.5-mL portions of cold 2 *M* aqueous sodium hydroxide solution, 1.0 mL of water, and 1.0 mL of saturated aqueous sodium chloride. Transfer the cloudy 1-bromobutane layer to a screw-cap centrifuge tube, and dry it over several microspatula-tips full of anhydrous sodium sulfate.★ Swirl the contents in the centrifuge tube occasionally for a period of 10–15 min until the crude 1-bromobutane is clear; add further small portions of anhydrous sodium sulfate if the solution appears cloudy. Using a Pasteur pipet, carefully transfer the crude 1-bromobutane to a tared, sample vial.

Analysis   Determine the weight of the 1-bromobutane isolated and compute the yield. Perform the alcoholic silver nitrate and sodium iodide/acetone classification tests on your product (Sec. 23.9). Obtain IR and ^{1}H NMR spectra of your starting material and product and compare them with those of authentic samples (Figs. 8.26 and 14.1–14.4).

Finishing Touches

Carefully dilute the *stillpot residue from the reaction* with water and then slowly combine this with the *water and the sodium hydroxide washes*. Neutralize the *combined aqueous mixture* with sodium carbonate, and flush the solution down the drain with excess water. The *sodium sulfate* used as the drying agent is contaminated with product, so place it in the container for halogenated solids. Place the *residue in the stillpot from the distillation in the Miniscale Procedure* into the container for halogenated liquids. Dilute the *silver nitrate test solution* with water and flush it down the drain unless instructed otherwise. Place the *sodium iodide/acetone test solution* in the container for halogenated liquids.

EXERCISES

1. In the procedure some water was added to the initial reaction mixture.

   a. How might the yield of 1-bromobutane be affected if the water were not added, and what product(s) would be favored?

   b. How might the yield of product be affected by adding twice as much water as is specified, while keeping the quantities of the other reagents the same?

2. In the purification process, the organic layer is washed sequentially with 2 M NaOH, water, and saturated aqueous sodium chloride. What is the purpose of each of these washes and why is *cold* 2 M NaOH recommended?

3. After the washes described in Exercise 2, the 1-bromobutane is treated with anhydrous sodium sulfate. Why is this done?

4. The final step of the purification process in the Miniscale Procedure involves a simple distillation. What impurities are removed by this distillation?

5. The reaction of 3-methyl-2-butanol with concentrated hydrobromic acid gives the two isomeric bromoalkanes shown.

$$(CH_3)_2CHCHCH_3 \xrightarrow{\text{HBr}} (CH_3)_2CHCHCH_3 + (CH_3)_2CCH_2CH_3$$

$$\underset{OH}{|} \qquad\qquad \underset{\underset{10\%}{Br}}{|} \qquad\qquad \underset{\underset{90\%}{Br}}{|}$$

   Using curved arrows to symbolize the flow of electrons, suggest two reasonable mechanisms for this reaction that account for the formation of each of the products. Indicate whether each is an S_N1 or an S_N2 reaction.

6. The alkyl bromide is dried with anhydrous sodium sulfate in this procedure. Could solid sodium hydroxide or potassium hydroxide be used for this purpose? Explain.

Refer to Tables 8.1–8.8 as needed for answering the following questions on spectroscopy.

7. Consider the spectral data for 1-butanol (Figs. 8.26, 14.1, and 14.2).

   a. In the functional group region of the IR spectrum, specify the absorption associated with the hydroxyl group, and explain why this peak is broad.

   b. In the ^{1}H NMR spectrum, assign the various resonances to the hydrogen nuclei responsible for them.

   c. If the NMR sample of 1-butanol in CDCl$_3$ were shaken with D$_2$O *prior* to taking the ^{1}H NMR spectrum, what differences might you expect to observe?

   d. For the ^{13}C NMR data, assign the various resonances to the carbon nuclei responsible for them.

8. Consider the NMR spectral data for 1-bromobutane (Fig. 14.4).

   a. In the ^{1}H NMR spectrum, assign the various resonances to the hydrogen nuclei responsible for them.

   b. For the ^{13}C NMR data, assign the various resonances to the carbon nuclei responsible for them.

9. Discuss the differences in the IR and NMR spectra of 1-butanol and 1-bromobutane that are consistent with the conversion of an alcohol into a bromoalkane in this procedure.

10. Discuss the differences observed in the ^{1}H and ^{13}C NMR spectra of 1-bromobutane, which appear in Figure 14.4a and b, with those of 1-chlorobutane, which appear in Figure 9.3a and b. To what factor do you attribute these differences?

Spectra

Starting Material and Product
The ^{1}H NMR spectrum of 1-butanol is provided in Figure 8.26.

FIGURE 14.1
IR spectrum of 1-butanol (neat).

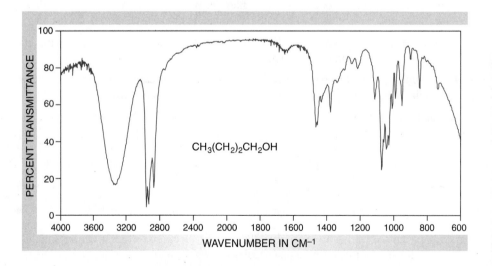

CH$_3$(CH$_2$)$_2$CH$_2$OH

FIGURE 14.2
NMR data for 1-butanol (CDCl$_3$).

CH$_3$(CH$_2$)$_2$ CH$_2$OH ^{13}C NMR data: δ 13.9, 19.2, 35.0, 62.2.

FIGURE 14.3

IR spectrum of 1-bromobutane (neat).

CH$_3$(CH$_2$)$_2$CH$_2$Br

(a) ^{1}H NMR spectrum (300 MHz).

FIGURE 14.4

NMR data for 1-bromobutane (CDCl$_3$).

CH$_3$(CH$_2$)$_2$CH$_2$Br

(b) ^{13}C NMR data: δ 13.2, 21.5, 33.1, 35.0.

14.5 Preparation of 2-Chloro-2-methylbutane: An S_N1 Reaction

Different reagents such as HX and PX$_3$ can be used to prepare alkyl halides from primary and secondary alcohols. However, because elimination reactions predominate when tertiary alcohols are treated with phosphorous trihalides, preparing tertiary alkyl halides from tertiary alcohols proceeds with good yields only if *concentrated* hydrogen halides, HX, are used. The reaction of 2-methyl-2-butanol with hydrochloric acid to produce 2-chloro-2-methylbutane as shown in Equation 14.17 illustrates this process.

$$
\underset{\text{2-Methyl-2-butanol}}{\text{CH}_3\text{CH}_2\underset{\underset{\text{CH}_3}{|}}{\overset{\overset{\text{CH}_3}{|}}{\text{C}}}\text{—OH}} + \text{HCl} \rightleftharpoons \underset{\text{2-Chloro-2-methylbutane}}{\text{CH}_3\text{CH}_2\underset{\underset{\text{CH}_3}{|}}{\overset{\overset{\text{CH}_3}{|}}{\text{C}}}\text{—Cl}} + \text{H}_2\text{O} \qquad \textbf{(14.17)}
$$

The conversion of tertiary alcohols to the corresponding tertiary alkyl halide using concentrated hydrogen halides proceeds via an S_N1 mechanism (Eqs. 14.18–14.20). The first step of the reaction involves protonation of the hydroxyl group of the alcohol by combination of a Lewis acid with a Lewis base (Eq. 14.18). In the second step, ionization occurs, and a molecule of water is lost (Eq. 14.19). Owing partially to steric hindrance at the tertiary carbon atom bearing the protonated hydroxyl group, attack of

$$
\underset{\underset{\text{CH}_3}{|}}{\overset{\overset{\text{CH}_3}{|}}{\text{CH}_3\text{CH}_2-\text{C}-\overset{H}{\underset{}{\text{O}}}}} + \text{H}^+ \underset{}{\overset{\text{fast}}{\rightleftharpoons}} \underset{\underset{\text{CH}_3}{|}}{\overset{\overset{\text{CH}_3}{|}}{\text{CH}_3\text{CH}_2-\text{C}-\overset{H}{\underset{H}{\overset{+}{\text{O}}}}}} \qquad (14.18)
$$

3

$$
\underset{\underset{\text{CH}_3}{|}}{\overset{\overset{\text{CH}_3}{|}}{\text{CH}_3\text{CH}_2-\text{C}-\overset{H}{\underset{H}{\overset{+}{\text{O}}}}}} \overset{\text{rds}}{\rightleftharpoons} \underset{\underset{\text{CH}_3}{|}}{\overset{\overset{\text{CH}_3}{|}}{\text{CH}_3\text{CH}_2-\text{C}^+}} + \text{H}_2\text{O} \qquad (14.19)
$$

Relatively stable
tertiary carbocation

chloride on **3** in an S_N2 process (Sec. 14.4) does not occur. This difference in reactivity between the oxonium ions derived from tertiary alcohols and those of secondary or primary alcohols reflects the relative stabilities of the three types of carbocations ($3° > 2° > 1°$) that would be formed upon the loss of a molecule of water. This step is the slow, rate-determining (rds), step in this sequence. In the final step of the reaction, chloride ion attacks the intermediate carbocation to give 2-chloro-2-methylbutane (Eq. 14.20).

$$
\underset{\underset{\text{CH}_3}{|}}{\overset{\overset{\text{CH}_3}{|}}{\text{CH}_3\text{CH}_2-\text{C}^+}} + :\ddot{\text{Cl}}:^- \overset{\text{fast}}{\rightleftharpoons} \underset{\underset{\text{CH}_3}{|}}{\overset{\overset{\text{CH}_3}{|}}{\text{CH}_3\text{CH}_2-\text{C}-\text{Cl}}} \qquad (14.20)
$$

The principal side reaction in S_N1 reactions of this type is E1 elimination, which results from the loss of a proton from the carbocation to give 2-methyl-1-butene, together with 2-methyl-2-butene (Eqs. 14.21 and 14.22) as discussed in more detail in

$$
\underset{\underset{\text{CH}_3}{|}}{\overset{\overset{\text{CH}_2-\text{H}}{|}}{\text{CH}_3\text{CH}_2-\text{C}^+}} + :\ddot{\text{Cl}}:^- \longrightarrow \text{CH}_3\text{CH}_2-\text{C}\overset{\overset{\text{CH}_2}{\diagup}}{\underset{\diagdown}{\text{CH}_3}} + \text{H}-\text{Cl} \qquad (14.21)
$$

2-Methyl-1-butene

$$
\underset{\underset{\underset{:\ddot{\text{Cl}}:^-}{}}{\overset{\text{H}}{|}}}{\overset{\overset{\text{CH}_3}{|}}{\text{CH}_3\text{CH}-\text{C}^+}} \longrightarrow \text{CH}_3\text{CH}=\text{C}\overset{\overset{\text{CH}_3}{\diagup}}{\underset{\diagdown}{\text{CH}_3}} + \text{H}-\text{Cl} \qquad (14.22)
$$

2-Methyl-2-butene

Section 10.3. However, under the reaction conditions of the present experiment, elimination is reversible through Markovnikov addition of HCl to the 2-methyl-1-butene or the 2-methyl-2-butene thus produced to give the desired product, 2-chloro-2-methylbutane (Eq. 14.23). A more extensive discussion of the ionic addition of hydrogen halides to alkenes is presented in Sections 10.4 and 10.5.

(14.23)

EXPERIMENTAL PROCEDURES

PREPARATION OF 2-CHLORO-2-METHYLBUTANE

Purpose: To demonstrate the conversion of a tertiary alcohol into a tertiary alkyl chloride using hydrochloric acid.

> **Safety Alert**
>
> If possible, wear latex gloves throughout the experiment because *concentrated* hydrochloric acid is being used. If any acid spills on your skin, wash it off with large volumes of water and then with dilute sodium bicarbonate solution.

Miniscale Procedure

Preparation Answer the Pre-Lab Exercises on page PL. 83. Review Sections 2.10, 2.13, and 2.21.

Apparatus A 25-mL round-bottom flask, separatory funnel, ice-water bath, apparatus for simple distillation, magnetic stirring, and *flameless* heating.

Setting Up Place 10 mL of 2-methyl-2-butanol and 25 mL of *concentrated* (12 *M*) hydrochloric acid in the separatory funnel.

2 methyl– 2– propanol

Reaction, Work-Up, and Isolation Swirl the contents of the separatory funnel gently *without* the stopper on the funnel to mix the reactants. After swirling the funnel for about 1 min, first stopper and then invert it; with the funnel inverted, release the excess pressure by carefully opening the stopcock. *Do not shake the funnel until the pressure has been equalized.* Now close the stopcock and shake the funnel for several minutes, with intermittent venting. Allow the mixture to separate into two distinct layers.

Separate the layers and determine which is the organic layer (Sec. 2.21). Wash the *Water is more density.* organic layer sequentially with 10-mL portions of saturated aqueous sodium chloride and *cold* saturated aqueous sodium bicarbonate. On initial addition of the bicarbonate solution, vigorous gas evolution will normally occur; gently swirl the *unstoppered* separatory funnel until the vigorous gas effervescence ceases. Stopper the funnel and invert it carefully; vent the funnel immediately to release gas pressure. Shake the funnel gently with *frequent* venting; then shake it vigorously, again with frequent venting. Separate the organic layer, and again wash it sequentially with 10-mL portions of water and saturated aqueous sodium chloride. Carefully remove the aqueous layer, transfer the 2-chloro-2-methylbutane to an Erlenmeyer flask, and dry the product over several spatula-tips full of anhydrous sodium sulfate.★ Swirl the flask occasionally for 10–15 min until the product is dry; add further small portions of anhydrous sodium sulfate if the solution appears cloudy.

Using a Pasteur or a filter-tip pipet, carefully transfer the crude product to the round-bottom flask, add a stirbar, and equip the flask for shortpath distillation. Distil the product, collecting the product in a tared receiver that is cooled in an ice-water bath. Because of the relatively small quantity of material, an accurate boiling point may be difficult to measure, and you should collect the fraction having a boiling point greater than about 75 °C (760 Torr).

Analysis   Determine the weight of the 2-chloro-2-methylbutane isolated and calculate the percent yield. Perform the alcoholic silver nitrate and sodium iodide/acetone classification tests on your product (Sec. 23.9). Obtain IR and ^{1}H NMR spectra of your starting material and product and compare them with those of authentic samples (Figs. 14.5–14.8).

Microscale Procedure

Preparation   Answer the Pre-Lab Exercises on page PL. 83. Review Sections 2.10, 2.13, and 2.21.

Apparatus   A 5-mL conical vial.

Setting Up   Place 1.0 mL of 2-methyl-2-butanol and 2.5 mL of *concentrated* (12 *M*) hydrochloric acid in the conical vial.

Reaction, Work-Up, and Isolation   Using a small glass rod, stir the contents of the conical vial gently to mix the reactants. After about 1 min of stirring, cap the vial, invert it, and return it to its original upright position; release the excess pressure by venting carefully. *Do not shake the vial until the pressure has been equalized.* Now replace the screw-cap and shake the vial for several minutes, with intermittent venting. Allow the mixture to separate into two distinct layers.

Separate the layers and determine which is the organic layer (Sec. 2.21). Wash the organic layer sequentially with 1-mL portions of saturated aqueous sodium chloride and *cold* saturated aqueous sodium bicarbonate. On initial addition of the bicarbonate solution, vigorous gas evolution will normally occur. Using a glass rod, gently stir the contents of the vial until the vigorous gas effervescence ceases. Cap the vial, invert it carefully, and return the vial to its original upright position; release the excess pressure by venting *carefully*. Shake the vial gently, with *frequent* venting; then shake it vigorously, again with frequent venting. Separate the organic layer, and again wash it sequentially with 1-mL portions of water and saturated sodium chloride solution. Carefully remove the aqueous layer.

Using a filter-tip pipet, transfer the 2-chloro-2-methylbutane to a centrifuge tube, and dry the product over several microspatula-tips full of anhydrous sodium sulfate.★ Swirl the tube occasionally for 10–15 min until the product is dry. Add several small portions of anhydrous sodium sulfate if the product does not become clear. Using a Pasteur or a filter-tip pipet, transfer the crude product to a tared, sample vial.

Analysis Determine the weight of the 2-chloro-2-methylbutane isolated and calculate the percent yield. Perform the alcoholic silver nitrate and sodium iodide/acetone classification tests on your product (Sec. 23.9). Obtain IR and ^{1}H NMR spectra of your starting material and product and compare them with those of authentic samples (Figs. 14.5–14.8).

Carefully dilute the *aqueous layer* from the first separation with water and then neutralize it with sodium carbonate. Combine this solution with the other *aqueous washes (water, saturated sodium bicarbonate, and saturated sodium chloride) and layers,* and flush them down the drain with excess water. The *sodium sulfate* used as the drying agent is contaminated with product, so place it in the container for halogenated solids. Pour the *residue in the stillpot from the Miniscale Procedure* into the container for halogenated liquids. Dilute the *silver nitrate test solution* with water and flush it down the drain unless instructed otherwise. Place the *sodium iodide/acetone test* solution in the container for halogenated liquids.

Finishing Touches

EXERCISES

1. Draw the structures of all the alcohols that are isomeric with 2-methyl-2-butanol. Arrange these alcohols in order of increasing reactivity toward concentrated hydrochloric acid. Which, if any, of these other alcohols would you expect to give a reasonable yield of the corresponding alkyl chloride under such reaction conditions?

2. The work-up procedure calls for washing the crude 2-chloro-2-methylbutane with *cold* sodium bicarbonate solution.

   a. What purpose does this wash serve?

   b. This washing procedure is accompanied by vigorous gas evolution, which increases the difficulty of handling and requires considerable caution. Alternatively, one might consider using a dilute solution of sodium hydroxide instead of the sodium bicarbonate. Discuss the relative advantages and disadvantages of using these two basic solutions in the work-up.

   c. On the basis of your answer to **b**, why were you instructed to use sodium bicarbonate, even though it is more difficult to handle?

3. On prolonged heating with concentrated hydrochloric acid, 2,2-dimethyl-1-propanol, a primary alcohol, reacts to give 2-chloro-2-methylbutane (Eq. 14.24). Provide a stepwise mechanism for this reaction, using curved arrows to symbolize the flow of electrons. Also account for the fact that the observed product is a tertiary alkyl chloride even though the starting alcohol is primary.

$$CH_3-\underset{\underset{CH_3}{|}}{\overset{\overset{CH_3}{|}}{C}}-CH_2-OH \xrightarrow{HCl} CH_3-\underset{\underset{Cl}{|}}{\overset{\overset{CH_3}{|}}{C}}-CH_2CH_3 + H_2O \qquad \textbf{(14.24)}$$

2,2-Dimethyl-1-propanol   2-Chloro-2-methylbutane

4. What is the purpose of the final wash of the purification process, in which the organic layer is washed with saturated sodium chloride?

5. The alkyl chloride is dried with anhydrous sodium sulfate in this procedure. Could solid sodium hydroxide or potassium hydroxide be used for this purpose? Explain your answer.

Refer to Tables 8.1–8.8 as needed for answering the following questions on spectroscopy.

6. Consider the spectral data for 2-methyl-2-butanol (Figs. 14.5 and 14.6).

   a. In the functional group region of the IR spectrum, identify the absorption associated with the hydroxyl group, and explain why this peak is broad.

   b. In the ^{1}H NMR spectrum, assign the various resonances to the hydrogen nuclei responsible for them.

   c. If the NMR sample of 2-methyl-2-butanol in CDCl$_3$ were shaken with D$_2$O *prior* to taking the ^{1}H NMR spectrum, what differences might you expect to observe?

   d. For the ^{13}C NMR data, assign the various resonances to the carbon nuclei responsible for them.

7. Consider the NMR spectral data for 2-chloro-2-methylbutane (Fig. 14.8).

   a. In the ^{1}H NMR spectrum, assign the various resonances to the hydrogen nuclei responsible for them.

   b. For the ^{13}C NMR data, assign the various resonances to the carbon nuclei responsible for them.

8. Discuss the differences in the IR and NMR spectra of 2-methyl-2-butanol and 2-chloro-2-methylbutane that are consistent with the conversion of an alcohol into an alkyl chloride in this procedure.

Spectra

Starting Material and Product

FIGURE 14.5

IR spectrum of 2-methyl-2-butanol (neat).

(a) ¹H NMR spectrum (300 MHz).

FIGURE 14.6
NMR data for 2-methyl-2-butanol
(CDCl₃).

(b) ¹³C NMR data: δ 8.7, 28.6, 36.5, 71.1.

FIGURE 14.7
IR spectrum of 2-chloro-
2-methylbutane (neat).

(a) ¹H NMR spectrum (300 MHz).

FIGURE 14.8
NMR data for 2-chloro-
2-methylbutane (CDCl₃).

(b) ¹³C NMR data: δ 9.5, 32.0, 39.0, 70.9.

Electrophilic Aromatic Substitution

15.1 Introduction

Electrophilic aromatic substitution is an important reaction in organic chemistry that allows the introduction of many different functional groups onto an aromatic ring. A general form of the reaction is given by Equation 15.1, where Ar–H is an **aromatic compound**, an **arene**, and E^+ represents an **electrophile** that replaces H on the ring. This equation is oversimplified because the electrophile is usually generated during the reaction, and a Lewis base assists in the removal of H^+.

$$Ar\text{—}H \ + \ E^+ \longrightarrow Ar\text{—}E \ + \ H^+ \qquad (15.1)$$

An   An   A substituted
arene   electrophile   arene

The rates of electrophilic aromatic substitutions are generally *second-order* overall, being *first order* in both the aromatic component *and* the electrophile, as shown in Equation 15.2. A general mechanism that is consistent with this kinetic expression is depicted in Equations 15.3–15.5.

$$\text{rate} = k_2\,[Ar\text{—}H][E^+] \qquad (15.2)$$

Step 1: Formation of electrophile:

$$E\text{—}Nu \xrightleftharpoons{\text{catalyst}} E^+ \ + \ Nu^- \qquad (15.3)$$

Step 2: Reaction of electrophile:

$$Ar\text{—}H \ + \ E^+ \xrightleftharpoons{\text{rds}} \left[\overset{+}{Ar} \overset{H}{\underset{E}{\diagdown}} \right] \qquad (15.4)$$

Step 3: Loss of proton to give product:

$$\left[\begin{array}{c} \overset{+}{Ar} \overset{H}{\diagup} \\ \diagdown E \end{array} \right] \xrightleftharpoons{\text{fast}} Ar\text{—}E + H^+ \qquad \textbf{(15.5)}$$

The electrophile is usually produced by the reaction between a catalyst and a compound containing a potential electrophile (Eq. 15.3). The second-order nature of the reaction arises from the step shown in Equation 15.4 in which one molecule each of arene and electrophile react to give the positively charged intermediate. The formation of this cation is the **rate-determining step (rds)** in the overall reaction; the subsequent deprotonation of the cation (Eq. 15.5) is fast. The bimolecular nature of the transition state for the rate-limiting step and the fact that an electrophile is involved in attacking the aromatic substrate classifies the reaction as S_E2 (**S**ubstitution **E**lectrophilic **Bi**molecular). Experiments involving four different such reactions are given in this chapter: Friedel-Crafts alkylation and acylation, nitration, and bromination.

15.2 Friedel-Crafts Alkylation of *p*-Xylene with 1-Bromopropane

The **Friedel-Crafts alkylation reaction** using alkyl halides (Eq. 15.6) is a classic example of electrophilic aromatic substitution (see the Historical Highlight at the end of this chapter). As a versatile method for directly attaching alkyl groups to aromatic rings, it is a process of great industrial importance. The three main limitations to this reaction as a synthetic tool are: (1) The substitution fails if the aromatic ring carries strongly **ring-deactivating groups**, such as NO_2, R_3N^+, C(O)R, and CN; (2) more than one alkyl group may be introduced onto the aromatic ring, a process termed polyalkylation; and (3) mixtures of isomeric products may be formed if the alkyl group rearranges. The first restriction is associated with the fact that aromatic rings bearing electron-withdrawing substituents are not sufficiently nucleophilic to react with the electrophiles generated under the reaction conditions. The second problem arises because addition of the first alkyl group activates the ring toward further substitution; this difficulty is minimized by using a large excess of the arene, as is done in our experiment. The final limitation results from structural rearrangements within the alkyl group, R.

$$\underset{\substack{\text{An arene}}}{Ar\text{—}H} + \underset{\substack{\text{An alkyl halide} \\ (X = \text{halogen})}}{R\text{—}X} \xrightleftharpoons{\text{catalyst}} \underset{\substack{\text{An alkylarene}}}{Ar\text{—}R} + HX \qquad \textbf{(15.6)}$$

A generally accepted mechanism for Friedel-Crafts alkylation using alkyl halides is illustrated in Equations 15.7–15.9. In the first step, the active electrophile, written as R^+ for the present purpose, is produced from an alkyl halide by reaction with a Lewis acid such as $AlCl_3$ (Eq. 15.7). The rate-determining step involves attack of the arene, which functions as a Lewis base, on the electrophile to yield a positively charged, resonance-stabilized **sigma complex** (Eq. 15.8). Finally, a base such as a halide ion from $XAlCl_3{}^-$ deprotonates this complex to give HX, reforming the aromatic system and regenerating the catalyst (Eq. 15.9).

$$R-X + AlCl_3 \rightleftharpoons R^+ + X\bar{A}lCl_3 \tag{15.7}$$

$$\tag{15.8}$$

σ-Complex

$$Cl_3\bar{A}l-X + H \cdots \longrightarrow^{\text{fast}} \quad + HX + AlCl_3 \tag{15.9}$$

Representing the electrophile as a carbocation, as in Equation 15.7, is reasonable when *secondary* and *tertiary* alkyl halides are used. However, this is probably *not* the case with the electrophile derived from a *primary* alkyl halide, where the reactive species is better written as a *polarized complex* such as $R^{\delta+}$- - - -X- - - -$Al^{\delta-}Cl_3$.

Molecular rearrangements involving hydride, H^-, and alkyl, R^-, shifts become possible when a carbocation or a polarized cation-like complex is an intermediate in a reaction. Thus, the aluminum chloride-promoted reaction of 1-bromopropane with benzene gives *n*-propylbenzene (**1**) and isopropylbenzene (**2**) in a ratio of about 1:2 (Eq. 15.10). As shown in Scheme 15.1, the formation of **2** is explained by a 1,2-hydride

$$\tag{15.10}$$

Benzene	1-Bromopropane	*n*-Propylbenzene	Isopropylbenzene
		1	**2**

rearrangement of the *primary carbocation-like* complex **3** to produce the *secondary carbocation* **4**. The formation of comparable amounts of **1** and **2** in the reaction means that the rate of addition of **3** to benzene followed by deprotonation to give **1** must be competitive with that for isomerization of **3** to the secondary ion **4**, which is the precursor of **2**. In mathematical terms, $k_2[\mathbf{3}][C_6H_6]$ must be similar to $k_1[\mathbf{3}]$, since these expressions measure the rates of formation of **1** and **2**, respectively.

SCHEME 15.1

The present experiment demonstrates Friedel-Crafts alkylation and provides a test of the following hypothesis: The amount of alkyl rearrangement that occurs during a Friedel-Crafts alkylation of a series of arenes depends on the *nucleophilicity* of the arene, Ar–R. This hypothesis is derived as follows. The rate constant, k_1, is for a *unimolecular* process of **3** that does *not* involve the arene (Scheme 15.1), so its value is *independent* of the nucleophilicity of the arene. In contrast, k_2 is the rate constant for a *bimolecular* reaction between **3** *and* the arene, so its value is *dependent* on the nucleophilicity of the arene being used. Substituents, R, that make the arene more nucleophilic yield higher values of the bimolecular rate constant k_2, thereby increasing the overall rate of the electrophilic substitution. Consequently, formation of the product derived from rearrangement becomes a relatively less important reaction pathway.

The importance of the nucleophilicity of the arene in determining the ratio of unrearranged products to rearranged products may be studied by comparing the Friedel-Crafts reactions of benzene and *p*-xylene with 1-bromopropane (Eq. 15.11). Because methyl groups activate the ring toward the S_E2 reaction and make k_2(*p*-xylene) > k_2(benzene), we expect from our hypothesis that the ratio of unrearranged product **5** to its rearranged isomer **6** to be *greater* than that of **1**:**2**. The experiment you will perform tests this prediction.

EXPERIMENTAL PROCEDURES

FRIEDEL-CRAFTS ALKYLATION OF *p*-XYLENE

Purpose: To demonstrate alkylation by electrophilic aromatic substitution and to assess the relationship between the extent of carbocation rearrangement and arene nucleophilicity.

Safety Alert

1. Anhydrous aluminum chloride is *extremely* hygroscopic and reacts rapidly with water, even the moisture on your hands, producing fumes of hydrogen chloride, which are highly corrosive. *Do not allow aluminum chloride to come in contact with your skin.* If it does, flush the affected area with copious amounts of water. *Minimize exposure of this chemical to the atmosphere!*

2. You may be provided with a weighed amount of aluminum chloride. If not, *quickly* weigh it into a dry vial in the hood and then transfer it to the reaction flask. The success of this experiment is highly dependent on the quality of the aluminum chloride that is used, so obtain it from a *freshly* opened bottle. In handling this chemical, be aware that it is a powdery solid that easily becomes airborne.

3. *p*-Xylene is flammable. Assemble the apparatus carefully and be sure that all joints are tightly mated. Have your instructor inspect your set-up before you begin the distillation and use *flameless* heating.

Miniscale Procedure

Preparation   Answer the Pre-Lab Exercises on page PL. 85. Review Sections 2.9, 2.10, 2.11, 2.13, 2.14, 2.23, and 2.27.

Apparatus   A 25- and a 50-mL round-bottom flask, drying tube, Claisen adapter, separatory funnel, gas trap, apparatus for simple distillation, fractional distillation, magnetic stirring, and *flameless* heating.

Setting Up   Equip the 50-mL round-bottom flask for simple distillation and add a stirbar and 15 mL of *p*-xylene. Distil the *p*-xylene into a receiver until the distillate is not cloudy; this should occur after no more than 3 mL of distillate has been obtained.

Allow the residual *p*-xylene in the stillpot to cool to room temperature, and equip the flask with a Claisen adapter, a water-cooled condenser, the gas trap specified by your instructor for removing fumes of hydrogen chloride, and a separatory funnel (Fig. 2.63b); fit the funnel with a drying tube rather than a stopper. *Briefly* remove the condenser from the reaction flask, *quickly* pour 0.7 g of *anhydrous* powdered aluminum chloride into the flask, and *immediately* reassemble the apparatus.

Using a graduated cylinder, measure 8.5 mL of 1-bromopropane and pour it into the funnel (stopcock *closed*!). Prepare an ice-water bath in case it is needed to cool the reaction mixture.

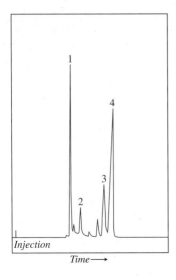

FIGURE 15.1

GLC analysis of reaction mixture from miniscale alkylation of *p*-xylene with 1-bromopropane. Numbers in parentheses are retention times in minutes and peak areas, respectively. Peak 1: solvent; peak 2: *p*-xylene (0.7, 14417); peak 3: isopropyl-*p*-xylene (1.8, 71721); peak 4: *n*-propyl-*p*-xylene (2.0, 229275). Column and conditions: 0.25-mm × 25-m AT-1 Helifex 100% dimethylsiloxane; 185 °C, 40 mL/min.

Addition and Work-Up   Add the 1-bromopropane dropwise to the stirred mixture of *p*-xylene and aluminum chloride. If the evolution of gas becomes too vigorous, reduce the rate of addition and raise the cooling bath to lower the temperature of the reaction mixture. The addition should take about 10 min. After completing the addition, continue stirring the reaction mixture for about 0.5 h. *Working at the hood* because of the hydrogen chloride that will be evolved, pour the mixture into a beaker containing about 10 g of crushed ice. Stir the mixture until all the ice has melted, pour it into a separatory funnel, and remove the aqueous layer. Pour the organic layer into an Erlenmeyer flask containing anhydrous sodium sulfate and swirl it for about 1 min; if the solution appears cloudy, add additional sodium sulfate with swirling.★

Isolation   Decant the dried solution into the 25-mL round-bottom flask containing a stirbar and equip the flask for fractional distillation using an *air-cooled* condenser. Make sure that all connections are tight and then distil. Collect the excess *p*-xylene and forerun in a tared, labeled receiver until the head temperature reaches 180 °C (760 Torr). Discontinue heating and allow any liquid in the column to drain into the stillpot.★ Refit the stillpot for shortpath distillation and resume the distillation, collecting any additional distillate boiling below 180 °C (760 Torr) in the original receiver. Use a second tared, labeled receiver to collect the fraction boiling between 180 and 207 °C (760 Torr). Isomers **5** and **6** have bp 204 °C and 196 °C (760 Torr), respectively.

Analysis   Weigh the two fractions and analyze each by GLC; a typical trace of the second fraction is provided in Figure 15.1. Use the GLC analysis to calculate the yield of propylxylenes obtained and their proportion to each other; assume that *p*-xylene and the propylxylenes have the same density and GLC response factor. Obtain IR and ^{1}H NMR spectra of your starting materials and products and compare them with those of authentic samples (Figs. 15.4–15.11).

Microscale Procedure

Preparation   Answer the Pre-Lab Exercises on page PL. 85. Review Sections 2.9, 2.11, 2.13, 2.17, and 2.23.

Apparatus   A 3- and a 5-mL conical vial, 1-mL syringe, Claisen adapter, gas trap, apparatus for distillation, magnetic stirring, and *flameless* heating.

FIGURE 15.2
Microscale apparatus for alkylation of *p*-xylene.

Setting Up   Equip the 5-mL vial for simple distillation and add a spinvane and 2 mL of *p*-xylene. Distil the *p*-xylene until the distillate is not cloudy; this should occur after no more than 0.5 mL of distillate has been obtained.

Allow the residual *p*-xylene in the vial to cool to room temperature. Replace the condenser on the vial with a Claisen adapter fitted with a rubber septum on the opening directly above the vial; on the other arm of the adapter put a water-cooled condenser equipped with the gas trap specified by your instructor for removing fumes of hydrogen chloride (Fig. 15.2). *Briefly* remove the vial from the adapter, *quickly* pour 0.1 g of *anhydrous* powdered aluminum chloride into it, and *immediately* reassemble the apparatus. Draw 0.8 mL of 1-bromopropane into the syringe and pierce the rubber septum with its needle.

Addition and Work-Up   Add the 1-bromopropane dropwise to the stirred mixture of *p*-xylene and aluminum chloride over a period of 5–10 min. If the evolution of gas becomes too vigorous, reduce the rate of addition. After completing the addition, continuing stirring the reaction mixture for about 20 min. *Working at the hood* because of the hydrogen chloride that will be evolved, pour the mixture into a beaker containing about 1 g of crushed ice. Stir the mixture until all the ice has melted, pour it into a centrifuge tube, and remove the aqueous layer. Add a microspatula-tip full of anhydrous sodium sulfate to the organic layer and swirl it for about 1 min; add additional sodium sulfate with swirling if the solution appears cloudy.★

Isolation   Using a filter-tip pipet, transfer the dried solution into the 3-mL conical vial containing a stirbar and equip the vial for distillation; with internal monitoring of the temperature. Make sure that all connections are tight and then distil. Collect the excess *p*-xylene and forerun until the bath temperature reaches 180 °C (760 Torr). Discontinue heating and allow the apparatus to cool for a few minutes.★ Transfer the distillate to a tared, labeled screw-cap vial and resume the distillation, collecting as one fraction all

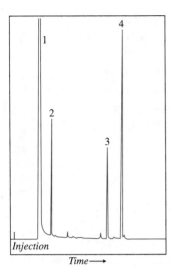

FIGURE 15.3

GLC analysis of reaction mixture from microscale alkylation of *p*-xylene with 1-bromopropane. (Numbers in parentheses are retention times in minutes and peak areas, respectively.) Peak 1: solvent; peak 2: *p*-xylene (1.8, 74565); peak 3: isopropyl-*p*-xylene (4.3, 92699); peak 4: *n*-propyl-*p*-xylene (5.0, 236258). Column and conditions: 0.25-mm × 25-m 100% dimethylpolysiloxane; 110 °C, 40 mL/min.

volatiles that distil with a bath temperature of up to 240 °C. Isomers **5** and **6** have bp 204 °C and 196 °C (760 Torr), respectively.

Analysis   Weigh the two fractions and analyze each by GLC; a typical trace of the second fraction is provided in Figure 15.3. Obtain IR and ^{1}H NMR spectra of your starting materials and products and compare them with those of authentic samples (Figs. 15.4–15.11).

Neutralize the *aqueous layer* from hydrolysis of the reaction mixture with sodium carbonate and flush it down the drain. Spread the *sodium sulfate* used for drying on a tray in the hood to allow evaporation of the volatiles; then put spent *drying agent* in the container for nontoxic solids. Place the *liquid collected as the low-boiling fraction* and the *pot residue* from the distillation in the container for nonhalogenated organic liquids.

Finishing Touches

EXERCISES

1. What purpose is served by distilling a small quantity of *p*-xylene prior to combining the reagents in this experiment? (*Hint:* What accounts for the cloudiness that normally appears in the initial stages of this distillation?)

2. What compound(s) might be present in the higher-boiling residue remaining in the stillpot after collecting the isomeric propylxylenes?

3. Calculate the molar ratio of 1-bromopropane to *p*-xylene used in the experiment. What would be the effect on the experimental results of using a ratio twice that actually used? Consider this in connection with your answer to Exercise 2.

4. Use either your own GLC analysis or that given in Figure 15.1 or 15.3 to calculate the yield of propylxylenes obtained and their proportion to each other; assume that *p*-xylene and the propylxylenes have the same densities and GLC response factor. Discuss whether these calculations support the hypothesis regarding the relative amounts of unrearranged product versus rearranged product to be expected from *p*-xylene as compared to benzene.

5. Account for the higher boiling point of **5** as compared to that of **6**. Are the boiling points consistent with the relative retention times of the two isomers as seen in the GLC trace of Figure 15.1 or 15.3? Explain.

6. Why do you think it is recommended that the distillation fraction containing the mixture of **5** and **6** be taken over such a wide temperature range?

7. On the assumption that relative cost is not an overriding consideration, give a reason why *p*-xylene was used in preference to *o*-xylene as the arene in this experiment.

8. Alkylation of toluene with 1-bromopropane gives a mixture of four isomeric propyltoluenes. Write formulas for these compounds, and predict the relative amounts in which they would be formed, based on the results of the alkylation of *p*-xylene.

Refer to Tables 8.1–8.8 as needed for answering the questions on spectroscopy.

9. Consider the spectral data for *p*-xylene (Figs. 15.4 and 15.5).

    a. In the functional group region of the IR spectrum, specify the absorption associated with the π-bonds of the aromatic ring. Indicate with what structural feature the strong band at about 800 cm^{-1} is associated.

    b. In the ^{1}H NMR spectrum, assign the various resonances to the hydrogen nuclei responsible for them.

    c. For the ^{13}C NMR data, assign the various resonances to the carbon nuclei responsible for them.

10. Consider the NMR spectral data for 1-bromopropane (Fig. 15.7).

    a. In the ^{1}H NMR spectrum, assign the various resonances to the hydrogen nuclei responsible for them.

    b. For the ^{13}C NMR data, assign the various resonances to the carbon nuclei responsible for them.

11. Consider the spectral data for *n*-propyl-*p*-xylene (Figs. 15.8 and 15.9).

    a. In the functional group region of the IR spectrum, specify the absorption associated with the π-bonds of the aromatic ring. Indicate with what structural feature the strong band at about 810 cm^{-1} is associated.

    b. In the ^{1}H NMR spectrum, assign the various resonances to the hydrogen nuclei responsible for them.

    c. For the ^{13}C NMR data, assign the various resonances to the carbon nuclei responsible for them.

12. Consider the spectral data for isopropyl-*p*-xylene (Figs. 15.10 and 15.11).

    a. In the functional group region of the IR spectrum, specify the absorption associated with the π-bonds of the aromatic ring. Indicate with what structural feature the strong band at about 800 cm^{-1} is associated.

    b. In the ^{1}H NMR spectrum, assign the various resonances to the hydrogen nuclei responsible for them.

    c. For the ^{13}C NMR data, assign the various resonances to the carbon nuclei responsible for them.

13. What differences in the IR and NMR spectra of *p*-xylene, **5**, and **6** are consistent with the incorporation of a C_3H_7 group on the ring in this experiment?

14. What differences in the IR and NMR spectra of *p*-xylene, **5**, and **6** are consistent with *mono*substitution rather than *di*substitution of *p*-xylene having occurred in the Friedel-Crafts alkylation you performed?

15. *Ipso* substitution, as given in the equation below, is sometimes observed in electrophilic aromatic substitution reactions. Explain how the data provided in Figures 15.8–15.11 are inconsistent with the structures of the *ipso* products shown.

Ipso substitution products

Starting Materials and Products

Spectra

FIGURE 15.4
IR spectrum of *p*-xylene (neat).

(a) ^1H NMR spectrum (300 MHz).

FIGURE 15.5
NMR data for *p*-xylene (CDCl$_3$).

(b) ^{13}C NMR data: δ 20.9, 129.0, 134.6.

FIGURE 15.6
IR spectrum of 1-bromopropane
(neat).

FIGURE 15.7
NMR data for 1-bromopropane
(CDCl₃).

(a) ^{1}H NMR spectrum (300 MHz).

(b) ^{13}C NMR data: δ 13.0, 26.2, 35.9.

FIGURE 15.8
IR specrum of *n*-propyl-*p*-xylene
(neat).

(a) ¹H NMR spectrum (300 MHz).

FIGURE 15.9
NMR data for *n*-propyl-*p*-xylene
(CDCl₃).

(b) ¹³C NMR data: δ 14.2, 18.8, 21.0, 23.5, 35.4, 126.4, 129.7, 129.9, 132.7, 135.1, 140.7.

FIGURE 15.10
IR spectrum of isopropyl-*p*-xylene
(neat).

(a) ¹H NMR spectrum (300 MHz).

FIGURE 15.11
NMR data for isopropyl-*p*-xylene
(CDCl₃).

(b) ¹³C NMR data: δ 18.8, 21.2, 23.2, 29.1, 125.4, 126.2, 130.1, 131.8, 135.4, 146.6.

15.3   Friedel-Crafts Acylation of *m*-Xylene

As described in Section 15.2 and in the Historical Highlight at the end of this chapter, Friedel and Crafts discovered that alkyl groups could be substituted onto the aromatic ring by reaction of arenes with alkyl halides in the presence of aluminum chloride, $AlCl_3$ (Eq. 15.6). The role of aluminum chloride, a strong Lewis acid, is to convert the alkyl halide into a reactive electrophilic intermediate, in the form of a carbocation or a highly polarized carbon-halogen bond. This electrophile then undergoes attack by an arene (Sec. 15.1), which functions as a Lewis base, resulting in aromatic substitution. They also explored the reaction of aluminum chloride with acid chlorides, **7**, and anhydrides, **8**, to produce **acylonium ions**, **9**, which function as electrophiles, just as alkyl carbocations do. Upon reaction of **9** with an arene, an acyl group is introduced onto the ring to provide an aryl ketone, **10** (Eq. 15.12). The overall reaction is named the **Friedel-Crafts acylation**.

The resonance structures in Equation 15.12 symbolize distribution of the positive charge in acylonium ions. This charge delocalization stabilizes the cation so that acylonium ions such as **9** do *not* undergo structural rearrangements as do alkyl carbocations (Sec. 15.2). The increased stability of acylonium ions also makes them less electrophilic and thus less reactive than alkyl carbocations. Nevertheless, cation **9** is sufficiently reactive to undergo the S_E2 reaction (Sec. 15.1) with arenes, *provided* the aromatic ring does not bear strongly deactivating groups such as NO_2, R_3N^+, $C(O)R$, and CN.

| **7** | **8** | **9** | **10** |
| An acid chloride | An acid anhydride | An acylonium ion | An aryl ketone |

(15.12)

11

An important difference between the Friedel-Crafts alkylation and acylation reactions is that the latter process requires use of a *stoichiometric* rather than a catalytic amount of aluminum chloride. This is because the product aryl ketones **10** undergo a Lewis acid-base reaction with aluminum chloride to form a strong 1:1 complex, **11**; the aluminum chloride involved in the complex no longer fosters formation of acylonium ions and in essence is "consumed" in the reaction. Thus, although generation of acylonium ions needed for the Friedel-Crafts acylation is catalytic in aluminum chloride, complexation with the product makes aluminum chloride a stoichiometric reagent.

The acylation performed in this section involves reaction of phthalic anhydride (**12**), a cyclic anhydride, with *m*-xylene (**13**) in the presence of aluminum chloride to afford 2-(2′,4′-dimethylbenzoyl)-benzoic acid (**15**) (Scheme 15.2), a ketoacid. The initial product of the acylation is **14**, in which one mole of aluminum chloride has reacted with the acid function to form the salt RCO_2AlCl_2, and a second mole is complexed to the carbonyl group. Adding ice and hydrochloric acid decomposes the complex to produce **15** and water-soluble aluminum salts.

SCHEME 15.2

| **12** | **13** | **14** |
| Phthalic anhydride | *m*-Xylene | |

15

2-(2′4′-Dimethylbenzoyl)-benzoic acid

The formation of **14** involves reaction of aluminum chloride with anhydride **12** to give the electrophile **16** (Eq. 15.13); this species then reacts with *m*-xylene (**13**). An issue in this step of the sequence is which of the three different positions of **13** undergoes attack by **16**. Because methyl groups are *o,p*-directors, reaction should occur preferentially at C-2 and C-4 of the ring. Steric hindrance impedes attack at C-2 so *m*-xylene reacts selectively at C-4 to give **14**, which provides the ketoacid **15** upon hydrolysis, as shown in Scheme 15.2.

12   **16**

(15.13)

EXPERIMENTAL PROCEDURES

FRIEDEL-CRAFTS ACYLATION OF *m*-XYLENE WITH PHTHALIC ANHYDRIDE

Purpose: To demonstrate acylation by electrophilic aromatic substitution.

> **Safety Alert**
>
> 1. Anhydrous aluminum chloride is *extremely* hygroscopic and reacts rapidly with water, even the moisture on your hands, producing fumes of hydrogen chloride, which are highly corrosive. *Do not allow aluminum chloride to come in contact with your skin.* If it does, flush the affected area with copious amounts of water. *Minimize exposure of this chemical to the atmosphere!*
> 2. You may be provided with a weighed amount of aluminum chloride. If not, *quickly* weigh it into a dry vial or test tube *in the hood* and then transfer it to the reaction flask. The success of this experiment is highly dependent on the quality of the aluminum chloride that is used, so obtain it from a *freshly* opened bottle. In handling this chemical, be aware that it is a powdery solid that easily becomes airborne.
> 3. *m*-Xylene is flammable. Assemble the apparatus carefully and be sure that all joints are tightly mated. Use *flameless* heating.

Miniscale Procedure

Preparation   Answer the Pre-Lab Exercises on page PL. 87. Review Sections 2.9, 2.10, 2.11, 2.13, 2.23, 2.25, and 2.27.

Apparatus   A dry 25-mL round-bottom flask, gas trap, ice-water bath, separatory funnel, apparatus for magnetic stirring, heating under reflux, vacuum filtration, and *flameless* heating.

Setting Up   *Working at the hood*, weigh 2 g of *anhydrous* aluminum chloride into a dry test tube or screw-cap vial and *immediately* close the vessel. Equip the round-bottom flask with a stirbar and a dry reflux condenser fitted with the gas trap specified by your instructor for removing fumes of hydrogen chloride. Place 1 g of phthalic anhydride and 6 mL of *m*-xylene in the flask, and cool the mixture to 0 °C in an ice-water bath.

Acylation   Remove the condenser from the flask, and, in one portion, *quickly* add the anhydrous aluminum chloride. *Immediately* replace the condenser and stir the reaction mixture for a minute or two while maintaining its temperature at 0 °C and then allow it to warm to room temperature. Initiation of the reaction will be evidenced by evolution of hydrogen chloride. If this does not occur, gently warm the flask until gas evolution begins. Be prepared to immerse the flask into an ice-water bath if the reaction becomes too vigorous. Continue gently heating the mixture with stirring until the reaction is proceeding smoothly, and then heat it under reflux for about 0.5 h.

Work-Up, Isolation and Purification Remove the flask from the apparatus and stopper it. *Working at the hood,* pour the cold reaction mixture onto 8–10 g of ice contained in a beaker, while stirring the mixture thoroughly. With continued vigorous stirring, slowly add 2 mL of *concentrated* hydrochloric acid and then 8–10 mL of ice-cold water to the beaker; if necessary, cool the mixture in an ice-water bath as necessary to keep it near room temperature during the additions. Add an additional 4 mL of ice-cold water with stirring and cool the resulting mixture to room temperature or below.★

Add 5 mL of diethyl ether to the beaker. With the aid of a stirring rod, scrape any solid from the neck and walls of the beaker and carefully break up any lumps at the bottom. Stir the mixture vigorously for several minutes to complete decomposition of the aluminum salt of the product, extraction of the organic product into diethyl ether, and dissolution of the inorganic aluminum salts.★

Transfer the two-phase mixture into a separatory funnel, rinse the beaker with a few milliliters of diethyl ether, and add the rinse to the funnel. Shake the mixture, frequently venting the funnel, and allow the layers to separate. A grayish, fluffy precipitate may appear below the organic layer. If so, before separating the layers, dissolve the precipitate by adding 5 mL of 6 *M* HCl and shaking the mixture for a few minutes, occasionally venting the funnel. Should any precipitate or grayish emulsion remain with the ethereal solution, remove it by vacuum filtration through a pad of a filter-aid. After separating the layers, extract the aqueous layer sequentially with two additional 10-mL portions of diethyl ether, venting the funnel occasionally during the extraction. Combine all of the ethereal solutions in an Erlenmeyer flask and add several spatula-tips full of anhydrous sodium sulfate, swirling the mixture occasionally over a period of about 15 min; if the solution appears cloudy, add additional sodium sulfate with swirling.★

Decant the ethereal solution into a 50-mL round-bottom flask and equip the flask for simple distillation. Concentrate the solution to about half its original volume. Alternatively, use rotary evaporation or other techniques (Sec. 2.29) to effect concentration. Transfer the concentrated solution to an Erlenmeyer flask and cool it to room temperature. Once crystallization has begun, further cool the solution in an ice-water bath and then collect **15** by vacuum filtration. Recrystallize it from 50% aqueous ethanol and air-dry the purified product.

Analysis Weigh your product and calculate the yield. Measure the melting point of **15**. Its reported melting point is 142.5–142.8 °C. Obtain IR and NMR spectra of your starting materials and product and compare them with those of authentic samples (Figs. 15.12–15.17).

Microscale Procedure

Preparation Answer the Pre-Lab Exercises on page PL. 87. Review Sections 2.9, 2.10, 2.11, 2.13, 2.23, 2.25, and 2.29.

Apparatus A dry 5-mL conical vial, gas trap, ice-water bath, 15-mL screw-cap centrifuge tube, apparatus for magnetic stirring, heating under reflux, vacuum and Craig-tube filtration, and *flameless* heating.

Setting Up *Working at the hood*, weigh 0.7 g of *anhydrous* aluminum chloride into a dry test tube or vial and *immediately* stopper the vessel. Equip the conical vial with a spinvane and a dry reflux condenser fitted with the gas trap specified by your instructor for removing fumes of hydrogen chloride. Add 0.3 g of phthalic anhydride and 1.8 mL of *m*-xylene to the vial and cool the mixture to 0 °C in an ice-water bath.

Acylation Remove the condenser from the conical vial, and, in one portion, quickly add the *anhydrous* aluminum chloride. *Immediately* replace the condenser, stir the reaction mixture for a minute or two, and then allow it to warm to room temperature. The start of the reaction will be evidenced by evolution of bubbles of hydrogen chloride. If this does not occur, gently warm the flask in a sand or oil bath until gas evolution begins. Be prepared to immerse the vial into an ice-water bath if the reaction becomes too vigorous. Continue heating the mixture gently with stirring until the reaction appears to be proceeding smoothly and then heat it under reflux for about 20 min. At the end of the reflux period, cool the reaction mixture to 0° with stirring.

Work-Up, Isolation, and Purification Remove the conical vial from the apparatus and stopper it. *Working at the hood,* pour the cold reaction mixture into a 15-mL screw-cap centrifuge tube or screw-cap vial containing 2 mL of ice-water and stir the mixture thoroughly. With continued vigorous stirring, slowly add 0.5 mL of *concentrated* hydrochloric acid and then 2 mL of ice-cold water. Cool the resulting mixture to room temperature or below.★

Add 1 mL of diethyl ether and break up any lumps of solid with the aid of a stirring rod. Cap the centrifuge tube and shake it vigorously for several minutes to complete decomposition of aluminum salts, extraction of the product into the organic phase, and dissolution of inorganic aluminum salts.★ Vent the system occasionally to relieve any pressure that may develop.

Using a Pasteur or filter-tip pipet, transfer the organic phase to a test tube or vial with a 15-mL capacity. Extract the aqueous layer sequentially with three 2-mL portions of diethyl ether, combining the organic layers with the original ethereal extract. Combine all of the ethereal solutions in an Erlenmeyer flask, and add several microspatula-tips full of anhydrous sodium sulfate, swirling the mixture occasionally over a period of about 15 min; if the solution appears cloudy, add additional sodium sulfate with swirling.

Decant about one-half of the ethereal solution into a 5-mL conical vial equipped for simple distillation and concentrate the solution to about half its original volume, removing the distillate from the Hickman stillhead as necessary. Allow the solution to cool below the boiling point, add the remainder of the dried ethereal extracts, and again concentrate the solution to about one-half its original volume. Alternatively, use rotary evaporation or other technique (Sec. 2.29) to effect concentration. Cool the concentrated solution to room temperature. Once crystallization has begun, further cool the solution in an ice-water bath, and then collect the product **15** by vacuum filtration. Recrystallize it from 50% aqueous ethanol and air-dry the purified product.

Analysis Weigh your product and calculate the yield. Measure the melting point of **15**. Its reported melting point is 142.5–142.8 °C. Obtain IR and NMR spectra of your starting materials and product and compare them with those of authentic samples (Figs. 15.12–15.17).

Finishing Touches

Discard the *grayish solid* obtained by gravity filtration in the container for nonhazardous solids. Pour the *diethyl ether* obtained by distillation in a container for nonhalogenated organic liquids. Flush the *filtrate from the recrystallization* and the *aqueous layer containing aluminum salts* down the drain.

EXERCISES

1. Explain why a catalytic amount of $AlCl_3$ is insufficient to promote the Friedel-Crafts reaction of **12** and **13**.

2. Why is loss of carbon monoxide (decarbonylation) from the acylonium ion **16** an unlikely reaction?

3. Why is it important that the apparatus used for the Friedel-Crafts acylation be dry?

4. Write the products of the reaction of $AlCl_3$ and excess H_2O.

5. What gas is evolved in the early stages of the reaction between *m*-xylene, phthalic anhydride, and aluminum chloride?

6. In which of the two solvents, water or ethanol, should the product first be dissolved for the mixed solvent recrystallization? Explain.

7. Why does the acylonium ion **16** react more rapidly with **13** than with **12**?

8. What evidence do you have from this experiment that substitution of phthalic acid for phthalic anhydride would not provide acylonium ion **16** required for the acylation reaction? (*Hint:* Write the product of reaction of aluminum chloride with one of the carboxylic acid groups of phthalic acid and compare it to the carboxylate **14**.)

Refer to Tables 8.1–8.8 as needed for answering the questions on spectroscopy.

9. Consider the spectral data for *m*-xylene (Figs. 15.12 and 15.13).

   a. In the functional group region of the IR spectrum, specify the absorption associated with the π-bonds of the aromatic ring. Indicate the bands in the finger print region that characterize the *meta* orientation of the methyl groups.

   b. In the ^1H NMR spectrum, assign the various resonances to the hydrogen nuclei responsible for them.

   c. For the ^{13}C NMR data, assign the various resonances to the carbon nuclei responsible for them.

   d. Explain how ^{13}C spectroscopy allows differentiation of the *meta* isomer from the *ortho* and *para* isomers.

10. Consider the spectral data for phthalic anhydride (Figs. 15.14 and 15.15).

   a. In the functional group region of the IR spectrum, specify the absorptions associated with the anhydride function and the aromatic ring.

   b. In the ^1H NMR spectrum, assign the various resonances to the hydrogen nuclei responsible for them.

   c. For the ^{13}C NMR data, assign the various resonances to the carbon nuclei responsible for them.

11. Consider the spectral data for 2-(2′,4′-dimethylbenzoyl) benzoic acid (Figs. 15.16 and 15.17).

   a. In the functional group region of the IR spectrum, specify the absorptions associated with the ketone carbonyl group, the aromatic rings, and the carboxylic acid function. Explain why the latter absorption is broad.

   b. In the ^1H NMR spectrum, assign the various resonances to the hydrogen nuclei responsible for them.

   c. For the ^{13}C NMR data, assign the various resonances to the carbon nuclei responsible for them.

12. What differences in the IR and NMR spectra of phthalic anhydride and 2-(2′,4′-dimethylbenzoyl)-benzoic acid are consistent with the conversion of the anhydride group to ketone and carboxylic acid groups in this experiment?

13. Explain how ^{1}H NMR and ^{13}C NMR spectra could be used to differentiate between **15** and the isomer of it shown below.

Spectra   *Starting Materials and Products*

FIGURE 15.12
IR spectrum of *m*-xylene (neat).

FIGURE 15.13
NMR data for *m*-xylene (CDCl$_3$).

(a) ^{1}H NMR spectrum (300 MHz).

(b) ^{13}C NMR data: δ 21.3, 126.2, 128.2, 130.0, 137.6.

FIGURE 15.14
IR spectrum of phthalic anhydride
(IR card).

(a) ^{1}H NMR spectrum (300 MHz).

FIGURE 15.15
NMR data for phthalic anhydride
(DMSO-d$_6$).

(b) ^{13}C NMR data: δ 125.3, 131.1, 136.1, 163.1.

FIGURE 15.16
IR spectrum of 2-
(2′,4′-dimethylbenzoyl)-benzoic
acid (IR card).

FIGURE 15.17
NMR data for
2-(2′,4′-dimethylbenzoyl)-benzoic
acid (DMSO-d$_6$).

(a) ^{1}H NMR spectrum (300 MHz).

(b) ^{13}C NMR data: δ 21.3, 21.4, 125.7, 128.0, 128.1, 129.4, 130.6, 131.8, 132.7, 132.8, 133.8, 140.3, 142.4, 143.9, 171.2, 198.4.

15.4 Nitration of Bromobenzene

Reaction of an aromatic compound such as benzene with a mixture of concentrated sulfuric and nitric acids introduces a nitro group on the ring by way of electrophilic aromatic substitution (Sec. 15.1) as depicted in Equation 15.14. The electrophilic species is the **nitronium ion**, NO_2^+, which is produced by reaction of sulfuric acid with nitric acid (Eq. 15.15); note that nitric acid, the *weaker* of these two strong mineral acids, is serving as a *base*, rather than an acid in the equilibrium!

$$\text{Benzene} \xrightarrow[\text{H}_2\text{SO}_4]{\text{HNO}_3} \text{Nitrobenzene} + \text{H}_2\text{O} \qquad (15.14)$$

$$\text{HÖ}-\text{NO}_2 \; \xrightleftharpoons \; \text{H}_2\text{O} + {}^+\text{NO}_2 \qquad (15.15)$$

Nitric acid   Nitronium
                                                       ion

The rate-determining step (rds) in the nitration reaction involves nucleophilic attack of the aromatic ring, a Lewis base, on the nitronium ion, a Lewis acid, to form the delocalized **σ-complex 17** (Eq. 15.16). In the final step of the reaction, a base such as water or bisulfate deprotonates this complex to regenerate the aromatic ring (Eq. 15.17).

(15.16)

17

σ-Complex

(15.17)

17

    The mechanism of the reaction of a nitronium ion with a *substituted* benzene such as bromobenzene (**18**), the arene used in this experiment, is similar to that with benzene. Now, however, *three* different products can be formed because of the presence of the bromine atom on the ring (Eq. 15.18). The activation energies for attack at the positions

(15.18)

| **18** | **19** | **20** | **21** |
| Bromobenzene | 2-Bromo-nitrobenzene | 4-Bromo-nitrobenzene | 3-Bromo-nitrobenzene |

ortho and *para* to the substituent are lower than that at the *meta* position. This is because the σ-complexes **22** and **23**, which are the precursors to the *ortho* and *para* products **19** and **20**, respectively, are more stable than complex **24**, which gives the *meta* compound **21**. The increased stability of **22** and **23** relative to **24** is associated with the participation of nonbonding electrons on the bromine atom in delocalizing the positive charge in the complexes **22** and **23**. Thus, like an alkyl group, a bromine atom is an example of an ***ortho-para director***, since it favors formation of these isomers over the *meta* isomer. For the nitration reaction the preference is found to be about 100 : 1.

22   **23**   **24**

    The *o*:*p* ratio itself is of interest because isolating pure disubstituted products requires separation of these isomers. Assuming that attack of the electrophile at the two positions occurs on a purely statistical basis, this ratio is predicted to be 2 : 1, because there are two *ortho* carbon atoms and only one *para* carbon atom in bromobenzene. This prediction neglects the possibility that the steric bulk and the inductive electron-withdrawing effect of the bromine atom might inhibit approach of the electrophile to the *ortho* relative to the *para* position. Indeed, these two factors appear to be important in determining the *o*:*p* ratio, which is approximately 1:2. Consequently, 2-bromonitrobenzene (**19**) is an important by-product in the preparation of 4-bromonitrobenzene (**20**), and this must be taken into consideration in the procedure used to isolate pure product.

Another type of by-product results if further nitration of **19** or **20** occurs. Considering the directing effects of the bromo and nitro substituents leads to the prediction that the major product from both isomers would be 1-bromo-2,4-dinitrobenzene (**25**), as shown in Equation 15.19. The possible formation of 1-bromo-2,6-dinitrobenzene (**26**) from **19** is less likely because of the steric and inductive effects of the bromo substituent. Fortunately it is relatively easy to suppress dinitration by performing the reaction at temperatures below 60 °C. Controlling the temperature suppresses dinitration because the strongly deactivating nitro substituent in **19** and **20** raises the activation energy for their nitration above that of **18**. Another factor decreasing the dinitration of **20** is its precipitation during the course of the reaction, thereby removing it from the nitrating medium.

$$\textbf{19 or 20} \quad \xrightarrow[\text{H}_2\text{SO}_4]{\text{HNO}_3} \quad \underset{\substack{\textbf{25}\\ \text{1-Bromo-2,4-dinitrobenzene}}}{\text{[structure]}} \quad + \quad \text{H}_2\text{O} \qquad (15.19)$$

26
1-Bromo-2,6-dinitrobenzene

Based on the preceding discussion, the major obstacle to isolating pure 4-bromonitrobenzene (**20**) is removal of 2-bromonitrobenzene (**19**). The strategy for doing this relies on the greater polarity of **19**, which makes this isomer more soluble in polar solvents than **20**. This is dramatically demonstrated by their relative solubilities in ethanol. At room temperature, the *ortho* isomer **19** is very soluble, whereas the *para* isomer is slightly soluble, dissolving to the extent of only 1.2 g/100 mL. This large difference in solubilities allows separation of the isomers by the technique of **fractional crystallization**. The mixture of products obtained from nitration of bromobenzene is dissolved in hot 95% ethanol, and the solution is allowed to cool. The less soluble *para* isomer selectively crystallizes from solution and is isolated by filtration. Concentration of the filtrate allows isolation of a second crop of **20**.

In a fractional crystallization it is commonly possible to induce crystallization of the more soluble component, in this case the *ortho* isomer **19**, once the less soluble material has been mostly removed. However, the low melting point of **19** means that this isomer is difficult to crystallize in the presence of impurities. Consequently, isolating this isomer involves column chromatography (Sec. 6.2).

EXPERIMENTAL PROCEDURES

NITRATION OF BROMOBENZENE

Purpose: To demonstrate nitration by electrophilic aromatic substitution.

Because *concentrated* sulfuric and nitric acids may cause severe chemical burns, *do not allow them to contact your skin.* Wear latex gloves when handling these reagents. Wipe off any drips and runs on the outside surface of reagent bottles and graduated cylinders *before* picking them up. Wash any affected area immediately and thoroughly with cold water, and apply 5% sodium bicarbonate solution.

Safety Alert

A · NITRATION

Miniscale Procedure

Preparation   Answer the Pre-Lab Exercises on page PL. 89. Review Sections 2.9, 2.10, 2.11, and 2.29.

Apparatus   A 25-mL round-bottom flask, Claisen adapter, thermometer, water-cooled condenser, ice-water bath, apparatus for magnetic stirring and *flameless* heating.

Setting Up   Prepare a solution of 4.0 mL of concentrated nitric acid and 4.0 mL of concentrated sulfuric acid in the round-bottom flask, and cool it to room temperature with a water bath. Equip the flask with a stirbar and a Claisen adapter fitted with the condenser and a thermometer that extends into the flask (Fig. 15.18).

Reaction and Work-Up   In portions of approximately 0.5 mL, add 4.5 mL of bromobenzene to the stirred mixture through the top of the condenser over a period of about 10 min. Do *not* allow the temperature of the reaction mixture to exceed 50–55 °C during the addition. Control the temperature by allowing more time between the addition of successive portions of bromobenzene and by cooling the reaction flask with an ice-water bath.

After the addition is complete and the exothermic reaction has subsided, heat the stirred mixture for 15 min, keeping its temperature below 60 °C.★ Cool the reaction mixture to room temperature, and then pour it carefully and with stirring into 40 mL of cold water contained in a beaker.

Isolation and Purification   Isolate the mixture of crude bromonitrobenzenes by vacuum filtration. Wash the filter cake thoroughly with cold water until the washes are neutral to pHydrion paper; allow the solid to drain under vacuum until nearly dry.★

Transfer the filter cake to an Erlenmeyer flask and recrystallize the crude product from 95% ethanol. Allow the residual solution to cool slowly to room temperature; then cool it to 0 °C in an ice-water bath. Isolate the crystalline product by vacuum filtration. Wash the product with a little *ice-cold* 95% ethanol, allowing the washes to drain into the filter flask with the mother liquors. Transfer the product to a watchglass or a piece of filter paper for air-drying.

Water out

Clamp

Water in

Clamp

Clamp

FIGURE 15.18
Apparatus for brominating nitrobenzene.

Concentrate the mother liquors to a volume of about 10 mL by simple distillation. Perform this operation in a hood or use one of the techniques described in Section 2.29 to prevent release of vapors into the laboratory. Allow the residual solution to cool to room temperature to produce a second crop of 4-bromonitrobenzene. Isolate it by vacuum filtration and, after air-drying, put it in a separate vial from the first crop.

Further concentrate the mother liquors from the second crop to a volume of 3–4 mL. The resulting oil contains crude 2-bromonitrobenzene. Separate the oil from the two-phase mixture by means of a Pasteur pipet and reserve it for chromatographic analysis (Parts B and C).

Analysis   Weigh both crops of product and calculate the yield. Measure the melting points of both crops. Obtain IR and NMR spectra of your starting material and product and compare them with those of authentic samples (Figs. 8.29, 15.19–15.22).

Microscale Procedure

Preparation   Answer the Pre-Lab Exercises on page PL. 89. Review Sections 2.9, 2.10, 2.11, and 2.29.

Apparatus   A 3-mL conical vial, water-cooled condenser, ice-water bath, apparatus for magnetic stirring, vacuum and Craig tube filtration, and *flameless* heating.

Setting Up   Prepare a solution of 0.5 mL of concentrated nitric acid and 0.5 mL of concentrated sulfuric acid in the vial, and cool it to room temperature with a water bath. Equip the vial with a spinvane and the condenser.

Reaction and Work-Up   In portions of approximately 0.1 mL, add 0.5 mL of bromobenzene to the stirred mixture through the top of the condenser over a period of about 5 min. Although the temperature of the reaction mixture may become *warm* to the touch, do not allow the reaction mixture to become *hot* to the touch during the addition. Control the temperature by allowing more time between the addition of successive portions of bromobenzene and by cooling the vial with an ice-water bath if necessary.

After the addition is complete and the exothermic reaction has subsided, heat the stirred reaction mixture for 15 min, keeping the bath temperature below 70 °C.★ Cool the vial to room temperature, and then pour the reaction mixture carefully and with stirring into 5 mL of cold water contained in a beaker.

Isolation and Purification   Isolate the mixture of crude bromonitrobenzenes by vacuum filtration. Wash the filter cake with small portions of cold water until the washes are neutral, and allow the solid to drain under vacuum until nearly dry.★

Transfer the filter cake to a Craig tube and recrystallize the crude product from 95% ethanol. Allow the solution to cool slowly to room temperature; then cool it to 0 °C in an ice-water bath. Isolate the crystalline product by vacuum or Craig tube filtration.

Concentrate the mother liquors to a volume of about 1 mL by simple distillation. Perform this operation in a hood or use one of the techniques described in Section 2.29 to prevent release of vapors into the laboratory. Allow the residual solution to cool to room temperature to produce a second crop of 4-bromonitrobenzene. Isolate it by Craig tube filtration and, after air-drying, put it in a separate vial from the first crop.

Further concentrate the mother liquors from the second crop to a volume of 3–4 mL. The resulting oil contains crude 2-bromonitrobenzene. Separate the oil from the two-phase mixture by means of a Pasteur pipet and reserve it for TLC analysis (Part C).

Analysis   Weigh both crops of product and calculate the yield. Measure the melting points of both crops. Obtain IR and NMR spectra of your starting material and product and compare them with those of authentic samples (Figs. 8.29, 15.19–15.22).

Neutralize the *aqueous filtrates* with sodium carbonate and flush them down the drain. Pour any *oil* obtained by concentration of the *ethanolic filtrate* into the container for halogenated liquids, *unless* the oil is to be subjected to chromatography.

Finishing Touches

B · THIN-LAYER CHROMATOGRAPHY

Preparation   Answer the Pre-Lab Exercises on page PL. 91. Review Section 6.3.

Apparatus   Capillary pipets, developing chamber.

Setting Up   In two small vials, separately prepare solutions of 4-bromonitrobenzene and of the oil containing 2-bromonitrobenzene in about 0.5 mL of dichloromethane. Apply spots of each solution to the TLC plate and allow the spots to dry.★

Chromatography and Analysis   Develop the chromatogram in a TLC chamber with 9:1 (v:v) hexane:ethyl acetate as the eluting solvent. When the solvent is within about 0.5 cm of the top of the plate, remove the developed chromatogram from the chamber, *quickly* mark the solvent front with a pencil, and allow the plate to dry.

Visualize the spots by placing the dry plate under an ultraviolet (uv) lamp or in a chamber whose atmosphere is saturated with iodine vapor. Calculate the R_f values of the spots observed, and identify them as either 2- or 4-bromonitrobenzene. A small orange spot may be observed very near the origin of the chromatogram; this spot is 1-bromo-2,4-dinitrobenzene.

Put the unused *eluant* in the container for nonhalogenated organic liquids. Discard the dry *TLC plates* in the container for halogenated organic compounds.

Finishing Touches

C · COLUMN CHROMATOGRAPHY

Miniscale Procedure

Preparation   Answer the Pre-Lab Exercises on page PL. 93. Review Sections 6.2 and 6.3.

Apparatus   Silica gel chromatography column, eight 13-mm × 100-mm test tubes, capillary pipets, developing chamber, *flameless* heating.

Setting Up   Obtain a 25- or 50-mL buret and 5 g of silica gel. Pack the column according to the procedure described in the experimental procedure of Section 6.2 with the exception that a 9:1 (v:v) solution of hexane and ethyl acetate rather than petroleum ether is to be used for packing the column. With a Pasteur pipet, apply a 0.2-g sample of the oil containing the 2-bromonitrobenzene to the head of the column, and rinse the inside of the buret with 1 mL of ethyl acetate. Open the stopcock and allow the liquid to drain just to the top of the sand. Close the stopcock and fill the buret with 9:1 (v:v) hexane:ethyl acetate.

Chromatography and Analysis   Elute the column until a total of 40 mL of the solvent has passed through the column. Do not allow the level of liquid to drain below the sand at the top of the column. Collect the eluate in 5- to 8-mL fractions in separate, labeled test tubes.

Analyze each fraction by TLC using 9:1 (v:v) hexane:ethyl acetate as the eluting solvent. With careful spotting it should be possible to analyze at least three different fractions on each 3-cm × 8-cm strip of a silica gel chromatogram. In order to introduce sufficient sample onto the strip, re-spot two or three times.

Combine the fractions containing only 2-bromonitrobenzene and then remove the solvent by simple distillation. Alternatively, rotary evaporation or one of the techniques described in Section 2.29 may be used. Characterize the residue obtained by either melting point determination, TLC, or spectroscopic analysis. 2-Bromonitrobenzene has a low reported melting point and thus may be isolated as an oil. Obtain IR and NMR spectra of your product and compare them with those of an authentic sample (Figs. 15.23 and 15.24).

Finishing Touches

After first allowing any residual solvent to evaporate from the silica gel in the hood, put the *silica gel* in the container for non-toxic solids. Pour all *eluates containing bromonitrobenzenes* into the container for halogenated organic compounds; put *all other eluates* and any unused eluant in the container for nonhalogenated solvents.

EXERCISES

1. Explain why 4-bromonitrobenzene (**20**) is *less* polar than 2-bromonitrobenzene (**19**).

2. Explain why the melting point of 4-bromonitrobenzene (**20**) is considerably higher than that of the 2-isomer **19**.

3. The pK_as of sulfuric and nitric acids are −3 (estimated value) and −1.3. Which of the two is the stronger acid?

4. **a.** Provide the resonance structures that contribute to the σ-complexes **23** and **24**.

   b. Use these resonance structures to explain why the formation of 3-bromonitrobenzene (**21**) is disfavored relative to 4-bromonitrobenzene.

5. Why does 4-bromonitrobenzene (**20**) have a larger R_f-value in the TLC analysis than does the 2-isomer **19**?

6. The $o:p$ ratio in the mononitration of bromobenzene has been reported to be 38:62. Use this ratio and the amount of 4-bromonitrobenzene actually isolated to estimate the experimental yield of mononitration in the reaction. What errors are there in using this method to calculate the extent of mononitration?

Refer to Tables 8.1–8.8 as needed for answering the questions on spectroscopy.

7. Consider the spectral data for bromobenzene (Figs. 15.19 and 15.20).

   a. In the IR spectrum, specify the absorption associated with the π-bonds of the aromatic ring. Indicate with what structural feature the strong absorptions at about 740 and 690 cm^{-1} is associated.

b. In the ^{1}H NMR spectrum, assign the various resonances to the hydrogen nuclei responsible for them.

c. For the ^{13}C NMR data, assign the various resonances to the carbon nuclei responsible for them.

8. Consider the spectral data for 4-bromonitrobenzene (Figs. 8.29, 15.21 and 15.22).

    a. In the functional group region of the IR spectrum, specify the absorption associated with the π-bonds of the aromatic ring and indicate with what structural feature the strong absorption at 825 cm^{-1} is associated.

    b. In the ^{1}H NMR spectrum, assign the various resonances to the hydrogen nuclei responsible for them.

    c. For the ^{13}C NMR data, assign the various resonances to the carbon nuclei responsible for them.

9. Consider the spectral data for 2-bromonitrobenzene (Figs. 15.23 and 15.24).

    a. In the functional group region of the IR spectrum, specify the absorption associated with the π-bonds of the aromatic ring and indicate with what structural feature the strong absorption at about 750 cm^{-1} is associated.

    b. In the ^{1}H NMR spectrum, assign the various resonances to the hydrogen nuclei responsible for them.

    c. For the ^{13}C NMR data, assign the various resonances to the carbon nuclei responsible for them.

10. What differences in the IR and NMR spectra of bromobenzene and 4-bromonitrobenzene are consistent with the introduction of a nitro group onto the ring in this experiment?

Starting Material and Products
    The ^{1}H NMR spectrum of 4-bromonitrobenzene is provided in Figure 8.29.

Spectra

FIGURE 15.19
IR spectrum of bromobenzene (neat).

FIGURE 15.20
NMR data for bromobenzene (CDCl$_3$).

(a) ^{1}H NMR spectrum (300 MHz).

(b) ^{13}C NMR data: δ 122.5, 126.7, 129.8, 131.4.

FIGURE 15.21
IR spectrum of 4-bromonitrobenzene (IR card).

FIGURE 15.22
NMR data for 4-bromonitrobenzene (CDCl$_3$).

^{13}C NMRdata: δ 124.9, 129.8, 132.5, 147.0.

FIGURE 15.23
IR spectrum of 2-bromonitrobenzene (neat).

(a) ^{1}H NMR spectrum (300 MHz).

FIGURE 15.24
NMR data for 2-bromonitrobenzene
(CDCl$_3$).

(b) ^{13}C NMR data: δ 114.1, 125.4, 128.2, 133.2, 134.8, 136.4.

15.5 Relative Rates of Electrophilic Aromatic Substitution

Electrophilic aromatic substitution reactions (Sec. 15.1) are among the best understood of all organic reactions. The *qualitative* aspects of the reactions that are discussed in textbooks include the effect substituents have on the reactivity of arenes toward electrophiles and the orientation, *ortho*, *meta*, or *para*, of their attack on the ring. Relatively little information is given about the *quantitative* differences in rates and reactivities of substituted aromatic compounds. This experiment provides a *semiquantitative* measure of the differences in reactivity of a series of arenes toward the bromonium ion, Br$^+$, to produce the corresponding aryl bromide (Eq. 15.20).

$$\underset{\substack{\text{An arene}}}{\text{Ar}-\text{H}} + \underset{\substack{\text{Bromine}\\\text{Red-brown}}}{\text{Br}_2} \xrightarrow{\text{HOAc}} \underset{\substack{\text{An aryl bromide}\\\text{Colorless}}}{\text{Ar}-\text{Br}} + \text{HBr} \qquad \textbf{(15.20)}$$

Studies of electrophilic substitutions on arenes are reported in which the experimental conditions allow a direct comparison of the relative reaction rates. For example, the relative reactivities of benzene and toluene toward halogenation, acetylation, sulfonation, nitration, and methylation have been determined. Electrophilic aromatic substitution was more rapid with toluene in all cases; bromination of toluene, for instance, is some 600 times faster than that of benzene. Such studies have led to the classification of substituents as **ring activators** or **deactivators**, depending on whether the substituted arene reacts *faster* or *slower* than benzene itself. Thus, the methyl group of toluene is a ring activator.

The effect substituents have on the relative rate of a specific type of electrophilic substitution may exceed several orders of magnitude, and the range of reactivity when comparing substituents may then be enormous. For example, in electrophilic bromination of arenes, phenol reacts 10^{12} times faster and nitrobenzene reacts 10^5 times *slower* than benzene. The hydroxy group is thus a powerful *ring activator*, and the nitro group is a potent *ring deactivator*. The range of difference in reactivity due to these two substituents is a remarkable 10^{17}!

The relative rates of bromination of the monosubstituted arenes shown below are evaluated in the following procedure. This particular reaction is selected because relative reaction rates can be determined visually by measuring the time required for disappearance of the characteristic reddish color of molecular bromine.

Phenol   Anisole   Diphenyl ether   Acetanilide   4-Bromophenol   1-Naphthol

EXPERIMENTAL PROCEDURE

RELATIVE RATES OF ELECTROPHILIC AROMATIC BROMINATION

Purpose: To evaluate the relative reactivities of different arenes in electrophilic aromatic bromination.

> **Safety Alert**
>
> **1.** *Bromine is a hazardous chemical.* Do not breathe its vapors or allow it to come into contact with your skin because it may produce severe chemical burns. Perform all operations involving the transfer of the pure liquid or its solutions in a hood, and wear latex gloves for these manipulations. Even though the solutions used in the experiments are dilute, take proper precautions. If you get bromine on your skin, wash the area *immediately* with soap and warm water and soak any affected area in 0.6 *M* sodium thiosulfate solution, for up to 3 h if the burn is particularly serious.
>
> **2.** Do not use your mouth to fill pipets when transferring *any* of the solutions in this experiment; this applies to both the bromine solutions and the acetic acid solutions of the substrates.

Note: It is convenient to work in pairs for this experiment.

Miniscale Procedure

Preparation   Answer the Pre-Lab Exercises on page PL. 95. Review Section 2.5.

Apparatus   A 1-L beaker, three calibrated Pasteur pipets, six 13-mm $\times$ 100-mm test tubes, and *flameless* heating.

Setting Up   As accurately as possible, calibrate several Pasteur pipets to contain 1.5 mL. In separate labeled test tubes, place 1.5 mL of a 0.2 *M* stock solution of each of the following substrates in 15 *M* (90%) acetic acid: (a) phenol, (b) anisole, (c) diphenyl ether, (d) acetanilide, (e) 4-bromophenol, and (f) 1-naphthol. Use special care to ensure that one stock solution is not contaminated by another during any transfer operations you perform.

Use the 1-L beaker to make a water bath and establish a bath temperature of 35 ± 2 °C. Suspend the test tubes partially in the water bath by looping a piece of copper wire around the neck of the test tube and over the rim of the beaker. Transfer about 25 mL of a solution containing 0.05 *M* bromine in 15 *M* acetic acid to an Erlenmeyer flask, and equilibrate this solution in the water bath for a few minutes.

Bromination   Add 1.5 mL of the bromine solution to a test tube containing one of the substrates. Make the addition *rapidly*, mix the solution *quickly*, and *record the exact time* of addition. Monitor the reaction mixture and note the elapsed time for the bromine color to become faint yellow or to disappear. If decoloration has not occurred within 5 min, which can happen because some substrates react slowly, proceed to a second substrate while waiting for the endpoint of the first to be reached. Record the reaction times. Repeat this procedure with each of the arenes, using the same endpoint color for each one.

To confirm your conclusions regarding the relative reactivities of the series of arenes, repeat the experiment at 0 °C in an ice-water bath, using only those substrates that were not clearly differentiated at 35 °C.

Analysis   On the basis of your observations, list the arenes in order of *decreasing* reactivity toward bromine.

Finishing Touches

Neutralize the combined *acetic acid solutions* by adding saturated aqueous sodium bicarbonate and filter the resulting mixture under vacuum. Flush the *filtrate* down the drain; put the *filter cake*, which contains aryl bromides, in the container for halogen-containing solid waste.

EXERCISES

1. Write the structures for the major monobromination products that would be formed from each of the substrates used. Explain your predictions.

2. Using inductive and/or resonance effects, explain the reactivity order you observed.

3. **a.** Determine the *initial* concentrations of bromine and of substrate arene in each reaction mixture.

   b. Which reagent, bromine or arene, is present in excess in each case?

   c. Explain why the experimental procedure calls for an excess of the specified reactant in **b** over the other.

   d. Discuss whether you would be able to determine the rates of reaction if the concentrations of arene and of bromine had been reversed over what is specified in the experimental procedure.

4. Considering only electronic factors, predict the reactivity toward electrophilic aromatic bromination of 3-methoxyphenol relative to benzene.

Refer to Tables 8.1–8.8 as needed for answering the questions on spectroscopy.

5. Consider the spectral data for phenol (Figs. 15.25 and 15.26).

    a. In the functional group region of the IR spectrum, specify with what structural feature the broad band centered at about 3280 cm^{-1} is associated, and explain why the band is broad.

    b. In the ^{1}H NMR spectrum, assign the various resonances to the hydrogen nuclei responsible for them.

    c. For the ^{13}C NMR data, assign the various resonances to the carbon nuclei responsible for them.

6. Consider the spectral data for anisole (Figs. 15.27 and 15.28).

    a. In the functional group region of the IR spectrum, specify the absorption associated with the π-bonds of the aromatic ring.

    b. In the ^{1}H NMR spectrum, assign the various resonances to the hydrogen nuclei responsible for them.

    c. For the ^{13}C NMR data, assign the various resonances to the carbon nuclei responsible for them.

7. Consider the spectral data for diphenyl ether (Figs. 15.29 and 15.30).

    a. In the functional group region of the IR spectrum, specify the absorption associated with the π-bonds and with the C–H bonds of the aromatic rings.

    b. In the ^{1}H NMR spectrum, assign the various resonances to the hydrogen nuclei responsible for them.

    c. For the ^{13}C NMR data, assign the various resonances to the carbon nuclei responsible for them.

8. Consider the spectral data for acetanilide (Figs. 8.6 and 15.31).

    a. In the functional group region of the IR spectrum, specify the absorptions associated with the N–H bond, with the carbon-oxygen double bond, and with the aromatic ring.

    b. In the ^{1}H NMR spectrum, assign as many as possible of the various resonances to the hydrogen nuclei responsible for them.

    c. For the ^{13}C NMR data, assign the various resonances to the carbon nuclei responsible for them.

9. Consider the spectral data for 4-bromophenol (Figs. 15.32 and 15.33).

    a. In the functional group region of the IR spectrum, specify the absorption associated with the π-bonds of the aromatic ring; indicate with what structural feature the broad band centered at about 3400 cm^{-1} is associated, and explain why the band is broad.

    b. In the ^{1}H NMR spectrum, assign the various resonances to the hydrogen nuclei responsible for them.

    c. For the ^{13}C NMR data, assign the various resonances to the carbon nuclei responsible for them.

10. Consider the spectral data for 1-naphthol (Figs. 15.34 and 15.35).

    a. In the functional group region of the IR spectrum, specify the absorption associated with the π-bonds of the aromatic ring. Indicate with what structural feature the broad band centered at about 3300 cm^{-1} is associated and explain why the band is broad.

b. In the ^{1}H NMR spectrum, assign the various resonances to the hydrogen nuclei responsible for them.

c. For the ^{13}C NMR data, assign the various resonances to the carbon nuclei responsible for them.

11. When the NMR solutions of phenol, 4-bromophenol, and 1-naphthol were each shaken with a drop of D_2O, the resonances assigned to the hydrogen of the hydroxyl group disappeared. Explain this observation.

Starting Materials and Products
  The IR spectrum of acetanilide is provided in Figure 8.6.

Spectra

FIGURE 15.25
IR spectrum of phenol (neat).

(a) ^{1}H NMR spectrum (300 MHz).

(b) ^{13}C NMR data: δ 115.4, 121.0, 129.7, 154.9.

FIGURE 15.26
NMR data for phenol (CDCl$_3$).

FIGURE 15.27
IR spectrum of anisole (neat).

FIGURE 15.28
NMR data for anisole (CDCl₃).

(a) ¹H NMR spectrum (300 MHz).

(b) ¹³C NMR data: δ 54.8, 114.1, 120.7, 129.5, 159.9.

FIGURE 15.29
IR spectrum of diphenyl ether
(IR card).

(a) ^{1}H NMR spectrum (300 MHz).

FIGURE 15.30
NMR data for diphenyl ether (CDCl$_3$).

(b) ^{13}C NMR data: δ 119.0, 123.2, 129.8, 157.6.

(a) ^{1}H NMR spectrum (300 MHz).

FIGURE 15.31
NMR data for acetanilide (CDCl$_3$).

(b) ^{13}C NMR data: δ 24.0, 119.7, 123.3, 128.5, 139.2, 168.7.

FIGURE 15.32
IR spectrum of 4-bromophenol (IR card).

FIGURE 15.33
NMR data for 4-bromophenol
(CDCl$_3$).

(a) ^{1}H NMR spectrum (300 MHz).

(b) ^{13}C NMR data: δ 113.2, 117.2, 132.5, 153.9.

FIGURE 15.34
IR spectrum of 1-naphthol
(IR card).

FIGURE 15.35
NMR data for 1-naphthol (CDCl$_3$).

(a) ^{1}H NMR spectrum (300 MHz).

(b) ^{13}C NMR data: δ 108.8, 119.8, 122.1, 124.9, 126.2, 127.6, 134.9, 152.4.

HISTORICAL HIGHLIGHT

Discovery of the Friedel-Crafts Reaction

The Friedel-Crafts reaction bears the names of two chemists, Charles Friedel, a Frenchman, and James Crafts, an American, who accidentally discovered this reaction in Friedel's laboratory in Paris in 1877. Friedel and Crafts quickly recognized the potential practical significance of their discovery, and immediately secured patents in France and England on procedures for preparing hydrocarbons and ketones using the reaction. Their judgment was accurate. Probably no other organic reaction has been of more practical value. Major industrial processes for producing high-octane gasoline, synthetic rubber, plastics, and synthetic detergents are representative applications of Friedel-Crafts chemistry. All of this came from an unexpected laboratory result and the brilliance of the two scientists who observed, interpreted, and extended it.

Crafts was born in 1839 in Boston. After graduating from Harvard University at the age of 19, he spent a year studying mining engineering, and then decided to go to Europe for further education. Crafts became fascinated with chemistry while studying metallurgy in Freiburg, Germany, and he subsequently secured positions in the laboratories of Bunsen in Heidelberg and Wurtz in Paris. It was in Wurtz's laboratory that he met Friedel and, owing to their similar scientific interests, the two began a collaboration in 1861.

Crafts returned to the United States in 1865 and, after a brief tenure as a mining inspector in Mexico and California, accepted a position as Professor of Chemistry at Cornell University shortly after this institution was founded. Three years later he moved to M.I.T. Because of poor health he decided to return to Europe in 1874 and resume his collaboration with Friedel in Wurtz's laboratory in Paris.

When Crafts left M.I.T., he had expected to return in a short time, presumably after recovering his health. Owing to the change of climate or, perhaps, the excitement of the discovery he shared with Friedel in 1877, his health improved dramatically, yet he remained in Paris for another 17 years. Friedel and Crafts conducted an energetic research collaboration and by 1888 had produced over 50 publications, including patents, related to the reactions of aluminum chloride with organic compounds.

What was the accidental research discovery that engraved the names of Friedel and Crafts in scientific and industrial annals? They were attempting to prepare 1-iodopentane by treating 1-chloropentane with aluminum and iodine. They observed that the reaction took an entirely different course from that expected, producing large amounts of hydrogen chloride and, unexpectedly, hydrocarbons. Upon further investigation, they found that using aluminum chloride in place of aluminum gave the same results. Other researchers had earlier reported somewhat similar results from reactions of organic chlorides with certain metals, but had not explained them or implicated the metal chloride as a reactant or catalyst. It was the work of Friedel and Crafts that first showed that the presence of the metal chloride was essential.

They subsequently performed a reaction of 1-chloropentane with aluminum chloride, using benzene as the solvent. Once again, they observed evolution of hydrogen chloride, but this time they found that 1-pentylbenzene (''amylbenzene'') was a major product! In describing their discovery to the Chemical Society of France in 1877, they reported, ''With a mixture of [alkyl] chloride and hydrocarbon [an arene], the formation is established, in good yield, of hydrocarbons from the residues of the hydrocarbon less H and from the chloride less Cl. It is thus that ethylbenzene, amylbenzene, benzophenone, etc., are obtained.'' A general representation of the transformation is shown in the following equation.

$$ \text{C}_6\text{H}_5\text{—H} + \text{R—Cl} \xrightarrow{\text{AlCl}_3} \text{C}_6\text{H}_5\text{—R} + \text{HCl} $$

Friedel and Crafts recognized that their unexpected result promised the possibility of synthesizing a wide

HISTORICAL HIGHLIGHT
Discovery of the Friedel-Crafts Reaction *(continued)*

variety of hydrocarbons and ketones, and they immediately proved this by experiment. In the ensuing years the voluminous research papers and patents that came from Friedel and Crafts established a whole new area of research and practice in organic chemistry and laid the foundation for some important modern industrial chemical processes.

Friedel-Crafts chemistry has touched our lives in many important but perhaps unrecognized ways. For example, Winston Churchill, in referring to the winning of the air war over Britain by the fighter pilots, once said ''Never in the field of human conflict was so much owed by so many to so few.'' What was not recognized by many at the time, perhaps even Churchill, was that the victory in the air war was due not only to the skill and daring of the British pilots, but to the superiority of their aviation gasoline over that of the Germans. It has been generally acknowledged that the German fighter planes were mechanically superior to the British planes, but their fuel was not. The gasoline used in the British planes was of higher quality because it contained toluene and other alkylated aromatic hydrocarbons, which provided higher octane ratings. Production of these types of hydrocarbons was a direct outgrowth of Friedel-Crafts chemistry.

Similarly, ''synthetic rubber'' was vital to the ground-war effort in World War II, after the Japanese cut off the Allies from sources of natural rubber in Southeast Asia. A synthetic rubber was developed in an amazingly short time in a remarkable display of cooperation between government officials and industrial and academic scientists. This rubber was a copolymer (Sec. 21.1) that contained styrene as a key component. Styrene is made from ethylbenzene, which is prepared industrially by the Friedel-Crafts reaction of benzene and ethylene, as shown below.

Styrene can also be homopolymerized to produce polystyrene (Sec. 21.2). This material is one of the most versatile plastics ever invented. For example, it can be molded into rigid cases for radios and batteries and formulated in a more flexible form, useful in the manufacture of toys and all kinds of containers. Polystyrene can also be produced in the form of a lightweight foam known as Styrofoam®. This foam is used for insulation in buildings and is molded into ice chests and disposable cups for hot and cold drinks.

The Friedel-Crafts reaction is a key part of the manufacture of many synthetic detergents, cleaning agents that have revolutionized the way we live today. We wash dishes and clothes with them, and, in contrast to soap, they work as well in hard water as in soft. A typical example of a biodegradable synthetic detergent is sodium dodecylbenzenesulfonate, in which the 12-carbon side chain is attached to the benzene ring by a Friedel-Crafts alkylation reaction, as illustrated here.

A dodecylbenzene
(other isomers also formed)

A sodium dodecylbenzene-
sulfonate

Ethylbenzene

Styrene

HISTORICAL HIGHLIGHT
Discovery of the Friedel-Crafts Reaction *(continued)*

These are just a few of the practical applications that have been developed in the years since Friedel and Crafts made their initial observation of an unexpected experimental result. By proper interpretation of the result, they have earned credit for a major discovery.

Relationship of Historical Highlight to Experiments

The experiments in Sections 15.2 and 15.3 illustrate the Friedel-Crafts alkylation and acylation of aromatic hydrocarbons, respectively. A complication of Friedel-Crafts reactions is apparent in the alkylation experiment, wherein rearrangements of the carbocations generated from the alkyl halide provide mixtures of substitution products. The acylation reaction of Section 15.3 provides an example of how a combination of electronic and steric effects can affect the orientation of electrophilic attack on an aromatic ring.

CHAPTER
16

Oxidation of Alcohols and Carbonyl Compounds

16.1 Introduction

The oxidation of organic compounds is an important class of reactions that involves interaction of an oxidizing agent, often symbolized as [O], with an organic substrate. Oxidations are frequently used to induce simple functional group transformations in synthetic organic chemistry. For example, the conversions of an alkene into a 1,2-diol (Eq. 16.1 and Sec. 10.4), of a primary or secondary alcohol into an aldehyde or ketone, respectively (Eq. 16.2), and of an aldehyde into a carboxylic acid (Eq. 16.3) are common laboratory reactions. In processes involving the degradation of organic molecules, oxidations may be used to cleave carbon-carbon double bonds, as illustrated by the ozonolysis of an alkene to give the corresponding carbonyl compounds (Eq. 16.4).

$$R^1 \overset{KMnO_4}{\longrightarrow} R^1 \text{ (1,2-diol with OH groups)} \qquad (16.1)$$

An alkene   A 1,2-diol

$$R^1\text{—}CH_2\text{—}OH \ \text{ or } \ R^1\text{—}\underset{R^2}{CH}\text{—}OH \ \xrightarrow{[O]} \ \underset{R^1 \quad H}{\overset{O}{C}} \ \text{ or } \ \underset{R^1 \quad R^2}{\overset{O}{C}} \qquad (16.2)$$

A primary   A secondary   An aldehyde   A ketone
alcohol   alcohol

$$\underset{R^1 \quad H}{\overset{O}{C}} \ \xrightarrow{[O]} \ \underset{R^1 \quad OH}{\overset{O}{C}} \qquad (16.3)$$

An aldehyde   A carboxylic acid

434

$$R^1\diagup\!\!\diagdown\begin{smallmatrix}R^2\\R^3\end{smallmatrix} \xrightarrow[\text{(2) Zn, HOAc}]{\text{(1) O}_3} R^1\!-\!CHO + \underset{R^2\quad R^3}{\overset{O}{\underset{\|}{C}}} \qquad\qquad \textbf{(16.4)}$$

An alkene   An aldehyde   A ketone

Indeed, the processes of life depend on oxidation of organic substrates, whereby metabolic energy is derived from the overall oxidation of carbohydrates, fats, and proteins to carbon dioxide and water, among other products, as illustrated in Equation 16.5. Potentially poisonous substances are commonly detoxified by biological oxidation to more benign substances. Thus, nicotine, which is toxic to humans if present in sufficiently high concentration, is oxidized in the liver to cotinine (Eq. 16.6), a substance having low toxicity.

$$\text{A carbohydrate} \xrightarrow{[O]} 6\,CO_2 + 6\,H_2O + \text{energy} \qquad\qquad \textbf{(16.5)}$$

Nicotine   Cotinine   **(16.6)**

Oxidation does *not* necessarily result in introducing oxygen into an organic molecule. In the most general sense, and in analogy to concepts of oxidation as applied to inorganic compounds, oxidation of an organic substance simply requires an increase in the **oxidation number**, or **oxidation state**, of carbon. There are several ways of defining the oxidation numbers of carbon atoms, but the one used here involves the following sequence.

1. *Select the carbon atom* whose oxidation number is to be defined.
2. *Assign oxidation numbers* to the atoms attached to this carbon atom using the following values:

   a. +1 for hydrogen
   b. −1 for halogen, nitrogen, oxygen and sulfur
   c. 0 for carbon

   In other words, with respect to the atoms bound to the carbon atom, those that are *more* electronegative are assigned a value of −1, whereas those that are *less* electronegative are given a value of +1.
3. *Sum the oxidation numbers* of these atoms. If a heteroatom (see 2b) is multiply bound to a carbon atom, multiply its oxidation number by the number of bonds linking it to the carbon atom involved.
4. The sum of the number from step 3 and the oxidation number of the carbon atom under consideration must equal the charge on the carbon atom, which is zero unless it bears a positive or negative charge.

Application of this method shows that the conversion of an alkene to the corresponding dibromide (Eq. 16.7) and that of an alkane to an alkyl halide (Eq. 16.8), reactions discussed in Section 10.6 and Chapter 9, respectively, are oxidations even though oxygen is not incorporated in these reactions. Determining the changes in the oxidation number of the various carbon atoms during the conversion of acetic acid to carbon dioxide and water (Eq. 16.9) illustrates how the method is used when heteroatoms are multiply bound to a carbon atom.

$$Cyclopentene \xrightarrow[CCl_4]{Br_2} \textit{trans}\text{-1,2-Dibromo-cyclopentane}$$

Net change: loss of 2 e^- (oxidation at carbon atom)

(16.7)

$$\overset{-3}{CH_3}\overset{}{CH_2}\!\!-\!\!H \xrightarrow{Cl_2} \overset{-1}{CH_3}\overset{}{CH_2}\!\!-\!\!Cl$$

Ethane   1-Chloroethane

(16.8)

$$\underset{-3}{CH_3}\!\!-\!\!\overset{\overset{\displaystyle O}{\|}}{\underset{+3}{C}}\!\!-\!\!OH \xrightarrow{[O]} 2\ \underset{+4}{O\!\!=\!\!C\!\!=\!\!O}$$

Acetic acid   Carbon dioxide

(16.9)

16.2   Preparation of Aldehydes and Ketones by Oxidation of Alcohols

Aldehydes and ketones have a central role in organic synthesis, and efficient methods for their preparation are of great importance. Such compounds are synthesized in a variety of ways, including hydration or hydroboration/oxidation of alkynes (Eqs. 16.10 and 16.11, respectively, and Chap. 11), and from carboxylic acids or their derivatives by reaction with organometallic reagents or reducing agents (Eqs. 16.12 and 16.13). The latter two reactions involve net **reduction** of the carbonyl carbon atom, whereas the transformations shown in Equations 16.10 and 16.11 are neither net oxidation nor reduction of the alkyne, as can be shown by determining the net change in oxidation state at the carbon atoms involved in the transformations.

$$R\!\!-\!\!C\!\!\equiv\!\!C\!\!-\!\!R(H) \xrightarrow[\substack{H_2SO_4 \\ H_2O}]{HgSO_4} R\!\!-\!\!\overset{\overset{\displaystyle O}{\|}}{C}\!\!-\!\!CH_2\!\!-\!\!R\ (H)$$

An alkyne   A ketone

(16.10)

$$R\!\!-\!\!C\!\!\equiv\!\!C\!\!-\!\!R(H) \xrightarrow[\substack{(2)\ H_2O_2/ \\ HO^-}]{(1)\ B_2H_6} R\!\!-\!\!CH_2\!\!-\!\!\overset{\overset{\displaystyle O}{\|}}{C}\!\!-\!\!R\ (H)$$

(16.11)

$$R^1-\overset{\overset{\textstyle O}{\|}}{C}-OH \xrightarrow[\text{(2) } H_3O^+]{\text{(1) } 2\ R^2Li} R^1-\overset{\overset{\textstyle O}{\|}}{C}-R^2 \qquad \textbf{(16.12)}$$

A carboxylic acid

$$R^1-\overset{\overset{\textstyle O}{\|}}{C}-OR \xrightarrow[\text{(2) } H_3O^+]{\text{(1) } (i\text{-Bu})_2AlH} R^1-\overset{\overset{\textstyle O}{\|}}{C}-H \qquad \textbf{(16.13)}$$

An ester

One of the most common synthetic methods for preparing aldehydes and ketones is the oxidation of primary and secondary alcohols, a reaction involving a two-electron change in the oxidation number of the functionalized carbon atom (Eq. 16.14). Chromic ion, Cr^{6+}, frequently in the form of chromic acid, H_2CrO_4, is a common oxidant, although other oxidants, such as halonium ion, X^+, as found in hypohalous acids, HOX, and permanganate ion, Mn^{7+}, as in potassium permanganate, $KMnO_4$, can be used.

$$R-\overset{\overset{\textstyle OH}{|}}{CH}-R(H) \xrightarrow[\substack{\text{or HOX} \\ \text{or } KMnO_4}]{H_2CrO_4} R-\overset{\overset{\textstyle O}{\|}}{C}-R(H) \qquad \textbf{(16.14)}$$

Oxidation number 0 (2° alcohol)   +2 (ketone)
−1 (1° alcohol)   +1 (aldehyde)

Chromic acid and potassium permanganate are excellent oxidizing agents that are widely used, but these oxidants are derived from heavy metals, a class of elements that are commonly toxic and therefore environmentally hazardous. The safe disposal of these metals and their derivatives is of considerable concern, and avoiding their use is desirable. Fortunately, an alternative to chromate and permanganate ions for oxidizing alcohols is available in the form of the hypohalites, XO– (X = halogen), which are made by reaction of the corresponding halogen with aqueous base (Eq. 16.15). The most familiar sources of this type of oxidizing agent are household bleach, which is an aqueous solution of sodium hypochlorite, NaOCl, and the chlorinating agent for swimming pools, which usually is calcium hypochlorite, $Ca(OCl)_2$, in granular or tablet form.

$$X-X + HO^- \rightleftharpoons X^- + XO-H \overset{HO^-}{\rightleftharpoons} XO^- + H_2O \qquad \textbf{(16.15)}$$

(X = halogen)   Hypohalous   Hypohalite
acid   ion

The mechanism by which hypohalites oxidize alcohols probably involves initial formation of an alkyl hypohalite (Eq. 16.16). This product arises from reaction of the alcohol with the hypohalous acid that is in equilibrium with hypohalite ion in the aqueous medium (Eq. 16.15). Base-promoted E2 elimination of the elements of H–X from the alkyl hypochlorite leads directly to either an aldehyde or ketone. The advantage of using hypohalite as an oxidant is immediately obvious upon review of Equation 16.16. The inorganic by-product derived from the oxidant is a halide salt, and this can be safely flushed down the drain.

$$R-\underset{\underset{H}{|}}{\overset{\overset{R}{|}}{\underset{0}{C}}}-\ddot{O}H + \underset{+1}{X}-OH \underset{-H_2O}{\rightleftharpoons} R-\underset{\underset{H}{|}}{\overset{\overset{R}{|}}{C}}-\overset{\frown}{O} \xrightarrow[{}^-OH]{} \underset{R'}{\overset{R}{>}}\underset{+2}{C}=O + \underset{-1}{X^-} \qquad (16.16)$$

Alkyl hypohalite   Halide ion

The stoichiometry of the reaction may be determined from changes in oxidation numbers in the reactants and products. During the oxidation, the carbon atom of the substrate alcohol undergoes a two-electron change in oxidation state, from 0 to +2 in the case of a secondary alcohol, while the halogen also undergoes a two-electron change in oxidation number from +1 to −1 (Eq. 16.16). Consequently a one-to-one ratio of the two reactants is required.

Some side reactions may complicate the oxidation of a primary alcohol to an aldehyde using hypohalous acid. For example, an aldehyde may undergo reaction with an additional equivalent of hypohalite to form a carboxylic acid (Eq. 16.17), a process that may be initiated by acid-catalyzed formation of the hydrate of the intermediate aldehyde as shown in Equation 16.18. Subsequent steps in the oxidation are then analogous to those for converting an alcohol to an aldehyde or ketone (Eq. 16.16). It may be difficult to suppress this further oxidation, so unless carboxylic acids are the desired products, the use of hypohalite as an oxidant is limited to the conversion of secondary alcohols to ketones.

$$R^l-\overset{\overset{O}{\|}}{C}-H + X-OH + H_2O \longrightarrow R^l-\overset{\overset{O}{\|}}{C}-OH \qquad (16.17)$$

An aldehyde   A carboxylic acid

$$R-\overset{\overset{O}{\|}}{C}-H \underset{}{\overset{H^+}{\rightleftharpoons}} R-\overset{\overset{\overset{H}{\underset{}{}}\!\!\overset{+}{O}}{\|}}{C}-H \overset{:\ddot{O}H_2}{\longrightarrow} R-\underset{\underset{+OH_2}{|}}{\overset{\overset{OH}{|}}{C}}-H \underset{}{\overset{-H^+}{\rightleftharpoons}} R-\underset{\underset{OH}{|}}{\overset{\overset{OH}{|}}{C}}-H \qquad (16.18)$$

An aldehyde   A hydrate

Another side reaction that may occur when hypohalite is used as an oxidizing agent is the α-halogenation of the desired carbonyl compound (Eq. 16.19). This process involves the rapid, *base-catalyzed* formation of an enolate ion, followed by its reaction with hypohalous acid or some other source of halonium ion, such as X_2 (X = halogen). The replacement of the α-hydrogens of a ketone or aldehyde is the basis for the **haloform test**, which is described in Section 23.7E. When oxidations of alcohols to give carbonyl compounds are conducted in acidic media, α-halogenation of the product is not a significant problem for two reasons: The rate of enolization of a ketone is slower in acidic than in basic media; and the *enols* that are produced in acid are less reactive than the enolates that are formed in base.

$$R-\overset{\overset{O}{\|}}{C}-\overset{\overset{H}{|}}{C}HR \underset{}{\overset{HO^-}{\rightleftharpoons}} R-\overset{\overset{O^-}{|}}{C}=CHR \longleftrightarrow R-\overset{\overset{O}{\|}}{C}-CHR \overset{X-OH}{\longrightarrow} R-\overset{\overset{O}{\|}}{C}-\overset{\overset{X}{|}}{C}HR$$

A carbonyl   An enolate   An α-halocarbonyl
compound   compound

$$(16.19)$$

In the experiments that follow, several representative oxidations of alcohols using hypochlorous acid will be performed. In the first experiment, the secondary alcohol cyclododecanol (**1**) is oxidized to cyclododecanone (**2**) using sodium hypochlorite or commercial household bleach (Eq. 16.20). In the second experiment, 4-chlorobenzyl alcohol (**3**) is oxidized directly to 4-chlorobenzoic acid (**4**) using calcium hypochlorite (16.21).

Because the concentration of sodium hypochlorite in bleach depends upon the age of the solution, it is important to monitor the oxidation of **1** with starch/iodide test paper to ensure that excess oxidant is present. The test paper should turn blue-black owing to the formation of the complex of starch and iodine that is produced upon oxidation of iodide ion by the hypochlorite ion. On the other hand, commercially available calcium hypochlorite is quite stable and may be stored at room temperature without decomposition. Therefore, the amount of oxidant used is easy to determine, but it is still good practice to ensure that an excess of oxidant is present by applying the starch/iodide test, especially if the reaction mixture is heated.

$$\underset{\substack{\textbf{1}\\ \text{Cyclododecanol}}}{} \xrightarrow[\text{CH}_3\text{CO}_2\text{H}]{\text{NaOCl, H}_2\text{O}} \underset{\substack{\textbf{2}\\ \text{Cyclododecanone}}}{} \qquad (16.20)$$

$$\underset{\substack{\textbf{3}\\ \text{4-Chlorobenzyl alcohol}}}{\text{Cl}\!-\!\!\!\bigcirc\!\!\!-\text{CH}_2\text{OH}} \xrightarrow[\text{CH}_3\text{CO}_2\text{H}]{\text{Ca(OCl)}_2, \text{H}_2\text{O}} \underset{\substack{\textbf{4}\\ \text{4-Chlorobenzoic acid}}}{\text{Cl}\!-\!\!\!\bigcirc\!\!\!-\text{CO}_2\text{H}} \quad (16.21)$$

EXPERIMENTAL PROCEDURES

OXIDATION OF ALCOHOLS

A · OXIDATION OF CYCLODODECANOL TO CYCLODODECANONE

Purpose: To demonstrate the oxidation of a secondary alcohol to the corresponding ketone.

Do not allow the solution of sodium hypochlorite to come in contact with your skin or eyes. If it does, flush the affected area immediately with copious amounts of water. Sodium hypochlorite will also bleach clothing.

Miniscale Procedure

Preparation   Answer the Pre-Lab Exercises on page PL. 97. Review Sections 2.9, 2.11, 2.13, 2.21, 2.22, and 2.29.

Apparatus   A 25-mL round-bottom flask, separatory funnel, apparatus for heating under reflux, magnetic stirring, simple distillation, and *flameless* heating.

Setting Up   Place 0.5 g of cyclododecanol, 1.2 mL of acetone, and 0.4 mL of glacial acetic acid in the round-bottom flask. Equip the flask with a stirbar and set up the apparatus for heating under reflux.

Oxidation   Stir the mixture and warm it to approximately 45 °C; maintain this temperature within ±5 °C throughout the course of the reaction. Using a Pasteur pipet, add 4.5 mL of commercial bleach (ca. 5.3% sodium hypochlorite) dropwise to the stirred mixture through the top of the condenser over a period of about 0.5 h. Upon completing the addition, stop stirring and heating the mixture so the layers may separate. Using a Pasteur pipet, remove a small portion of the aqueous layer, and place a drop or two of this solution on a dampened piece of starch/iodide test paper to determine whether sufficient hypochlorite has been added. The indicator paper immediately turns blue-black in color if sufficient bleach has been added. If this color does not develop, add an additional 0.4 mL of bleach to the reaction mixture. Stir the resulting mixture with heating for 2–3 min and repeat the test for excess hypochlorite. Add additional 0.4-mL portions of bleach until a positive test for oxidant is observed. Then stir the reaction mixture with heating for an additional 10 min and retest for hypochlorite. If the test is negative, add a final 0.4 mL of bleach. Whether this last test is positive or negative, stir the mixture with heating for 10 min more to complete the reaction.

Work-Up   Allow the reaction mixture to cool to room temperature, and transfer it to a separatory funnel using a Pasteur pipet. Rinse the round-bottom flask with 5 mL of diethyl ether, and use a filter-tip pipet to transfer this wash to the separatory funnel. Shake the two-phase mixture, and separate the layers. Extract the aqueous layer with an additional 5-mL portion of diethyl ether, and add this extract to the original one. Wash the combined organic extracts with 5 mL of saturated sodium bicarbonate. Before shaking this mixture, swirl the *unstoppered* funnel until the evolution of carbon dioxide ceases. Shake the mixture, venting the funnel frequently to relieve any pressure that might develop. Wash the organic solution sequentially with 5-mL portions of saturated aqueous sodium bisulfite and saturated aqueous sodium chloride. Transfer the organic solution to an Erlenmeyer flask, and dry it over several spatula-tips full of anhydrous sodium sulfate.★ Swirl the flask occasionally for a period of 10–15 min to facilitate drying; add further small portions of anhydrous sodium sulfate if the solution does not become clear.

Isolation and Purification   Using a filter-tip pipet, transfer the dried ethereal solution to a *tared* 25-mL round-bottom flask, and equip it for simple distillation. Remove the diethyl ether by simple distillation. Alternatively, rotary evaporation or other techniques (Sec. 2.29) may be used to concentrate the solution. The final traces of solvent may be removed by attaching the flask to a vacuum source, gently swirling the contents as the vacuum is applied.★ The oil that is initially formed after removal of the solvents should solidify. Recrystallize the cyclododecanone from aqueous methanol.

Analysis   Weigh the flask and calculate the yield of solid cyclododecanone. Determine the melting point of the product. Prepare the semicarbazone (mp 218–219 °C) or oxime (mp 131–132 °C) according to the procedures given in Sections 23.7G and 23.7H. If necessary, recrystallize the derivatives from methanol. Obtain IR and ^{1}H NMR spectra of your starting material and product and compare them with those of authentic samples (Figs. 16.1–16.4).

Flush all *aqueous solutions* down the drain. Spread the *sodium sulfate* on a tray in the hood to evaporate residual solvent from it and then put the used drying agent in the nonhazardous solid waste container. Place the *diethyl ether* recovered by distillation in the container for nonhalogenated organic solvents.

Finishing Touches

EXERCISES

1. Prove that no net oxidation occurs at the carbon atoms in the transformations of Equations 16.10 and 16.11 by comparing the oxidation numbers for carbon in the reactants and products.

2. What is the function of sodium bisulfite in the procedure for isolating cyclododecanone?

3. Would cyclododecanone be expected to give a positive iodoform test (Sec. 23.7E)? Explain.

4. Determine whether the conversion of iodide ion to iodine is an oxidation or a reduction.

Refer to Tables 8.1–8.8 as needed for answering the questions on spectroscopy.

5. Consider the spectral data for cyclododecanol (Figs. 16.1 and 16.2).

   a. In the IR spectrum, there is a broad absorption at about 3250 cm^{-1}. What functional group is responsible for this absorption, and why is the absorption broad?

   b. In the ^{1}H NMR spectrum, assign the various resonances to the hydrogen nuclei responsible for them.

   c. For the ^{13}C NMR data, assign the various resonances to the carbon nuclei responsible for them.

6. Consider the spectral data for cyclododecanone (Figs. 16.3 and 16.4).

   a. In the functional group region of the IR spectrum, specify the absorption associated with the carbon-oxygen double bond.

   b. In the ^{1}H NMR spectrum, assign the various resonances to the hydrogen nuclei responsible for them.

   c. For the ^{13}C NMR data, assign the various resonances to the carbon nuclei responsible for them.

7. Discuss the differences observed in the IR and NMR spectra of cyclododecanol and cyclododecanone that are consistent with the conversion of an alcohol to a ketone in this experiment.

Spectra   *Starting Material and Product*

FIGURE 16.1
IR spectrum of cyclododecanol (IR card).

FIGURE 16.2
NMR data for cyclododecanol (CDCl₃).

(a) ¹H NMR spectrum (300 MHz).

(b) ¹³C NMR data: δ 21.0, 23.5, 24.0, 24.4, 32.5, 69.1.

FIGURE 16.3
IR spectrum of cyclododecanone (IR card).

(a) ^{1}H NMR spectrum (300 MHz).

FIGURE 16.4
NMR data for cyclododecanone (CDCl$_3$).

(b) ^{13}C NMR data: δ 22.7, 24.4, 24.9, 40.2, 211.4.

B • OXIDATION OF 4-CHLOROBENZYL ALCOHOL TO 4-CHLOROBENZOIC ACID

Purpose: To demonstrate the oxidation of an alcohol to the corresponding carboxylic acid.

Safety
Alert

Do not allow the solution of calcium hypochlorite to come in contact with your skin or eyes. If it does, flush the affected area immediately with copious amounts of water. Calcium hypochlorite will also bleach clothing.

Miniscale Procedure

Preparation   Answer the Pre-Lab Exercises on page PL. 99. Review Sections 2.9, 2.11, 2.17, and 2.21.

Apparatus   A 50-mL round-bottom flask, separatory funnel, apparatus for magnetic stirring, vacuum filtration, and *flameless* heating.

Setting Up   Weigh 2.4 g of commercial calcium hypochlorite (65%) and transfer it to the round-bottom flask containing a stirbar. Add 20 mL of water to the flask and stir the mixture while adding 2 mL of glacial acetic acid dropwise. Weigh 0.5 g of 4-chlorobenzyl alcohol and add it to a small Erlenmeyer flask containing 5 mL of acetonitrile; swirl the contents to dissolve the alcohol. Set up the apparatus for heating under reflux.

Oxidation   Using a Pasteur pipet, transfer the solution of 4-chlorobenzyl alcohol in acetonitrile dropwise to the stirred solution of calcium hypochlorite through the top of the condenser. Stir the mixture vigorously for 1.5–2 h at 45–50 °C; *do not heat above 50 °C*. Remove a small aliquot of the reaction mixture using a Pasteur pipet, and place a drop or two of this solution on a dampened piece of starch/iodide paper to determine whether sufficient hypochlorite has been added. The indicator paper immediately turns blue-black in color if excess oxidizing agent is present. If this color does not develop, add an additional 150 mg of commercial calcium hypochlorite to the reaction mixture. Whether or not additional calcium hypochlorite is required, continue stirring the mixture for an additional 1 h at 45–50° C, and then cool it to room temperature.★

Work-Up, Isolation, and Purification   Transfer the cooled reaction mixture to the separatory funnel and add 10 mL of water. Rinse the reaction flask with 10 mL of diethyl ether, and transfer this rinse to the separatory funnel. Shake the funnel and separate the layers. Extract the aqueous layer sequentially with two additional 10-mL portions of diethyl ether. Extract the *combined* diethyl ether layers with two separate 10-mL portions of saturated aqueous sodium bicarbonate. Slowly add *concentrated* hydrochloric acid dropwise to the *combined* bicarbonate washings until the solution is slightly acidic (pH = 3). Collect the precipitate by vacuum filtration, wash the filter cake with 5–10 mL of water, and air-dry the crude product. 4-Chlorobenzoic acid may be recrystallized from methanol.

Analysis   Weigh the product and calculate the percent yield. Determine the melting point of the recrystallized 4-chlorobenzoic acid. Obtain IR and ¹H NMR spectra of your starting material and product and compare them with those of authentic samples (Figs. 16.5–16.8).

Microscale Procedure

Preparation   Answer the Pre-Lab Exercises on page PL. 99. Review Sections 2.7, 2.9, 2.11, 2.17, and 2.21.

Apparatus   A 5-mL conical vial, apparatus for magnetic stirring, vacuum filtration, Craig tube filtration, and *flameless* heating.

Setting Up   Weigh 0.5 g of commercial calcium hypochlorite (65%) and transfer it to the conical vial containing a spinvane. Add 3.0 mL of water to the vial and stir the mixture while adding 0.3 mL of glacial acetic acid dropwise. Weigh 0.1 g of 4-chlorobenzyl alcohol and add it to a small test tube containing 1 mL of acetonitrile; swirl the contents to dissolve the alcohol. Set up the apparatus for heating under reflux.

Oxidation   Using a Pasteur pipet, transfer the solution of 4-chlorobenzyl alcohol in acetonitrile dropwise to the stirred solution of calcium hypochlorite through the top of the condenser. Stir the mixture vigorously for 1.5–2 h at 45–50 °C; *do not heat above 50 °C*. Remove a small aliquot of the reaction mixture using a Pasteur pipet, and place a drop or two of this solution on a dampened piece of starch/iodide paper to determine whether sufficient hypochlorite has been added. The indicator paper immediately turns blue-black in color if there is excess oxidizing agent present. If this color does not develop, add an additional 50 mg of commercial calcium hypochlorite to the reaction mixture. Whether or not additional calcium hypochlorite is required, continue stirring the mixture for an additional 1 h and then cool it to room temperature.★

Work-Up, Isolation, and Purification   Transfer the cooled reaction mixture to a screw-cap centrifuge tube and add 1.5 mL of water. Rinse the reaction vial with 2 mL of diethyl ether, and add this rinse to the centrifuge tube. Cap the centrifuge tube, and shake the contents to mix the layers. Separate the layers and extract the aqueous layer sequentially with two additional 2-mL portions of diethyl ether. Extract the *combined* diethyl ether layers with two separate 2-mL portions of saturated aqueous sodium bicarbonate. Slowly add *concentrated* hydrochloric acid dropwise to the *combined* bicarbonate washings until the solution is slightly acidic (pH = 3). Collect the precipitate by vacuum filtration, wash the filter cake with 1–2 mL of water, and air-dry the crude product. 4-Chlorobenzoic acid may be recrystallized from methanol in a Craig tube.

Analysis   Weigh the product and calculate the percent yield. Determine the melting point of the recrystallized 4-chlorobenzoic acid. Obtain IR and ^{1}H NMR spectra of your starting material and product and compare them with those of authentic samples (Figs. 16.5–16.8).

Neutralize the *combined aqueous solutions* with solid sodium carbonate and flush them down the drain. Because there may be some residual starting alcohol or intermediate aldehyde in the *diethyl ether extract*, transfer it to the container for halogenated organic liquids.

Finishing Touches

EXERCISES

1. Prove that no net oxidation occurs at the carbon atoms in the transformations of Equations 16.10 and 16.11 by comparing the oxidation numbers for carbon in the reactants and products.

2. Determine whether the conversion of iodide ion to molecular iodine is an oxidation or a reduction.

Refer to Tables 8.1–8.8 as needed for answering the following questions on spectroscopy.

3. Consider the spectral data for 4-chlorobenzyl alcohol (Figs. 16.5 and 16.6).

    a. In the IR spectrum, there is a broad absorption at about 3400 cm^{-1}. What functional group is responsible for this absorption, and why is the absorption broad?

    b. In the ^{1}H NMR spectrum, assign the various resonances to the hydrogen nuclei responsible for them.

    c. For the ^{13}C NMR data, assign the various resonances to the carbon nuclei responsible for them.

4. Consider the spectral data for 4-chlorobenzoic acid (Figs. 16.7 and 16.8).

    a. In the functional group region of the IR spectrum, specify the absorptions associated with the carbonyl component of the carboxyl group. What functional group is responsible for the broad absorption in the region of about 2800 cm^{-1}, and why is it broad?

    b. In the ^{1}H NMR spectrum, assign the various resonances to the hydrogen nuclei responsible for them.

    c. For the ^{13}C NMR data, assign the various resonances to the carbon nuclei responsible for them.

5. Discuss the differences observed in the IR and NMR spectra of 4-chlorobenzyl alcohol and 4-chlorobenzoic acid that are consistent with the oxidation of an alcohol to a carboxylic acid rather than to an aldehyde in this experiment.

Spectra   *Starting Material and Product*

FIGURE 16.5
IR spectrum of 4-chlorobenzyl alcohol (IR card).

(a) ^{1}H NMR spectrum (300 MHz).

FIGURE 16.6
NMR data for 4-chlorobenzyl alcohol (CDCl$_3$).

(b) ^{13}C NMR data: δ 64.3, 128.2, 128.6, 133.2, 139.2.

FIGURE 16.7
IR spectrum of 4-chlorobenzoic acid (IR card).

(a) ^{1}H NMR spectrum (300 MHz).

FIGURE 16.8
NMR data for 4-chlorobenzoic acid (DMSO-d$_6$).

(b) ^{13}C NMR data: δ 128.5, 129.8, 131.1, 138.3, 166.8.

447

16.3   Base-Catalyzed Oxidation-Reduction of Aldehydes: The Cannizzaro Reaction

Aldehydes having no hydrogens on the **α-carbon atom**, the carbon atom adjacent to the carbonyl group, undergo mutual oxidation and reduction in the presence of strong alkali (Eq. 16.22). In contrast, aldehydes with hydrogen atoms on the **α-carbon atom** preferentially undergo other types of base-promoted reactions such as the **aldol condensation** (Eq. 16.23), which is described in Section 18.3. The mutual oxidation-reduction, called the **Cannizzaro reaction**, is a consequence of the fact that an aldehyde is intermediate in oxidation state between an alcohol and a carboxylic acid. It may be converted into either one by a decrease or gain, respectively, of two in the oxidation number (Sec. 16.1) of its carbonyl carbon atom, as indicated in Equation 16.22.

$$2 \; R\text{—}\underset{R}{\overset{R}{\underset{|}{\overset{|}{C}}}}\text{—}\overset{O}{\overset{\parallel}{C}}\text{—}H \xrightarrow[\text{(2) } H_3O^+]{\text{(1) } HO^-} R\text{—}\underset{R}{\overset{R}{\underset{|}{\overset{|}{C}}}}\text{—}\underset{H}{\overset{O\text{—}H}{\underset{|}{\overset{|}{C}}}}\text{—}H \; + \; R\text{—}\underset{R}{\overset{R}{\underset{|}{\overset{|}{C}}}}\text{—}\overset{O}{\overset{\parallel}{C}}\text{—}OH \qquad (16.22)$$

An aldehyde   An alcohol   A carboxylic acid

$$2 \; R\text{—}\underset{H}{\overset{H}{\underset{|}{\overset{|}{C}}}}\text{—}\overset{O}{\overset{\parallel}{C}}\text{—}H \xrightarrow[\text{(2) } H_3O^+]{\text{(1) } HO^-} R\text{—}\underset{H}{\overset{H}{\underset{|}{\overset{|}{C}}}}\text{—}\underset{H}{\overset{OH}{\underset{|}{\overset{|}{C}}}}\text{—}\underset{R}{\overset{}{\underset{|}{CH}}}\text{—}\overset{O}{\overset{\parallel}{C}}\text{—}H \qquad (16.23)$$

An aldehyde   An aldol
                           (β-hydroxycarbonyl compound)

The accepted mechanism of the reaction follows logically from the ease with which nucleophiles add to the carbonyl group, particularly that of an aldehyde. The first step is attack of hydroxide ion on the carbonyl group to give a tetrahedral intermediate (Eq. 16.24); this is followed by transfer of a hydride ion to the carbonyl group of another aldehyde function (Eq. 16.25). The step depicted in Equation 16.25 is the one in which both oxidation and reduction occur, as is shown by determining the oxidation numbers of the carbon atoms involved. The remaining steps (Eqs. 16.26 and 16.27) simply illustrate the types of acid-base chemistry expected to occur in the strongly basic medium in which the reaction is performed. Summing of Equations 16.24–16.27 gives Equation 16.22.

$$\underset{R}{\overset{O}{\overset{\parallel}{C}}}\diagdown_H \; + \; {}^-OH \longrightarrow R\text{—}\underset{OH}{\overset{O^-}{\underset{|}{\overset{|}{C}}}}\text{—}H \qquad (16.24)$$

$$R\text{—}\underset{OH}{\overset{O^-}{\underset{|}{\overset{|}{C}}}}\text{—}H \; + \; \underset{R}{\overset{O}{\overset{\parallel}{C}}}\diagdown_H \longrightarrow \underset{R}{\overset{O}{\overset{\parallel}{C}}}\diagdown_{OH} \; + \; R\text{—}\underset{H}{\overset{O^-}{\underset{|}{\overset{|}{C}}}}\text{—}H \qquad (16.25)$$

$$R-\overset{\displaystyle O}{\overset{\|}{C}}-\overset{\curvearrowleft}{O}-H \; + \; \overset{\curvearrowleft}{^-OH} \longrightarrow R-\overset{\displaystyle O}{\overset{\|}{C}}-O^- \; + \; H-OH \qquad \textbf{(16.26)}$$

$$R-\overset{\displaystyle O^-}{\underset{\displaystyle H}{\overset{|}{C}}}-H \; + \; H\overset{\curvearrowleft}{-OH} \longrightarrow R-\overset{\displaystyle O-H}{\underset{\displaystyle H}{\overset{|}{C}}}-H \; + \; ^-OH \qquad \textbf{(16.27)}$$

Aromatic aldehydes, formaldehyde, and trisubstituted acetaldehydes all undergo the Cannizzaro reaction. In the experiment of this section, 4-chlorobenzaldehyde, an aromatic aldehyde, is converted to 4-chlorobenzyl alcohol and potassium 4-chloro-benzoate, which gives 4-chlorobenzoic acid upon acidification of the reaction mixture (Eq. 16.28).

| 4-Chlorobenzaldehyde | 4-Chlorobenzyl alcohol | Potassium 4-chlorobenzoate | 4-Chlorobenzoic acid |

(16.28)

EXPERIMENTAL PROCEDURES

BASE-CATALYZED OXIDATION-REDUCTION OF ALDEHYDES BY THE CANNIZZARO REACTION

Purpose: To demonstrate the oxidation-reduction of an aromatic aldehyde to give the corresponding alcohol and carboxylic acid salt by a Cannizzaro reaction.

> **Safety Alert**
>
> The concentrated solution of potassium hydroxide is *highly* corrosive and caustic. *Do not allow it to come into contact with the skin.* Should it do so, flood the affected area immediately with water and then thoroughly rinse it with 1% acetic acid. Wear latex gloves when preparing and transferring solutions of potassium hydroxide.

Miniscale Procedure

Preparation   Answer the Pre-Lab Exercises on page PL. 101. Review Sections 2.9, 2.11, 2.13, 2.17, 2.21, 2.22, and 2.29.

Apparatus   A 25-mL round-bottom flask, thermometer, separatory funnel, apparatus for heating under reflux, magnetic stirring, simple distillation, vacuum filtration, and *flameless* heating.

Setting Up   Dissolve 5.0 g of potassium hydroxide in 5 mL of distilled water contained in an Erlenmeyer flask. Weigh 1.0 g of 4-chlorobenzaldehyde and transfer it to the round-bottom flask containing a stirbar. Add 2.5 mL of methanol and stir to dissolve the 4-chlorobenzaldehyde. Using a Pasteur pipet, transfer 1.5 mL of the 50% aqueous potassium hydroxide solution to the round-bottom flask; do not get any of this solution on the ground-glass joint. Assemble the apparatus for heating under reflux; be sure that the joint of the flask is well-greased so that it does not freeze.

Reaction and Work-Up   Place the round-bottom flask in a heating bath that has been preheated to about 75 °C, and heat the reaction mixture with stirring at this temperature for 1.5 h. Remove the flask from the bath, cool the mixture to room temperature, and add 15 mL of water. Transfer the mixture to a separatory funnel. Extract the aqueous mixture with three separate 5-mL portions of dichloromethane; shake the funnel *gently* to avoid forming an emulsion. Test the lower layer to verify that it is the organic layer, and combine all of the organic extracts. *Save both the organic and the aqueous layers as each must be worked up separately to isolate the products.*

Isolation and Purification   Wash the combined *dichloromethane extracts* with two separate 5-mL portions of saturated aqueous sodium chloride. Dry the organic layer over several spatula-tips full of anhydrous sodium sulfate.★ Swirl the flask occasionally for a period of 10–15 min to facilitate drying; add further small portions of anhydrous sodium sulfate if the solution does not become clear. Filter or decant the dried solution into a clean, tared round-bottom flask, rinse the sodium sulfate with about 1 mL of fresh dichloromethane, and transfer the rinse to the round-bottom flask. Remove the dichloromethane by simple distillation. Alternatively, rotary evaporation or other techniques (Sec. 2.29) may be used to concentrate the solution. The final traces of solvent may be removed by attaching the flask to a vacuum source, gently swirling the contents as the vacuum is applied.★ The crude 4-chlorobenzyl alcohol may be recrystallized by mixed solvent techniques from acetone and hexane (Sec. 3.2). Air-dry the purified product.

Cool the *aqueous phase* saved from the initial dichloromethane extraction in an ice-water bath. Acidify the solution by slowly adding 2.5 mL of *concentrated* hydrochloric acid. If the solution is not acidic (pH = 3), continue adding concentrated hydrochloric acid *dropwise* until it is. Collect the precipitate by vacuum filtration, wash the filter cake with 5–10 mL of water, and air-dry the product. The crude 4-chlorobenzoic acid may be recrystallized from methanol.

Analysis   Weigh the products and calculate percent yields. Determine the melting points of the recrystallized products. Obtain IR and ¹H NMR spectra of your starting material and product and compare them with those of authentic samples (Figs. 16.5–16.10).

Microscale Procedure

Preparation   Answer the Pre-Lab Exercises on page PL. 101. Review Sections 2.9, 2.11, 2.13, 2.17, 2.21, 2.22, and 2.29.

Apparatus   A 5-mL conical vial, thermometer, screw-cap centrifuge tubes, apparatus for heating under reflux, magnetic stirring, simple distillation, vacuum filtration, Craig tube filtration, and *flameless* heating.

Setting Up   Dissolve 1.0 g of potassium hydroxide in 1 mL of distilled water contained in a test tube. Weigh 200 mg of 4-chlorobenzaldehyde and transfer it to the 5-mL conical vial containing a spinvane. Add 0.5 mL of methanol and stir to dissolve the 4-chlorobenzaldehyde. Using a Pasteur pipet, transfer 0.3 mL of the 50% aqueous potassium hydroxide solution to the vial; do not get any of this solution on the ground-glass joint. Assemble the apparatus for heating under reflux; be sure that the joint of the vial is well-greased so it does not freeze.

Reaction and Work-Up   Place the base of the 5-mL conical vial in a heating bath that has been preheated to about 75 °C, and heat the reaction mixture with stirring at this temperature for 1.5 h. Remove the vial from the bath, cool the mixture to room temperature, and add 2.5 mL of water. Extract the aqueous mixture with three separate 0.7-mL portions of dichloromethane. For each extraction, you should cap the vial and shake it gently to avoid forming an emulsion; carefully vent the vial by slowly loosening the cap. Remove the lower layer using a Pasteur pipet and verify that it is the organic layer; combine all of the organic extracts in a centrifuge tube. *Save both the organic and the aqueous layers as each must be worked up separately to isolate the products.*

Isolation and Purification   Wash the combined *dichloromethane extracts* with two separate 0.5-mL portions of saturated aqueous sodium chloride. Transfer the dichloromethane layer to a dry centrifuge tube and add several microspatula-tips full of anhydrous sodium sulfate.★ Swirl the tube occasionally for a period of 10–15 min to facilitate drying; add further small portions of anhydrous sodium sulfate if the solution does not become clear. Using a filter-tip pipet, transfer the dried solution into a clean 5-mL conical vial, rinse the sodium sulfate with about 0.2 mL of dichloromethane, and transfer the rinse to the conical vial. Remove the dichloromethane by simple distillation. Alternatively, rotary evaporation or other techniques (Sec. 2.29) may be used to concentrate the solution. The final traces of solvent may be removed by attaching the vial to a vacuum source, gently swirling the contents as the vacuum is applied.★ The crude 4-chlorobenzyl alcohol may be recrystallized from 4% acetone in hexane in a Craig tube. Air-dry the purified product.

Cool the *aqueous phase* saved from the initial dichloromethane extraction in an ice-water bath. Acidify the solution by slowly adding 0.5 mL of *concentrated* hydrochloric acid. If the solution is not acidic (pH = 3), continue adding concentrated hydrochloric acid *dropwise* until it is. Collect the precipitate by vacuum filtration, wash the filter cake with 1–2 mL of water, and air-dry the product. The crude 4-chlorobenzoic acid may be recrystallized from methanol in a Craig tube.

Analysis   Weigh the products and calculate the percent yields. Determine the melting points of the recrystallized products. Obtain IR and ^{1}H NMR spectra of your starting material and product and compare them with those of authentic samples (Figs. 16.5–16.10).

Flush the *saturated sodium chloride solution* used during the extraction process to wash the dichloromethane down the drain. Neutralize the *aqueous filtrate* from isolation of 4-chlorobenzoic acid with solid sodium carbonate and flush it down the drain. Place the *dichloromethane* obtained by distillation in the container for halogenated organic liquids. Dispose of the *sodium sulfate* in the container for nontoxic solids after allowing residual solvent to evaporate from it in the hood.

Finishing Touches

EXERCISES

1. Determine the oxidation number for the carbonyl carbon in 4-chlorobenzaldehyde. Then confirm that oxidation and reduction of the aldehyde group occurs in the Cannizzaro reaction by calculating the oxidation numbers for the corresponding carbon atoms in 4-chlorobenzyl alcohol and 4-chlorobenzoic acid.

2. After separating the organic and aqueous layers, the dichloromethane layer is first washed with two portions of saturated aqueous sodium chloride prior to drying with anhydrous sodium sulfate. What is the purpose of these washes?

3. When the Cannizzaro reaction is performed on 4-chlorobenzaldehyde in D_2O solution, no deuterium is found on the benzylic carbon atom in the 4-chlorobenzyl alcohol formed. How does this support the mechanism given in Equations 16.24–16.27?

4. What is the solid that is formed after 4-chlorobenzaldehyde has been allowed to react with aqueous potassium hydroxide?

5. The Cannizzaro reaction occurs much more slowly in dilute than in concentrated potassium hydroxide solution. Why is this?

6. How would 2,2-dimethylpropanal react with potassium hydroxide solution under the conditions of this experiment?

7. A *crossed* Cannizzaro reaction involving two different aldehydes may be performed if neither of the aldehydes contains α-hydrogens. Write an equation for the crossed Cannizzaro reaction of a mixture of 4-chlorobenzaldehyde and formaldehyde with concentrated potassium hydroxide solution, followed by acidification of the reaction mixture. Show all organic products.

Refer to Tables 8.1–8.8 as needed for answering the following questions on spectroscopy.

8. Consider the spectral data for 4-chlorobenzaldehyde (Figs. 16.9 and 16.10).

   a. In the functional group region of the IR spectrum, specify the absorptions associated with the carbonyl group and with the aromatic ring.

   b. In the ^1H NMR spectrum, assign the various resonances to the hydrogen nuclei responsible for them.

   c. For the ^{13}C NMR data, assign the various resonances to the carbon nuclei responsible for them.

9. Consider the spectral data for 4-chlorobenzyl alcohol (Figs. 16.5 and 16.6).

   a. In the functional group region of the IR spectrum, there is a broad absorption at about 3400 cm^{-1}. What functional group is responsible for this absorption, and why is it broad?

   b. In the ^1H NMR spectrum, assign the various resonances to the hydrogen nuclei responsible for them.

   c. For the ^{13}C NMR data, assign the various resonances to the carbon nuclei responsible for them.

10. Consider the spectral data for 4-chlorobenzoic acid (Figs. 16.7 and 16.8).

    a. In the functional group region of the IR spectrum, specify the absorptions associated with the carbonyl component of the carboxyl group. What functional group is responsible for the broad absorption in the region of about 2800 cm^{-1}, and why is the absorption broad?

b. In the ^{1}H NMR spectrum, assign the various resonances to the hydrogen nuclei responsible for them.

c. For the ^{13}C NMR data, assign the various resonances to the carbon nuclei responsible for them.

11. Discuss the differences observed in the IR and NMR spectra of 4-chlorobenzaldehyde, 4-chlorobenzyl alcohol, and 4-chlorobenzoic acid that are consistent with the formation of the two products from 4-chlorobenzaldehyde by the Cannizzaro reaction.

Spectra

Starting Material and Products

The IR and NMR spectra of 4-chlorobenzyl alcohol are given in Figures 16.5 and 16.6, respectively; those of 4-chlorobenzoic acid are presented in Figures 16.7 and 16.8, respectively.

FIGURE 16.9
IR spectrum of 4-chlorobenzaldehyde (IR card).

(a) ^{1}H NMR spectrum (300 MHz).

(b) ^{13}C NMR data: δ 129.4, 130.8, 134.9, 140.7, 190.5.

FIGURE 16.10
NMR data for 4-chlorobenzaldehyde (CDCl$_3$).

CHAPTER 17

Reduction Reactions of Double Bonds

Alkenes, Carbonyl Compounds, and Imines

17.1 Introduction

The general term **reduction** refers to those reactions in which hydrogen or electrons are added to a functional group present in a reactant; it is the *opposite* of **oxidation**, which involves the addition of oxygen or the net loss of electrons from a substrate. In organic chemistry, *reduction results in a decrease in the oxidation number of carbon and hence a net gain of hydrogen or electrons by the carbon atom.* For example, from the discussion in Section 16.1, we know that the **oxidation number** of each carbon atom in ethylene, $CH_2{=}CH_2$, is -2; however, in ethane, CH_3CH_3, the oxidation number is -3. Thus, conversion of ethylene to ethane involves a net gain of $1\ e^-$ by *each* carbon atom and is therefore a reduction.

Organic functional groups containing double and triple bonds undergo reduction by the *net* addition of the elements of one or two molecules of hydrogen across the π-bond(s). The starting compounds in these processes are termed **unsaturated**, whereas the final products are **saturated** with respect to hydrogen, since they contain only single bonds and can absorb no more hydrogen. Equations 17.1–17.4 illustrate some types of reduction reactions that are commonly performed by organic chemists. The symbol [**H**] is generally used to represent the **catalytic** and **chemical** methods that are used to effect reduction.

Catalytic reduction involves adding hydrogen to a carbon-carbon or carbon-heteroatom double or triple bond in the presence of metal **catalysts** such as platinum, palladium, nickel, or rhodium. The selection of the best catalyst for a specific reduction is sometimes difficult, but this problem is typically solved by analogy with similar reactions that have been reported in the literature.

Chemical methods of reduction include the use of hydride-donating reagents such as lithium aluminum hydride, $LiAlH_4$, and sodium borohydride, $NaBH_4$, and electron-transfer processes such as dissolving metal reductions. In addition to the functional groups shown in Equations 17.1–17.4, many other functional groups containing π-

454

bonds can be reduced to form products containing only single σ-bonds. These include carboxylic acids and their derivatives (esters, amides, anhydrides, and nitriles), nitro groups, and aromatic rings.

$$\text{An alkene} \quad\xrightarrow{[H]}\quad \text{An alkane} \tag{17.1}$$

$$\text{An alkyne} \quad\xrightarrow[\text{1 mol}]{[H]}\quad \text{An alkene} \quad\xrightarrow[\text{1 mol}]{[H]}\quad \text{An alkane} \tag{17.2}$$

$$\text{A carbonyl compound} \quad\xrightarrow{[H]}\quad \text{An alcohol} \tag{17.3}$$

$$\text{An imine} \quad\xrightarrow{[H]}\quad \text{An amine} \tag{17.4}$$

Catalytic hydrogenation is widely used to reduce the π-bonds in a variety of functional groups. Analytical methods have been developed in which catalytic hydrogenations are performed on a small scale, and the amount of hydrogen consumed and the heat evolved are accurately measured. Using such techniques, the number of double and/or triple bonds in a compound is easily ascertained, as are the relative stabilities of isomeric alkenes and alkynes. Catalytic hydrogenation is frequently used in synthetic organic chemistry to prepare novel compounds that may be biologically active or exhibit interesting structural or physical properties. The catalytic hydrogenation of a compound containing a carbon-carbon double bond is performed in the experimental procedure in Section 17.2.

The **chemical reduction** of a number of organic functional groups is commonly effected using **metal hydride** reducing agents, which have different reactivities toward specific functional groups. This property is of considerable practical importance because one of the challenges in contemporary synthetic organic chemistry is developing new reducing reagents that react *selectively* with only one type of functional group in the presence of other reducible groups. For example, lithium aluminum hydride, $LiAlH_4$, is *not* a selective reducing agent but rather is a very reactive hydride donor that reduces aldehydes, ketones, and esters with equal ease. Sodium borohydride, $NaBH_4$, on the other hand, is less reactive and hence more selective; it reduces aldehydes and ketones but *not* esters. Not only is chemical selectivity a concern in choosing the appropriate reagent, but experimental factors such as ease of handling and safety often play important roles. Thus, if both lithium aluminum hydride and sodium borohydride could be used to effect a specific reduction, the latter is preferred because it is safer and easier to handle. Section 17.3 includes experiments for several chemical reductions using sodium borohydride.

The nitro group, $-NO_2$, is reduced to the amino group, $-NH_2$, by catalytic hydrogenation or by using lithium aluminum hydride. However, this conversion is also accomplished by an alternative method of chemical reduction known as **dissolving metal reduction**. These reductions are typically performed either using finely divided iron or tin with hydrochloric acid or an alkaline earth metal such as lithium, sodium, and potassium in liquid ammonia. The reduction of nitrobenzene, $C_6H_5NO_2$, to aniline, $C_6H_5NH_2$, using tin and hydrochloric acid (Sec. 20.2) is an example of such a reduction.

17.2   Catalytic Hydrogenation of the Carbon-Carbon Double Bond

Catalytic hydrogenation of alkenes belongs to the general class of reactions known as **addition reactions**, which were introduced in Section 10.4. In these reactions, one mole of hydrogen is *stereospecifically* added to the *same* side of the carbon-carbon π-bond, as exemplified by the hydrogenation of 1,2-dimethylcyclopentene (**1**) to give *cis*-1,2-dimethylcyclopentane (**2**) as shown in Equation 17.5. Such processes are termed *syn*-**additions**. Although the detailed mechanism by which this reaction occurs is not fully understood, it is generally believed that hydrogen gas is adsorbed on the surface of the finely divided metal catalyst to produce hydrogen atoms. The π-electrons of the double bond also complex with the surface of the catalyst, and the hydrogen atoms are transferred from the surface to the carbon atoms to form the saturated product. Catalytic hydrogenation is the method of choice for reducing alkenes to alkanes and alkynes to *cis*-alkenes.

$$(17.5)$$

1	2
1,2-Dimethylcyclopentene	*cis*-1,2-Dimethylcyclopentane

Catalytic hydrogenation may be accomplished in several ways. Generally, the appropriate metal catalyst is suspended in a solution containing the reactant in a suitable solvent, and the resulting mixture is stirred or shaken under an atmosphere of hydrogen gas at pressures ranging from 1 to 60 psi. Hydrogen, a highly flammable gas, is normally purchased in pressurized gas cylinders, and special apparatus and precautions must be used to deliver the hydrogen from the cylinder to the apparatus. It is more convenient in simple experiments to generate hydrogen gas *in situ*. Thus, the combination of sodium borohydride and concentrated hydrochloric acid provides a ready source of hydrogen via the reaction of H^+ from HCl with the hydride ions, $H:^-$, contained in $NaBH_4$ (Eq. 17.6). The catalyst, platinum on carbon, may be prepared in the same flask by reduction of chloroplatinic acid, H_2PtCl_6, with hydrogen in the presence of decolorizing carbon, which serves as a solid support for the finely divided metallic platinum produced (Eq. 17.7).

$$NaBH_4 \ + \ HCl \ + \ 3 \ H_2O \ \ 4 \ H_2 \ + \ B(OH)_3 \ + \ NaCl \qquad (17.6)$$

$$H_2PtCl_6 \ + \ 2 \ H_2 \ \ Pt° \ + \ 6 \ HCl \qquad (17.7)$$

The experiment that follows involves the conversion of 4-cyclohexene-*cis*-1,2-dicarboxylic acid (**3**) to cyclohexane-*cis*-1,2-dicarboxylic acid (**4**), as shown in Equation 17.8. The catalytic hydrogenation of **3** is selective, because the carboxylic acid groups, −CO₂H, are *not* reduced under the reaction conditions. Moreover, the relative stereochemistry of the two carboxyl groups remains unchanged, even though the *trans* diacid is more stable. The starting material for this reduction may be readily prepared by the hydrolysis of 4-cyclohexene-*cis*-1,2-dicarboxylic acid anhydride, which is the product of the Diels-Alder reaction between butadiene and maleic anhydride, according to the procedure outlined in Section 12.3, Part C1.

$$\xrightarrow{\text{H}_2,\ \text{Pt(C)}}$$

(17.8)

3	**4**
4-Cyclohexene-*cis*- 1,2-dicarboxylic acid	Cyclohexane-*cis*- 1,2-dicarboxylic acid

---

EXPERIMENTAL PROCEDURES

HYDROGENATION OF 4-CYCLOHEXENE-*cis*-1,2-DICARBOXYLIC ACID

Purpose: To demonstrate the reduction of an alkene by catalytic hydrogenation.

> **Safety Alert**
>
> 1. *Hydrogen gas is extremely flammable.* Use no flames in this experiment, and make certain that there are no open flames nearby.
> 2. The platinum catalyst prepared in this experiment may be *pyrophoric* and combust spontaneously in air. Do *not* allow it to become dry!
> 3. Use *flameless* heating when removing the diethyl ether.
> 4. Wear latex gloves throughout the experiment to avoid skin contamination by the solutions you are using. If you do get these solutions on your hands, wash them thoroughly with water. In the case of contact with the sodium borohydride solution, rinse the affected areas with 1% acetic acid solution. In the case of acid burns, apply a paste of sodium bicarbonate to the area for a few minutes and then rinse with large amounts of water.

Miniscale Procedure

Preparation   Answer the Pre-Lab Exercises on page PL. 103. Review Sections 2.9, 2.11, 2.13, 2.17, 2.21, and 2.29.

Apparatus   A 50-mL filter flask, a 2-mL plastic syringe, separatory funnel, a 50-mL round-bottom flask, apparatus for magnetic stirring, simple distillation, vacuum filtration, and *flameless* heating.

Setting Up   Prepare a reaction vessel for hydrogenation by using rubber bands or wire to attach a balloon to the sidearm of the filter flask. Place a stirbar in the flask. Prepare a 1 *M* aqueous solution of sodium borohydride by dissolving 0.2 g of sodium borohydride in 5 mL of 1% aqueous sodium hydroxide. Place 5 mL of water, 0.5 mL of a 5% solution of chloroplatinic acid ($H_2PtCl_6 \cdot 6\ H_2O$) and 0.2 g of decolorizing carbon in the reaction flask and slowly add 1.5 mL of the 1 *M* sodium borohydride solution with stirring. Stir the resulting slurry for 5 min to allow the catalyst to form. During this time, dissolve 0.5 g of 4-cyclohexene-*cis*-1,2-dicarboxylic acid (Sec. 12.3, Part C1) in 5 mL of water by heating. Pour 2 mL of *concentrated* hydrochloric acid into the reaction flask containing the catalyst, and add the hot aqueous solution of the diacid to the flask. Seal the flask with a rubber septum that is securely wired in place, and stir the contents.

Reaction   Draw 0.8 mL of the 1 *M* sodium borohydride solution into the syringe, carefully push the needle of the syringe through the rubber septum, and inject the solution dropwise into the stirred solution of reactants in the filter flask. If the balloon on the sidearm of the flask begins to inflate *rapidly*, add the solution of sodium borohydride more slowly. When the syringe is empty, remove it from the flask, refill it with an additional 0.8 mL of 1 *M* sodium borohydride solution, and add this second portion of sodium borohydride to the stirred reaction mixture; the balloon should inflate, indicating a positive pressure of hydrogen in the system. Remove the syringe from the rubber septum, and continue stirring for 5 min. Then heat the flask at 70–80 °C for 1.5–2 h with continued stirring to complete the reaction.

Work-Up and Isolation   Release the pressure from the reaction flask by pushing a syringe needle through the rubber septum. *Do not allow the reaction mixture to cool.* Remove the catalyst by vacuum filtration of the *hot* reaction mixture using a Hirsch funnel, and *immediately* treat the catalyst according to the protocol given in the FINISHING TOUCHES section.★ Cool the filtrate and transfer it to the separatory funnel. Rinse the filter flask with 15 mL of diethyl ether to dissolve the solid product that may have separated upon cooling, and add this rinse to the separatory funnel. Based upon the amount of water used in the experiment, calculate the weight of sodium chloride that would be needed to saturate the filtrate and add this amount of solid sodium chloride to the separatory funnel. Shake the separatory funnel, venting as necessary to relieve pressure. Allow the layers to separate; any insoluble solid should be sodium chloride. Separate the layers, and extract the aqueous layer sequentially with two 10-mL portions of diethyl ether. Dry the *combined* ethereal extracts by swirling them over several spatula-tips full of anhydrous sodium sulfate.★ Swirl the flask occasionally for a period of 10–15 min to facilitate drying; add further small portions of anhydrous sodium sulfate if the solution remains cloudy.

Carefully decant the solution into a tared round-bottom flask containing a stirbar. Equip the flask for simple distillation and remove the diethyl ether by simple distillation. Alternatively, rotary evaporation or other techniques (Sec. 2.29) may be used to concentrate the solution. The final traces of solvent may be removed by attaching the flask to a vacuum source, gently swirling the contents as the vacuum is applied. The product remains as a solid residue in the flask.★ Air-dry the crude product and determine its weight and melting point.

Purification The crude product may be recrystallized from a *minimum* amount of water as follows. Carefully scrape the bulk of the crude diacid from the round-bottom flask into a *small* Erlenmeyer flask. Add *no more* than 1 mL of water to the round-bottom flask, heat to dissolve any residual diacid, and pour the hot solution into the Erlenmeyer flask. Heat the aqueous mixture to boiling; if necessary, *slowly* add water *dropwise* to the boiling mixture to dissolve all the diacid. Do *not* add more than an additional 1 mL of water, as the diacid is rather soluble in water. Determine the pH of the aqueous solution to ascertain whether hydrogenation has affected the acidic nature of the molecule. When all of the diacid has dissolved, add 2 drops of *concentrated* hydrochloric acid to decrease its solubility. Allow the mixture to cool to room temperature, then place the flask in an ice-water bath to effect complete crystallization of the product. Collect the product by vacuum filtration and air-dry it completely.★ The reported melting point of cyclohexane-*cis*-1,2-dicarboxylic acid is 192–193 °C.

Analysis Weigh the product and calculate the percent yield; determine its melting point. Perform the qualitative tests for unsaturation using the bromine in dichloromethane (Sec. 23.8A) and the Baeyer tests (Sec. 23.8B) on your product, and record your observations in your notebook. Obtain IR and ^{1}H NMR spectra of your starting material and product and compare them with those of authentic samples (Figs. 12.11, 12.12, 17.1, and 17.2).

| | Finishing Touches |

Wet the *catalyst* and the *filter paper* immediately with water, and remove them from the Hirsch funnel; place them in the container reserved for recovered catalyst, which should be kept wet with water at all times. Neutralize the *combined aqueous filtrates and washes* with sodium carbonate, and then flush them down the drain with excess water. Place the *recovered ethereal distillate* in the container for nonhalogenated organic solvents. After allowing the diethyl ether to evaporate from the *sodium sulfate* on a tray in the hood, place the used drying agent in the container for nonhazardous solids. Place the *dichloromethane solution* from the bromine test for unsaturation in a container for halogenated organic liquids; put the *manganese dioxide* from the Baeyer test for unsaturation in a container for heavy metals.

EXERCISES

1. Why is the sodium borohydride solution prepared in basic rather than acidic aqueous media?

2. Why does adding hydrochloric acid to an aqueous solution of a carboxylic acid decrease its solubility in water?

3. What is the purpose of swirling the ethereal extracts over sodium sulfate?

4. Determine the oxidation numbers (Sec. 16.1) of the carbon atoms undergoing change in the following reactions, and indicate the net gain or loss of electrons in each.

   a. $CH_3Cl \rightarrow CH_4$

   b. $CH_3CHO \rightarrow CH_3CH_2OH$

   c. acetylene $\rightarrow$ ethylene

   d. $CH_3CO_2H \rightarrow CH_3CHO$

   e. $CH_3CH{=}NH \rightarrow CH_3CH_2NH_2$

   f. $H_2C{=}CH_2 \rightarrow BrCH_2CH_2Br$

5. Draw the structure of the product of catalytic hydrogenation of

    a. 1,2-dimethylcyclohexene   **c.** *trans*-2,3-dideutero-2-butene

    b. *cis*-2,3-dideutero-2-butene   **d.** *cis*-3,4-dimethyl-3-hexene

Refer to Tables 8.1–8.8 as needed for answering the questions on spectroscopy.

6. Consider the spectral data for 4-cyclohexene-*cis*-1,2-dicarboxylic acid (Figs. 12.11 and 12.12).

    a. In the functional group region of the IR spectrum, specify the absorptions associated with the hydroxyl and the carbonyl components of the carboxyl group.

    b. In the ^{1}H NMR spectrum, assign the various resonances to the hydrogen nuclei responsible for them.

    c. For the ^{13}C NMR data, assign the various resonances to the carbon nuclei responsible for them.

7. Consider the spectral data for cyclohexane-*cis*-1,2-dicarboxylic acid (Figs. 17.1 and 17.2).

    a. In the functional group region of the IR spectrum, specify the absorptions associated with the hydroxyl and the carbonyl components of the carboxyl group.

    b. In the ^{1}H NMR spectrum, assign the various resonances to the hydrogen nuclei responsible for them.

    c. For the ^{13}C NMR data, assign the various resonances to the carbon nuclei responsible for them.

8. Discuss the differences observed in the IR and NMR spectra of 4-cyclohexene-*cis*-1,2-dicarboxylic acid and cyclohexane-*cis*-1,2-dicarboxylic acid that are consistent with the reduction of the carbon-carbon double bond in this experiment.

Spectra

Starting Material and Product

    The IR and NMR spectra of 4-cyclohexene-cis-1,2-dicarboxylic acid are given in Figures 12.1 and 12.2, respectively.

FIGURE 17.1

IR spectrum of cyclohexane-*cis*-1,2-dicarboxylic acid (IR card).

(a) ^{1}H NMR spectrum (300 MHz).

FIGURE 17.2
NMR data for cyclohexane-*cis*-1,2-
dicarboxylic acid (DMSO-d$_6$).

(b) ^{13}C NMR data: δ 23.8, 26.3, 42.3, 175.6.

17.3 Reduction of Imines; Preparation of Amines

Imines are readily formed by the condensation of primary amines with carbonyl compounds such as aldehydes and ketones (Sec. 23.7) according to Equation 17.9. The stability of an imine depends upon the nature of the substituents, R^1–R^3, on the nitrogen and carbon atoms of the imine functional group. For example, imines that bear only alkyl groups are less stable toward hydrolysis and polymerization than those having one or more aromatic substituents on the carbon-nitrogen double bond; the most stable imines are those having only aryl substituents.

$$\underset{\substack{\text{An aldehyde or}\\\text{a ketone}}}{\overset{R^1}{\underset{R^2}{>}}\!\!=\!\!O} \ + \ \underset{\text{A primary amine}}{R^3\!\!-\!\!NH_2} \ \longrightarrow \ \underset{\text{An imine}}{\overset{R^1}{\underset{R^2}{>}}\!\!=\!\!N\!\!-\!\!R^3} \ + \ H_2O \qquad (17.9)$$

Imines undergo addition of one mole of hydrogen in the presence of a catalyst such as palladium, platinum or nickel to produce secondary amines (Eq. 17.10). This process is analogous to the hydrogenation of alkenes to alkanes and of carbonyl compounds to alcohols.

$$\underset{\text{An imine}}{\overset{R^1}{\underset{R^2}{>}}\!\!=\!\!N\!\!-\!\!R^3} \ + \ H_2 \ \xrightarrow{\text{Catalyst}} \ \underset{\text{A secondary amine}}{\overset{R^1 \quad H}{\underset{R^2 \quad NR^3}{\diagdown\!\diagup}}\!\!-\!\!H} \qquad (17.10)$$

Other functional groups such as C=C, C≡C, N=O (as in NO$_2$), or C—X, undergo catalytic hydrogenation under the same conditions required to reduce imines.

This limitation is readily overcome by using a *selective* metal hydride, such as sodium borohydride, $NaBH_4$, to reduce the imine, as depicted in Equation 17.11. Sodium borohydride reacts with methanol, which is frequently used as the solvent, but it reacts much faster with the imine. Thus, when performing small-scale reactions, it is generally more convenient to use an excess of $NaBH_4$ to allow for its reaction with the solvent than to use a less-reactive solvent in which this reagent is less soluble.

$$\underset{\text{An imine}}{\overset{R^1}{\underset{R^2}{>}}=N-R^3} \xrightarrow[\text{(2) } H_2O]{\text{(1) } NaBH_4, CH_3OH} \underset{\text{A secondary amine}}{\overset{R^1 \quad H}{\underset{R^2}{>}}\underset{\underset{H}{|}}{C}\overset{}{N}-R^3} \qquad \textbf{(17.11)}$$

In this section, you will prepare *N*-cinnamyl-*m*-nitroaniline (**9**) by a sequence beginning with the condensation of cinnamaldehyde (**5**) with *m*-nitroaniline (**6**), followed by reduction of the intermediate imine **7** with sodium borohydride, as shown in Equations 17.12–17.14. The formation of the imine is reversible, but the reaction is driven to completion by **azeotropic distillation**. Because cyclohexane and water form a minimum-boiling azeotrope, the water generated by the condensation of **5** and **6** is continuously removed by distilling the cyclohexane-water azeotrope throughout the course of the reaction.

$$C_6H_5CH{=}CH{-}\underset{\underset{H}{|}}{C}{=}O + H_2N{-} \underset{\underset{NO_2}{}}{\bigcirc} \xrightleftharpoons{\text{Cyclohexane}} C_6H_5CH{=}CH{-}\underset{\underset{H}{|}}{C}{=}N{-}\underset{\underset{NO_2}{}}{\bigcirc} + H_2O \qquad \textbf{(17.12)}$$

$$\qquad\quad \textbf{5} \qquad\qquad\qquad \textbf{6} \qquad\qquad\qquad\qquad\qquad \textbf{7}$$
$$\text{Cinnamaldehyde} \qquad \textit{m}\text{-Nitroaniline} \qquad\qquad \textit{N}\text{-Cinnamylidene-}\textit{m}\text{-nitroaniline}$$

$$4\,C_6H_5CH{=}CH{-}\underset{\underset{H}{|}}{C}{=}N{-}\underset{\underset{NO_2}{}}{\bigcirc} + NaBH_4 \longrightarrow \left(C_6H_5CH{=}CH{-}\overset{\overset{H}{|}}{C}H{-}\underset{\underset{O_2N}{}}{N}{-}\underset{\underset{\quad}{}}{\bigcirc} \right)_4 B^- Na^+ \qquad \textbf{(17.13)}$$

$$\qquad\qquad\qquad\qquad\qquad \textbf{7} \qquad\qquad\qquad\qquad\qquad\qquad\qquad\qquad \textbf{8}$$

$$\left(C_6H_5CH{=}CH{-}CH_2{-}\underset{\underset{O_2N}{}}{N}{-}\underset{\underset{\quad}{}}{\bigcirc} \right)_4 B^- Na^+ + 3\,H_2O \longrightarrow 4\,C_6H_5CH{=}CH{-}CH_2{-}\underset{\underset{O_2N}{}}{NH}{-}\underset{\underset{\quad}{}}{\bigcirc} + NaH_2BO_3 \qquad \textbf{(17.14)}$$

$$\qquad\qquad\qquad \textbf{8} \qquad\qquad\qquad\qquad\qquad\qquad\qquad\qquad\qquad \textbf{9}$$
$$\qquad\qquad\qquad\qquad\qquad\qquad\qquad\qquad\qquad \textit{N}\text{-Cinnamyl-}\textit{m}\text{-nitroaniline}$$

The reaction of sodium borohydride with the imine **7** (Eq. 17.13) is analogous to the addition of sodium borohydride (Sec. 17.4) or a Grignard reagent (Secs. 19.1 and 19.3) to a carbonyl compound. A nucleophilic hydride ion (H:⁻) is transferred from the borohydride anion (BH_4^-) to the electrophilic carbon atom of the carbon-nitrogen double bond, and the electron-deficient boron atom becomes attached to nitrogen. All four hydrogens of the borohydride anion may be transferred to the imine carbon to produce the organoborate anion **8**, which is subsequently decomposed with water to yield the secondary amine **9** as shown in Equation 17.14. This experiment nicely illustrates the selectivity of sodium borohydride, which reduces the intermediate imine **7** but not the carbon-carbon double bond, the nitro group, or the benzene rings that are also present.

EXPERIMENTAL PROCEDURES

FORMATION AND REDUCTION OF *N*-CINNAMYLIDENE-*m*-NITROANILINE

Purpose: To demonstrate the reductive alkylation of an amine by the sequence of forming and reducing an imine.

> **Safety Alert**
>
> 1. Avoid contact of methanolic solutions of sodium borohydride with your skin; they are *highly caustic*. If possible, wear latex gloves when handling these solutions. If these solutions accidentally get on your skin, wash the affected area first with a 1% acetic acid solution and then with copious quantities of water.
> 2. Do *not* stopper flasks containing methanolic sodium borohydride; the solution slowly evolves hydrogen gas, and a dangerous buildup of pressure could occur in a stoppered flask.
> 3. Cyclohexane is flammable; use *flameless* heating in this experiment.

Miniscale Procedure

Preparation   Answer the Pre-Lab Exercises on page PL. 105. Review Sections 2.9, 2.11, 2.13, 2.17, and 2.22.

Apparatus   A 25-mL round-bottom flask, apparatus for heating under reflux, simple distillation, magnetic stirring, vacuum filtration, and *flameless* heating.

Setting Up   Place 0.6 g of *freshly distilled* cinnamaldehyde, 0.6 g of *m*-nitroaniline, 5 mL of cyclohexane, and a stirbar in the round-bottom flask. Assemble the apparatus for simple distillation.

Imine Formation   Heat the reaction flask with stirring and distil until most of the cyclohexane is removed; about 4 mL of distillate should be obtained in approximately 5–10 min. *Discontinue* heating when the distillation rate decreases. Add another 5 mL of cyclohexane and resume the distillation. Stop heating when the rate of distillation again decreases and remove the heat source.

Isolation and Characterization of the Imine (Optional)   Take about 0.2 mL of the residual liquid from the stillpot with a Pasteur pipet. Transfer this aliquot to a small test tube, and add 1 mL of methanol. Swirl the tube to effect solution and place the test tube in an ice-water bath. Collect any crystals that separate, air-dry them, and determine their melting point. The crude *N*-cinnamylidene-*m*-nitroaniline may be purified by recrystallization from a small volume of methanol. Pure *N*-cinnamylidene-*m*-nitroaniline has a melting point of 92–93 °C.

Reduction   Add 4 mL of methanol to the stillpot, stir to dissolve the crude imine, and set up the apparatus for heating under reflux. Prepare a solution of 0.15 g of sodium borohydride in 3 mL of methanol. Because sodium borohydride reacts slowly with methanol to evolve hydrogen gas, this solution should be prepared in an *unstoppered* vessel *immediately* before use. Using a Pasteur pipet or a syringe, transfer the methanolic solution of sodium borohydride dropwise through the top of the reflux condenser to the stirred solution of the imine; the addition should be completed within 1 min. Try to add the solution *directly* into the flask without touching the walls of the condenser. After all of the borohydride solution is added, heat the reaction mixture under reflux for 15 min.

Work-Up, Isolation, and Purification   Cool the reaction mixture to room temperature and add 10 mL of water.★ Stir the mixture for 10–15 min. Collect the crystals that separate and wash them with water.★ Recrystallize the crude *N*-cinnamyl-*m*-nitroaniline from 95% ethanol. Air-dry the product.

Analysis   Weigh the product and calculate the percent yield; determine its melting point. The melting point of pure *N*-cinnamyl-*m*-nitroaniline is 106–107 °C. Obtain IR and ^{1}H NMR spectra of your starting materials, the intermediate imine if isolated, and the product and compare them with those of authentic samples (Figs. 17.3–17.10).

Microscale Procedure

Preparation   Answer the Pre-Lab Exercises on page PL. 105. Review Sections 2.9, 2.11, 2.13, 2.17, and 2.22.

Apparatus   A 5-mL conical vial, 1-mL syringe, apparatus for heating under reflux, simple distillation, vacuum filtration, magnetic stirring, Craig tube filtration, and *flameless* heating.

Setting Up   Place 120 mg of *freshly distilled* cinnamaldehyde, 120 mg of *m*-nitroaniline, 1 mL of cyclohexane, and a spinvane in the conical vial. Assemble the apparatus for simple distillation.

Imine Formation   Heat the reaction vial with stirring and distil until most of the cyclohexane is removed; this should require about 3–5 min. *Discontinue* heating when the distillation rate decreases. Add another 1 mL of cyclohexane and resume the distillation; stop heating when the rate of distillation again decreases and remove the heat source.

Reduction   Add 0.8 mL of methanol to the conical vial, stir to dissolve the crude imine, and set up the apparatus for heating under reflux. Prepare a solution of 30 mg of sodium borohydride in 0.6 mL of methanol. Because sodium borohydride slowly reacts with methanol to evolve hydrogen gas, this solution should be prepared in an *unstoppered* vessel *immediately* before use. Using a Pasteur pipet or a syringe, transfer the methanolic solution of sodium borohydride dropwise through the top of the reflux condenser to the stirred solution of the imine; the addition should be completed within 1 min. Try to add the solution *directly* into the vial without touching the walls of the condenser. After all of the borohydride solution is added, heat the reaction mixture under reflux for 15 min.

Work-Up, Isolation, and Purification   Cool the reaction mixture to room temperature and add 2 mL of water.★ Stir the mixture for 10–15 min. Collect the crystals that separate and wash them with water.★ Recrystallize the crude *N*-cinnamyl-*m*-nitroaniline from 95% ethanol in a Craig tube. Air-dry the product.

Analysis   Weigh the product and calculate the percent yield; determine its melting point. The melting point of pure *N*-cinnamyl-*m*-nitroaniline is 106–107 °C. Obtain IR and ^{1}H NMR spectra of your starting materials and product and compare them with those of authentic samples (Figs. 17.3–17.6, 17.9, and 17.10).

Pour the *cyclohexane distillates* in the container for nonhalogenated organic solvents. Dilute the *aqueous methanolic filtrate* with water, neutralize it with acetic acid to destroy any excess sodium borohydride, and flush the mixture down the drain with excess water.

Finishing Touches

EXERCISES

1. What causes the turbidity in the distillate collected while heating the carbonyl compound and primary amine in cyclohexane solution?

2. Why was the use of *dry* cyclohexane not specified?

3. Determine the molar ratio of NaBH$_4$ to imine that you actually used in the experiment. Why is it necessary to use a greater molar ratio than theoretical?

4. After the reaction between sodium borohydride and the imine is complete, the reaction mixture is treated with water to produce the desired secondary amine. Explain this reaction by indicating the source of the hydrogen that ends up on the nitrogen atom.

5. Although sodium borohydride is fairly unreactive toward methanol, adding a mineral acid to this solution results in the rapid destruction of the NaBH$_4$. Explain.

6. Suppose that each of the following pairs of compounds were subjected to the reactions provided in the experimental procedures. Draw the structures of the intermediate imines that would be formed, as well as those of the final amine product.

   a. cinnamaldehyde and aniline

   b. benzaldehyde and *m*-nitroaniline

   c. benzaldehyde and aniline

   d. cyclohexanone and aniline

7. Propose a mechanism to show how imine **7** can be hydrolyzed by water at pH 6 to produce **5** and **6**. Symbolize the flow of electrons with curved arrows.

8. Explain why aliphatic imines are much less stable than those having at least one aromatic group attached to the carbon-nitrogen double bond.

Refer to Tables 8.1–8.8 as needed for answering the following questions on spectroscopy.

9. Consider the spectral data for cinnamaldehyde (Figs. 17.3 and 17.4).

   a. In the functional group region of the IR spectrum, specify the absorptions associated with the α, β-unsaturated aldehyde and with the aromatic ring.

   b. In the ^{1}H NMR spectrum, assign the various resonances to the hydrogen nuclei responsible for them.

   c. For the ^{13}C NMR data, assign the various resonances to the carbon nuclei responsible for them.

10. Consider the spectral data for *m*-nitroaniline (Figs. 17.5 and 17.6).

    a. In the functional group region of the IR spectrum, specify the absorptions associated with the nitro group and the aromatic ring. There is a sharp absorption at about 3400 cm^{-1}. What functional group is responsible for this absorption?

    b. In the ^{1}H NMR spectrum, assign the various resonances to the hydrogen nuclei responsible for them.

    c. For the ^{13}C NMR data, assign the various resonances to the carbon nuclei responsible for them.

11. Consider the spectral data for *N*-cinnamylidene-*m*-nitroaniline (Figs. 17.7 and 17.8).

    a. In the functional group region of the IR spectrum, specify the absorptions associated with the nitro group, the carbon-nitrogen double bond, and the carbon-carbon π-bonds of the aromatic ring and alkene.

    b. In the ^{1}H NMR spectrum, assign the various resonances to the hydrogen nuclei responsible for them.

    c. For the ^{13}C NMR data, assign the various resonances to the carbon nuclei responsible for them.

12. Consider the spectral data for *N*-cinnamyl-*m*-nitroaniline (Figs. 17.9 and 17.10).

    a. In the functional group region of the IR spectrum, specify the absorptions associated with the amino group, the nitro group, and the aromatic ring.

    b. In the ^{1}H NMR spectrum, assign the various resonances to the hydrogen nuclei responsible for them.

    c. For the ^{13}C NMR data, assign the various resonances to the carbon nuclei responsible for them.

13. Discuss the differences observed in the IR and NMR spectra of cinnamaldehyde and *N*-cinnamylidene-*m*-nitroaniline that are consistent with the formation of the carbon-nitrogen double bond of the imine in this experiment.

14. Discuss the differences observed in the IR and NMR spectra of *N*-cinnamylidene-*m*-nitroaniline and *N*-cinnamyl-*m*-nitroaniline that are consistent with the *selective* reduction of the carbon-nitrogen double bond in this experiment.

Starting Materials and Products

Spectra

FIGURE 17.3
IR spectrum of cinnamaldehyde (neat).

(a) ^{1}H NMR spectrum (300 MHz).

FIGURE 17.4
NMR data for cinnamaldehyde (CDCl$_3$).

(b) ^{13}C NMR data: δ 128.5, 129.0, 131.1, 134.1, 152.3, 193.2.

FIGURE 17.5
IR spectrum of *m*-nitroaniline (IR card).

FIGURE 17.6
NMR data for *m*-nitroaniline
(CDCl₃).

(a) ¹H NMR spectrum (300 MHz).

(b) ¹³C NMR data: δ 107.8, 110.5, 120.2, 129.6, 149.1, 149.8.

FIGURE 17.7
IR spectrum of *N*-cinnamylidene-*m*-nitroaniline (IR card).

FIGURE 17.8
NMR data for *N*-cinnamylidene-*m*-nitroaniline (CDCl₃).

(a) ¹H NMR spectrum (300 MHz).

(b) ¹³C NMR data: δ 114.8, 120.1, 127.4, 128.6, 129.5, 129.8, 129.9, 130.2, 134.8, 145.7, 148.6, 152.6, 163.4.

FIGURE 17.9
IR spectrum of *N*-cinnamyl-*m*-nitroaniline (IR card).

(a) ¹H NMR spectrum (300 MHz).

FIGURE 17.10
NMR data for *N*-cinnamyl-*m*-nitroaniline (CDCl₃).

(b) ¹³C NMR data: δ 45.7, 106.5, 111.9, 118.8, 125.4, 126.3, 127.7, 128.6, 129.6, 132.1, 136.4, 148.7, 149.3.

17.4 Reduction of Carbonyl Compounds; Preparation of Alcohols

Carbonyl compounds are commonly reduced to alcohols by catalytic hydrogenation or with metal hydrides. When applied to aldehydes, reduction, which is generally indicated by the symbol [H], provides a convenient route to primary alcohols (Eq. 17.15), whereas the reduction of ketones gives secondary alcohols (Eq. 17.16). Although catalytic hydrogenation of carbonyl groups is frequently the method of choice in industrial processes, lithium aluminum hydride, sodium borohydride, and their derivatives are generally used in the research laboratory. Sodium borohydride may be used in alcoholic and even aqueous solutions, because it reacts much more rapidly with the carbonyl group than with the solvent. On the other hand, lithium aluminum hydride reacts rapidly with protic solvents, and it *must* be used in *anhydrous* ethereal solvents such as diethyl ether or tetrahydrofuran.

$$R-\overset{\overset{\displaystyle O}{\parallel}}{C}-H \xrightarrow{\text{[H]}} R-CH_2OH \qquad (17.15)$$

An aldehyde   A primary alcohol

R = alkyl or aryl

$$R^1-\overset{\overset{\displaystyle O}{\parallel}}{C}-R^2 \xrightarrow{\text{[H]}} R^1-\underset{\underset{\displaystyle R^2}{\mid}}{C}HOH \qquad (17.16)$$

A ketone   A secondary alcohol

R^1 and R^2 = alkyl or aryl

In this experiment, you will examine the reduction of fluorenone (**10**) by sodium borohydride to give fluorenol (**12**), as shown in Equation 17.17. This reaction is mechanistically analogous to the reduction of imines with sodium borohydride (Sec. 17.3) and involves the transfer of hydride ion (H:$^-$) from BH_4^- to the electrophilic carbonyl carbon with concomitant transfer of the electron-deficient boron atom to the carbonyl oxygen. Theoretically, all four of the hydrogen atoms attached to boron may be transferred in this way to produce the intermediate borate salt **11**, which is decomposed upon addition of water and acid to yield fluorenol (**12**).

10	**11**	**12**
Fluorenone		Fluorenol

$$(17.17)$$

EXPERIMENTAL PROCEDURES

REDUCTION OF FLUORENONE

Purpose: To demonstrate the reduction of a ketone to an alcohol using sodium borohydride.

Safety Alert

1. Avoid contact of methanolic solutions of sodium borohydride with your skin; they are *highly caustic*. If possible, wear latex gloves when handling these solutions. If these solutions accidentally get on your skin, wash the affected area first with a 1% acetic acid solution and then with copious quantities of water.
2. Do *not* stopper flasks containing methanolic sodium borohydride; the solution slowly evolves hydrogen gas, and a dangerous buildup of pressure could occur in a stoppered flask.

Miniscale Procedure

Preparation   Answer the Pre-Lab Exercises on page PL. 107. Review Sections 2.9 and 2.17.

Apparatus   A 25-mL Erlenmeyer flask, apparatus for vacuum filtration, and *flameless* heating.

Setting Up   Add 0.6 g of fluorenone to the Erlenmeyer flask containing 6 mL of methanol, and swirl the flask with slight warming to dissolve the ketone. Allow the solution to cool to room temperature. *Quickly* weigh 0.05 g of sodium borohydride into a dry test tube and stopper the test tube *immediately* to avoid undue exposure of the hygroscopic reagent to atmospheric moisture.

Reduction   Add the sodium borohydride in *one* portion to the solution of fluorenone in methanol, and swirl the mixture vigorously to dissolve the reagent. After all of the sodium borohydride dissolves, allow the solution to stand at room temperature for 20 min; swirl the solution occasionally. If the solution does not become colorless during this time, add an additional small portion of sodium borohydride with swirling to complete the reaction.★

Work-Up   Add 2 mL of 3 *M* sulfuric acid to the reaction mixture. Heat the contents in the flask gently and intermittently for 5–10 min. Stir the mixture occasionally with a glass rod to help dissolve the solid; maintain the internal temperature just below the reflux point to *minimize the loss of solvent*. If all of the solids do not dissolve, *gradually* add methanol in about 0.5- to 1-mL portions with continued heating until a solution is obtained. When all of the precipitated solids redissolve, allow the solution to cool to room temperature, and then in an ice-water bath for 10–15 min.

Isolation and Purification   Collect the solid product by vacuum filtration, *wash it thoroughly* with water until the filtrate is *neutral*, and air-dry the product.★ Recrystallize the fluorenol by a mixed solvent recrystallization (Sec. 3.2) using methanol and water. After cooling the solution first to room temperature and then in an ice-water bath for 10–15 min, collect and air-dry the crystals.

Analysis   Weigh the product and calculate the percent yield; determine its melting point. Obtain IR and ^{1}H NMR spectra of your starting material and product and compare them with those of authentic samples (Figs. 17.11–17.14).

Microscale Procedure

Preparation   Answer the Pre-Lab Exercises on page PL. 107. Review Sections 2.9 and 2.17.

Apparatus   A 5-mL conical vial, apparatus for magnetic stirring, heating under reflux, vacuum filtration, Craig tube filtration, and *flameless* heating.

Setting Up   Add 100 mg of fluorenone to the conical vial containing a spinvane and 1 mL of methanol, and stir the contents of the vial with slight warming to dissolve the ketone. Allow the solution to cool to room temperature. *Quickly* weigh 10 mg of sodium borohydride into a dry test tube and stopper the test tube *immediately* to avoid undue exposure of the hygroscopic reagent to atmospheric moisture.

Reduction   Add the sodium borohydride in *one* portion to the stirred solution of fluorenone in methanol, and then stir the mixture rapidly to dissolve the reagent. After all of the sodium borohydride dissolves, continue stirring at room temperature for 20 min. If the solution does not become colorless during this time, add an additional small portion of sodium borohydride to complete the reaction.★

Work-Up   Add 0.35 mL of 3 *M* sulfuric acid to the reaction mixture. Set up the apparatus for heating under reflux, and heat the vial gently with stirring for 5–10 min to dissolve the solid; adjust the internal temperature to maintain a *gentle* reflux. If all of the solids do not dissolve, *gradually* add methanol *dropwise* to the vial with continued heating and stirring until a solution is obtained. When all of the precipitated solids redissolve, allow the solution to cool to room temperature, and then in an ice-water bath for 10–15 min.

Isolation and Purification   Collect the solid product by vacuum filtration, *wash it thoroughly* with water until the filtrate is *neutral*, and air-dry the product.★ Recrystallize the fluorenol by a mixed solvent recrystallization (Sec. 3.2) using methanol and water in a Craig tube. After cooling the solution first to room temperature and then in an ice-water bath for 10–15 min, collect and air-dry the crystals.

Analysis   Weigh the product and calculate the percent yield; determine its melting point. Obtain IR and ^{1}H NMR spectra of your starting material and product and compare them with those of authentic samples (Figs. 17.11–17.14).

Finishing Touches
Dilute the combined *aqueous methanol filtrates* with water, neutralize the resulting solution with sodium carbonate, and flush the mixture down the drain with excess water.

EXERCISES

1. Determine the molar ratio of NaBH$_4$ to fluorenone that you used in the experiment. Why is it necessary to use a greater molar ratio than theoretical?

2. After the reaction between sodium borohydride and the ketone is complete, the reaction mixture is treated with water and acid to produce the desired secondary alcohol. Explain this reaction by indicating the source of the hydrogen that ends up on the oxygen atom.

3. Sodium borohydride is fairly unreactive toward methanol, but adding a mineral acid to this solution results in the rapid destruction of the NaBH$_4$. Explain.

4. Suggest a structure for the white precipitate formed in the reaction of fluorenone with sodium borohydride.

5. What gas is evolved when sulfuric acid is added to the reaction mixture?

6. Draw the structure of the product that results from complete reduction of the following compounds by NaBH$_4$.

    a. cyclohexanone   **b.** 3-cyclohexen-1-one   **c.** 1,4-butanedial

    d. 4-oxohexanal   **e.** acetophenone

7. Draw the structure of the product that would be formed from allowing each of the compounds in Exercise 6 to react with excess hydrogen gas in the presence of a nickel catalyst.

Refer to Tables 8.1–8.8 as needed for answering the following questions on spectroscopy.

8. Consider the spectral data for fluorenone (Figs. 17.11 and 17.12).

   a. In the functional group region of the IR spectrum, specify the absorptions associated with the carbonyl group and with the aromatic rings.

   b. In the ^{1}H NMR spectrum, assign the various resonances to the hydrogen nuclei responsible for them.

   c. For the ^{13}C NMR data, assign the various resonances to the carbon nuclei responsible for them.

9. Consider the spectral data for fluorenol (Figs. 17.13 and 17.14).

   a. In the functional group region of the IR spectrum, specify the functional group that is associated with the broad absorption at about 3100–3250 cm^{-1}; why is this absorption broad?

   b. In the ^{1}H NMR spectrum, assign the various resonances to the hydrogen nuclei responsible for them.

   c. For the ^{13}C NMR data, assign the various resonances to the carbon nuclei responsible for them.

10. Discuss the differences observed in the IR and NMR spectra of fluorenone and fluorenol that are consistent with the reduction of the ketone functional group in this experiment.

Starting Material and Product

Spectra

FIGURE 17.11
IR spectrum of fluorenone (IR card).

FIGURE 17.12
NMR data for fluorenone (CDCl$_3$).

(a) ^{1}H NMR spectrum (300 MHz).

(b) ^{13}C NMR data: δ 120.1, 123.8, 128.8, 133.9, 134.4, 144.1, 193.1.

FIGURE 17.13
IR spectrum of fluorenol (IR card).

WAVENUMBER IN CM^{-1}

FIGURE 17.14
NMR data for fluorenol (CDCl$_3$).

(a) ^{1}H NMR spectrum (300 MHz).

(b) ^{13}C NMR data: δ 73.8, 119.6, 125.0, 127.2, 128.2, 139.5, 146.8.

17.5 Enzymatic Reduction: A Chiral Alcohol from a Ketone

In modern synthetic organic chemistry, it is often necessary to use optically pure starting materials to prepare a target molecule, such as a drug, in enantiomerically pure form. However, reduction, [H], of a **prochiral** ketone **13**, which bears two *different* alkyl or aryl residues on the carbonyl carbon atom, by catalytic hydrogenation or with hydride reducing agents such as sodium borohydride or lithium aluminum hydride gives equal amounts of the chiral alcohol **14** and its enantiomer *ent*-**14** (Eq. 17.18). This **racemate** is produced because these *achiral* reducing reagents attack *both* faces *a* and *b* of the planar prochiral carbonyl function with *equal* probability.

$$(17.18)$$

However, if the reducing agent is chiral, it is possible to obtain the resulting chiral alcohol as an optically pure substance. For example, the reduction of prochiral ketones by catalytic hydrogenation in the presence of chiral catalysts or by the use of chiral hydride reducing agents may produce secondary alcohols with high levels of **optical purity**. There are also certain enzymes, which are polypeptides composed of L-amino acids and thus have chiral active sites, that can perform such **enantioselective** reductions.

When a chiral product is formed from an achiral starting material, the **optical yield** or optical purity, not just the chemical yield, of the product is important. Optical purity is commonly evaluated by calculating the **enantiomeric excess (ee)**. For example, reduction of a prochiral ketone **13** with sodium borohydride produces 50% (*R*)- and 50% (*S*)-alcohol **14** and *ent*-**14**; the ee for this process is 0%. If 90% of the (*S*)- and 10% of the (*R*)-isomer were produced by the reaction, the ee would be 80%.

In this experiment, methyl acetoacetate (**15**) is reduced to methyl (*S*)-(+)-3-hydroxybutanoate (**16**), using one of the reducing enzymes found in baker's yeast (Eq. 17.19). The enzyme is one of many that are involved in the metabolism of D-glucose to ethanol. Enantiomeric excesses ranging from 70–97% have been reported for this reaction, with higher optical purities being reported when oxygen is excluded during fermentation. Such fermentations are termed **anaerobic**. The hydroxy ester **16** is a building block that has been used as a starting material in a number of syntheses of optically pure natural products.

$$(17.19)$$

15
Methyl acetoacetate

16
Methyl (*S*)-(+)-3-hydroxybutanoate

EXPERIMENTAL PROCEDURES

ENZYMATIC REDUCTION OF METHYL ACETOACETATE

Purpose: To demonstrate the asymmetric reduction of a ketone.

Miniscale Procedure

Preparation   Answer the Pre-Lab Exercises on page PL. 109. Review Sections 2.11, 2.17, 2.21, and 2.29.

Apparatus   A 250-mL Erlenmeyer flask, separatory funnel, apparatus for magnetic stirring, anaerobic fermentation, vacuum filtration, simple distillation, and *flameless* heating.

Setting Up   Dissolve 40 g of sucrose and 0.25 g of disodium hydrogen phosphate, which buffers the mixture to maintain an optimal pH, in 75 mL of warm (40 °C) tap water contained in Erlenmeyer flask. Add a stirbar and one packet (8 g) of dry, *active* baker's yeast, and assemble the apparatus for anaerobic fermentation shown in Figure 17.15. The 3% barium hydroxide solution should be protected from atmospheric carbon dioxide by a layer of mineral oil. A precipitate of barium carbonate will then form as carbon dioxide is produced during the course of the fermentation. Stir the mixture *vigorously* for about 1 h in a warm location to suspend all of the yeast and to initiate the fermentation.

Reduction   Remove the stopper in the Erlenmeyer flask, and add 2.5 mL of methyl acetoacetate. Replace the stopper and continue to stir the contents of the flask *vigorously*, under anaerobic conditions, and in a warm place, ideally at 30–35 °C, for at least 48 h although one week is better.★

Work-Up and Isolation   Prepare a uniform bed of a filter-aid by filtering a slurry made from 10 g of filter-aid in 50 mL of water through a 10-cm Büchner funnel containing a piece of filter paper; do not allow the bed to dry completely. Discard the filtrate according to the procedure described in FINISHING TOUCHES. Add another 10 g of filter-aid to the reaction mixture containing the yeast cells, and swirl the mixture thoroughly to mix the filter-aid with the cells. Allow the solids to settle for a few minutes. Then *carefully* decant the clear supernatant fluid through the bed of filter-

One-hole rubber stopper

Glass tubing

500-mL Erlenmeyer flask

Stirbar

Mineral oil

Ba(OH)$_2$

FIGURE 17.15   Apparatus for the anaerobic fermentation of methyl acetoacetate.

aid while applying *gentle* suction to remove the suspended yeast cells. After filtering the supernatant liquid, pour the residual solids and liquid through the bed of filter-aid. Because the tiny yeast particles tend to clog the filter-bed, it will be necessary to increase the vacuum during the filtration.

Wash the filter-aid with 25 mL of water and saturate the combined filtrates with solid sodium chloride. Extract the resulting solution three times with 25-mL portions of dichloromethane by *gently* shaking the separatory funnel to mix the layers. Do *not* shake the funnel too vigorously, because a bad emulsion will form. If this occurs, pass the contents of the separatory funnel through a *fresh* bed of filter-aid by vacuum filtration; this will remove most of the fine particles that cause the emulsion. Wash the bed with 15 mL of dichloromethane. Transfer this filtrate to the separatory funnel and continue the extractions; any fine emulsion that remains may be included with the aqueous phase. Dry the combined dichloromethane layers with several spatula-tips full of anhydrous sodium sulfate.★ Swirl the flask occasionally for a period of 10–15 min to facilitate drying; add further small portions of anhydrous sodium sulfate if the solution remains cloudy.

Decant the dry dichloromethane solution into a tared distilling flask, and remove most of the dichloromethane by simple distillation. Alternatively, rotary evaporation or other techniques (Sec. 2.29) may be used to concentrate the solution.★ Cool the stillpot to room temperature and remove the last traces of solvent by attaching the flask to a vacuum source, gently swirling the contents as the vacuum is applied.★

Analysis   Weigh the product and determine the percent yield of crude methyl (*S*)-(+)-3-hydroxybutanoate. Use the ferric ion test (Sec. 23.12C) to ascertain whether starting ketoester remains; record the result in your notebook. Also analyze the product by TLC using dichloromethane as the eluting solvent. If starting material is present, the optical yield of methyl (*S*)-(+)-3-hydroxybutanoate as measured by specific rotation will be lower. Obtain IR and ^{1}H NMR spectra of your starting material and product and compare them with those of authentic samples (Figs. 17.16–17.19).

Finishing Touches

Flush the *filtrate* from preparing the filter-aid bed down the drain. Dilute the *aqueous layer* with water and flush it down the drain. Dry the *filter-aid and yeast cells* and place them in the container for nonhazardous solids. After the dichloromethane evaporates from the *sodium sulfate* in the hood, place the solid in the container for nonhazardous solids. Place the *recovered dichloromethane* that was used as the extraction solvent and as the eluting solvent for the TLC in the container for halogenated organic solvents.

EXERCISES

1. Why is it necessary to saturate the aqueous mixture with sodium chloride before extracting the product with dichloromethane?

2. Based upon the stereochemistry of the reduction of **15**, predict the structure of the alcohols obtained by reducing the ketones **a**–**c** with baker's yeast.

a. (structure: $CH_3COCH_2CO_2C_2H_5$)

b. (structure: ethyl group–CO–phenyl ketone)

c. (structure: CH_3COCH_2–O–P(OCH$_3$)$_2$ with P=O)

Refer to Tables 8.1–8.8 as needed for answering the questions on spectroscopy.

3. Consider the spectral data for methyl acetoacetate (Figs. 17.16 and 17.17).

   a. In the functional group region of the IR spectrum, specify the absorptions associated with the two carbonyl groups.

   b. In the ^{1}H NMR spectrum, assign the various resonances to the hydrogen nuclei responsible for them.

   c. For the ^{13}C NMR data, assign the various resonances to the carbon nuclei responsible for them.

4. Consider the spectral data for methyl (S)-($+$)-3-hydroxybutanoate (Figs. 17.18 and 17.19).

   a. In the functional group region of the IR spectrum, specify the absorptions associated with the hydroxyl and ester groups. Why is the absorption at about 3200–3600 cm^{-1} broad?

   b. In the ^{1}H NMR spectrum, assign the various resonances to the hydrogen nuclei responsible for them.

   c. For the ^{13}C NMR data, assign the various resonances to the carbon nuclei responsible for them.

5. Discuss the differences observed in the IR and NMR spectra of methyl acetoacetate and methyl (S)-($+$)-3-hydroxybutanoate that are consistent with the reduction of the carbonyl group in this experiment.

Spectra   *Starting Material and Product*

FIGURE 17.16
IR spectrum of methyl acetoacetate (neat).

(a) ¹H NMR spectrum (300 MHz).

(b) ¹³C NMR data: δ 29.9, 49.8, 52.1, 168.0, 200.8.

FIGURE 17.17
NMR data for methyl acetoacetate (CDCl₃).

FIGURE 17.18
IR spectrum of methyl (*S*)-(+)-3-hydroxybutanoate (neat).

(a) ¹H NMR spectrum (300 MHz).

(b) ¹³C NMR data: δ 22.6, 42.8, 51.7, 64.3, 173.2.

FIGURE 17.19
NMR data for methyl (*S*)-(+)-3-hydroxybutanoate (CDCl₃).

17.6   Determining Optical Purity

There are several ways to determine the optical purity of a compound obtained as the product of an asymmetric synthesis. The simplest involves measuring the optical rotation using a polarimeter (Sec. 7.5). The reported specific rotation, $[\alpha]_D^{25}$, of pure methyl (S)-(+)-3-hydroxybutanoate is +38.5° (c = 1.80, chloroform), but the specific rotation of the product you obtain might range from +27° to +36°. For example, a measured specific rotation of 32.7° (c = 1.3, chloroform) for the product of this enzymatic reduction would correspond to an ee of 85%.

When only small quantities of a material are available, the use of ¹H NMR and a chiral shift reagent is a more accurate method for determining optical purity. In an *achiral* environment, the ¹H NMR spectra of methyl (S)-(+)-3-hydroxybutanoate and methyl (R)-(−)-3-hydroxybutanoate are *identical*, and each corresponds to that shown in Figure 17.19a. However, a Lewis acid having chiral ligands can form *diastereomeric complexes* with the two enantiomers, and these complexes often give different ¹H NMR spectra. In the present case, the Lewis acid is the chiral shift reagent *tris*[3-(heptafluoropropylhydroxymethylene)-(+)-camphorato]europium (III) or Eu(hfc)₃ (**17**); the europium atom coordinates with the hydroxyl group in methyl 3-hydroxybutanoate and its camphor-derived ligand to form the diastereomeric complexes, each of which gives a different ¹H NMR spectrum.

17

An important property of europium shift reagents is that they spread the peaks in the ¹H NMR spectrum over a wider range, generally moving all of the peaks downfield. The magnitude of the shift of each peak depends on the concentration of the shift reagent and the proximity to the europium ion of the protons corresponding to that peak. The complexes that are formed between **17** and methyl (S)-(+)-3-hydroxybutanoate and methyl (R)-(+)-3-hydroxybutanoate are *diastereomeric*, so the peaks for the protons on one enantiomer move downfield to different extents than the corresponding peaks for the other. Thus, the two diastereomeric complexes may be easily distinguished in the ¹H NMR spectrum, as is shown in Figure 17.20. It is then possible to integrate the peak areas corresponding to each of the enantiomers and determine the enantiomeric excess. Determining the ee using this NMR method gives results that are accurate to within approximately 2–3%.

In this experiment, you will determine the enantiomeric excess of the product that you obtained upon asymmetric reduction of methyl acetoacetate with baker's yeast using the chiral europium chiral shift reagent **17**.

EXPERIMENTAL PROCEDURE

DETERMINING OPTICAL PURITY OF
METHYL (*S*)-(+)-3-HYDROXYBUTANOATE

Purpose: To determine the optical purity of methyl (*S*)-(+)-3-hydroxybutanoate by ^{1}H NMR spectroscopy.

Procedure

Preparation   Answer the Pre-Lab Exercises on page PL. 111.

Apparatus   Two NMR tubes.

Preparing the Samples   Dissolve 25–30 mg of racemic methyl 3-hydroxybutanoate in 0.75 mL of deuterochloroform and transfer the solution to an NMR tube using a Pasteur pipet. Dissolve 25–30 mg of enantiomerically enriched methyl (*S*)-(+)-3-hydroxybutanoate, prepared by the enzymatic reduction in the first part of this experiment, in 0.75 mL of deuterochloroform and transfer the solution to another NMR tube, using a Pasteur pipet.

Measurements   Measure the ^{1}H NMR spectrum of the racemic methyl 3-hydroxybutanoate. Add 20 mg of *tris*[3-(heptafluoropropylhydroxymethylene)-(+)-camphorato]europium shift reagent to this solution, shake the mixture to effect dissolution, and allow the solution to stand for about 20 min. Again measure the ^{1}H NMR spectrum. The peaks of primary interest for determining the optical purity are the methoxy hydrogens of the ester. Continue the procedure of adding shift reagent and measuring the spectrum until these peaks are separated and accurate integrals may be obtained as shown in Figure 17.20; no more than about 80 mg of shift reagent should be required.

Now perform the same series of measurements on the sample of enantiomerically enriched methyl (*S*)-(+)-3-hydroxybutanoate. A typical spectrum is shown in Figure 17.21. Determine the ratio of the two enantiomers in both of the samples by comparing the heights of the singlets that correspond to the methoxy protons of the ester, and calculate the percentages of each enantiomer in the two samples based upon this ratio. Calculate the enantiomeric excess (ee) of methyl (*S*)-(+)-3-hydroxybutanoate.

Place the *solution of deuterochloroform and shift reagent* in the container for halogenated organic liquids.

Finishing Touches

EXERCISES

1. The ^{1}H NMR spectrum in Figure 17.21 is of a mixture of the chiral shift reagent **17** and a sample of methyl (*S*)-(+)-3-hydroxybutanoate that was produced by the enzymatic reduction of methyl acetoacetate using baker's yeast. Determine the enantiomeric excess of the methyl (*S*)-(+)-3-hydroxybutanoate contained in this sample.

2. Why is it necessary to obtain a ^{1}H NMR spectrum of *racemic* methyl 3-hydroxybutanoate in the presence of **17**?

3. Which of the following compounds will form diastereomeric complexes with the chiral europium shift reagent **17**?

a. CH_3—C(=O)—CH(CH_3)—$CO_2C_2H_5$

b. OH

c. OH

d. OH

e. NHMe

f.

Spectra   *Starting Materials & Products*

FIGURE 17.20
^{1}H NMR spectrum (300 MHz, CDCl_3) of racemic methyl 3-hydroxybutanoate in the presence of chiral shift reagent **17**.

FIGURE 17.21
^{1}H NMR spectrum (300 MHz, CDCl_3) of enriched methyl (S)-(+)-3-hydroxybutanoate in the presence of chiral shift reagent **17**.

Reactions of Carbonyl Compounds

The **carbonyl group**, **1**, is a rich source of many important reactions in organic chemistry, with two fundamental properties of this functionality being primarily responsible for its diverse chemistry. The first is the polarization of the carbon-oxygen π-bond, owing to the relatively high electronegativity of the oxygen atom. In terms of resonance theory, this polarization arises from the contribution of the dipolar resonance structure **1b** to the resonance hybrid for this functional group.

1a   **1b**

Consequently, the carbonyl group undergoes a variety of reactions in which the electrophilic carbonyl carbon atom is attacked by **nucleophiles**, which are Lewis bases; the oxygen atom, in turn, reacts with **electrophiles** or **Lewis acids**. The net effect is *addition* of a reagent Nu–E across the π-bond of the carbonyl function. The sequence by which the elements of Nu–E add to the carbonyl group varies according to the particular reagents and reaction conditions used, but in a general sense follow the pathway of Equation 18.1 in basic media and that of Equation 18.2 in acidic media.

$$\text{(18.1)}$$

$$ \text{(18.2)} $$

The second property of a carbonyl moiety is to increase the acidity of the **α-hydrogen atoms**, which are the hydrogens on any carbon atoms attached to the carbonyl group. The enhanced acidity of these α-hydrogens means that the **α-carbon atoms** can become **nucleophilic**. This nucleophilicity arises in one of two general ways: through deprotonation to form an **enolate ion** (Eq. 18.3), or by keto–enol equilibration, called **tautomerization**, to give an **enol** (Eq. 18.4). As shown in Equation 18.3, an enolate, which is the resonance hybrid of the two contributing resonance structures **2a** and **2b**, can react with electrophiles, E^+, at an α-carbon atom to give net substitution of the electrophile for an α-hydrogen atom. A similar result attends the reaction of an enol **3** with an electrophile, E^+ (Eq. 18.4).

$$ \text{(18.3)} $$

2a   **2b**

An enolate

$$ \text{(18.4)} $$

3

An enol

The experiments described in this chapter illustrate some of the representative reactions of carbonyl compounds according to the principles outlined in Equations 18.1–18.4. For example, you may study reactions in which the electrophilic carbonyl carbon atom of an aldehyde undergoes attack by a carbon nucleophile that is stabilized by an electron-withdrawing group such as a phosphorus-containing substituent, as found in the **Wittig reaction** and the **Wadsworth-Emmons modification** (Sec. 18.2), or a carbonyl function, as found in the **aldol condensation** (Sec. 18.3). The **1,4-** or **conjugate addition** of the enol form of an aldehyde to an α,β-unsaturated ketone to give a dicarbonyl compound that cyclizes by an *intramolecular* aldol reaction is explored in Section 18.4.

18.2 The Wittig and Related Reactions

As noted in the preceding section, the ability of the carbonyl group to increase the acidity of the α-hydrogens makes it possible to generate enolate ions, carbon nucleophiles that are stabilized by resonance (Eq. 18.3). Certain phosphorus-containing groups are also known to increase the acidity of hydrogens bound to the same carbon atom, and deprotonation of such compounds gives rise to another type of carbon nucleophile. For example, the hydrogens on the carbon atom attached to the phosphorus atom of the phosphonium salt **4** exhibit enhanced acidity, so treatment of **4** with a strong base produces the nucleophilic species **5**, which is called a phosphorus **ylide** because of the dipolar character reflected in resonance structure **5a** (Eq. 18.5). This ylide is further stabilized by contribution of the resonance structure **5b**.

$$(C_6H_5)_3\overset{+}{P}\!-\!\overset{H}{\underset{|}{\underset{C}{C}}}HC_6H_5 \; Cl^- \longrightarrow \left[(C_6H_5)_3\overset{+}{P}\!-\!\overset{-}{C}HC_6H_5 \longleftrightarrow (C_6H_5)_3P\!=\!CHC_6H_5\right] \qquad (18.5)$$

| **4** | **5a** | **5b** |

Benzyltriphenylphosphonium
chloride

Benzylidenetriphenylphosphorane

Phosphorus ylides like **5** are frequently referred to as **Wittig reagents**, in recognition of their inventor, Georg Wittig, who showed that they react with aldehydes and ketones to produce alkenes. (See the Historical Highlight at the end of this chapter.) Wittig won the Nobel Prize in Chemistry in 1979 for this discovery, which is illustrated in Scheme 18.1 for the synthesis of (Z)-stilbene (**9**) and (E)-stilbene (**10**) from the reaction of the ylide **5** with benzaldehyde (**6**). The reaction commences with the nucleophilic addition of the carbon atom of the ylide **5** to the electrophilic carbonyl carbon of **6** to give the zwitterion **7** as a mixture of two diastereoisomers. The zwitterion **7** then collapses to the oxaphosphetane **8**, which then undergoes fragmentation to give a mixture of the isomeric alkenes **9** and **10**, together with triphenylphosphine oxide (**11**).

SCHEME 18.1

5a   6   7

Benzaldehyde

8

An oxaphosphetane

9   **10**   **11**

(Z)-Stilbene   (E)-Stilbene   Triphenylphosphine oxide

The net effect of the Wittig reaction is conversion of a carbon-oxygen double bond to a carbon-carbon double bond. This transformation represents a general method for preparing alkenes that has two important advantages over other methods for preparing this functional group: (1) the carbonyl group is replaced *specifically* by a carbon-carbon double bond, without forming isomeric alkenes having the π-bond at other positions; (2) the reactions are carried out under mild conditions.

The Wittig reaction typically involves highly reactive phosphoranes and is therefore not normally suited for the introductory laboratory. For example, the very strong bases, such as *n*-butyllithium, that are normally used to produce the intermediate ylides, require special handling techniques, and it is generally necessary to perform all experimental operations under a dry, inert atmosphere. However, it has recently been shown that the Wittig reaction of **5** with aromatic aldehydes such as benzaldehyde can be performed under **phase-transfer** conditions, in which the phosphonium salt serves as the phase-transfer catalyst (Sec. 10.5). This simple expedient will be used in one of the experimental procedures at the end of this section to prepare a mixture of (*Z*)-stilbene (**9**) and (*E*)-stilbene (**10**). You will be able to determine the ratio of **9** and **10** by examining an ^{1}H NMR spectrum of the mixture. It is also possible to isomerize (*Z*)-stilbene to (*E*)-stilbene by irradiation of a solution of (*Z*)-stilbene containing a small amount of molecular iodine (Eq. 18.6).

$$\text{(18.6)}$$

9	10
(*Z*)-Stilbene	(*E*)-Stilbene

A variant of the Wittig reaction is the **Wadsworth-Emmons modification** and involves using a **phosphonate ester** in place of a phosphonium salt. Like the phase-transfer Wittig reaction just discussed, these reactions may be easily performed in the undergraduate laboratory. In one of the procedures that follow, the phosphonate ester **12** is deprotonated with potassium *tert*-butoxide in the polar, *aprotic* solvent *N,N*-dimethylformamide, $(CH_3)_2NCHO$ (DMF), to provide the nucleophilic phosphonate anion **13**, which is stabilized by resonance (Eq. 18.7).

$$\text{(18.7)}$$

12	*tert*-Butoxide	**13a** **13b**
Diethyl benzylphosphonate		Potassium diethyl benzylphosphate

As shown in Scheme 18.2, the anion **13** then reacts with benzaldehyde to give (*E*)-stilbene (**10**) by a sequence of reactions similar to that observed for the reaction of the ylide **5** with **6** in Scheme 18.1. The intermediate adduct **14** undergoes cyclization to form **15**, which then fragments to give **10** and potassium diethyl phosphate (**16**). Because **16** and the solvent DMF are both water-soluble, the reaction may be easily worked up by adding water to the mixture and isolating the product by vacuum filtration.

SCHEME 18.2

13a   **6**   **14**

15

10   **16**
Potassium diethyl phosphate

EXPERIMENTAL PROCEDURES

WITTIG AND WADSWORTH-EMMONS REACTIONS

A · SYNTHESIS OF (Z)- AND (E)-STILBENES BY A WITTIG REACTION

Purpose: To demonstrate the synthesis of a mixture of alkenes by a phase-transfer Wittig reaction and the geometric isomerization of an alkene.

> **Safety Alert**
>
> The 50% (by weight) aqueous sodium hydroxide solution is *highly* caustic. *Do not allow it to come in contact with your skin.* If this should happen, wash the affected area with dilute acetic acid and then copious amounts of water. Wear latex gloves when transferring the solution.

Miniscale Procedure

Preparation   Answer the Pre-Lab Exercises on page PL. 113. Review Sections 2.9, 2.11, 2.13, 2.17, 2.21, 2.22, and 2.29.

Apparatus   A 25- and a 50-mL round-bottom flask, separatory funnel, apparatus for heating under reflux, magnetic stirring, simple distillation, vacuum filtration, and *flameless* heating.

Setting Up   Prepare a 50% (by weight) sodium hydroxide solution by dissolving 5.0 g of sodium hydroxide in 5 mL of water. The mixture may be heated gently to facilitate dissolution, but be sure to cool the solution to room temperature before proceeding. Add 3.8 g of benzyltriphenylphosphonium chloride and 1 mL of benzaldehyde to the 50-mL round-bottom flask containing 10 mL of dichloromethane and a stirbar. Set up the apparatus for heating under reflux with magnetic stirring. Heat the mixture with stirring until it refluxes *gently*.

Reaction   Using a Pasteur pipet, add 5 mL of the 50% aqueous sodium hydroxide solution dropwise through the top of the reflux condenser to the solution, which should be stirred as *vigorously* as possible to ensure adequate mixing of the phases. Try to add the solution *directly* into the flask without touching the walls of the condenser. After the addition is complete, continue stirring under gentle reflux for 30 min.

Work-Up and Isolation of Isomeric Stilbenes   Allow the reaction mixture to cool to room temperature and transfer it to a separatory funnel. Rinse the round-bottom flask with a 5-mL portion of dichloromethane and transfer the wash to the separatory funnel. Separate the layers and wash the organic layer sequentially with 10 mL of water and 15 mL of saturated aqueous sodium bisulfite. Finally, wash the organic layer with 10-mL portions of water until the pH of the wash is neutral. Transfer the organic solution to a dry Erlenmeyer flask and add several spatula-tips full of anhydrous sodium sulfate. Occasionally swirl the mixture over a period of 10–15 min to facilitate drying; add further small portions of anhydrous sodium sulfate if the solution remains cloudy.★

To determine the ratio of (*Z*)- and (*E*)-stilbene in the *crude* reaction mixture by ^{1}H NMR, use a filter-tip pipet to transfer a 0.5-mL portion of the dry organic solution to a small round-bottom flask, and evaporate the dichloromethane by one of the techniques described in Section 2.29. The final traces of solvent may be removed by attaching the flask to a vacuum source, gently swirling the contents as vacuum is applied.★

Isomerization   Filter or decant the remainder of the dry dichloromethane solution into the 25-mL round-bottom flask containing a stirbar, rinse the sodium sulfate with about 1 mL of fresh dichloromethane, and transfer this wash to the round-bottom flask. Add about 75 mg of iodine to the dichloromethane solution, fit the flask with a reflux condenser, and irradiate the solution with stirring for 1 h with a 150-W light bulb.★

Work-Up and Isolation of (E)-Stilbene   Decant the solution into a separatory funnel and wash the dichloromethane solution with a 5-mL portion of saturated aqueous sodium bisulfite; shake the mixture vigorously to decolorize the dichloromethane layer. Wash the organic solution with a 5-mL portion of saturated aqueous sodium chloride, transfer it to a dry Erlenmeyer flask and add several spatula-tips full of anhydrous sodium sulfate. Occasionally swirl the mixture over a period of 10–15 min to facilitate drying; add further small portions of anhydrous sodium sulfate if the solution remains cloudy.★

Filter or decant the dried solution into a tared round-bottom flask and equip it for simple distillation. Remove the dichloromethane by simple distillation. Alternatively, rotary evaporation or other techniques (Sec. 2.29) may be used to concentrate the solution. The final traces of solvent may be removed by attaching the flask to a vacuum source, gently swirling the contents as vacuum is applied.★

Remove about 50 mg of the residue for analysis by ^{1}H NMR. Dissolve the remaining thick, cloudy residue in 10–12 mL of hot 95% ethanol. Cool the resulting solution first to room temperature, and then in an ice-water bath for 15–20 min to complete crystallization. Collect the (*E*)-stilbene that precipitates and air-dry it. Recrystallize the crude (*E*)-stilbene from 95% ethanol.

Analysis   Weigh the product and calculate the percent yield; determine its melting point. Perform the tests for unsaturation on the product using bromine in dichloromethane (Sec. 23.8A) and the Baeyer test (Sec. 23.8B). Obtain IR and ^{1}H NMR spectra of your starting materials, the final product, and the crude product mixtures before and after irradiation with iodine; compare them with those of authentic samples (Figs. 10.34, 10.35, 18.1–18.4).

Microscale Procedure

Preparation   Answer the Pre-Lab Exercises on page PL. 113. Review Sections 2.9, 2.11, 2.13, 2.17, 2.21, 2.22, and 2.29.

Apparatus   A 5-mL conical vial, screw-cap centrifuge tube, apparatus for heating under reflux, magnetic stirring, simple distillation, vacuum filtration, Craig tube filtration, and *flameless* heating.

Setting Up   Prepare a 50% (by weight) sodium hydroxide solution by dissolving 1.0 g of sodium hydroxide in 1 mL of water. The mixture may be heated gently to facilitate dissolution, but be sure to cool the solution to room temperature before proceeding. Add 0.76 g of benzyltriphenylphosphonium chloride and 0.2 mL of benzaldehyde to the conical vial containing 2 mL of dichloromethane and a spinvane. Set up the apparatus for heating under reflux with magnetic stirring. Heat the solution with stirring until it refluxes *gently*.

Wittig Reaction   Using a Pasteur pipet, add 1 mL of the 50% aqueous sodium hydroxide solution dropwise through the top of the reflux condenser to the solution, which should be stirred as *vigorously* as possible to ensure adequate mixing of the phases. Try to add the solution *directly* into the flask without touching the walls of the condenser. After the addition is complete, continue stirring under *gentle* reflux for 30 min.

Work-Up and Isolation of Isomeric Stilbenes   Allow the reaction mixture to cool to room temperature and transfer it to a screw-cap centrifuge tube. Rinse the conical vial with 0.5 mL of dichloromethane and transfer the rinse to the centrifuge tube. Separate the layers and wash the organic layer sequentially with 1.5 mL of water and then 2.5 mL of saturated aqueous sodium bisulfite. Finally, wash the organic layer with 1.5-mL portions of water until the pH of the wash is neutral. Transfer the organic solution to a dry centrifuge tube, and add several microspatula-tips full of anhydrous sodium sulfate. Occasionally shake the mixture over a period of 10–15 min to facilitate drying; add further small portions of anhydrous sodium sulfate if the solution remains cloudy.★

Isomerization   Using a filter-tip pipet, transfer the dried dichloromethane solution into the 5-mL conical vial equipped with a spinvane. Rinse the sodium sulfate with 0.5 mL of dichloromethane and transfer the rinse to the conical vial. Add a *very small* crystal of iodine to the dichloromethane solution and irradiate the solution with stirring for 1 h with a 100-W light bulb.★

Work-Up and Isolation of (E)-Stilbene   Remove the spinvane, add 1 mL of saturated aqueous sodium bisulfite, and shake the mixture vigorously to decolorize the dichloromethane layer. Separate the layers and wash the organic layer with 1 mL of saturated aqueous sodium chloride. Transfer the organic solution to a dry centrifuge tube and add several microspatula-tips full of anhydrous sodium sulfate. Occasionally swirl the mixture over a period of 10–15 min to facilitate drying; add further small portions of anhydrous sodium sulfate if the solution remains cloudy.★

Using a filter-tip pipet, transfer the dried dichloromethane solution into a clean, dry 5-mL conical vial equipped with a spinvane. Remove the dichloromethane by simple distillation; withdraw solvent from the Hickman stillhead as needed. Alternatively, rotary evaporation or other techniques (Sec. 2.29) may be used to concentrate the solution. The final traces of solvents may be removed by attaching the vial to a vacuum source.★

Remove about 25 mg of the residue for analysis by ^{1}H NMR. Dissolve the remaining thick, cloudy residue in 2 mL of hot 95% ethanol. Cool the resulting solution first to room temperature, and then in an ice-water bath for 15-20 min to complete crystallization. Collect the (E)-stilbene that precipitates and air-dry it. Recrystallize the crude (E)-stilbene from 95% ethanol in a Craig tube.

Analysis   Weigh the product and calculate the percent yield; determine its melting point. Perform the tests for unsaturation on the product using bromine in dichloromethane (Sec. 23.8A) and the Baeyer test (Sec. 23.8B). Obtain IR and ^{1}H NMR spectra of your starting materials, the final product, and the crude product mixtures after irradiation with iodine; compare them with those of authentic samples (Figs. 10.34, 10.35, 18.1–18.4).

Finishing Touches

Neutralize the combined *aqueous layers and washes* with dilute hydrochloric acid and flush them down the drain. Place the *recovered dichloromethane* and the *dichloromethane solution* from the bromine test for unsaturation in a container for halogenated organic liquids; put the *manganese dioxide* from the Baeyer test for unsaturation in a container for heavy metals. Flush the *ethanolic filtrate* down the drain.

B • SYNTHESIS OF (E)-STILBENE BY THE WADSWORTH-EMMONS REACTION

Purpose: To demonstrate the synthesis of (E)-stilbene by the Wadsworth-Emmons modification of the Wittig reaction.

Safety Alert

Solutions of potassium *tert*-butoxide are *highly* caustic. *Do not allow them to come in contact with your skin.* If this should happen, wash the affected area with dilute acetic acid and then copious amounts of water. Wear latex gloves when transferring the solution.

Miniscale Procedure

Preparation   Answer the Pre-Lab Exercises on page PL. 115. Review Sections 2.10, 2.11, 2.17, 2.27, and 2.28.

Apparatus   A 25-mL round-bottom flask, Claisen adapter, rubber septum, drying tube, two 1-mL syringes and a 5-mL syringe with needles, cold-water bath, apparatus for magnetic stirring, vacuum filtration, and *flameless* heating.

Setting Up   Dry the glass parts of the syringes in an oven at 110 °C for 15 min. Equip the round-bottom flask with a stirbar and the Claisen adapter. Fit the rubber septum onto the joint of the Claisen adapter that is directly *in line* with the lower male joint (see Fig. 10.42, p. 322). Using a small flame or a heat gun, gently heat the flask and Claisen adapter to remove traces of water from the apparatus; do *not* overheat the joint with the septum. After the flask has been heated for a few minutes, discontinue the heating, attach the drying tube to the *sidearm* of the adapter, and allow the apparatus to cool to room temperature. Remove the glass syringes from the oven and allow them to cool to room temperature, preferably in a desiccator.

Draw 5 mL of a 1-*M* solution of potassium *tert*-butoxide in dry *N,N*-dimethylform-amide that is obtained from your instructor into the 5-mL syringe. Pierce the rubber septum with the needle of the syringe, being careful not to press down on the plunger, and add the solution directly to the round-bottom flask. Cool the flask below 20 °C in a cold-water bath. Using a 1-mL syringe, transfer 1 mL of diethyl benzylphosphonate through the rubber septum into the stirred solution of potassium *tert*-butoxide, and continue stirring the solution for 5 min.

Reaction   Using a 1-mL syringe, slowly add 0.5 mL of benzaldehyde through the rubber septum while continuing to stir the cooled solution. Once the addition is complete, remove the apparatus from the cooling bath, and stir the mixture at room temperature for 30 min.★

Work-Up and Isolation   Remove the Claisen adapter from the flask and add 5 mL of water with stirring. Collect the precipitated product by vacuum filtration, wash it with cold 1:1 methanol-water, and air-dry it. Recrystallize the crude (*E*)-stilbene from 95% ethanol.

Analysis   Weigh the product and calculate the percent yield; determine its melting point. Perform the tests for unsaturation on the product using bromine in dichloromethane (Sec. 23.8A) and the Baeyer test (Sec. 23.8B). Obtain IR and ¹H NMR spectra of your starting materials, and compare them with those of authentic samples (Figs. 10.34, 10.35, 18.1–18.4). In the ¹H NMR spectrum of the crude product, determine whether there is any (*Z*)-stilbene present; if so calculate the isomeric ratio.

Microscale Procedures

Preparation   Answer the Pre-Lab Exercises on page PL. 115. Review Sections 2.10, 2.11, 2.17, 2.27, and 2.28.

Apparatus   A 5-mL conical vial, Claisen adapter, rubber septum, drying tube, three 1-mL syringes with needles, cold-water bath, apparatus for magnetic stirring, vacuum filtration, Craig tube filtration, and *flameless* heating.

Setting Up   Dry the conical vial, the Claisen adapter, and the glass parts of the syringes in an oven at 110 °C for 15 min. Remove the vial and adapter from the oven, and after they have cooled, preferably in a desiccator, place a spinvane in the vial, and attach the Claisen adapter. Fit the rubber septum onto the joint of the Claisen adapter that is directly *in line* with the lower male joint, and attach the drying tube to the *sidearm* of the adapter (see Fig. 10.43, p. 323). Remove the glass syringes from the oven and allow them to cool to room temperature, preferably in a desiccator.

Draw 1 mL of a 1 *M* solution of potassium *tert*-butoxide in dry *N*,*N*-dimethylform-amide that is obtained from your instructor into one of the 1-mL syringes. Pierce the rubber septum with the needle of the syringe, being careful not to press down on the plunger, and add the solution directly to the vial. Cool the vial below 20 °C in a cold-water bath. Using another 1-mL syringe, transfer 0.2 mL of diethyl benzylphosphonate through the rubber septum into the stirred solution of potassium *tert*-butoxide, and continue stirring the solution for 5 min.

Reaction   Using a 1-mL syringe, slowly add 0.1 mL of benzaldehyde through the rubber septum while continuing to stir the cooled solution. Once the addition is complete, remove the apparatus from the cooling bath, and stir the mixture at room temperature for 30 min.★

Work-Up and Isolation   Remove the Claisen adapter from the vial and add 1 mL of water with stirring. Collect the precipitated product by vacuum filtration, wash it with cold 1:1 methanol-water, and air-dry it. Recrystallize the crude (*E*)-stilbene from 95% ethanol in a Craig tube.

Analysis   Weigh the product and calculate the percent yield; determine its melting point. Perform the tests for unsaturation on the product using bromine in dichloro-methane (Sec. 23.8A) and the Baeyer test (Sec. 23.8B). Obtain IR and ^{1}H NMR spectra of your starting materials, and compare them with those of authentic samples (Figs. 10.34, 10.35, 18.1–18.4). In the ^{1}H NMR spectrum of the crude product, determine whether there is any (*Z*)-stilbene present; if so calculate the isomeric ratio.

Finishing Touches   Place the *recovered dichloromethane* and the *dichloromethane solution* from the bromine test for unsaturation in a container for halogenated organic liquids; put the *manganese dioxide* from the Baeyer test for unsaturation in a container for heavy metals. Flush all *aqueous filtrates* down the drain.

EXERCISES

Refer to Tables 8.1–8.8 as needed for answering the questions on spectroscopy.

General Questions

1. Compare the mechanism of aldol addition (Sec. 18.3) to that of the Wittig synthesis, pointing out similarities and differences.

2. Write equations for the preparation of the following alkenes by the Wittig reaction or the Wadsworth-Emmons modification of the Wittig reaction. Start with any carbonyl compound and Wittig or Wadsworth-Emmons reagent together with any other organic or inorganic reagents that you require.

 a. $C_6H_5CH{=}C(CH_3)C_6H_5$

 b. $CH_2{=}CH{-}CH{=}CH{-}C_6H_5$

 c. $(CH_3)_2C{=}CH{-}CO_2C_2H_5$

3. Explain why you would expect an anion of type **5** to be more stable if one of the R groups is cyano, $C{\equiv}N$, rather than alkyl.

4. Why should the aldehydes used as starting materials in Wittig syntheses be free of contamination by carboxylic acids?

5. Consider the spectral data for benzaldehyde (Figs. 18.1 and 18.2).

 a. In the functional group region of the IR spectrum, specify the absorptions associated with the carbonyl group and the aromatic ring.

 b. In the ^1H NMR spectrum, assign the various resonances to the hydrogen nuclei responsible for them.

 c. For the ^{13}C NMR data, assign the various resonances to the carbon nuclei responsible for them.

6. Consider the spectral data for (*E*)-stilbene (Figs. 10.34 and 10.35).

 a. In the functional group region of the IR spectrum, specify the absorptions associated with the aromatic rings.

 b. In the ^1H NMR spectrum, assign the various resonances to the hydrogen nuclei responsible for them.

 c. For the ^{13}C NMR data, assign the various resonances to the carbon nuclei responsible for them.

7. Consider the spectral data for (*Z*)-stilbene, (Figs. 18.3 and 18.4).

 a. In the functional group region of the IR spectrum, specify the absorptions associated with the aromatic rings.

 b. In the ^1H NMR spectrum, assign the various resonances to the hydrogen nuclei responsible for them.

 c. For the ^{13}C NMR data, assign the various resonances to the carbon nuclei responsible for them.

8. Discuss the differences observed in the IR and NMR spectra of (*Z*)-stilbene and (*E*)-stilbene that enable you to distinguish between the two isomers. What differences in the IR and NMR spectra of benzaldehyde, (*Z*)-stilbene, and (*E*)-stilbene are consistent with the conversion of benzaldehyde to a mixture of the isomeric stilbenes in this experiment?

Questions for Part A

9. What is the purpose of washing the crude reaction mixture obtained *immediately* after the Wittig reaction with saturated sodium bisulfite?

10. Propose a mechanism for the isomerization of (Z)-stilbene into (E)-stilbene using a trace of iodine and light (Eq. 18.6). (*Hint:* The iodine-iodine bond undergoes homolysis upon irradiation with light; see also Section 7.3.)

11. In the ^{1}H NMR spectra of the crude product before *and* after irradiation with iodine, determine the approximate ratio of (Z)-stilbene and (E)-stilbene.

12. Consider the NMR spectral data for benzyltriphenylphosphonium chloride (Fig. 18.6). Note that ^{31}P has $I_z = \frac{1}{2}$ and therefore couples with neighboring hydrogen and carbon and carbon nuclei having $I_z = \frac{1}{2}$.

    a. In the ^{1}H NMR spectrum, assign the various resonances to the hydrogen nuclei responsible for them.

    b. For the ^{13}C NMR data, assign the various resonances to the carbon nuclei responsible for them.

Questions for Part B

13. Why should the exposure of the solution of potassium *tert*-butoxide to the atmosphere be minimized?

14. *N,N*-Dimethylformamide (DMF) is very water-soluble. Why?

15. Consider the NMR spectral data for diethyl benzylphosphonate (Fig. 18.8). Note that ^{31}P has $I_z = \frac{1}{2}$ and therefore couples with neighboring hydrogen and carbon and carbon nuclei having $I_z = \frac{1}{2}$.

    a. In the ^{1}H NMR spectrum, assign the various resonances to the hydrogen nuclei responsible for them.

    b. For the ^{13}C NMR data, assign the various resonances to the carbon nuclei responsible for them.

Spectra

Starting Materials and Products

   The IR and NMR spectra of (E)-stilbene are provided in Figures 10.34 and 10.35, respectively.

FIGURE 18.1

IR spectrum of benzaldehyde (neat).

(a) ^{1}H NMR spectrum (300 MHz).

FIGURE 18.2
NMR data for benzaldeyde
(CDCl$_3$).

(b) ^{13}C NMR data: δ 129.0, 129.7, 134.4, 136.6, 192.0.

FIGURE 18.3
IR spectrum of (Z)-stilbene (neat).

(a) ^{1}H NMR spectrum (300 MHz).

FIGURE 18.4
NMR data for (Z)-stilbene (CDCl$_3$).

(b) ^{13}C NMR data: δ 127.3, 128.4, 129.1, 130.5, 137.5.

FIGURE 18.5
IR spectrum of
benzyltriphenylphosphonium
chloride (IR card).

FIGURE 18.6
NMR data for
benzyltriphenylphosphonium
chloride (CDCl$_3$).

(a) ^{1}H NMR spectrum (300 MHz).

(b) ^{13}C NMR data (^{31}P-coupled): δ 29.8, 30.6, 116.6, 118.0, 126.8, 127.0, 127.9, 127.9, 128.3, 128.4, 129.6, 129.8, 131.0, 131.1, 133.8, 133.9, 134.6, 134.6.

FIGURE 18.7
IR spectrum of diethyl
benzylphosphonate (neat).

(a) ^{1}H NMR spectrum (300 MHz).

FIGURE 18.8
NMR data for diethyl
benzylphosphonate (CDCl$_3$).

(b) ^{13}C NMR data (^{31}P-coupled): δ 15.9, 16.0, 32.2, 34.4, 61.5, 61.6, 126.3, 126.4, 128.0, 128.1, 129.3, 129.4, 131.1, 131.3.

18.3 Reactions of Stabilized Carbanions Derived from Carbonyl Compounds

The presence of the carbonyl function in a molecule allows the α-carbon atom to become nucleophilic by forming an **enolate ion** (Eq. 18.3) or an **enol** (Eq. 18.4). The nucleophilic property of an enol is the basis for the experiment in Section 18.4. Enolate ions, which are considerably more reactive nucleophiles than enols, are the subject of this section.

Formation of enolate ions (Eq. 18.3) involves the use of a base, and the strength of the base required will depend on the **pK$_a$** of the hydrogen atom α to the carbonyl carbon atom. The pK$_a$ is defined as the negative log of K$_a$, the equilibrium constant for the ionization shown in Equation 18.8. Thus, pK$_a$ = $-$log K$_a$. Water is the usual reference solvent for reporting values of pK$_a$, which are known to be solvent-dependent. Experimentally, the **pK$_a$-values** of aldehydes and ketones are in the range 18–20. This is a remarkably low value, when compared to those for saturated hydrocarbons, which are in excess of 50!

$$\text{O} \quad \underset{\alpha}{\overset{\text{O}}{\|}} \text{H}_{(aq)} \xrightleftharpoons{\text{K}_a} \text{H}^{+}_{(aq)} + \underset{\textbf{17}}{\overset{\text{O}}{\|}}_{(aq)} \qquad (18.8)$$

17
Enolate ion

The dramatic acidifying effect of the carbonyl group is due mainly to its ability to delocalize, and therefore to stabilize, the negative charge in the enolate ion, as reflected in **17** (Eq. 18.8), which is the **resonance hybrid** of the **resonance structures 3a** and **3b** (Sec. 18.1). Because the pK$_a$-values of aldehydes and ketones fall in the range of water (pK$_a$ 15.7) and alcohols (pK$_a$ 15.5–18), it is possible to generate enolate ions using anions such as hydroxide or alkoxide (Eq. 18.3, B$^-$ = HO$^-$ and RO$^-$, respectively); these are the bases utilized in the experimental procedures of this section.

ALDOL CONDENSATIONS

An important general reaction of enolate ions involves **nucleophilic addition** to the electrophilic carbonyl carbon atom of the aldehyde or ketone from which the enolate is derived. A dimeric anion **18** results, which may then be neutralized by abstraction of a proton to produce a β-hydroxycarbonyl compound, **19** (Eq. 18.9). If the reaction is performed in hydroxylic solvents such as water or an alcohol, the source of the proton may be the solvent, whose deprotonation will regenerate the base required for forming the enolate ion. The overall process is thus *catalytic* in the base that is used.

$$\text{(18.9)}$$

| **3b** | **3a** | | **18** | **19** |

Enolate ion

An aldol
(β-Hydroxycarbonyl compound)

The β-hydroxycarbonyl compound **19** is termed an **aldol**, and the reaction leading to its formation is referred to as an **aldol addition**. The origin of the word "aldol" is understood by realizing that use of an aldehyde in the reaction provides a dimerization product **19** containing an *ald*ehyde function and an **alco*hol* group**. The isolation of such addition products depends on the reaction conditions used, because dehydration to an α,β-unsaturated carbonyl compound **20** may occur *if* **19** still has a hydrogen atom on the α-carbon atom that served as the nucleophile in the addition reaction (Eq. 18.10). The overall reaction described by Equations 18.9 and 18.10 is called an **aldol condensation**.

$$\text{(18.10)}$$

19

20

α,β-Unsaturated
compound

The experiment described in this section is an example of a **mixed** or **crossed aldol condensation**. This term refers to cases in which two *different* carbonyl compounds are the reactants. Under certain circumstances, such reactions are synthetically practical, selectively producing a single major condensation product. For example, a ketone may condense selectively with an aldehyde because self-addition of two molecules of the ketone does not readily occur. This is because the carbonyl group of ketones is sterically and electronically not as susceptible to nucleophilic attack as is the carbonyl function of aldehydes.

You will perform the crossed aldol condensation of acetophenone (**21**) with *p*-anisaldehyde (4-methoxybenzaldehyde, **23**) to give *trans-p*-anisalacetophenone (**25**) according to the sequence outlined in Scheme 18.3. For steric reasons, the enolate ion **22** reacts preferentially with **23** rather than **21**, leading to the aldol **24**. This product then dehydrates in the presence of base, in the manner outlined in Equation 18.10, to yield the condensation product **25**. Part of the thermodynamic driving force for this dehydration is associated with forming a new carbon-bond π-bond that is **conjugated** with the aromatic ring as well as with the carbonyl function. As a general rule, formation of an extended conjugated system, as occurs here, increases the stability of a molecule.

SCHEME 18.3

21
Acetophenone

22

23
p-Anisaldehyde

24

25
trans-p-Anisalacetophenone

EXPERIMENTAL PROCEDURES

SYNTHESIS OF *trans-p*-ANISALACETOPHENONE

Purpose: To demonstrate the synthesis of an α, β-unsaturated carbonyl compound by a crossed aldol condensation.

Safety Alert

The 50% (by weight) aqueous sodium hydroxide solution is *highly caustic. Do not allow it to come in contact with your skin.* If this should happen, wash the affected area with dilute acetic acid and then copious amounts of water. Wear latex gloves when transferring the solution.

Miniscale Procedure

Preparation   Answer the Pre-Lab Exercises on page PL. 117. Review Sections 2.9, 2.10, 2.11, and 2.17.

Apparatus   A 13-mm × 100-mm test tube, two 1-mL syringes, apparatus for vacuum filtration and *flameless* heating.

Setting Up   Prepare a 50% (by weight) sodium hydroxide solution by dissolving 1.0 g sodium hydroxide in 1 mL of water. The mixture may be heated gently to facilitate dissolution, but be sure to cool the solution to room temperature before proceeding. Place 1.0 mL of *p*-anisaldehyde and 1.0 mL of acetophenone in the test tube and add 3.0 mL of 95% ethanol. Shake the test tube gently to mix and dissolve the reactants.

Aldol Reaction, Work-Up, and Isolation   Using a Pasteur pipet, transfer 5 drops of the 50% sodium hydroxide solution into the ethanolic solution of the carbonyl compounds, shake the mixture for a minute or two until a homogeneous solution results, and allow it to stand with occasional shaking at room temperature for 15 min.★ Cool the reaction mixture in an ice-water bath. If crystals do not form, scratch at the liquid-air interface to induce crystallization. Collect the product by vacuum filtration, wash the product with 1-2 mL of *cold* 95% ethanol, and air-dry the crystals.★ Recrystallize the crude *trans-p*-anisalacetophenone from methanol.

Analysis   Weigh the product and calculate the percent yield; determine its melting point. Prepare the 2,4-dinitrophenylhydrazones of the product and of the two starting materials according to the procedure in Section 23.7F; compare the melting points of these derivatives. Perform the tests for unsaturation on the product using bromine in dichloromethane (Sec. 23.8A) and the Baeyer test (Sec. 23.8B). Obtain IR and ^{1}H NMR spectra of your starting materials and product and compare them with those of authentic samples (Figs. 18.9–18.14).

Microscale Procedure

Preparation   Answer the Pre-Lab Exercises on page PL. 117. Review Sections 2.9, 2.10, 2.11, and 2.17.

Apparatus   A 3-mL conical vial, two 1-mL syringes, apparatus for magnetic stirring, vacuum filtration, Craig tube filtration, and *flameless* heating.

Setting Up   Prepare a 50% (by weight) sodium hydroxide solution by dissolving 0.5 g of sodium hydroxide in 0.5 mL of water. The mixture may be heated gently to facilitate dissolution, but be sure to cool the solution to room temperature before proceeding. Place 0.2 mL of *p*-anisaldehyde and 0.2 mL of acetophenone in the conical vial containing a spinvane. Add 0.7 mL of 95% ethanol and stir the contents of the vial to mix and dissolve the reactants.

Aldol Reaction, Work-Up, and Isolation   Using a Pasteur pipet, transfer 2 drops of the 50% NaOH solution into the ethanolic solution of the carbonyl compounds and stir the reaction mixture at room temperature for 15 min.★ Cool the reaction mixture in an ice-water bath. If crystals do not form, scratch at the liquid-air interface to induce crystallization. Collect the product by vacuum filtration, wash the product with about 5–10 drops of *cold* 95% ethanol, and air-dry the crystals.★ Recrystallize the crude *trans-p*-anisalacetophenone from methanol in a Craig tube.

Analysis   Weigh the product and calculate the percent yield; determine its melting point. Prepare the 2,4-dinitrophenylhydrazones of the product and the two starting materials according to the procedure in Section 23.7F; compare the melting points of these derivatives. Perform the tests for unsaturation on the product using bromine in dichloromethane (Sec. 23.8A) and the Baeyer test (Sec. 23.8B). Obtain IR and ^{1}H NMR spectra of your starting materials and product and compare them with those of authentic samples (Figs. 18.9–18.14).

Neutralize the *ethanolic filtrates* with dilute hydrochloric acid and flush them down the drain. Place the *dichloromethane solution* from the bromine test for unsaturation in a container for halogenated organic liquids.

Finishing Touches

EXERCISES

1. Compute the equilibrium constant, K_{eq}, for the reaction of equimolar amounts of acetophenone, $C_6H_5COCH_3$, and hydroxide ion to generate the enolate ion. The pK_a-values of the ketone and of water are 19.0 and 15.7, respectively.

2. Explain why the main reaction between acetophenone and *p*-anisaldehyde is the mixed aldol reaction rather than (a) self-condensation of acetophenone or (b) the Cannizzaro reaction (Sec. 16.3) of *p*-anisaldehyde.

3. Explain why *trans-p*-anisalacetophenone would be expected to be more stable than the corresponding *cis* isomer.

4. α,β-Unsaturated carbonyl compounds such as *trans-p*-anisalacetophenone (**25**) undergo reactions that are typical of both carbonyl compounds and alkenes. Illustrate this by showing the reaction of **25** with bromine (Sec. 10.6) and 2,4-dinitrophenylhydrazine (Sec. 23.7F) to give the corresponding products.

5. The addition of bromine to *trans-p*-anisalacetophenone (**25**) gives a single racemic dibromide.

    a. Using curved arrows to symbolize the flow of electrons and suitable stereochemical drawings, show the mechanism of this reaction (*Hint:* See Sec. 10.6).

    b. How does the mechanism you proposed in part **a** support the formation of the racemic form of a single dibromide?

Refer to Tables 8.1–8.8 as needed for answering the questions on spectroscopy.

6. Consider the spectral data for acetophenone (Figs. 18.9 and 18.10).

    a. In the functional group region of the IR spectrum, specify the absorptions associated with the carbonyl group and the aromatic ring.

    b. In the ^{1}H NMR spectrum, assign the various resonances to the hydrogen nuclei responsible for them.

    c. For the ^{13}C NMR data, assign the various resonances to the carbon nuclei responsible for them.

7. Consider the spectral data for *p*-anisaldehyde (Figs. 18.11 and 18.12).

    a. In the functional group region of the IR spectrum, specify the absorptions associated with the carbonyl group and the aromatic ring. Also specify the absorption in the fingerprint region that is associated with the *para* substitution on the aromatic ring.

    b. In the ^{1}H NMR spectrum, assign the various resonances to the hydrogen nuclei responsible for them.

    c. For the ^{13}C NMR data, assign the various resonances to the carbon nuclei responsible for them.

8. Consider the spectral data for *trans-p*-anisalacetophenone (Figs. 18.13 and 18.14).

    a. In the functional group region of the IR spectrum, specify the absorptions associated with the unsaturated ketone group and the aromatic rings. Also specify the absorptions in the fingerprint region that are associated with the *para* substitution on the aromatic ring and the *trans* carbon-carbon double bond.

    b. In the ^{1}H NMR spectrum, assign the various resonances to the hydrogen nuclei responsible for them.

    c. For the ^{13}C NMR data, assign the various resonances to the carbon nuclei responsible for them.

9. Discuss the differences observed in the IR and NMR spectra of acetophenone, *p*-anisaldehyde, and *trans-p*-anisalacetophenone that are consistent with the crossed aldol condensation occurring in this experiment.

Spectra   *Starting Materials and Product*

FIGURE 18.9
IR spectrum of acetophenone (neat).

(a) ^{1}H NMR spectrum (300 MHz).

(b) ^{13}C NMR data: δ 26.3, 128.3, 128.8, 133.0, 137.3, 197.4.

FIGURE 18.10
NMR data for acetophenone (CDCl$_3$).

FIGURE 18.11
IR spectrum of *p*-anisaldehyde (neat).

(a) ^{1}H NMR spectrum (300 MHz).

(b) ^{13}C NMR data: δ 55.5, 114.5, 130.2, 131.9, 164.6, 190.5.

FIGURE 18.12
NMR data for *p*-anisaldehyde (CDCl$_3$).

FIGURE 18.13
IR spectrum of *trans-p-*anisalacetophenone (IR card).

FIGURE 18.14
NMR data for *trans-p-*anisalacetophenone (CDCl$_3$).

(a) ^{1}H NMR spectrum (300 MHz).

(b) ^{13}C NMR data: δ 55.3, 114.4, 119.7, 127.6, 128.4, 128.5, 130.2, 132.5, 138.5, 144.6, 161.7, 190.4.

18.4   Conjugate Addition to an α,β-Unsaturated Ketone

A carbon atom alpha to a carbonyl function may become nucleophilic either by forming an *enolate* (Eq. 18.3) or an *enol* (Eq. 18.4). In the crossed aldol reaction described in the previous section (Scheme 18.3), the nucleophilic character of the carbon atom *alpha* to a ketone function was provided by enolate formation. However, in the experiment that follows, nucleophilicity at the α-carbon atom is promoted by enol formation. The overall reaction is the *acid-catalyzed* conversion of 2-methylpropanal (**26**) and 3-buten-2-one (**27**) into 4,4-dimethyl-2-cyclohexen-1-one (**29**), as shown in Equation 18.11. In the first stage of the reaction, the enol form of **26** serves as the nucleophile for the **1,4-** or **conjugate addition** to the α,β-unsaturated ketone **27** to give **28**. Then in the second stage, an enol is the nucleophile in the cyclization of **28** by an **intramolecular aldol condensation** that is acid-catalyzed rather than base-catalyzed. The conjugate addition to give **28** is an example of a **Michael-type addition**, whereas the combination of this addition followed by an aldol reaction to form the six-membered ring of **29** is called the **Robinson annulation**. The names honor the chemists who discovered the two types of reactions.

(18.11)

26 2-Methylpropanal (isobutyraldehyde)

27 3-Buten-2-one (methyl vinyl ketone)

28 2,2-Dimethylhexanal-5-one

29 4,4-Dimethyl-2-cyclohexen-1-one

A plausible overall mechanism for the reactions leading to **29** is shown in Scheme 18.4. 2-Methylpropanal (**26**) first undergoes acid-catalyzed tautomerization to its enol form **30**, in which the α-carbon atom is now nucleophilic. Protonation of the 3-buten-2-one (**27**) converts it into the delocalized cation **31**, which is more electrophilic than the unprotonated form. Because **30** is a weak nucleophile, relative to an enolate ion, the formation of **31** facilitates the next stage of the reaction, which is creating a new carbon-carbon bond between **26** and **27** to give the enol **32**. The *bold-faced* atoms in structure **32** show that the α-C–H bond of **26** has added in a conjugate, or 1,4- manner to **27**. Acid-catalyzed tautomerization of **32** leads to the thermodynamically more stable keto form **28**.

Although **28** is potentially isolable, it is not observed under our reaction conditions; rather it undergoes an acid-catalyzed *intramolecular aldol addition*. This process is initiated by the acid-catalyzed tautomerization of **28** to a different enol form, **33**. Activating the remaining carbonyl function by protonation sets the stage for cyclization to give the β-hydroxycarbonyl compound **34** after deprotonation. The fact that **28** undergoes efficient cyclization rather than intermolecular condensation with the enolic form of 2-methylpropanal or other potential nucleophiles is mainly attributable to *entropic* considerations that tend to favor *intra*molecular processes over *inter*molecular ones. In the final stage of the reaction, acid-catalyzed dehydration of **34** to give the desired **29** occurs; the mechanism of this reaction is left as an exercise at the end of this section.

SCHEME 18.4

26   30   31

27   32

Tautomerization

33   28

34

A β-hydroxycarbonyl compound

The dehydration of **34** to **29** is a reversible process that is driven to completion by removing the water from the reaction mixture. This is conveniently done by using

FIGURE 18.15
Dean-Stark trap for removal of water by azeotropic distillation.

FIGURE 18.16
Apparatus for reaction of 2-methylpropanal with 3-buten-2-one.

toluene as the solvent for the reaction; water and toluene form an **azeotrope**, so azeotropic distillation allows the continuous separation of water as dehydration occurs. To minimize the amount of solvent that is required for distillations of this type, a Dean-Stark trap (Fig. 18.15) is commonly used. Although such traps are not usually available in the introductory laboratory, an operational equivalent may be devised by assembling the apparatus in such a way that water, but not toluene, is prevented from returning to the reaction flask (Fig. 18.16).

The choice of an aldehyde, such as 2-methylpropanal (**26**) that has only *one* α-hydrogen atom, is important to the success of this crossed aldol condensation because its self-addition is of no consequence. For example, although aldehyde **26** undergoes facile self-dimerization to **35** (Eq. 18.12), this process is *reversible* under the reaction conditions used. Moreover, unlike the aldol product **34**, **35** is incapable of dehydrating to give an α,β-unsaturated product and water, which could be removed to drive the equilibrium and lead to self-dimerization product. It is left to you to consider the possibility of self-condensation of 3-buten-2-one (**27**) and of the reaction of enol **30** with the carbon atom of the protonated carbonyl function in **31** (see Exercises 6 and 7 at the end of this section).

$$2\ (CH_3)_2\overset{\overset{\displaystyle O}{\|}}{\underset{\underset{\displaystyle H}{|}}{C}}{\!-\!}H \;\xrightleftharpoons{H_3O^+}\; (CH_3)_2CH\overset{\overset{\displaystyle H}{|}}{\underset{}{C}}{\!-\!}\overset{\overset{\displaystyle OH}{|}}{\underset{\underset{\displaystyle CH_3}{|}}{C}}{\!-\!}CHO \qquad (18.12)$$

$$\underset{\textbf{26}}{} \qquad\qquad \underset{\substack{\textbf{35}\\ \text{2,2,4-Trimethyl-3-}\\ \text{hydroxypentanal}}}{}$$

EXPERIMENTAL PROCEDURES

PREPARATION OF 4,4-DIMETHYL-2-CYCLOHEXEN-1-ONE

Purpose: To demonstrate the preparation of a substituted cyclohexenone via sequential conjugate addition and an intramolecular aldol condensation.

Safety Alert	
Safety Alert	1. Because volatile and flammable solvents are used in this experiment, *flameless* heating should be used for heating the reaction under reflux and for distilling the product. However, if it is necessary to use a small microburner to distil the product, be sure that no one is using flammable solvents in the laboratory. 2. Be certain that all joints in the apparatus are properly lubricated and firmly mated in order to keep vapors from escaping. 3. 2-Methylpropanal and 3-buten-2-one are highly volatile irritants with penetrating odors. Do *not* breathe their vapors. If possible, they should be handled in a hood and measured using syringes to minimize their release into the laboratory.

Miniscale Procedure

Preparation   Answer the Pre-Lab Exercises on page PL. 119. Review Sections 2.9, 2.11, 2.13, 2.21, 2.22, 2.29, and 6.4.

Apparatus   A 125-mL and a 25-mL round-bottom flask, two 5-mL syringes, stillhead, reflux condenser, separatory funnel, microburner, apparatus for magnetic stirring, simple distillation, and *flameless* heating.

Setting Up   Equip the 125-mL round-bottom flask with a stirbar and add 0.1 g of 2-naphthalenesulfonic acid and 25 mL of toluene. Fit the stillhead with a rubber septum or stopper firmly seated in the male joint that is normally used for connection to a condenser. Secure the rubber septum or stopper to the joint with wire. Working at the hood, use two different syringes to measure 5.0 mL of 2-methylpropanal and 3.5 mL of 3-buten-2-one into an Erlenmeyer flask containing 10 mL of toluene. Transfer this solution quickly, but carefully to avoid spillage, to the round-bottom flask. Attach the stillhead to the flask and fit the stillhead with a reflux condenser. Finish assembling the apparatus shown in Figure 18.16 for heating under reflux with azeotropic removal of water. Tilt the flask at an angle of about 30° from the vertical so that the condensate will drip into the sidearm of the distillation head.

Robinson Annulation   Bring the solution to a brisk reflux with stirring and continue heating under reflux for 2–2.5 h. The reflux ring must rise above the sidearm of the stillhead so that, as the reaction proceeds, water collects in the plugged sidearm. Following the period of reflux, allow the reaction mixture to cool to room temperature. Carefully disconnect the apparatus, *making certain that the collected water is not spilled back into the reaction flask.* Pour the water and solvent present in the sidearm into a small graduated cylinder to determine what percentage of the theoretical amount of water has been collected.★

Work-Up and Isolation   Transfer the reaction mixture to a separatory funnel and wash it with a 20-mL portion of saturated aqueous sodium bicarbonate. Transfer the organic solution to a dry Erlenmeyer flask and add several spatula-tips full of anhydrous sodium sulfate. Occasionally shake the mixture over a period of 10–15 min to facilitate drying; add further small portions of anhydrous sodium sulfate if the solution remains cloudy.★

Filter or decant the dry organic solution into a dry 125-mL round-bottom flask and equip the flask for simple distillation. Carefully (*caution*: foaming) remove the solvent and any unreacted starting materials by simple distillation; collect a single fraction boiling up to 110–115 °C (760 Torr). Alternatively, rotary evaporation or other techniques (Sec. 2.29) may be used to concentrate the solution. The final traces of solvent may be removed by attaching the flask to a vacuum source, gently swirling the contents as the vacuum is applied.★

Purification   Using a Pasteur pipet, transfer the *cooled* pot residue to a 25-mL round-bottom flask containing a stirbar. Rinse the original stillpot with 1–2 mL of diethyl ether and add the rinse to the 25-mL round-bottom flask. Remove the diethyl ether by attaching the flask to a vacuum source, gently swirling the contents as the vacuum is applied. Fit this flask for a shortpath distillation. Insulate the top of the distillation flask and stillhead with cotton or glasswool wrapped with aluminum foil. Using a micro-burner, distil the 4,4-dimethyl-2-cyclohexen-1-one into a *tared* receiver. Collect it as a single fraction boiling above 130 °C (760 Torr) and record the temperature range. Note that the head temperature may never reach the reported boiling point, owing to the high boiling temperature and the small amount of sample being distilled. Terminate heating when the dark pot residue becomes viscous and starts to evolve fumes.

Analysis   Weigh the product and calculate the percent yield. The product may be contaminated with residual toluene, as illustrated in Figures 18.17 and 18.18, the amount of which will be dependent upon the efficiency with which toluene was removed in the first stage of the distillation. Determine the purity of your product by GLC or ^{1}H NMR spectroscopic analysis. Prepare a 2,4-dinitrophenylhydrazone derivative from a portion of your product according to the procedure in Section 23.7F; this derivative may be recrystallized from 95% ethanol. Obtain IR and ^{1}H NMR spectra of your starting materials and product and compare them with those of authentic samples (Figs. 18.19–18.24).

Microscale Procedure

Preparation   Answer the Pre-Lab Exercises on page PL. 119. Review Sections 2.9, 2.11, 2.13, 2.21, 2.22, 2.25, 2.29, and 6.4.

Apparatus   A 5-mL conical vial, two 1-mL syringes, microburner, apparatus for magnetic stirring, simple distillation, and *flameless* heating.

Setting Up   Equip the conical vial with a spinvane, and add 10 mg of 2-naphthalene-sulfonic acid and 2.5 mL of toluene. Working at the hood, use two different syringes to measure first 0.35 mL of 3-buten-2-one and then 0.50 mL of 2-methylpropanal directly into the conical vial. Set up the apparatus for simple distillation. Using a Pasteur pipet, fill the well of the Hickman stillhead with toluene, being careful not to allow any toluene to flow back into the conical vial.

Robinson Annulation   Bring the solution to a brisk reflux with stirring and continue heating under reflux for 2–2.5 h. It is important that the reflux ring rise *into the bulb*

of the Hickman stillhead to remove the water from the reaction. Allow the reaction mixture to cool to room temperature.★

Work-Up and Isolation   Remove the spinvane and add 0.5 mL of saturated aqueous sodium bicarbonate to the conical vial. Cap the vial, invert it, and return it to its original upright position; release the pressure by *carefully* venting the screw-cap. Now replace the cap and shake the vial with intermittent venting to relieve pressure. Separate the layers and transfer the organic layer to a dry, screw-cap centrifuge tube. Add several microspatula-tips full of anhydrous sodium sulfate. Shake the centrifuge tube occasionally for a period of 10–15 min to facilitate drying; add further small portions of anhydrous sodium sulfate if the solution remains cloudy.★

Using a Pasteur pipet, transfer the organic solution to a dry 5-mL conical vial and equip the vial for simple distillation. Carefully (*caution*: foaming) remove the solvent by simple distillation; withdraw solvent from the Hickman stillhead as needed. Alternatively, rotary evaporation or other techniques (Sec. 2.29) may be used to concentrate the solution. The final traces of solvents may be removed by attaching the vial to a vacuum source, gently swirling the contents as the vacuum is applied.★

Purification (Optional)   Remove the source of heat and dissemble the apparatus. Clean and dry the Hickman stillhead. Reassemble the apparatus for simple distillation. Insulate the upper portion of the conical vial and the lower portion of the stillhead with cotton or glasswool wrapped with aluminum foil. Using a microburner, distil the 4,4-dimethyl-2-cyclohexen-1-one. Collect it as a single fraction. Terminate heating when the dark pot residue becomes viscous and starts to evolve fumes. Transfer the product to a tared sample vial.

Analysis   Weigh the product and calculate the percent yield. The product may be contaminated with residual toluene, as illustrated in Figures 18.17 and 18.18, the amount of which will be dependent upon the efficiency with which toluene was removed. Determine the purity of your product by GLC or ^{1}H NMR spectroscopic analysis. Prepare a 2,4-dinitrophenylhydrazone derivative of a portion of your product according to the procedure in Section 23.7F; this derivative may be recrystallized from 95% ethanol. Obtain IR and ^{1}H NMR spectra of your starting materials and product and compare them with those of authentic samples (Figs. 18.19–18.24).

Finishing Touches   Place the *recovered toluene* that was obtained from the azeotropic distillation and from removal from the reaction mixture into the container for nonhalogenated organic liquids. Flush the *sodium bicarbonate wash* down the drain. Use a small amount of *recovered toluene* to rinse the *pot residue* from the simple distillation of the product into the container for nonhalogenated organic liquids. After residual solvent has evaporated from the *sodium sulfate* on a tray in the hood, place the drying agent in a container for nonhazardous waste.

EXERCISES

1. Although benzene could be used as the solvent for this experiment, toluene is the solvent of choice because it is not carcinogenic. Furthermore, the reaction appears to go faster in toluene than in benzene, as judged by the rate at which water is formed. How might you account for this rate acceleration?

2. Why is it important that water *not* return to the reaction vessel during the reflux period?

3. Product **36** could potentially be produced by aldol condensation of **28** but is not. Provide a stepwise reaction mechanism for the possible formation of **36** in the presence of an acid catalyst. Explain why this alternate product is not observed.

36

4. a. Figure 18.23 is the GLC trace of a sample of product obtained in this experiment. Calculate the ratio of toluene to **29** that is present, assuming that the response factor of the detector (Sec. 6.4.) is the same for the two compounds. Note that the GLC column used in this analysis separates substances on the basis of their relative boiling points.

  b. Figure 18.24 is the ^{1}H NMR spectrum of this same sample. Calculate the ratio of toluene to **29** that is present according to this method of analysis.

  c. Account for the discrepancy, if any, between the two ratios that you obtain.

  d. The distillate from which the samples for these analyses were drawn weighed a total of 2.8 g. Based on the ^{1}H NMR analysis of the mixture, calculate how many grams of the desired **29** were present in the distillate. Calculate the yield of the reaction, assuming that it was run on the scale described in the Miniscale Procedure of this section.

5. Attempted acid-catalyzed reaction of propanal, CH_3CH_2CHO, with 3-buten-2-one (**27**) fails to give a good yield of the desired product, **37**. Explain why this might be.

27	**37**
Propanal	4-Methyl-2-cyclohexen-1-one

6. Write a product of self-condensation of 3-buten-2-one (**27**) and explain why this process might be less favorable than the observed reaction between **26** and **27**.

7. Write the product expected to result from attack of enol **30** on the carbonyl carbon atom of **31**. Explain why formation of this product should be thermodynamically *less* favorable than formation of **28**.

8. Why is protonation of the carbonyl group in **33** likely to be involved in the conversion of **33** to **34**?

9. Write a stepwise reaction mechanism for the acid-catalyzed dehydration of **34** to **29**.

Refer to Tables 8.1–8.8 as needed for answering the following questions on spectroscopy.

10. Consider the spectral data for 2-methylpropanal (Figs. 18.17 and 18.18).

  a. In the functional group region of the IR spectrum, specify the absorptions associated with the carbonyl component and the carbon-hydrogen bond of the aldehyde group.

b. In the ^{1}H NMR spectrum, assign the various resonances to the hydrogen nuclei responsible for them.

c. For the ^{13}C NMR data, assign the various resonances to the carbon nuclei responsible for them.

11. Consider the spectral data for 3-buten-2-one (Figs. 18.19 and 18.20).

a. In the functional group region of the IR spectrum, specify the absorptions associated with the carbonyl group and the carbon-carbon double bond. Also specify the absorption in the fingerprint region that is characteristic for the terminal vinyl group.

b. In the ^{1}H NMR spectrum, assign the various resonances to the hydrogen nuclei responsible for them.

c. For the ^{13}C NMR data, assign the various resonances to the carbon nuclei responsible for them.

12. Consider the spectral data for 4,4-dimethyl-2-cyclohexen-1-one (Figs. 18.21 and 18.22).

a. In the functional group region of the IR spectrum, specify the absorptions associated with the carbonyl group and the carbon-carbon double bond.

b. In the ^{1}H NMR spectrum, assign the various resonances to the hydrogen nuclei responsible for them.

c. For the ^{13}C NMR data, assign the various resonances to the carbon nuclei responsible for them.

13. Discuss the differences observed in the IR and NMR spectra of 2-methylpropanal, 3-buten-2-one, and 4,4-dimethyl-2-cyclohexen-1-one that are consistent with the formation of the latter by the Robinson annulation in this experiment.

Spectra   *Starting Materials and Product*

FIGURE 18.17
IR spectrum of 2-methylpropanal (neat).

(a) ^{1}H NMR spectrum (300 MHz).

FIGURE 18.18
NMR data for 2-methylpropanal
(CDCl$_3$).

(b) ^{13}C NMR data: δ 15.5, 41.2, 204.7.

FIGURE 18.19
IR spectrum of 3-buten-2-one
(neat).

(a) ^{1}H NMR spectrum (300 MHz).

FIGURE 18.20
NMR data for 3-buten-2-one
(CDCl$_3$).

(b) ^{13}C NMR data: δ 26.3, 128.8, 137.4, 198.7.

FIGURE 18.21

IR spectrum of 4,4-dimethyl-
2-cyclohexen-1-one (neat).

FIGURE 18.22

NMR data for 4,4-dimethyl-
2-cyclohexen-1-one (CDCl₃).

(a) ¹H NMR spectrum (300 MHz).

(b) ¹³C NMR data: δ 27.7, 32.8, 34.4, 36.1, 126.8, 159.6, 199.0.

FIGURE 18.23

GLC analysis of reaction mixture
for Exercise **4a**. The retention times
in minutes and the peak areas,
respectively, are: peak 1: 1.7,
10250; peak 2: 1.8, 37513.

514

FIGURE 18.24
^{1}H NMR spectrum (300 MHz) of 4,4-dimethyl-2-cyclohexen-1-one and toluene for Exercise **4b**.

HISTORICAL HIGHLIGHT

The Wittig Reaction

In the early 1950s, the synthesis of alkenes with complete control of the position of the carbon-carbon double bond was not possible. Mixtures of alkenes, sometimes having rearranged carbon frameworks, were often formed upon base-induced dehydrohalogenation of alkenes and acid-catalyzed dehydration of alcohols (Secs. 102. and 10.3). This situation was soon to change, however, because in 1953 Georg Wittig observed the unexpected formation of 1,1-diphenylethylene from the reaction of methylenetriphenylphosphorane with benzophenone. At the time Wittig and one of his students, Georg Geissler, were studying the chemistry of pentavalent organophosphorus compounds, and they published this unexpected result as a small part of a paper summarizing other aspects of the chemistry of these organophosphorus compounds. Because Wittig and Geissler were unaware of a related reaction that had been reported 30 years earlier by Staudinger, they were astonished by their result. Nevertheless, Wittig recognized the importance of the discovery. He and another student, Ullrich Schöllkopf, almost immediately published a second paper on the reactions of methylenetriphenylphosphorane with a variety of aldehydes and ketones to give alkenes under very mild conditions and with no isomerization. For the first time, it was possible to prepare alkenes with pre-dictable structure, because the carbon-carbon double bond that was produced by the reaction *always* replaced the carbon-oxygen bond of the carbonyl-containing reagent.

$$(C_6H_5)_3P = CH_2 + O = \hspace{2cm} \longrightarrow \hspace{0.5cm} H_2C =$$

Methylenetriphenyl-phosphorane
An ylide

Benzo-phenone

1,1-Diphenyl-ethylene

Soon the reaction was being widely used in research laboratories throughout the world, and today the Wittig reaction and variants thereof stand as some of the most important reactions in organic chemistry for the efficient synthesis of a diverse array of substituted alkenes. In research laboratories, the Wittig synthesis of alkenes is commonly applied to the preparation of small quantities of alkenes, but it may also be carried out on a large scale in industry where it has been used to prepare tonnage quantities of vitamin A!

HISTORICAL HIGHLIGHT
The Wittig Reaction (continued)

Vitamin A methyl ester

Vitamin A

Georg Friedrich Wittig was born in Berlin in 1897. Early in his life he became keenly interested in science, music, and mountain climbing, and he pursued these passions for his entire life. He began his studies in chemistry at the University of Tübingen, but his education was interrupted by his military service during World War I. After the war, he finished his studies at the University of Marburg, where he also obtained his Ph.D. degree. He remained at Marburg until 1932, when he went to the University of Braunschweig and then to the University of Freiburg. In 1944, he moved to Tübingen as Professor and head of the Institute of Chemistry. Finally he moved in 1956 to the University of Heidelberg as head of the department where he became Professor Emeritus in 1967. He died in 1987 at the age of 90.

Wittig was involved in many areas of research including the concept of ring strain, valence tautomerism, diradicals, the formation and reactions of dehydrobenzenes, carbanions, and organophosphorus chemistry. His interest in the latter two subjects gave rise to his work with the **ylides** that were formed upon deprotonation of alkyltriphenylphosphonium salts. In due course, this research resulted in his discovery of the Wittig synthesis of alkenes, work for which he shared the Nobel Prize in Chemistry in 1979 with Herbert Brown of Purdue University. His studies of carbanions also led to important discoveries with metalated derivatives of Schiff bases or imines, which are termed metalloenamines; these reactive intermediates have proven to be very useful in directed aldol reactions (Sec. 18.3) and for effecting the monoalkylations of aldehydes and ketones. But it is for his serendipitous discovery of the reaction that bears his name that Wittig is best remembered.

Relationship of Historical Highlight to Experiments

The Wittig synthesis of alkenes and variations of it are now regarded as one of the classical reactions of organic chemistry. Such reactions are of enormous importance because they produce alkenes of predictable structure under mild conditions. In the experiments in this chapter, you will have an opportunity not only to perform a reaction that earned its discoverer a Nobel Prize in Chemistry (Sec. 18.2) but also to explore the Wadsworth-Emmons reaction, a valuable modification of the original reaction.

CHAPTER 19

Organometallic Chemistry

19.1 | Introduction

Organometallic compounds are substances that contain carbon-metal bonds, and they may be generally represented by the structure **1**, in which the metal, **M**, may be Li, Na, Mg, Cu, Hg, Pd, or other transition elements. Organomagnesium compounds **1**, M = MgBr, which are commonly called **Grignard reagents**, were the first organometallic substances to be extensively studied, and they are still among the most important of all organometallic reagents. (See Historical Highlight at the end of this chapter.) The polarization of the bond between the carbon atom and the electropositive metal in these reagents renders the carbon atom electron-rich, or Lewis basic, and the carbon atom bears a partial negative charge, δ^-, as shown. Thus, one of the characteristic properties of organometallic reagents **1** is that the carbon atom serves as a **nucleophile** in chemical reactions. In contrast, when a carbon atom is bonded to more electronegative elements such as the halogen atom in the alkyl halide **2** (X = Cl, Br, or I) or oxygen atom in the carbonyl compound **3**, it is electron-deficient or Lewis acidic and possesses a partial positive charge. Such carbon atoms then serve as **electrophiles** in chemical reactions.

$$
\begin{array}{ccc}
\overset{R^1}{\underset{R^3}{R^2-\overset{\delta^-}{C}-M^{\delta^+}}} & \overset{R^1}{\underset{R^3}{R^2-\overset{\delta^+}{C}-X^{\delta^-}}} & \overset{R^1}{\underset{R^2}{C}=O^{\delta^-}}^{\delta^+}
\end{array}
$$

1	2	3
An organometallic reagent	An alkyl halide	A carbonyl compound

Because of their nucleophilic character, organometallic compounds are widely utilized as reagents in reactions that produce new carbon-carbon bonds. For example, two typical reactions of organometallic reagents **1** with carbon electrophiles such as alkyl halides **2** and carbonyl compounds **3** are illustrated by the general transformations shown in Equations 19.1 and 19.2. In each of these reactions, the nucleophilic carbon atom of one reactant becomes attached to the electrophilic carbon atom of the other reactant with the resulting formation of a new carbon-carbon bond. Thus, like many bond-forming processes, these reactions may be viewed in the simple context of *combinations of Lewis bases with Lewis acids.*

$$R^1 \overset{\delta^-}{C}H_2 - \overset{\delta^+}{M} + \overset{\delta^-}{X} - \overset{\delta^+}{C}H_2R^2 \longrightarrow R^1CH_2 - CH_2R^2 + MX \quad \text{(19.1)}$$

An organometallic
reagent   An alkyl halide   An alkane

$$R^1 \overset{\delta^-}{C}H_2 - \overset{\delta^+}{M} + \overset{R^2}{\underset{R^3}{\overset{\delta^+}{C}}} \overset{\delta^-}{=} O \longrightarrow R^1CH_2 - \overset{R^2}{\underset{R^3}{C}} - O^- M^+ \xrightarrow{H_3O^+} R^1CH_2 - \overset{R^2}{\underset{R^3}{C}} - OH \quad \text{(19.2)}$$

An organometallic   A carbonyl
reagent   compound   An alcohol

The following discussions focus on the preparation and reactions of one important class of organometallic compounds, the Grignard reagents **1** (M = Mg). However, many of the principles that are presented may be applied to the chemistry of other organometallic reagents. The experiments that may be performed in this chapter involve the preparation of Grignard reagents from aryl and alkyl bromides, followed by their typical reactions with (1) an ester to produce a tertiary alcohol, (2) carbon dioxide to produce a carboxylic acid, and (3) an aldehyde to produce a secondary alcohol.

19.2 Grignard Reagents: Preparation

REACTIONS OF ORGANIC HALIDES WITH MAGNESIUM METAL

Grignard reagents, R–MgX or Ar–MgX, are typically prepared by the reaction of an alkyl halide, R–X, or an aryl halide, Ar–X, with magnesium metal in an *anhydrous* ethereal solvent (Eq.19.3); the reagent dissolves as it is formed. It is noteworthy that in this process *carbon is transformed from an electrophilic center in the starting material, R–X or Ar–X, into a nucleophilic center in the product, R–MgX or Ar–MgX.*

$$\overset{\delta^+}{R} - \overset{\delta^-}{X} \quad \text{or} \quad \overset{\delta^+}{Ar} - \overset{\delta^-}{X} + Mg^\circ \xrightarrow[\text{or THF (solvent)}]{\text{Dry ether}} \overset{\delta^-}{R} - \overset{\delta^+}{MgX} \quad \text{or} \quad \overset{\delta^-}{Ar} - \overset{\delta^+}{MgX} \quad \text{(19.3)}$$

An alkyl   An aryl   An alkyl   An aryl
halide   halide   magnesium halide   magnesium halide

Although it is customary to represent the Grignard reagent by the formula R–MgX or Ar–MgX, the structure of the organometallic species in solution is complex and not well understood. For example, with alkyl magnesium halides, there is an equilibrium between RMgX, R_2Mg, and MgX_2 that depends on the solvent and the nature of the alkyl group. Moreover, the various organometallic species form aggregates in solution.

Grignard reagents are readily prepared from alkyl and aryl chlorides, bromides, and iodides; Grignard reagents derived from organofluorides are rarely used. The ease of formation of Grignard reagents from alkyl halides follows the order R–I > R–Br > R–Cl; aryl halides are less reactive than their alkyl counterparts, with aryl bromides and chlorides being comparable in reactivity.

In the experiments in this chapter, the Grignard reagents derived from bromobenzene (**4**) and 1-bromobutane (**6**) are prepared in an ethereal solvent according to Equations 19.4 and 19.5. The preparation of these reagents theoretically requires equivalent amounts of the organic halide and magnesium, but a slight excess of magnesium is normally used experimentally.

$$\underset{\substack{\textbf{4}\\ \text{Bromobenzene}}}{\text{C}_6\text{H}_5\text{Br}} + \text{Mg} \xrightarrow{\text{diethyl ether}} \underset{\substack{\textbf{5}\\ \text{Phenylmagnesium bromide}}}{\text{C}_6\text{H}_5\text{MgBr}} \qquad \textbf{(19.4)}$$

$$\underset{\substack{\textbf{6}\\ \text{1-Bromobutane}}}{\text{CH}_3\text{CH}_2\text{CH}_2\text{CH}_2{-}\text{Br}} + \text{Mg} \xrightarrow{\text{diethyl ether}} \underset{\substack{\textbf{7}\\ \text{1-Butylmagnesium bromide}}}{\text{CH}_3\text{CH}_2\text{CH}_2\text{CH}_2{-}\text{MgBr}} \qquad \textbf{(19.5)}$$

Use of an ethereal solvent is critical for the efficient preparation of the Grignard reagent because the basic oxygen atom of the ether complexes with the electropositive magnesium atom of $R(Ar){-}MgX$ to stabilize the organometallic species. The ethereal solvents most commonly used in this reaction are diethyl ether, $(C_2H_5)_2O$, and tetrahydrofuran (THF), $(CH_2)_4O$. Diethyl ether is often the solvent of choice because it is less expensive, may be purchased in anhydrous form, and is easily removed from the reaction mixture owing to its low boiling point (bp 36 °C, 760 Torr). Tetrahydrofuran is a stronger Lewis base than diethyl ether, and it also has better solvating ability; it may be used when the Grignard reagent does not form readily in diethyl ether.

Caution is necessary when using anhydrous diethyl ether that has been stored in metal containers. If these containers have been opened so that the diethyl ether has been exposed to oxygen, **hydroperoxides**, $CH_3CH(OOH)OCH_2CH_3$, may be present. These peroxides are *explosive*, and large volumes of anhydrous ether suspected to contain peroxides should *not* be evaporated to dryness. Moreover, emptied containers should be *thoroughly* rinsed with water to remove any peroxides before being discarded. Peroxides in anhydrous diethyl ether or tetrahydrofuran may be conveniently destroyed by distillation from alkali metals such as sodium or potassium metal or from lithium aluminum hydride, $LiAlH_4$. The presence of peroxides in diethyl ether and tetrahydrofuran may be detected by placing a drop on a piece of *moistened* starch/iodide test paper. If the paper turns dark violet, owing to formation of a starch-iodine complex, then the ether contains peroxides.

The magnesium metal used for the preparation of the Grignard reagent is normally in the form of turnings—thin shavings that have a high surface area relative to chunks of the metal. This type of magnesium is generally suitable for the preparation of most Grignard reagents. However, if the turnings have been repeatedly exposed to oxygen, their surface may be dull owing to a coating of magnesium oxide, which decreases the reactivity of the turnings. Magnesuim ribbon may also be used *after* its oxide coating has been removed by scraping with the edge of a spatula.

Initiating the reaction between the organic halide and magnesium may sometimes be difficult, especially with unreactive halides. In these cases, the reaction may often be initiated by adding a small crystal of iodine, I_2, to the reaction mixture. The iodine facilitates the reaction either by activating the metal by removing some of its oxide coating, or by converting a small amount of an alkyl halide to the corresponding iodide, which is more reactive toward magnesium. Alternatively, a small amount of

1,2-dibromoethane can be added to the reaction mixture to activate the magnesium, presumably by removing the oxide coating. 1,2-Dibromoethane is very reactive toward magnesium metal, giving ethylene and magnesium bromide (Eq. 19.6).

$$Br-CH_2CH_2-Br + Mg° \xrightarrow[\text{Dry } (C_2H_5)_2O]{} CH_2{=}CH_2 + MgBr_2 \qquad (19.6)$$

1,2-Dibromoethane   Ethylene
                                           (Ethene)

The formation of a Grignard reagent is an **exothermic** process. Since side reactions may occur if the reaction is allowed to proceed uncontrolled, it is important to regulate the rate at which the reaction proceeds. Grignard reactions often require an **induction period**, so it is *important* to be sure that the reaction of the alkyl or aryl halide with the magnesium has initiated *before* adding large quantities of the halide. Once the reaction is in progress, the halide should be added *dropwise and slowly* to the stirred suspension of the magnesium metal in the dry ethereal solvent; the halide may be added either neat or as a solution in the ethereal solvent. Such slow addition maintains a low concentration of the halide, and the rate of the reaction and the evolution of heat may thus be easily controlled. Because it is ultimately transferred to the cooling water in the condenser, the heat generated in exothermic reactions is also removed by the vaporization and refluxing of the solvent. The rate of the reaction must be controlled so that the refluxing solvent remains in the condenser. Should it appear that the reaction is getting out of control, as evidenced by vapors escaping from the top of the condenser, the reaction mixture should be immediately cooled with an ice-water bath. Indeed, it is prudent always to have such a cooling bath prepared when performing an exothermic reaction.

SIDE REACTIONS

Although the formation of Grignard reagents is quite general, side reactions may be encountered. For example, as the **conjugate bases** of the exceedingly weak organic acids R–H (R = aryl and alkyl), Grignard reagents are *strong* bases that react rapidly with water according to Equation 19.7. The overall result is the destruction of the Grignard reagent and the formation of the corresponding hydrocarbon, RH, and a basic magnesium salt. Grignard reagents also cannot be prepared when carboxyl (CO_2H), hydroxyl (OH), or amino (NH_2) groups are present in the alkyl or aryl halide; the acidic hydrogens of these functional groups will simply protonate the highly basic carbon atom of the Grignard reagent as it forms.

$$R-MgX + H_2O \longrightarrow R-H + HO-Mg-X \qquad (19.7)$$

The reactivity of Grignard reagents with water (Eq. 19.7) dictates that all reagents, solvents, and apparatus used for their preparation must be *thoroughly* dry. Consequently, *anhydrous* ethereal solvents, which typically contain less than 0.01% water, must be used. Anhydrous diethyl ether rapidly absorbs atmospheric moisture, so if a container of diethyl ether has been opened a number of times over a period of several days, the ether will not be suitable. Only freshly opened cans should be used, and the cans should always be tightly sealed *immediately* after the necessary volume of ether has been removed. These operations will also minimize the formation of dangerous peroxides as mentioned previously.

Other side reactions that may occur during formation of a Grignard reagent are shown in Equations 19.8–19.10, but these can be minimized by taking certain precautions. Reaction of the Grignard reagent with oxygen and carbon dioxide (Eqs. 19.8 and 19.9) is typically avoided in the research laboratory by performing the reaction under an inert atmosphere, such as nitrogen (N_2) or argon. However, in the undergraduate laboratory, this precaution is not essential because diethyl ether, which is used as the solvent, has a very high vapor pressure that effectively excludes most of the air from the reaction vessel.

The coupling reaction (Eq. 19.10) is an example of a Wurtz-type reaction. Although this can be a useful process for preparing symmetrical hydrocarbons R–R, it is generally desirable to minimize this side reaction by using dilute solutions, thereby avoiding high localized concentrations of the alkyl halide. This is accomplished by efficient stirring and by slowly adding the halide to the suspension of magnesium in the ethereal solvent.

$$RMgX + O_2 \longrightarrow R-O-O^- \overset{+}{M}gX \qquad (19.8)$$

$$RMgX + CO_2 \longrightarrow RCO_2^- \overset{+}{M}gX \qquad (19.9)$$

$$RMgX + RX \longrightarrow R-R + MgX_2 \qquad (19.10)$$

In the experiments that follow, phenylmagnesium bromide (**5**) and 1-butylmagnesium bromide (**7**) are prepared according to Equations 19.4 and 19.5, respectively. The most important side reaction in these experiments involves the Wurtz-type coupling of the Grignard reagent with the organic halide. For example, during the preparation of phenylmagnesium bromide, small quantities of biphenyl (**8**) are formed according to Equation 19.11; however, the presence of **8** does not interfere with the subsequent reactions of phenylmagnesium bromide. Although coupling may also occur during the preparation of 1-butylmagnesium bromide, the *n*-octane that is produced is volatile and easily removed.

$$(19.11)$$

5   **4**   **8**
                                    Biphenyl

19.3 Grignard Reagents: Reactions

REACTION OF PHENYLMAGNESIUM BROMIDE WITH METHYL BENZOATE

The aryl carbon atom bearing the magnesium in phenylmagnesium bromide is nucleophilic and reacts readily with electrophiles (Sec. 19.1). In the experiment that follows, two equivalents of phenylmagnesium bromide (**5**, Eq. 19.4) react with an ester, methyl benzoate (**9**), in an exothermic reaction to give triphenylmethanol (**11**) via the sequence of reactions depicted in Scheme 19.1. In the first step of the reaction, nucleophilic attack of phenylmagnesium bromide on the electrophilic carbon atom of the ester group of **9** gives benzophenone (**10**) as an intermediate. Subsequent reaction of **10** with an additional mole of **5** then produces an alkoxide salt, which is converted to **11** upon protonation with acid. The reaction is worked up with acid rather than water alone to avoid precipitation of the basic magnesium salt, HOMgX.

SCHEME 19.1

9   **5**
Methyl benzoate

10   **5**   **11**
Benzophenone   Triphenylmethanol

The principal organic products present after the aqueous work-up are the desired triphenylmethanol (**11**), benzene from any unreacted **5**, and biphenyl (**8**). Fortunately, it is possible to separate **11** and **8** easily owing to their relative solubilities in nonpolar hydrocarbon solvents. Biphenyl is considerably more soluble in cyclohexane than triphenylmethanol, so recrystallization of the crude product mixture gives pure triphenylmethanol.

REACTION OF PHENYLMAGNESIUM BROMIDE WITH CARBON DIOXIDE

Nucleophilic addition of phenylmagnesium bromide (**5**, Eq. 19.4) to carbon dioxide produces an intermediate carboxylate salt, which may be converted into benzoic acid (**12**) by acidification of the reaction mixture, as shown in Equation 19.12. This process is accompanied by side reactions that lead to the formation of benzophenone (**10**) and/or triphenylmethanol (**11**), as shown in Equation 19.13. However, by controlling the reaction conditions, formation of these by-products can be minimized. The bromomagnesium salt of benzoic acid is only slightly soluble in diethyl ether, so when the reaction

5   **12**
                                                                            Benzoic acid

(19.12)

is performed in this solvent, the salt precipitates from solution and cannot undergo further reaction. A large excess of carbon dioxide is also used to increase the likelihood that **5** will react with carbon dioxide rather than with the magnesium salt of benzoic acid. If the phenylmagnesium bromide is added slowly *to* the Dry Ice, carbon dioxide is always present in excess, thus favoring the desired reaction. Finally, when Dry Ice is used as the source of carbon dioxide, the temperature of the reaction is maintained at −78 °C until all of the Grignard reagent has been consumed. At this low temperature, the reaction of phenylmagnesium bromide with the bromomagnesium salt of benzoic acid is slow.

$$C_6H_5-C\overset{O}{\underset{O^-\overset{+}{M}gBr}{\diagdown}} \xrightarrow{C_6H_5-MgBr} C_6H_5-C\overset{O}{\underset{C_6H_5}{\diagdown}} \xrightarrow{C_6H_5-MgBr} \qquad (19.13)$$

10

$$\underset{C_6H_5}{\overset{\overset{+}{O^-}\overset{+}{M}gBr}{\underset{|}{C_6H_5-C-C_6H_5}}} \xrightarrow{H_3O^+} \underset{C_6H_5}{\overset{OH}{\underset{|}{C_6H_5-C-C_6H_5}}}$$

11

After the addition of the Grignard reagent to carbon dioxide is complete, the excess Dry Ice is allowed to evaporate, and the mixture is acidified. Extraction of the aqueous mixture with diethyl ether then gives a solution containing benzoic acid, benzophenone, triphenylmethanol, benzene, and biphenyl. The benzoic acid is readily separated from the neutral side products in the crude mixture by extracting the solution with dilute aqueous sodium hydroxide, whereby benzoic acid is converted to its water-soluble sodium salt; the neutral by-products remain in the organic layer. Cautious acidification of the aqueous layer regenerates the benzoic acid, which can be purified by recrystallization.

REACTION OF 1-BUTYLMAGNESIUM BROMIDE WITH 2-METHYLPROPANAL

1-Butylmagnesium bromide (**7**, Eq. 19.5) is also readily prepared, and its reaction with 2-methylpropanal (**13**) provides an excellent example of the Grignard synthesis of secondary alcohols (Eq. 19.14). In this reaction, the Grignard reagent **7** adds to 2-methylpropanal to give the magnesium salt of 2-methyl-3-heptanol, and the alcohol **14** may be isolated after acidification of the mixture.

$$CH_3CH_2CH_2CH_2\overset{\delta^-}{\underset{}{-}}\overset{\delta^+}{MgBr} + \overset{\overset{\delta^-}{O}}{\underset{H}{\overset{\delta^+}{C}-CHCH_3}} \longrightarrow \qquad (19.14)$$

7
1-Butylmagnesium bromide   **13**
2-Methylpropanal

$$\underset{CH_3}{\overset{\overset{+}{O^-}\overset{+}{M}gBr}{\underset{|}{CH_3(CH_2)_3-CH-CHCH_3}}} \xrightarrow{H_3O^+} \underset{CH_3}{\overset{OH}{\underset{|}{CH_3(CH_2)_3-CH-CHCH_3}}}$$

14
2-Methyl-3-heptanol

19.4 Special Experimental Techniques

Some of the side reactions that are encountered during the preparation of Grignard reagents and during their reaction with electrophiles can be controlled by prudent choice of solvent and reaction temperature. Another experimental technique that can be used to minimize undesired side reactions involves the **order of addition** of the Grignard reagent and the electrophilic reactant. For example, the electrophiles methyl benzoate and 2-methylpropanal are added *to* a solution of the appropriate Grignard reagent, which is a highly reactive nucleophile, to prepare triphenylmethanol (**11**) and 2-methyl-3-heptanol (**14**), respectively. This mode of addition is designated as **normal addition**. On the other hand, in the preparation of benzoic acid, the nucleophilic Grignard reagent is added *to* the carbon dioxide, the electrophile. This procedure is called **inverse addition**.

Whether an addition is performed according to the *normal* or *inverse* mode is dictated by the specific nature of the reactants and the potential side reactions that may occur. For example, if carbon dioxide gas were bubbled into a solution of phenylmagnesium bromide (**5**) to prepare benzoic acid (normal addition), the by-products **10** and/or **11** (Eq. 19.13) would become significant, perhaps predominant. There would always be an excess of the Grignard reagent **5** present relative to carbon dioxide, so further reaction between **5** and the bromomagnesium salt of benzoic acid would be favored. In the preparation of triphenylmethanol (**11**), the combination of the Grignard reagent with methyl benzoate could be done equally well in either the normal or inverse fashion, since two equivalents of **5** are necessary to complete the reaction, and the magnesium alkoxide salt of **11** is unreactive toward **5**. Inverse addition is less convenient in this case, however, because it requires transferring the Grignard reagent **5** from one flask to a dropping funnel and then addition of **5** to the reaction flask containing the methyl benzoate. Similarly, the reaction of 1-butylmagnesium bromide with 2-methylpropanal (Eq. 19.13) may be performed using either normal or inverse addition, but normal addition is again more convenient.

EXPERIMENTAL PROCEDURES

PREPARATION AND REACTIONS OF GRIGNARD REAGENTS

A · PREPARATION OF GRIGNARD REAGENTS

Purpose: To develop techniques required to prepare Grignard reagents from aryl and alkyl halides.

Safety Alert

1. Diethyl ether is *extremely* flammable and volatile, and its vapors can easily travel several feet along the bench top or the floor and then be ignited. Consequently, *be certain there are no open flames anywhere in the laboratory* whenever you are working with ether. Use a *flameless* heating source *whenever* heating is required.

2. The *anhydrous* diethyl ether used in this experiment is contained in metal cans, and the screw or plastic cap should *always* be in place when the can is not in use to prevent evaporation, absorption of moisture and oxygen, and accidental fires.
3. *Open containers of diethyl ether must not be kept at your laboratory bench or stored in your laboratory drawer.* Estimate the total volume of ether you will need and measure it in the hood into a container that is *loosely* stoppered before returning to your bench.
4. You should use ovens to dry your glassware if possible. However, if drying ovens are not available and it is necessary to dry the glass apparatus with a flame or a heat gun, be *certain* that no one in the laboratory is working with diethyl ether. *Consult with your instructor before using any open flame.* Avoid excessive heating in the vicinity of the ring seals in the condenser and near the stopcock in the addition funnel, particularly if the stopcock is made of plastic or Teflon.
5. Lubricate all ground-glass joints in the apparatus carefully, and mate them tightly to prevent the escape of diethyl ether during the reaction.
6. On the small scale of these experiments, the exothermic formation and reaction of Grignard reagents rarely causes a problem. Nevertheless, it is still good laboratory practice to have an ice-water bath ready if the reaction proceeds too rapidly, as evidenced by an *excessively* rapid rate of reflux *and* the emission of vapors from the top of the condenser.

Miniscale Procedure

Preparation   Answer the Pre-Lab Exercises on page PL. 121. Review Sections 2.9, 2.10, 2.11, 2.22, 2.27, and 2.28.

Apparatus   A 50-mL round-bottom flask, 5-mL syringe, separatory funnel with a ground-glass joint, condenser, Claisen adapter, drying tube, ice-water bath, and apparatus for magnetic stirring and *flameless* heating.

Setting Up   Weigh 0.5 g of magnesium turnings that have been freshly crushed with a spatula into the round-bottom flask and add a stirbar. Place this flask and its contents, the separatory funnel, condenser, Claisen adapter, and drying tube in an oven at 110 °C for at least 30 min. If the separatory funnel has a plastic or Teflon stopcock and stopper, do *not* place the stopcock, its plastic retaining nut, and the stopper in the oven as they may melt or soften. Using gloves or tongs, remove the glassware from the oven. After the glassware is cool enough to handle, lubricate all the joints and quickly assemble the apparatus shown in Figure 2.63b. Attach the drying tube to the top of the condenser and place the stopper and stopcock in the separatory funnel. Allow the apparatus to cool to room temperature.

Optional Measures If an oven is not available, it will be necessary to dry the apparatus with a microburner or a heat gun. Assemble the apparatus as described above. *Be sure that no one in the laboratory is working with diethyl ether*, and then dry the assembled apparatus. Do not overheat any plastic parts of the apparatus. Allow the apparatus to cool to room temperature.

Verify that there are *NO flames* in the laboratory before continuing. Add 5 mL of *anhydrous* diethyl ether to the round-bottom flask *through* the separatory funnel; *close the stopcock.* Be sure that water is running through the condenser. Prepare a solution of *either* 2.4 mL of bromobenzene *or* 2.5 mL of 1-bromobutane in 5 mL of anhydrous diethyl ether in a small, *dry* Erlenmeyer flask. Swirl the solution to achieve homogeneity and transfer this solution to the separatory funnel.

Reaction   Add a 0.5-mL portion of the aryl or alkyl halide-ether solution from the separatory funnel onto the magnesium turnings and stir the resulting mixture. If small bubbles form at the surface of the magnesium turnings or if the mixture becomes slightly cloudy or chalky, the reaction has started. The flask should become *slightly* warm. If the reaction has started, *disregard* the optional instructions in the next paragraph.

> **Optional Measures** If the reaction does not start spontaneously, warm the mixture gently for several minutes; observe whether the mixture becomes slightly cloudy or chalky. If it does not, then obtain one or two additional magnesium turnings and crush them *thoroughly* with a heavy spatula or the end of a clamp. Remove the separatory funnel just long enough to add these broken pieces of magnesium to the flask, and quickly replace the funnel. The clean, unoxidized surfaces of magnesium that are exposed should aid in initiating the reaction. If the reaction still has not started after an additional 3–5 min of warming, consult your instructor. The best remedy at this point is to add a small crystal of iodine or 2–3 drops of 1,2-dibromoethane to the mixture. Alternatively, a small amount of the preformed Grignard reagent may be added.

Once the reaction has started, *gently* heat the reaction mixture so the solvent re-fluxes smoothly. Add another 5-mL portion of anhydrous diethyl ether to the reaction mixture through the top of the *condenser* and continue heating and stirring until the solvent is again refluxing. The rest of the halide-ether solution should now be added *dropwise* to the stirred reaction mixture at a rate that is just fast enough to maintain a gentle reflux. If the reaction becomes too vigorous, reduce the rate of adding the ethereal solution of aryl or alkyl halide and, if necessary, discontinue heating the flask. If the spontaneous boiling of the mixture slows, increase the rate of addition slightly. If the rate of reflux still does not increase, heat the mixture as necessary to maintain gentle reflux during the remainder of addition. *It is important that reflux be maintained throughout the addition of the halide-ether solution.* The addition should take about 5–10 min. Upon completing the addition, continue heating the mixture under gentle reflux for 15 min. If necessary add anhydrous diethyl ether to the reaction flask so that there is no less than about 15 mL of solution. At the end of the reaction, the solution normally has a tan to brown, chalky appearance, and most of the magnesium has dis-appeared, although residual bits of metal usually remain. Discontinue heating and allow the mixture to cool to room temperature.

Use the Grignard reagent as soon as possible after its preparation. Phenylmagne-sium bromide is used in Parts B and C that follow, whereas 1-butylmagnesium bromide is used in Part D.

Microscale Procedure

Preparation   Answer the Pre-Lab Exercises on page PL. 121. Review Sections 2.9, 2.10, 2.11, 2.22, 2.27, and 2.28.

Apparatus A 3-mL and 5-mL conical vial, screw-cap centrifuge tube, two 1-mL plastic (or glass) syringes, Pasteur pipet with 0.5- and 1.0-mL calibration marks, Claisen adapter, drying tube, ice-water bath, and apparatus for magnetic stirring and *flameless* heating.

Setting Up Dry the conical vials, Claisen adapter, centrifuge tube, drying tube, and calibrated Pasteur pipet in an oven at 110 °C for at least 30 min. Do *not* place any plastic connectors or rubber *O*-rings in the oven as they may melt or soften. Using gloves or tongs, remove the glassware from the oven. Cool the glassware in a desiccator if possible. Lubricate the joints and assemble the apparatus shown in Figure 19.1 by adding a spinvane to the 5-mL conical vial and then fitting the Claisen adapter to the vial. Finally, place a rubber septum and the drying tube on the Claisen adapter.

Optional Measures If an oven is not available, it will be necessary to dry the apparatus with a microburner or a heat gun. Assemble the apparatus as described above. *Be sure that no one in the laboratory is working with diethyl ether*, and then dry the assembled apparatus. Be careful heating the conical vial as this heavy-walled vessel can easily crack from thermal shock. Also do not overheat any plastic parts of the apparatus. Allow the apparatus to cool to room temperature.

Verify that there are *NO flames* in the laboratory before continuing. Transfer about 3–4 mL of anhydrous diethyl ether to a screw-cap centrifuge tube, and cap the tube. Use this during the remainder of the experiment whenever anhydrous diethyl ether is required. Weigh 0.05 g of magnesium turnings that have been freshly crushed with a heavy spatula or the end of a clamp. Remove the rubber septum and transfer the turnings to the reaction vial. Add one *small* crystal of iodine to the vial and replace the rubber septum. Add 0.2 mL of anhydrous diethyl ether to the vial using a dry 1-mL syringe inserted through the rubber septum. Prepare a solution of 0.24 mL of bromobenzene in 0.5 mL of anhydrous diethyl ether in a dry 3-mL conical vial. Swirl the solution to achieve homogeneity.

FIGURE 19.1 Microscale apparatus for preparing and reacting Grignard reagents.

Reaction   Stir the contents of the conical vial while warming gently. Draw the ethereal solution of bromobenzene into the plastic syringe and cap the conical vial that contained the bromobenzene solution. Insert the syringe needle through the rubber septum, and add a 0.1-mL portion of this solution onto the magnesium turnings; stir the mixture. *Hold* the plunger of the syringe to control the amount of solution added. If small bubbles form at the surface of the magnesium turnings or if the mixture becomes slightly cloudy or chalky, the reaction has started.

> **Optional Measures** If the reaction does not start spontaneously, consult your instructor. The best remedy at this point is to continue warming the vial and add an additional crystal of iodine or a small drop of 1,2-dibromoethane to the mixture. You may also add a small sample of phenylmagnesium bromide from a stock solution.

Once the reaction has started, continue heating the reaction mixture gently so the solvent refluxes slightly. Using a second plastic syringe inserted through the rubber septum, add another 0.5-mL portion of anhydrous diethyl ether to the reaction mixture, and continue heating and stirring until the solvent is again refluxing. The rest of the bromobenzene-ether solution should now be added *dropwise* to the stirred reaction mixture at a rate that is just fast enough to maintain a gentle reflux. If the reaction becomes too vigorous, reduce the rate of adding the ethereal solution of aryl halide and, if necessary, discontinue heating the vial. If the spontaneous boiling of the mixture slows, increase the rate of addition slightly. If the rate of reflux still does not increase, increase heating and maintain gentle reflux during the remainder of addition. *It is important that reflux be maintained throughout the addition of the bromobenzene-ether solution.* The addition should take about 3–5 min. Upon completion of the addition, place about 0.5 mL of anhydrous diethyl ether in the vial that contained the bromobenzene solution. Draw this solution into a syringe and add it in one portion to the reaction mixture. Continue heating the mixture with stirring under gentle reflux for 15 min. If necessary, add anhydrous diethyl ether to maintain a level *no lower* than the 2-mL mark on the conical vial. At the end of the reaction, the solution normally has a tan to brown, chalky appearance, and most of the magnesium has disappeared, although residual bits of metal usually remain. Discontinue heating and allow the mixture to cool to room temperature.

Use the Grignard reagent as soon as possible after its preparation following one of the procedures given in Parts B and C that follow.

B · PREPARATION OF TRIPHENYLMETHANOL

Purpose: To demonstrate the preparation of a tertiary alcohol by the reaction of a Grignard reagent with an ester.

 Safety Alert   Review the Safety Alert for Part A of these experiments.

Miniscale Procedure

Preparation   Answer the Pre-Lab Exercises on page PL. 123. Review Sections 2.10, 2.11, 2.13, 2.17, 2.21, 2.22, and 2.29.

Apparatus   Glass apparatus from Part A of these experiments, separatory funnel, ice-water bath, and apparatus for magnetic stirring, simple distillation, vacuum filtration, and *flameless* heating.

Setting Up   While the reaction mixture for the preparation of phenylmagnesium bromide (Part A) is cooling to room temperature, dissolve 1.2 mL of methyl benzoate in about 5 mL of *anhydrous* diethyl ether, and place this solution in the separatory funnel with the *stopcock closed*. Cool the reaction flask containing the phenylmagnesium bromide in the ice-water bath.

Reaction   Begin the *slow, dropwise* addition of the ethereal solution of methyl benzoate to the *stirred* solution of phenylmagnesium bromide. This reaction is *exothermic*, so you should control the rate of reaction by adjusting the rate of addition *and* by occasionally cooling the reaction flask as needed with the ice-water bath. The ring of condensate should be allowed to rise no more than one-third of the way up the reflux condenser. A white solid may form during the reaction, and this is normal. After the addition is complete and the exothermic reaction subsides, you may complete the reaction in one of two ways. Consult with your instructor to determine whether you should (1) heat the reaction mixture at reflux for 30 min or (2) stopper the flask after cooling the contents to room temperature and place it in the *hood* until the next laboratory period (no reflux required).★

Work-Up, Isolation, and Purification   Place about 10 mL of cold 6 *M* sulfuric acid and about 5–10 g of crushed ice in a beaker. If the reaction mixture solidified upon cooling, add a small quantity of solvent-grade diethyl ether to the reaction flask. Pour the reaction mixture gradually with stirring into the ice-acid mixture. Rinse the round-bottom flask with 2–3 mL of solvent-grade diethyl ether and add this wash to the beaker. Continue stirring until the heterogeneous mixture is completely free of undissolved solids. It may be necessary to add a small portion of solvent-grade diethyl ether to dissolve all the organic material; the total volume of ether should be about 15–20 mL. Verify that the aqueous layer is acidic; if it is not, add cold 6 *M* sulfuric acid dropwise until the layer is acidic. If necessary, sequentially add 2–3-mL portions of solvent-grade diethyl ether and then water to dissolve all of the solids.

Transfer the entire mixture to a separatory funnel. Shake the funnel vigorously with venting to relieve pressure; separate the aqueous layer.★ Wash the organic layer sequentially with about 5 mL of 3 *M* sulfuric acid, two 5-mL portions of saturated aqueous sodium bicarbonate (*vent!*), and finally with one 5-mL portion of saturated sodium chloride solution. Dry the organic layer using several spatula-tips full of anhydrous sodium sulfate. Swirl the flask occasionally for a period of 10–15 min to facilitate drying; add further small portions of anhydrous sodium sulfate if the solution remains cloudy.★

Filter or decant the solution into a 50-mL round-bottom flask and equip the flask for simple distillation. Remove the diethyl ether by simple distillation. Alternatively, rotary evaporation or other techniques (Sec. 2.29) may be used to concentrate the solution. The final traces of solvent may be removed by attaching the flask to a vacuum source, gently swirling the contents as the vacuum is applied. After the crude solid residue has dried, determine its melting range, which may be wide.★

Purify the triphenylmethanol by dissolving it in a *minimum* amount of boiling cyclohexane (ca. 10 mL/g product). Perform this operation at the hood, or use a funnel that is attached to a vacuum source and inverted over the flask (Fig. 2.68b). Once all the material is in solution, evaporate the solvent *slowly* until small crystals of triphenylmethanol start to form. Allow the crystallization to continue at room temperature and then in an ice-water bath until no more crystals form. Isolate the product by vacuum filtration and air-dry it.

Analysis   Weigh the triphenylmethanol and calculate the percent yield; determine its melting point. Obtain IR and ^{1}H NMR spectra of your starting materials and product and compare them with those of authentic samples (Figs. 15.12, 15.13, and 19.2–19.5).

Microscale Procedure

Preparation   Answer the Pre-Lab Exercises on page PL. 123. Review Sections 2.10, 2.11, 2.13, 2.17, 2.21, 2.22, and 2.29.

Apparatus   Glass apparatus from Part A of these experiments, 3-mL conical vial, two screw-cap centrifuge tubes, 1-mL plastic syringe, Pasteur pipet with 0.5- and 1.0-mL calibration marks, ice-water bath, and apparatus for magnetic stirring, simple distillation, vacuum filtration, Craig tube filtration, and *flameless* heating.

Setting Up   While the reaction mixture for the preparation of phenylmagnesium bromide (Part A) is cooling to room temperature, dissolve 120 μL of methyl benzoate in 0.5 mL of anhydrous diethyl ether contained in the 3-mL conical vial. Cool the conical vial containing the phenylmagnesium bromide in the ice-water bath.

Reaction   Draw the ethereal solution of methyl benzoate into the syringe and recap the vial. Insert the syringe needle through the rubber septum and begin the *slow, drop-wise* addition of the ethereal solution of methyl benzoate to the *stirred* solution of phenylmagnesium bromide. This reaction is *exothermic*, so you should control the rate of reaction by adjusting the rate of addition *and* by occasionally cooling the reaction vial with the ice-water bath if necessary. The reflux ring should not be allowed to rise into the Claisen adapter. A white solid may form during the reaction, and this is normal. Upon completion of the addition, place about 0.2–0.3 mL of anhydrous diethyl ether in the vial that contained the methyl benzoate solution. Draw this solution into a syringe and add it in one portion to the reaction mixture. After the exothermic reaction subsides, you may complete the reaction in one of two ways. Consult with your instructor to determine whether you should (1) heat the reaction mixture at *gentle* reflux for 30 min or (2) cap the vial after cooling the contents to room temperature and place it in the *hood* until the next laboratory period (no reflux required). The mixture may solidify on cooling.★

Work-Up, Isolation, and Purification   Remove the Claisen adapter and drying tube, and place the conical reaction vial in an ice-water bath. Using a Pasteur pipet, *slowly* add 1.5 mL of cold 3 *M* sulfuric acid dropwise to neutralize the reaction mixture. *Be careful* as the addition of acid may be accompanied by frothing. Use a small glass stirring rod to break up the solid during this addition. You may need to cap the vial tightly and shake it with venting to facilitate dissolution of the solids. Continue shaking the vial until the heterogeneous mixture is free of undissolved solids. You should obtain two distinct layers as the solid gradually dissolves. It may be necessary to add solvent-grade diethyl ether to dissolve all the organic material; the total volume of ether should be about 2.0–2.5 mL. Verify that the aqueous layer is acidic; if it is not, add cold 3 *M* sulfuric acid dropwise until the layer is acidic.

Optional Measures If some solid material remains after the addition of aqueous acid is complete, the entire mixture may be transferred to a screw-cap centrifuge tube. Sequentially add solvent-grade diethyl ether and water in 0.5-mL portions to the centrifuge tube. After each addition, cap the tube and shake it to dissolve the solids. Continue the extraction as described, using appropriately sized conical vials or screw-cap centrifuge tubes.

Remove the spinvane with forceps and, holding it above the vial, rinse it with several drops of solvent-grade ether. Cap the vial tightly, invert it gently several times with venting to mix the layers, and allow the layers to separate. Using a Pasteur pipet, transfer the lower aqueous layer to a conical vial. Wash this layer with one 0.5-mL portion of solvent-grade ether and combine this ethereal layer with the first.★ Wash the combined organic layers sequentially with a 0.5-mL portion of 3 *M* sulfuric acid, two 0.5-mL portions of saturated aqueous sodium bicarbonate (*vent!*), and finally with one 0.5-mL portion of saturated aqueous sodium chloride. Using a clean filter-tip pipet, transfer the ether layer to a screw-cap centrifuge tube containing several microspatula-tips full of anhydrous sodium sulfate. Swirl the tube occasionally for a period of 10–15 min to facilitate drying; add further small portions of anhydrous sodium sulfate if the solution remains cloudy.★

With a filter-tip pipet, transfer the dried organic solution into a clean 5.0-mL conical vial, rinse the sodium sulfate with about 0.2 mL of diethyl ether and add this rinse to the vial. Equip the vial for simple distillation. Remove the solvent by simple distillation; withdraw the distillate from the Hickman stillhead as needed. Alternatively, rotary evaporation or other techniques (Sec. 2.29) may be used to concentrate the solution. The final traces of solvents may be removed by attaching the vial to a vacuum source. After the crude solid residue has dried, determine its melting range, which may be wide.★

Purify the triphenylmethanol by dissolving it in a *minimum* amount of boiling cyclohexane (ca. 1 mL/0.1 g product) in a Craig tube. Perform this operation at the hood. Once all the material is in solution, evaporate the solvent *slowly* until small crystals of triphenylmethanol start to form. Allow the crystallization to continue at room temperature and then in an ice-water bath until no more crystals form. Isolate and air-dry the product.

Analysis   Weigh the triphenylmethanol and calculate the percent yield; determine its melting point. Obtain IR and ^{1}H NMR spectra of your starting materials and product and compare them with those of authentic samples (Figs. 15.12, 15.13, and 19.2–19.5).

Dilute the *combined aqueous layers and washes* with water, neutralize the solution if necessary, and flush it down the drain with excess water. Place the *ether distillate* and the *cyclohexane mother liquor* in the container for nonhalogenated organic solvents. Spread the *calcium chloride* from the drying tube and the *sodium sulfate* on a tray in the hood and, after the ether has evaporated, place them and the *filter paper* in the container for nonhazardous solids.

Finishing Touches

C · PREPARATION OF BENZOIC ACID

Purpose: To demonstrate the preparation of a carboxylic acid by the reaction of a Grignard reagent with carbon dioxide.

Safety Alert

1. Review the Safety Alert for Part A of these experiments.
2. Use tongs or gloves to handle the solid carbon dioxide (Dry Ice) because contact with the skin can cause severe frostbite. Crush the Dry Ice by wrapping larger pieces in a clean dry towel and striking them with a mallet or a wooden block. Use the Dry Ice *immediately* after crushing to minimize contact with atmospheric moisture.

Miniscale Procedure

Preparation   Answer the Pre-Lab Exercises on page PL. 125. Review Sections 2.17 and 2.21.

Apparatus   Glass apparatus from Part A of these experiments, separatory funnel, apparatus for vacuum filtration and *flameless* heating.

Setting Up   After the reaction mixture for the preparation of phenylmagnesium bromide (Part A) has cooled to room temperature, place about 10 g of coarsely crushed Dry Ice in a 125-mL Erlenmeyer flask. The Dry Ice should be crushed *immediately* before use and protected as much as possible from atmospheric moisture.

Reaction   Slowly pour the phenylmagnesium bromide solution from Part A onto the Dry Ice with vigorous swirling; the mixture normally becomes rather viscous. Rinse the original round-bottom reaction flask with 2–3 mL of anhydrous diethyl ether and add this wash to the Erlenmeyer flask. Continue swirling the contents of the flask for a few minutes. After the addition is complete, cover the flask with a small watch glass or piece of filter paper, and allow the excess carbon dioxide to sublime by letting the flask stand in the hood, properly labeled with your name, until the next laboratory period.★

You may expedite the removal of the excess carbon dioxide by shaking or swirling the flask while warming it *very slightly* in a warm-water bath or by adding small amounts of warm water to the reaction mixture and then shaking or swirling the flask. Because both of these methods may cause a *sudden* evolution of carbon dioxide gas and splash the contents of the flask onto the floor or bench top, be careful if you use one of them. The flask should *never* be stoppered.

Work-Up, Isolation, and Purification   After the excess Dry Ice has completely sublimed, most of the ether will have also evaporated, so add 20 mL of solvent-grade diethyl ether to the reaction flask. Prepare a mixture from about 5 g of ice and 10 mL of cold 3 M sulfuric acid in an Erlenmeyer flask. Add this acid to the reaction mixture slowly to avoid excessive foaming. If the ether evaporates appreciably during this operation, more solvent-grade diethyl ether should be added so that the total volume of ether is about 20–25 mL. Verify that the aqueous layer is acidic; if it is not, add cold 3 M sulfuric acid dropwise until the layer is acidic.

Swirl the mixture and transfer it to a separatory funnel. Rinse the flask with a small portion of diethyl ether and add the rinse to the separatory funnel. Shake the funnel cautiously with venting and separate the layers. Extract the aqueous layer with a 10-mL portion of solvent-grade diethyl ether and add this to the original ether layer. Extract the *combined* ethereal fractions sequentially with two 10-mL portions of a 1 *M* solution of sodium hydroxide; vent the funnel frequently during the extractions. Transfer the two alkaline extracts to an Erlenmeyer flask and slowly add 6 *M* hydrochloric acid until precipitation of the benzoic acid is complete and the aqueous mixture is *acidic*. Cool the solution in an ice-water bath, isolate the solid benzoic acid by vacuum filtration, wash it with water, and air-dry it.★ Recrystallize the benzoic acid from water.

Analysis   Weigh the benzoic acid and calculate the percent yield; determine its melting point. Obtain IR and ^{1}H NMR spectra of your starting material and product and compare them with those of authentic samples (Figs. 15.12, 15.13, 19.6, and 19.7).

Microscale Procedure

Preparation   Answer the Pre-Lab Exercises on page PL. 125. Review Sections 2.17 and 2.21.

Apparatus   Glass apparatus from Part A of these experiments, a 3-mL and a 5-mL conical vial, screw-cap centrifuge tube, Pasteur pipet with 0.5- and 1.0-mL calibration marks, apparatus for vacuum filtration, and Craig tube filtration.

Setting Up   After the reaction mixture for the preparation of phenylmagnesium bromide (Part A) has cooled to room temperature, place about 1 g of coarsely crushed Dry Ice in a small beaker. The Dry Ice should be crushed immediately before use and protected as much as possible from atmospheric moisture.

Reaction   Using a filter-tip pipet, slowly transfer the phenylmagnesium bromide solution from Part A onto the Dry Ice. Stir the mixture with a glass rod during the addition, as the mixture normally becomes rather viscous. Rinse the conical vial with 0.5 mL of anhydrous diethyl ether and add this wash to the beaker. After the addition is complete, cover the beaker with a small watchglass or piece of filter paper, and allow the excess carbon dioxide to sublime by letting the beaker stand in the hood, properly labeled with your name, until the next laboratory period.★

You may expedite the removal of the excess carbon dioxide by stirring the contents of the beaker while warming it *very slightly* in a warm-water bath or by adding small amounts of warm water to the reaction mixture with stirring. Because both of these methods may cause a *sudden* evolution of carbon dioxide gas and splash the contents of the beaker onto the floor or bench top, be careful using them.

Work-Up, Isolation, and Purification   After the excess Dry Ice has completely sublimed, most of the ether will have also evaporated, so add 2.5 mL of solvent-grade diethyl ether. Then *slowly* add 1.5 mL of cold 3 *M* sulfuric acid to the reaction mixture to avoid excessive foaming. If the ether evaporates appreciably during this operation, more solvent-grade diethyl ether should be added so that the total volume of ether is about 2.0–2.5 mL. Verify that the aqueous layer is acidic; if it is not, add cold 3 *M* sulfuric acid dropwise until the layer is acidic.

Optional Measures If some solid material remains after the addition of aqueous acid is complete, the entire mixture may be transferred to a screw-cap centrifuge tube. Sequentially add solvent-grade diethyl ether and water in 0.5-mL portions to the centrifuge tube. After each addition, cap the tube and shake it to dissolve the solids. Continue the extraction as described using appropriately sized conical vials or screw-cap centrifuge tubes.

Using a Pasteur pipet, transfer the entire mixture to a 5-mL conical vial. Rinse the beaker with a small portion of solvent-grade diethyl ether and add the rinse to the conical vial.★ Cap the vial tightly, invert it gently several times with venting to mix the layers thoroughly, and allow the layers to separate. Transfer the aqueous layer to a 3-mL conical vial, extract the aqueous layer with a 0.5-mL portion of solvent-grade diethyl ether, and add this extract to the original ether layer. Extract the *combined* ethereal fractions sequentially with two 1-mL portions of a 1 *M* solution of sodium hydroxide, venting the vial frequently during the extractions. Transfer the two alkaline extracts to a small Erlenmeyer flask and warm the flask for about 5 min to remove the ether that is dissolved in the aqueous phase; a saturated solution of diethyl ether in water contains 7% diethyl ether. Cool the alkaline solution, and slowly add 6 *M* hydrochloric acid until precipitation of the benzoic acid is complete and the aqueous mixture is acidic. Cool the solution in an ice-water bath, isolate the solid benzoic acid by vacuum filtration and air-dry the sample.★ Recrystallize the benzoic acid from water in a Craig tube.

Analysis   Weigh the benzoic acid and calculate the percent yield; determine its melting point. Obtain IR and ^{1}H NMR spectra of your starting material and product and compare them with those of authentic samples (Figs. 15.12, 15.13, 19.6, and 19.7).

Finishing Touches

Dilute the *combined aqueous layers and washes* with water, neutralize the solution if necessary, and flush it down the drain with excess water. Place the *combined ethereal layers* in the container for nonhalogenated organic solvents. Spread the *calcium chloride* from the drying tube on a tray in the hood to allow the ether to evaporate; then place it and the *filter paper* in the container for nonhazardous solids.

D · PREPARATION OF 2-METHYL-3-HEPTANOL

Purpose: To demonstrate the preparation of a secondary alcohol by the reaction of a Grignard reagent with an aldehyde.

Safety Alert   Review the Safety Alert for Part A of these experiments.

Miniscale Procedure

Preparation   Answer the Pre-Lab Exercises on page PL. 127. Review Sections 2.10, 2.11, 2.13, 2.17, 2.21, and 2.29.

Apparatus   Glass apparatus from Part A of these experiments, a 25-mL round-bottom flask, separatory funnel, ice-water bath, and apparatus for magnetic stirring, simple distillation, shortpath distillation, and *flameless* heating.

Setting Up   While the reaction mixture for the preparation of 1-butylmagnesium bromide (Part A) is cooling to room temperature, dissolve 1.8 mL of *freshly distilled* 2-methylpropanal in 5 mL of *anhydrous* diethyl ether, and place this solution in the separatory funnel with the *stopcock closed*. Cool the reaction flask containing the 1-butylmagnesium bromide in an ice-water bath and begin stirring.

Reaction   Over a period of about 5 min and in a *dropwise* manner, add the ethereal solution of 2-methylpropanal to the stirred solution of 1-butylmagnesium bromide. Control the resulting exothermic reaction by adjusting the rate of addition so that the ring of condensate rises no more than one-third of the way up the reflux condenser. After completing the addition, allow the reaction mixture to stand for about 15 min.★ The reaction mixture may be stored, loosely stoppered, *in the hood* until the next laboratory period.

Work-Up, Isolation, and Purification   Place about 10 mL of cold 6 *M* sulfuric acid and 5–10 g of crushed ice in a beaker. Pour the reaction mixture *slowly* with stirring into the ice-acid mixture. After the addition is complete, transfer the cold mixture, which may contain some precipitate, to a separatory funnel and shake it gently. The precipitate should dissolve. Separate the layers. Extract the aqueous layer sequentially with two 5-mL portions of solvent-grade diethyl ether, and add these extracts to the main ethereal layer. Transfer the combined ethereal extracts into the separatory funnel. Venting the funnel frequently to relieve pressure, wash the solution sequentially with 5 mL of saturated aqueous sodium bisulfite, two 5-mL portions of saturated aqueous sodium bicarbonate, and finally with 5 mL of saturated aqueous sodium chloride. Dry the ethereal solution over several spatula-tips full of anhydrous sodium sulfate.★ Swirl the flask occasionally for a period of 10–15 min to facilitate drying; add further small portions of anhydrous sodium sulfate if the solution remains cloudy.

Filter or carefully decant the dried organic solution into a round-bottom flask equipped for simple distillation, and remove the diethyl ether by simple distillation. Alternatively, rotary evaporation or other techniques (Sec. 2.29) may be used to concentrate the solution. The final traces of solvent may be removed by attaching the flask to a vacuum source, gently swirling the contents as the vacuum is applied.★

Using a Pasteur pipet, transfer the cooled pot residue to a 25-mL round-bottom flask containing a stirbar. Rinse the original stillpot with 1–2 mL of diethyl ether and add the rinse to the 25-mL round-bottom flask. Remove the remaining diethyl ether by attaching the flask to a vacuum source, gently swirling the contents as the vacuum is applied. Fit this flask for a shortpath distillation. Insulate the top of the distilling flask and the stillhead with cotton or glasswool wrapped with aluminum foil, and distil the 2-methyl-3-heptanol into a *tared* receiver. Collect all distillate that boils above 130 °C (760 Torr). Note that the head temperature may never reach the reported boiling point, owing to the high boiling temperature and the small amount of sample being distilled. Terminate heating when the dark pot residue becomes viscous and starts to evolve fumes.

Analysis   Weigh the product and calculate the yield. Obtain IR and ^{1}H NMR spectra of your starting materials and product and compare them with those of authentic samples (Figs. 14.3, 14.4, 18.17, 18.18, 19.8, and 19.9).

Dilute the *combined aqueous layers and washes* with water, neutralize the solution if necessary, and flush it down the drain with excess water. Place the *ether distillate* in the container for nonhalogenated organic solvents. Spread the *calcium chloride* from the drying tube and the *sodium sulfate* on a tray in the hood, and then place them in the container for nonhazardous solids after the volatiles have evaporated.

Finishing Touches

EXERCISES

Refer to Tables 8.1–8.8 as needed for answering the questions on spectroscopy.

General Questions

1. Arrange the following compounds in order of increasing reactivity toward attack of a Grignard reagent at the carbonyl carbon atom: methyl benzoate, benzoic acid, benzaldehyde, and benzophenone. Explain the basis for your decision, making use of mechanisms where needed.

2. What is (are) the product(s) of reaction of each of the carbonyl-containing compounds in Exercise 1 with *excess* Grignard reagent, RMgBr?

3. How might primary, secondary, and tertiary alcohols be prepared from a Grignard reagent and a suitable carbonyl-containing compound? Write chemical reactions for these preparations using any starting materials you wish; indicate stoichiometry where important.

Questions for Part A

4. Why were you cautioned not to heat excessively in the vicinity of plastic parts, rubber *O*-rings, and the ring seals of the condenser when *flame*-drying the apparatus for this experiment?

5. Ethanol is often present in solvent-grade diethyl ether. If this grade rather than anhydrous were used, what effect would the ethanol have on the formation of the Grignard reagent?

6. What is the purpose of adding an additional portion of ether after the Grignard reaction is initiated and before the addition of halide is continued?

7. Give a plausible, three-dimensional structure for the complex $RMgBr \cdot 2 (C_2H_5)_2O$. How do you think the molecules of diethyl ether are bound to the Grignard reagent?

8. What problems might be encountered if bromocyclohexane were used as an additive to help initiate the formation of phenylmagnesium bromide and 1-butylmagnesium bromide? How does the use of 1,2-dibromoethane avoid such experimental difficulties?

9. The reaction of 5-bromo-1-pentanol with magnesium in anhydrous ether did not give the expected Grignard reagent, 5-hydroxy-1-pentylmagnesium bromide. Provide an explanation for the failure of this reaction. What organic product is formed instead?

$$H{-}OCH_2(CH_2)_3CH_2{-}Br \; + \; Mg° \xrightarrow[\text{Dry Et}_2O]{} H{-}OCH_2(CH_2)_3CH_2{-}MgBr$$

  5-Bromo-1-pentanol   5-hydroxy-1-pentylmagnesium bromide

10. Why is it unwise to allow the solution of the Grignard reagent to remain exposed to air, even if it is protected from moisture by drying tubes?

Questions for Part B

11. Why is it unwise to begin addition of the solution of methyl benzoate to the Grignard reagent before the latter has cooled to room temperature and then placed in an ice-water bath?

12. Why should anhydrous rather than solvent-grade diethyl ether be used to prepare the solution of methyl benzoate that is added to the Grignard reagent?

13. What is the solid that forms during the addition of the ester to the Grignard reagent?

14. Why is it necessary to acidify the mixture obtained after the reaction of methyl benzoate with phenylmagnesium bromide?

15. Cyclohexane is used as the recrystallization solvent to purify the triphenylmethanol by removing the biphenyl impurity. Why is this a better choice of solvent than isopropyl alcohol?

16. Comment on the use of steam distillation (Sec. 4.4) as a possible alternative procedure for purifying crude triphenylmethanol. Consider what possible starting materials, products, and by-products might be present, and indicate which of these should steam-distil and which should not. Would this method of purification yield pure triphenylmethanol? Give your reasoning.

17. Consider the spectral data for bromobenzene (Figs. 15.12 and 15.13).

    a. In the functional group region of the IR spectrum, specify the absorptions associated with the aromatic ring.

    b. In the ^{1}H NMR spectrum, assign the various resonances to the hydrogen nuclei responsible for them.

    c. For the ^{13}C NMR data, assign the various resonances to the carbon nuclei responsible for them.

18. Consider the spectral data for methyl benzoate (Figs. 19.2 and 19.3).

    a. In the functional group region of the IR spectrum, identify the absorptions associated with the ester functional group and the aromatic ring.

    b. In the ^{1}H NMR spectrum, assign the various resonances to the hydrogen nuclei responsible for them.

    c. For the ^{13}C NMR data, assign the various resonances to the carbon nuclei responsible for them.

19. Consider the spectral data for triphenylmethanol (Figs. 19.4 and 19.5).

    a. In the functional group region of the IR spectrum, specify the absorptions due to the aromatic ring. There is a broad absorption at about 3450 cm^{-1}. What functional group is responsible for this absorption and why is the absorption broad?

    b. In the ^{1}H NMR spectrum, assign the various resonances to the hydrogen nuclei responsible for them.

    c. For the ^{13}C NMR data, assign the various resonances to the carbon nuclei responsible for them.

20. Discuss the differences observed in the IR and NMR spectra of methyl benzoate and triphenylmethanol that are consistent with the conversion of an ester into a tertiary alcohol in this experiment.

Questions for Part C

21. Why is it necessary to acidify the mixture obtained after the reaction of phenylmagnesium bromide with carbon dioxide?

22. What is the purpose of using diethyl ether to extract the aqueous mixture obtained after acidification of the reaction? What products are in the aqueous and organic layers?

23. What function does extracting the ethereal solution of organic products with aqueous base have in the purification of benzoic acid? What remains in the ether layer?

24. The yield of benzoic acid obtained when only enough acid is added to the aqueous solution of sodium benzoate to bring the pH to 7 is lower than that obtained if the pH is brought below 5. Explain.

25. Consider the spectral data for bromobenzene (Figs. 15.12 and 15.13).

    a. In the functional group region of the IR spectrum, specify the absorptions associated with the aromatic ring.

    b. In the ^{1}H NMR spectrum, assign the various resonances to the hydrogen nuclei responsible for them.

    c. For the ^{13}C NMR data, assign the various resonances to the carbon nuclei responsible for them.

26. Consider the spectral data for benzoic acid (Figs. 19.6 and 19.7).

    a. In the functional group region of the IR spectrum, specify the absorptions associated with the carbonyl component of the carboxyl group and the aromatic ring. There is a broad absorption in the region of about 2800–3050 cm^{-1}. What functional group is responsible for this absorption and why is the absorption broad?

    b. In the ^{1}H NMR spectrum, assign the various resonances to the hydrogen nuclei responsible for them.

    c. For the ^{13}C NMR data, assign the various resonances to the carbon nuclei responsible for them.

27. Discuss the differences observed in the IR and NMR spectra of bromobenzene and benzoic acid that are consistent with the conversion of bromobenzene into benzoic acid in this experiment.

Questions for Part D

28. Given that commercially available aldehydes are usually produced by oxidative methods, what contaminant(s) might be present in the 2-methylpropanal that could lead to a decrease in yield?

29. Why is it unwise to begin addition of the solution of 2-methylpropanal to the Grignard reagent before the latter has cooled to room temperature and then placed in an ice-water bath?

30. What is the solid that forms upon reaction of 1-butylmagnesium bromide with 2-methylpropanal?

31. Why is it necessary to acidify the mixture obtained after the reaction of 1-butyl-magnesium bromide with 2-methylpropanal?

32. The work-up in this reaction calls for successive washes of an ethereal solution of the product with aqueous sodium bisulfite, sodium bicarbonate, and sodium chloride. What is the purpose of each of these steps?

33. Consider the NMR spectral data for 1-bromobutane (Fig 14.4).

    a. In the ^{1}H NMR spectrum, assign the various resonances to the hydrogen nuclei responsible for them.

    b. For the ^{13}C NMR data, assign the various resonances to the carbon nuclei responsible for them.

34. Consider the spectral data for 2-methylpropanal (Figs. 18.17 and 18.18).

    a. In the functional group region of the IR spectrum, specify the absorptions associated the carbonyl component and the carbon-hydrogen bond of the aldehyde group.

    b. In the ^{1}H NMR spectrum, assign the various resonances to the hydrogen nuclei responsible for them.

    c. For the ^{13}C NMR data, assign the various resonances to the carbon nuclei responsible for them.

35. Consider the spectral data for 2-methyl-3-heptanol (Figs. 19.8 and 19.9).

    a. In the IR spectrum, specify the absorptions associated with the hydroxyl group.

    b. In the ^{1}H NMR spectrum, assign the various resonances to the hydrogen nuclei responsible for them.

    c. For the ^{13}C NMR data, assign the various resonances to the carbon nuclei responsible for them. Explain why there are eight peaks in the ^{13}C NMR spectrum.

36. Discuss the differences observed in the IR and NMR spectra of 2-methylpropanal and 2-methyl-3-heptanol that are consistent with the conversion of an aldehyde into a secondary alcohol in this experiment.

37. The IR spectrum of 2-methyl-3-heptanol prepared by the procedure of Part D sometimes shows a band at 1720 cm^{-1}. Give a possible source for this absorption.

Starting Materials and Products

The IR and NMR spectra of bromobenzene are given in Figures 15.12 and 15.13, respectively. The complete ^{13}C NMR spectrum of methyl benzoate is shown in Figure 8.46. The IR and NMR spectra for 1-bromobutane are presented in Figures 14.3 and 14.4, respectively, and those for 2-methylpropanal are given in Figures 18.17 and 18.18, respectively.

Spectra

FIGURE 19.2
IR spectrum of methyl benzoate (neat).

FIGURE 19.3
NMR data for methyl benzoate
(CDCl₃).

(a) ¹H NMR spectrum (300 MHz).

(b) ¹³C NMR data: δ 51.8, 128.5, 129.7, 130.5, 132.9, 166.8 (see Figure 8.46 for spectrum).

FIGURE 19.4
IR spectrum of triphenylmethanol
(IR card).

FIGURE 19.5
NMR data for triphenylmethanol
(CDCl₃).

(a) ¹H NMR spectrum (300 MHz).

(b) ¹³C NMR data: δ 82.0, 127.2, 127.9, 146.9.

FIGURE 19.6
IR spectrum of benzoic acid
(IR card).

(a) ^{1}H NMR spectrum (300 MHz).

FIGURE 19.7
NMR data for benzoic acid
(CDCl$_3$).

(b) ^{13}C NMR data: δ 128.5, 129.5, 130.3, 133.8, 172.7.

FIGURE 19.8
IR spectrum of 2-methyl-3-heptanol
(neat).

FIGURE 19.9

NMR data for 2-methyl-3-heptanol (CDCl$_3$).

(a) ^{1}H NMR spectrum (300 MHz).

(b) ^{13}C NMR data: δ 14.2, 17.5, 19.3, 23.1, 28.6, 33.8, 34.1, 76.6.

■■■■■■■■■■ H I S T O R I C A L   H I G H L I G H T ■■■■■■■■■■

Grignard and the Beginnings of Modern Organometallic Chemistry

François Auguste Victor Grignard was born in Cherbourg, France in 1871. He began his graduate studies at the University of Lyons under Barbier, one of the pioneers in developing the chemistry of compounds that contain a carbon-metal bond. In 1899 Barbier had discovered that magnesium could be used in place of zinc to promote the reaction of methyl iodide with a ketone to give a tertiary alcohol. This represented a useful advance in organometallic chemistry because the zinc reagents that were being used at the time were pyrophoric and difficult to handle. Based upon this exciting finding, Barbier suggested to the young Grignard that he conduct further exploratory studies of the reactions of magnesium with alkyl halides for his doctoral research project. Every young chemist should be given such a golden opportunity and Grignard made the most of his!

Grignard quickly discovered that a wide variety of alkyl and aryl halides reacted readily with magnesium in anhydrous ether to give the corresponding alkyl- and arylmagnesium halides that are now known as Grignard reagents. His initial results were reported in 1900, and he published seven additional papers within the next year on the preparation of organomagnesium compounds and their reactions with carbonyl compounds to give alcohols. After receiving his Ph.D. degree in 1901, Grignard continued his research at the University of Lyons and then the University of Nancy where he explored the scope and limitations of the reactions of Grignard reagents. During this time, he discovered that the reactions that now bear his name could be used to prepare primary, secondary, and tertiary alcohols, ketones, esters, and carboxylic acids.

The discovery and development of the chemistry of Grignard reactions were expecially important for applications to the synthesis of complex molecules because a new carbon-carbon bond is formed by combining two simpler building blocks, namely an alkyl or aryl halide and a carbonyl compound. The products of Grignard reactions also contain functionality that may then be used in subsequent transformations to fashion even larger molecules.

Grignard was awarded the Nobel Prize in Chemistry in 1912 for his extensive work with organomagnesium compounds. He returned to the

HISTORICAL HIGHLIGHT
Grignard and the Beginnings of Modern Organometallic Chemistry *(continued)*

University of Lyons in 1919 where he succeeded Barbier as head of the Department and continued his research. Grignard also conducted an extensive research program in the field of terpenoid natural products, ozonolysis, aldol reactions, and catalytic hydrogenation and dehydrogenation. However, it is for his development of the chemistry of Grignard reagents that he is best remembered. Indeed, at the time of his death in 1935, more than 6000 papers had appeared in which applications of Grignard reactions were described.

Relationship of Historical Highlight to Experiments

Grignard reagents belong to the larger class of substances known as organometallic compounds. The reactivity of these compounds is determined by the polarized nature of the carbon-metal bond in which carbon bears a partial negative charge. In the experiments in this chapter, you will conduct some of the same reactions that Grignard performed while a graduate student. For example, you will find that the reaction of an alkyl or aryl halide with magnesium in anhydrous ether occurs rapidly. You will also discover that these reagents are highly reactive toward electrophiles and that the carbon atom of the reagent serves as a nucleophile in its chemical reactions. Although you will not win a Nobel Prize for your observations in these experiments, there are other reactions that remain to be discovered. Perhaps your training in the organic laboratory will enable you to find one of these.

CHAPTER 20

Multistep Organic Synthesis

The synthesis of complex molecules from simpler starting materials is one of the most important aspects of modern organic chemistry. In the pharmaceutical and agricultural industries, large quantities of biologically active molecules, sometimes having very complex structures and numerous functional groups, must be prepared from commercially available compounds in the most efficient and cost-effective fashion. This often poses considerable challenges, and it is usually necessary to marshal the combined skills of expert synthetic and mechanistic organic chemists to solve the problems.

There are presently routes to biologically important molecules that require more than 20 or 30 sequential reactions or steps; however the ''arithmetic demon'' haunts these multistep syntheses because the yield in each step is not quantitative. For example, even a five-step synthesis in which each step occurs in a respectable 80% yield gives an overall yield of only 33% [i.e., $(0.8)^5 \times 100 = 32.8\%$]. To offset this cumulative effect, the organic chemist must find the shortest and most efficient pathways to the targeted molecule. Each reaction must be carefully studied to minimize side reactions, and a thorough knowledge of the mechanism of each reaction is critical to success. Multistep syntheses that exceed 8–10 steps are economically feasible only for end products that are biologically highly active and cannot be obtained readily from natural sources.

The *optimal route* to a particular compound *is not always the shortest*, as is illustrated by the simple problem of preparing 2-chloropropane on an industrial scale, as shown in Scheme 20.1. Although 2-chloropropane is produced from propane in one step by direct chlorination, it must be separated from the 1-chloropropane formed concurrently; this is *not* a simple separation. However, if the mixture of 1- and 2-chloropropane is treated with a base, both isomers undergo dehydrochlorination to give propene, and subsequent addition of hydrogen chloride produces 2-chloropropane as the *sole* product. In this case a three-step process is the preferred way to obtain pure 2-chloropropane. As this example demonstrates, an important consideration in designing a synthesis is the ease of separating the desired product from unwanted by-products and purifying it. Other considerations include the availability and cost of starting materials, simplicity of equipment and instrumentation, energy costs, activity of catalysts, selectivity of reactions in polyfunctional molecules, and stereochemical control.

544

SCHEME 20.1

$$CH_3\overset{\overset{\displaystyle H}{|}}{CH}-CH_2-H + Cl_2 \xrightarrow{\text{Heat or } h\nu} \left\{ \begin{array}{c} \overset{\overset{\displaystyle Cl}{|}}{CH_3CHCH_3} \\ \text{2-Chloropropane} \\ + \\ CH_3CH_2CH_2-Cl \\ \text{1-Chloropropane} \end{array} \right\} \xrightarrow[\text{Base}]{-HCl}$$

$$CH_3CH{=}CH_2 \xrightarrow{\text{HCl}} CH_3\overset{\overset{\displaystyle Cl}{|}}{CH}CH_3$$

2-Chloropropane

Because so many variables must be considered in planning the synthesis of complex molecules, it is not surprising that organic chemists sometimes use computers to help design and analyze multistep syntheses. Computers can handle an enormous amount of information, and it may eventually be possible to design optimal sequences of reaction steps using sophisticated computer programs. In considering this fascinating prospect, however, it must be remembered that the information is provided to the computer by humans, specifically chemists, who will be vital to the success of such a project. Chemists still have at least one significant competitive advantage over computers—they are more creative!

Appreciation of and insight into some of the problems that may be encountered in planning a multistep synthesis is gained from preparing a compound that requires a fairly small number of steps. In this chapter, the syntheses of sulfanilamide (Sec. 20.2) and 1-bromo-3-chloro-5-iodobenzene (Sec. 20.3) are used to illustrate the fundamental principles. In these experiments, the product of one reaction is used in a subsequent step, and if you are to be successful, you *must* use good experimental technique. You may either isolate and purify the intermediates by distillation or recrystallization or you may use the crude material directly in the next step. This is a choice the practicing synthetic organic chemist constantly faces. In general, extensive purification of each intermediate in a sequence is avoided *if* the impurities can eventually be removed *and* their presence does not interfere with the course of the desired reactions. However, even if the entire quantity of an intermediate is not purified, it is good scientific practice to purify a small sample of it for *complete* characterization by spectroscopic (Chap. 8) and physical (bp, mp, etc.) methods.

20.2 Sulfanilamide: Discovery and Synthesis of the First Antibiotic

An important area of research in the pharmaceutical industry is the discovery and development of new orally active **antibiotics** to treat bacterial infections. Commonly used antibiotics fall into several important classes including **β-lactams**, the **macrolides**, the **tetracyclines**, the **aminocyclitols**, and the **sulfa drugs**, (Fig. 20.1). Each antibiotic has a different profile of biological activity, and all of them are used in modern medicine to treat various infections.

FIGURE 20.1
Classes of antibiotics.

Ampicillin
A β-lactam antibiotic

Terramycin
A tetracycline antibiotic

Erythromycin A
A macrolide antibiotic

Spectinomycin
An aminocyclitol antibiotic

Sulfamethoxazole
A sulfa drug

BACKGROUND

The broad-spectrum antibacterial activity of sulfanilamide (**1**) was first revealed in the mid-1930s by serendipity, as are many important discoveries in science, and an interesting account of this discovery may be found in the Historical Highlight at the end of this chapter. Over one thousand derivatives of sulfanilamide have been synthesized and tested as potential antibiotics, and some of these are still used today. The mode of action of the sulfa drugs is another interesting story because it provides some insights into strategies that might be generally exploited to design biologically active compounds.

Early in the development of sulfa drugs as antibiotics, it was found that *p*-aminobenzoic acid (PABA, **2**), which is also used in sunscreens and sunblocks, inhibits the antibacterial action of sulfanilamide. Since *p*-aminobenzoic acid and sulfanilamide are structurally similar, this discovery led to the speculation that the two compounds competed with each other in some biological process that was essential to bacterial growth. This speculation was eventually supported by experimentation. *p*-Aminobenzoic acid is used by bacteria in the synthesis of the essential enzyme cofactor folic acid (**3**). When sulfanilamide is present, it successfully competes with *p*-aminobenzoic acid for the active site in the enzyme that incorporates *p*-aminobenzoic acid into folic acid. By functioning as a competitive inhibitor of this enzyme, sulfanilamide blocks the biosynthesis of folic acid, and without folic acid, the bacteria cannot grow. On the other hand, animal cells cannot synthesize folic acid, which is an **essential vitamin**, and it therefore must be part of the diet. Because only bacteria rely on the biosynthesis of folic acid from *p*-aminobenzoic acid, the sulfa compounds are "ideal" drugs as they kill only the bacteria and not the animal host. Of course, they are not truly ideal, since some people have allergic reactions to sulfa drugs and because bacteria develop resistance to them over time.

1	2	3
Sulfanilamide	*p*-Aminobenzoic acid (PABA)	Folic acid

The preparation of sulfanilamide (**1**) from benzene (**4**) (Scheme 20.2) serves as an excellent example of a multistep synthesis that produces a biologically active, nonnatural molecule. Owing to the toxicity of benzene, the experimental sequence provided here will commence with the second step, the reduction of nitrobenzene (**5**), which is considerably less toxic. Each of the intermediate compounds in the synthesis may either be isolated or used directly in the subsequent step without further purification. A discussion of the reactions involved in each step follows.

SCHEME 20.2

REDUCTION OF AROMATIC NITRO COMPOUNDS: PREPARATION OF ANILINE (6)

Aromatic nitro compounds, which are readily prepared by electrophilic aromatic nitration (Sec. 15.1), are **reduced** to the corresponding aromatic amines by a variety of methods. Chemical reduction is commonly used in the research laboratory, but the most important commercial method is catalytic hydrogenation. Although the precise mechanism of the reduction of nitro compounds is not fully understood, the isolation of various intermediates, some of which are shown in Equation 20.1, suggests that the reaction is stepwise. The symbol [**H**] over the arrows symbolizes reduction without specifying the reagent. The product that is ultimately isolated from the reduction of an aromatic nitro compound can be controlled to a significant degree by the nature of the reducing agent. For example, the reduction of nitrobenzene (**5**) with zinc metal and ammonium chloride gives only *N*-phenylhydroxylamine (**11**), whereas the use of tin metal and hydrochloric acid gives aniline (**6**) in excellent yield.

5	**10**	**11**	**6**
Nitrobenzene	Nitrosobenzene	*N*-Phenylhydroxylamine	Aniline

The reduction of nitrobenzene to aniline is a typical oxidation-reduction reaction in which tin metal, Sn^0, is oxidized to stannic chloride, $SnCl_4$; hydrochloric acid serves as the source of protons. The mechanism of this reaction is outlined in some detail in Scheme 20.3. Generally, the reduction occurs by a sequence of steps in which an electron is first transferred from a tin atom to the organic substrate to give an intermediate radical ion that is then protonated. The oxygen atoms on the nitro group are eventually removed as water molecules. It is left as an exercise to write a *balanced* equation for the overall reaction and to provide a mechanism for the reduction of *N*-phenylhydroxylamine (**11**) into aniline (see Exercises 8 and 9 at the end of this section).

SCHEME 20.3

$$C_6H_5-\overset{..}{\underset{.}{N}}{}^+ + e^- \longrightarrow C_6H_5-\overset{..}{\underset{..}{N}}-OH \xrightarrow{H^+} C_6H_5-\overset{\overset{H}{|}}{N}-OH + \xrightarrow[3\ H^+]{Sn}$$

$$\underset{\textbf{11}}{}$$

$$C_6H_5-\overset{+}{N}H_3 \xrightarrow{NaOH} C_6H_5-NH_2$$

$$\underset{\textbf{6}}{}$$

The procedure for isolating and purifying the aniline formed in this experiment represents an excellent example of how the physical and chemical properties of an individual component in a mixture of organic substances can be exploited to isolate it without using chromatographic techniques. For example, after completing the procedure for the chemical reduction of nitrobenzene, it is necessary to remove aniline from its principal impurities, which are unchanged nitrobenzene and two by-products, benzidine (**12**), and 4-aminophenol (**13**). Steam distillation (Secs. 2.16 and 4.4) is a good technique for separating volatile organic compounds from nonvolatile organic and inorganic substances. However, the reaction mixture is *acidic*, and it must be made *basic* prior to performing the steam distillation so the aniline is present as the free base and the 4-aminophenol is converted to its *water-soluble* sodium phenoxide salt. The aniline and nitrobenzene are then removed from the reaction mixture by steam distillation, and the nonvolatile salt of 4-aminophenol, together with the nonvolatile benzidine, remains in the aqueous phase.

$$H_2N-\underset{\textbf{12}}{\underset{Benzidine}{\bigcirc\!\!-\!\!\bigcirc}}-NH_2 \qquad HO-\underset{\textbf{13}}{\underset{4\text{-Aminophenol}}{\bigcirc}}-NH_2$$

To separate aniline from nitrobenzene, the aqueous steam distillate is acidified with hydrochloric acid to convert aniline to its water-soluble hydrochloride salt **14**. The acidic solution is then extracted with diethyl ether to remove nitrobenzene. The aqueous layer, which contains **14**, is made basic, and the aniline that results is isolated from the mixture by extraction with diethyl ether.

$$\underset{\underset{\text{Anilinium hydrochloride}}{\textbf{14}}}{\bigcirc\!\!-\!\!\overset{+}{N}H_3\ Cl^-}$$

PROTECTING GROUPS: PREPARATION OF ACETANILIDE (7)

The amino function on aniline is a highly electron-donating group that *activates* the aromatic ring toward electrophilic aromatic substitution reactions (Chap. 15). However, if an aryl amino group is protonated or complexed with a Lewis acid, the resulting ammonium group *deactivates* the ring toward subsequent electrophilic aromatic substitution. Electrophilic reagents or other functional groups that may be present on the ring can also react directly with the amino group. Thus, because of its basic and nucleophilic properties, it is necessary to **protect** the amino group on aniline to ensure that the desired reaction, rather than a deleterious side reaction, will take place. In general, placing a **protecting group** on a reactive functional group alters the chemical properties of that group, thereby rendering it unreactive toward certain reagents. The protecting group must be carefully selected, however, because it must be stable to the reaction conditions employed in the various transformations, *and* it must be removable under conditions that do not affect other functional groups in the molecule.

In our synthesis of sulfanilamide (**1**), adding a sulfonamido, $-SO_2NH_2$, group to the aromatic ring requires the use of a protecting group. This is necessary because creating a sulfonamido function on an arene, Ar–H, commonly involves sequential chlorosulfonation, an electrophilic aromatic substitution reaction, followed by treatment of the intermediate aryl sulfonyl chloride, Ar–SO$_2$Cl with ammonia to give the requisite sulfonamide, Ar–SO$_2$NH$_2$ (Eq. 20.2). If aniline itself were treated with chlorosulfonic acid, the amino group could react with either hydrogen chloride or SO$_3$, which are generated *in situ*, to provide ammonium salts such as **14** or **15**; these groups would then deactivate the ring toward electrophilic aromatic substitution. Moreover, a chlorosulfonyl group cannot be generated in the presence of an amino group since the chlorosulfonyl group of one molecule could react with the amino group of another, thereby forming a *polymeric* material containing sulfonamide linkages.

$$Ar\!-\!H \xrightarrow[\text{(2) NH}_3]{\text{(1) ClSO}_3\text{H}} Ar\!-\!SO_2NH_2 \qquad\qquad \textbf{(20.2)}$$

$$\text{An arene} \qquad\qquad\qquad \text{An aryl sulfonamide}$$

15

Because the presence of the free amino group of aniline would interfere with the preparation of the arylsulfonyl chloride, it is necessary to *protect* the amino group as an amide by acetylation. The free amine is regenerated by removing the acetyl group via controlled hydrolysis *after* the chlorosulfonyl moiety has been introduced onto the aromatic ring and converted into the sulfonamido group. The selective hydrolysis of the amido group in the presence of the sulfonamido group is possible because the sulfur atom of the sulfonamido group is *tetrahedral* and hence more sterically hindered toward nucleophilic attack by water than the *trigonal* carbonyl carbon of **9** (Scheme 20.2).

Aniline (**6**) is converted to acetanilide (**7**) by acetylation according to the sequence shown in Equation 20.3. In this reaction, aniline is first converted to the hydrochloride salt **14**. Acetic anhydride and sodium acetate are then added to give a mixture in which sodium acetate and **14** are in equilibrium with acetic acid and aniline. As free aniline is produced by this acid-base reaction, it rapidly undergoes acetylation by acetic anhydride to give acetanilide, which is isolated by filtration and purified by recrystallization from water. This method for acetylation of amines is general, and the yields are usually high.

$$+ \ CH_3CO_2H \ + \ NaCl \qquad \textbf{(20.3)}$$

6   **14**   **7**

A possible side reaction in this procedure is diacetylation (Eq. 20.4). Although the acetylation of aniline in pure acetic anhydride gives substantial quantities of the diacetyl compound **16**, this side reaction is minimized in aqueous solution.

$$\text{(20.4)}$$

Preparation

CHLOROSULFONATION: OF 4-ACETAMIDOBENZENESULFONYL CHLORIDE (8)

A simple one-step reaction can be used to introduce the sulfonyl group, $-SO_2Cl$, *para* to the N-acetyl group of acetanilide by a process called chlorosulfonation, (Eq. 20.5). Although an electron-donating group can direct an incoming electrophile to either the *ortho* or *para* position, the acetamido group orients the incoming group predominantly *para*, presumably as a consequence of the steric bulk of the acetamido group; virtually none of the *ortho* isomer is observed. The reaction is known to proceed through the intermediate sulfonic acid **17**, which is converted to the sulfonyl chloride **8** on further reaction with chlorosulfonic acid. At least two equivalents of chlorosulfonic acid per

$$\text{(20.5)}$$

4-Acetamidobenzene-
sulfonic acid

equivalent of acetanilide (**7**) are thus required for complete reaction. The attacking electrophile is probably SO_3, which is generated *in situ* from chlorosulfonic acid according to Equation 20.6.

$$ClSO_3H \rightleftharpoons SO_3 + HCl \qquad \text{(20.6)}$$

To isolate the 4-acetamidobenzenesulfonyl chloride, the reaction mixture is poured into ice water, and the product is obtained as a white precipitate. The water rapidly hydrolyzes the excess chlorosulfonic acid (Eq. 20.7). In general, sulfonyl chlorides are much less reactive toward water than are carboxylic acid chlorides, but it is unwise

$$ClSO_3H + H_2O \longrightarrow HCl + H_2SO_4 \qquad \text{(20.7)}$$

to expose them to water for extended periods of time because they hydrolyze slowly to give the corresponding sulfonic acids according to Equation 20.8. Nevertheless, it is unnecessary to dry or purify the 4-acetamidobenzenesulfonyl chloride (**8**) in this sequence; rather, **8** is simply treated directly with aqueous ammonia.

$$NHCOCH_3 \quad + H_2O \longrightarrow NHCOCH_3 \quad + HCl \qquad (20.8)$$

AMMONOLYSIS AND HYDROLYSIS: PREPARATION OF SULFANILAMIDE (1)

The preparation of sulfanilamide (**1**) involves treatment of 4-acetamidobenzenesulfonyl chloride (**8**) with an excess of aqueous ammonia to give 4-acetamidobenzenesulfonamide (**9**), followed by selective hydrolytic removal of the *N*-acetyl protecting group (Eq. 20.9). To avoid hydrolysis of the sulfonyl chloride group (Eq. 20.8) before its reaction with ammonia, it is *imperative* to treat **8** with ammonia *immediately* after its isolation in the previous experiment. If the conversion of acetanilide (**7**) to **9** is not completed within a *single* laboratory period, the overall yield for the sequence is significantly reduced.

Hydrolysis of the acetamido moiety of **9** may be effected using either aqueous acid or aqueous base without affecting the sulfonamido group, which hydrolyzes slowly; an acid-catalyzed hydrolysis is performed in this experiment. Because the amine group will form a hydrochloride salt under the conditions of the hydrolysis, it is necessary to neutralize the solution with a base such as sodium carbonate in order to isolate sulfanilamide, which may be purified by recrystallization from water.

$$(20.9)$$

EXPERIMENTAL PROCEDURES

SYNTHESIS OF SULFANILAMIDE

Purpose: To demonstrate principles of multi-step synthesis, protecting groups, and electrophilic aromatic substitution.

Note   Throughout the sequence of reactions, quantities of reagents must be adjusted according to the amount of starting material that is available from the previous reaction; you should *not* run the reactions on a larger scale than is indicated, however. If the amount of your starting material varies from that indicated, do *not* change the reaction times unless directed to do so by your instructor. You should be able to obtain sufficient quantities of product from each reaction to complete the entire sequence starting with the amount of acetanilide provided in the first step.

A · PREPARATION OF ANILINE

Safety Alert

1. Nitrobenzene is toxic; avoid inhaling its vapors or allowing it to come into contact with your skin. Wear latex gloves when transferring this chemical.
2. *Concentrated* hydrochloric acid can cause burns if it comes in contact with the skin. Should contact occur, flood the affected area with water and rinse it thoroughly with dilute aqueous sodium bicarbonate solution. Wear latex gloves when transferring this acid.
3. The solution of 12 *M* sodium hydroxide is highly caustic and can cause burns and loss of hair; avoid contact of this solution with your skin. Should contact occur, flood the affected area with water and rinse it thoroughly with dilute aqueous acetic acid solution. Wear latex gloves when transferring this solution.
4. Use *flameless* heating when concentrating ethereal solutions.

Miniscale Procedure

Preparation   Answer the Pre-Lab Exercises on page PL. 129. Review Sections 2.9, 2.10, 2.11, 2.13, 2.16, 2.21, 2.22, 2.25, and 2.29.

Apparatus   A 250-mL round-bottom flask, thermometer, ice-water bath, separatory funnel, apparatus for heating under reflux, magnetic stirring, steam distillation, simple distillation, and *flameless* heating.

Setting Up   Add 13.1 g of finely divided tin powder and 5.2 mL of nitrobenzene to the round-bottom flask equipped with a stirbar.

Reduction   Add 28 mL of *concentrated* hydrochloric acid to the flask, insert a thermometer, and stir the contents of the flask to mix the three-phase system. Monitor the temperature, and cool the flask as necessary in an ice-water bath to maintain the temperature below 60 °C; the initial exothermic portion of the reaction should be complete after about 15 min. Attach a reflux condenser and heat the contents of the flask under reflux for about 20 min with continued stirring. During this time the color due to the intermediate reduction product and any droplets of nitrobenzene in the condenser should disappear.

Work-Up   When the reflux period is complete, cool the acidic solution in an ice-water bath, and slowly add 50 mL of 12 *M* sodium hydroxide solution directly to the reaction mixture. Check the pH of the mixture to ensure that it is basic; if not, add additional base. Equip the flask for steam distillation, and steam-distil the mixture until the condensate no longer appears oily, although it may still be slightly cloudy; about 35–40 mL of distillate should be collected.★

Add 4.5 mL of *concentrated* hydrochloric acid to the distillate and transfer the mixture to a separatory funnel. Check the mixture to be certain it is acidic, and

then extract it sequentially with two 15-mL portions of diethyl ether to remove any unreacted nitrobenzene. Transfer the *aqueous* layer to an Erlenmeyer flask; cool the contents of the flask in an ice-water bath, and make the aqueous solution *basic* by slowly adding a *minimum* volume of 12 *M* sodium hydroxide solution. Saturate the aqueous solution with solid sodium chloride, cool the mixture to room temperature, and transfer it to a clean separatory funnel. Extract the aqueous mixture sequentially with two 15-mL portions of diethyl ether, using the first portion to rinse the flask in which the neutralization was done; separate the aqueous layer from the organic layer as thoroughly as possible each time. Dry the *combined* organic extracts over anhydrous sodium sulfate.★ Decant the solution into a round-bottom flask and remove most of the diethyl ether by simple distillation. Alternatively, rotary evaporation or other techniques (Sec. 2.29) may be used to concentrate the solution. The final traces of solvents may be removed by connecting the cool stillpot to a vacuum source.

The aniline remaining in the stillpot may be used directly in the next step of the sequence. Alternatively, you may purify the aniline by shortpath distillation. Wrap the stillhead with a layer of glasswool and then aluminum foil to minimize heat losses and facilitate the distillation of aniline. Collect three fractions having approximate boiling ranges (760 Torr) of 35–90 °C, 90–180 °C, and 180–185 °C, respectively.

Analysis   Weigh the product and calculate its yield. Obtain IR and ^{1}H NMR spectra of your starting material and product and compare them with those of authentic samples (Figs. 20.2–20.5). Pure aniline is colorless but may darken immediately following distillation, owing to air-oxidation.

Finishing Touches

Vacuum-filter the *pot residue from the steam distillation* through a bed of filter-aid, and place the *tin salts* together with the *filter-aid* and *filter paper* in the container for nonhazardous solids. Neutralize the *filtrate* and the other *aqueous layers* with dilute hydrochloric acid and flush them down the drain with excess water. Pour the combined *ethereal extracts* and *distillate* into a container for nonhalogenated organic liquids. Put the *sodium sulfate* on a tray in the hood and, after the ether evaporates, pour it into the container for nonhazardous solids.

B · PREPARATION OF ACETANILIDE

Miniscale Procedure

Preparation   Answer the Pre-Lab Exercises on page PL. 131. Review Sections 2.9, 2.10, 2.11, 2.17, and 2.26.

Apparatus   A 250-mL Erlenmeyer flask, thermometer, ice-water bath, apparatus for magnetic stirring, vacuum filtration, and *flameless* heating.

Setting Up   Add 100 mL of 0.4 *N* hydrochloric acid followed by 3.6 mL of aniline to the Erlenmeyer flask equipped with a stirbar. Stir the mixture to effect solution and warm it to about 50 °C. Prepare a solution containing 6.0 g of sodium acetate trihydrate dissolved in 20 mL of water and, in a *separate* container, measure out 4.4 mL of acetic anhydride.

Reaction and Isolation   Add the acetic anhydride in *one* portion to the warm solution of anilinium hydrochloride. Stir the resulting solution *vigorously* and then add the solution of sodium acetate *immediately and in one portion*. Cool the reaction mixture to 5 °C in an ice-water bath, and continue to stir the mixture until the crystalline product completely precipitates. Collect the acetanilide by vacuum filtration, wash it with a small portion of ice-cold water, and air-dry it.★ If impure or slightly colored acetanilide is obtained, recrystallize it from a minimum volume of hot water, using decolorizing carbon to give a colorless product.

Analysis   Weigh the product and calculate its yield; determine its melting point. Obtain IR and ^{1}H NMR spectra of your starting material product and compare them with those of an authentic sample (Figs. 8.6 and 15.31).

After neutralizing the *aqueous filtrate* with sodium carbonate, flush it and the *mother liquor* from crystallization down the drain with a large excess of water.

Finishing Touches

C · PREPARATION OF 4-ACETAMIDOBENZENESULFONYL CHLORIDE

Note   *This entire procedure as well as the next one, up to the first stopping point, must be completed in a single laboratory period.*

Safety Alert

1. If possible, perform this reaction in a hood. Otherwise, use the gas trap prescribed by your instructor to prevent escape of hydrogen chloride, SO_2, and SO_3 into the laboratory.
2. Chlorosulfonic acid is *highly corrosive* and may cause serious burns if it is allowed to come in contact with your skin. Wear latex gloves when handling or transferring this reagent, and use *extreme* care when working with it. Should any chlorosulfonic acid come in contact with your skin, *immediately* flood the affected area with cold water and then rinse it with 5% sodium bicarbonate solution.
3. Chlorosulfonic acid reacts *violently* with water. Open containers of chlorosulfonic acid will fume upon exposure to atmospheric moisture owing to the reaction of chlorosulfonic acid with water to give HCl and SO_3, both of which are noxious and corrosive gases. Measure and transfer this acid *only in the hood*. Several graduated cylinders should be kept in the hood for all students to use for measuring the volume of chlorosulfonic acid needed.
4. To destroy residual chlorosulfonic acid in graduated cylinders and other glassware that has contained it, add cracked ice to the

glassware *in the hood* and let the glassware remain there until the ice has melted. Then rinse the apparatus with large amounts of cold water.

5. Make sure your glassware is dry, lubricate the ground-glass joints of the apparatus carefully, and be certain that the joints are intimately mated. Otherwise, noxious gases will escape.

Miniscale Procedure

Preparation   Answer the Pre-Lab Exercises on page PL. 133. Review Sections 2.9, 2.10, 2.11, 2.17, 2.23, and 2.28.

Apparatus   A 50-mL round-bottom flask, Claisen adapter, gas trap, separatory funnel with ground-glass joint, thermometer, ice-water bath, apparatus for magnetic stirring, vacuum filtration, and *flameless* heating.

Setting Up   Oven-dry the glass apparatus prior to assembling it. Because an *airtight* seal is required for all connections, *carefully grease all joints* when assembling the apparatus. Place 2.7 g of *dry* acetanilide in the round-bottom flask, and equip the flask with a stirbar and the Claisen adapter. Fit the *sidearm* of the adapter with the gas trap specified by your instructor; one example is illustrated in Figure 2.65a. *Working at the hood*, measure 8.0 mL of chlorosulfonic acid into a small separatory funnel having a standard-taper ground-glass joint; be certain that the stopcock of the funnel is firmly seated and closed prior to this transfer. Stopper the funnel and place it on the straight arm of the Claisen adapter above the reaction flask so that the chlorosulfonic acid will drip *directly* onto the acetanilide. Cool the flask in a water bath that is maintained at 10–15 °C; do *not* allow the bath temperature to go below 10 °C.

Reaction   Open the stopcock of the funnel fully so that the chlorosulfonic acid is added as rapidly as possible to the flask containing the acetanilide; it may be necessary to *briefly* lift the stopper on the funnel to equalize the pressure in the system if the flow of acid becomes erratic or slow. When the addition has been completed, stir the mixture to hasten dissolution of the acetanilide; maintain the temperature of the water bath below 20 °C. After most of the solid has dissolved and the initial reaction has subsided, remove the cooling bath and allow the reaction mixture to warm to room temperature with stirring. To complete the reaction, warm the mixture at about 70–80 °C with continued stirring until there is no longer any increase in the rate of gas evolution; about 10–20 min of heating will be required.

Work-Up and Isolation   Cool the mixture to room temperature or slightly below using an ice-water bath. *Working at the hood*, place about 150 g of crushed ice in a beaker, and pour the reaction mixture slowly onto the ice with stirring; stir the mixture thoroughly with a glass rod during the addition. *Be careful not to add the mixture too quickly to avoid splattering the chlorosulfonic acid*. Rinse the reaction flask with a little ice-water and transfer this to the beaker; the remainder of the work-up procedure may be performed at the bench. The precipitate that forms is crude 4-acetamidobenzenesulfonyl chloride and may be white to pink in color. If it is allowed to stand without stirring, a hard mass may form; any lumps that form should be thoroughly broken up with a stirring rod. Collect the crude material by vacuum filtration. Wash the solid with 15-mL portions of ice water *until the filtrate is neutral*; press the filter cake as dry as possible with a clean cork. The crude sulfonyl chloride should be used *immediately* in the following procedure.

Purification   Purify about 50 mg of the product by dissolving it in a minimum amount of boiling dichloromethane in a test tube. Using a Pasteur pipet, remove the upper, aqueous layer as quickly as possible; use care in separating the layers. Cool the organic layer, collect the purified product by vacuum filtration, and air-dry it.

Analysis   Determine the melting point of the recrystallized solid. Obtain its IR and ¹H NMR spectra and compare them with those of an authentic sample (Figs. 20.6 and 20.7).

Neutralize the combined *aqueous solutions* with sodium carbonate and flush them down the drain with a large excess of water. Cover any spilled droplets of *chlorosulfonic acid* with sodium carbonate and flush the resulting powder down the drain. If a gas trap that contains a drying agent has been used, put the *drying agent* in the container for nonhazardous solids.

Finishing Touches

D • PREPARATION OF 4-ACETAMIDOBENZENESULFONAMIDE

Safety Alert

1. Ammonia is evolved in this procedure. If possible, perform this reaction in a hood. Otherwise, use an inverted funnel attached to a vacuum source (Fig. 2.68b) to minimize escape of vapors of ammonia into the laboratory.
2. Concentrated ammonium hydroxide is caustic and may cause burns if it is allowed to come in contact with your skin. Wear latex gloves when handling or transferring this reagent, and use care when working with it. Should any of the liquid come in contact with your skin, *immediately* flood the affected area with cold water.

Miniscale Procedure

Preparation   Answer the Pre-Lab Exercises on page PL. 135. Review Sections 2.9, 2.10, 2.11, and 2.17.

Apparatus   A 125-mL Erlenmeyer flask, ice-water bath, apparatus for vacuum filtration and *flameless* heating.

Reaction, Work-Up and Isolation   Transfer the crude 4-acetamidobenzenesulfonyl chloride obtained in the previous experiment to the Erlenmeyer flask, and add 15 mL of concentrated (28%) ammonium hydroxide. A rapid exothermic reaction may occur if the crude 4-acetamidobenzenesulfonyl chloride contains acidic contaminants that were not removed by the aqueous washings in the previous step. Use a stirring rod to break up any lumps of solid that may remain; the reaction mixture should be thick but homogeneous. Heat the mixture at 70–80 °C for about 0.5 h. Cool the reaction mixture in an ice-water bath and collect the product by vacuum filtration. Wash the crystals with cold water and air-dry them.★

Analysis   Weigh the product and calculate its yield; determine its melting point. Obtain IR and ^{1}H NMR spectra of your product and compare them with those of an authentic sample (Figs. 20.8 and 20.9).

Finishing Touches	**N**eutralize the combined *aqueous solutions and filtrates* with 10% hydrochloric acid and flush them down the drain with a large excess of water.

E · PREPARATION OF SULFANILAMIDE

> **Safety Alert**   Concentrated hydrochloric acid is corrosive and may cause serious burns if it is allowed to come in contact with your skin. Wear latex gloves when handling or transferring this reagent and use care when working with it. Should any acid come in contact with your skin, *immediately* flood the affected area with cold water and then rinse it with 5% sodium bicarbonate solution.

Miniscale Procedure

Preparation   Answer the Pre-Lab Exercises on page PL. 137. Review Sections 2.9, 2.10, 2.11, 2.17, and 2.22.

Apparatus   A 50-mL round-bottom flask, ice-water bath, apparatus for heating under reflux, magnetic stirring, vacuum filtration, and *flameless* heating

Setting Up   Prepare a solution of *dilute* hydrochloric acid by mixing equal volumes of *concentrated* hydrochloric acid and water; add the acid *to* the water. Equip the round-bottom flask with a stirbar, and add the 4-acetamidobenzenesulfonamide and an amount of *dilute* hydrochloric acid that is about *twice* the weight of the crude 4-acetamido-benzenesulfonamide. Assemble the apparatus for heating under reflux.

Reaction, Work-Up, and Isolation   Heat the stirred reaction mixture under gentle reflux for 45 min and then allow the mixture to cool to room temperature. If any solid, which is unreacted starting material, appears on cooling, reheat the mixture under reflux for an additional 15 min. Add an equal volume of water to the cooled solution and transfer the resulting mixture to a beaker. Neutralize the excess hydrochloric acid by adding *small* quantities of *solid* sodium carbonate until the solution is slightly alkaline.

A precipitate should form during neutralization; scratching at the air-liquid interface may be necessary to initiate crystallization. Cool the mixture in an ice-water bath to complete the precipitation of product.★ Collect the crystals by vacuum filtration, wash them with a small amount of cold water, and air-dry them.

Purification   Purify the crude product by recrystallization from the minimum volume of hot water. If necessary, decolorize the hot solution and use a preheated funnel for the hot-filtration step to prevent premature crystallization of the product in the funnel. Cool the aqueous solution of sulfanilamide in an ice-water bath, isolate the crystalline product by vacuum filtration and air-dry it.★

Analysis Weigh the product and calculate its yield; determine its melting point. Test the solubility of sulfanilamide in 1.5 M hydrochloric acid solution and in 1.5 M sodium hydroxide solution. Obtain IR and ^{1}H NMR spectra of your product and compare them with those of an authentic sample (Figs. 20.10 and 20.11).

Neutralize the combined *aqueous solutions and filtrates* with sodium carbonate and flush them down the drain with a large excess of water.

Finishing Touches

EXERCISES

Refer to Tables 8.1–8.8 as needed for answering the questions on spectroscopy.

General Questions

1. Outline a possible synthesis of the sulfanilamide derivative, **18**, using benzene as the only source of an aromatic ring. Use any needed aliphatic or inorganic reagents.

$$H_2N-\bigcirc-SO_2NH-\bigcirc$$

18

2. Two hypothetical sequences have been developed for converting an arbitrary compound A into E in the same overall yield.

$$A \xrightarrow{25\%} B \xrightarrow{49\%} C \xrightarrow{60\%} D \xrightarrow{57\%} E$$

$$A \xrightarrow{58\%} B \xrightarrow{57\%} C \xrightarrow{51\%} D \xrightarrow{30\%} E$$

a. What is the overall yield for each sequence?

b. As the production manager responsible for selecting the more economical of the two routes, what factor(s) would you consider in reaching a decision? Assume that no new capital investment would be required for either sequence.

3. A student proposed the following alternate sequence of reactions for preparing **1** from sulfanilic acid. Discuss this modified approach, commenting on any possible side reactions that might be encountered in either of the two steps. Propose a modification of this sequence that could be used to prepare sulfanilamide.

Sulfanilic acid   *p*-Aminosulfonyl chloride

Questions for Part A

4. Outline in a flow diagram the procedure for the purification of aniline. Indicate the importance of each step in the procedure, and give reasons for doing the steam distillation with a *basic* solution. What is the purpose of performing the subsequent extractions from aqueous acid and then base? Write the equation(s) for the reactions that occur when base and acid are added.

5. What gas is evolved upon adding concentrated hydrochloric acid to tin powder in this reduction?

6. Suppose a student decided that a suitable stopping point in the procedure might be immediately after the initially exothermic reaction between tin and hydrochloric acid. What is the potential problem with storing the reaction mixture at this stage?

7. Why is sodium chloride added to the steam distillate *after* it has been made basic but *before* extraction with diethyl ether?

8. Write the balanced half-reactions for the reduction of nitrobenzene (**5**) into aniline (**6**) and for the oxidation of tin metal into stannic ion, Sn^{4+}. Sum these equations and write the balanced equation for the overall process.

9. Write a mechanism for the reduction of *N*-phenylhydroxylamine (**11**) into aniline (**6**). Show all electron and proton transfer steps, using curved arrows to symbolize the flow of electrons.

10. Consider the spectral data for nitrobenzene (Figs. 20.2 and 20.3).

    a. In the functional group region of the IR spectrum, specify the absorptions associated with the nitro group and with the aromatic ring.

    b. In the ^{1}H NMR spectrum, assign the various resonances to the hydrogen nuclei responsible for them.

    c. For the ^{13}C NMR data, assign the various resonances to the carbon nuclei responsible for them.

11. Consider the spectral data for aniline (Figs. 20.4 and 20.5).

    a. In the functional group region of the IR spectrum, specify the absorptions associated with the amino group and with the aromatic ring.

    b. In the ^{1}H NMR spectrum, assign the various resonances to the hydrogen nuclei responsible for them.

    c. For the ^{13}C NMR data, assign the various resonances to the carbon nuclei responsible for them.

12. Discuss the differences observed in the IR and NMR spectra of nitrobenzene and aniline that are consistent with the formation of the latter in this procedure.

Questions for Part B

13. Why should the acetic anhydride not be allowed to stay in contact with the aqueous solution of anilinium hydrochloride (**14**) for an extended period of time before the solution of sodium acetate is added?

14. Why is aqueous sodium acetate preferred to aqueous sodium hydroxide for the conversion of anilinium hydrochloride (**14**) to aniline?

15. Why is aniline (**6**) soluble in aqueous hydrochloric acid whereas acetanilide (**7**) is not?

16. Give a stepwise reaction mechanism for the reaction of aniline (**6**) with acetic anhydride; use curved arrows to symbolize the flow of electrons.

17. Consider the spectral data for acetanilide (Figs. 8.6 and 15.31).

    a. In the functional group region of the IR spectrum, specify the absorptions associated with the nitrogen-hydrogen bonds, with the carbon-oxygen double bond, and with the aromatic ring.

    b. In the ^{1}H NMR spectrum, assign the various resonances to the hydrogen nuclei responsible for them.

    c. For the ^{13}C NMR data, assign the various resonances to the carbon nuclei responsible for them.

18. Discuss the differences observed in the IR and NMR spectra of aniline and acetanilide that are consistent with the formation of the latter in this procedure.

Questions for Part C

19. Explain why 4-acetamidobenzenesulfonyl chloride (**8**) is much *less* susceptible to hydrolysis of the acid chloride function than is 4-acetamidobenzoyl chloride.

$$CH_3COHN-\langle\ \rangle-COCl$$
4-Acetamidobenzoyl chloride

20. What materials, organic or inorganic, may contaminate the crude sulfonyl chloride **8** prepared in this reaction? Which of them are likely to react with the ammonia used in the next reaction step of the sequence?

21. Write a stepwise mechanism for the conversion of acetanilide (**7**) into 4-acetamidobenzenesulfonic acid (**17**); use curved arrows to symbolize the flow of electrons.

22. Write the structure of the material containing sulfonamide linkages that might be obtained upon polymerization of 4-aminobenzenesulfonyl chloride.

$$H_2N-\langle\ \rangle-SO_2Cl$$
4-Aminobenzenesulfonyl chloride

23. Consider the spectral data for 4-acetamidobenzenesulfonyl chloride (Figs. 20.6 and 20.7).

    a. In the functional group region of the IR spectrum, specify the absorptions associated with the nitrogen-hydrogen bond, with the carbon-oxygen double bond, and with the aromatic ring.

    b. In the ^{1}H NMR spectrum, assign the various resonances to the hydrogen nuclei responsible for them.

    c. For the ^{13}C NMR data, assign the various resonances to the carbon nuclei responsible for them.

24. Discuss the differences observed in the IR and NMR spectra of acetanilide and 4-acetamidobenzenesulfonyl chloride that are consistent with the formation of the latter in this procedure.

Questions for Part D

25. What acids might be present in the crude sulfonyl chloride that would cause an exothermic reaction with ammonia?

26. Why would you expect the yield of 4-acetamidobenzenesulfonamide (**9**) to be lowered if the crude sulfonyl chloride **8** were not combined with ammonia until the laboratory period following its preparation?

27. Concentrated ammonia contains 72% water, yet its reaction with 4-acetamidobenzenesulfonyl chloride (**8**) produces 4-acetamidobenzenesulfonamide (**9**) rather than 4-acetamidobenzenesulfonic acid (**17**). Why is ammonia more reactive than water toward **9**?

28. Consider the spectral data for 4-acetamidobenzenesulfonamide (Figs. 20.8 and 20.9).

    a. In the functional group region of the IR spectrum, specify the absorptions associated with the nitrogen-hydrogen bonds, with the carbon-oxygen double bond, and with the aromatic ring.

    b. In the ^{1}H NMR spectrum, assign the various resonances to the hydrogen nuclei responsible for them.

    c. For the ^{13}C NMR data, assign the various resonances to the carbon nuclei responsible for them.

29. Discuss the differences observed in the IR and NMR spectra of 4-acetamidobenzenesulfonyl chloride and 4-acetamidobenzenesulfonamide that are consistent with the formation of the latter in this procedure.

Questions for Part E

30. In the preparation of sulfanilamide (**1**) from 4-acetamidobenzenesulfonamide (**9**), only the acetamido group is hydrolyzed. Explain this difference in reactivity of the acetamido and sulfonamido groups toward aqueous acid.

31. Following hydrolysis of 4-acetamidobenzenesulfonamide (**9**) with aqueous acid, the reaction mixture is homogeneous, whereas 4-acetamidobenzenesulfonamide is insoluble in aqueous acid. Explain the change in solubility that occurs as a result of the hydrolysis.

32. Explain the results obtained when the solubility of sulfanilamide was determined in 1.5 *M* hydrochloric acid and in 1.5 *M* sodium hydroxide. Write equations for any reaction(s) that occurred.

33. Why might there be a preference for using solid sodium carbonate instead of solid sodium hydroxide for basifying the acidic hydrolysis solution obtained in this experiment?

34. What would be observed if 4-acetamidobenzenesulfonamide were subjected to vigorous hydrolysis conditions, such as concentrated hydrochloric acid and heat for a long period of time? Write an equation for the reaction that might occur.

35. Calculate the overall yield of sulfanilamide obtained in the sequence of reactions that you performed.

36. Consider the spectral data for sulfanilamide (Figs. 20.10 and 20.11).

    a. In the functional group region of the IR spectrum, specify the absorptions associated with the nitrogen-hydrogen bonds and with the aromatic ring.

b. In the ^{1}H NMR spectrum, assign the various resonances to the hydrogen nuclei responsible for them.

c. For the ^{13}C NMR data, assign the various resonances to the carbon nuclei responsible for them.

37. Discuss the differences observed in the IR and NMR spectra of 4-acetamidobenzenesulfonyl chloride and 4-acetamidobenzenesulfonamide that are consistent with the formation of the latter in this procedure.

Starting Materials and Products
The IR and NMR spectra of acetanilide are provided in Figures 8.6 and 15.31, respectively.

Spectra

FIGURE 20.2
IR spectrum of nitrobenzene (neat).

(a) ^{1}H NMR spectrum: (300 MHz).

(b) ^{13}C NMR data: δ 123.5, 129.5, 134.8, 148.3.

FIGURE 20.3
NMR data for nitrobenzene
($CDCl_3$).

FIGURE 20.4

IR spectrum of aniline (neat).

FIGURE 20.5

NMR data for aniline (CDCl$_3$).

(a) ^{1}H NMR spectrum: (300 MHz).

(b) ^{13}C NMR data: δ 115.1, 118.2, 129.2, 146.7.

FIGURE 20.6

IR spectrum of
4-acetamidobenzenesulfonyl
chloride (IR card).

(a) ^{1}H NMR spectrum: (300 MHz).

(b) ^{13}C NMR data: δ 25.3, 119.7, 127.5, 141.6, 141.7, 170.3.

FIGURE 20.7
NMR data for
4-acetamidobenzenesulfonyl
chloride (DMSO-d_6).

FIGURE 20.8
IR spectrum of
4-acetamidobenzenesulfonamide
(IR card).

(a) ^{1}H NMR spectrum: (300 MHz).

(b) ^{13}C NMR data: δ 24.2, 118.5, 126.7, 138.1, 142.3, 169.0.

FIGURE 20.9
NMR data for
4-acetamidobenzenesulfonamide
(DMSO-d_6).

FIGURE 20.10
IR spectrum of sulfanilamide
(IR card).

FIGURE 20.11
NMR data for sulfanilamide
(DMSO-d_6).

(a) ^{1}H NMR spectrum: (300 MHz).

(b) ^{13}C NMR data: δ 113.1, 127.6, 130.2, 151.6.

20.3 Synthesis of 1-Bromo-3-chloro-5-iodobenzene

The preparation of 1-bromo-3-chloro-5-iodobenzene (**24**), as illustrated in Scheme 20.4, is a fascinating excursion into the role of **protecting groups** and **substituent effects** as important factors in multi-step synthesis. The sequence used here involves a sequence of **electrophilic aromatic substitutions** and illustrates two important general strategies in synthesis: (1) Protecting a functionality to prevent undesired side reactions and (2) using substituents to control the regiochemical course of a reaction. In the present example, the first principle is represented by the conversion of the *amino* group of **6** into the *amido* group of **7**, and the second is illustrated by the directing and ring-activating properties of the amino and amido functions. The yields of the individual steps are good to high, ranging from 60–96%, an important consideration in designing multi-step syntheses, as noted in Section 20.1.

Although benzene (**4**) is a potential starting point of our synthesis (Scheme 20.4), it is an alleged carcinogen, so the sequence of reactions begins with nitrobenzene (**5**). Discussion of and experimental procedures for the reduction of nitrobenzene and the conversion of aniline (**6**) to acetanilide (**7**) are contained in Section 20.2, so the present description starts with the bromination of acetanilide.

SCHEME 20.4

ACETYLATION AND BROMINATION: PREPARATION OF 4-BROMOACETANILIDE (21)

The introduction of bromine by electrophilic aromatic substitution is not performed directly on aniline (**6**) because the amino group of **6** so greatly activates the ring toward electrophiles that even mild brominating agents such as an aqueous solution of bromine lead to trisubstitution (Eq. 20.10). Acetylating the amino group of aniline moderates the activating effect of the non-bonded pair of electrons on the nitrogen atom. Like the amino group, the acetamido group, $CH_3CONH—$, of acetanilide (**7**) is a ring activator, so selective monobromination of **7** to give **19** occurs under mild conditions, such as bromine in acetic acid *without* a Lewis acid catalyst. The acetamido group is *ortho*- and *para*-directing, so both 2-bromo- and 4-bromoacetanilide are produced. However, the steric bulk of this substituent hinders attack at the 2-position, so formation of the 4-bromo isomer is strongly favored (Eq. 20.11).

2,4,6-Tribromoaniline

(20.10)

7   19   2-Bromoacetanilide

major   minor

(20.11)

As a general rule, separating *ortho-* and *para-*isomers from a mixture containing both is fairly simple. Because *para-*isomers are more structurally symmetrical than *ortho-*isomers, they are *less* soluble in a given solvent as noted in Section 15.4. 4-Bromoacetanilide (**19**), for example, is separable from the 2-bromo isomer by a single recrystallization of the crude mixture from methanol. A second consequence of the greater molecular symmetry of *para-*isomers is that they typically have a higher melting point than their *ortho* relatives. In the case of the *ortho-* and *para-*isomers of bromo-nitrobenzene, this difference is approximately 70 °C.

CHLORINATION: PREPARATION OF 4-BROMO-2-CHLOROACETANILIDE (20)

4-Bromoacetanilide is slightly less reactive toward further electrophilic substitution than acetanilide, but it is still sufficiently reactive to be chlorinated under mild conditions *without* a strong Lewis acid catalyst. Thus, molecular chlorine dissolved in acetic acid chlorinates 4-bromoacetanilide to produce 4-bromo-2-chloroacetanilide (**20**) as shown in Equation 20.12. The formation of the 2- rather than the 3-chloro isomer demonstrates the more powerful directing effect of the acetamido group as compared to the bromo substituent.

19   20

(20.12)

To avoid the hazards associated with using chlorine gas, this reagent may be generated *in situ* by reaction of concentrated hydrochloric acid and sodium chlorate ($NaClO_3$); an oxidation-reduction process occurs by which chloride ion is oxidized to chlorine (Eq. 20.13) and chlorate ion is reduced to chlorine (Eq. 20.14); balancing of these equations is left to the student. In addition to the benefits of ease and safety in handling provided by this method, it allows for preparation of precise amounts of molecular chlorine.

$$Cl^- = 1/2\ Cl_2 + e^- \tag{20.13}$$

$$6H^+ + ClO_3^- + 5e^- = 1/2\ Cl_2 + 3H_2O \tag{20.14}$$

HYDROLYSIS: PREPARATION OF 4-BROMO-2-CHLORO-6-IODOANILINE (22)

The acetamido group of 4-bromo-2-chloroacetanilide (**20**) is the most powerful director of the three substituents on the aromatic ring, so further electrophilic substitution will preferentially occur at the 6-position of the molecule. However, attempted iodination of **20** fails because of the deactivating effects of the halogen substituents and because the iodonium ion, I^+, is a very unreactive electrophile compared to bromonium, Br^+, or chloronium, Cl^+, ions. The solution to this problem is to regenerate the amino group, a more powerful ring activator, by hydrolysis of the amide (Eq. 20.15). This reaction is effected using a solution of concentrated hydrochloric acid in ethanol.

$$\tag{20.15}$$

The iodinating agent in our procedure is iodine monochloride, ICl. Because iodine is the less electronegative of the two halogens, the reagent is polarized to make the iodine atom electrophilic, and it may dissociate to give Cl^- and I^+. The electrophile for the reaction is thus either polarized molecular I–Cl or iodonium ion (Eq. 20.16).

$$\tag{20.16}$$

DIAZOTIZATION AND PROTONOLYSIS: PREPARATION OF 1-BROMO-3-CHLORO-5-IODOBENZENE (24)

The final step in the synthesis is the removal of the amino group from 2-chloro-4-bromo-6-iodoaniline (**22**). After the amino group has been diazotized with nitrous acid at 0 °C to produce **23**, the diazo group may be removed by treatment with absolute ethanol or hypophosphorous acid, H_3PO_2, as depicted in Equation 20.17. Although the latter is the more general reagent for replacing a diazonium group by a hydrogen atom, ethanol works well with certain types of aromatic compounds, in particular those containing halogen, and will be used in our procedure. The transformation presumably involves an oxidation-reduction reaction in which ethanol is oxidized to acetaldehyde or acetic acid. The process is mechanistically ill-defined, but donation of a hydride ion to reduce the diazo compound is a likely possibility (Eq. 20.18).

(20.17)

(20.18)

A potentially important side reaction that might occur is dissociation of the diazonium salt **23** to a carbocation and molecular nitrogen. The carbocation could then react with ethanol to give an aryl ethyl ether as a by-product (Eq. 20.19). Fortunately, ether formation is a minor process in the present reduction, and any ether that is produced is removed by recrystallization of the product.

(20.19)

OVERVIEW

The logic underlying the specific sequence of steps for making **24** is summarized as follows. Introducing the amino group as a director on the benzene ring provides a way to control the locations at which the various halogen substituents appeared. When the directive effects of the amino group are no longer needed, this functionality is replaced with a hydrogen atom, the final step of the present synthesis. Before introducing a halogen atom onto the ring, the ring-activating power of the amino group is lessened by converting it to an acetamido group, which makes monohalogenation possible. The reason that the bromo substituent is added first is that the acetamido group preferentially directs an electrophile to the *para* position. The chloro substituent was introduced next because chlorination of 4-bromoacetanilide is possible, whereas iodination is not. If the amino function had been regenerated *prior* to chlorinating and iodinating the ring, undesired polyhalogenation would have occurred. The iodo group is added last because the aromatic ring must be highly activated for this reaction to succeed, and there is only one activated position remaining where substitution can occur, making polyiodination improbable.

EXPERIMENTAL PROCEDURES

SYNTHESIS OF 1-BROMO-3-CHLORO-5-IODOBENZENE

Purpose: To demonstrate principles of multi-step synthesis, protecting groups, and electrophilic aromatic substitution.

Note   Throughout the sequence of reactions, quantities of reagents must be adjusted according to the amount of starting material that is available from the previous reaction; you should *not* run the reactions on a larger scale than is indicated, however. If the amount of your starting material varies from that indicated, do *not* change the reaction *times* unless directed to do so by your instructor. You should be able to obtain sufficient quantities of product from each reaction to complete the entire sequence starting with the amount of acetanilide provided in the first step.

A · PREPARATION OF ANILINE AND ACETANILIDE

See Section 20.2 for Pre-Lab Exercises and Experimental Procedures for the preparation of nitrobenzene, aniline, and acetanilide, and Figures 8.6, 15.31, and 20.2–20.5 for their spectra.

B · PREPARATION OF 4-BROMOACETANILIDE

Safety Alert

1. Bromine is a hazardous chemical. Do not breathe its vapors or allow it to come into contact with the skin because it may cause serious chemical burns. Perform all operations involving the transfer of solutions of bromine at a hood, and wear latex gloves when handling bromine or solutions of it. If you get bromine on your skin, immediately wash the affected area with soap and warm water and soak

the skin in 0.6 *M* sodium thiosulfate solution for up to 3 h if the burn is particularly serious.

2. Do *not* use acetone to rinse glassware containing residual bromine! This prevents formation of α-bromoacetone, a severe *lachrymator*. First rinse the glassware with aqueous sodium bisulfite.

3. *Glacial* acetic acid is a corrosive liquid. Wear latex gloves when handling this chemical and thoroughly wash with large amounts of water any areas of your skin that may come in contact with it.

Miniscale Procedure

Preparation   Answer the Pre-Lab Exercises on page PL. 139. Review Sections 2.9, 2.11, and 2.17.

Apparatus   A 250-mL round-bottom flask, apparatus for magnetic stirring, vacuum filtration, and *flameless* heating.

Setting Up   Working at the hood, prepare a solution of 3.2 mL of bromine in 6 mL of *glacial* acetic acid in an Erlenmeyer flask. Dissolve 8.1 g of acetanilide in 30 mL of glacial acetic acid in the round-bottom flask containing a stirbar, and rapidly stir the solution.

Bromination, Work-Up, and Isolation   Using a Pasteur pipet, add the bromine solution to the stirred solution of acetanilide over a period of 1–2 min; continue stirring the reaction mixture for about 10 min after completing the addition. Slowly add 100 mL of ice-cold water with stirring, and then add just enough ice-cold saturated aqueous sodium bisulfite to discharge the color of the mixture that results. Cool the mixture in an ice-water bath and then collect the product by vacuum filtration. If the crude product appears yellow, wash it with aqueous sodium bisulfite. In any case wash it well with cold water and press it as dry as possible on the filter. The product may be recrystallized from methanol.★

Analysis   Weigh the product and calculate its yield; determine its melting point. Obtain IR and ^{1}H NMR spectra of your starting material and product and compare them with those of authentic samples (Figs. 8.6, 15.31, 20.12, and 20.13).

Microscale Procedure

Preparation   Answer the Pre-Lab Exercises on page PL. 139. Review Sections 2.9, 2.11, and 2.17.

Apparatus   A 5-mL conical vial, apparatus for magnetic stirring, vacuum filtration, and *flameless* heating.

Setting Up   Working at the hood, prepare a solution of 0.15 mL of bromine and 0.3 mL of *glacial* acetic acid in a test tube. Prepare a solution of 375 mg of acetanilide and 3 mL of *glacial* acetic acid in the conical vial equipped with a spinvane. Stir the solution rapidly.

Bromination, Work-Up, and Isolation   In one portion, add the bromine solution to the stirred mixture and continue stirring for about 10 min at room temperature. Then pour the reaction mixture into 25 mL of ice-cold water contained in a beaker that is equipped with a stirbar. Add just enough ice-cold saturated aqueous sodium bisulfite to discharge the color of the mixture. Stir the resulting mixture in an ice-water bath for 15 min and isolate the precipitated product by vacuum filtration.★ If the crude product appears yellow, wash it with aqueous sodium bisulfite. In any case wash it well with cold water and press it as dry as possible on the filter. The product may be recrystalized from methanol.★

Analysis   Weigh the product and calculate its yield; determine its melting point. Obtain IR and ^{1}H NMR spectra of your starting material and product and compare them with those of authentic samples (Figs. 8.6, 15.31, 20.12, and 20.13).

Finishing Touches

If the *filtrate* and *washings* contain residual bromine, as evidenced by a yellow color, add saturated aqueous sodium bisulfite until the solution is colorless. Flush the *decolorized solution* down the drain.

C · PREPARATION OF 4-BROMO-2-CHLOROACETANILIDE

Safety Alert

1. Chlorine gas is evolved in this procedure. To minimize exposure to it, perform this reaction in the hood if possible. Alternatively, invert a funnel over the reaction flask and attach the funnel to the water aspirator to remove the gaseous chlorine (Fig. 2.68b).
2. *Concentrated* hydrochloric acid and *glacial* acetic acid are corrosive liquids. Wear latex gloves when handling these chemicals and thoroughly wash any areas of your skin that may come in contact with them.

Miniscale Procedure

Preparation   Answer the Pre-Lab Exercises on page PL. 141. Review Sections 2.9, 2.10, 2.11, and 2.17.

Apparatus   A 250-mL round-bottom flask, ice-water bath, apparatus for magnetic stirring, vacuum filtration, and *flameless* heating.

Setting Up   Suspend 10.7 g of 4-bromoacetanilide in a magnetically stirred solution of 23 mL of *concentrated* hydrochloric acid and 28 mL of *glacial* acetic acid contained in the round-bottom flask. Heat the mixture gently with stirring until it becomes homogeneous, and then cool the solution to 0 °C. Prepare a solution of 2.8 g of sodium chlorate in about 7 mL of water.

Chlorination, Work-Up, and Isolation   Using a Pasteur pipet, add the aqueous sodium chlorate solution dropwise to the stirred cold mixture over a period of about 5 min;

some chlorine gas is evolved. After completing the addition, stir the reaction mixture at room temperature for 1 h and then collect the precipitate by vacuum filtration.★ Wash the crude product *thoroughly* with ice-cold water until the washings are neutral or nearly so. The product may be recrystallized from methanol.

Analysis   Weigh the product and calculate its yield; determine its melting point. Obtain its IR and ^{1}H NMR spectra and compare them with those of an authentic sample (Figs. 20.14 and 20.15).

Microscale Procedure

Preparation   Answer the Pre-Lab Exercises on page PL. 141. Review Sections 2.9, 2.10, 2.11, and 2.17.

Apparatus   A 5-mL conical vial, ice-water bath, apparatus for magnetic stirring, vacuum filtration, and *flameless* heating.

Setting Up   Suspend 500 mg of 4-bromoacetanilide in a magnetically stirred solution of 1 mL of *concentrated* hydrochloric acid and 1.3 mL of *glacial* acetic acid contained in the conical vial. If necessary heat the mixture gently with stirring until it becomes homogeneous and then cool the solution to 0 °C. Prepare a solution of 150 mg of sodium chlorate in about 0.5 mL of water.

Chlorination, Work-Up, and Isolation   Using a Pasteur pipet, add the aqueous sodium chlorate solution dropwise to the stirred cold mixture over a period of about 2 min; some chlorine gas is evolved. After completing the addition, stir the reaction mixture at room temperature for 1 h and then collect the precipitate by vacuum filtration.★ Wash the crude product *thoroughly* with ice-cold water until the washings are neutral or nearly so. The product may be recrystallized from methanol.

Analysis   Weigh the product and calculate its yield; determine its melting point. Obtain its IR and ^{1}H NMR spectra and compare them with those of an authentic sample (Figs. 20.14 and 20.15).

Finishing Touches   If the *filtrate* and *washings* contain residual chlorine, as evidenced by formation of a blue-black color when a drop of the liquid is placed on starch/iodide paper, add saturated aqueous sodium bisulfite until the solution gives a negative test to starch/iodide paper. Flush the *decolorized solution* down the drain.

D · PREPARATION OF 4-BROMO-2-CHLOROANILINE

Safety Alert   *Concentrated* hydrochloric acid is a corrosive liquid and 14 *N* aqueous sodium hydroxide is highly caustic. Wear latex gloves when handling these chemicals and thoroughly wash any areas of your skin that may come in contact with them.

Miniscale Procedure

Preparation   Answer the Pre-Lab Exercises on page PL. 143. Review Sections 2.9, 2.11, 2.17, and 2.22.

Apparatus   A 250-mL round-bottom flask, apparatus for heating under reflux, magnetic stirring, vacuum filtration, and *flameless* heating.

Setting Up   Mix 11.2 g of crude 4-bromo-2-chloroacetanilide with 20 mL of 95% ethanol and 13 mL of *concentrated* hydrochloric acid in the flask. Add a stirbar and assemble the apparatus for heating under reflux.

Hydrolysis and Isolation   Heat the mixture under reflux for *ca.* 0.5 h. At the end of this period, add 90 mL of *hot* water, swirl the mixture to dissolve any solids completely and pour the hot solution onto about 150 g of ice contained in a beaker. To the resulting mixture slowly add 12 mL of 14 *N* sodium hydroxide solution, stirring the mixture well during the addition. Add more ice if it has all melted before completing addition of the aqueous base. Check the pH of the reaction mixture to ensure that it is basic; if not, add more aqueous base. Isolate the solid by vacuum filtration, wash it thoroughly with ice-cold water, and press it as dry as possible on the filter.★

Purification   The crude product may be recrystallized by first dissolving it in boiling methanol, allowing the solution to cool to below 40 °C, and adding cold water dropwise until crystallization begins. Should the product begin to oil out rather than crystallize, add a few drops of methanol to dissolve the oil. Cool the mixture thoroughly in an ice-water bath and collect the product by vacuum filtration, washing it with a *small* volume of ice-cold 1:1 methanol-water.

Analysis   Weigh the product and calculate its yield; determine its melting point. Obtain its IR and ¹H NMR spectra and compare them with those of an authentic sample (Figs. 20.16 and 20.17).

Microscale Procedure

Preparation   Answer the Pre-Lab Exercises on page PL. 143. Review Sections 2.9, 2.11, 2.17, and 2.22.

Apparatus   A 5-mL conical vial, apparatus for heating under reflux, magnetic stirring, vacuum filtration, Craig tube filtration, and *flameless* heating

Setting Up   Mix 450 mg of crude 4-bromo-2-chloroacetanilide with 1.5–2 mL of 95% ethanol and 0.5 mL of *concentrated* hydrochloric acid in the conical vial. Add a spinvane and assemble the apparatus for heating under reflux.

Hydrolysis and Isolation   Heat the mixture under reflux for *ca.* 0.5 h. At the end of this period, add 2 mL of *hot* water, stir the mixture to dissolve any solids completely, and pour the hot solution onto 3–4 g of ice contained in a beaker. To the resulting mixture slowly add 0.5 mL of 14 *N* sodium hydroxide solution, stirring the mixture well during the addition. Add more ice if it has all melted before completing addition of the aqueous base. Check the pH of the reaction mixture to ensure that it is basic; if not, add more aqueous base. Isolate the solid by vacuum filtration, and thoroughly dry it.★

Purification   The crude product may be recrystallized by first dissolving it in boiling methanol, allowing the solution to cool to below 40 °C, and adding cold water dropwise until crystallization begins. Should the product begin to oil out rather than crystallize, add a few drops of methanol to dissolve the oil. Cool the mixture thoroughly in an ice-water bath and collect the product by Craig tube or vacuum filtration, washing it with a *few drops* of ice-cold 1 : 1 methanol-water.

Analysis   Weigh the product and calculate its yield; determine its melting point. Obtain its IR and ^{1}H NMR spectra and compare them with those of an authentic sample (Figs. 20.16 and 20.17).

Finishing Touches

Neutralize the *filtrate* and flush it down the drain.

E • PREPARATION OF 4-BROMO-2-CHLORO-6-IODOANILINE

Safety Alert

   Glacial acetic acid is a corrosive liquid, can cause burns, and is harmful if inhaled. Wear latex gloves when handling this chemical and thoroughly wash with water any areas of your skin that may come in contact with it.

Miniscale Procedure

Preparation   Answer the Pre-Lab Exercises on page PL. 145. Review Sections 2.9, 2.10, 2.11, and 2.17.

Apparatus   A 250-mL round-bottom flask, thermometer, ice-water bath, apparatus for magnetic stirring, vacuum filtration, and *flameless* heating.

Setting Up   With magnetic stirring, dissolve 2.5 g of recrystallized 4-bromo-2-chloroaniline in 40 mL of *glacial* acetic acid in the round-bottom flask, and add 10 mL of water to the mixture. In an Erlenmeyer flask, prepare a solution of 2.5 g of technical iodine monochloride in 10 mL of *glacial* acetic acid.

Iodination, Work-Up, and Isolation   Over a period of 8–10 min, add the iodine monochloride solution to the stirred solution of the aniline, heat the resulting mixture to 90 °C with continued stirring, and then allow the solution to cool to about 50 °C. Add just enough saturated aqueous sodium bisulfite to turn the color of the mixture bright yellow; record the volume of sodium bisulfite solution required. Dilute the reaction mixture with enough extra water such that the *combined* volume of the aqueous sodium bisulfite *plus* the volume of added water is about 15 mL. Cool the reaction mixture in an ice-water bath, and isolate the crude product that results by vacuum filtration. Wash the crystals with a small amount of ice-cold 33% aqueous acetic acid and then at least 50 mL of ice-cold water. Press out as much water as possible on the filter.★

Purification   Recrystallize the crude product from acetic acid-water as follows. Mix the product with *glacial* acetic acid in the ratio of about 20 mL of glacial acetic acid per gram of product. Heat the mixture and slowly add 5 mL of water per gram of product to the mixture as it is heating. Terminate heating once dissolution has occurred, and allow the solution to cool slowly to room temperature. Complete the crystallization by cooling the mixture in an ice-water bath and isolate the product by vacuum filtration.

Analysis   Weigh the product and calculate its yield; determine its melting point. Obtain its IR and ^{1}H NMR spectra and compare them with those of an authentic sample (Figs. 20.18 and 20.19).

Microscale Procedure

Preparation   Answer the Pre-Lab Exercises on page PL. 145. Review Sections 2.9, 2.10, 2.11, and 2.17.

Apparatus   A 25-mL Erlenmeyer flask, thermometer, ice-water bath, apparatus for magnetic stirring, vacuum filtration, and *flameless* heating.

Setting Up   With magnetic stirring dissolve 250 mg of recrystallized 4-bromo-2-chloroaniline in 4 mL of *glacial* acetic acid in the Erlenmeyer flask, and add 1 mL of water to the mixture. In a test tube, prepare a solution of 250 mg of technical iodine monochloride in 1 mL of *glacial* acetic acid.

Iodination, Work-Up, and Isolation   Over a period of 3–5 min, add the iodine monochloride solution to the stirred solution of the aniline. With continued stirring, heat the resulting mixture to 90 °C, and then allow the solution to cool to about 50 °C. Add just enough saturated aqueous sodium bisulfite to turn the color of the mixture bright yellow; record the volume of sodium bisulfite solution required. Dilute the reaction mixture with enough extra water such that the *combined* volume of the aqueous sodium bisulfite *plus* the volume of added water is about 3 mL. Cool the reaction mixture in an ice-water bath and isolate the crude product that results by vacuum filtration. Wash the crystals with a small amount of ice-cold 33% aqueous acetic acid and then with at least 5 mL of ice-cold water.★

Purification   Using a test tube, recrystallize the crude product from acetic acid-water as follows. Mix the product with *glacial* acetic acid in the ratio of about 2 mL of glacial acetic acid per 100 mg of product. Heat the mixture and slowly add 0.5 mL of water per 100 mg of product to the mixture as it is heating. Terminate heating once dissolution has occurred, and allow the solution to cool slowly to room temperature. Complete the crystallization by cooling the mixture in an ice-water bath and isolate the product by vacuum filtration.

Analysis   Weigh the product and calculate its yield; determine its melting point. Obtain its IR and ^{1}H NMR spectra and compare them with those of an authentic sample (Figs. 20.18 and 20.19).

Flush the *aqueous filtrates* down the drain. Use solid sodium carbonate to neutralize the *filtrate* from the recrystallization before flushing it down the drain.

Finishing Touches

F · PREPARATION OF 4-BROMO-2-CHLORO-6-IODOBENZENE

Miniscale Procedure

Preparation   Answer the Pre-Lab Exercises on page PL. 147. Review Sections 2.9,
2.10, 2.11, 2.13, 2.16, 2.17, 2.22, and 2.29.

Apparatus   A 250-mL and a 50-mL round-bottom flask, thermometer, ice-water bath,
apparatus for heating under reflux, magnetic stirring, steam distillation, vacuum filtration, simple distillation, and *flameless* heating.

Setting Up   Suspend 2 g of 4-bromo-2-chloro-6-iodoaniline in about 10 mL of absolute
ethanol in the 250-mL round-bottom flask containing a stirbar. While stirring the mixture, add 4.0 mL of *concentrated* sulfuric acid dropwise. Equip the flask for heating
under reflux.

Diazotization, Isolation, and Isolation   Add 0.7 g of powdered sodium nitrite in small
portions through the top of the condenser. When the addition is complete, use a few
drops of *absolute* ethanol to rinse any solid that may stick to the walls of the condenser
into the flask, and then heat the mixture under gentle reflux for about 10 min.
Add 125 mL of hot water to the flask and equip it for steam distillation. Steam-
distil the mixture, collecting about 50 mL of distillate, which should be clear
and not contain any organic product. The desired product should form as a solid
in the condenser, so monitor the condenser closely to ensure that it does not become plugged during the distillation. Should the condenser become clogged with
solid, open it by running hot water or a slow stream of steam through the condenser
jacket.

Upon completing the distillation, extract the distillate with two 15-mL portions of
dichloromethane and put the extract in the 50-mL round-bottom flask. Use a few milliliters of dichloromethane to rinse any solid in the condenser into the flask containing
the dichloromethane extract.★ Remove the dichloromethane by simple distillation;
alternatively, use rotary evaporation or other technique (Sec. 2.29) to remove the
solvent. Recrystallize the residue from methanol and isolate the product by vacuum
filtration.

Analysis   Weigh the product and calculate its yield; determine its melting point. Obtain
its IR and ^{1}H NMR spectra and compare them with those of an authentic sample (Figs.
20.20 and 20.21).

Microscale Procedure

Preparation   Answer the Pre-Lab Exercises on page PL. 145. Review Sections 2.9,
2.10, 2.11, 2.13, 2.16, 2.17, 2.22, and 2.29.

Apparatus A 5-mL conical vial, ice-water bath, apparatus for heating under reflux, magnetic stirring, simple distillation, Craig tube filtration, and *flameless* heating.

Setting Up Suspend 200 mg of 4-bromo-2-chloro-6-iodoaniline in about 1.5 mL of absolute ethanol in the conical vial containing a spinvane. While stirring the mixture, add 0.4 mL of *concentrated* sulfuric acid dropwise.

Diazotization, Isolation, and Purification Add 0.07 g of powdered sodium nitrite in small portions to the vial. When the addition is complete, use a drop or two of *absolute* ethanol to rinse any solid that may stick to the sides of the vial into the solution. Equip the vial with a condenser and heat the mixture under gentle reflux for about 10 min. Add 2 mL of hot water to the vial and equip it for simple distillation. Steam-distil the mixture, until about 1 mL of liquid remains in the vial. Remove distillate from the Hickman stillhead as necessary and place it in a conical vessel.

Upon completing the distillation, combine all distillates in the conical vessel. Rinse any solid contained in the Hickman stillhead into the vessel, using 1–2 mL of dichloromethane, and separate the layers. Extract the aqueous layer once with 2 mL of dichloromethane and combine the dichloromethane solutions.★ Remove the dichloromethane by simple distillation; alternatively, use rotary evaporation or other technique (Sec. 2.29). Recrystallize the residue from methanol in a Craig tube and isolate the product by Craig tube filtration.

Analysis Weigh the product and calculate its yield; determine its melting point. Obtain its IR and ^{1}H NMR spectra and compare them with those of authentic samples (Figs. 20.20 and 20.21).

Finishing Touches

Neutralize the *pot residue* from the steam distillation of the *reaction mixture* and flush the resulting *aqueous layer* and that obtained from the *steam distillate* down the drain. Pour the *organic layer* from the *pot residue*, the recovered *dichloromethane*, and the *filtrate* from recrystallization into the container for halogenated compounds.

EXERCISES

Refer to Tables 8.1–8.8 as needed for answering the questions on spectroscopy.

General Questions

1. Using the concepts developed in the synthesis of 1-bromo-3-chloro-5-iodobenzene, outline synthetic procedures, starting from acetanilide, for preparing each of the following compounds:

    a. 1,3,5-Tribromobenzene

    b. 2-Bromo-4-chloro-6-iodophenol

    c. 2-Bromo-4,6-dichloroaniline

    d. 3-Bromotoluene

2. Why is the amino group a more powerful ring activator than the amido group?

3. In the sequence used, which steps demonstrate that the amido group is a more powerful *ortho-*, *para*-director than a bromo or chloro substituent?

4. Hydroxylic solvents, such as water or low molecular-weight alcohols, are often used to purify amides such as the acetanilides produced in this experiment. Briefly explain why these are used in preference to hydrocarbon solvents like petroleum ether.

Questions for Part A

See questions 4–18 of Section 20.2.

Questions for Part B

5. Explain why electrophilic bromination of acetanilide produces 4- rather than 2-bromoacetanilide as the major product.

6. Compared to Br_2/HOAc, Br_2/$FeBr_3$ is a more powerful combination of reagents for effecting electrophilic aromatic bromination. What difficulty might arise if this combination were used for brominating acetanilide?

7. Account for the color change you observed upon reaction of bromine with acetanilide.

8. Using curved arrows to symbolize the flow of electrons, provide a stepwise mechanism for the bromination of acetanilide with Br_2/HOAc.

9. Consider the spectral data for 4-bromoacetanilide (Figs. 20.12 and 20.13).

   a. In the functional group region of the IR spectrum, specify the absorptions associated with the nitrogen-hydrogen bond, carbon-oxygen double bond, and the aromatic ring.

   b. In the ^{1}H NMR spectrum, assign the various resonances to the hydrogen nuclei responsible for them.

   c. For the ^{13}C NMR data, assign the various resonances to the carbon nuclei responsible for them.

10. How many resonances would you expect for the ^{1}H-decoupled ^{13}C spectrum of 2-bromoacetanilide?

11. Discuss the differences observed in the IR and NMR spectra of acetanilide and 4-bromoacetanilide that are consistent with the formation of the latter in this procedure.

Questions for Part C

12. How might acetic acid interact with molecular chlorine to make it more electrophilic in character?

13. Would 4-bromoacetanilide or 4-bromo-2-chloroacetanilide be more reactive toward electrophilic substitution? Explain.

14. 4-Bromo-3-chloroacetanilide is a possible product from chlorination of 4-bromoacetanilide. Why is this isomer *not* produced to any significant extent?

15. Consider the spectral data for 4-bromo-2-chloroacetanilide (Figs. 20.14 and 20.15).

   a. In the functional group region of the IR spectrum, specify the absorptions associated with the nitrogen-hydrogen bond, the carbon-oxygen double bond, and the aromatic ring.

b. In the ^{1}H NMR spectrum, assign the various resonances to the hydrogen nuclei responsible for them.

c. For the ^{13}C NMR data, assign the various resonances to the carbon nuclei responsible for them.

16. Discuss the differences observed in the IR and NMR spectra of 4-bromoacetanilide and 4-bromo-2-chloroacetanilide that are consistent with the formation of the latter in this procedure.

Questions for Part D

17. What is the white precipitate that forms during the hydrolysis of 4-bromo-2-chloroacetanilide?

18. Why must aqueous base be added in order to isolate 4-bromo-2-chloroaniline from the aqueous acidic mixture?

19. Using curved arrows to symbolize the flow of electrons, provide a stepwise mechanism for the acid-catalyzed hydrolysis of 4-bromo-2-chloroacetanilide.

20. Consider the spectral data for 4-bromo-2-chloroaniline (Figs. 20.16 and 20.17).

a. In the functional group region of the IR spectrum, specify the absorptions associated with the amino group and the aromatic ring.

b. In the ^{1}H NMR spectrum, assign the various resonances to the hydrogen nuclei responsible for them.

c. For the ^{13}C NMR data, assign the various resonances to the carbon nuclei responsible for them.

21. Discuss the differences observed in the IR and NMR spectra of 4-bromo-2-chloroacetanilide and 4-bromo-2-chloroaniline that are consistent with the formation of the latter in this procedure.

Questions for Part E

22. a. Using curved arrows to symbolize the flow of electrons, provide a stepwise mechanism for the reaction that occurs between iodine monochloride and 4-bromo-2-chloroaniline.

b. Suppose that this same reaction were carried out using bromine monochloride, BrCl. What electrophilic substitution reaction might occur were this reagent, rather than iodine monochloride, allowed to react with 4-bromo-2-chloroaniline? Explain.

23. Iodonium ion, I^+, is a less reactive electrophile than bromonium ion, Br^+. Explain why.

24. The pK_a of acetic acid is about 4.8, whereas that of 4-bromo-2-chloro-6-iodoanilinium ion is estimated to be about 3.

a. Using these values, calculate K_{eq} for the acid–base reaction given below.

b. Given that the anilinium ion should be soluble in polar media such as the combination of HOAc and H_2O used in recrystallizing 4-bromo-2-chloro-6-

iodoaniline (**24**), explain how your calculated K_{eq} is consistent with the fact that **24** precipitates from aqueous acetic acid.

$$CH_3CO_2H + Br-\text{(ring)}-NH_2 \rightleftharpoons CH_3CO_2^- + Br-\text{(ring)}-\overset{+}{N}H_3$$

24

4-Bromo-2-chloro-6-
iodoanilinium ion

25. Consider the spectral data for 4-bromo-2-chloro-6-iodoaniline (Figs. 20.18 and 20.19).

   a. In the functional group region of the IR spectrum, specify the absorptions associated with the amino group and the aromatic ring.

   b. In the ^{1}H NMR spectrum, assign the various resonances to the hydrogen nuclei responsible for them.

   c. For the ^{13}C NMR data, assign the various resonances to the carbon nuclei responsible for them.

26. Discuss the differences observed in the IR and NMR spectra of 4-bromo-2-chloroaniline and 4-bromo-2-chloro-6-iodoaniline that are consistent with the formation of the latter in this procedure.

Questions for Part F

27. Using curved arrows to symbolize the flow of electrons, provide a stepwise mechanism for the conversion of 4-bromo-2-chloro-6-iodoaniline (**24**) to the diazonium salt **25**.

28. The reduction of the diazonium salt **25** may involve transfer of hydrogen from ethanol, which is oxidized to acetaldehyde. Propose an experiment that would allow you to determine whether the hydrogen comes from the hydroxyl group or from C-1 of ethanol.

29. Calculate the overall yield of 1-bromo-3-chloro-5-iodobenzene obtained in the sequence of reactions that you performed.

30. Consider the spectral data for 4-bromo-2-chloro-6-iodoaniline (Figs. 20.20 and 20.21).

   a. In the functional group region of the IR spectrum, specify the absorptions associated with the aromatic ring.

   b. In the ^{1}H NMR spectrum, assign as many as possible of the various resonances to the hydrogen nuclei responsible for them.

   c. For the ^{13}C NMR data, assign the various resonances to the carbon nuclei responsible for them.

31. Discuss the differences observed in the IR and NMR spectra of 4-bromo-2-chloro-6-iodoaniline and 4-bromo-2-chloro-6-iodobenzene that are consistent with the formation of the latter in this procedure.

Spectra

Starting Materials and Products
    The IR and NMR spectra of acetanilide are provided in Figures 8.6 and 15.31, respectively.

FIGURE 20.12
IR spectrum of 4-bromoacetanilide (IR card).

(a) ^{1}H NMR spectrum: (300 MHz).

FIGURE 20.13
NMR data for 4-bromoacetanilide (DMSO-d$_6$).

(b) ^{13}C NMR data: δ 25.1, 115.5, 121.9, 132.5, 139.7, 169.5.

FIGURE 20.14
IR spectrum of 4-bromo-2-chloroacetanilide (IR card).

FIGURE 20.15

NMR data for 4-bromo-2-chloroacetanilide (CDCl₃).

(a) ¹H NMR spectrum: (300 MHz).

(b) ¹³C NMR data: δ 24.8, 116.2, 122.6, 123.2, 130.7, 131.3, 133.7, 168.2.

FIGURE 20.16

IR spectrum of 4-bromo-2-chloroaniline (IR card).

FIGURE 20.17

NMR data for 4-bromo-2-chloroaniline (CDCl₃).

(a) ¹H NMR spectrum: (300 MHz).

(b) ¹³C NMR data: δ 109.2, 116.8, 119.8, 130.4, 131.5, 142.1.

FIGURE 20.18
IR spectrum of 4-bromo-2-chloro-6-iodoaniline (IR card).

(a) ^{1}H NMR spectrum: (300 MHz).

FIGURE 20.19
NMR data for 4-bromo-2-chloro-6-iodoaniline (CDCl$_3$).

(b) ^{13}C NMR data: δ 84.0, 109.7, 118.6, 132.4, 140.0, 143.2.

FIGURE 20.20
IR spectrum of 1-bromo-3-chloro-5-iodobenzene (IR card).

FIGURE 20.21

NMR data for 1-bromo-3-chloro-5-iodobenzene (CDCl$_3$).

(a) ^{1}H NMR spectrum: (300 MHz).

(b) ^{13}C NMR data: δ 94.7, 123.7, 131.6, 136.3, 136.5, 138.7.

■■■■■■■ H I S T O R I C A L   H I G H L I G H T ■■■■■■■

Discovery of Sulfa Drugs

The sulfanilamides were the first antibacterial drugs invented by chemists, and a fascinating story underlies the discovery of the medicinal properties of these compounds. Although many scientists played important parts in this discovery, a key individual in the effort was Gerhard Domagk. Born in Lagow, Germany, in 1895, Domagk attended the University ofKiel intending to become a doctor, but World War I interrupted his medical studies. Following the Armistice in 1918, he reentered the University of Kiel and earned his medical degree in 1921. After a brief career in academia, he moved to I. G. Farbenindustrie (I.G.F.), the German dye cartel, where his responsibility was testing the pharmacological properties of the new dyes being synthesized by chemists.

At the time Domagk joined I.G.F., there were no antibacterials. This presented a serious health problem because bacteria were known to be the agents that caused pneumonia, meningitis, gonorrhea, and streptococcic and staphylococcic infections. The I.G.F. team of chemists and pharmacologists set out to find compounds that would kill these microbes without harming their animal or human hosts, and

they formulated a plan to determine whether certain dyes might be bactericidal. Their strategy evolved from the observation that particular dyes, specifically those containing a sulfonamido group, seemed to be particularly "fast," or tightly bound, to wool fabrics, indicating their affinity for the protein molecules comprising wool. Because bacteria are proteinaceous in nature, the researchers reasoned that the dyes might fasten to the bacteria in such a way as to inhibit or kill them selectively. As we shall see, this simple hypothesis was partially correct: The sulfonamido group was indeed essential, but the part of the molecule that made the substance function as a dye was irrelevant to its effectiveness as a bactericide.

One dye Domagk tested on laboratory mice and rabbits infected with streptococci was called Prontosil. This compound was found to be strongly disinfective against these bacteria and could be tolerated by the animals in large doses with no ill effects. This discovery of the bactericidal effect of Prontosil in animals was probably made in early 1932; I.G.F. applied for a patent in December of that year.

HISTORICAL HIGHLIGHT
Discovery of Sulfa Drugs *(continued)*

Clinical tests on human patients apparently began soon after this, but the record is confused. Some accounts say that before any other tests had been made on humans, Domagk gave a dose of Prontosil in desperation to his deathly ill young daughter, who had developed a serious streptococcal infection following a needle prick; the girl then made a rapid recovery. Others report that the first clinical test was on a ten-month-old boy who was dying of staphylococcal septicemia. His doctor, R. Förster, was a friend of Domagk's superior at I.G.F. and through him learned about a red dye (Prontosil) that was miraculously effective in animals against streptococci. Since the baby was close to death anyway and feeling that there was nothing to lose if the dye was not effective against staphylococci, Förster gave the child two doses of the red dye; complete recovery rapidly followed.

Regardless of which of these two stories is correct, or whether both are true, it was widely recognized by the middle 1930s that the discovery of the bactericidal properties of Prontosil was a medical miracle, for which Domagk was awarded the Nobel Prize in Physiology or Medicine in 1939. There were other important developments in the years between 1933 and 1939, however. Domagk did not publish the results of his tests of Prontosil on animal infections until February 1935, more than two years after the work was done. Learning of his results, the Trefouels, a wife and husband team at the Pasteur Institute in Paris, were prompted to test the bactericidal properties of several compounds, all of which were "azo" dyes closely related in chemical structure to Prontosil. The feature common to their dyes was the sulfonamido portion, but the other parts of the molecules differed significantly. Remarkably, they found that the antibacterial properties of these dyes were virtually identical with those of Prontosil.

This finding led to an explanation of a puzzling fact about Prontosil: It was ineffective against bacteria *in vitro* but was strongly effective *in vivo*. Apparently, a metabolic process within animals was necessary to make the sulfonamide dyes antibacterial. The Trefouels reasoned that the dye is broken into two parts in animals and only the sulfonamido portion is effective as an antibacterial. To prove this, they synthesized the sulfonamido component of Prontosil, which was the known compound *p*-aminobenzenesulfonamide, or sulfanilamide, and found it to be as effective as Prontosil against bacterial infections. Comparison of the formulas of Prontosil and sulfanilamide makes it clear that cleavage of Prontosil at the azo double bond affords the skeleton of sulfanilamide. This cleavage occurs biochemically when Prontosil is injected or imbibed, and the sulfanilamide so produced is the actual antibacterial agent. The original hypothesis that sulfonamido dyes would be bactericidal was thus partly misconceived, in that only the sulfonamido part of the dye molecule kills microbes; the fact that it was a part of a dye molecule was incidental.

Prontosil

Sulfanilamide

Interestingly, the Trefouels' observations made the patent on Prontosil filed by I.G.F. useless. Sulfanilamide had been synthesized and patented many years before as a dye intermediate, but the patent had expired by the time the substance was found to be a potent bactericide. Moreover, their findings led to clinical trials of sulfanilamide in France, England, and the United States, all of which were highly successful. One case that gave great publicity to the new drug was the use of Prontosil to save the life of Franklin D. Roosevelt, Jr., son of the president. In

HISTORICAL HIGHLIGHT
Discovery of Sulfa Drugs *(continued)*

1936, young Roosevelt was dying from a strepto-coccic infection when his mother convinced a doctor to administer Prontosil, which saved his life.

By 1947 over 5,000 sulfonamides related to sul-fanilamide had been prepared and tested for their ef-ficacy as antibacterials. Although not all were effec-tive, some were found to be better than sulfanilamide against certain infections. Of the thousands of com-pounds prepared and tested, the active ones are al-most always those in which the only variation in structure is a change in the group of atoms attached to the nitrogen atom of the sulfonamido moiety.

We mentioned that Domagk was awarded the No-bel Prize in 1939. This is not quite correct. He was selected for the prize in that year but did not actually receive it until many years later. When he received notice of the award in October 1939, Domagk sent a letter of acceptance, but a second letter declining it was received in Stockholm in November. The second letter was the result of pressure from the Nazis: Do-magk was by this time in the hands of the Gestapo. In 1947, Domagk was finally able to visit Stockholm, deliver the Nobel lecture, and receive the medal and diploma—but not the prize money, which by then had reverted to the Nobel Foundation.

Relationship of Historical Highlight to Experiments

The experiments performed in this chapter illustrate some of the various strategies that are the keys to-sucessful multistep syntheses. In such processes, the product on one reaction serves as the starting mate-rial for the next, so it is necessary to develop efficient procedures that afford products of sufficient purity that they may be used directly in the next step with minimal purification. The preparation of the antibi-otic sulfanilamide from the readily available starting material nitrobenzene (Sec. 20.2) represents the use of multistep synthesis to produce a biologically im-portant anti-bacterial agent. That of 1-bromo-3-chloro-5-iodobenzene (Sec. 20.3) shows how a poly-substituted aromatic compound can be selectively formed in a multistep sequence.

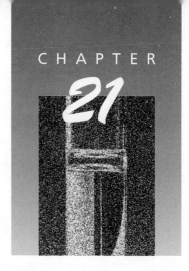

CHAPTER
21

Polymers

Introduction

Polymers are a class of molecules characterized by their high molecular weights, which range from the thousands to the hundreds of thousands, and by the presence of simple repeating structural units called **monomers**. Because of their large size, polymers are often referred to as **macromolecules**. A polymer comprised of a *single* recurring monomer, M, is termed a **homopolymer**, whereas one containing at least two structurally distinct monomeric units, M_1 and M_2, distributed at random in the molecule is called a **copolymer**. These two classes of macromolecules are represented by **1** and **2**, respectively.

$$R—M—(M)_n—M—R \qquad R—M_1—(M_1—M_2)_n—M_2—R$$
$$\textbf{1} \qquad\qquad\qquad\qquad \textbf{2}$$
Homopolymer   Copolymer

Polymers are found in nature and may also be produced by laboratory synthesis. Important examples of naturally occurring macromolecules, or **biopolymers**, are **proteins**, **polysaccharides**, **terpenes**, and **nucleic acids**. General representations of these substances are provided by structures **3–6**, respectively, in which the monomeric subunits of an α-amino acid, **3**, a pyranose, **4**, an isoprene, **5**, and a ribonucleotide phosphate, **6**, are seen. Synthetic, or ''manmade,'' polymers are represented by the myriad of **plastics**, **elastomers**, and **fibers** that are commonplace in contemporary society.

3   **4**   **5**   **6**

(Base = purine or pyrimidine)

589

Two primary methods of polymerization are commonly used to convert monomers into synthetic polymers. In the older literature, these techniques are referred to as **addition** and **condensation polymerization**, but because of ambiguities in these terms, the preferred names now are **chain-reaction polymerization** and **step-growth polymerization**, respectively. As discussed in Sections 21.2 and 21.3, the major distinction between these two types of polymerization is the general mechanism by which the polymer forms.

The term *addition polymerization* arose because such polymers are produced by combining a large number of monomer molecules by addition reactions. For example, the self-addition of thousands of ethylene molecules yields polyethylene (**7**), a homopolymer (Eq. 21.1). Another homopolymer is polystyrene (**9**), which is formed by self-addition of styrene (**8**) as shown in Equation 21.2. It is appropriate to note that representation of the molecular formula of the polymer as essentially n times that of the monomer, as shown in these two equations, is only a slight over-simplification of the actual formula. Because n is such a large number, inclusion of the elemental compositions of the end-groups, R, that appear at the termini of the polymeric chain makes an insignificant change in the molecular formula in comparison with the rest of the molecule.

$$n\ CH_2{=}CH_2 \xrightarrow[\text{and/or heat}]{\text{catalyst}} \ {+}CH_2{-}CH_2{)}_n \qquad (21.1)$$

7

$n = 10{,}000{-}30{,}000$

Ethylene   Polyethylene

$$(21.2)$$

8   **9**

Styrene   Polystyrene

Some other common addition polymers have trade names that do not indicate their structure. For example, Teflon is a homopolymer of tetrafluoroethylene (**10**), and Plexiglas^{TM} is a polymer of methyl methacrylate (**11**).

$CF_2{=}CF_2$ $CH_2{=}C\big<^{CO_2CH_3}_{CH_3}$

10   **11**

Tetrafluoroethylene   Methyl methacrylate

Copolymers are produced from a *mixture* of monomers, as was noted earlier. For example, saran (**14**), a widely used plastic film, is made by polymerizing a mixture of vinyl chloride (**12**) and vinylidene chloride (**13**), as depicted in Equation 21.3. Note that the abbreviated formula of **14** is not meant to imply that the two monomeric units appear as a sequence of two distinct blocks, each of which individually represents a homopolymer. Although such **block copolymers** can be produced by special techniques, copolymers usually have the two monomers distributed randomly along the chain, as in **15**.

$$n\ CH_2\!\!=\!\!CH \underset{Cl}{|} + m\ CH_2\!\!=\!\!CCl_2 \longrightarrow \left(\!\!-CH_2\!\!-\!\!CH\underset{Cl}{|}\right)_{\!n}\!\!\left(CH_2\!\!-\!\!CCl_2\!\!- \right)_{\!m} \qquad (21.3)$$

$$\begin{array}{ccc} \mathbf{12} & \mathbf{13} & \mathbf{14} \\ \text{Vinyl chloride} & \text{Vinylidene chloride} & \text{Saran} \end{array}$$

$$R\!-\!M_1\!-\!(M_2\!-\!M_2\!-\!M_1\!-\!M_2\!-\!M_1\!-\!M_1\!-\!M_1)_n\!-\!M_2\!-\!R$$

$$\mathbf{15}$$

The following sections contain discussions of the mechanisms of chain-reaction and step-growth polymerization. The associated experimental procedures illustrate the preparation of polymers derived from each type of polymerization.

21.2   Chain-Reaction Polymerization

As the name implies, *chain-reaction polymerization* is a chain reaction in which the initiator may be a cation, anion, or free radical. An example of a **cationic polymerization** is found in the polymerization of isobutylene in the presence of protic or Lewis acid catalysts to give polyisobutylene (**16**) as depicted in Equation 21.4. The conversion of acrylonitrile to polyacrylonitrile (**17**) using sodium amide, a strong base, represents **anionic polymerization** (Eq. 21.5).

$$(CH_3)_2C\!\!=\!\!CH_2 \xrightarrow{\ H^+\ } (CH_3)_2\overset{+}{C}\!\!-\!\!CH_3 \xrightarrow{\ H_2C=C(CH_3)_2\ } (CH_3)_3C\!\!-\!\!CH_2\overset{CH_3}{\underset{CH_3}{\overset{|}{\underset{|}{C}}}}{}^{+}$$

Isobutylene

$$\Big\downarrow\ n\ H_2C\!\!=\!\!C(CH_3)_2 \qquad\qquad (21.4)$$

$$(CH_3)_3C\!\!-\!\!\left(CH_2\overset{CH_3}{\underset{CH_3}{\overset{|}{\underset{|}{C}}}} \right)_{\!n}\!\!-\!\!CH_2\overset{|}{\underset{|}{C}}\!\!-\!\!R$$

16
Polyisobutylene

$$CH_2\!\!=\!\!CHCN \xrightarrow{\ :H_2N^-\ } CH_2\!\!-\!\!\bar{C}HCN \xrightarrow{\ CH_2=CHCN\ } H_2NCH_2\overset{CN}{\overset{|}{CH}}\!\!-\!\!CH_2\overset{CN}{\overset{|}{\underset{}{CH}}}$$

$$\underset{NH_2}{}$$

Acrylonitrile

$$\Big\downarrow\ n\ CH_2\!\!=\!\!CHCN \qquad\qquad (21.5)$$

$$H_2N\!\!-\!\!\left(CH_2\overset{CN}{\overset{|}{CH}} \right)_{\!n}\!\!-\!\!CH_2\overset{CN}{\overset{|}{CH}}\!\!-\!\!R$$

17
Polyacrylonitrile

Free-radical polymerization is a widely used method to induce chain-reaction polymerization, and its mechanistic course is parallel to that of the free-radical halogenation of hydrocarbons (Sec. 9.2). In the experiment in this section, you will perform the free-radical polymerization of styrene to give polystyrene.

The reaction is started by the thermal decomposition of an *initiator*, which in our experiment is *tert*-butyl peroxybenzoate (**18**); this compound produces the free radicals **19** and **20** when heated (Eq. 21.6). If In$^\bullet$ stands for one or both of these free radicals,

$$
\begin{array}{ccc}
& \text{heat} & \\
\text{18} & \xrightarrow{\quad} & (\text{CH}_3)_3\text{C}\!-\!\text{O}^\bullet \ + \ {}^\bullet\!\underset{\text{O}}{\overset{\text{O}}{\text{C}}}\!-\!\text{Ph} \qquad\qquad (21.6)
\end{array}
$$

18 **19** **20**

tert-Butyl peroxybenzoate

the course of the polymerization may be illustrated as shown in Equations 21.7–21.10. Equation 21.7 indicates the function of the free radicals in **initiating** the polymerization. Equation 21.8 represents the **propagation** of the growing polymer chain. Equations 21.9 and 21.10 show possible **termination** processes. In Equation 21.9 the free-radical end of one growing polymer chain abstracts a hydrogen atom from the carbon atom next to the end of another polymer radical to produce the unsaturated and saturated polymer molecules **22** and **23**, respectively, in a process termed **disproportionation**. For the termination reaction illustrated by Equation 21.10, Rad$^\bullet$ may be one of the initiating radicals, In$^\bullet$, or another growing polymer chain.

$$\text{In}^\bullet + \text{CH}_2\!\!=\!\!\underset{\underset{\text{C}_6\text{H}_5}{|}}{\text{CH}} \longrightarrow \text{In}\!-\!\text{CH}_2\!-\!\underset{\underset{\text{C}_6\text{H}_5}{|}}{\overset{\bullet}{\text{CH}}} \qquad\qquad (21.7)$$

8

$$\text{InCH}_2\!-\!\underset{\underset{\text{C}_6\text{H}_5}{|}}{\overset{\bullet}{\text{CH}}} + \text{CH}_2\!\!=\!\!\underset{\underset{\text{C}_6\text{H}_5}{|}}{\text{CH}} \longrightarrow \text{InCH}_2\underset{\underset{\text{C}_6\text{H}_5}{|}}{\text{CH}}\!-\!\text{CH}_2\underset{\underset{\text{C}_6\text{H}_5}{|}}{\overset{\bullet}{\text{CH}}} \qquad (21.8a)$$

21

$$\underset{\textbf{21}}{\text{InCH}_2\underset{\underset{\text{C}_6\text{H}_5}{|}}{\text{CH}}\!-\!\text{CH}_2\underset{\underset{\text{C}_6\text{H}_5}{|}}{\overset{\bullet}{\text{CH}}}} + n\ \text{CH}_2\!\!=\!\!\underset{\underset{\text{C}_6\text{H}_5}{|}}{\text{CH}} \longrightarrow \text{InCH}_2\underset{\underset{\text{C}_6\text{H}_5}{|}}{\text{CH}}\!-\!(\text{CH}_2\underset{\underset{\text{C}_6\text{H}_5}{|}}{\text{CH}})_n\!-\!\text{CH}_2\underset{\underset{\text{C}_6\text{H}_5}{|}}{\overset{\bullet}{\text{CH}}} \quad (21.8b)$$

$$\underset{\underset{\text{C}_6\text{H}_5}{|}}{\overset{\bullet}{\text{CH}}}\!\!-\!\!\overset{\overset{\text{H}}{|}}{\text{CHR}} + \underset{\underset{\text{C}_6\text{H}_5}{|}}{\overset{\bullet}{\text{CH}}}\!-\!\text{CH}_2\text{R} \longrightarrow \underset{\underset{\text{C}_6\text{H}_5}{|}}{\text{CH}}\!\!=\!\!\text{CHR} + \underset{\underset{\text{C}_6\text{H}_5}{|}}{\overset{\overset{\text{H}}{|}}{\text{CH}}}\!-\!\text{CH}_2\text{R} \qquad (21.9)$$

22 **23**

$$\text{R} = -\!(\text{CH}_2\underset{\underset{\text{C}_6\text{H}_5}{|}}{\text{CH}})_n\!-$$

$$\text{InCH}_2\text{CH}\underset{\underset{C_6H_5}{|}}{}\text{—(CH}_2\text{CH)}_n\underset{\underset{C_6H_5}{|}}{}\text{—CH}_2\overset{\bullet}{\underset{\underset{C_6H_5}{|}}{\text{CH}}} + \text{Rad}^\bullet \longrightarrow \text{InCH}_2\text{CH}\underset{\underset{C_6H_5}{|}}{}\text{—(CH}_2\text{CH)}_n\underset{\underset{C_6H_5}{|}}{}\text{—CH}_2\text{CH}\underset{\underset{C_6H_5}{|}}{}\text{—Rad} \quad (21.10)$$

The commercially available styrene used in our experiments contains *tert*-butyl-catechol (**24**), a phenol that stabilizes styrene by functioning as a **radical scavenger**. The catechol does this by donating a hydrogen atom to reactive free radicals to convert them into nonradical products as shown in Equation 21.11. The resulting phenoxy radical **25** that is produced is relatively unreactive as an initiator of a free-radical chain reaction. The stabilizer is necessary to prevent premature polymerization of styrene during storage or shipment, because it is so readily polymerized by traces of substances such as atmospheric oxygen.

$$(CH_3)_3C\text{—}\underset{\text{OH}}{\overset{}{\bigcirc}}\text{—O} \quad + \quad \overset{H}{} \quad + \quad R^\bullet \longrightarrow (CH_3)_3C\text{—}\underset{\text{OH}}{\overset{}{\bigcirc}}\text{—O}^\bullet + RH \quad (21.11)$$

24
tert-Butycatechol

25

In the following procedure, directions are given for producing polystyrene in the form of an amorphous solid, a film, and a clear glass.

EXPERIMENTAL PROCEDURES

PREPARATION OF POLYSTYRENE

Purpose: To demonstrate the synthesis of polystyrene by free radical polymerization.

Safety Alert

The free-radical initiator, *tert*-butyl peroxybenzoate, is a safe material to use in this experiment because it decomposes at a moderate rate when heated. Nonetheless, do not heat this catalyst excessively when performing the polymerization.

Miniscale Procedure

Preparation   Answer the Pre-Lab Exercises on page PL. 149. Review Sections 2.9, 2.17, 2.19, 2.21, and 2.22.

Apparatus   A separatory funnel, small soft-glass test tube, 25-mL round-bottom flask, microburner, apparatus for magnetic stirring, heating under reflux, and *flameless* heating.

A. Removal of the Inhibitor from Commercial Styrene

Place about 10 mL of commercial styrene in a small separatory funnel and add 4 mL of 3 M sodium hydroxide and 15 mL of water. Shake the mixture thoroughly, allow the layers to separate, and withdraw the aqueous layer. Wash the organic layer sequentially with two 8-mL portions of water, carefully separating the aqueous layers after each wash. Dry the styrene by pouring it into a small Erlenmeyer flask containing a little anhydrous calcium chloride and then swirling the flask. Allow the mixture to stand for 5–10 min, decant the liquid from the drying agent, and use the dried styrene in the following experiments.

Finishing Touches

Allow the volatiles to evaporate from the *calcium chloride* by placing it on a tray in the hood; then discard it in the container for nontoxic solids. Neutralize the *aqueous layers* before flushing them down the drain.

B. Polymerization of Pure Styrene

Place about 2–3 mL of dry styrene in a small soft-glass test tube, and add 2 or 3 drops of *tert*-butyl peroxybenzoate. Clamp the test tube in a *vertical* position over a wire gauze, insert a thermometer so that its bulb is in the liquid, and heat the styrene and catalyst with a *small* burner flame. When the temperature reaches 140 °C, remove the flame temporarily. If boiling stops, resume heating to maintain gentle boiling. The exothermicity of the polymerization increases the rate of formation of free radicals by thermal decomposition of the initiator, and this in turn increases the rate of polymerization. Thus be watchful for a rapid increase in the rate of boiling, and remove the flame if the refluxing liquid nears the top of the test tube.

After the onset of polymerization the temperature should rise to 180–190 °C, much above the boiling point of styrene. The viscosity of the liquid will increase rapidly during this time. As soon as the temperature begins to decrease, remove the thermometer and pour the polystyrene onto a watchglass. Do not touch the thermometer *before* the temperature decreases because movement of the thermometer in the boiling liquid might cause a sudden "bump," which could throw hot liquid out of the tube. Note the formation of fibers as the thermometer is pulled out of the polymer. The rate of solidification of the polystyrene depends on the amount of catalyst used, the temperature, and the length of time the mixture is heated.

C. Solution Polymerization of Styrene

Place about 2 mL of dry styrene and 5 mL of xylene in a 25-mL round-bottom flask and add 7 drops of *tert*-butyl peroxybenzoate from a Pasteur pipet. Assemble the apparatus for heating under reflux and heat the mixture under reflux for 20 min. Cool the solution to room temperature; then pour about *half* of it into 25 mL of methanol. Collect the white precipitate of polystyrene that forms by decantation or, if decantation is not practical, by vacuum filtration. Resuspend the polystyrene in fresh methanol and stir it vigorously; collect the polystyrene by filtration and allow it to dry in the hood.

Pour the remaining *half* of the polystyrene solution onto a watchglass or the bottom of a large inverted beaker, and allow the solvent to evaporate. A clear film of polystyrene should form.

Place the *filtrate* containing a mixture of xylene and methanol in the container for nonhalogenated organic liquids. Flush the *methanolic filtrate* obtained after resuspension of the polystyrene down the drain.

EXERCISES

1. *tert*-Butylcatechol (**24**) is capable of reacting with *two* equivalents of a radical, R•, to produce two moles of RH and a stable non-radical oxidation product of **24**. Propose a structure for this product and write a stepwise reaction mechanism for its formation. Use curved "fishhook" arrows to symbolize flow of electrons.

2. The use of phenols such as *tert*-butylcatechol as free-radical scavengers is based on the fact that phenolic hydrogens are readily abstracted by radicals, producing relatively stable phenoxyl radicals, which interrupt chain processes of oxidation and polymerization. Alcohols such as cyclohexanol, on the other hand, do *not* function as radical scavengers. Explain why the two types of molecules differ in their abilities to donate a hydrogen atom to a radical, R•.

3. Write an equation for the reaction involved in the removal of *tert*-butylcatechol from styrene by extraction with sodium hydroxide.

4. Why is the polymerization of styrene an exothermic reaction? Explain in terms of a calculation based on the following equation using these bond dissociation energies: PhCH(R)—H, 83 kcal/mol; CH_2=CHPh, 53 kcal/mol (π-component only); $PhCH_2CH_2$—CH(CH$_3$)Ph, 73 kcal/mol.

$$PhCH-H \ + \ CH_2=CH \ \longrightarrow \ PhCH-CH_2CHPh$$
$$\quad \ | \qquad\qquad\quad | \qquad\qquad\qquad | \qquad\quad |$$
$$\quad CH_3 \qquad\qquad Ph \qquad\qquad\quad CH_3 \quad\ H$$

5. Explain why polystyrene is soluble in xylene but insoluble in methanol.

6. What effect would using a smaller proportion of catalyst to styrene have on the average molecular weight of polystyrene?

7. In principle, radicals could add to styrene at the carbon atom bearing the phenyl group rather than the other one, yet they do not. Explain the basis of this selectivity for the addition reaction.

Refer to Tables 8.1–8.8 as needed for answering the questions on spectroscopy.

8. Consider the spectral data for styrene (Figs. 21.1 and 21.2).

   a. In the functional group region of the IR spectrum, specify the absorptions associated with the carbon-carbon double bond and the aromatic ring. Also specify the absorptions in the fingerprint region that are characteristic for the terminal vinyl group.

   b. In the ${}^1H$ NMR spectrum, assign the various resonances to the hydrogen nuclei responsible for them.

   c. For the ${}^{13}C$ NMR data, assign the various resonances to the carbon nuclei responsible for them.

9. Discuss the differences in the IR spectrum of styrene and polystyrene that are consistent with the loss of the vinyl group during the polymerization of styrene in this experiment.

Spectra

Starting Material and Product
The IR spectrum of polystyrene is provided as Figure 8.8.

FIGURE 21.1
IR spectrum of styrene (neat).

FIGURE 21.2
NMR data for styrene (CDCl₃).

(a) ¹H NMR spectrum (300 MHz).

(b) ¹³C NMR data: δ 113.5, 126.2, 127.8, 128.5, 137.0, 137.7.

Step-growth polymerization typically involves the reaction between two different, *difunctionalized* monomers. Both functionalities of each monomer react, and this leads to the formation of polymers. For example, a diacid such as terephthalic acid (**26**) can react with a diol such as ethylene glycol (**27**) in the presence of an acid catalyst to produce a polyester, as shown in Equation 21.12.

$$n\ HO_2C\text{---}\langle\bigcirc\rangle\text{---}CO_2H \xrightarrow[n\ HOCH_2CH_2OH]{H^+} \qquad (21.12)$$

26   **27**

Terephthalic acid   Ethylene glycol

$$H\text{---}\left(\!\!\begin{array}{c} O \\ \| \\ O\text{---}C\text{---}\langle\bigcirc\rangle\text{---}C\text{---}O\text{---}CH_2CH_2 \\ \| \\ O \end{array}\!\!\right)_{\!n}\!\!\text{---}OH\ +\ (2n-1)\ H_2O$$

A polyester

Chain growth is initiated by the acid-catalyzed reaction of one carboxyl group of the diacid with one hydroxy function of the diol to provide an ester and a molecule of water. The free carboxyl group or hydroxyl group of the resulting dimer then reacts with an appropriate functional group in another monomer or dimer, and the process is repeated in such *steps* until all of the monomers are converted into dimers, trimers, tetramers, and, eventually, polymers. The formation of new intermolecular bonds as the polymer grows involves elimination of a small molecule, water, which led to this type of polymerization being called *condensation polymerization*. Step-growth polymerization processes are much slower than chain-reaction processes. Because they also typically have higher activation energies, heating is often required to attain satisfactory rates of polymerization. Finally, condensation polymers generally have lower average molecular weights than polymers formed by chain-reaction polymerization.

Polyamides are one type of useful polymer that is produced by a step-growth process, and a variety of such polymers are preparable from various diacids and diamines. Nylon-6,6™ was the first commercially successful polyamide and is derived from the two monomers, hexanedioic acid (**28**) and 1,6-hexanediamine (**29**), as shown in Equation 21.13. Its trade name reflects the presence of six carbon atoms in each of the monomeric units that comprise the polymer. Of interest regarding the commercial importance of the nylons is the fact that the tremendous financial success enjoyed by E. I. Du Pont and Company from these types of polymers stems from the firm's patent on a method to draw the substance into fibers rather than from a patent on the molecular composition of the polymer itself. (See the Historical Highlight at the end of this chapter.)

$$n \; HO-\overset{\overset{\displaystyle O}{\|}}{C}(CH_2)_4\overset{\overset{\displaystyle O}{\|}}{C}-OH \; + \; n \; H_2N(CH_2)_6NH_2 \longrightarrow$$

28   **29**

Hexanedioic acid   1,6-Hexanediamine

(21.13)

$$H-\left(O-\overset{\overset{\displaystyle O}{\|}}{C}(CH_2)_4\overset{\overset{\displaystyle O}{\|}}{C}-HN(CH_2)_6NH\right)_n H \; + \; (2n-1) \; H_2O$$

Nylon-6,6

In the typical industrial process for preparing polyamides, equimolar amounts of the diacid and diamine are mixed to give a salt, which is then heated to high temperature under vacuum to eliminate the water. The polymer so produced has a molecular weight of about 10,000 and a melting point of about 250 °C. Fibers may be spun from melted polymer, and if the fibers are stretched to several times their original length, they become very strong. This "cold-drawing" orients the polymer molecules parallel to one another so that hydrogen bonds form between C—O and N—H groups on adjacent polymer chains, as shown in **30**, greatly increasing the strength of the fibers. The strength of the fibers of silk, a well-known biopolymer involving protein molecules, is ascribed to the same stabilizing factor.

30

The preparation of Nylon-6,10™, rather than Nylon-6,6, has been chosen to illustrate step-growth polymerization for the present experiment. This polyamide is commercially produced from decanedioic acid (**31**) and 1,6-hexanediamine (**29**), as shown in Equation 21.14. To facilitate forming the polyamide under simple laboratory conditions, however, the diacid chloride of **31** is used because it is more reactive toward diamine **29**. Using the diacid chloride means that the small molecule eliminated in this step-growth polymerization is hydrogen chloride rather than water (Eq. 21.15). Sodium carbonate is added to neutralize the acid formed to prevent consumption of the expensive diamine via an acid-base reaction. If the base were not added, an excess of diamine would be required for complete polymerization.

$$n \text{ HO}-\overset{\overset{\text{O}}{\|}}{\text{C}}(\text{CH}_2)_8\overset{\overset{\text{O}}{\|}}{\text{C}}-\text{OH} + n \text{ H}_2\text{N}(\text{CH}_2)_6\text{NH}_2$$

31   **29**

Decanedioic acid

$$\text{H}-\left(\text{O}-\overset{\overset{\text{O}}{\|}}{\text{C}}(\text{CH}_2)_8\overset{\overset{\text{O}}{\|}}{\text{C}}-\text{HN}(\text{CH}_2)_6\text{NH}\right)_n\text{H} + (2n-1) \text{ H}_2\text{O}$$ (21.14)

Nylon-6,10

$$\overset{\overset{\text{O}}{\|}}{\text{ClC}}(\text{CH}_2)_8\overset{\overset{\text{O}}{\|}}{\text{CCl}} + \text{H}_2\text{NCH}_2-\text{R} \longrightarrow \overset{\overset{\text{O}}{\|}}{\text{ClC}}(\text{CH}_2)_8\overset{\overset{\text{O}}{\|}}{\text{C}}-\text{NHCH}_2-\text{R}$$ (21.15)

Decanedioyl dichloride
(Sebacoyl chloride)

The reactivity of the diacid chloride toward nucleophilic acyl substitution (Eq. 21.15) allows this polymerization to be performed under mild conditions. When a solution of the diacid chloride in a water-immiscible solvent is brought into contact with an aqueous solution of the aliphatic diamine, a film of high molecular-weight polymer instantly forms at the interface between the two solutions. The film is thin but strong, and can be pulled from the interface, where it is immediately and continuously replaced by additional polymer. In this way a long cord or rope of polyamide can be produced, much as a magician pulls a string of silk handkerchiefs out of a top hat. When this experiment was first described by two Du Pont chemists, they characterized it as the "Nylon Rope Trick." It does seem to be almost magic that a polymer can attain an average molecular weight in the range 5000 to 20,000 in a fraction of a second!

To perform this experiment properly, the necessary equipment must be assembled so the polymer rope can be pulled from the reaction zone rapidly. A convenient way to do this is illustrated in Figure 21.3. A can, preferably with a diameter of 10 cm or more, makes a good drum on which to wind the polymer. After puncturing the can in the center of each end, a wooden or metal rod is passed through the center holes to make an axle for the drum. The rod is supported horizontally by clamps attached to ring stands in the usual way.

The circumference of the drum should be measured and a reference point should be marked on the drum so that an estimate of the length of the nylon rope can be made by counting the revolutions made as the rope is produced. A length of 6 m or more can usually be obtained with the procedure described here.

The dry polymer obtained in this experiment does not appear to have the properties expected of a nylon; it is fragile and of low density. However, fibers produced from it are much more dense, and have the appearance and strength more characteristic of a typical polyamide. You will make fibers and films from the dry polymer so you can compare the properties of the different forms of the polymer.

FIGURE 21.3
Apparatus for the ''Nylon Rope Trick.''

Aqueous solution

Dichloromethane solution

EXPERIMENTAL PROCEDURES

PREPARATION OF NYLON-6,10

Purpose: To demonstrate step-growth polymerization by the synthesis of Nylon-6,10.

Safety Alert	1. If Pasteur pipets are used instead of syringes to measure the reactants, use a rubber bulb to draw up the liquid.
	2. Until it has been washed free of solvent and reagents, do not handle the polymer rope with your bare hands more than is necessary. Use latex gloves, tongs, or forceps to manipulate it. If you touch the crude polymer, immediately wash your hands with soap and warm water.
	3. If formic acid is used to form a film, do not let it get on your skin because it causes deep skin burns that are not immediately apparent. If the acid does accidentally come in contact with your skin, wash the affected area immediately with 5% sodium bicarbonate solution and then with copious amounts of water.

Miniscale Procedure

Preparation Answer the Pre-Lab Exercises on page PL. 151. Discuss the experiment with your lab partner.

Apparatus A 250-mL beaker, a 5-mL syringe, separatory funnel, apparatus for the "Nylon Rope Trick."

Setting Up Measure 2 mL of decanedioyl dichloride into a 250-mL beaker using the syringe. *The size of the beaker is important:* In smaller beakers the polymer tends to stick to the walls, whereas in larger beakers poor "ropes" are obtained unless larger amounts of reagents are used. Dissolve the decanedioyl dichloride in 100 mL of dichloromethane. Place 1.1 g of crystalline 1,6-hexanediamine, or 1.3 mL of a commercially available 80–95% aqueous solution, in a separatory funnel, and add 2.0 g of sodium carbonate and 50 mL of water. Gently shake the mixture to dissolve both reactants. Arrange the drum onto which the polymer is to be wound at a height such that the beaker containing the decanedioyl dichloride solution can be placed on the lab bench about 40 cm beneath and slightly in front of the drum (Fig. 21.3). Support the separatory funnel containing the other reagents so the lower tip of the funnel is centered no more than a centimeter above the surface of the dichloromethane solution of the decanedioyl dichloride.

Reaction Open the stopcock of the separatory funnel slightly so the aqueous solution runs *slowly and gently* onto the surface of the organic solution. A film of polymer will form immediately at the interface of the two solutions. Use long forceps or tongs to grasp the *center* of the polymer film and pull the rope that forms up to the front of the drum, loop it over the drum, and rotate the drum away from you to wind the rope onto the drum. For the first turn or two it may be necessary for you to use your fingers to secure the rope to the drum. Continue to rotate the drum and rapidly wind the nylon rope onto the drum until the reactants are consumed, remembering to count the revolutions of the drum as you wind.

Work-Up and Isolation Replace the beaker with a large dish or pan containing about 200 mL of 50% aqueous ethanol, and unwind the nylon rope into the wash solution. After stirring the mixture gently, decant the wash solution, and transfer the polymer to a filter on a Büchner funnel. Press the polymer as dry as possible, and then place it in your desk to dry until the next laboratory period. When the nylon is thoroughly dry, weigh it and calculate the yield. Note the decrease in the bulk of polymer on drying.

Film Formation To produce a film of Nylon-6,10, dissolve the dry polymer in about 10 times its weight of 90–100% formic acid (*Caution*: see the **SAFETY ALERT**, item 3) by stirring the mixture at *room temperature*; heating to achieve dissolution degrades the polymer. Spread the viscous solution on a glass plate. Leave the plate *in a hood* until the next laboratory period to allow evaporation of the formic acid.

Fiber Formation Form fibers by carefully melting the polymer in a metal spoon or spatula with gentle heating over a very small burner flame or a hot plate, and then drawing fibers from the melt with a small glass rod. If necessary, combine your polymer with that of several students to provide enough material to be melted and drawn successfully. Do not heat the polymer much above the melting temperature because it becomes discolored and charred.

**Finishing
Touches**

After the rope has been drawn, stir the remaining reaction mixture thoroughly until no more *polymer* forms. Isolate any additional polymer and, after thoroughly washing it with water, put it in the container for nonhazardous organic solids. Separate the *dichloromethane and aqueous layers* of the reaction mixture. Pour the *dichloromethane* into the container for halogenated organic liquids. Flush the *aqueous layer and all aqueous solutions* down the drain.

EXERCISES

1. Write an equation for the formation of the salt produced from one molecule of hexanedioic acid and two molecules of 1,6-hexanediamine.

2. Why is sodium carbonate used in the reaction to prepare Nylon-6,10?

3. Using full structural formulas, draw a typical portion of a Nylon-6,6 molecule; that is, expand a portion of the formula given in Equation 21.13. Show at least two hexanedioic acid units and two 1,6-hexanediamine units.

4. Draw formulas that illustrate the hydrogen bonding that may exist between two polyamide molecules after fibers have been ''cold-drawn.''

5. Nylons undergo *de*polymerization when heated in aqueous acid. Propose a reaction mechanism that accounts for this fact, using curved arrows to symbolize the flow of electrons.

6. Nylon-6 is produced from caprolactam by adding a small amount of aqueous base and then heating the mixture to about 270 °C.

Caprolactam

   a. Draw a representative portion of the polyamide molecule.

   b. Suggest a mechanism for the polymerization, using curved arrows to symbolize the flow of electrons, and indicate whether it is of the chain-reaction or step-growth type.

7. There is a large decrease in the bulk of the rope of Nylon-6,10 upon drying. Explain this observation.

Refer to Tables 8.1.–8.8 as needed for answering the questions on spectroscopy.

8. Consider the spectral data for decanedioyl dichloride (Figs. 21.4 and 21.5).

   a. In the functional group region of the IR spectrum, specify the absorptions associated with the carbonyl component of the acid chloride.

   b. In the ^{1}H NMR spectrum, assign the various resonances to the hydrogen nuclei responsible for them.

   c. For the ^{13}C NMR data, assign the various resonances to the carbon nuclei responsible for them.

9. Consider the spectral data for 1,6-hexanediamine (Figs. 21.6 and 21.7).

   a. In the IR spectrum, specify the absorptions associated with the amino group.

   b. In the ^{1}H NMR spectrum, assign the various resonances to the hydrogen nuclei responsible for them.

   c. For the ^{13}C NMR data, assign the various resonances to the carbon nuclei responsible for them.

Starting Materials and Products

FIGURE 21.4
IR spectrum of decanedioyl
dichloride (IR card)

(a) ^{1}H NMR spectrum (300 MHz).

FIGURE 21.5
NMR data for decanedioyl
dichloride (CDCl$_3$).

(b) ^{13}C NMR data: δ 24.7, 27.9, 28.4, 46.7, 173.2.

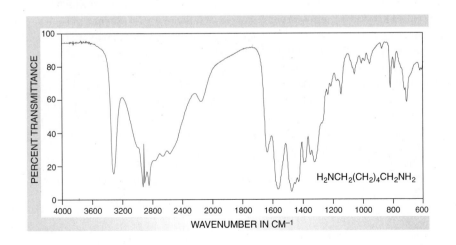

FIGURE 21.6
IR spectrum of 1,6-hexanediamine
(neat).

FIGURE 21.7
NMR data for 1,6-hexanediamine (CDCl$_3$).

(a) ^{1}H NMR spectrum (300 MHz).

H$_2$NCH$_2$(CH$_2$)$_4$CH$_2$NH$_2$

(b) ^{13}C NMR data: δ 26.0, 33.0, 41.4.

H I S T O R I C A L   H I G H L I G H T

Discovery of Polyethylene and Nylon

Polyethylene, or "polythene," as its British inventors called it, and the nylons are two types of polymers that have particularly interesting histories of discovery that bear repeating here. Both stories illustrate the role of serendipity in scientific achievements of great importance.

Polyethylene was discovered accidentally by British chemists at Imperial Chemicals Industries (I.C.I.) as an unexpected result of experiments on chemical reactions at very high pressures. In 1933, a reaction of benzaldehyde and ethylene at 170 °C and 1400 atmospheres gave no adducts involving the two reagents and was considered a complete failure. However, an observant chemist noticed a thin layer of "white waxy solid" on the walls of the reaction vessel used for the experiment. This was recognized as a polymer of ethylene, but additional experiments with ethylene alone to produce the same polymer only resulted in violent decompositions that destroyed the equipment.

Two years elapsed before better and stronger equipment was available for further experimentation. When ethylene was heated to 180 °C in this new

equipment, the pressure in the apparatus dropped unexpectedly, so more ethylene was pumped in. Then, when the reaction vessel was opened, the I.C.I. chemists found a large amount of white powdery solid, which was the long-sought polyethylene. Because they knew that the polymerization could not account for all of the pressure drop that had been observed, they suspected a leak in one of the joints of the apparatus. This idea led to the proposal that the polymerization had been catalyzed by oxygen in the air that had leaked into the apparatus, and this hypothesis was confirmed by experiments in which air was intentionally included with the ethylene. Oxygen can act as a radical initiator and catalyze the polymerization by a chain-reaction mechanism analogous to Equations 21.7–21.10.

The polyethylene produced by the oxygen-catalyzed, high-pressure, high-temperature process developed by I.C.I. in the mid-1930s was ideal for many applications, including insulation of radar equipment, where it was used to great advantage by the Allies in World War II. Referring to the contribution radar made to naval operations, the British

HISTORICAL HIGHLIGHT
Discovery of Polyethylene and Nylon *(continued)*

Commander-in-Chief said it enabled the Home Fleet to "find, fix, fight, and finish the Scharnhorst (the pride of Hitler's navy)."

The group of polymers called nylons was first produced in 1939 as a textile material for women's hose and other garments, but with the onset of World War II and the involvement of the U. S. by 1941, nylon was taken off the domestic market because it was found to be the best available material for military parachutes.

The first nylon to be produced industrially was Nylon-6,6. The remarkable fact about the discovery of this polymer is not how it was first prepared from the two monomers, but how it was first prepared in a form suitable for a textile fiber. This depended on invention of the "cold-drawing process," and this technique was discovered almost completely by accident, as we shall see.

Wallace Hume Carothers was brought to Du Pont to direct its new basic chemical research program because his colleagues at the University of Illinois and Harvard University recommended him as the most brilliant organic chemist they knew. Carothers initiated a program aimed at understanding the composition of the high polymers of Nature such as cellulose, silk, and rubber, and of producing synthetic materials like them. By 1934 his group had contributed valuable fundamental knowledge in these areas, but Carothers had just about decided that their efforts to produce a synthetic fiber like silk was a failure. It was a shrewd observation made during some "horseplay" among Carothers' chemists in the laboratory that turned this failure to compete with Nature into the enormous success ultimately advertised at the 1939 New York World's Fair as "Nylon, the Synthetic Silk Made from Coal, Air, and Water!"

The Carothers group had learned how to make Nylon-6,6, but even though this polyamide had a molecular structure similar to that of silk, they had "put it on the back shelf" without patent protection because the polymer did not have the tensile strength of silk, a necessary criterion for a good textile fiber. The group continued its research by investigating the polyester series, polymers which were more soluble, easier to handle, and thus simpler to work with in the laboratory. It was while working with one of these softer materials that Julian Hill noted that if he gathered a small ball of such a polymer on the end of a glass stirring rod and drew a thread out of the mass, the thread of polymer so produced became very silky in appearance. This attracted his attention and that of the others working with him, and it is reported that one day while Carothers was in downtown Wilmington, Hill and his cohorts tried to see how far they could stretch one of these samples. One chemist put a little ball of the polymer on a stirring rod and a second chemist touched a glass rod to the polymer ball and then ran down the hall to see how far he could stretch the thread of polymer. While doing this they noticed not only the silky appearance of the extended strands, but also their increased strength. They soon realized that this additional strength might be the result from some special orientation of the polymer molecules produced by the stretching procedure.

Because the polyesters they were working with at that time had melting points too low for use in textile products, a deficiency that has since been removed, the researchers returned to the polyamides (nylons) that had earlier been put aside. They soon found that these polymers, too, could be "cold-drawn" to increase their tensile strength so much that they could be made into excellent textiles. Filaments, gears, and other molded objects could also be made from the strong polymer produced by cold-drawing.

The alignment of the long polyamide molecules in a manner that produces extensive intermolecular hydrogen-bonding (**30**, Sec. 21.3) binds the individual polymer molecules together in much the same way that separate strands in a rope, when twisted together, form a cable. This association of linear polymer molecules through hydrogen bonding is responsible for the greatly increased strength of the nylon fibers. We believe that the same principle accounts for the strength of silk fibers; the natural polyamide molecules of silk are oriented in such a way that hydrogen bonds hold the individual molecules together. Interestingly, the silkworms accomplish the equivalent of "cold-drawing" as they extrude the viscous silk filaments to produce cocoons!

HISTORICAL HIGHLIGHT
Discovery of Polyethylene and Nylon *(continued)*

*Relationship of Historical Highlight
to Experiments*

The experiments in this chapter represent examples of chain-reaction polymerization to produce polystyrene (Sec. 21.2) and step-growth polymerization (Sec. 21.3) to yield a nylon. The procedures of Section 21.2 provide an opportunity to observe formation of the same polymer in three physically different forms, whereas that of Section 21.3 illustrates how strong hydrogen-bonded fibers can result from the ''cold-drawing'' technique patented by Du Pont.

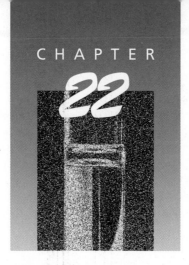

Carbohydrates

22.1 Introduction

Carbohydrates are one of the important classes of natural compounds that are found in living systems. The term carbohydrate originates from the fact that many, but not all, members of this class have the general molecular formula $C_nH_{2n}O_n$ and thus were once considered to be "hydrates of carbon." The simplest carbohydrates are either polyhydroxy aldehydes, which have the general structure **1** and are referred to as **aldoses**, or polyhydroxy ketones, which have the general structure **2** and are called **ketoses**. Such simple carbohydrates are termed **monosaccharides**. Two monosaccharides may be combined with the loss of one molecule of water to form a **disaccharide**. **Polysaccharides** and other complex carbohydrates are then produced by the condensation of more monosaccharide subunits with the loss of a molecule of water for each additional monosaccharide. Depending upon its constitution, hydrolysis of a polysaccharide yields either a single monosaccharide or a mixture of monosaccharides. In this chapter, some of the fundamental chemical and physical properties of carbohydrates will be investigated using simple mono- and disaccharides; several classical qualitative tests to characterize and classify carbohydrates will also be performed.

$$
\begin{array}{cc}
\text{CHO} & \text{CH}_2\text{OH} \\
| & | \\
(\text{CHOH})_n & \text{C}=\text{O} \\
| & | \\
\text{CH}_2\text{OH} & (\text{CHOH})_n \\
& | \\
& \text{CH}_2\text{OH} \\
\mathbf{1} & \mathbf{2} \\
\text{An aldose} & \text{A ketose}
\end{array}
$$

22.2 Monosaccharides: General Principles

The monosaccharide D-glucose, whose chemistry is representative of all aldoses containing four or more carbon atoms, exists predominantly in the two **pyranose** forms **4** and **5**, which are six-membered **hemiacetals** formed from the acyclic form **3** by the reversible, intramolecular addition of the δ-hydroxyl group to the free aldehyde function (Eq. 22.1). In the cyclic forms **4** and **5**, the ring carbon that is derived from the carbonyl group is referred to as the **anomeric carbon** atom. The **specific rotation**, $[\alpha]_D^{25}$ (Sec. 7.5), of α-D-(+)-glucose (**4**) is $+112°$, whereas that of the β-anomer **5** is $+19°$. When crystals of either pure **4** or pure **5** are dissolved in water, the $[\alpha]_D^{25}$ changes to an equilibrium value of $+52.7°$. This process is termed **mutarotation**. At equilibrium in water, the α- and β-forms are present in the ratio of $36:64$; only about 0.03% of D-glucose is in the acyclic form **3**.

4	**3**	**5**	(22.1)
α-D-Glucopyranose	D-Glucose	β-D-Glucopyranose	

Another common monosaccharide is D-fructose (**6**), a ketose. In aqueous solution, D-fructose also undergoes mutarotation to produce a complex equilibrium mixture of the acyclic form **6** ($<1\%$), the five-membered hemiacetals **7** (31%) and **8** (9%), which are called **furanoses**, and the cyclic pyranoses **9** (57%) and **10** (3%); this mixture exhibits an $[\alpha]_D^{25} = -92°$ (Scheme 22.1).

SCHEME 22.1

7	**9**
β-D-Fructofuranose	β-D-Fructopyranose
	6
	D-Fructose
8	**10**
α-D-Fructofuranose	α-D-Fructopyranose

The ready oxidation of the aldehyde group of an aldose to a carboxylic acid function forms the basis of a number of useful qualitative tests for classifying a carbohydrate as a **reducing sugar**. Ketoses also yield positive tests for reducing sugars even though no aldehyde group is present. The positive test for ketoses arises because the α-hydroxy keto group of the open form of a ketose undergoes base-catalyzed **tautomerization** to give an **enediol** that is protonated to provide a pair of epimeric aldehydes differing in configuration at C-2. These transformations are outlined in the form of partial structures in Equation 22.2. Since the tests for reducing sugars are performed under basic conditions that allow the equilibria of Equation 22.2 to be established, *all* known *monosaccharides* are reducing sugars.

$$\tag{22.2}$$

A ketose   An enediol   An aldose   An aldose

22.3 Disaccharides: Hydrolysis of Sucrose

Disaccharides are *O*-**glycosides** in which the hydroxyl group attached to the anomeric carbon atom of one monosaccharide is replaced with a hydroxyl group of another monosaccharide, thereby forming a cyclic **acetal**. For example, maltose (**11**) is a disaccharide in which the anomeric hydroxyl group of one α-D-glucopyranose has been replaced with the C(4)-hydroxyl group of another D-glucopyranose subunit. Because the other D-glucopyranose ring of **11** is in equilibrium with the open-chain isomer **12** having a free aldehyde function (Eq. 22.3), maltose is a reducing sugar. Indeed, any disaccharide in which *one* of the rings is a hemiacetal or hemiketal is a *reducing sugar* (Sec. 22.2), since the cyclic hemiacetal or hemiketal moiety is in equilibrium with the open-chain isomer in which the aldehyde or the α-hydroxyketone function can be oxidized.

$$\tag{22.3}$$

11
Maltose

12

Disaccharides in which *both* rings are in the **acetal** or **ketal** form are *not* reducing sugars because they *cannot* be in equilibrium with an aldehydo or keto form under neutral or basic conditions. For example, sucrose (**13**), a common foodstuff, is a **nonreducing** disaccharide because the glycosidic linkage between the two monosaccharide subunits is formed between the two anomeric carbon atoms, thereby incorporating the potential aldehyde and α-hydroxyketone functions into cyclic acetal and ketal groups. Thus sucrose, $[\alpha]_D^{25} = +66.5°$, does not undergo mutarotation under either *neutral* or *alkaline* conditions. However, sucrose hydrolyzes under acidic conditions to give the reducing sugars D-glucose and D-fructose (Eq. 22.4). The 1:1 mixture of D-glucose and D-fructose that is obtained upon hydrolysis of sucrose is called **invert sugar** because the specific rotation of the mixture is dominated by the negative rotation of D-fructose.

$$\xrightarrow{\text{H}_3\text{O}^+} \text{D-Glucose} + \text{D-Fructose} \qquad (22.4)$$

13
Sucrose

Fructose is the sweetest common sugar, being about twice as sweet as sucrose; consequently, invert sugar is sweeter than sucrose. The enzyme **invertase**, which bees use in making honey, accomplishes the same chemical result as does the acid-catalyzed hydrolysis of sucrose.

In the experiment in this section, you will examine the acid-catalyzed hydrolysis of sucrose to give a mixture of D-glucose and D-fructose and monitor the consequent change in the specific rotation that occurs using a polarimeter.

EXPERIMENTAL PROCEDURES

HYDROLYSIS OF SUCROSE

Purpose: To demonstrate the acid-catalyzed hydrolysis of a disaccharide into its component monosaccharides.

Safety Alert   Wear latex gloves when measuring the *concentrated* hydrochloric acid. If any acid spills on your skin, wash it off with large volumes of water and then with dilute sodium bicarbonate solution to neutralize any residual acid.

Miniscale Procedure

Preparation   Answer the Pre-Lab Exercises on page PL. 153. Review Sections 2.9, 2.11, and 2.22.

Apparatus   A 100-mL round-bottom flask, separatory funnel, 50-mL volumetric flask, polarimeter, apparatus for magnetic stirring, heating under reflux, and *flameless* heating.

Setting Up and Reaction   Accurately weigh about 7.5 g of sucrose and place it in the round-bottom flask containing a stirbar. Add about 40 mL of water, swirl the contents of the flask to effect solution, and add about 0.5 mL of *concentrated* hydrochloric acid. Assemble the apparatus for heating under reflux. Heat the solution at reflux for about 2 h.

Determining the Specific Rotation of Sucrose   While the solution of sucrose in aqueous acid is being heated under reflux, determine the specific rotation of sucrose according to the general directions that follow, together with any specific directions of your instructor. Carefully fill the sample tube for the polarimeter with water; *be certain that no air bubbles are trapped in it.* Place the sample tube in the polarimeter and determine the blank reading for the solvent. Record the blank rotation, the temperature, and the length of the tube. Empty and carefully dry the tube. Then accurately weigh 5–10 g of sucrose to the nearest 0.05 g, and transfer the sample *quantitatively* to the volumetric flask. Fill the flask to within several milliliters of the volumetric mark, tightly stopper the flask, and shake the flask thoroughly. Carefully fill the flask to the mark and again shake the flask thoroughly. Transfer a portion of this solution to the polarimeter tube, measure the rotation, and record this value.

Analysis   At the end of the period of reflux, cool the reaction mixture to room temperature and transfer the solution to a clean volumetric flask. Use small amounts of water to rinse the round-bottom flask, and add the rinses to the flask to effect quantitative transfer; fill the flask to the volumetric mark. Using the polarimeter as before, determine the specific rotation of the product mixture from the hydrolysis of sucrose. Compare this value with the specific rotation of pure sucrose determined earlier.

Neutralize the *aqueous solutions* with sodium carbonate and flush them down the drain.

Finishing Touches

EXERCISES

1. Explain the change in sign of the optical rotation that occurs when sucrose undergoes hydrolysis.

2. When determining the rotation, why is it important that no air bubbles be present in the polarimeter tube?

3. Calculate the specific rotation of *invert* sugar from the known rotations of the equilibrium mixtures of the anomers of D-glucose and D-fructose. How does this number compare with that determined experimentally?

4. The specific rotation of invert sugar and of a racemic mixture represent average rotations produced by the molecules in solution. Why is the rotation of the racemic mixture 0°, whereas that of invert sugar is not?

5. Using curved arrows to symbolize the flow of electrons, write a stepwise reaction mechanism for the acid-catalyzed hydrolysis of sucrose to D-glucose and D-fructose.

6. Specify which of the structures in Scheme 22.1 are α-anomers and which are β-anomers.

7. Using curved arrows to symbolize the flow of electrons, write a stepwise reaction mechanism for acid-catalyzed isomerization of **9** to **6**, followed by cyclization to produce a five-membered ring as in **7**.

Determining the complete structure of an unknown monosaccharide was a formidable challenge to early organic chemists (see the Historical Highlight at the end of this chapter). Many carbohydrates, particularly when impure, have a tendency to form syrups rather than crystallize from solution, and this sometimes makes it difficult to obtain pure compounds for characterization and identification. The number and nature of functional groups present in carbohydrates further exacerbates the problem of assigning their structures.

Some of these difficulties are illustrated by considering the example of D-(+)-glucose, whose nature in solution is shown in Equation 22.1. The first step toward solving the structure of D-(+)-glucose was to identify D-(+)-glucose as a derivative of the 2,3,4,5,6-pentahydroxyhexanal **14**. Because D-(+)-glucose exists primarily as a cyclic hemiacetal at equilibrium, another issue involved determining the ring size. Namely, did D-(+)-glucose exist in the form of a *furanose*, **15**, or a *pyranose*, **16**? Finally, it was necessary to assign the relative and absolute configuration to each of the stereo-centers (Sec. 7.1), including the anomeric carbon atom (Sec. 22.2). Despite these tremendous experimental challenges, the complete structures of many monosaccharides, some containing up to nine carbon atoms, are now known. It is amazing that much of this structural work was performed *prior* to the availability of the modern spectroscopic methods discussed in Chapter 8.

The structural elucidation of polysaccharides represents an even greater challenge. The individual monosaccharide subunits that constitute the unknown polysaccharide must first be identified. Then the ring size and position in the polysaccharide sequence must be elucidated for each monosaccharide. Finally, the nature of the glycosidic linkages that form the polysaccharide backbone must be defined. Toward this end, it is necessary to establish which hydroxyl group on one monosaccharide is involved in the formation of the acetal or ketal that forms the glycosidic bond to the adjoining monosaccharide; the stereochemistry at this anomeric center must also be determined.

CLASSIFICATION TESTS

A number of useful qualitative chemical tests have been invented to obtain information for unknown carbohydrates. Some of these tests are used to classify such molecules according to their structural type. In the experiments that follow, you will use such tests to identify certain structural features that are found in mono- and polysaccharides. Information derived from these experiments can provide significant information that may be used to prove the structure of an unknown carbohydrate.

Tollens' Test. Tollens' test is designed to distinguish between aldehydes and ketones. A complete discussion of this test together with an experimental procedure is included with the classification tests for aldehydes in Section 23.7C.

Benedict's Test. This test is based upon the fact that cupric ion will oxidize aliphatic aldehydes, including α-hydroxyaldehydes, but not aromatic aldehydes. The reagent used in this test is a solution of cupric sulfate, sodium citrate, and sodium carbonate. The citrate ion forms a complex with the Cu(II) ion so that $Cu(OH)_2$ does not precipitate from the basic solution. A positive test for the presence of the aliphatic aldehyde group in an aldose is evidenced by the formation of a red precipitate of cuprous oxide, Cu_2O (Eq. 22.5). A yellow precipitate is occasionally observed as a positive test. This yellow solid has apparently not been characterized, but its formation seems to depend on the amount of oxidizing agent present.

$$\text{RCHO} + 2\ \text{Cu}^{2+} + 5\ \text{HO}^- \xrightarrow{\text{Citrate}} \text{RCO}_2^- + \underset{\text{Brick red}}{\text{Cu}_2\text{O}} + 3\ \text{H}_2\text{O} \qquad (22.5)$$

Barfoed's Test for Monosaccharides. Like Tollens' and Benedict's tests, this test depends on the *reducing* properties of the saccharides. However, because of the specific conditions employed for the test, it is possible to distinguish between monosaccharides and disaccharides. The test reagent consists of an aqueous solution of cupric acetate and acetic acid. Thus, in contrast to the previous tests, the reaction is carried out under *acidic* conditions. A positive test for monosaccharides is indicated by the formation of the brick-red precipitate of Cu_2O within *two* or *three* minutes. For reasons that are not understood, disaccharides require a longer time, providing the precipitate only after about ten minutes or more. Nonreducing sugars such as sucrose undergo slow hydrolysis under the acidic aqueous conditions of the test, and therefore they give a precipitate after *extended* time.

EXPERIMENTAL PROCEDURES

CLASSIFICATION TESTS FOR CARBOHYDRATES

Purpose: To demonstrate the classification tests for carbohydrates.

Microscale Procedures

Preparation   Answer the Pre-Lab Exercises on page PL. 155.

Apparatus   A number of 13-mm $\times$ 100-mm test tubes.

Setting Up   Obtain the carbohydrates that you will use in the following classification tests from your instructor. Perform each test on each carbohydrate and enter your observations and conclusions in your notebook.

Tollens' Test   *Before performing this test, read the* SAFETY ALERT *and* FINISHING TOUCHES *sections in Section 23.7C.* Perform Tollens' test according to the procedure outlined in that section using about 0.1 g of the carbohydrate in a *clean* glass test tube and about 1 mL of Tollens' reagent. The formation of a silver mirror or a black precipitate constitutes a positive test.

Benedict's Test   Place about 0.1 g of the carbohydrate in a test tube and add 2 mL of water. Stir the mixture to effect solution and add 2–3 mL of the stock solution of Benedict's reagent. Heat the solution to boiling. The formation of a yellow to red precipitate is a positive test for aliphatic aldehydes and α-hydroxyaldehydes; the precipitate appears green when viewed in the blue solution of the reagent. For comparison, perform the test simultaneously on an unknown, on glucose, and on sucrose and record the results.

Finishing Touches | Remove any precipitated *cuprous oxide* by vacuum filtration and place it in the container for nonhazardous solids. Flush the *filtrate* down the drain with an excess of water.

Barfoed's Test   Place 3 mL of Barfoed's reagent and 1 mL of a 1% solution of the carbohydrate in a test tube; place the test tube in a beaker of boiling water for *5 min*. Remove the test tube and cool it under running water. A red precipitate of cuprous oxide is a positive test. To see the precipitate, it may be necessary to view the tube against a dark background in good light. For comparison, run the test simultaneously on the unknown, on glucose, and on lactose, and record the results.

Finishing Touches | Remove any precipitated *cuprous oxide* by vacuum filtration and place it in the container for nonhazardous solids. Flush the *filtrate* down the drain with an excess of water.

FORMATION OF OSAZONES

Carbohydrates that can exist in solution in an acyclic form with a free aldehyde or ketone function react with *three* equivalents of phenylhydrazine to form bright yellow crystalline derivatives called **phenylosazones**, together with aniline and ammonia as the other products of the reaction. For example, α- and β-D-glucopyranoses, **4** and **5**, are in dynamic equilibrium in aqueous solution with the ring-opened form **3** (Sec. 22.2), which, because of its free aldehyde group, will react with phenylhydrazine to form the phenylosazone **17**, as shown in Equation 22.6. Since phenylosazones are readily identified by either their melting points or temperatures of thermal decomposition, these derivatives may be used to characterize carbohydrates.

$$
\begin{array}{c}
\text{CHO} \\
| \\
\text{H}-\text{C}-\text{OH} \\
| \\
\text{HO}-\text{C}-\text{H} \\
| \\
\text{H}-\text{C}-\text{OH} \\
| \\
\text{H}-\text{C}-\text{OH} \\
| \\
\text{CH}_2\text{OH} \\
\mathbf{3}
\end{array}
\;+\; 3\ \text{C}_6\text{H}_5\text{NHNH}_2
\;\xrightarrow{\text{H}_3\text{O}^+}\;
\begin{array}{c}
\text{CH}=\text{NNHC}_6\text{H}_5 \\
| \\
\text{C}=\text{NNHC}_6\text{H}_5 \\
| \\
\text{HO}-\text{C}-\text{H} \\
| \\
\text{H}-\text{C}-\text{OH} \\
| \\
\text{H}-\text{C}-\text{OH} \\
| \\
\text{CH}_2\text{OH} \\
\mathbf{17}
\end{array}
\;+\; \text{C}_6\text{H}_5\text{NH}_2 \;+\; \text{NH}_3
\qquad (22.6)
$$

D-Glucose · Phenylhydrazine · Phenylosazone of D-glucose

The accepted mechanism for the formation of phenylosazones is presented in Scheme 22.2. Following the initial formation of the phenylhydrazone **18** from the aldose, an internal oxidation-reduction occurs by tautomerization of *two* hydrogens from

C-2 to the hydrazone moiety to give the ketone **19**. The newly formed carbonyl group then condenses with a second equivalent of phenylhydrazine to give **20**, which tautomerizes to **21**. After the 1,4-elimination of aniline to produce either **22** or **23**, a third equivalent of phenylhydrazine condenses with the imine group of **22** or **23** to yield the phenylosazone **24** and ammonia. Although it may appear that **24** should undergo further reaction by intramolecular oxidation-reduction between the secondary alcohol group at C-3 and the hydrazone group at C-2, the reaction stops at this point, so that only two phenylhydrazine units are introduced. The formation of the intramolecular hydrogen bond shown in **24** limits the reaction to the first two carbon atoms of the chain. Thus, when 1-methylphenylhydrazine [$H_2NNMeC_6H_5$] is used in place of phenylhydrazine, the reaction readily proceeds down the chain at least as far as C-5.

SCHEME 22.2

EXPERIMENTAL PROCEDURES

FORMATION OF OSAZONES

Purpose: To demonstrate the preparation of osazones.

Safety
Alert

Phenylhydrazine is toxic; avoid contact with it. If you spill some on your skin, rinse it off thoroughly with household bleach, dilute acetic acid, and then water.

Microscale Procedure

Preparation   Answer the Pre-Lab Exercises on page PL. 157. Review Sections 2.9 and 2.17.

Apparatus   Test tubes, hot water bath, apparatus for vacuum filtration.

Setting Up   Heat a large beaker of water to boiling. In each of three test tubes, *separately* dissolve 0.2-g portions of D-glucose, D-fructose, and sucrose in 4 mL of water. In a fourth test tube place 4 mL of the solution that was used for rotation experiments for the hydrolysis of sucrose (Sec. 22.3). Add 0.5 mL of saturated sodium bisulfite solution to each test tube to suppress oxidation of the phenylhydrazine during the reaction; this avoids contaminating the phenylosazone with tarry by-products.

Reactions   To each of the test tubes, add *either* 0.6 mL of glacial acetic acid, 0.6 g of sodium acetate, and 0.4 g of phenylhydrazine *or* 0.6 g of sodium acetate and 0.6 g of phenylhydrazine hydrochloride. Stir the solutions thoroughly, place the test tubes in the beaker of boiling water, and discontinue heating. Allow the test tubes to remain in the hot-water bath for 30 min. Remove the test tubes from the hot-water bath and cool the contents to room temperature.★

Isolation and Purification   Collect any precipitates by vacuum filtration.★ Recrystallize each of the phenylosazones formed from ethanol-water.

Analysis   Determine the melting points of the purified phenylosazones. Make a mixture containing equal parts of the two phenylosazones obtained from the pure carbohydrates, and determine the melting point of this mixture. Because the melting points of phenylosazones may depend on the rate of heating, perform these determinations simultaneously. Make a mixture containing equal parts of the phenylosazone obtained from the hydrolysis of sucrose and each of the phenylosazones obtained from the pure carbohydrates and determine their melting points as well. Record your observations.

Finishing Touches

Neutralize the *filtrates* obtained from isolating the crude phenylosazones with a small amount of sodium carbonate, and add 2 mL of laundry bleach (5.3% aqueous sodium hypochlorite) for every 1 mL of the filtrate. Heat the mixture at 45–50 °C for 2 h to oxidize and decompose any phenylhydrazine that is present. Cool the mixture and then flush it down the drain with excess water. Flush the *filtrates from the recrystallizations* down the drain.

EXERCISES

1. Which of the pure carbohydrates formed phenylosazones?

2. What conclusions can be drawn from the results of the mixed melting-point determination on the phenylosazones formed from the pure carbohydrates?

3. What conclusions can be drawn from the results of the mixed melting-point determinations on the phenylosazones formed from the pure carbohydrates and that obtained from the hydrolysis product of sucrose?

4. D-Glucose and D-mannose give the same phenylosazone. Explain.

5. Draw the structure of the product that would be expected upon treatment of D-glucose with a large excess of 1-methylphenylhydrazine.

■ H I S T O R I C A L   H I G H L I G H T ■

Emil Fischer: Carbohydrate Chemist Extraordinaire

One of the great pioneers in the development of carbohydrate chemistry was Emil Fischer, who was born near Bonn, Germany, in 1852. Although he went to the University of Strasbourg to study chemistry under Friedrich Kekulé, he actually obtained his Ph.D. degree under Adolf von Baeyer in 1874. He moved with von Baeyer to Munich, where he received his *habilitation* in 1878. He was then appointed as an *ausserordentlicher*, or assistant professor, in 1879. Afterwards he held appointments as Professor of Chemistry at Erlangen (1882) and Würzburg (1885) before succeeding von Hoffmann at the prestigious Chemical Institute of the University of Berlin. His original laboratories and office in Berlin are now a part of the Humboldt University, which has a special exhibit of some of his library, writings, and laboratory equipment.

In 1875, Fischer synthesized phenylhydrazine, $C_6H_5NHNH_2$, for the first time. Although the preparation of a new chemical entity for the first time must itself have been an exhilarating experience, the discovery of its utility in carbohydrate and carbonyl chemistry quickly overshadowed its mere creation. Fischer found that phenylhydrazine reacted with certain sugars to form *osazones*. He not only studied the mechanism of the formation of osazones, but he was able to determine the structures of the osazones of a large number of sugars and to use these derivatives as the basis for developing techniques for the systematic classification of sugars. Fischer was the first to synthesize glucose, an achievement for which he was awarded the Davy medal of the Royal Society in 1890. Shortly thereafter, he prepared unnatural sugars such as the nonoses. Synthetic organic chemistry had come a long way since the preparation of urea, H_2NCONH_2, by Wöhler in 1828!

Fischer applied the concept of the tetrahedral carbon atom, which had been proposed in 1874–1875 by van't Hoff and Le Bel, to assigning the relative stereochemistries of *all eight* of the isomeric hexoses. The solution to this difficult problem eluded Fischer for some time, until, on the advice of Victor Meyer, he built molecular models of the different sugars. Having done so, the complex stereochemical relationships became apparent. It was at this time that he developed the protocol for representing stereochemical relationships in sugars via planar structures. The result was *Fischer projections*, which are used to this day. The elucidation of the structures of glucose and other known sugars is regarded as Fischer's most important contribution to science and earned him the Nobel Prize in Chemistry in 1902; he was only the second chemist to receive this award.

Fischer is best known for his work in carbohydrate chemistry, but he was also active in the fields of photosynthesis and protein and heterocyclic chemistry. He demonstrated that amino acids were the basic building blocks in proteins. His ingenious contribution of the "lock and key" hypothesis of how enzymes selectively bind substrates of complementary shapes evolved from his work with the proteins known as glucosidases. For example, he found that α-D-methyl glucoside was easily hydrolyzed by a yeast extract termed invertin but not by a preparation of emulsin isolated from almonds. Conversely, β-D-methyl glucoside was hydrolyzed by emulsin, but not by invertin. This work helped lay the foundation of modern biochemistry. Fischer felt strongly that the great chemical secrets of life would only be unveiled by a cooperative effort in which chemical methods were applied to solving biological problems, a point he made in a seminal lecture to the Faraday Society in 1907.

α-D-Methyl glucoside β-D-Methyl glucoside

Fischer's interest in other heterocyclic molecules found in nature led to the first synthesis of caffeine, which is the stimulant in coffee and tea. The purine ring system found in caffeine also comprises the ring systems of the heterocyclic bases adenine and quanine that are found in nucleic acids. He also developed a general synthesis of indole, the heterocyclic ring-containing component in the amino acid tryptophan, and the method is known today as the Fischer indole synthesis.

Caffeine

Adenine

Guanine

Indole

L-Tryptophan

While Fischer must have taken satisfaction and joy from his scientific accomplishments, his personal life ended in tragedy. Before moving to Berlin, he married Agnes Gerlach with whom he had three sons. His wife died at the age of 33, and two of his sons were killed in World War I. By the end of the war, Fischer himself suffered from mercury and phenylhydrazine poisoning and advanced intestinal cancer. Depressed by his own poor health, the losses of his wife and two sons, and the socioeconomic conditions of post-war Germany, he committed suicide in 1919. However, his legacy as one of the great bio-organic chemists in the history of the science lives to this day.

Relationship of Historical Highlight to Experiments

Fischer was the first to synthesize and determine the structures of a number of the sugars used in the experiments in this section. He studied the hydrolysis of glucosides and disaccharides in much the same way you may if you perform the hydrolysis of sucrose (Sec. 22.3). He was also the first to synthesize phenylhydrazine, which you may use to form the same osazones he did in his pioneering work in characterizing sugars (Sec. 22.4). You can share his excitement as you observe the formation of the solid osazones from the combination of different monosaccharides with excess phenylhydrazine.

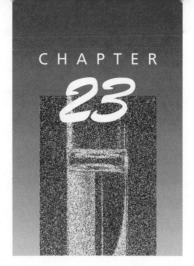

C H A P T E R

23

Identifying Organic
Compounds

One of the greatest challenges to the chemist is identifying the substances that are obtained from chemical reactions or isolated from natural sources. Although structure elucidation may be difficult and time-consuming, the availability of modern spectroscopic techniques (Chap. 8) in combination with chemical methods has greatly facilitated this aspect of experimental organic chemistry. Some of the techniques and approaches that are used to accomplish this goal are described in this chapter.

A successful systematic scheme of qualitative analysis for identifying organic compounds was developed early in the twentieth century. This scheme, together with some more recent modifications, is termed **classical qualitative organic analysis**, and is the basis for most textbook discussions on the subject (Ref. 1 at the end of Section 23.5, for example).

In recent years, the development of chromatographic methods of separation (Chap. 6) and structural analysis by spectroscopic techniques (Chap. 8) have revolutionized the laboratory practice of organic chemistry. Nevertheless, interest in classical qualitative organic analysis remains high because it is an effective and interesting means of understanding fundamental organic chemistry. Consequently, this chapter contains an outline of this classical scheme, and the experimental procedures associated with it, that is adequate for obtaining a good understanding of the approach. The scheme now is rarely used as the sole means of elucidating structures because a wealth of information is readily available from spectroscopic techniques. Thus, an introduction to identification and structure determination based on data from modern spectroscopic methods is included in Section 23.5. The advantages of the combined use of the classical and instrumental methods are also described in that section.

23.2   Overview of Organic Analysis

An overview of a systematic procedure that can be used to identify an unknown pure organic compound serves to introduce the classical scheme for organic analysis. The first step is to ensure that the compound is pure, and this can be done in several ways. In the case of a liquid, gas-liquid chromatography (GLC, Sec. 6.4) may be used to demonstrate that only one component is present; a pure compound produces only a single peak in a gas chromatogram, assuming that no decomposition occurs under the conditions of the analysis. If enough liquid is available, it can be distilled (Chap. 4). A 1–2 °C boiling-point range implies that the compound is pure, although such a narrow range is also consistent with the distillation of an **azeotrope**. However, GLC provides better evidence for purity because of its ability to detect low levels of impurities. The purity of a solid substance can be ascertained from its melting point (Sec. 3.3). A 1–2 °C melting-point range usually indicates a pure compound except in the rare instance of a **eutectic mixture** (Sec. 3.3). Impure liquids can be purified by simple or fractional distillation (Secs. 4.2 and 4.3) and by column or preparative GLC (Secs. 6.2 and 6.4), whereas solids can be purified by recrystallization (Sec. 3.2), by sublimation (Sec. 2.20), or by column or thin-layer chromatography (Secs. 6.2 and 6.3). The purification and identification of gases is not included in this text because of the difficulties in handling them and because they are not commonly encountered in organic chemistry.

Once purity is established, various physical properties are determined. The melting point of a solid or boiling point of a liquid is considered essential. Occasionally, the density and/or refractive index of a liquid may be useful, and for certain compounds, either liquid or solid, the specific rotation can be determined if a substance is optically active.

Establishing what elements other than carbon and hydrogen are present is critical for ultimately identifying the compound, and techniques for such **elemental analysis** are described later. Molecular weight, as determined by cryoscopic techniques or mass spectrometry, or percentage by weight composition of the elements present, also provides important data. The solubility of the unknown compound in water, in dilute acids and bases, or in various organic solvents may signal the presence or absence of various functional groups.

Perhaps the most important step in identifying an unknown substance is determining the functional group(s) that may be present, and IR and NMR spectroscopy (Secs. 8.2 and 8.3) are now commonly used for this purpose. Before the development of these spectroscopic methods as routine experimental techniques, functional group determination involved performing qualitative chemical tests for each possible group. Although spectral analyses may not provide an *unequivocal* answer about the presence of certain functional groups, they at least permit narrowing the possibilities to a small number, so only one or two chemical tests are then needed to complete the identification of the functionalities present.

Final assignment of a structure to the ''unknown'' compound is achieved by one of several procedures. The classical method involves the chemical conversion of the substance into a *solid* derivative. The success of this technique depends on the availability of information about the unknown and its various derivatives. Of prime importance is knowledge of the melting or boiling points of possible candidates for the unknown as well as the melting points of solid derivatives. Many tabulations of organic compounds are available for this purpose, and references to two of them are provided at the end of Section 23.5.

The identification of the compound can also be completed by thoughtful analysis of the IR, ^{1}H, and ^{13}C NMR spectra of the compound. However, unequivocal proof of structure based *solely* upon NMR spectra is usually not possible unless the spectra are being used to compare the "unknown" with a number of known compounds; demonstrating that the IR and NMR spectra of an unknown and of a known compound are *identical* will suffice to prove that the substances are identical.

The following sections contain descriptions of the stepwise procedures that may be used to identify an unknown compound using classical methods alone or a combination of classical and spectroscopic methods. As a reminder, these procedures should *not* be performed unless the compound is pure. Since it is possible that you may be given a mixture of unknown compounds to identify, a procedure is provided in Section 23.4 for separating a mixture into its individual components so that each one can be identified.

23.3 Classical Qualitative Analysis Procedure to Identify a Pure Compound

The *classical system of qualitative organic analysis* consists of six steps, each of which is discussed in the following subsections. The first four steps, which can be carried out in any order, should be completed *before* performing the qualitative tests for functional groups. The final step must always be the preparation of one or more solid derivatives.

1. Preliminary examination of physical and chemical characteristics
2. Determining physical constants
3. Elemental analysis to determine the presence of elements other than carbon, hydrogen, and oxygen
4. Solubility tests in water, dilute acid, and dilute base
5. Functional group analysis using classification tests
6. Derivatization

It is a tribute to the power of the system that one can identify an unknown organic compound with certainty, even though it may be one of several million known compounds. With the exception of a few general guidelines, there are no rigid directions to be followed. You must rely on good judgment and initiative in selecting a course of attack on the unknown, and it is particularly important to observe and consider each experimental result. Negative results may be as useful as positive ones in the quest to reveal the identity of an unknown.

PRELIMINARY EXAMINATION

If it is carried out intelligently, the **preliminary examination** may provide more information with less effort than any other part of the procedure. The simple observation that the unknown is a *crystalline* solid, for example, eliminates from consideration a large fraction of all organic compounds because many are liquids at room temperature. The **color** is also informative: Most pure organic compounds are white or colorless. A brown color is often characteristic of small amounts of impurities; for example, aromatic amines and phenols quickly become discolored by the formation of trace amounts of highly colored air-oxidation products. Color in a pure organic compound is usually attributable to conjugated double bonds.

The **odor** of many organic compounds, particularly those of lower molecular weight, is highly distinctive. A conscious effort should be made to learn and recognize the odors that are characteristic of several classes of compounds such as the alcohols, esters, ketones, and aliphatic and aromatic hydrocarbons. The odors of certain compounds demand respect, even when they are encountered in small amounts and at considerable distance; for example, the unpleasant odors of thiols (mercaptans), isonitriles, and higher carboxylic acids and diamines cannot be described definitively, but they are recognizable once encountered. *Be extremely cautious* in smelling unknowns, because some compounds are not only disagreeable but also possibly toxic. Large amounts of organic vapors should *never* be inhaled.

The **ignition test** involves a procedure in which a drop or two of a liquid or about 50 mg of a solid is heated gently on a small spatula or crucible cover with a microburner flame. Whether a solid melts at low temperature or only upon heating more strongly is then noted. The flammability and the nature of any flame from the sample are also recorded. A yellow, sooty flame is indicative of an aromatic or a highly unsaturated aliphatic compound; a yellow but non-sooty flame is characteristic of aliphatic hydrocarbons. The oxygen content of a substance makes its flame more colorless or blue; a high oxygen content lowers or prevents flammability, as does halogen content. The unmistakable and unpleasant odor of sulfur dioxide indicates the presence of sulfur in the compound. If a white, nonvolatile residue is left after ignition, a drop of water is added and the resulting aqueous solution is tested with litmus or pHydrion paper; a metallic salt is indicated if the solution is alkaline.

PHYSICAL CONSTANTS

If the unknown is a solid, its **melting point** is measured by the capillary tube method (Sec. 2.7) An observed melting-point range of more than 2–3 °C indicates that the sample should be recrystallized (Sec. 3.1).

For an unknown that is a liquid, the **boiling point** is determined by the micro boiling-point technique (Sec. 2.8). An indefinite or irreproducible boiling point or discoloration or inhomogeneity of the unknown requires that the sample be distilled (Secs. 2.13 and 2.15); the boiling point is obtained during the distillation.

Other physical constants that *may* be of use for liquids are the **refractive index** and the **density**. Consult with your instructor about the advisability of making these measurements.

ELEMENTAL ANALYSIS

The technique of **elemental analysis** involves determining *which* elements may be present in a compound. The halogens, sulfur, oxygen, phosphorus, and nitrogen are the elements other than carbon and hydrogen that are most commonly found in organic molecules. Although there is no simple way to test for the presence of oxygen, it is fairly easy to determine the presence of the other heteroatoms, and the appropriate procedures are provided for the halogens, sulfur, and nitrogen.

The basis of these procedures is as follows. Because the bonding found in organic compounds is principally covalent, there rarely are direct methods analogous to those applicable to ionic inorganic compounds for determining the presence of the aforementioned elements. However, the covalent bonds between carbon and these heteroatoms may be broken by heating an organic compound with sodium metal. This process, called **sodium fusion**, results in the formation of inorganic ions involving these elements if they are present in the original compound; thus, the products are halide ions, X^-, from halogens, sulfide ion, S^{2-}, from sulfur, and cyanide ion, CN^-, from nitrogen. After the organic compound has been heated with sodium metal, the residue is *cautiously* treated with *distilled* water to destroy the excess sodium and to dissolve the inorganic ions that have been formed. The ***fusion aqueous solution***, designated as **FAQS** can then be analyzed for the presence of halide, sulfide, and cyanide ions.

The presence of halide is determined by first acidifying a portion of the FAQS with dilute nitric acid and boiling the solution *in the hood* to remove any sulfide or cyanide ions as hydrogen sulfide or hydrogen cyanide, respectively; sulfide and cyanide must be removed because they interfere with the test for halogens. Silver nitrate solution is then added, and the formation of a precipitate of silver halide signals the presence of halide in the FAQS (Eq. 23.1).

$$Ag^+ + X^- \longrightarrow \underline{AgX} \qquad\qquad \textbf{(23.1)}$$
A silver halide

The color of the precipitate provides a *tentative* indication of which halogen is present in the unknown: AgCl is *white* but turns *purple* on exposure to light, AgBr is *light yellow*, and AgI is *dark yellow*. *Definitive* identification is made by standard inorganic qualitative analysis procedures or by means of thin-layer chromatography (TLC, Sec. 6.3).

Sulfur is detected by taking a second portion of the FAQS and carefully acidifying it. Any sulfide ion that is present will be converted to H_2S gas, which forms a dark precipitate of PbS upon contact with a strip of paper saturated with lead acetate solution (Eq. 23.2).

$$Pb(OAc)_2 + H_2S \longrightarrow \underline{\underset{Black}{PbS \text{ (solid)}}} + 2\,HOAc \qquad \textbf{(23.2)}$$

| Lead acetate | Hydrogen sulfide | Lead sulfide | Acetic acid |

The presence of nitrogen in the unknown is shown by the following qualitative test. A third portion of the FAQS is carefully acidified, and ferrous ion, Fe^{2+}, and ferric ion, Fe^{3+}, are added; this converts the cyanide ion into potassium ferric ferrocyanide, which precipitates as an intense blue solid called **Prussian blue** (Eq. 23.3).

$$4\,CN^- + Fe^{2+} \longrightarrow Fe(CN)_6^{4-} \xrightarrow[Fe^{3+}]{K^+ \text{ and}} \underline{KFeFe(CN)_6} \qquad \textbf{(23.3)}$$
Potassium ferric ferrocyanide
(Prussian blue)

EXPERIMENTAL PROCEDURES

ELEMENTAL ANALYSIS

> **Safety Alert**
>
> 1. Sodium fusion involves heating sodium metal or a sodium-lead alloy to a high temperature and then adding the organic compound. Use extreme care when performing both the heating and addition.
> 2. Perform the sodium fusion in the hood if possible.
> 3. Use a Pyrex™ test tube and check it for cracks or other imperfections before performing the sodium fusion.
> 4. If pure sodium metal is used, hydrolyze the residue *very* carefully as directed, because any excess metal reacts *vigorously* with alcohol or water.
> 5. Be careful when handling the test tube after the fusion is complete; remember that it may still be hot.
> 6. *Throughout this procedure, point the mouth of the test tube away from yourself and your neighbors*; the organic material may burst into flame when it contacts the hot metal, or it may react so violently that hot materials are splattered from the test tube.

A · SODIUM FUSION

Sodium fusion may be executed using either sodium metal or a commercially available sodium-lead alloy, which contains nine parts lead and one part sodium. For safety reasons, we recommend using the sodium-lead alloy method because the alloy is easier to handle than is sodium metal and poses less potential danger during hydrolysis.

Sodium-Lead Alloy Method

Asbestos or bare metal

Sodium vapor

Na/Pb pellet

1–2 cm

FIGURE 23.1
Apparatus for sodium fusion.

Support a small Pyrex test tube in a *vertical* position using a clamp whose jaws have either an asbestos liner or none at all; the jaws must *not* be lined with materials such as rubber or plastic (Fig. 23.1). Place a 0.5-g sample of sodium-lead alloy in the test tube and heat the alloy with a flame until it melts and vapors of sodium rise 1–2 cm in the test tube. Do *not* heat the test tube to redness. Then remove the flame and immediately add 2–3 drops of a liquid sample or about 10 mg of a solid sample to the hot alloy, being careful not to allow the sample to touch the sides of the hot test tube during the addition. If there is no visible reaction, heat the fusion mixture gently to initiate the reaction, then discontinue the heating and allow the reaction to subside. Next, heat the test tube to redness for a minute or two, and then *let it cool* to room temperature. Add 3 mL of *distilled water* to the cooled reaction mixture, and heat the mixture gently for a few minutes to complete hydrolysis. Decant or filter the solution. If the latter option is used, wash the filter paper with about 2 mL of water. Otherwise simply dilute the decanted solution with about 2 mL of water. Label the filtrate FAQS and use it in the appropriate tests for detecting sulfur, nitrogen, and the halogens.

Put the *lead pellet* that remains after hydrolysis of the reaction mixture in the container for heavy metals.

Sodium Metal Method

Support a small Pyrex test tube in a *vertical* position using a clamp whose jaws have either an asbestos liner or none at all; the jaws must *not* be lined with materials such as rubber or plastic (Fig. 23.1). Place a clean cube of sodium metal about 3–4 mm on an edge in the tube and heat it gently with a microburner flame until the sodium melts and the vapors rise 1–2 cm in the test tube. Do *not* heat the test tube to redness. Then remove the flame and immediately add 2 or 3 drops of a liquid or about 10 mg of a solid sample to the hot test tube, being careful not to allow the sample to touch the sides of the hot test tube during the addition. A brief flash of fire may be observed; this is normal. Reheat the bottom of the tube, remove the flame and add a second portion of the unknown organic compound in the same amount as for the first addition. Now heat the bottom of the tube until it is a dull red color; then remove the flame and allow the tube to cool to room temperature. Carefully add about 1 mL of 95% ethanol dropwise to the cooled reaction mixture to decompose the excess sodium; stir the contents of the tube with a stirring rod. After a few minutes, add another 1 mL of ethanol with stirring. After the reaction has subsided, apply gentle heat to boil the ethanol; either work at the hood or use an inverted funnel connected to a vacuum source (Fig. 2.69b) over the mouth of the test tube to keep the vapors of ethanol from entering the room. When the ethanol has been removed, allow the tube to cool and add about 10 mL of *distilled water*; stir the mixture and pour it into a small beaker. Rinse the tube with an additional 5 mL of distilled water, and combine the rinse with the main solution. The total amount of water should be 15 to 20 mL. Boil the aqueous mixture briefly, filter it, and label the filtrate FAQS; use this filtrate in the tests for sulfur, nitrogen, and the halogens.

B • QUALITATIVE ANALYSIS FOR HALOGENS, SULFUR, AND NITROGEN

Halogens.   Acidify about 2 mL of the FAQS by dropwise addition of 6 *M* nitric acid. Working at the hood, boil the solution gently for 2–3 min to expel any hydrogen sulfide or cyanide that may be present. Cool the solution and then add several drops of 0.3 *M* aqueous silver nitrate solution to it. A *heavy precipitate* of silver halide indicates the presence of chlorine, bromine, or iodine in the original organic compound. A faint turbidity of the solution should *not* be interpreted as a positive test.

Tentative identification of the particular halogen is made on the basis of color. *Definitive* identification must be made by standard inorganic qualitative procedures (Reference 1 at the end of Section 23.5) or by means of TLC (Sec. 6.2), using the following procedure.

Obtain a 3-cm × 10-cm strip of fluorescent silica gel chromatogram sheet. About 1 cm from one end, place four equivalently spaced spots as follows. Using a capillary, spot the FAQS at the left side of the sheet. Because this solution is likely to be relatively dilute in halide ion, several reapplications of it may be required. Allow the spot to dry following each application, a process that is hastened by blowing on the plate. Do not permit the spot to broaden to a diameter greater than about 2 mm. Next, and in order, spot samples of 1 *M* potassium chloride, 1 *M* potassium bromide, and 1 *M* potassium iodide. Develop the plate in a solvent mixture of acetone, 1-butanol, concentrated ammonium hydroxide, and water in the volume ratio of $13:4:2:1$. Following development, allow the plate to air-dry, and, working at the hood, spray the plate lightly with an indicator spray prepared by dissolving 1 g of silver nitrate in 2 mL of water and adding this solution to 100 mL of methanol containing 0.1 g of fluorescein and 1 mL of concentrated ammonium hydroxide. Allow the now-yellow strip to dry, and then irradiate it for several minutes with a long-wavelength ultraviolet lamp (366 nm). Compare the spots formed from the test solution with those formed from the solutions of known halides. Be aware that iodide gives two spots according to this procedure.

Sulfur.   Acidify a 1–2 mL sample of the FAQS with acetic acid, and add a few drops of 0.15 M lead acetate solution. A black precipitate of PbS signifies the presence of sulfur in the original organic compound.

Nitrogen.   Determine the pH of a 1-mL sample of the FAQS with pHydrion paper. The pH should be about 13. If it is above 13, add a *small* drop of 3 M sulfuric acid to bring the pH down to about 13. If the pH of the fusion solution is below 13, add a *small* drop of 6 M NaOH to bring the pH up to about 13. Add 2 drops each of a saturated solution of ferrous ammonium sulfate and of 5 M potassium fluoride. Boil the mixture gently for about 30 sec, cool it, and add 2 drops of 5% ferric chloride solution. Then carefully add 3 M sulfuric acid dropwise to the mixture until the precipitate of iron hydroxide *just* dissolves. Avoid an excess of acid. At this point the appearance of the deep-blue color of potassium ferric ferrocyanide (Prussian blue) indicates the presence of nitrogen in the original organic compound. If the solution is green or blue-green, filter it; a blue color remaining on the filter paper is a weak but nonetheless positive test for nitrogen.

Finishing Touches

Transfer precipitated *silver halides* to a container labeled for them so that the silver can be recovered. Put the precipitated *lead sulfide* in a container for heavy metals. Any *FAQS* that remains may be flushed down the drain, as can the *solution* from *the test for nitrogen*.

SOLUBILITY TESTS

The solubility of an organic compound in water, dilute acid, or dilute base can provide useful, but not definitive, information about the presence or absence of certain functional groups. Note that the assignment of an unknown to a formal solubility class is rather arbitrary because a large number of compounds exhibit *borderline* behavior. It is recommended that the solubility tests be done in the order presented here.

Water. Test the solubility of the unknown in water. For the present purposes, a compound is defined as soluble if it dissolves to the extent of 3 g in 100 mL of water, or more practically, 30 mg in 1 mL of water. As a general rule, few organic compounds exhibit appreciable water solubility as defined here.

   Several structural features of the unknown can be deduced if it is water-soluble. It must be of low molecular weight and will usually contain no more than four to five carbon atoms, unless it is polyfunctional. It must contain a polar group that will hydrogen-bond with water, such as the hydroxy group of an alcohol or a carboxylic acid, the amino functionality of an amine, or the carbonyl group of aldehydes or ketones.

Esters, amides, and nitriles dissolve to a lesser extent, and acid chlorides or anhydrides react with water rather than simply dissolving in it. On the other hand, alkanes, alkenes, alkynes, and alkyl halides are water-insoluble.

If the unknown is water-soluble as defined above, test its aqueous solution with pHydrion paper. If the solution is acidic, the unknown is likely to be a low molecular-weight carboxylic acid such as acetic acid. If the solution is basic, a low molecular weight organic base such as diethylamine is possible. A neutral solution suggests the presence of a neutral polar compound such as an alcohol or a ketone.

The borderline for water solubility of *monofunctional* organic compounds is most commonly at or near the member of the homologous series containing five carbon atoms. Thus, butanoic acid is soluble, pentanoic acid is borderline, and hexanoic acid is insoluble in water; similarly, 1-butanol is soluble, 1-pentanol is borderline, and 1-hexanol is insoluble in this solvent. There are exceptions to this generalization, however. For example, if a molecule is spherically shaped, it can contain a larger number of carbon atoms and still remain soluble in water. A case in point is that 2-methyl-2-butanol (*tert*-pentyl alcohol) is water-soluble to the extent of 12.5 g/100 mL of water even though it contains five carbon atoms. The increased solubility is because the molecular surface areas of spherical molecules are less than those of non-spherical ones, and this decreases the **hydrophobicity** of the molecule.

Solubility in Aqueous Acid and Base. If an unknown is insoluble in water, test its solubility first in sodium hydroxide, then in sodium bicarbonate, and finally in hydrochloric acid. Solubility in one or more of these acids and bases is defined in terms of the compound being *more soluble in base or acid than in water* and reflects the presence of an acidic or basic functional group in the water-insoluble unknown compound.

In each of the following solubility tests, *shake the unknown with the test reagent at room temperature*. If it does not dissolve, warm the mixture for several minutes in a hot-water bath and continue shaking it. If the substance still appears not to dissolve, decant or filter the liquid from the undissolved sample and carefully *neutralize* the filtrate; observing a precipitate or turbidity is indicative of greater solubility in the aqueous acid or base than in water itself. The importance of bringing the filtrate to pH 7 arises because an unknown may show enhanced solubility in both acidic and basic solutions if it contains *both* basic and acidic functional groups. Assessing these possibilities necessitates performing *all* the solubility tests.

1. *Sodium Hydroxide.* If the compound is water-insoluble, test its solubility in 1.5 *M* NaOH solution. Carboxylic acids, which are strong acids, and phenols, which are weak acids, dissolve in sodium hydroxide because they are converted into their water-soluble sodium salts (Eq. 23.4). An unknown that is more soluble in NaOH than in water may be either a phenol or a carboxylic acid, and it must be tested for solubility in the weaker base, 0.6 *M* NaHCO$_3$, which may permit distinction between these two functional groups (Part 2, below). If the unknown does not exhibit solubility in NaOH, its solubility in NaHCO$_3$ need not be tried; rather, it should next be tested for solubility in 1.5 *M* HCl.

$$\text{Carboxylic acids:} \quad \text{RCO}_2\text{H} \xrightarrow{\text{NaOH}} \text{RCO}_2^- \, \text{Na}^+ \qquad \textbf{(23.4a)}$$

A carboxylic acid   A sodium carboxylate
Water-insoluble   *Water-soluble*

$$\text{Phenols:} \quad \text{ArOH} \xrightarrow{\text{NaOH}} \text{ArO}^- \, \text{Na}^+ \qquad \textbf{(23.4b)}$$

A phenol   A sodium phenoxide
Water-insoluble   *Water-soluble*

2. *Sodium Bicarbonate.* An unknown that is soluble in dilute NaOH solution should also be tested for its solubility in 0.6 *M* NaHCO$_3$. If it is soluble, the *tentative* conclusion is that a carboxylic acid group is present, owing to the formation of the water-soluble sodium salt (Eq. 23.5a); phenols are normally not deprotonated in this medium (Eq. 23.5b). Dissolution should be accompanied by effervescence resulting from decomposition of the carbonic acid, H$_2$CO$_3$, formed from reaction of bicarbonate with the carboxylic acid, to carbon dioxide and water (Eq. 23.5a).

$$\text{Carboxylic acids:} \quad \text{RCO}_2\text{H} \xrightarrow{\text{NaHCO}_3} \text{RCO}_2^-\text{Na}^+ + \text{H}_2\text{CO}_3 \longrightarrow \text{CO}_2 + \text{H}_2\text{O} \qquad \textbf{(23.5a)}$$

A carboxylic acid   A sodium carboxylate
Water-insoluble   *Water-soluble*

$$\text{Phenols:} \qquad \text{ArOH} \xrightarrow{\text{NaHCO}_3} \text{No reaction} \qquad \textbf{(23.5b)}$$

A phenol
Water-insoluble

Carboxylic acids are *usually* soluble in NaHCO$_3$ *and* in NaOH, whereas phenols usually dissolve *only* in NaOH. However, caution must be used in making *definitive* conclusions about the presence of a carboxylic acid or phenol based upon solubility in NaHCO$_3$. For example, a phenol containing one or more strong electron-withdrawing substituents, such as a nitro group, can be as acidic as a carboxylic acid and thus may form a water-soluble sodium salt by reaction with NaHCO$_3$. Similarly, the salt of a carboxylic acid of high molecular weight may *not* be completely soluble in the aqueous medium; nevertheless, evolution of CO$_2$ should still be observable when the acid comes in contact with the base.

3. *Hydrochloric Acid.* The solubility of the unknown in 1.5 *M* hydrochloric acid should be determined. If the unknown is soluble, the presence of an amino group in the compound is indicated because amines are organic bases that react with dilute acids to form ammonium salts that are usually water-soluble (Eq. 23.6). However, this solubility test does *not* permit the distinction between weak and strong organic bases.

$$\text{RNH}_2 \xrightarrow{\text{HCl}} \text{RNH}_3^+ \text{Cl}^- \qquad \textbf{(23.6)}$$

An amine   An alkylammonium chloride
Water-insoluble   *Water-soluble*

4. *Concentrated Sulfuric Acid.* Many compounds that are too weakly basic or acidic to dissolve in dilute aqueous acid or base will dissolve in or react with concentrated H$_2$SO$_4$. Such solubility is often accompanied by the observation of a dark solution or the formation of a precipitate; any detectable reaction such as evolution of a gas or formation of precipitate is considered "solubility" in concentrated H$_2$SO$_4$. This behavior can usually be attributed to the presence of carbon-carbon π-bonds, oxygen, nitrogen, or sulfur in the unknown. The solubility is usually due to reaction of one of these functional groups with the concentrated acid, which results in the formation of a salt that is soluble in the reagent. For example, an alkene adds the elements of sulfuric acid to form an alkyl hydrogen sulfate (Eq. 23.7) that is soluble in the acid, and an oxygen-containing compound becomes protonated in concentrated acid to form a soluble oxonium salt (Eq. 23.8). Substances that exhibit this solubility behavior are termed "neutral" compounds.

$$R_2C{=}CR_2 + H_2SO_4 \longrightarrow \underset{\underset{H \quad OSO_3H}{|\quad\;\;|}}{R_2C{-}CR_2} \qquad \textbf{(23.7)}$$

An alkene   Sulfuric acid   An alkyl sulfonate

$$R_2C{=}O \text{ or } ROR \xrightarrow{\;H_2SO_4\;} R_2C{=}\overset{+}{O}H\;HSO_4^- \text{ or } R_2\overset{+}{O}H\;HSO_4^- \qquad \textbf{(23.8)}$$

An aldehyde
or ketone   An ether   Oxonium ions

Compounds that are soluble in dilute HCl or neutral water-insoluble compounds containing N or S need not be tested for their solubility in concentrated H_2SO_4 because they will invariably dissolve in or react with it. Note that solubility or insolubility of an unknown in concentrated H_2SO_4 does *not* yield a great deal of evidence for the presence or absence of any *specific* group, whereas the solubility of a compound in dilute HCl or NaOH or $NaHCO_3$ provides valuable information about the type of functional group present.

Assuming that a compound contains only *one* functional group, the following scheme classifies compounds according to their solubility in acid or base. This picture may be changed dramatically if a compound contains several polar functional groups that cause its solubility properties to be different from those of a monofunctional compound or if an acidic or basic compound contains one or more strongly electron-withdrawing groups.

Acidic Compounds Soluble in NaOH and NaHCO₃
     Carboxylic acids
     Phenols
Acidic Compounds Soluble in NaOH but not in NaHCO₃
     Phenols
Basic Compounds Soluble in HCl
     Amines
Neutral Compounds Soluble in Concentrated H₂SO₄
     Carbonyl compounds (aldehydes and ketones)
     Unsaturated compounds (alkenes and alkynes)
     Alcohols
     Esters
     Amides
     Nitriles
     Nitro compounds
     Ethers
Neutral Compounds Insoluble in Cold Concentrated H₂SO₄
     Alkyl halides
     Aryl halides
     Aromatic hydrocarbons

After determining the physical constants, elemental analysis, and solubility properties, provide a **preliminary report** of these findings to your instructor, who may give advice regarding the validity of these observations. This protocol serves to minimize unnecessary loss of time in finding the structure of the unknown.

CLASSIFICATION TESTS

The next step in identifying an unknown is to determine which functional groups are present. The classical scheme involves performing a number of chemical tests on a substance, each of which is specific for a type of functional group. These tests can normally be done quickly and are designed so that the observation of a color change or the formation of a precipitate indicates the presence of a particular functional group. The results of these tests usually allow the assignment of the unknown to a structural class such as alkene, aldehyde, ketone, or ester, for example. The following factors should be considered when performing qualitative classification tests for functional groups.

1. A compound may contain more than one functional group, so the complete series of tests must be performed unless you have been told that the compound is mono-functional.
2. Careful attention is required when the functional group tests are performed. Record *all* observations, such as the formation and color of any solid produced as a result of a test.
3. Some of the color-forming tests occur for several different functional groups, and although the *expected* colors are given in the experimental procedures, the *observed* color may be affected by the presence of other functional groups.
4. It is of utmost importance to perform a qualitative test on *both the unknown and a known compound that contains the group being tested*. Some functional groups may appear to give only a slightly positive test, and it is helpful to determine how a compound known to contain a given functional group behaves under the conditions of the test being performed. It is most efficient and reliable to do the tests on standards at the same time as on the unknown. In this manner, inconclusive positive tests may be interpreted correctly. Because aliphatic compounds are sometimes more reactive than aromatic ones, it is wise to perform a test on both of these types of standards along with the unknown.
5. The results obtained from the elemental analysis and the solubility tests can be used in deciding which functional group tests should be performed initially and which should not be done at all. The following examples illustrate the use of the preliminary work in making these decisions:

    a. A classification test for an amine should be applied first if a compound is found to be soluble in dilute hydrochloric acid and to contain nitrogen.
    b. The test for a phenol should be performed on an unknown that is soluble in dilute sodium hydroxide but insoluble in dilute sodium bicarbonate.
    c. The tests for alkyl or aryl halides should be omitted if the elemental analysis indicates the absence of halogen.
    d. The tests for amines, amides, nitriles, and nitro compounds need not be performed if nitrogen is absent, as shown by the elemental analysis.

6. A logical approach should be followed in deciding which tests are needed. The result obtained from one test, whether positive or negative, often has a bearing on which additional tests should be done. A random ''hit-or-miss'' approach is wasteful of time and often leads to erroneous results. Another error commonly made by beginners in qualitative organic analysis is to omit the tests for functional groups and immediately attempt preparation of a derivative. This tactic has a very low probability for success; for example, trying to make a derivative of a ketone is certain to fail if the unknown is actually an alcohol. The different classification tests should

be performed until the nature of the functional group(s) present in the unknown is defined as completely as possible! This will minimize unproductive efforts at derivatization of the compound.

Classification tests for most of the common functional groups are presented in the following sections.

Neutral Compounds

Alcohols	Section 23.11
Aldehydes	Section 23.7
Alkenes	Section 23.8
Alkyl halides	Section 23.9
Alkynes	Section 23.8
Amides	Section 23.18
Aromatic hydrocarbons	Section 23.10
Aryl halides	Section 23.10
Esters	Section 23.16
Ketones	Section 23.7
Nitriles	Section 23.17
Nitro compounds	Section 23.15

Acidic Compounds

Carboxylic acids	Section 23.13
Phenols	Section 23.12

Basic Compounds

Amines	Section 23.14

DERIVATIZATION

It was mentioned earlier that the classical approach to structure elucidation usually involves converting an ''unknown'' liquid or solid into a second compound that is a solid, the latter being called a **derivative** of the first compound. It is better to prepare a derivative that is a solid rather than a liquid because solids can be obtained in pure form by recrystallization and because the melting point of a solid can be determined on a small quantity of material, whereas a larger amount of a liquid derivative would be required in order to determine its boiling point.

Two solid derivatives of an unknown compound should be prepared in order to double-check its identity. The melting points of the derivatives, along with the melting point or boiling point of the unknown compound, usually serve to identify the unknown completely. However, the success of this type of identification depends upon the availability of tables listing the melting points and/or boiling points of known compounds and the melting points of suitable derivatives. Extensive listings of compounds are found in the references at the end of Section 23.5 and abbreviated compilations are available in the Instructor's Manual for this textbook. These tabulations are by no means comprehensive, and many other compounds that have been identified on the basis of derivatives appear in the scientific literature (Chap. 24).

The warning given in this section under the heading ''Classification Tests'' is repeated here: Do *not* proceed directly to the preparation of a derivative merely on the basis of a hunch about the class of compound to which your unknown belongs. Rather, make certain of the type of functional group present by obtaining one *or more* positive classification tests *before* attempting the preparation of any derivative.

The following example illustrates how classification tests and preparation of derivatives are used to identify an unknown compound.

Preliminary analysis	Colorless liquid with pleasant odor
Physical constant	bp 119–120 °C (760 Torr)
Elemental analysis	No X, S, or N
Solubility tests	Slightly soluble in water; no increased solubility in dilute HCl, NaOH, or NaHCO$_3$; soluble in concentrated H$_2$SO$_4$.
Preliminary report	Observations confirmed by instructor
Classification tests	Negative test for aldehyde, alkene, alkyne; positive test for ketone, positive test for methyl ketone, negative tests for all other functional groups

The experimental error inherent in determining melting or boiling points means it is prudent to consider compounds melting or boiling within about 5 °C of the observed melting or boiling point as possibilities for the unknown. Applying this principle to the entries in Table 23.1 of selected liquid ketones leads to compounds **1–6** as possibilities.

From the experimental facts that were given, some of these compounds can be eliminated as possibilities for the following reasons: **4** because it contains a halogen, **5** because it contains a carbon-carbon double bond, and **6** because it is not a methyl ketone. Now suppose that two derivatives, the 2,4-dinitrophenylhydrazone (Sec. 23.7A) and the semicarbazone (Sec. 23.7F), were prepared from the unknown and found to melt at 93–95 °C and 69–71 °C, respectively. Examining Table 23.1 reveals that the derivatives of only one of the ketones under consideration, namely 3-methyl-2-pentanone (**3**), melt at these temperatures. Hence, the identity of the unknown is deduced. Although it is possible that other ketones with similar boiling points *may* exist and may *not* be listed in Table 23.1, it is highly unlikely that any of these will give two derivatives with the same melting points as those of the derivatives obtained from **3**. This emphasizes the desirability of preparing *two* derivatives.

The preceding analysis indicates how positive *and* negative information can be utilized in determining the identity of an unknown substance. Of particular note is the importance of the functional group classification tests, which must be done carefully and thoroughly in order to exclude the possible presence of all groups other than the keto group in the unknown.

Table 23.1	**Liquid Ketones with Boiling Points 115–125 °C (760 Torr).**			
Name and Number	Structure	BP (°C)	2,4-DNP (°C)	Oxime (°C)
1-Methoxy-2-propanone (**1**)	CH$_3$OCH$_2$C(=O)CH$_3$	115		163
4-Methyl-2-pentanone (**2**)	(CH$_3$)$_2$CHCH$_2$C(=O)CH$_3$	117	135	95
3-Methyl-2-pentanone (**3**)	CH$_3$CH$_2$CH(CH$_3$)C(=O)CH$_3$	118	95	71
Chloroacetone (**4**)	ClCH$_2$C(=O)CH$_3$	119	150	125
3-Penten-2-one (**5**)	(CH$_3$)$_2$CH=CHC(=O)CH$_3$	122	142	155
2,4-Dimethyl-3-pentanone (**6**)	(CH$_3$)$_2$CHC(=O)CH(CH$_3$)$_2$	124	160	188

23.4 Separating Mixtures of Organic Compounds

The preceding section contains information identifying a *pure* organic compound. However, when a chemist is faced with the problem of identifying an organic compound, it is seldom pure; rather, it is often contaminated with by-products or starting materials if it has been synthesized in the laboratory. Modern methods of separation, particularly chromatographic techniques (Chap. 6), make the isolation of a pure compound easier than it once was, but one must not lose sight of the importance of classical techniques of separation and purification, which are treated in detail in Chapters 3–5.

The common basis of the procedures most often used to separate mixtures of organic compounds is the difference in **polarity** that exists or may be induced in the components of the mixture. This difference in polarity is exploited in nearly all the separation techniques, including distillation, recrystallization, extraction, and chromatography. The greatest differences in polarity, which make for the simplest separations, are those that exist between salts and nonpolar organic compounds. Whenever one or more of the components of a mixture can be converted to a salt, it can be separated easily and efficiently from the nonpolar components by distillation (Chap. 4) or extraction (Chap. 5).

When learning the techniques of qualitative organic analysis, you may be given a *mixture* of unknown compounds, each of which is to be identified. Before this can be done, each component of the mixture must be obtained in pure form. The general approach shown in Figure 23.2 illustrates how this may be accomplished using principles based on differences in polarity and functionality. For this scheme to be applicable, each component of the mixture must have a low solubility in water and must not undergo appreciable hydrolysis by reaction with dilute acids or bases at room temperature. The procedure is based primarily on partitioning compounds of significantly different polarities between diethyl ether and water and separating these liquid layers in a separatory funnel. The underlying concept of this process involves extraction, the theory of which is discussed in Section 5.2.

Assuming that a mixture contains a carboxylic acid, a phenol, an amine, and a neutral compound, each of which is water-insoluble, the separation is initiated by dissolving the mixture in a suitable organic solvent such as diethyl ether. The ethereal solution of the mixture is first extracted with sodium bicarbonate solution, which removes the carboxylic acid by converting it to its water-soluble sodium salt. This extraction is followed by one with sodium hydroxide solution, which removes the water-soluble sodium salt of the phenol. Finally, the ethereal solution of the mixture is treated with hydrochloric acid, which reacts with the amine, converting it into a water-soluble ammonium salt. The ethereal solution that remains after removal of the aqueous solution contains the neutral compound.

Note that each extraction is performed on the same ethereal solution that originally contained all the components of the mixture, and the *sequence* of extraction, namely, $NaHCO_3$, then NaOH, and finally HCl, is extremely important. The bases and acid each remove one type of organic compound from the mixture and leave the neutral compound in the other layer when the extractions have been completed. Each of the basic and acidic extracts is subsequently treated with acid or base to liberate the carboxylic acid, phenol, and amine from its salt; each of these compounds is then removed from the respective aqueous solutions by extraction with ether or by collecting the solid by vacuum filtration.

Many different layers and solutions are obtained in the experimental procedure that follows. This can lead to much confusion unless the flasks containing each liquid are

FIGURE 23.2

General scheme for separating a simple mixture of water-insoluble compounds.

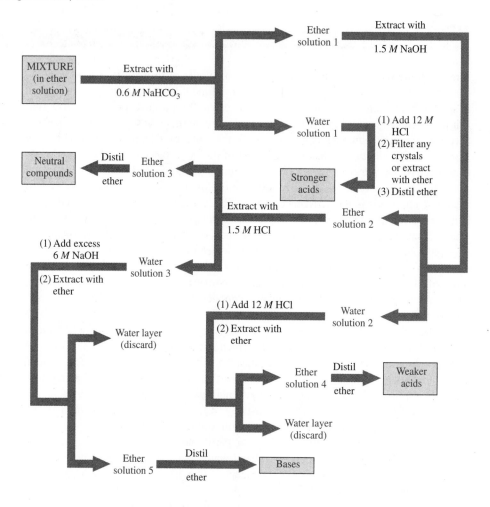

clearly labeled as to the identity of their contents. Moreover, *it is prudent to retain the flasks containing all layers and solutions until it is certain that they are no longer needed.*

EXPERIMENTAL PROCEDURE

SEPARATING MIXTURES ON THE BASIS OF SOLUBILITY

Safety Alert   Diethyl ether is removed from solutes after several extraction steps in this procedure. Use *flameless* heating to remove the ether.

The following procedure, based on the scheme of Figure 23.2, may be used to separate the components of a mixture consisting of three or four compounds, 1–2 g of each, of the types described. It is important to understand that this generalized procedure may not give complete separation of all compounds of even the limited types for which it is intended. The separated products must be tested for purity by the usual methods of melting- or boiling-point determinations and, if possible, by GLC or TLC. Before attempting identification of the individual compounds by any of the classical or modern instrumental methods described later, the samples should be purified by recrystallization, distillation, or chromatography.

Stir or swirl the mixture (3–8 g) with 10 mL of diethyl ether at room temperature. If necessary, collect any solid material that does not dissolve by gravity filtration. Extract the ethereal solution with 4-mL portions of 0.6 M NaHCO$_3$ solution until the aqueous extract remains slightly basic. *Caution:* Vent the separatory funnel quickly after the first mixing because of possible build-up of pressure from carbon dioxide. This solution, **water solution 1** of Figure 23.2, should contain the sodium salt of any carboxylic acid present in the mixture; the ether layer, **ether solution 1**, should contain all other components of the mixture of unknowns. Regenerate the free organic acid by careful acidification of **water solution 1** to pH 2–3 with 12 M hydrochloric acid. If a solid acid separates, collect it by vacuum filtration. Otherwise, extract the aqueous acid solution with several 4-mL portions of diethyl ether, dry the ethereal solution over anhydrous sodium sulfate, and remove the ether by one of the techniques described in Section 2.29.

Extract **ether solution 1** with two 4-mL portions of 1.5 M NaOH solution to remove a phenol or any other weak acid from the solution and to give **ether solution 2**. Acidify the combined aqueous extracts, **water solution 2**, with 12 M hydrochloric acid, and extract the regenerated weak acid with several 4-mL portions of ether to yield **ether solution 4**. If **amphoteric** compounds were included in the unknown mixture, they would be carried through to the water layer separated from **ether solution 4**. Remove the ether from this solution as already described, and isolate the weak acid that remains as the residue.

Extract **ether solution 2** with one or more 4-mL portions of 1.5 M hydrochloric acid until the aqueous extract, **water solution 3**, remains acidic. The ether layer, designated as **ether solution 3** in Figure 23.2, should contain any neutral organic compounds; recover these by removing the ether as described previously.

Add sufficient 5 M NaOH solution to **water solution 3** to make the solution strongly basic; then extract it with several 4-mL portions of diethyl ether. The combined ether extracts, **ether solution 5**, should contain any organic base present in the original unknown mixture; isolate the base by removing the ether as described previously.

Transfer any recovered *diethyl ether* in the container for nonhalogenated organic liquids. Neutralize all *basic or acidic aqueous solutions,* and flush them down the drain. Place the *filter paper* containing any solids that were not soluble in diethyl ether in the container for nonhazardous solids.

Finishing Touches

23.5 | Applying Spectroscopic Methods to Qualitative Organic Analysis

A major limitation inherent to the classical system of qualitative organic analysis is that only *known* compounds can be identified. The research chemist is constantly faced with the task of identifying *new* substances. Although information about the type of compound can be derived from this classical approach, the complete identification of a new organic compound traditionally required a combination of degradation and synthesis in order to achieve a correlation with a known substance; this was usually a lengthy and laborious task. The advent of spectroscopic techniques changed this picture dramatically. New and unknown compounds may be identified quickly and with certainty by using a combination of spectroscopic methods such as those described in Chapter 8. A number of examples could be cited of structural elucidations that were completed in a matter of days or weeks for molecules of greater complexity than those of compounds that were the lifework of several of the great nineteenth-century and early twentieth-century organic chemists.

The modern approach to identifying organic compounds typically involves a combination of spectroscopic and classical methods. Ideally, students should be introduced to the application of spectroscopy in organic chemistry by using the instruments that produce the spectra, but this is not feasible in many instances because some of these instruments are expensive. The next best alternative is to have access to the spectra of typical known compounds and then be provided with the spectra of "unknowns" to identify. This textbook contains over 300 IR and NMR spectra of starting materials and of the products obtained from various preparative experiments, and the CD-ROM accompanying the textbook has the ^{1}H NMR spectra of these compounds. A careful study of these spectra, aided by the material presented in Chapter 8 and by additional discussion by your instructor, should enable meaningful use of IR and NMR spectra for identifying unknown organic compounds.

The spectral data serve to complement or supplement the "wet" qualitative classification tests and in many instances substitute for these tests entirely. For example, a strong IR absorption in the 1690–1760 cm^{-1} region is just as indicative of the presence of a carbonyl group as is formation of a 2,4-dinitrophenylhydrazone, and absorptions in the δ 6.0–8.5 region of ^{1}H NMR spectrum are a more reliable indication of the presence of an aromatic ring than a color test with CHCl$_3$ and AlCl$_3$.

It must be emphasized that spectroscopic analysis also requires *careful interpretation of the data*. Although some problems can be solved quickly and uniquely by modern spectroscopy, others require the intelligent application of both the modern methods and the classical methods. The remainder of this section illustrates the use of spectroscopy in structure determination.

Because the primary use of IR spectroscopy is for identifying functional groups, the observation of certain IR absorptions provides information about the presence of particular functional groups in a compound. Conversely, the absence of certain peaks may be useful in *excluding* the possibility of specific functional groups. For example, the appearance of a strong IR band in the 1650–1760 cm^{-1} region is strong evidence that a substance contains the carbon-oxygen double bond characteristic of an aldehyde, a ketone, a carboxylic acid, or a derivative of a carboxylic acid. On the other hand, the absence of such an absorption is interpreted to mean that the compound does not contain this functional group. Furthermore, a strong band in the 3500–3650 cm^{-1} region is indicative that a substance is an alcohol, a phenol, a carboxylic acid, a 1° or 2° amine or amide, whereas the complete absence of such an absorption excludes these functional groups from further consideration.

However, caution must be used in drawing conclusions based on the absence of expected absorption bands. As noted in Section 8.2, absorption of IR energy by a molecule requires a change in dipole moment as a result of the molecular vibration. This means that symmetrically substituted alkynes do *not* have an IR band for the stretching mode of the triple bond, and even disubstituted alkynes in which the two substituents are similar may have only a weak absorption in this region of the spectrum. For similar reasons, the absence of an absorption in the carbon-carbon double-bond stretching region does not necessarily exclude the presence of this functional group, because the nature of substitution at the double bond may render the functionality inactive in the IR spectrum. Finally, because certain structural features in a molecule cause small shifts in the expected absorptions of some functional groups, IR spectroscopy does not always provide a unique answer about the presence of a particular group. However, it is very useful in limiting the possibilities to a small number of functional groups, and the uncertainty may then be resolved by performing just a few qualitative tests in the laboratory.

NMR spectra generally do not permit direct observation of specific functional groups, as noted in Section 8.3, but they do provide indirect evidence regarding the presence or absence of some groups. The three major features of ^{1}H NMR spectra (Sec. 8.3) that are useful for identifying a compound are the chemical shift, the splitting pattern, and the relative abundance of each type of hydrogen, as determined by the peak areas. The chemical shifts for carbon atoms in the ^{13}C NMR spectrum are also valuable for defining the nature of functional groups in unknown substances.

The research chemist usually obtains IR and NMR spectra of an unknown substance even before obtaining an elemental analysis or performing solubility tests because these types of spectra are easily and quickly measured on a small amount of sample. Only after analyzing them does the researcher undertake other experimental approaches for determining the structure. This process can save many hours of unnecessary laboratory work. Other useful information about a compound may be obtained from mass spectrometry, which provides the molecular weight of the compound and, if high-resolution data are available, its elemental composition.

The following example illustrates applying spectral analysis to identifying an unknown compound.

Example

The unknown **X** has the molecular formula $C_9H_{10}O_2$ and provides the spectra shown in Figure 23.3. Analysis of the molecular formula results in an **Index of Hydrogen Deficiency (IHD)** or degree of unsaturation for the compound of *five*, and thus mandates the presence of rings and/or multiple bonds. Examining the IR spectrum reveals two intense absorptions in the functional group region. Reference to Tables 8.1 and 8.2 shows that the band at about 1700 cm^{-1} is consistent with the presence of a carbonyl group, whereas the broad absorption ranging from 3400–2300 cm^{-1} is characteristic of the hydroxyl moiety of a carboxylic acid function: the O–H stretch of an alcohol is not as broad as that of a carboxylic acid. Further evidence for the acid functionality is found in the absorption at 1320 cm^{-1}, which may be assigned to C–O bond stretching. Stretching vibrations associated with the carbon-hydrogen bonds that must be present in the unknown appear to be buried in the hydroxyl absorption, so their absence should be of no concern. Finally, the presence at 1500 cm^{-1} of a sharp band of intermediate intensity may signify an aromatic ring, a possibility that seems reasonable in light of the value of IHD.

The ^{13}C NMR spectrum shows that there are seven magnetically different carbon atoms in **X**. The resonance at δ 179.5 has a chemical shift appropriate for the carbonyl carbon atom of a carboxylic function (Table 8.5), whereas those at δ 30.5 and 35.5 are clearly associated with sp^3- rather than sp^2- or sp-hybridized carbon atoms. The resonances appearing in the range of δ 126–140 have chemical shifts expected for sp^2-hybridized carbon atoms. Because there are only four such resonances and these must account for six carbon atoms, two of the peaks must each represent two magnetically identical carbon atoms. This requirement is met by proposing that **X** contains either a

FIGURE 23.3   IR and ^{1}H NMR spectra of unknown compound **X**.

monosubstituted or a *p*-disubstituted benzene ring, **7** or **8**, respectively. Given the conclusions reached earlier about the nature of the other three carbon atoms present, only structures **9–11** are possible. Which of these is actually the unknown compound could be determined by calculating the expected ^{13}C chemical shifts of each of the possibilities, using the data in Tables 8.6–8.8, and comparing the results with those observed experimentally. However, we shall use the data available from the ^{1}H NMR to differentiate among the three possibilities.

R^I	R^I	$CH_2CH_2CO_2H$	CH_2CO_2H	CO_2H
	R^2		CH_3	CH_2CH_3
7	**8**	**9**	**10**	**11**

The ^{1}H NMR spectrum reveals that there are *four* sets of chemically nonequivalent protons in the unknown. These are in the relative ratio of 1:5:2:2 when going upfield.

This is also the absolute ratio because there are only 10 hydrogen atoms in **X**. The appearance of a resonance at very low field, δ 11.9, confirms the earlier conclusion that a carboxylic acid function is present (see Table 8.3). The fact that the broad singlet at δ 7.2, which is in the range of chemical shifts for aromatic protons, integrates for *five* hydrogens means that the ring is *monosubstituted*. Consequently, the unknown compound must have structure **9**, in which the pair of two-proton multiplets centered at δ 2.8 correspond to the two heterotopic sets of methylene protons.

This example provides a logical approach to the use of a molecular formula and spectral data in deducing the structure of an unknown compound. The interplay of the information available from IR, ^{1}H NMR, and ^{13}C NMR spectra is critical to the process of structural elucidation, and it is commonly necessary to use more than one spectral technique before an assignment is possible. It cannot be overemphasized that *all* of the spectral data available must be consistent with the proposed structure of the sample. The existence of even one inconsistency between the data and the proposed structure renders that structure unlikely, unless the inconsistency is produced by an extraneous factor such as an impurity in the sample or a malfunction in the instrumentation used to obtain the spectra.

A second example illustrates the complementarity between spectral analysis and classical qualitative analysis for determining the structure of an unknown compound.

Example

The unknown **Y** is a liquid and the following information is available:

Preliminary analysis	Colorless liquid with pleasant odor
Physical constant	bp 143–145 °C (760 Torr)
Elemental analysis	No X, S, or N
Solubility tests	Water-soluble to give a neutral solution
Preliminary report	Observations confirmed by instructor
Spectra	IR and ^{1}H NMR spectra are shown in Figure 23.4.
Classification tests	Positive test for ester, negative tests for all other functional groups

Analysis

Owing to its water-solubility, **Y** most likely contains oxygen-bearing polar functional groups; it also probably contains a relatively small number of carbon atoms because compounds of more than five or six carbon atoms are usually water-insoluble. It is immediately evident from the IR spectrum shown in Figure 23.4 that the liquid contains an ester group because of the strong absorption peaks at about 1750 and 1230 cm^{-1}; the former peak is due to C—O stretching and the latter to the C—O—C bond (Table 8.1). Although the 1750 cm^{-1} band could indicate a five-membered cyclic ketone, this and other possibilities are negated by the presence of numerous peaks in the ^{1}H NMR spectrum. The presence of an ester function was confirmed by obtaining a positive qualitative test for this group (Sec. 23.16).

The ^{1}H NMR spectrum has four different resonances: a multiplet centered at δ 4.2, a second multiplet centered at δ 3.6, a singlet at δ 3.4, and another singlet at δ 2.1, and their relative areas are 2:2:3:3, respectively. This means that *at least* four chemically different carbon atoms must be present in the molecule. In fact, reference to the ^{13}C NMR data shows that *five* chemically distinct carbon atoms are contained in **Y**. The *relative* abundances of different types of protons are likely to correspond to the *absolute* ones, because the water-solubility of the unknown signals that **Y** is likely to contain fewer than seven carbon atoms. Consequently, the two singlets at δ 2.1 and 3.4, each with relative integrations of three, must represent two rather than some larger

FIGURE 23.4   IR and ^{1}H NMR spectra of unknown compound **Y**.

number of methyl groups, and the two groups have no nearest neighbors. The higher-field peak at δ 2.1 is characteristic of a methyl ketone or acetate (Table 8.3), the latter being consistent with the deduction provided by the IR spectrum and the qualitative test that **Y** is an ester. The second methyl peak at δ 3.4 is shifted downfield, as would be expected if the methyl group were bonded to oxygen. This peak is too far upfield for a methyl ester, because methyl esters normally show the methyl absorption at about δ 3.7–4.1. However, it is within the δ 3.3–4.0 range frequently observed for aliphatic α-hydrogens of an alcohol or ether. Because neither spectrum shows any evidence for a hydroxy group, the compound probably contains a methoxy group, and the presence of an aliphatic ether is consistent with the C–O–C absorption observed at 1050 cm^{-1} in the IR spectrum. The multiplets centered at δ 4.2 and 3.6 each integrate for two hydrogens and show mutual spin-spin coupling. This pattern is diagnostic of chemically nonequivalent, adjacent methylene groups such as X–CH$_2$CH$_2$–Y. The downfield absorption for the CH$_2$ groups indicates that each is bonded to oxygen.

Table 23.2 is a listing of some liquid esters, and only a single structure, 2-methoxyethyl acetate (**15**) appears to be consistent with all of the spectral observations for **Y**. Confirmation of this structural assignment is possible by applying the classical tech-

Table 23.2	**Liquid Esters with Boiling Points 140–150 °C (760 Torr).**		
Name and Number	Structure	BP (°C)	α-Naphthylurethane (°C)
3-Methylbutyl acetate (**12**)	$CH_3C(=O)OCH_2CH_2CH(CH_3)_2$	142	82
Methyl lactate (**13**)	$CH_3CH(OH)C(=O)OCH_3$	145	124
Ethyl chloroacetate (**14**)	$ClCH_2C(=O)OCH_2CH_3$	145	79
2-Methoxyethyl acetate (**15**)	$CH_3C(=O)OCH_2CH_2OCH_3$	145	113
Ethyl valerate(**16**)	$CH_3(CH_2)_3C(=O)OCH_2CH_3$	146	79
Ethyl α-chloropropionate (**17**)	$CH_3CH(Cl)C(=O)OCH_2CH_3$	146	79
Diisopropyl carbonate (**18**)	$(CH_3)_2CHOC(=O)OCH(CH_3)_2$	147	106
Pentyl acetate (**19**)	$CH_3C(=O)O(CH_2)_4CH_3$	149	68

niques of derivatization. In this instance, hydrolyzing the suspected ester **15** (Sec. 23.16), should produce acetic acid and 2-methoxyethanol, both of which could be separately characterized through a solid derivative. The alcohol, for example, could be converted into an α-naphthylurethane derivative (Sec. 23.11D). Finding that this derivative has a melting point of 111–113 °C, a range corresponding to that expected for the α-naphthylurethane of 2-methoxyethanol (Table 23.2), verifies that the unknown **Y** contains the 2-methoxyethoxy, $CH_3OCH_2CH_2O-$, fragment, as required in **15**.

This example of proving the structure of an unknown should provide an appreciation of the complementary aspects of spectral analysis and the "wet" classification scheme, including the use of physical properties. Although the spectra were not in themselves specifically definitive in the assignment of the structures of the unknown, they provided data regarding the possible presence of certain functional groups, which could be confirmed by performing the appropriate classification tests. Thus, the spectral analyses helped minimize the effort and time spent in performing numerous classification tests, most of which would have been negative. The final decision on the structure of the unknown came from preparing a solid derivative whose melting point could be compared with that of the known compound. Of course, when solving the structure of an unknown, it is often possible to find the IR and NMR spectra of the suspected compounds in one of the catalogs of spectra given in the references in Chapter 8. If the reported spectra are *identical* to those obtained for the unknown, the identity of the unknown is confirmed.

REFERENCES

1. Shriner, R. L.; Fuson, R. C.; Curtin, D. Y.; Morrill, T. C. *The Systematic Identification of Organic Compounds*, 6th ed., John Wiley & Sons, New York, **1980**.
2. *Handbook of Tables for Organic Compound Identification*, 3rd ed., Z. Rapoport, ed., Chemical Rubber Company, Cleveland, OH, **1967**. Provides physical properties and derivatives for more than 4000 compounds, arranged according to functional groups.

23.6 Qualitative Classification Tests and Preparation of Derivatives

This section contains discussions about and experimental procedures for the qualitative classification tests and preparation of derivatives for compounds having most of the commonly encountered functional groups. Proper interpretation of the information from the qualitative tests is *crucial* for successfully applying the principles of Section 23.3

to structure elucidation. Consequently, you should exercise great care not only when performing the tests but also in analyzing their results. In the end, you may find solving the structures of unknowns the most intellectually challenging, yet most satisfying experience you have in the introductory organic laboratory.

23.7   Aldehydes and Ketones

CLASSIFICATION TESTS

A. 2,4-Dinitrophenylhydrazine. Arylhydrazines ($ArNHNH_2$), which include phenyl-hydrazine, 4-nitrophenylhydrazine, and 2,4-dinitrophenylhydrazine, are commonly used to make crystalline derivatives of carbonyl compounds. The formation of a 2,4-dinitrophenylhydrazone **3** is represented by Equation 23.9, and the mechanism of this transformation typifies that followed by a number of compounds, RNH_2, that may be considered as derivatives of ammonia. Thus, the overall reaction involves initial acid-catalyzed addition of the elements of N—H across the carbonyl π-bond to afford a tetrahedral intermediate, which subsequently dehydrates to the product (Eq. 23.9).

The arylhydrazines are valuable reagents for both classifying and forming derivatives because the solid products of these test reactions may be used as derivatives of the aldehyde or ketone; 2,4-dinitrophenylhydrazine is particularly useful in this regard.

The arylhydrazines will give a positive test for either an aldehyde or a ketone. Schiff's and Tollens' tests given in Parts B and C, respectively, provide methods for distinguishing between these two types of compounds.

$$R_2C{=}O \;+\; H_2NNH{-}Ar \;\underset{}{\overset{H_3O^+}{\rightleftharpoons}}\; R_2C\!\!\begin{array}{c}OH\\ \\ NH\\ | \\ NHAr\end{array} \;\rightleftharpoons\; R_2C{=}NNHAr \;+\; H_2O \qquad (23.9)$$

An aldehyde 2,4-Dinitrophenyl- **3**
or ketone hydrazine A 2,4-dinitrophenylhydrazone

$$Ar =$$

EXPERIMENTAL PROCEDURE

2,4-DINITROPHENYLHYDRAZINE TEST FOR ALDEHYDES AND KETONES

Safety Alert   Concentrated sulfuric acid may produce severe chemical burns and 2,4-dinitrophenylhydrazine is a possible mutagen and may cause skin sensitization. Wear latex gloves when preparing and handling solutions involving these reagents. If this acid or the ethanolic solution of 2,4-dinitrophenylhydrazine comes in contact with your skin, thoroughly wash the affected area with water. Also rinse the area with 5% sodium bicarbonate solution.

If the test reagent is not supplied, prepare it by dissolving 0.2 g of 2,4-dinitrophenyl-hydrazine in 1 mL of concentrated sulfuric acid. Add this solution, with stirring, to 1.5 mL of water and 5 mL of 95% ethanol. Stir the solution vigorously and then filter it to remove any undissolved solids.

Dissolve 1 or 2 drops of a liquid, or about 100 mg of a solid, in 2 mL of 95% ethanol, and add this solution to 2 mL of the 2,4-dinitrophenylhydrazine reagent. Shake the mixture vigorously; if a precipitate does not form immediately, let the solution stand for 15 min.

If more crystals are desired for a melting-point determination, dissolve 200–500 mg of the carbonyl compound in 20 mL of 95% ethanol, and add this solution to 15 mL of the reagent. Recrystallize the product from aqueous ethanol.

Finishing Touches

Neutralize and then filter any excess *2,4-dinitrophenylhydrazine solution*. Put the filter cake in the container for nonhazardous solids. Flush the *filtrate*, as well as *filtrates* obtained from recrystallization, down the drain.

B. Schiff's Test. A common method for distinguishing between aldehydes and ketones is Schiff's test. Aldehydes give positive tests, whereas ketones do not.

Schiff's reagent is an aqueous solution formed by combining *p*-rosaniline hydro-chloride (basic fuchsin), sodium bisulfite, and concentrated hydrochloric acid (Eq. 23.10). The reagent reacts with aldehydes to produce an imine, also called a Schiff base, that is magenta or purple in color (Eq. 23.11). The mechanism for forming the addition product presumably involves nucleophilic attack of sulfur on the aldehydic carbonyl group.

Color development typically occurs in no more than 10 minutes and ketones do not produce a corresponding color change. However, because the reagent may produce a light blue color in the presence of other functional groups, it is important to perform this test with known compounds such as benzaldehyde and acetone.

p-Rosaniline hydrochloride
Fuchsia (purple)

(1) $NaHSO_3$
(2) HCl

Schiff's reagent
Colorless

(23.10)

$$\text{Schiff's reagent} + \text{RCH}{=}\text{O} \longrightarrow$$

An aldehyde   An addition product
                               Magenta or purple

(23.11)

EXPERIMENTAL PROCEDURE

SCHIFF'S TEST FOR ALDEHYDES

Safety Alert

p-Rosaniline hydrochloride is a cancer suspect agent and a mutagen. Wear latex gloves to prevent Schiff's reagent from contacting your skin. If you get the reagent on your skin, wash the area immediately with warm water.

If Schiff's reagent is not available as a stock solution, prepare it by dissolving 0.01 g of *p*-rosaniline hydrochloride in 10 mL of water, adding 0.4 mL of saturated aqueous sodium bisulfite, and allowing the resulting solution to stand for 1 h. Then add 0.2 mL of concentrated hydrochloric acid to complete the preparation.

Place 1 mL of Schiff's reagent in a small test tube and add 1 drop of the unknown. Compare any color changes with those observed when a known aldehyde and a known ketone are subjected to the same test.

Finishing Touches

Neutralize the *acidic solutions* by adding solid sodium carbonate and pour the resulting solution into a container for hazardous aqueous waste.

C. Tollens' Test. Another method for distinguishing between aldehydes and ketones is Tollens' test. A positive test indicates the presence of an aldehyde function, whereas no reaction occurs with ketones. Tollens' reagent consists of silver-ammonia complex, $Ag(NH_3)_2^+$, in an ammonia solution. This reagent oxidizes both aliphatic and aromatic aldehydes to the corresponding carboxylic acids; silver ion is reduced to elemental silver, which is deposited as a silver mirror on the glass wall of a *clean* test tube. Thus, the formation of the silver mirror or of a precipitate is considered a positive test. Equation 23.12 shows the reaction that occurs.

$$\text{RCHO} + 2\,\text{Ag(NH}_3)_2^+ + 2\,\text{HO}^- \longrightarrow 2\,\underline{\text{Ag}} + \text{RCO}_2^-\,\text{NH}_4^+ + \text{H}_2\text{O} + 3\,\text{NH}_3 \qquad (23.12)$$

An aldehyde   Metallic silver

Similar tests for aldehydes make use of Fehling's and Benedict's reagents, which contain complex tartrate and citrate salts, respectively, of cupric ion as the oxidizing agents. With these reagents a positive test is the formation of a brick-red precipitate of cuprous oxide (Cu_2O), which forms when Cu^{2+} is reduced to Cu^+ by the aldehyde. These two tests are more useful in distinguishing between aliphatic and aromatic aldehydes, because the aliphatic compounds give a faster reaction. They have been used widely to detect reducing sugars (Sec. 22.4), whereas Tollens' test is used to distinguish between aldehydes, both aliphatic and aromatic, and ketones.

EXPERIMENTAL PROCEDURE

TOLLEN'S TEST FOR ALDEHYDES

> **Safety Alert**
>
> 1. Avoid spilling Tollens' reagent on your skin because staining may result. If you do come into contact with the solution, wash the affected area immediately with copious amounts of water.
> 2. *Do not store unused Tollens' reagent because it decomposes on standing and yields silver azide, AgN₃, an explosive precipitate.* Follow the directions under FINISHING TOUCHES for proper disposal of the solution.

Tollens' reagent must be prepared at the time of use by combining two other solutions, *A* and *B*, which should be available as stock solutions. If not, make up solution *A* by dissolving 0.25 g of silver nitrate in 4.3 mL of distilled water and solution *B* by dissolving 0.3 g of potassium hydroxide in 4.2 mL of distilled water.

Prepare the test reagent according to the following directions. Add concentrated ammonium hydroxide solution dropwise to 3 mL of solution *A* until the initial brown precipitate begins to clear. The solution should be grayish and almost clear. Then add 3 mL of solution *B* to this mixture, and again add concentrated ammonium hydroxide dropwise until the solution is almost clear.

To carry out the test, add 0.5 mL of the reagent to 3 drops or 50–100 mg of the unknown compound; the formation of a silver mirror or black precipitate constitutes a positive test. The silver deposits in the form of a mirror only on a *clean* glass surface. A black precipitate, although not as aesthetically pleasing, still constitutes a positive test. If no reaction occurs at room temperature, warm the solution slightly in a beaker of warm water.

Finishing Touches

Remove the *silver mirrors* from the test tube with nitric acid, and pour the resulting *solution* into a beaker containing unused *Tollens' reagent*, *ammonium hydroxide*, and *sodium hydroxide*. Acidify this mixture with nitric acid to destroy the unreacted Tollens' reagent. Neutralize the solution with sodium carbonate and add saturated sodium chloride solution to precipitate silver chloride. Collect the silver chloride by vacuum filtration and place the *filter cake* in a container for recovered silver halides; flush the filtrate down the drain.

D. Chromic Acid Test. Another method for distinguishing between aldehydes and ketones is the chromic acid test. Because chromic acid is unstable when stored for extended periods of time, the test reagent is prepared as needed by dissolving chromic anhydride in sulfuric acid (Eq. 23.13) and used in acetone solution. This reagent oxidizes primary and secondary alcohols (Eqs. 23.14 and 23.15, respectively) and all aldehydes (Eq. 23.14); it gives no visible reaction with ketones and tertiary alcohols (Eqs. 23.15 and 23.16, respectively) under the test conditions. A distinctive color change from the orange-red of Cr^{6+} to the green of Cr^{3+} occurs as oxidation of the organic substrate proceeds. The stoichiometry of oxidation of an alcohol to an aldehyde or ketone is depicted in Equation 23.17, where it can be seen that three moles of sulfuric acid are consumed in the process.

$$CrO_3 + H_2O \xrightarrow[\text{Orange-red}]{H_2CrO_4} H_2CrO_4 \tag{23.13}$$

$$\underset{\text{1° alcohol}}{RCH_2OH} \xrightarrow[\text{Orange-red}]{H_2CrO_4} [RCHO] \xrightarrow[\text{Orange-red}]{H_2CrO_4} RCO_2H + \textit{Green} \text{ solution or precipitate} \tag{23.14}$$

$$\underset{\text{A 2° alcohol}}{R_2CHOH} \xrightarrow[\text{Orange-red}]{H_2CrO_4} R_2CO + \textit{Green} \text{ solution} \xrightarrow{H_2CrO_4} \text{No visible reaction} \atop \text{or precipitate} \tag{23.15}$$

$$R_3COH \xrightarrow{H_2CrO_4} \text{No visible reaction} \tag{23.16}$$

$$3 \ R^2 \overset{\overset{\text{OH}}{|}}{-CH-} R^1 + 2\,H_2CrO_4 + 3\,H_2SO_4 \longrightarrow 3\ R^2 \overset{\overset{\text{O}}{\|}}{-C-} R^1 + Cr_2(SO_4)_3 + 8\,H_2O \tag{23.17}$$

Thus, the chromic acid reagent gives a clear-cut distinction between primary and secondary alcohols and aldehydes on the one hand and tertiary alcohols and ketones on the other. Aldehydes may be distinguished from primary and secondary alcohols by means of Schiff's, Tollens', Benedict's (Sec. 22.4), and Fehling's tests, and lower-molecular-weight primary and secondary alcohols may be differentiated on the basis of their rates of reaction with concentrated hydrochloric acid containing zinc chloride—the Lucas reagent (Sec. 23.11B).

EXPERIMENTAL PROCEDURES

CHROMIC ACID TEST FOR ALDEHYDES AND 1° AND 2° ALCOHOLS

Safety Alert

1. When preparing and handling solutions of chromic acid, wear latex gloves to keep the acids from contacting your skin. Chromic acid causes unsightly stains on your hands for several days, and may

cause severe chemical burns. Wash the affected area thoroughly with warm water if the oxidant comes in contact with the skin. Also rinse the area with 5% sodium bicarbonate solution.

2. The preparation of chromic acid requires diluting a paste of chromic anhydride and concentrated sulfuric acid with water. Be certain to add the water *slowly* to the acid and swirl the mixture to ensure continuous mixing. Swirling keeps the denser sulfuric acid from layering at the bottom of the flask and avoids possible splattering because of the heat generated when the two layers are suddenly mixed.

If the chromic acid reagent is not available, prepare it as follows. Add 1 g of chromic anhydride (CrO_3) to 1 mL of concentrated sulfuric acid and stir the mixture until a smooth paste is obtained. Dilute the paste *cautiously* with 3 mL of distilled water, and stir this mixture until a clear orange solution is obtained.

Dissolve 1 drop of a liquid or about 10 mg of a solid alcohol or carbonyl compound in 1 mL of *reagent grade* acetone. Add 1 drop of the acidic chromic anhydride reagent to the acetone solution, and shake the tube to mix the contents. A positive reaction is indicated by disappearance of the orange color of the reagent and the formation of a green or blue-green precipitate or emulsion.

Primary and secondary alcohols and aliphatic aldehydes give a positive test within 5 sec. Aromatic aldehydes require 30 to 45 sec. Color changes occurring after about 1 min should not be interpreted as positive tests; other functional groups such as ethers and esters may slowly hydrolyze under the conditions of the test, releasing alcohols that in turn provide "false-positive" tests. Tertiary alcohols and ketones produce no visible change in several minutes. Phenols and aromatic amines give dark precipitates, as do aromatic aldehydes and benzylic alcohols having hydroxyl or amino groups on the aromatic ring.

Finishing Touches

Add sodium sulfite to the *aqueous solution* of chromium salts in order to destroy excess Cr^{6+}. Make the solution slightly basic, to form chromium hydroxide, and isolate this salt by vacuum filtration through a bed of a filter-aid. Place the *filter paper* and the *filter cake* in the container for heavy metals; flush the *filtrate* down the drain.

E. Iodoform Test. When an aldehyde or ketone that has α-hydrogens is treated with a halogen in basic medium, halogenation occurs on the α-carbon (Eq. 23.18). This reaction involves the formation of an enolate ion (Eq. 18.3), which subsequently reacts

$$R_2C-\overset{O}{\overset{\|}{C}}\underset{R}{\overset{}{}} \xrightarrow{\text{Base, } X_2} R_2C-\overset{O}{\overset{\|}{C}}\underset{R}{\overset{}{}} \qquad (23.18)$$
$$\underset{H}{\overset{|}{}} \qquad\qquad \underset{X}{\overset{|}{}}$$

An aldehyde or
ketone

with halogen to produce the substitution product. In the case of acetaldehyde or a methyl ketone, all three of the α-hydrogens of the methyl group are replaced by halogen to give **20** (Eq. 23.19), which then reacts with excess base to give products **21** and **22** (Eq. 23.20).

$$\underset{\substack{\text{An aldehyde or}\\ \text{ketone}}}{\text{H}_3\text{C}-\text{C(=O)R}} \xrightarrow{\text{NaOH, X}_2} \underset{\textbf{20}}{\text{X}_3\text{C}-\text{C(=O)R}} \qquad (23.19)$$

$$\underset{\textbf{20}}{\text{X}_3\text{C}-\text{C(=O)}-\text{R}} \xrightarrow{\text{NaOH}} \underset{\textbf{21}}{\text{CHX}_3} + \underset{\textbf{22}}{\text{RCO}_2^- \text{Na}^+} \qquad (23.20)$$

A haloform

Although chlorine, bromine, and iodine all react in this manner, the qualitative test for a methyl ketone uses iodine because it is safer to handle and because the product is *iodoform*, CHI_3, a highly *insoluble* crystalline *yellow* solid that is readily observed and identified on the basis of its characteristic odor and melting point. It should be pointed out that a positive iodoform test involves two experimental observations: (1) the disappearance of the characteristic red-brown color of iodine as the test reagent is added to the compound being analyzed, *and* (2) the formation of a yellow precipitate of iodoform. The reason both observations are required is that *all* aldehydes and ketones containing α-hydrogens react with iodine in base to decolorize the test reagent. Only methyl ketones, after being trisubstituted with iodine on the methyl group, react with base to produce iodoform and the salt of a carboxylic acid.

A positive test also occurs for alcohols having a hydroxymethyl ($-\text{CHOHCH}_3$) functionality, as in ethanol and isopropyl alcohol, and for acetaldehyde, which is analogous to a methyl ketone in having three α-hydrogens. The reason that a hydroxymethyl compound gives a positive iodoform test is that the combination of iodine and sodium hydroxide produces sodium hypoiodite (Eq. 23.21) which is a mild oxidant (Sec. 16.2). Consequently, the test reagent is able to oxidize a hydroxymethyl function to a methyl ketone, and the methyl ketone then reacts with iodine in the presence of base to give iodoform. It is thus always necessary to consider this possibility before making a final decision regarding the functional group responsible for a positive test. The ambiguity in interpreting this test is resolved by determining whether a keto group is present or absent by using another qualitative test, such as that involving 2,4-dinitrophenylhydrazine (Eq. 23.9).

$$\text{I}_2 + 2\,\text{NaOH} \longrightarrow \text{NaI} + \text{NaOI} + \text{H}_2\text{O} \qquad (23.21)$$

The iodoform test is sometimes called the **hypoiodite test** because sodium hypoiodite is formed according to Equation 23.21. In more general terms, the reactions just described can be called the **haloform test** or the **sodium hypohalite test** when the specific halogen is not specified.

EXPERIMENTAL PROCEDURE

IODOFORM TEST

> **Safety Alert**
>
> Iodine is a hazardous chemical. Do not breathe its vapors and wear latex gloves to prevent this reagent or solutions of it from contacting your skin because it may cause serious chemical burns. It you get iodine on your skin, wash the area quickly with warm water and soak the area in 0.6 M sodium thiosulfate solution for up to 3 h if the burn is particularly serious.

If the iodine reagent is not available as a stock solution, prepare it by dissolving 1 g of iodine in a solution of 2 g of potassium iodide in 8 mL of water. The potassium iodide is added to increase the solubility of iodine in water by formation of potassium triiodide, KI_3 (Eq. 23.22).

$$I_2 + KI \longrightarrow KI_3 \qquad (23.22)$$

If the substance is water-soluble, dissolve 2 to 3 drops of a liquid or an estimated 50 mg of a solid in 2 mL of water in a small test tube, add 2 mL of 3 M sodium hydroxide, and then slowly add 3 mL of iodine solution. In a positive test the brown color disappears, and yellow iodoform separates. If the substance tested is insoluble in water, dissolve it in 2 mL of dioxane, proceed as above, and dilute the mixture with 10 mL of water *after* addition of the iodine solution.

Iodoform is recognizable by its yellow color and, more definitively, by its melting point, 118–119 °C. Isolate this product either by vacuum filtration of the test mixture or by the following sequence: Add 2 mL of dichloromethane to the mixture, shake the stoppered test tube to extract the iodoform into the small lower layer, withdraw the clear part of this layer with a Pasteur pipet, and evaporate it in a small test tube or on a watchglass with a steam bath in the hood. Whichever mode of isolation is used, recrystallize the crude solid from methanol-water.

Finishing Touches

Place the residual *aqueous solutions* in the container for halogenated liquids. Put the *iodoform* in the container for halogenated solids. Flush the *filtrate* down the drain.

DERIVATIVES

Two of the most useful solid derivatives of aldehydes and ketones are the **2,4-dinitrophenylhydrazone** and the **semicarbazone**. The **oxime** is also sometimes useful, but it often forms as an oil rather than a solid.

F. 2,4-Dinitrophenylhydrazone. The qualitative test for aldehydes and ketones is described in Section 23.7A, and the solid that forms may be isolated and purified as indicated in that section.

G. Semicarbazone. Semicarbazide reacts with aldehydes and ketones to produce a derivative that is called a semicarbazone (Eq. 23.23), a reaction that is discussed in Chapter 13. Because semicarbazide is unstable as the free base, it is usually stored in the form of its hydrochloric acid salt. In the procedure that follows, the free base is liberated from the salt by addition of sodium acetate.

$$
\underset{\substack{\text{An aldehyde} \\ \text{or ketone}}}{\overset{\displaystyle O}{\underset{R^1}{\overset{\|}{C}}}{}_{R^2}} + \underset{\text{Semicarbazide}}{H_2N-NH\overset{\displaystyle O}{\overset{\|}{C}}NH_2} \xrightarrow{H^+} \underset{\text{A semicarbazone}}{\overset{\displaystyle N}{\underset{R^1}{\overset{NHCNH_2}{\overset{\|}{C}}}}{}_{R^2}}
$$

(23.23)

PREPARATION OF SEMICARBAZONES

> **Safety Alert**   Semicarbazide and its hydrochloride may have carcinogenic, teratogenic, and mutagenic effects. Should either of these substances contact your skin, wash the affected area with copious amounts of water.

Dissolve 0.5 g of semicarbazide hydrochloride and 0.8 g of sodium acetate in 5 mL of water in a test tube, and then add about 0.5 mL of the carbonyl compound. Stopper and shake the tube vigorously, remove the stopper, and place the test tube in a beaker of boiling water. Discontinue heating the water, and allow the test tube to cool to room temperature in the beaker of water. Transfer the test tube to an ice-water bath, and if crystals have not formed, scratch the side of the tube with a glass rod at the interface between the liquid and air to induce crystallization. Recrystallize the semicarbazone from water or aqueous ethanol.

If the carbonyl compound is insoluble in water, dissolve it in 5 mL of ethanol. Add water until the solution becomes turbid, then add a little ethanol until the turbidity disappears. Add the semicarbazide hydrochloride and sodium acetate, and continue as above from this point.

Finishing Touches   Flush *excess reagent* and *filtrates* down the drain.

H. Oximes. Hydroxylamine reacts with aldehydes or ketones to yield an oxime (Eq. 23.24). Hydroxylamine is usually stored as the hydrochloric acid salt because it is not stable as the free base; it is liberated from its salt by addition of base. The use of an oxime as a derivative has the limitation that such compounds frequently are *not* crystalline solids.

$$\underset{\substack{\text{An aldehyde} \\ \text{or ketone}}}{\underset{R^1}{\overset{O}{\underset{\|}{C}}}\overset{}{R^2}} + \underset{\text{Hydroxylamine}}{H_2N-OH} \xrightarrow{H^+} \underset{\substack{\text{An oxime}}}{\underset{R^1}{\overset{N^{OH}}{\underset{\|}{C}}}R^2} + H_2O \qquad (23.24)$$

EXPERIMENTAL PROCEDURES

PREPARATION OF OXIMES

> **Safety Alert**
>
> Hydroxylamine and its hydrochloride may have mutagenic effects. Should either of these substances contact your skin, wash the affected area with copious amounts of water.

Dissolve 0.5 g of hydroxylamine hydrochloride in 5 mL of water and 3 mL of 3 M sodium hydroxide solution, and add 0.5 g of the aldehyde or ketone. If the carbonyl compound is insoluble in water, add just enough ethanol to give a clear solution. Warm the mixture on a steam bath or boiling-water bath for 10 min; then cool it in an ice-water bath. If crystals do not form immediately, scratch the side of the tube with a glass rod at the air-liquid interface to induce crystallization. Recrystallize the oxime from water or aqueous ethanol.

In some cases the use of 3 mL of pyridine and 3 mL of absolute ethanol in place of the 3 mL of 3 M sodium hydroxide solution and 5 mL of water is more effective. A longer heating period is often necessary. After the heating is finished, pour the mixture into an evaporating dish and remove the solvent with a current of air in a hood. Grind the solid residue with 3–4 mL of cold water and filter the mixture. Recrystallize the oxime from water or aqueous ethanol.

Flush *excess reagent* and *filtrates* down the drain.

Finishing Touches

23.8 Alkenes and Alkynes

CLASSIFICATION TESTS

Two common types of unsaturated compounds are alkenes and alkynes, characterized by the carbon-carbon double and triple bonds, respectively, as the functional group. The two common qualitative tests for unsaturation are the reaction of the compounds with *bromine in dichloromethane* and with *potassium permanganate*. In both cases a positive test is denoted by decoloration of the reagent. There are no simple direct ways to prepare solid derivatives of unsaturated aliphatic compounds having no other functional groups.

A. Bromine in Dichloromethane. Bromine adds to the carbon-carbon double bond of alkenes to produce dibromoalkanes (Eq. 23.25) and reacts with alkynes to produce tetrabromoalkanes (Eq. 23.26). When this reaction occurs, molecular bromine is rapidly consumed, and its characteristic dark red-orange color disappears if bromine is not added in excess. The nearly *instantaneous* disappearance of the bromine color is a positive test for unsaturation. The test is ambiguous, however, because some alkenes do not react with bromine, and some react very slowly. In the case of a negative test, the Baeyer procedure using potassium permanganate (Sec. 23.8B) should be performed.

$$R_2C{=}CR_2 + Br_2 \xrightarrow{CH_2Cl_2} R_2\overset{\displaystyle Br}{\underset{\displaystyle Br}{C}}{-}CR_2 \tag{23.25}$$

An alkene   Bromine   A dibromide
                 Red-orange   *Colorless*

$$RC{\equiv}CR + Br_2 \xrightarrow{CH_2Cl_2} R\overset{\displaystyle Br}{\underset{\displaystyle Br}{C}}{-}\overset{\displaystyle Br}{\underset{\displaystyle Br}{C}}R \tag{23.26}$$

An alkyne   A tetrabromide
                                   Colorless

More details about the reaction of bromine in dichloromethane with alkenes are found in Section 10.4.

EXPERIMENTAL PROCEDURE

BROMINE TEST FOR UNSATURATION

Safety Alert

1. *Bromine is a hazardous chemical.* Do not breathe its vapors or allow it to come into contact with your skin because it may cause serious chemical burns. Perform all operations involving the transfer of solutions of bromine at a hood; wear latex gloves. If you get bromine on your skin, wash the area quickly with warm water and soak the area in 0.6 *M* sodium thiosulfate solution for up to 3 h if the burn is particularly serious.

2. Do *not* use acetone to rinse glassware containing residual amounts of bromine. This prevents formation of α-bromoacetone, a severe lachrymator. Rather, follow the procedure described in FINISHING TOUCHES.

Dissolve 50 mg or 1 or 2 drops of the unknown in dichloromethane and add to this solution 0.1 M bromine in dichloromethane dropwise until a light orange color just persists. *Rapid* disappearance of the bromine color to give a colorless solution is a positive test for unsaturation.

Decolorize any *solutions* in which the color of *bromine* is visible by dropwise addition of cyclohexene; then discard the resulting mixtures in the container for halogenated organic liquids.

Finishing Touches

B. Potassium Permanganate. A second qualitative test for unsaturation, the Baeyer test, depends on the ability of potassium permanganate to oxidize the carbon-carbon double bond to give alkanediols (Eq. 23.27) or the carbon-carbon triple bond to yield carboxylic acids (Eq. 23.28). The permanganate is destroyed in the reaction, and a brown precipitate of MnO_2 is produced. The disappearance of the characteristic purple color of the permanganate ion is a positive test for unsaturation. However, care is needed in interpreting this test, because compounds containing certain other functional groups, such as aldehydes, also decolorize permanganate ion because they undergo oxidation.

$$R_2C{=}CR_2 + MnO_4^- \xrightarrow{H_2O} \underset{\substack{\text{OH OH}\\|\;\;\;|}}{R_2C{-}CR_2} + MnO_2 \qquad (23.27)$$

An alkene	Permanganate	A glycol	Manganese dioxide
	Purple	*Colorless*	*Brown*

$$R^1C{\equiv}CR^2 + MnO_4^- \xrightarrow{H_2O} R^1CO_2H + R^2CO_2H + MnO_2 \qquad (23.28)$$

An alkyne	Permanganate	Carboxylic acids	Manganese dioxide
	Purple	*Colorless*	*Brown*

EXPERIMENTAL PROCEDURE

BAEYER TEST FOR UNSATURATION

Safety Alert

Wear latex gloves to prevent solutions of potassium permanganate from contacting your skin. These solutions cause unsightly stains on your hands for several days. If this oxidant comes in contact with the skin, wash the affected area thoroughly with warm water.

Dissolve 1 or 2 drops of the unknown in 2 mL of 95% ethanol, and then add 0.1 *M* KMnO$_4$ solution dropwise; note the results. Count the number of drops added before the permanganate color persists. For a **blank determination**, count the number of drops of aqueous permanganate that can be added to 2 mL of 95% ethanol before the color persists. A significant difference in the number of drops required in the two cases is a positive test for unsaturation.

Finishing Touches

Filter all *solutions* to remove manganese dioxide. Place the *filter paper* and the *filter cake* containing manganese salts in the container for heavy metals. Flush all *filtrates* down the drain.

23.9   Alkyl Halides

CLASSIFICATION TESTS

Qualitative tests for alkyl halides are useful in deciding whether the compound in question is a primary, secondary, or tertiary halide. In general it is quite difficult to prepare solid derivatives of alkyl halides, so this discussion is limited to two qualitative tests: (a) reaction with **alcoholic silver nitrate** solution, and (b) reaction with **sodium iodide in acetone**.

A. Alcoholic Silver Nitrate. If a compound is known to contain bromine, chlorine, or iodine, information concerning the environment of the halogen is obtained from reacting the substance with alcoholic silver nitrate. The overall reaction is shown in Equation 23.29.

$$\underset{\substack{\text{An alkyl} \\ \text{halide}}}{\text{RX}} + \underset{\substack{\text{Silver} \\ \text{nitrate}}}{\text{AgNO}_3} \xrightarrow{\text{ethanol}} \underset{\substack{\text{A silver} \\ \text{halide}}}{\text{AgX}} + \underset{\substack{\text{An alkyl} \\ \text{nitrate}}}{\text{RONO}_2} \qquad (23.29)$$

Such a reaction mechanistically will be of the S$_N$1 type. As asserted in Section 14.2, tertiary halides are more reactive in an S$_N$1 reaction than are secondary halides, which are in turn more reactive than primary halides. Thus, the rate of precipitation of silver halide in this test is expected to correspond to the ease of ionization of the alkyl halide to a carbocation, that is, primary < secondary < tertiary. From a practical standpoint, these differences are best determined by testing authentic samples of primary, secondary, and tertiary halides with silver nitrate in separate test tubes and comparing the results.

Alkyl bromides and iodides react more rapidly than chlorides, and the latter may even require warming to produce a reaction in a reasonable time-period. Aryl halides are *unreactive* toward the test reagent, as in general are any vinyl or alkynyl halides. Allylic and benzylic halides, even when primary, show reactivities as great as or greater than tertiary halides because of resonance stabilization of the resulting allyl or benzyl carbocations.

EXPERIMENTAL PROCEDURE

SILVER NITRATE TEST FOR ALKYL HALIDES

Safety Alert

Avoid spilling alcoholic silver nitrate on your skin because staining may result. If you do come into contact with the solution, wash the affected area immediately with copious amounts of water.

Add 1 drop of the alkyl halide to 2 mL of a 0.1 *M* solution of silver nitrate in 95% ethanol. If no reaction is observed within 5 min at room temperature, warm the mixture in a beaker of boiling water and observe any change. Note the color of any precipitates: Silver chloride is white, but turns purple on exposure to light, silver bromide is pale yellow, and silver iodide is yellow. If there is any precipitate, add several drops of 1 *M* nitric acid solution to it, and note any changes; the silver halides are insoluble in acid. To determine expected reactivities, test known primary, secondary, and tertiary halides in this manner. If possible, use alkyl iodides, bromides, and chlorides so that differences in halogen reactivity can also be observed.

Finishing Touches

Neutralize and then filter all test mixtures and place the *filter cake* in a container for recovered silver halides. Flush the *filtrate* down the drain.

B. Sodium Iodide in Acetone. Another method for distinguishing between primary, secondary, and tertiary halides makes use of sodium iodide dissolved in acetone. This test complements the alcoholic silver nitrate test, and when these two tests are used together, it is possible to determine the gross structure of the attached alkyl group with reasonable accuracy.

The basis of this test is that both sodium chloride and sodium bromide are *not* very soluble in acetone, whereas sodium iodide is. Mechanistically, the reactions occurring are S_N2 substitutions in which iodide ion is the nucleophile (Eqs. 23.30 and 23.31); the order of reactivity is primary > secondary > tertiary.

$$RCl + NaI \xrightarrow{\text{acetone}} RI + \underline{NaCl} \tag{23.30}$$

$$RBr + NaI \xrightarrow{\text{acetone}} RI + \underline{NaBr} \tag{23.31}$$

With the test reagent, *primary bromides* give a precipitate of sodium bromide in about 3 min at room temperature, whereas the *primary* and *secondary chlorides* must be heated to about 50 °C before reaction occurs. *Secondary* and *tertiary bromides* react at 50 °C, but the *tertiary chlorides* fail to react in a reasonable time. This test is necessarily limited to bromides and chlorides.

EXPERIMENTAL PROCEDURE

SODIUM IODIDE TEST FOR ALKYL CHLORIDES AND BROMIDES

Place 1 mL of the sodium iodide–acetone test solution in a test tube, and add 2 drops of the chloro or bromo compound. If the compound is a solid, dissolve about 50 mg of it in a minimum volume of acetone and add this solution *to* the reagent. Shake the test tube, and allow it to stand for 3 min at room temperature. Note whether a precipitate forms; if no change occurs after 3 min, warm the mixture in a beaker of water at 50 °C. After 6 min of heating, cool the solution to room temperature and note whether any precipitate forms. Occasionally a precipitate forms immediately after combination of the reagents; this represents a positive test *only* if the precipitate remains after the mixture is shaken and allowed to stand for 3 min.

   Carry out this reaction with a series of primary, secondary, and tertiary chlorides and bromides. Note in all cases the differences in reactivity as evidenced by the rate of formation of sodium bromide or sodium chloride.

Finishing Touches   Place all *solutions* and *solids* involved in this test in the container for halogenated liquids, unless instructed to do otherwise.

23.10   Aromatic Hydrocarbons and Aryl Halides

CLASSIFICATION TEST

A. Chloroform and Aluminum Chloride. This test for the presence of an aromatic ring should be performed only on compounds that have been shown to be *insoluble* in concentrated sulfuric acid (Sec. 23.3). The test involves the Friedel-Crafts reaction (Sec. 15.2) between an aromatic compound and chloroform in the presence of anhydrous aluminum chloride as a catalyst. The colors produced in this type of reaction are often quite characteristic for certain aromatic compounds, whereas aliphatic compounds give little or no color with this test. Some typical examples are tabulated below. Often these colors change with time and ultimately yield brown-colored solutions.

Type of Compound	Color
Benzene and homologs	Orange to red
Aryl halides	Orange to red
Naphthalene	Blue
Biphenyl	Purple

   The chemistry of the test involves a series of alkylation reactions (Sec. 15.2); for benzene, the ultimate product is triphenylmethane (Eq. 23.32).

$$3 \ C_6H_6 \ + \ CHCl_3 \ \xrightarrow{\ AlCl_3\ } \ (C_6H_5)_3CH \ + \ 3 \ HCl \qquad (23.32)$$

Benzene   Chloroform   Triphenylmethane

The colors arise because species such as triphenylmethyl cations, $(C_6H_5)_3C^+$, form and remain in the solution as their $AlCl_4^-$ salts; ions of this sort are highly colored owing to the extensive delocalization of charge that is possible throughout the three aromatic rings.

The test is significant if positive, but a negative test does *not* rule out an aromatic structure; some compounds bearing electron-withdrawing groups are so unreactive that they do not readily undergo Friedel-Crafts alkylation reactions.

Positive tests for aryl halides are difficult to obtain directly, and some of the best evidence for their presence involves indirect methods. Elemental analysis indicates the presence of halogen. If *both* the silver nitrate and sodium iodide–acetone tests are negative, the compound is most likely a vinyl or an aromatic halide, both of which are very unreactive toward silver nitrate and sodium iodide. Distinction between a vinyl and an aromatic halide can be made by means of the aluminum chloride–chloroform test.

3–4 cm $AlCl_3$

FIGURE 23.5
Apparatus for subliming aluminum chloride.

EXPERIMENTAL PROCEDURE

FRIEDEL-CRAFTS REACTION FOR DETECTING ARENES

Safety Alert

Anhydrous aluminum chloride reacts vigorously with water, even the moisture on your hands, producing fumes of hydrogen chloride that are highly corrosive if inhaled. Do not allow aluminum chloride to come in contact with your skin. If it does, flush the affected area with copious amounts of water. Because aluminum chloride is a powdery solid that can easily become airborne, weigh and transfer it into the reaction flask *in the hood*.

In a *dry* Pyrex test tube held almost horizontally, heat about 100 mg of *anhydrous* aluminum chloride until the material has sublimed to 3 or 4 cm above the bottom of the tube (Fig. 23.5). Allow the tube to cool until it is almost comfortable to touch; then add about 20 mg of a solid, or 1 drop of a liquid down the side of the tube, followed by 2 or 3 drops of chloroform. The appearance of a bright color ranging from red to blue where the sample and chloroform come in contact with the aluminum chloride is a positive indication of an aromatic ring.

Finishing Touches

Cool the *contents of the test tube* in an ice-water bath and cautiously add a few milliliters of cold water to hydrolyze the aluminum salts. Flush the resulting *aqueous layer* down the drain and place the *organic layer* in the container for halogenated organic liquids.

DERIVATIVES

Two types of derivatives are used to characterize aromatic hydrocarbons and aryl halides. These are prepared by nitration (Eq. 23.33) and side-chain oxidation (Eq. 23.34).

$$\underset{\text{An aromatic hydrocarbon}}{\boxed{R}} \xrightarrow[\text{H}_2\text{SO}_4]{\text{HNO}_3} \underset{\text{A nitroaromatic}}{\boxed{R}-\text{NO}_2} \tag{23.33}$$

$$\underset{\substack{\text{An alkylated or}\\\text{acylated aromatic hydrocarbon}}}{\overset{\displaystyle \text{R(H)}}{\underset{\displaystyle \text{H}}{\text{ArC}-\text{R}}}} \xrightarrow[\text{or H}_2\text{CrO}_4]{\text{KMnO}_4/\text{base}} \underset{\text{A benzoic acid}}{\overset{\displaystyle\text{O}}{\text{Ar}\overset{\|}{\text{C}}-\text{OH}}} \tag{23.34}$$

B. Nitration. Some of the best solid derivatives of aryl halides are mono- and dinitration products. Two general procedures can be used for nitration.

EXPERIMENTAL PROCEDURES

PREPARATION OF NITROARENES

> **Safety Alert**
>
> 1. Use care whenever performing a nitration, whether it is with a known or an unknown substance, because many compounds react vigorously under typical nitration conditions.
> 2. Concentrated or fuming sulfuric and nitric acids are used in this procedure. These acids are very corrosive; handle them with care and wear latex gloves. If the acids contact your skin, immediately wash the affected area with copious amounts of water, followed by 5% sodium bicarbonate solution.

Method A.   *Note:* This method yields 1,3-dinitrobenzene from benzene or nitrobenzene and the 4-nitro derivative from chloro- or bromobenzene, benzyl chloride, or toluene; dinitro derivatives are obtained from phenol, acetanilide, naphthalene, and biphenyl.

Carry out the reaction in a large test tube or small Erlenmeyer flask. Add about 0.5 g of the compound to 2 mL of concentrated sulfuric acid. Add 2 mL of concentrated nitric acid dropwise to this mixture; stopper the container and shake the mixture after each addition. After the addition is complete, heat the mixture at 45 °C on a water bath for about 5 min. Then pour the reaction mixture onto 15 g of ice, and collect the precipitate by vacuum filtration. Recrystallize the solid from aqueous ethanol.

Method B.   *Note:* This method is the better one to use for nitrating halogenated benzenes because dinitration occurs and these compounds have higher melting points and are easier to purify than are the mononitration products obtained from *Method A*. The xylenes, mesitylene, and pseudocumene yield trinitro compounds using the present method.

Follow the procedure used for Method *A*, except use 2 mL of *fuming* nitric acid in place of the *concentrated* nitric acid and warm the mixture with a steam bath for 10 min. If little or no nitration occurs, substitute *fuming* sulfuric acid for the *concentrated* sulfuric acid. *Carry out this reaction in a hood.*

Neutralize all *acidic solutions* before flushing them down the drain.

Finishing
Touches

C. Side-Chain Oxidation. A second method for characterizing aromatic hydrocarbons involves oxidation of alkyl side chains to carboxylic acid groups, using either alkaline potassium permanganate or chromic acid solution (Eq. 23.34). Carbon-carbon as well as carbon-hydrogen bonds are cleaved in this process. Moreover, if more than one alkyl side chain is present, each will be oxidized to a carboxyl group, giving a polycarboxylic acid (Eq. 23.35). Identifying the acid provides information regarding the location of the alkyl side chains on the aromatic ring, and the acid itself, which usually will be a solid, serves as a derivative of the unknown.

$$
\begin{array}{c}
\underset{\text{(1,2-, 1,3-, or 1,4-)}}{R-\!\!\!\bigcirc\!\!\!-R} \xrightarrow[\substack{\text{or} \\ H_2CrO_4}]{\substack{(1)\ KMnO_4/HO^- \\ (2)\ \text{Neutralize}}} \underset{\text{(1,2-, 1,3-, or 1,4-)}}{HO_2C-\!\!\!\bigcirc\!\!\!-CO_2H}
\end{array}
\qquad (23.35)
$$

The mechanism of this type of oxidation is not well understood, but may involve benzyl radicals generated by action of the oxidant, followed by conversion to the benzylic alcohol (Eq. 23.36). The alcohol then undergoes further oxidation.

$$
\underset{\substack{\text{An aromatic} \\ \text{hydrocarbon}}}{\bigcirc\!\!-CHR_2} \xrightarrow[\text{or } Cr^{6+}]{Mn^{7+}} \underset{\substack{\text{A benzylic} \\ \text{alcohol}}}{\bigcirc\!\!-CR_2OH} \xrightarrow{\text{oxidant}} \underset{\substack{\text{An aromatic} \\ \text{aldehyde or ketone}}}{\bigcirc\!\!-C(O)R} \longrightarrow \underset{\substack{\text{An aromatic} \\ \text{carboxylic acid}}}{\bigcirc\!\!-C(O)OH}
\qquad (23.36)
$$

EXPERIMENTAL PROCEDURES

SIDE-CHAIN OXIDATION OF ARENES

1. Wear latex gloves to keep the acids from contacting your skin when you prepare and handle solutions of potassium permanganate or chromic acid. These solutions cause unsightly stains on your skin for several days, and the chromic acid–sulfuric acid solution may cause severe chemical burns. Wash the affected area thoroughly with

copious amounts of water if these oxidants come in contact with the skin. In the case of chromic acid, also rinse the area with 5% sodium bicarbonate solution.

2. The preparation of chromic acid requires diluting sulfuric acid with water. Be certain to add the acid *slowly* to the water and swirl the container to ensure continuous mixing. The dissolution of sulfuric acid in water generates heat, and when the acid is added to water, the heat is dispersed through warming of the water. Swirling prevents the layering of the denser sulfuric acid at the bottom of the flask and the attendant possibility that hot acid will be splattered by the steam generated when the two layers are suddenly mixed later by agitation.

1. Permanganate Method.   Add 0.5 g of the compound to a round-bottom flask containing a stirbar and a solution prepared from 80 mL of water and 2 g of potassium permanganate. Add 0.5 mL of 3 *M* sodium hydroxide solution, and heat the mixture under reflux until the purple color of permanganate ion has disappeared; this will normally take 0.5–3 h. At the end of the reflux period, cool the mixture and carefully acidify it with 3 *M* sulfuric acid. Now heat the mixture for an additional 30 min and cool it again; remove excess brown manganese dioxide, if any, by addition of sodium bisulfite solution. The bisulfite reduces the manganese dioxide to manganous ion, which is water-soluble. Collect the solid acid that remains by vacuum filtration. Recrystallize the acid from toluene or aqueous ethanol. If little or no solid acid is formed, this may be because the acid is somewhat water-soluble. In this case, extract the aqueous layer with small portions of diethyl ether or dichloromethane, dry the organic extracts over anhydrous sodium sulfate, decant the solution, and remove the organic solvent by means of a steam bath in the hood. Recrystallize the acid that remains. In this particular method, the presence of base during the oxidation often means that some silicic acid, derived from the glass of the reaction vessel, forms on acidification. Thus, purify the carboxylic acid by recrystallization prior to determining the melting point.

Finishing Touches

Place the *aqueous filtrate* containing manganese salts in the container for heavy metals. Flush all *ethanolic filtrates* down the drain. Put *filtrates* involving other organic solvents in the container for halogenated or nonhalogenated liquids, as appropriate.

2. Chromic Acid Method.   Dissolve 1 g of sodium dichromate in 3 mL of water contained in a round-bottom flask, add 0.5 g of the compound to be oxidized and a stirbar. Carefully add 1.5 mL of concentrated sulfuric acid to the mixture with mixing and cooling. Attach a reflux condenser to the flask, and heat the mixture gently until a reaction ensues; as soon as reaction begins, remove the heating source and cool the mixture if necessary to control the rate of boiling. After spontaneous boiling subsides, heat the mixture under reflux for 2 h. Pour the reaction mixture into 5 mL of water, and collect the precipitate by filtration. Transfer the solid into a flask, add 3 mL of 2 *M* sulfuric acid, and then warm the mixture on a steam cone with stirring. Cool the mixture, collect the precipitate, and wash it with about 3 mL of cold water. Dissolve the residue in 3 mL of 1.5 *M* sodium hydroxide solution and filter the solution. Add the filtrate, with stirring, to 6 mL of 2 *M* sulfuric acid. Collect the new precipitate, wash it with cold water, and recrystallize the carboxylic acid from either toluene or aqueous ethanol.

Finishing Touches

Add sodium sulfite to the *aqueous filtrate* of chromium salts in order to destroy excess Cr^{6+}. Make the solution slightly basic, to form chromium hydroxide, and isolate this salt by vacuum filtration through a layer of filter-aid. Place the *filter paper* and the *filter cake* in the container for heavy metals; flush the *filtrate* down the drain. Neutralize all other *aqueous filtrates* before flushing them down the drain. Put any *filtrates* containing toluene in the container for nonhalogenated organic liquids.

23.11 Alcohols

CLASSIFICATION TESTS

The tests for the presence of a hydroxy group not only detect this functionality but may also indicate whether it is attached to a primary, secondary, or tertiary carbon atom.

A. Chromic Acid in Acetone. This test may be used to detect the presence of a hydroxy group, provided that it is shown previously that the molecule does *not* contain an aldehyde function. The reactions and experimental procedures for this test are given in Section 23.7C. Chromic acid does *not* distinguish between primary and secondary alcohols because they *both* give a positive test; tertiary alcohols give a negative test.

B. Lucas Test. This test is used to distinguish among primary, secondary, and tertiary alcohols. The reagent is a mixture of concentrated hydrochloric acid and zinc chloride, which converts alcohols to the corresponding alkyl chlorides. With this reagent, primary alcohols give no appreciable reaction (Eq. 23.37), secondary alcohols react more rapidly (Eq. 23.38), and tertiary alcohols react very rapidly (Eq. 23.39). A positive test depends on the fact that the alcohol is soluble in the reagent, whereas the alkyl chloride is not; thus the formation of a second layer or an emulsion constitutes a positive test. The *solubility of the alcohol in the reagent places limitations on the utility of the test*, and in general only monofunctional alcohols with six or fewer carbon atoms, as well as polyfunctional alcohols, may be used successfully.

$$RCH_2\text{---}OH + HCl \xrightarrow{ZnCl_2} \text{No reaction} \qquad (23.37)$$
A 1° alcohol

$$R_2CH\text{---}OH + HCl \xrightarrow[\text{slow}]{ZnCl_2} R_2CH\text{---}Cl + H_2O \qquad (23.38)$$
A 2° alcohol   A 2° alkyl halide

$$R_3C\text{---}OH + HCl \xrightarrow[\text{fast}]{ZnCl_2} R_3C\text{---}Cl + H_2O \qquad (23.39)$$
A 3° alcohol   A 3° alkyl halide

The similarity should be noted between this reaction and the nucleophilic displacement reactions between alcohols and hydrohalic acids as discussed in Sections 14.3 and 14.4. In the Lucas test, the presence of zinc chloride, a Lewis acid, greatly increases the reactivity of alcohols toward hydrochloric acid.

EXPERIMENTAL PROCEDURE

LUCAS TEST FOR ALCOHOLS

> **Safety Alert**   The solution of hydrochloric acid and zinc chloride is highly acidic and may produce severe chemical burns. Wear latex gloves when handling solutions of these compounds. If this solution comes in contact with your skin, thoroughly wash the affected area with water. Also rinse the area with 5% sodium bicarbonate solution.

Add 5 mL of the hydrochloric acid–zinc chloride reagent (Lucas reagent) to about 0.5 mL of the compound in a test tube. Stopper the tube and shake it; allow the mixture to stand at room temperature. Try this test with known primary, secondary, and tertiary alcohols, and note the *time* required for the formation of an alkyl chloride, which appears either as a second layer or as an emulsion. Repeat the test with an unknown, and compare the result with the results from the knowns.

Finishing Touches   Use a Pasteur pipet to separate the layer of alkyl chloride, if formed, from the test reagent. Put the *alkyl chloride* in the container for halogenated liquids. Neutralize the *aqueous solutions* before flushing them down the drain.

C. Ceric Nitrate Test. Although this test is used primarily for detecting phenols, it can also be applied as a qualitative test for alcohols. Discussion about and experimental procedures for this test are given in Section 23.12.

DERIVATIVES

Two common derivatives of alcohols are urethanes and the benzoate esters; the former are best for primary and secondary alcohols, whereas the latter are useful for all types of alcohols.

D. Urethanes. When an alcohol is allowed to react with an aryl substituted isocyanate, ArN=C=O, addition of the alcohol to the carbon-nitrogen double bond occurs to give a urethane (Eq. 23.40). Some commonly used isocyanates are α-naphthyl (1-naphthyl), 4-nitrophenyl, and phenyl.

$$ArN{=}C{=}O \ + \ ROH \ \longrightarrow \ \underset{\text{A urethane}}{\underset{\text{ArNH}\diagup\overset{\displaystyle C}{}\diagdown\text{OR}}{}} \qquad (23.40)$$

$$\underset{\text{An aryl isocyanate}}{ArN{=}C{=}O} \ + \ \underset{\text{An alcohol}}{ROH}$$

A major side reaction involves hydrolysis of the isocyanate to an amine, which then reacts with more isocyanate to give a disubstituted urea (Eqs. 23.41 and 23.42). Because of their symmetry, these ureas are high-melting, and their presence makes purification of the desired urethane difficult. Therefore, take precautions to ensure that the alcohol is *anhydrous* when using this procedure. The method works best for water-insoluble alcohols because they are more easily obtained in anhydrous form.

$$\underset{\text{An aryl isocyanate}}{ArN{=}C{=}O} \ + \ H_2O \ \longrightarrow \ \underset{\substack{\text{A carbamic acid}\\\text{(unstable)}}}{ArNH\diagup\overset{\displaystyle C}{}\diagdown OH} \ \xrightarrow{-CO_2} \ \underset{\text{An arylamine}}{ArNH_2} \qquad (23.41)$$

$$\underset{\text{An aryl isocyanate}}{ArN{=}C{=}O} \ + \ \underset{\text{An arylamine}}{ArNH_2} \ \longrightarrow \ \underset{\text{A disubstituted urea}}{ArNH\diagup\overset{\displaystyle C}{}\diagdown NHAr} \qquad (23.42)$$

This type of derivative is also useful for phenols; the procedure given here has been generalized so that it can be used for both alcohols and phenols. Other derivatives of phenols are provided in Section 23.12.

EXPERIMENTAL PROCEDURE

PREPARATION OF URETHANES

Aryl isocyanates such as those used in the preparation of these derivatives are toxic. Take normal precautions in handling them. Immediately wash with soap and warm water any areas of skin that come in contact with isocyanates.

Safety Alert

Place 1 g of the *anhydrous* alcohol or phenol in a round-bottom flask, and add 0.5 mL of phenyl isocyanate or α-naphthyl isocyanate; be sure to recap the bottle of isocyanate tightly to minimize exposure of the reagent to atmospheric moisture. If you are preparing the derivative of a phenol, also add 2 or 3 drops of dry pyridine as a catalyst. Affix a drying tube to the flask and warm the reaction mixture for 5 min. Cool the mixture in an ice-water bath, and scratch the mixture with a stirring rod at the air-liquid interface to induce crystallization. Recrystallize the crude derivative from petroleum ether. *Note:* 1,3-Di(α-naphthyl)urea has mp 293 °C, and 1,3-diphenylurea (carbanilide) has mp 237 °C; if your product has one of these melting points, repeat the preparation, taking greater care to maintain anhydrous conditions.

Finishing Touches

Flush any *filtrates* that contain the starting isocyanate down the drain. Pour the *filtrates* from the recrystallization of the product into the container for nonhalogenated organic liquids. Put the *descicant* used in the drying tube in the container for nonhazardous solids.

E. 3,5-Dinitrobenzoates. The reaction between 3,5-dinitrobenzoyl chloride and an alcohol gives the corresponding ester (Eq. 23.43). This method is useful for primary, secondary, and tertiary alcohols, especially those that are water-soluble and thus are likely to contain traces of water.

$$\text{ROH} \ + \ \underset{\substack{\text{3,5-Dinitrobenzoyl}\\\text{chloride}}}{\text{Cl—C}} \xrightarrow{\text{pyridine}} \underset{\text{A 3,5-dinitrobenzoate}}{\text{RO—C}} \ + \ \text{HCl} \qquad \textbf{(23.43)}$$

An alcohol

EXPERIMENTAL PROCEDURES

PREPARATION OF 3,5-DINITROBENZOATES

Safety Alert

1. 3,5-Dinitrobenzoyl chloride releases hydrochloric acid upon hydrolysis, so avoid contact of this reagent with your skin. If contact occurs, wash the area with water and rinse it with 5% sodium bicarbonate solution. Wear latex gloves when handling this reagent.
2. Thionyl chloride is a lachrymator. Measure this material in a hood and do not inhale its vapors. Wear latex gloves when handling this reagent.

Method A. *Note:* 3,5-Dinitrobenzoyl chloride is reactive toward water; it should be used *immediately* after weighing. Take care to minimize its exposure to air and to keep the bottle tightly closed.

Mix 0.6 mL of the alcohol with about 0.2 g of 3,5-dinitrobenzoyl chloride and 0.2 mL of pyridine in a small round-bottom flask equipped with a stirbar and a reflux condenser. Gently heat the mixture under reflux for 30 min, although 15 min is sufficient for a primary alcohol. Cool the solution and add about 5 mL of 0.6 *M* aqueous sodium bicarbonate solution. Cool this solution in an ice-water bath, and collect the crude crystalline product. Recrystallize the product from aqueous ethanol, using a minimum volume of solvent.

Finishing Touches

Flush all *filtrates* down the drain.

Method B. Perform this reaction in the hood, if possible, or use one of the gas traps shown in Figure 2.65. Add 0.5 g of 3,5-dinitrobenzoic acid, 1.5 mL of thionyl chloride, and 1 drop of pyridine to a small round-bottom flask equipped with a stirbar and a reflux condenser. Heat the mixture under reflux until the acid has dissolved and then for an additional 10 min. The total reflux time should be about 30 min.

Equip the flask for vacuum distillation. Cool the receiving flask with an ice-salt bath, and attach the vacuum adapter to an aspirator by means of a safety trap such as that shown in Figure 2.40a. Evacuate the system and distil the excess thionyl chloride. When the thionyl chloride has been removed, cautiously release the vacuum. Tightly stopper the flask containing thionyl chloride. To the residue in the stillpot, which is 3,5-dinitrobenzoyl chloride, add 1 mL of the alcohol and 0.3 mL of pyridine. Fit the flask with a reflux condenser bearing a calcium chloride drying tube. Proceed with the period of reflux according to the procedure of *Method A.*

Finishing Touches

Flush all *filtrates* down the drain with copious amounts of water. Discard the excess *thionyl chloride* either by pouring it slowly down a drain *in the hood* while running water or by putting it in a container for recovered thionyl chloride. Spread the *dessicant* used in the drying tube on a tray in the hood and discard it in the container for nonhazardous solids after the volatiles are removed.

23.12 Phenols

CLASSIFICATION TESTS

Several tests can be used to detect the presence of a phenolic hydroxy group: (a) bromine water, (b) ceric ammonium nitrate reagent, and (c) ferric chloride solution. In addition to these, solubility tests give a preliminary indication of a phenol because phenols are soluble in 1.5 *M* sodium hydroxide solution, but generally are insoluble in 0.6 *M* sodium bicarbonate solution. Care must be exercised in interpreting solubility data, however, because phenols containing highly electronegative groups are stronger acids and *may* be soluble in 0.6 *M* sodium bicarbonate; examples are 2,4,6-tribromophenol and 2,4-dinitrophenol.

A. Bromine Water. Phenols usually are highly reactive toward electrophilic substitution and consequently are brominated readily by bromine water, as illustrated in Equation 23.44. The rate of bromination is much greater in water than in nonpolar organic solvents. Water, being more polar, increases the ionization of bromine and thus enhances the rate of ionic bromination. Although hydrogen bromide is liberated, it is not observed when water is used as the solvent.

$$\text{Phenol} + 3\,Br_2 \xrightarrow{\;H_2O\;} \text{2,4,6-Tribromophenol} + 3\,Br_2 \qquad (23.44)$$

Phenols are so reactive toward electrophilic aromatic substitution that all unsubstituted positions *ortho* and *para* to the hydroxy group are brominated. The brominated compounds so formed are often solids and may be used as derivatives (Sec. 23.12E). Aniline and substituted anilines are also very reactive toward bromine and react analogously; however, solubility tests are normally used to distinguish between anilines and phenols.

EXPERIMENTAL PROCEDURE

BROMINE WATER TEST FOR PHENOLS

Safety Alert

1. *Bromine is a hazardous chemical.* Do not breathe its vapors or allow it to contact your skin because it may cause serious chemical burns. Perform all operations involving the transfer of solutions of bromine in a hood; wear latex gloves when handling this chemical. If you get bromine on your skin, wash the area immediately with warm water and soak the skin in 0.6 *M* sodium thiosulfate solution for up to 3 h if the burn is particularly serious.
2. Do *not* use acetone to rinse glassware containing residual bromine. This prevents formation of α-bromoacetone, a severe lachrymator. Rather, follow the procedure described in FINISHING TOUCHES.

Prepare a 1% aqueous solution of the unknown. If necessary, add dilute sodium hydroxide solution dropwise to effect solution of the phenol. Add a saturated solution of bromine in water dropwise to this solution; continue that addition until the bromine color persists. Note how much bromine water is used and perform this experiment on phenol and aniline for purposes of comparison.

Finishing Touches

Add solid sodium thiosulfate to all *test solutions* to destroy residual bromine, then neutralize and filter them. Flush the *filtrates* down the drain and place the *filter paper* in the container for halogenated organic compounds.

B. Ceric Nitrate Test. Alcohols are capable of replacing nitrate ions in complex cerate anions, resulting in a change in color of the solution from yellow to red (Eq. 23.45). Phenols are oxidized by this reagent to give a brown to greenish-brown precipitate in aqueous solution; a red-to-brown solution is produced in 1,4-dioxane. Because of solubility problems, the alcohol or phenol should have no more than ten carbon atoms for this test. Aromatic amines are also oxidized by the reagent to give a colored solution or precipitate; consequently the presence of this functionality must be excluded before concluding that a positive test signals an alcohol or phenol moiety.

$$\underset{\substack{\text{Ceric ammonium nitrate}\\ \textit{Yellow}}}{(NH_4)_2Ce(NO_3)_6} + \underset{\substack{\text{An alcohol}}}{ROH} \longrightarrow \underset{\substack{\text{An alkoxy ceric}\\ \text{ammonium nitrate}\\ \textit{Red}}}{(NH_4)_2\overset{\overset{\displaystyle OR}{|}}{Ce}(NO_3)_5} + HNO_3 \quad \textbf{(23.45)}$$

EXPERIMENTAL PROCEDURE

CERIC NITRATE TEST FOR ALCOHOLS AND PHENOLS

Dissolve about 20 mg of a solid or 1 drop of a liquid unknown in 1–2 mL of water, and add 0.5 mL of the ceric ammonium nitrate reagent; shake the mixture and note the color. If the unknown is insoluble in water, dissolve it in 1 mL of 1,4-dioxane, and proceed as before.

Finishing Touches

Flush the *test solutions* down the drain.

C. Ferric Chloride Test. Most phenols and enols react with ferric chloride to give colored complexes. The colors vary, depending on the nature of the phenol or enol and also on the solvent, concentration, and time of observation. Some phenols not giving coloration in aqueous or alcoholic solution do so in chloroform solution, especially after addition of a drop of pyridine. The nature of the colored complexes is still uncertain; they may be ferric phenoxide salts that absorb visible light. The production of a color is typical of phenols and enols; however, many of them do *not* give colors, so a negative ferric chloride test must *not* be taken as significant without supporting information such as that available from the ceric ammonium nitrate and bromine water tests.

FERRIC CHLORIDE TEST FOR PHENOLS AND ENOLS

Dissolve 30 to 50 mg of the unknown compound in 1–2 mL of water, or a mixture of water and 95% ethanol if the compound is not water-soluble, and add several drops of a 0.2 M aqueous solution of ferric chloride. Most phenols produce red, blue, purple, or green coloration; enols give red, violet, or tan coloration.

Finishing Touches

Flush all *test solutions* down the drain.

DERIVATIVES

Two useful solid derivatives of phenols are α-naphthylurethanes and bromo compounds.

D. Urethanes. The preparation of urethanes is discussed in Section 23.11D. Although either a phenyl- or naphthyl-substituted urethane could be prepared, the majority of the derivatives reported are the α-naphthylurethanes, and they are generally the urethanes of choice.

E. Bromophenols. The high reactivity of phenols toward electrophilic aromatic bromination is discussed in Section 23.12A. This reaction can be used to prepare bromophenols, which may serve as solid derivatives.

PREPARATION OF BROMOPHENOLS

Safety Alert

1. *Bromine is a hazardous chemical.* Do not breathe its vapors or allow it to contact your skin because it may cause serious chemical burns. Perform all operations involving the transfer of solutions of bromine in a hood; wear latex gloves when handling this chemical. If you get bromine on your skin, wash the area immediately with warm water and soak the skin in 0.6 M sodium thiosulfate solution for up to 3 h if the burn is particularly serious.
2. Do *not* use acetone to rinse glassware containing residual bromine. This prevents formation of α-bromacetone, a severe lachrymator. Rather, follow the procedure described in FINISHING TOUCHES.

A stock solution for bromination in this preparation should be supplied; if not, prepare it by dissolving 1 g of potassium bromide in 6 mL of water and adding 1 g of bromine. Dissolve 0.5 g of the phenolic compound in water or 95% ethanol, and add the bromine-containing solution to it *dropwise*. Continue the addition until the reaction mixture begins to develop a yellow color, indicating the presence of excess bromine. Let the mixture stand for about 5 min; if the yellow coloration begins to fade, add another drop or two of the solution of bromine. Add 5 mL of water to the mixture and then a few drops of 0.5 *M* sodium bisulfite solution to destroy the excess bromine. Shake the mixture vigorously, and remove the solid derivative by vacuum filtration. If necessary, neutralize the solution with concentrated hydrochloric acid to promote precipitation of the derivative. Purify the solid by recrystallization from 95% ethanol or aqueous ethanol.

Flush all *filtrates* down the drain.

Finishing Touches

23.13 Carboxylic Acids

CLASSIFICATION TEST

One of the best qualitative tests for the carboxylic acid group is solubility in basic solutions. Carboxylic acids are soluble both in 1.5 *M* sodium hydroxide solution *and* in 0.6 *M* sodium bicarbonate solution, from which they are regenerated by acidification. Solubility properties are discussed in Section 23.3.

The relatively high acidity of carboxylic acids enables ready determination of the **equivalent weight** or **neutralization equivalent** of the acid by titration with standard base. The equivalent weight of an acid is that weight, in grams, of acid that reacts with one equivalent of base. As an example, suppose that 0.1000 g of an unknown acid requires 16.90 mL of 0.1000 *N* sodium hydroxide solution to be titrated to a phenolphthalein end point. This means that 0.1000 g of the acid corresponds to (16.90 mL)(0.1000 equivalent/1000 mL) or 0.0016901 equivalent of the acid, or that one equivalent of the acid weighs 0.1000/0.00169 or 59.201 g. Thus the following expression applies:

$$equivalent\ weight = \frac{grams\ of\ acid}{(volume\ of\ base\ consumed\ in\ liters)(N)}$$

where *N* is the *normality* of the standard base.

Because each carboxylic acid function in a molecule is titrated with base, the equivalent weight corresponds to the molecular weight of the acid divided by *n*, where *n* is the number of acid functions present in the molecule. Thus, for the example given, the molecular weight is 59.2 for a single acid function, 118.4 for two, and 177.6 for three. If the molecular weight of an unknown compound is known, then the number of acid groups in the molecule is calculated by dividing the molecular weight by the equivalent weight. Hence, if the molecular weight of the unknown compound is 118 and its equivalent weight is 59.2, the unknown must have *two* titratable acid functions.

EXPERIMENTAL PROCEDURE

DETERMINATION OF EQUIVALENT WEIGHT OF AN ACID

Accurately weigh about 0.2 g of the acid and dissolve it in 50 to 100 mL of water or 95% ethanol or a mixture of the two. It may be necessary to warm the mixture to dissolve the compound completely. Using phenolphthalein as the indicator, titrate the solution with a *standardized* sodium hydroxide solution having a concentration of about 0.1 M (Fig. 23.6). From the data obtained, calculate the equivalent weight.

Finishing Touches   Flush the *aqueous solution* down the drain.

DERIVATIVES

Three satisfactory solid derivatives of carboxylic acids are amides (Eq. 23.46), anilides (Eq. 23.47), and *p*-toluidides (Eq. 23.48). These derivatives are prepared by treating the corresponding acid chlorides with ammonia, aniline, or *p*-toluidine. The amides are generally less satisfactory derivatives than the other two because they tend to be more soluble in water and thus are harder to isolate. The acid chlorides are most conveniently prepared from the acid, or its salt, and thionyl chloride (Eq. 23.49).

$$RCOCl + 2\ NH_3 \xrightarrow{\text{cold}} RCONH_2 + NH_4Cl \qquad (23.46)$$

An acid
chloride   An amide

$$RCOCl + 2\ H_2NC_6H_5 \longrightarrow RCONHC_6H_5 + C_6H_5NH_3^+\ Cl^- \qquad (23.47)$$

Aniline   An anilide

$$RCOCl + 2\ H_2NAr \longrightarrow RCONHAr + Ar\overset{+}{N}H_3\ Cl^- \qquad (23.48)$$

p-Toluidine   A *p*-toluidide

$$Ar = \text{—}\langle \text{benzene ring} \rangle\text{—CH}_3$$

FIGURE 23.6
Apparatus for quantitative titration.

$$RCO_2H + SOCl_2 \longrightarrow RCOCl + SO_2 + HCl \qquad (23.49)$$

A carboxylic   Thionyl   An acid   Sulfur
acid   chloride   chloride   dioxide

EXPERIMENTAL PROCEDURES

PREPARATION OF AMIDES

1. Thionyl chloride is a lachrymator. Measure this material in a hood and do not inhale its vapors. Wear latex gloves when handling this chemical.
2. Acid chlorides release hydrochloric acid upon hydrolysis, so avoid contact of such reagents with your skin. If contact occurs, wash the area with water and rinse it with 5% sodium bicarbonate solution. Wear latex gloves when handling acid chlorides.

To prepare the acid chloride from a carboxylic acid, place the acid in a small round-bottom flask with 1 mL of thionyl chloride and 5 drops of *N,N*-dimethylformamide (DMF). Add a stirbar and attach a calcium chloride drying tube *directly* to the flask. Warm the flask at 60–65 °C with stirring *in the hood*. Bubbling or fuming usually begins shortly after the addition of the DMF. The reaction is sufficiently complete when the bubbling greatly slows, which typically occurs within 30 min. Use the mixture containing the acid chloride to make the amide, anilides, or *p*-toluidide according to the procedures that follow.

A. Amides. Use 0.5 g of the acid to prepare the acid chloride. *At the hood* pour the mixture containing the acid chloride and unchanged thionyl chloride into 15 mL of *ice-cold*, concentrated ammonium hydroxide solution. Be very careful when performing this addition because the reaction is quite exothermic. Collect the precipitated amide derivative by vacuum filtration, and recrystallize the crude product from water or aqueous ethanol.

Neutralize all *filtrates* and then flush them down the drain.

Finishing Touches

B. Anilides and p-Toluidides. Use 0.5 g of the carboxylic acid to prepare the acid chloride. After the heating period, cool the reaction mixture to room temperature. Dissolve 0.5 g of either aniline or *p*-toluidine in 15 mL of cyclohexane; slight warming may be necessary to effect complete solution. Pour the acid chloride into the cyclohexane solution of the amine, and heat the resulting mixture for 2–3 min. A heavy white precipitate of the amine hydrochloride forms; isolate this precipitate by vacuum filtration and set it aside, but do *not* discard it. Transfer the filtrate to a separatory funnel and wash it sequentially with 3-mL portions of water, 1.5 *M* HCl, 1.5 *M* NaOH, and again with water. In some cases precipitation may occur in the organic layer during one or more of these washings. If this occurs, warm the solution gently with a warm-water bath to redissolve the precipitate.

Following the sequential washings, remove the cyclohexane from the organic layer either by distillation or using one of the other techniques described in Section 2.29. Recrystallize the derivative of the carboxylic acid from aqueous ethanol.

If little residue remains after evaporating the cyclohexane, dissolve the precipitate that was removed earlier in about 10 mL of water. Stir the mixture and then remove any undissolved solid by vacuum filtration; any such solid is the desired derivative, which became entrapped in the precipitated amine hydrochloride. Combine this material with the residue obtained from cyclohexane.

Finishing Touches

Neutralize all *filtrates* and then flush them down the drain. Put any recovered *cyclohexane* into the container for nonhalogenated organic liquids.

23.14 Amines

CLASSIFICATION TESTS

Two common qualitative tests for amines are the Hinsberg test and the nitrous acid test. The nitrous acid test is not included here because the *N*-nitroso derivatives of some secondary amines are carcinogenic. The risk of producing an as-yet-unrecognized carcinogenic material in this procedure outweighs any possible benefit of a test that can also be misleading and difficult to interpret. The modified sodium nitroprusside test has been included as an alternative.

A. Hinsberg Test. The reaction between primary or secondary amines and benzenesulfonyl chloride (Eqs. 23.50 and 23.52, respectively) yields the corresponding substituted benzenesulfonamide. The reaction is performed in excess aqueous base; if the amine is primary, the sulfonamide, which has an acidic amido hydrogen, is converted by base (Eq. 23.51) to the potassium salt, which is normally soluble in the medium. With few exceptions, which are discussed in the next paragraph, primary amines react with benzenesulfonyl chloride to provide *homogeneous* reaction mixtures. Acidifying this solution regenerates the insoluble primary benzenesulfonamide. On the other hand, the benzenesulfonamides of secondary amines bear no acidic amido hydrogens and thus typically are insoluble in *both* acid and base. Therefore, secondary amines react to yield *heterogeneous* reaction mixtures, with formation of either an oily organic layer or a solid precipitate.

$$R\ddot{N}H_2 + \langle\!\!\!\!\bigcirc\!\!\!\!\rangle - SO_2Cl \xrightarrow{KOH} \langle\!\!\!\!\bigcirc\!\!\!\!\rangle - SO_2\ddot{N}HR + KCl + H_2O \qquad (23.50)$$

A 1° amine   Benzenesulfonyl chloride

A benzenesulfonamide
Insoluble in water

$$\langle\!\!\!\!\bigcirc\!\!\!\!\rangle - SO_2\overset{H}{\underset{|}{N}}R \underset[\text{HCl}]{\overset{\text{excess KOH}}{\overset{\longrightarrow}{\underset{\longleftarrow}{\text{excess}}}}} \langle\!\!\!\!\bigcirc\!\!\!\!\rangle - SO_2\bar{N}R\ K^+ + H_2O \qquad (23.51)$$

A potassium benzenesulfonamide
Soluble in water

$$R_2\ddot{N}H + \left\langle\bigcirc\right\rangle - SO_2Cl \xrightarrow{KOH} \left\langle\bigcirc\right\rangle - SO_2\dot{N}R + KCl + H_2O \quad \textbf{(23.52)}$$

A 2° amine   Benzenesulfonyl   A benzenesulfonamide
           chloride   *Insoluble in water*

$\downarrow$ excess KOH

No reaction

The distinction between primary and secondary amines thus depends on the different solubility properties of their benzenesulfonamide derivatives. However, the potassium salts of *certain* primary sulfonamides are not completely soluble in basic solution. Examples usually involve primary amines of higher molecular weight and those having cyclic alkyl groups. To avoid confusion and possible misassignment of a primary amine as secondary, the basic solution is separated from the oil or solid and acidified. The formation of an oil or a precipitate indicates that the derivative is partially soluble and that the amine is primary. It is important not to overacidify the solution because this may precipitate certain side products, resulting in an ambiguous test. The original oil or solid should be tested for solubility in water and acid to substantiate the test for a primary or a secondary amine.

Tertiary amines behave somewhat differently. Under the conditions of the Hinsberg test, the processes shown in Equation 23.53 typically provide for converting benzenesulfonyl chloride to potassium benzenesulfonate with recovery of the tertiary amine. Tertiary amines are nearly always insoluble in the aqueous potassium hydroxide solution, so the test mixture remains heterogeneous. Proof that the nonaqueous layer is a tertiary amine can generally be obtained simply by noting the relative densities of the oil layer and the aqueous test solution: Tertiary amines are less dense than the solution, whereas benzenesulfonamides are generally more dense. Further support for the conclusion that the oil is a tertiary amine is available from testing its solubility in aqueous acid; solubility usually indicates a tertiary amine.

$$R_3\ddot{N} + \left\langle\bigcirc\right\rangle - SO_2Cl \longrightarrow \left\langle\bigcirc\right\rangle - SO_2 - \overset{+}{N}R_3\ Cl^-$$
23

A 3° amine   Benzenesulfonyl
           chloride

$\downarrow$ KOH

$$\left\langle\bigcirc\right\rangle - SO_3^-K^+ + \ddot{N}R_3 + H_2O \quad \textbf{(23.53)}$$
Potassium benzenesulfonate

$$\left\langle\bigcirc\right\rangle - SO_2\overset{+}{N}R_3Cl^- \xrightarrow{R_3\ddot{N}} \left\langle\bigcirc\right\rangle - SO_2\ddot{N}R_2 + NR_4^+\ Cl^- \quad \textbf{(23.54)}$$
23   A quaternary
            A benzenesulfonamide   ammonium chloride

The procedure for the Hinsberg test must be followed as closely as possible because it is designed to minimize complications that arise because of side reactions that may occur between tertiary amines and benzenesulfonyl chloride. For example, **23**, the initial adduct between the chloride and the amine, is subject to an S_N2 reaction with another molecule of amine to produce the benzenesulfonamide of a *secondary* amine and a quaternary ammonium salt (Eq. 23.54). Although this process is normally unimportant, particularly when excess amine is avoided, the observation of *small* amounts of an insoluble product formed by this pathway may erroneously cause designation of a tertiary amine as secondary.

Formation of adduct **23** is usually less of a problem with tertiary *aryl*amines because of their lesser nucleophilicity and lower solubility as compared to *trialkyl*amines. Consequently, the competing hydrolysis of benzenesulfonyl chloride by hydroxide ion (Eq. 23.55) allows recovery of most of the amine. However, this class of amines is often subject to other side reactions that produce a complex mixture of mainly insoluble products (Eq. 23.56). The ambiguity caused by side reactions of tertiary arylamines is minimized by keeping the reaction time short and the temperature low.

$$\langle \text{Ph} \rangle - SO_2Cl \xrightarrow{\text{KOH}} \langle \text{Ph} \rangle - SO_3^- \ K^+ + KCl + H_2O \qquad (23.55)$$

$$\xrightarrow[\text{ArNR}_2]{} \text{Complex mixture including} \quad \langle \text{Ph} \rangle - SO_2 - \ddot{N}RAr \qquad (23.56)$$

A further complication with tertiary amines is that they often contain quantities of secondary amines as impurities. If it is not possible to obtain a reliable boiling point or if the amine is not carefully distilled, such contaminants may be present and lead to formation of small quantities of precipitate, making the test results ambiguous.

To summarize the discussion, tertiary amines may produce small amounts of insoluble products if the concentration of the amine in the test solution is too high and if the reaction time is too long. If the directions of the procedure are followed and *care is taken not to interpret small amounts of insoluble product as a positive test for secondary amines*, the Hinsberg test may be used with confidence to designate an amine as primary, secondary, or tertiary.

EXPERIMENTAL PROCEDURE

HINSBERG TEST FOR AMINES

Safety Alert Benzenesulfonyl chloride is a lachrymator and produces strong acids upon hydrolysis. Wear latex gloves when handling this reagent, measure it out in a hood, and avoid inhaling its vapors. Should this reagent come in contact with your skin, wash the area with water and rinse it with 5% sodium bicarbonate solution.

Mix 5 mL of 2 *M* aqueous potassium hydroxide, 0.2 mL (5 drops) or 0.2 g of the amine, and 0.7 mL (15 drops) of benzenesulfonyl chloride in a test tube. Stopper the tube and shake the mixture *vigorously*, with cooling if necessary, until the odor of benzenesulfonyl chloride is gone (*caution!*). In even the slowest case this should take no more than about 5 min. Test the solution to see that the mixture is still basic; if it is not, add sufficient 2 *M* potassium hydroxide solution dropwise until it is.

If the mixture forms two layers or a precipitate, note the relative densities, and separate the oil or solid by decantation or filtration. Test any oil for solubility in 0.6 *M* hydrochloric acid. The sulfonamide of a secondary amine is insoluble, whereas an amine is at least partially soluble. A solubility test indicating the presence of an amine could be caused by either a tertiary amine or secondary amines that for steric reasons react very slowly with benzenesulfonyl chloride.

Test any solid for solubility in water and in dilute acid. The potassium salt of a water-insoluble sulfonamide is usually soluble in water; acidifying the salt regenerates the sulfonamide, which is insoluble in aqueous acid. A solid sulfonamide of a secondary amine is insoluble in both water and acid. Acidify the solution from the original reaction mixture to pH 4 as signaled by pHydrion paper or a few drops of Congo red indicator solution; the formation of a precipitate or oil indicates a primary amine.

If the original mixture from the Hinsberg test does not form two layers, a primary amine is indicated. Acidify the solution to pH 4; a sulfonamide of a primary amine will either separate as an oil or precipitate as a solid.

Finishing Touches

Neutralize all *test solutions*. Isolate any solids by filtration and any oils with the aid of a Pasteur pipet. Flush the *filtrates* down the drain and put any *filter cakes* or *oils* in the container for nonhalogenated organic compounds.

B. Sodium Nitroprusside Tests. Two color tests to distinguish primary and secondary *aliphatic* amines have been available for many years, although they have not been widely used. More recently, these tests have been extended to primary, secondary, and tertiary *aromatic* amines. No attempt is made here to explain the complex color-forming reactions that occur. However, they most likely involve the reaction of the amine with either acetone, the Ramini test, or acetaldehyde, the Simon test, and the interaction of the resulting products with sodium nitroprusside to form colored complexes.

To apply these tests to an unknown amine, the *conventional* Ramini or Simon tests are first performed. These will give positive results in the cases of primary and secondary aliphatic amines. If these tests are negative and an aromatic amine is suspected, the *modified* versions of these tests are then performed. Reference to Figure 23.7 is helpful for interpreting the results of these tests.

	1° Aliphatic	2° Aliphatic	1° Aromatic	2° Aromatic	3° Aromatic
Ramini	Deep red	Deep red			
Simon	Pale yellow to red-brown	Deep blue			
Modified Ramini			Orange-red to red-brown	Orange-red to red-brown	Green
Modified Simon			Orange-red to red-brown	Purple	Usually green

FIGURE 23.7
Colors formed in the Ramini and the Simon tests.

EXPERIMENTAL PROCEDURES

RAMINI AND SIMON TESTS FOR AMINES

Conventional Tests.   Prepare the sodium nitroprusside reagent for use in both the conventional Ramini and Simon tests by dissolving 0.4 g of sodium nitroprusside ($Na_2[Fe(NO)(CN)_5] \cdot 2 H_2O$ in 10 mL of 50% aqueous methanol.

Ramini Test.   To 1 mL of the sodium nitroprusside reagent, add 1 mL of water, 0.2 mL (5 drops) of acetone, and about 30 mg of an amine. In most cases the characteristic colors given in Figure 23.7 appear in a few seconds, although in some instances up to about 2 min may be necessary.

Simon Test.   To 1 mL of the sodium nitroprusside reagent, add 1 mL of water, 0.2 mL (5 drops) of 2.5 *M* aqueous acetaldehyde solution, and about 30 mg of an amine. As in the Ramini test, color formation normally occurs in a few seconds, although times up to 2 min are occasionally necessary.

Modified Tests.   Prepare the reagent for use in both the *modified* Ramini and Simon tests by dissolving 0.4 g of sodium nitroprusside in a solution containing 8 mL of dimethyl sulfoxide and 2 mL of water.

Modified Ramini Test.   To 1 mL of the *modified* sodium nitroprusside reagent, sequentially add 1 mL of saturated aqueous zinc chloride solution, 0.2 mL (5 drops) of acetone, and about 30 mg of an amine. Primary and secondary aromatic amines provide orange-red to red-brown colors within a period of a few seconds to 5 min. Tertiary aromatic amines give a color that changes from orange-red to green over a period of about 5 min.

Modified Simon Test.   To 1 mL of the *modified* sodium nitroprusside reagent, sequentially add 1 mL of saturated aqueous zinc chloride solution, 0.2 mL (5 drops) of 2.5 *M* aqueous acetaldehyde solution, and about 30 mg of an amine. Primary aromatic amines give an orange-red to red-brown color within 5 min; secondary aromatic amines give a color changing from red to purple within 5 min; tertiary aromatic amines give a color that changes from orange-red to green over a period of 5 min.

Finishing Touches | Flush the *test solutions* down the drain.

DERIVATIVES

Suitable derivatives of primary and secondary amines are benzamides and benzenesulfonamides (Eq. 23.57 and 23.58, respectively).

$$R\ddot{N}H_2 \ \text{ or } \ R_2\ddot{N}H \ + \ C_6H_5COCl \xrightarrow{\text{Pyridine}} \ C_6H_5\overset{O}{\overset{\|}{C}}\text{—}\ddot{N}HR \ \text{ or } \ C_6H_5\overset{O}{\overset{\|}{C}}\text{—}\ddot{N}R_2 \qquad (23.57)$$
A 1° amine   A 2° amine   Benzoyl chloride   A benzamide

$$\text{R\ddot{N}H}_2 \text{ or } \text{R}_2\ddot{N}\text{H} + \text{C}_6\text{H}_5\text{SO}_2\text{Cl} \longrightarrow \text{C}_6\text{H}_5\text{SO}_2\text{—\ddot{N}HR} \text{ or } \text{C}_6\text{H}_5\text{SO}_2\text{—\ddot{N}R}_2 \quad \textbf{(23.58)}$$

A 1° amine   A 2° amine   Benzenesulfonyl chloride   A benzenesulfonamide

These transformations are satisfactory for derivatizing most primary and secondary amines, but tertiary amines do not undergo such reactions. However, such amines form salts that constitute solid derivatives. Thus a useful crystalline salt is formed upon reaction with methyl iodide to afford a methiodide (Eq. 23.59).

$$\text{R}_3\text{N:} + \text{CH}_3\text{I} \longrightarrow \text{R}_3\overset{+}{\text{N}}\text{CH}_3\text{ I}^- \quad \textbf{(23.59)}$$

A 3° amide   Methyl iodine   A methiodide

EXPERIMENTAL PROCEDURES

PREPARATION OF BENZAMIDES, BENZENESULFONAMIDES, AND METHIODIDES

Safety Alert

Benzoyl chloride is a lachrymator and produces hydrochloric and benzoic acids upon hydrolysis. Wear latex gloves when handling this reagent, measure it out in a hood, and avoid inhaling its vapors. Should this reagent come in contact with your skin, wash the area with water and rinse it with 5% sodium bicarbonate solution.

A. Benzamides.   In a small round-bottom flask containing a stirbar dissolve 0.3 g of the amine in 3 mL of dry pyridine. *Slowly* add 0.3 mL of benzoyl chloride to this solution. Affix a drying tube to the flask and heat the reaction mixture to 60–70 °C for 30 min; then pour the mixture into 25 mL with stirring at water. If the solid derivative precipitates at this time, isolate it by vacuum filtration, and when it is nearly dry, dissolve it in 10 mL of diethyl ether. If no precipitate forms, extract the aqueous mixture twice with 5-mL portions of diethyl ether. Combine the extracts. Wash the ethereal solution sequentially with 5-mL portions of water, 1.5 *M* HCl, and 0.6 *M* sodium bicarbonate solution. Dry the ethereal layer over anhydrous sodium sulfate, filter or decant the dried solution, and remove the diethyl ether by one of the techniques described in Section 2.29. Recrystallize the solid derivative from one of the following solvents: cyclohexane-hexane mixtures, cyclohexane-ethyl acetate mixtures, 95% ethanol, or aqueous ethanol.

Finishing Touches

Transfer any recovered *diethyl ether* into the container for nonhalogenated organic liquids. Flush all *aqueous* and *ethanolic solutions* down the drain, and pour any other *filtrates* containing organic solvents into the container for nonhalogenated liquids. Spread the descicant used in the drying tube on a tray in the hood and discard it in the container for nonhazardous solids after the volatiles are removed.

B. Benzenesulfonamides.   The method for preparing the benzenesulfonamides is discussed under the Hinsberg Test (Sec. 23.14A). The derivatives may be prepared using that method, but sufficient amounts of material should be used so that the final product can be purified by recrystallization from 95% ethanol. If the derivative is obtained as an oil, it *may* be induced to crystallize by scratching at the liquid-liquid or air-liquid interface with a stirring rod. If the oil cannot be made to crystallize, separate it, dissolve it in a minimum quantity of hot ethanol, and allow the solution to cool. Note that some amines do not give *solid* benzenesulfonamide derivatives.

Safety Alert	Methyl iodide is a powerful alkylating agent. Avoid inhaling its vapors or contact of this reagent with your skin. Wear latex gloves when handling it.

C. Methiodides.   Mix 0.3 g of the amine with 0.3 mL of methyl iodide in a test tube and warm the mixture for several minutes. Cool the test tube in an ice-water bath. If necessary, scratch at the air-liquid interface with a glass rod to induce crystallization. Isolate the product by vacuum filtration and purify it by recrystallization from absolute ethanol or methanol or from ethyl acetate.

Finishing Touches	Put the *filtrate* from the reaction mixture in the container for halogenated liquids and flush all other *filtrates* down the drain.

23.15   Nitro Compounds

CLASSIFICATION TEST

Ferrous Hydroxide Test. Organic compounds that are oxidizing agents will oxidize ferrous hydroxide (blue) to ferric hydroxide (brown). The most common organic compounds that function in this way are aliphatic and aromatic *nitro* compounds, which are reduced to amines by the reaction (Eq. 23.60). Other less common types of compounds that give the same test are nitroso compounds, hydroxylamines, alkyl nitrates, alkyl nitrites, and quinones.

$$RNO_2 + Fe(OH)_2 + 4\ H_2O \longrightarrow RNH_2 + 6\ Fe(OH)_3 \qquad (23.60)$$

A nitro compound   Ferrous hydroxide *Blue*   A 1° amine   Ferric hydroxide *Brown*

EXPERIMENTAL PROCEDURE

HYDROXIDE TEST FOR NITRO COMPOUNDS

In a small test tube, mix about 20 mg of a solid or 1 drop of a liquid unknown with 1.5 mL of freshly prepared 5% ferrous ammonium sulfate solution. Add 1 drop of 3 *M* sulfuric acid and 1 mL of 2 *M* potassium hydroxide in *methanol*. Stopper the tube immediately and shake it. A positive test is indicated by the blue precipitate turning rust-brown within 1 min. A slight darkening or greenish coloration of the blue precipitate is *not* considered a positive test.

Finishing Touches

Filter the *test mixture* and flush the *filtrates* down the drain. Put the *filter cake* in the container for nonhazardous solids.

DERIVATIVES

Two different types of derivatives of nitro compounds may be prepared. Aromatic nitro compounds can be di- and trinitrated with nitric acid and sulfuric acid. Discussion of and procedures for nitration are given in Section 23.10.

The other method for preparation of a derivative may be utilized for both aliphatic and aromatic nitro compounds. This involves the reduction of the nitro compound to the corresponding primary amine (Eq. 23.61), followed by conversion of the amine to a benzamide or benzenesulfonamide, as described in Section 23.14. The reduction is most often carried out with tin and hydrochloric acid.

$$\underset{\text{A nitro compound}}{\text{RNO}_2 \text{ or ArNO}_2} \xrightarrow[\text{(2) NaOH}]{\text{(1) Sn, HCl}} \underset{\text{A 1° amine}}{\text{R}\ddot{\text{N}}\text{H}_2 \text{ or Ar}\ddot{\text{N}}\text{H}_2} \qquad \textbf{(23.61)}$$

EXPERIMENTAL PROCEDURE

REDUCTION OF NITRO COMPOUNDS

Combine 1 g of the nitro compound and 2 g of granulated tin in a small round-bottom flask containing a stirbar. Attach a reflux condenser, and add 20 mL of 3 *M* hydrochloric acid in small portions to the stirred mixture through the condenser. If the nitro compound is insoluble in the aqueous medium, add 5 mL of 95% ethanol to increase its solubility.

After the addition is complete, warm the mixture with stirring for 10 min at 70–80 °C. Decant the warm, homogeneous solution into 10 mL of water, and add sufficient 12 *M* sodium hydroxide solution to dissolve the tin hydroxide completely. Extract the basic solution with several 10-mL portions of diethyl ether. Dry the ethereal solution over potassium hydroxide pellets, decant the liquid, and remove the ether by one of the techniques described in Section 2.29.

The residue contains the primary amine. Convert it to one of the derivatives described in Section 23.14.

Finishing Touches

Transfer any recovered *diethyl ether* into the container for nonhalogenated organic liquids. Neutralize the *aqueous solution* before flushing it down the drain.

23.16   Esters

CLASSIFICATION TESTS

Two tests for the ester functionality are the hydroxylamine test, a color reaction, and the saponification equivalent. The latter procedure can provide *quantitative* information regarding the number of ester groups present.

A. Hydroxylamine. This test for an ester group involves the use of hydroxylamine and ferric chloride. The former converts the ester to a hydroxamic acid, which then complexes with Fe^{3+} to give a colored species (Eqs. 23.62 and 23.63).

$$R^1 \overset{\overset{O}{\|}}{-C} -OR^2 \; + \; H_2\ddot{N}OH \; \longrightarrow \; R^1 \overset{\overset{O}{\|}}{-C} -\ddot{N}HOH \; + \; R^2OH \qquad (23.62)$$

An ester   Hydroxylamine   A hydroxamic acid   An alcohol

$$R^1 \overset{\overset{O}{\|}}{-C} -\ddot{N}HOH \; + \; FeCl_3 \; \longrightarrow \; \left[R^1 -C \begin{smallmatrix} O \\ \\ \ddot{N}-O \\ | \\ H \end{smallmatrix} \right]_3 Fe \; + \; 3\,HCl \qquad (23.63)$$

An iron complex
Colored

All carboxylic acid esters, including polyesters and lactones, give magenta colors that vary in intensity depending on other structural features of the molecule. Acid chlorides and anhydrides also give positive tests. Formic acid produces a red color, but other free acids give negative tests. Primary or secondary aliphatic nitro compounds give a positive test because ferric chloride reacts with the *aci* form, which is equivalent to the enol form of a ketone, that is present in basic solution. Most imides also provide positive tests. Some amides, but not all, produce light magenta coloration, whereas most nitriles give a negative test. A modification of the following procedure provides a positive test for nitriles and amides, and the details of it are provided in Section 23.17.

EXPERIMENTAL PROCEDURE

HYDROXYLAMINE TEST FOR ESTERS

Hydroxylamine and its hydrochloride may have mutagenic effects. Wear latex gloves when handling them. Should either of these substances contact your skin, wash the affected area with copious amounts of water.

Before the final test is performed, it is necessary to run a preliminary test as a "blank" or "control" because some compounds produce a color in the absence of hydroxylamine hydrochloride even though they do *not* contain an ester linkage.

Preliminary Test.   Mix 1 mL of 95% ethanol and 50–100 mg of the compound to be tested, and add 1 mL of 1 *M* hydrochloric acid. Note the color that is produced when 1 drop of 0.6 *M* aqueous ferric chloride solution is added. If the color is orange, red, blue, or violet, the following test for the ester group does *not* apply and *cannot* be used.

Final Test.   Mix 40–50 mg of the unknown, 1 mL of 0.5 *M* hydroxylamine hydrochloride in 95% ethanol, and 0.2 mL of 6 *M* sodium hydroxide in a test tube. Heat the mixture to boiling, and after cooling it slightly, add 2 mL of hydrochloric acid. If the solution is cloudy, add about 2 mL more of 95% ethanol. Add 1 drop of 0.6 *M* ferric chloride, and observe any color. If needed, add more ferric chloride solution until the color persists. Compare the color obtained here with that from the preliminary test. If the color is burgundy or magenta, as compared to the yellow color in the preliminary experiment, the presence of an ester group is indicated.

Neutralize the *aqueous solutions* before flushing them down the drain.

Finishing Touches

B. Saponification Equivalent. It is possible to carry out the hydrolysis of an ester with alkali in a *quantitative* manner so that the **saponification equivalent, SE**, results. This value is analogous to the equivalent weight of an acid (Sec. 23.13) in that it is the molecular weight of the ester divided by the number of ester functions in the molecule. Therefore the SE is the number of grams of ester required to react with one gram-equivalent of alkali.

The SE of an ester is determined by hydrolyzing a weighed amount of the ester with standardized alkali and then titrating the excess alkali to a phenolphthalein endpoint using standardized hydrochloric acid. The SE is then calculated as follows:

$$Saponification\ equivalent = \frac{grams\ of\ ester}{equivalents\ of\ alkali\ consumed}$$

$$= \frac{grams\ of\ ester}{(volume\ of\ alkali\ in\ liters)(N) - (volume\ of\ acid\ in\ liters)(N')}$$

where N is the normality of the standard base and N' is the normality of the standard acid.

EXPERIMENTAL PROCEDURE

DETERMINATION OF SAPONIFICATION EQUIVALENT

Dissolve approximately 3 g of potassium hydroxide in 60 mL of 95% ethanol. Allow the small amount of insoluble material, if any, to settle to the bottom of the container, and by decantation fill a 50-mL buret with the clear supernatant. Measure 25.0 mL of the alcoholic solution into each of two round-bottom flasks. Quantitatively transfer an *accurately* weighed 0.3- to 0.4-g sample of pure, dry ester into one of the flasks; the other basic solution is used for a blank determination. Equip each flask with a stirbar and a reflux condenser.

Heat the stirred solutions in both flasks under gentle reflux for 1 h. When the flasks have cooled, rinse each condenser with about 10 mL of distilled water, catching the rinse water in the flask. Add a drop or two of phenolphthalein, and separately titrate the solutions in each flask with *standardized* hydrochloric acid that is approximately 0.5 *M* in concentration.

The difference in the volumes of hydrochloric acid required to neutralize the base in the flask containing the sample and in the flask containing the blank corresponds to the amount of potassium hydroxide that reacted with the ester. The volume difference, in milliliters, multiplied by the molarity of the hydrochloric acid equals the number of *milli*moles of potassium hydroxide consumed. Using the titration data, calculate the saponification equivalent of the unknown ester.

The ester may not completely saponify in the allotted time, as evidenced by a *non*-homogeneous solution; if this is the case, heat the mixture under reflux for a longer period of time (2–4 h). Higher temperatures may be required in some cases; if so, diethylene glycol must be used as a solvent *in place of* the original 60 mL of 95% ethanol.

Finishing Touches

Flush the *neutralized solution* down the drain.

DERIVATIVES

To characterize an ester completely, it is necessary to prepare solid derivatives of both the acid *and* the alcohol components. However, isolating both of these components in pure form so that suitable derivatives can be prepared may present problems. One way to do the isolation is to perform the ester hydrolysis in base (Eq. 23.64) in a high-boiling solvent. If the alcohol is low-boiling, it is distilled from the reaction mixture and characterized. The acid that remains in the mixture as a carboxylate salt is obtained upon acidification of the solution. Derivatives of alcohols and acids are discussed in Sections 23.11 and 23.13, respectively.

$$R^1CO_2R^2 + HO^- \longrightarrow R^1CO_2^- + R^2OH \xrightarrow{H_3O^+} R^1CO_2H \qquad (23.64)$$

An ester   A carboxylate   An   A carboxylic
                        salt   alcohol   acid

EXPERIMENTAL PROCEDURE

BASE-PROMOTED HYDROLYSIS OF ESTERS

Mix 3 mL of diethylene glycol, 0.6 g (2 pellets) of potassium hydroxide, and 10 drops of water in a small round-bottom flask equipped with a stirbar. Heat the mixture until the solution is homogeneous, and then cool it to room temperature. Add 1 mL of the ester and equip the flask for heating under reflux. Reheat the mixture to reflux with stirring; after the ester layer dissolves (3–5 min), allow the solution to cool. Equip the flask for simple distillation, and heat the solution strongly so that the alcohol distils; all but high-boiling alcohols can be removed by direct distillation. The distillate, which should be an alcohol in a fairly pure and dry state, is used directly for the preparation of a solid derivative (Sec. 23.17).

The residue that remains after distillation contains the salt of the carboxylic acid. Add 10 mL of water to the residue and mix the two thoroughly. Acidify the resulting solution with 6 M sulfuric acid. Allow the mixture to stand, and collect any crystals by vacuum filtration. If crystals do not form, extract the aqueous acidic solution with small portions of diethyl ether or dichloromethane, dry the organic solution over anhydrous sodium sulfate, decant the liquid, and remove the solvent by distillation or by evaporation (in the hood). Use the residual acid to prepare a derivative (Sec. 23.19).

Flush the *pot residue* from the distillation of the alcohol down the drain. Neutralize the *aqueous filtrate* and also flush it down the drain. Transfer any *organic solvent* recovered by distillation to the appropriate container, that for nonhalogenated organic liquids in the case of *diethyl ether* and the one for halogenated organic liquids if *dichloromethane* was used.

Finishing Touches

23.17 Nitriles

CLASSIFICATION TESTS

A. Hydroxylamine. A qualitative test used for nitriles is similar to the hydroxylamine test for esters (Sec. 23.22A). Nitriles, as well as amides, typically give a colored solution on treatment with hydroxylamine and ferric chloride (Eq. 23.65).

$$R-C{\equiv}N + H_2NOH \longrightarrow 3\ R-\overset{\overset{\displaystyle NH}{\|}}{C}-NHOH \xrightarrow{FeCl_3} \left[R_1-C\overset{\diagup NH}{\underset{\diagdown N-O}{}} \right]_3 Fe + 3\ HCl \quad (23.65)$$

$$\qquad\qquad\qquad\qquad\qquad\qquad\qquad\qquad\qquad\qquad\qquad\qquad\qquad H$$

A nitrile Hydroxylamine   An iron complex
                                          Colored

EXPERIMENTAL PROCEDURE

HYDROXYLAMINE TEST FOR NITRILES

Safety Alert	Hydroxylamine and its hydrochloride may have mutagenic effects. Wear latex gloves when handling them. Should either of these substances contact your skin, wash the affected area with copious amounts of water.

In a large test tube, prepare a mixture consisting of 2 mL of 1 *M* hydroxylamine hydrochloride in propylene glycol, 30–50 mg of the compound that has been dissolved in a minimum amount of propylene glycol, and 1 mL of 1 *M* potassium hydroxide. Heat the mixture to boiling for 2 min, and then cool it to room temperature; add 0.5–1.0 mL of a 0.5 *M alcoholic* ferric chloride solution. A red-to-violet color is a positive test. Yellow colors are negative, and brown colors and precipitates are neither positive nor negative.

Finishing Touches	**F**lush the *solutions* down the drain.

DERIVATIVES

B. Hydrolysis. On hydrolysis in either acidic or basic solution, nitriles are ultimately converted to the corresponding carboxylic acids (Eqs. 23.66 and 23.67). It is then possible to prepare a derivative of the acid using methods provided in Section 23.19.

$$\textit{Basic hydrolysis}: \quad RC\equiv N \ + \ NaOH \ \xrightarrow{H_2O} \ RCO_2^-Na^+ \ + \ NH_3 \quad \textbf{(23.66)}$$
$$\underset{\text{A nitrile}}{} \qquad\qquad\qquad \underset{\text{A sodium carboxylate}}{}$$

$$\textit{Acidic hydrolysis}: \quad RC\equiv N \ \xrightarrow[H_2SO_4]{H_2O} \ RCONH_2 \ \xrightarrow[H_2SO_4]{H_2O} \ RCO_2H \ + \ \overset{+}{N}H_4 \quad \textbf{(23.67)}$$
$$\underset{\text{A nitrile}}{} \qquad\qquad \underset{\text{An amide}}{} \qquad\qquad \underset{\substack{\text{A carboxylic}\\\text{acid}}}{}$$

EXPERIMENTAL PROCEDURES

HYDROLYSIS OF NITRILES

Safety Alert	The aqueous sodium hydroxide solution is caustic. Wear latex gloves when handling it. Should it contact the skin, wash the area thoroughly with water.

Basic Hydrolysis. In a small Erlenmeyer flask, mix 10 mL of 3 *M* sodium hydroxide solution and 1 g of the nitrile. Heat the mixture to boiling, and note the odor of ammonia, or hold a piece of moist pHydrion paper over the container and note the color change. After the mixture is homogeneous, cool and then acidify it. If the acid solidifies, collect the crystals by vacuum filtration. If it is a liquid, extract the acidic solution with small portions of diethyl ether, dry and decant the ethereal solution, and then remove the solvent by one of the techniques described in Section 2.29. The residue is the acid. Prepare a suitable derivative of the acid (Sec. 23.13).

Neutralize all *aqueous solutions* before flushing them down the drain. Pour any *diethyl ether* isolated into the container for nonhalogenated organic liquids.

Finishing Touches

Acidic Hydrolysis. In a small Erlenmeyer flask, treat 1 g of the nitrile with 10 mL of concentrated sulfuric acid or concentrated hydrochloric acid, and warm the mixture to 50 °C for about 30 min. Dilute the mixture by addition of 20 mL of water. *Caution:* add the mixture slowly *to* the water if sulfuric acid has been used. Heat the mixture under gentle reflux for 30 min to 2 h and then allow it to cool. The acid usually forms a separate layer. If the acid solidifies upon cooling, collect it by vacuum filtration. If it is a liquid, extract the acidic mixture with small portions of diethyl ether, dry and decant the ethereal solution, and remove the solvent by one of the techniques described in Section 2.29. Prepare suitable derivatives of the acid (Sec. 23.13).

Neutralize the *aqueous solution* before flushing it down the drain. Pour any *diethyl ether* isolated into the container for nonhalogenated organic liquids.

Finishing Touches

23.18   Amides

CLASSIFICATION TEST

A. Hydroxylamine. A qualitative test for an amide group (Eq. 23.68) is the same as that for an ester, as given in Section 23.16A. Follow the experimental procedure provided there. The colors observed with amides are the same as those with esters.

An amide	Hydroxylamine	A hydroxamic acid	An iron complex
			Colored

$$R-\overset{\overset{\textstyle O}{\|}}{C}-NH_2 + H_2NOH \longrightarrow R-\overset{\overset{\textstyle O}{\|}}{C}-NHOH \xrightarrow{FeCl_3} \left[R-C \underset{N-O}{\overset{O}{\diagdown}} \right]_3 Fe + 3\ HCl \qquad (23.68)$$

DERIVATIVES

B. Hydrolysis. Amides may be hydrolyzed under acidic or basic conditions, giving an amine and a carboxylic acid (Eq. 23.69). In the instance of unsubstituted amides, ammonia is liberated. Substituted amides provide a substituted amine as a product. In those cases it is necessary to classify the amine as being primary or secondary and to prepare derivatives of both the acid (Sec. 23.13) and the amine (Sec. 23.14).

$$R^1CONR_2^2 \xrightarrow[\text{HO}^-]{\text{H}_3\text{O}^+\text{ or}} R^1CO_2H + HNR_2^2 \qquad (23.69)$$

An amide   A carboxylic
(R = H, alkyl, aryl)   acid   An amine

EXPERIMENTAL PROCEDURE

BASE-PROMOTED HYDROLYSIS OF AMIDES

Safety Alert   The aqueous sodium hydroxide solution is caustic. Wear latex gloves when handling it. Should it contact the skin, wash the area with water.

Hydrolysis.   Equip a small round-bottom flask containing 10 mL of 3 *M* sodium hydroxide solution for shortpath distillation into a receiver containing a few milliliters of dilute hydrochloric acid. Heat the mixture to boiling and note changes, if any, in the appearance of the contents of the receiver. After the mixture is homogeneous, allow the solution to cool to room temperature.

Neutralize the acidic solution in the receiver, perform the Hinsberg test (Sec. 23.14A), and prepare derivatives of the amine (Sec. 23.14). If the amine is *not* volatile enough to distil, and thus is not found in the receiver, use small portions of diethyl ether to extract it from the aqueous layer contained in the stillpot. Dry the combined ethereal extracts over potassium hydroxide pellets, decant the solution, and then remove the solvent by one of the techniques described in Section 2.29. Prepare derivatives of the amine from the residue remaining after solvent removal (Sec. 23.14).

Acidify the alkaline solution from which the amine has been extracted and isolate the carboxylic acid by either vacuum filtration or by extraction with diethyl ether followed by drying over sodium sulfate and removal of solvent as before. Characterize this acid by preparing suitable solid derivatives (Sec. 23.13).

Finishing Touches   Neutralize all *aqueous solutions* before flushing them down the drain. Pour any *diethyl ether* isolated into the container for nonhalogenated organic liquids.

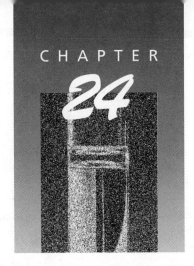

CHAPTER
24

The Literature of Organic Chemistry

This chapter is designed to assist those interested in obtaining additional information on different aspects of synthetic, physical, and theoretical organic chemistry. The chapter is not intended to be a comprehensive guide to the literature of organic chemistry. Rather, it should provide the reader with a general knowledge of the important literature sources of organic chemistry that may be used as valuable lead references to initiate a specific search.

24.1 Classification of the Literature

For common applications, most of the literature of organic chemistry can be divided into nine major categories: (1) primary research journals, (2) review journals, (3) encyclopedias and dictionaries, (4) abstract journals, (5) advanced textbooks, (6) monographs, (7) general multivolume references, (8) general reference works on synthetic procedures and techniques, and (9) catalogs of physical data. In this section, selected examples from each of these classes are given, along with brief explanatory notes. In Section 24.2, two examples are given to illustrate how the literature may be used to find information about a specific organic compound.

PRIMARY RESEARCH JOURNALS

Primary research journals are the ultimate source of most of the information in organic chemistry; these journals publish original research results in several formats. **Articles** are full papers that provide historical discussions together with a presentation of the important findings, conclusions, and experimental details for preparing new compounds. **Communications** are short articles, sometimes with brief experimental details, that are restricted to a single topic or important discovery; such articles must be especially timely and of general interest to the chemical community. Some journals also publish short papers called **Notes**, that contain limited discussions and experimental details. Some of the more useful journals for the practicing organic chemist and a brief description of their contents follow.

A. *Angewandte Chemie, International Edition in English.* Concurrent translation of the German journal *Angewandte Chemie.* This is an outstanding journal that publishes critical reviews of selected topics and communications covering all areas of chemistry.

B. *Biochemistry.* Articles and communications in biochemistry, but many of the contributions tend toward the biological side of organic chemistry.

C. *Bulletin of the Chemical Society of Japan.* Articles and notes in English covering all areas of chemistry.

D. *Canadian Journal of Chemistry.* Articles in English and French in all areas of chemistry.

E. *Chemische Berichte.* Excellent journal with articles and notes in German and English covering organic and organometallic chemistry; one of the two major German journals in organic chemistry. Effective 1997 will be *Chemische Berichte/Recueil* as European journal covering inorganic and organometallic chemistry.

F. *Chemistry Letters.* Communications in English covering all areas of organic chemistry.

G. *Helvetica Chimica Acta.* Articles and notes in English, French, or German covering all areas of organic chemistry.

H. *Heterocycles.* Reviews, communications, and articles in all areas of heterocyclic chemistry.

I. *Journal of the American Chemical Society.* Articles and communications covering all areas of chemistry. Generally, the articles and communications appearing in this journal, which is one of the foremost chemical journals in the world, tend to have broader interest to chemists of various disciplines.

J. *Journal of the Chemical Society, Perkin Transactions.* Divided into two sections: I. Organic and bio-organic chemistry. II. Physical organic chemistry. Primarily articles but a few communications, in all areas of organic and bio-organic chemistry. Most communications are published in a separate journal titled *Chemical Communications.*

K. *Journal of Medicinal Chemistry.* Articles, communications, and notes in English covering the preparation of new organic compounds having biological activity.

L. *Journal of Organic Chemistry.* Articles, communications, and notes in English covering all areas of organic chemistry.

M. *Liebigs Annalen.* Excellent journal with articles and notes in German and English covering organic chemistry; one of the two major German journals in organic chemistry. Effective 1997 will be *Leibigs Annalen/Recueil* as a European journal covering organic chemistry.

N. *Synlett.* Articles and communications in the general area of synthetic organic chemistry.

O. *Synthesis.* Reviews, articles, and communications in the general area of synthetic organic chemistry, mostly in English but some in German and French.

P. *Tetrahedron.* One of the best international journals dedicated to publishing articles, reviews and ''symposia in print'' in the general areas of organic and bio-organic chemistry; most articles are in English, but some are in German and French.

Q. *Tetrahedron: Asymmetry.* Reviews, articles, and communications in the specialized area of asymmetric synthesis and methods.

R. *Tetrahedron Letters.* Perhaps the best international journal dedicated to publishing brief two- or four-page communications in English, French, or German in all areas of organic and bio-organic chemistry.

REVIEW JOURNALS

Review journals publish longer articles covering selected topics in various areas of chemistry. These articles do not present new research, but they do reference the original work in the primary journals. Some review journals publish reviews in all areas of chemistry, while others cover only specific areas. In addition to the review journals listed below, a number of the primary research journals listed in the previous subsection publish excellent reviews. These include: *Angewandte Chemie, International Edition in English, Heterocycles, Synthesis,* and *Tetrahedron.*

A. *Accounts of Chemical Research.* Provides concise reviews of areas of active research in all areas of chemistry.

B. *Annual Reports on the Progress of Chemistry.* A series of timely reports in areas of general interest and importance; since 1968 this has been divided into sections, and Section B is devoted to organic chemistry. Supplemented by a series of *Specialist Periodical Reports*, which cover specific topics such as alkaloids, peptides, carbohydrates, nuclear magnetic resonance, and organometallic compounds.

C. *Chemical Reviews.* A review journal that covers all areas of chemistry and is published eight times per year by the American Chemical Society.

D. *Chemical Society Reviews.* A review journal with broad scope published quarterly by the Royal Chemical Society (London); succeeded *Quarterly Reviews.*

ENCYCLOPEDIAS AND DICTIONARIES

Encyclopedias and dictionaries of organic chemistry provide specific information such as physical properties and uses of known compounds. Although all of these are useful, the most exhaustive compilation of such data is found in Beilstein.

A. *Beilstein's Handbuch der Organischen Chemie,* which was first published in 1881–1883 in two volumes, is perhaps the most complete reference work in any branch of science. In contrast to other collections of physical and chemical data, those included in the *Beilstein* have been critically evaluated and checked for internal consistency by scientists. The Fourth Edition (1918–1938) contains data on and references for the 140,000 organic compounds known in 1909. The base set of the 4th edition is known as the **Hauptwerk**, abbreviated H. It is continued by a series of supplements called **Ergänzungswerke**, abbreviated E. The main work, H, and the first four supplements, E I–E IV, are in German, but the fifth supplement, E V, is published in English, and is not yet complete. The first supplement (E I) covers the literature for the years 1910–1919, and the second supplement (E II) covers the years 1920–1929. Supplements three (E III) and four (E IV) cover the years 1930–1959, and the fifth supplement (E V) covers heterocyclics from 1960–1979; Volumes 17–27 of Supplementary Series III and IV, covering heterocyclic compounds, were published jointly and are known as E III/IV. Thus, the series of the *Beilstein* handbook have the following coverage:

H	Vol. 1–27	up to 1909
E I	Vol. 1–27	1910–1919
E II	Vol. 1–27	1920–1929
E III	Vol. 1–16	1930–1949
E III/IV	Vol. 17–27	1930–1959
E IV	Vol. 1–16	1950–1959
E V	Vol. 17–27	1960–1979

The size and complexity of *Beilstein* can make it a daunting work to consult, so an understanding of how it is organized will help make it more usable. *Beilstein* is arranged by functional classes into a series of systems, and a compound is always treated in the same system, no matter in what supplement it may appear. This feature of the set greatly facilitates location of information; once a **system number** of a compound has been obtained, it is easy to trace it through the whole set.

There is a combined molecular formula (**Formelregister**), and compound-name (**Sachregister**) index for the main set and the first two supplements (H, E I–E II). In addition, molecular formula and compound name indexes are available for all volumes of the **Hauptwerk** and the **Ergänzungswerke** E I–E IV. Because the compound name indexes are in German, the molecular formula index generally provides the easiest access to *Beilstein*. A cumulative *Centennial Index* in English was published in 1991 that allows access by both substance name and formula to all compounds in *Beilstein* for the main work, H, and the supplements E I–E IV, thereby covering the period 1779–1959.

There are several guides and a brief German-English dictionary designed specifically for use with *Beilstein*. The computer program SANDRA (Structure and Reference Analyzer) can be used to search for a compound in *Beilstein*. SANDRA allows searching by the structure or substructure of a substance and predicts the system number of the requested substance, even if the substance is not actually included in *Beilstein*.

The *Beilstein Online* database is available in a computerized format through the STN and Dialog online vendors. *Beilstein Online* includes information in English from the printed Handbook, 1779–1959, as well as a large amount of data from 1960 onward. The online file is searchable in many ways, including Chemical Abstracts' Registry Number, chemical substructure, physical properties, and by other parameters not indexed in the print format. While some users may be able to search the online file without help, to make the fullest, most cost-effective use of the online file, it is best to consult a specialist for assistance. *Beilstein's Current Facts,* a computer-based information system that uses CD-ROM technology, complements the information system. Updated quarterly, it allows access to new data on previously known substances and provides basic data on new compounds.

In 1995 *Beilstein* launched the *Crossfire* system, which provides full access to the database of over six million compounds and five million reactions, through a dedicated client-server system that is installed locally or accessed via the Internet. In addition to having a more powerful graphical interface than can be found on STN or Dialog, *Crossfire* has the advantage of not charging a user by the time spent online. It seems likely that this kind of network access to large structure files will eventually supplant the pay-as-you-go model that has prevailed for some years.

B. *CRC Atlas of Spectral Data and Physical Constants for Organic Compounds*, Grasselli, J. G.; Ritchey, W. M., eds. Chemical Rubber Co., Cleveland, OH, 1975. Six volumes. Gives data on some 21,000 organic compounds.

C. *CRC Handbook of Chemistry and Physics*, annual editions, CRC Press, Boca Raton, FL. Gives physical properties and Beilstein references for about 15,000 organic compounds.

D. *Dictionary of Organic Compounds*, 6th ed., Buckingham, J., ed. Chapman and Hall, New York, 1996. This edition is in nine volumes. Volumes 1–6 contain the data for the compounds, Volume 7 is a name index with cross-references, Volume 8 contains a molecular-formula index, and Volume 9 is a Chemical Abstracts Service registry number index.

E. *Handbook of Tables for Identification of Organic Compounds*, 3rd ed., Rapoport, Z., ed. Chemical Rubber Co., Cleveland, OH, 1967. Gives physical properties and derivatives for more than 4000 compounds, arranged according to functional groups.

F. *Lange's Handbook of Chemistry*, 14th ed., Dean, J. A., ed. McGraw-Hill, New York, 1992. Gives physical properties for about 6500 organic compounds.

G. *Merck Index of Chemicals and Drugs*, 12th ed., Merck and Co., Rahway, NJ, 1996. Gives a concise summary of the physical and biological properties of more than 10,000 compounds, with some literature references. Organization is alphabetical by name, synonyms and trade names; contains cross-index and index of Chemical Abstracts Service registry numbers.

ABSTRACT JOURNALS

Abstract journals provide concise summaries of original research articles in journals and listings of reviews and books. Although abstracts are a fast way to locate chemical information, they are always incomplete and sometimes even misleading, so they should not be relied upon as the final source. A complete literature search should always include reference to the primary research publication.

A. *Chemical Abstracts.* This publication is one of the most important reference sources that indexes the original literature. It began in 1907 and annually publishes abstracts for over half a million documents that are drawn from about 9,000 journals as well as books, conferences, dissertations, and patents. *Chemical Abstracts* provides by far the most complete coverage of the chemical literature after 1940. *Chemical Abstracts* is published weekly, and at the end of each six-month period, indexes by author, general subject, chemical substance, patent number, and molecular formula appear. Prior to 1956, these indexes were cumulated every 10 years, but they are now cumulated at five-year intervals.

Effective use of *Chemical Abstracts,* especially the chemical substance indexes, requires a basic understanding of the nomenclature systems used currently and in the past. Current CA nomenclature practice is summarized in the 1987–1991 *Index Guide*, which is a listing of major indexing terms and cross references. The *Guide* also contains the indexing rules and procedures currently used. Supplementing the regular indexes is the *Ring Systems Handbook*, which contains nomenclature information on organic ring systems, allowing a user to determine the **CA Index Name** and **registry number** for over 100,000 ring and cage structures. In *Chemical Abstracts*, a unique registry number is assigned to each chemical substance described in the literature, and these numbers are widely used throughout the chemical literature. Articles that describe the preparation or the reactions of a particular substance may be easily located by simply using the registry number of that substance. The *Registry Number Handbook* is a comprehensive numerical listing of all registry numbers assigned to compounds since 1965. If one has a registry number, it is easy to determine the CA Index Name for that compound in order to search in the Chemical Substance Indexes.

It is possible to search *Chemical Abstracts* online from several different vendors. *CAS Online* is a database service from a vendor called STN that provides access to several related files. The **CA** file is the equivalent to the printed *Chemical Abstracts* from 1967 to the present. The **Registry** file is the world's largest database of chemical structures, over 15 million records for chemicals, polymers, biological compounds, and other materials. The **CAOLD** file is a small retrospective database that covers the 1957–66 period, but it is searchable only by registry number and is incomplete. *CAPreviews* offers early access to the most current literature that has not yet appeared in *Chemical Abstracts*. These resources are usually available in most college and university libraries as well as at most chemical companies.

There are many advantages to using a computer to search *Chemical Abstracts*. Computer searches using CAS Online may be tailored to find information by author's name, subject keywords, the name, registry number, or formula of a particular substance, and a chemical structure or substructure. Different terms and key words may be linked by Boolean operators to focus the search so that small sets of relevant information may be quickly found. Because there is a charge for using CAS Online, it is advisable to obtain training and assistance from a search specialist before beginning online searching. That assistance will improve the quality and the cost-effectiveness of the search.

In addition to *Chemical Abstracts*, many other databases provide access to other specialized types of literature, including medicine, patents, polymers, toxicology, and pharmacy. There are also databases that offer the entire text of articles from some chemistry journals. Many scientific publishers are exploring the possibility of providing full text access to primary journals on the Internet's Worldwide Web, a readily usable medium that is growing very quickly. The realm of electronic access to scientific information is undergoing profound and rapid change. Contact the librarian or information specialist at your institution to find out about the newest developments in this area.

B. *Chemisches Zentralblatt* (1830–1969) is the second major abstract journal and was published in German. It predates *Chemical Abstracts* by more than 70 years, and prior to 1940 it was more complete and more reliable than *Chemical Abstracts*. For this reason, in an exhaustive search for an organic compound in the years before 1940, it is advisable to consult the collective formula indexes of *Chemisches Zentralblatt*.

C. *Science Citation Index* is a unique reference source because it enables one to search forward from a particular literature reference. For example, to find all of the applications of a reaction or method that was described in a specific paper, one simply looks up the appropriate literature reference for that article in *Science Citation Index*, and all papers published subsequently that referenced the original article in a footnote are listed. It is now available in print, CD-ROM, and online formats.

ADVANCED TEXTBOOKS

Advanced textbooks in organic chemistry provide useful information for students who are interested in a more sophisticated or advanced treatment of the information found in typical undergraduate organic textbooks. Furthermore, material regarding a variety of interesting aspects of organic chemistry that are not covered in standard textbooks may be found. A selected listing includes:

A. Carey, F. A.; Sundberg, R. M. *Advanced Organic Chemistry; Part A: Structure and Mechanisms;* and *Part B: Reactions and Synthesis*, 3rd ed., Plenum, New York, 1990. An excellent survey of reactions and their applications.

B. Carruthers, W. *Some Modern Methods of Organic Synthesis*, 3rd. ed., Cambridge University Press, Cambridge, UK, 1986. Selected reactions used in synthetic organic chemistry are discussed in detail.

C. House, H. O. *Modern Synthetic Reactions*, 2nd ed., W. A. Benjamin, Menlo Park, CA, 1972. The general classes of reactions used for reduction, for oxidation, and for the formation of new carbon-carbon bonds are surveyed in terms of their scope, limitations, stereochemistry, and mechanisms.

D. Lowry, T. H.; Richardson, K. S. *Mechanism and Theory in Organic Chemistry*, 3rd ed., Harper and Row, New York, 1987. An excellent treatment of mechanistic and theoretical aspects of organic chemistry.

E. March, J. *Advanced Organic Chemistry: Reactions, Mechanisms, and Structures*, 4th ed., Wiley-Interscience, New York, 1992. Advanced text with broad coverage and many references to the original literature.

F. Smith, M. B. *Organic Synthesis*, McGraw-Hill, Inc., New York, 1994. The general classes of reactions used for reduction, for oxidation, and for the formation of new carbon-carbon bonds are surveyed in terms of their scope, limitations, stereochemistry, and mechanisms. Strategies for synthesis are also discussed.

MONOGRAPHS

Numerous monographs contain reviews of selected topics in organic and related areas of chemistry. A partial listing follows:

A. Collman, J. P.; Hegedus, L. S.; Norton, J. R.; Finke, R. G. *Principles and Applications of Organotransition Metal Chemistry*, University Science Books, Mill Valley, CA, 1987. An excellent survey of organometallic chemistry.

B. Eliel, E. L.; Allinger, N. L.; Angyal, S. J.; Morrison, G. A. *Conformational Analysis*, Interscience, New York, 1965. An important reference work covering all major aspects of conformational analysis.

C. Eliel, E. L.; Wilen, S. H.; Mander, L. N. *Stereochemistry of Organic Compounds*, Wiley-Interscience, New York, 1994. An excellent treatise on all aspects of stereochemistry.

D. Green, T. W.; Wuts, P. G. M. *Protective Groups in Organic Synthesis*, 2nd ed., Wiley-Interscience, New York, 1991. An extensive compilation of methods for protection and deprotection of various functional groups.

E. Larock, R. C. *Comprehensive Organic Transformations: A Guide to Functional Group Preparation,* VCH Publishers, New York, 1989. An exhaustive compilation of reactions of functional groups.

F. Paquette, L. A. *Principles of Modern Heterocyclic Chemistry*, W. A. Benjamin, New York, 1968. An introduction to the basic principles of heterocyclic chemistry.

G. Perrin, D. D.; Armarego, W. L. F.; Perrin, D. R. *Purification of Laboratory Chemicals*, 3rd ed., Pergamon Press, Oxford, UK, 1988. Complete procedures for purifying organic solvents and reagents.

H. Turro, N. J. *Modern Molecular Photochemistry*, Benjamin/Cummings, Menlo Park, CA, 1978. A treatise on the applications of photochemistry in organic chemistry.

GENERAL, MULTIVOLUME REFERENCES

General, multivolume references provide a comprehensive review of various topical and general aspects of organic chemistry. A partial listing follows.

A. *Asymmetric Synthesis*, Morrison, J. D., ed., Vol. 1–5, Academic Press, New York, 1983–1985. Series dedicated to critical review of modern techniques of organic asymmetric synthesis.

    B. *Comprehensive Heterocyclic Chemistry II*, Katritzky, A. R.; Rees, C. W.; Scriven, E. F. V., eds., Pergamon Press, Oxford, 1996. An 11-volume treatise covering the reactions, structure, synthesis, and uses of heterocyclic compounds through 1995.

    C. *Comprehensive Organic Chemistry*, Barton, D. H. R.; Ollis, W. D., eds., Pergamon Press, Oxford, UK, 1978. A six-volume treatise on the synthesis and reactions of organic compounds.

    D. *Comprehensive Organic Synthesis*, Trost, B. M.; Fleming, I., eds., Pergamon Press, Oxford, UK, 1991. A nine-volume treatise with index of all aspects of synthetic organic chemistry including carbon-carbon bond formation, heteroatom manipulation, and oxidation and reduction.

    E. *Comprehensive Organometallic Chemistry II*, Wilkinson, G., ed., Pergamon Press, Oxford, UK, 1996. A 14-volume treatise covering the synthesis, reactions, and structure of organometallic compounds through 1994.

    F. *Comprehensive Organic Functional Group Transformations*, Katritzky, A. R.; Meth-Cohn, O.; Rees, C. W., eds., Pergamon, Oxford, UK, 1995. A seven-volume series that comprehensively reviews transformations of functional groups.

    G. *Encyclopedia of Reagents for Organic Synthesis*, Paquette, L. A., ed., John Wiley & Sons, Inc. New York, 1995. An eight-volume series that comprehensively reviews reagents used in synthetic organic chemistry.

    H. *Chemistry of Functional Groups*, Patai, S., senior ed., Wiley-Interscience, New York. Series devoted to an in-depth review of each of the functional groups.

    I. *Methoden der Organischen Chemie (Houben-Weyl)*, 1952–present, Georg Thieme Verlag, Stuttgart. As of 1996 consisted of 20 multipart volumes with supplements reviewing all classes of organic compounds; originally in German but some recent volumes are in English.

    J. *Rodd's Chemistry of Carbon Compounds*, 2nd ed., Coffey, S., ed., Elsevier, Amsterdam; 1964–present. A comprehensive survey of all classes of organic compounds, giving properties and syntheses for many individual compounds; consists of four volumes in 30 parts, with supplements to maintain currency.

REFERENCE WORKS ON SYNTHETIC PROCEDURES AND TECHNIQUES

It is often necessary to survey the scope and limitations of a reaction, a synthetic method, a reagent, or a technique to determine whether it may be applied to solving a specific problem. To facilitate access to this information, a number of reference works are available that contain reviews of reactions, reagents, techniques, and methods in organic chemistry; often these are serial in nature.

    A. *Annual Reports in Organic Synthesis*, Academic Press, New York, 1970–present. An annual review of synthetically useful reactions.

    B. *Compendium of Organic Synthetic Methods*, Vol. 1–8, John Wiley & Sons, New York, 1971–1995. These volumes present in outline form possible interconversions among the major functional groups of organic compounds. References to the primary literature sources are given.

    C. *Organic Functional Group Preparations*, 2nd ed., Sandler, W.; Karo, W., eds., Vol. 1, 1983; Vol. 2, 1986, Academic Press, New York. A listing of different methods for preparing all functional groups.

    D. *Organic Reactions*, John Wiley & Sons, New York, 1942–present, 48 volumes through 1996. Each volume contains from 5 to 12 chapters, each of which deals with an organic reaction of wide applicability. Typical experimental procedures are given in detail, and extensive tables of examples with references are given. Each volume contains a cumulative author and chapter-title index.

E. *Organic Syntheses*, John Wiley & Sons, New York, 1932–present, 74 volumes through 1996. Every ten volumes have been collected, indexed, and published as *Collective Volumes I–VI*, through Volume 59. Beginning with the *Collective Volume VII*, which contains Volumes 60–64, the collective volumes are compiled every five years. Detailed directions for synthesis of more than 1000 compounds are given. The procedures have all been thoroughly checked by independent investigators before publication. Many of the methods are general and can be applied to the synthesis of related compounds other than those described. The collective volumes contain indexes of formulas, names, types of reaction, types of compounds, purification of solvents and reagents, and illustrations of special apparatus. A cumulative index to *Collective Volumes I–VIII* was published in 1995, and an excellent reaction guide has been published that provides a compilation of all the reactions in *Collective Volumes I–VII* and the annual Volumes 65–68.

F. *Synthetic Methods of Organic Chemistry*, Theilheimer, W.; Finch, A. F., eds., Interscience, New York, Karger, Basel, 1948–present, 50 volumes through 1996. Emphasis is upon functional group interconversions and chemistry and general reactions that may be used to construct carbon-carbon and carbon-heteroatom bonds. The reactions are classified by symbols that are arranged systematically. There are cumulative indexes.

G. *Techniques of Organic Chemistry*, 3rd ed., Weissberger, A., ed., Interscience, New York, 1959–present; some volumes have been revised. Title broadened in 1970 to *Techniques of Chemistry*. Examples of useful volumes are *Elucidation of Organic Structures*, Vol. 4, parts 1–3; *Investigation of Rates and Mechanisms of Reactions*, Vol. 6., parts 1 and 2; *Separation and Purification*, Vol. 12; and *Thin Layer Chromatography*, Vol. 14.

H. *Vogel's Elementary Practical Organic Chemistry*, 3rd ed., Vol. 1, Smith, B. V.; Waldron, N. M., eds., Longman, London, 1979. Good coverage of basic experimental techniques with a selection of contemporary laboratory preparations.

I. *Vogel's Textbook of Practical Organic Chemistry*, 4th ed., Furniss, B. S., et al., eds., Longman, London, 1978. For the more experienced researcher; furnishes additional tips on technique and a wide range of preparations.

J. *Survey of Organic Syntheses*, Buehler, C. A.; Pearson, D. E., Wiley-Interscience, New York, Vol. 1, 1970; Vol. 2, 1972. This extensive two-volume work covers the principal methods of synthesizing the main types of organic compounds. The limitations of the reactions, the preferred reagents, the newer solvents, and experimental conditions are considered.

K. *Fieser and Fieser's Reagents for Organic Synthesis*, Wiley-Interscience, New York, 1967–present, 17 volumes as of 1996. These volumes list some 8000 reagents and solvents, described in terms of methods of preparation or source, purification, and use in typical reactions. Ample references to original literature are given.

L. *Preparative Organic Chemistry*, Hilgetag, G; Martini, A., John Wiley & Sons, New York, 1972. An extensive compilation of reactions and methods for interconverting functional groups and forming carbon-carbon bonds.

CATALOGS OF PHYSICAL DATA

Identifying a compound that has been obtained from a chemical reaction or isolated from a natural source or the environment is a commonly encountered problem in organic chemistry. The spectroscopic characteristics of a sample provide important clues to the identity of the substance. Collections of spectral data of known compounds provide an important source of information that may be used either to identify known compounds or to assist in determining the structure of an unknown substance. Some useful compilations of spectral data are available.

A. ^{1}H NMR Spectra

Aldrich Library of NMR Spectra, 2nd ed., Pouchert, C. J., ed., Aldrich Chemical Company, Milwaukee, WI, 1983; two volumes. A collection of about 37,000 spectra.

Aldrich Library of ^{13}C and ^{1}H NMR Spectra, Pouchert, C. J.; Behnke, J., eds., Aldrich Chemical Company, Milwaukee, WI, 1993; three volumes. A collection of about 12,000 spectra.

Nuclear Magnetic Resonance Spectra, Sadtler Research Laboratories, Philadelphia. Proton NMR spectra of more than 64,000 compounds have been published as of 1996, with 1000 being added annually. Peaks are assigned to the hydrogen nuclei responsible for the absorptions, and integration of the signals is shown on many of the spectra.

B. ^{13}C NMR Spectra

Atlas of Carbon-13 NMR Data, Breitmaier, E.; Haas, G.; Voelter, W., Heyden, Philadelphia, 1979; two volumes. Tabular data on 3017 compounds, with chemical shifts for ^{13}C given and ^{1}H–^{13}C multiplicities indicated.

Carbon-13 NMR Nuclear Magnetic Resonance Spectra, Sadtler Research Laboratories, Philadelphia. Compilation of 42,000 proton-decoupled spectra as of 1996.

C. IR Spectra

Sadtler Standard Infrared Prism Spectra, Vol. 1–123, Sadtler Research Laboratories, Philadelphia, PA. Compilation of 91,000 spectra, as of 1996.

Sadtler Standard Infrared Grating Spectra, Vol. 1–123, Sadtler Research Laboratories, Philadelphia, PA. Compilation of 91,000 spectra as of 1996.

Aldrich Library of Infrared Spectra, 3rd ed., Pouchert, C. J., ed., Aldrich Chemical Company, Milwaukee, WI, 1981. A collection of about 12,000 spectra.

Aldrich Library of FT-IR Infrared Spectra, Vol. 1–3, Pouchert, C. J., ed., Aldrich Chemical Co., Milwaukee, WI, 1985–1989; three volumes.

Sigma Library of FT-IR Spectra, Keller, R. J., ed., Sigma Chemical Company, St. Louis, MO, 1986; two volumes. A collection of 10,400 spectra.

24.2   Use of the Literature of Organic Chemistry

The literature outline given in this chapter may be used in a variety of ways, according to the aims and needs of different courses and the library facilities available. Even in those cases where the pressure of time and/or lack of facilities preclude the use of literature beyond the pages of this textbook itself, this chapter is still valuable to the student who decides to go further in the study of organic chemistry.

In many organic laboratory courses, instructors make part of the experimentation open-ended—encouraging the students to plan and carry out experiments with some independence. This is a desirable objective, but there is an element of risk if the experimental procedure has not been carefully checked; *procedures found in the literature are not always easily reproduced*! One of the best sources of different experiments is

Organic Syntheses; although the experiments are typically reported on a large scale, they may often be easily scaled down. The experiments that occasionally appear in the *Journal of Chemical Education* also deserve mention.

The synthesis of known and unknown compounds is a task commonly encountered by the research organic chemist. In the course of this work, it is often necessary to determine whether a particular compound has been previously prepared; if it has been, then the questions of when, how, and by whom arise. If the compound has not been reported, then the organic chemist searches for similar compounds that might have been prepared, because the same methods might apply to the synthesis of the substance of interest. The examples that follow will provide some guidelines to performing literature searches to solve such problems, and the exercises at the end pose many types of problems that are routinely encountered in research.

WHAT IS MUSTARD GAS?

"Mustard gas" is one of the names that has been applied to the compound $ClCH_2CH_2$—S—CH_2CH_2Cl. Find the answers to the following questions: (1) Has this compound been synthesized or isolated? (2) By whom? (3) When? (4) Where can the most recent information on this compound be found?

First, write the molecular formula as $C_4H_8Cl_2S$, and look in *Beilstein's* General Formelregister, Zweites Ergänzungswerk. On page 65 is the entry "*β,β'*-Dichlordiäthylsulfid, Senfgas **1**, 349, I 174, II 348, 940," and just below it the entry "*α,α'*-Dichlordiäthylsulfid (**1**) II 685." The first entry is recognized as that pertaining to the subject compound. The references are to page 349 in Volume 1 of the main work, page 175 in Volume 1 of the first supplement, and to pages 348 and 940 in Volume 1 of the second supplement. Although the third supplement was published after the general index, by noting the system number of the subject compound, which is 23, one may locate it on page 1382 of the third supplement using the index in the appropriate volume.

Referring to page 349 in Volume 1 of the main work, one finds "*β,β'*-Dichlorodiäthylsulfid $C_4H_8Cl_2S$ = $(CH_2Cl \cdot CH_2)_2$S. B. Aus Thiodiglykol S $(CH_2 \cdot CH_2 \cdot OH)_2$ und PCl_3 (V. Meyer, B. **19**, 3260)." This translates: "B. = Bildung, Formation. From thiodiglycol and PCl_3 (V. Meyer, *Berichte*, **19**, 3260)." Looking up the reference in the *Berichte der Deutschen Chemischen Gesellschaft* (Vol. 19, p. 3260, published in 1886) reveals that Victor Meyer first prepared **1** in two steps as illustrated in Equations 24.1 and 24.2.

$$2 \; Cl\diagup\!\!\diagdown\!\!\diagup^{OH} \xrightarrow{K_2S} Cl\diagup\!\!\diagdown\!\!\diagup^{S}\diagdown\!\!\diagup^{OH} \qquad (24.1)$$

$$HO\diagup\!\!\diagdown^{S}\diagup\!\!\diagdown^{OH} \xrightarrow{PCl_3} Cl\diagup\!\!\diagdown^{S}\diagup\!\!\diagdown Cl \qquad (24.2)$$

Thiodiglycol   Mustard gas

The entry in the main work of *Beilstein* (Vol. 1, p. 349) gives in six lines the boiling point and solubility properties of **1** and one chemical reaction. The final two words are *Sehr giftig*, "very poisonous"—a terse commentary on a material that was much feared as a lethal military weapon in World War II, but was never used. Apparently the compound does not adversely affect everyone who is exposed to it, because Meyer noted in his *Berichte* article that his laboratory assistant developed skin eruptions and eye inflammation after preparing the compound; however, Meyer himself suffered no ill effects even though he took no precautions in handling the chemical. In the first supplement on page 175, 15 lines are devoted to "β,β'-diäthylsulfid," and in the second supplement (p. 348), five and one-half pages appear, indicating the increased interest in this compound during the period 1920–1929.

A search of *Chemical Abstracts* may be begun by using the *Chemical Abstracts* Collective Formula Index for 1920–1946. Thus, under the molecular formula $C_4H_8Cl_2S$ is found the entry "(See also Sulfide, bis(chloroethyl)) Sulfide, 1-chloroethyl 2-chloroethyl, **25**:2114[8]," wherein the reference is to volume 25, page 2114. Since "Sulfide, 1-chloroethyl 2-chloroethyl" is not the compound of interest, we look in the *Chemical Abstracts* Decennial Subject Index, 1917–1926, for the entry "Sulfide bis (β-chloroethyl)," which is followed by the names *mustard gas*, *yperite* and by six general references and two columns of more specific references beginning with "absorption by skin, mechanism of, **14**: 300[4]" and ending with "toxicity and skin-irritant effects of, **15**: 1943[6]."

The *Chemical Abstracts* Collective Subject Indexes can be used for more recent years, but *one must be on guard for changes in nomenclature*. Consequently, it is usually advantageous to use formula indexes first. For example, upon examining the January–June 1972 Formula Index, under $C_4H_8Cl_2S$ we find "Ethane, 1,1'-thiobis[2-chloro-" as the name for our compound, with seven references to abstracts. Under this name in the *Chemical Substance Index* for the same period, we find the same seven references, but with specific subject headings such as "DNA, cross-linking induced by, 95344w."

IS A COMPOUND OF INTEREST KNOWN?

A second example of a search demonstrates how *Beilstein* may be used to find compounds reported after 1929, the last year covered by the Formula Index of Volume 29. The search for a method of preparing 3-(2-furyl)-1-(3-nitrophenyl)propenone (**2**) is illustrative of the approach.

If one looks for this compound in the *Beilstein* Formelregister (Vol. 29), under the molecular formula $C_{13}H_9NO_4$, no name fitting the structure is found, indicating that this compound was not reported prior to 1929. However, the analogous compound without the nitro group is listed under the formula $C_{13}H_{10}O_2$ in the Formelregister, with the notation *17* 353; II 377, which means that it is described in Vol. 17 of the Hauptwerk on p. 353 and in the Zweites Ergänzungswerk on p. 377. Looking up these entries, one finds that the system number for this compound is 2467. In Vol. 17 of the combined Drittes und Viertes Ergänzungswerk (E III/IV), one then finds that compounds having System Number 2467 are in Part 6 of Vol. 17. In the formula index of Part 6 under $C_{13}H_9NO_4$, one finds the name of the compound sought, followed by the page number 5262, where a description of the preparation and properties of this compound is given. Alternatively, the Gesamtregister Formelregister for Volumes 17 and 18 directs one to the same page in Volume 17.

2

As mentioned earlier, *Chemical Abstracts* Collective Formula Indexes may conveniently be used for locating compounds appearing in the literature after 1929. The 1920–1946 Index has no entry for the compound located in *Beilstein* as described earlier, but the 1947–1956 Index lists under the formula $C_{13}H_9NO_4$ the name "Acrylophenone, 3-(2-furyl)-3′-nitro", which may, *or may not*, be recognized as another name for the compound. Both *Beilstein* E III/IV, *17* and *Chem. Abstr.*, **1949**, *43*, 3429g refer to the primary source, the *Am. Chem. Soc.*, **1949**, *71*, 612, where D. L. Turner describes the first preparation of this compound.

This example illustrates the advantage of using the formula indexes, of *Beilstein* and/or *Chemical Abstracts* for locating an organic compound in the literature. Several different names may be used for the same compound, but the formula will be more distinctive, and one can be on the lookout for different ways of naming the compound.

EXERCISES

1. Find the melting points of the following crystalline derivatives. (**a**) 2,4-dinitrophenylhydrazone of trichloroacetaldehyde; (**b**) semicarbazone of 3-methylcyclohexanone; (**c**) 3,5-dinitrobenzoate of 1,3-dichloro-2-propanol; (**d**) amide of 2-methyl-3-phenylpropanoic acid; (**e**) benzamide of 4-fluoroaniline.

2. Locate an article or a chapter on each of the following types of organic reactions: (**a**) the aldol reaction; (**b**) the Wittig reaction; (**c**) reactions of diazoacetic esters with unsaturated compounds; (**d**) hydration of alkenes and alkynes through hydroboration; (**e**) metalation with organolithium compounds; (**f**) reactions of lithium dialkylcuprates; (**g**) asymmetric synthesis of amino acids.

3. Give a literature reference for a practical synthetic procedure for each of the following compounds: (**a**) 1,2-dibromocyclohexane; (**b**) α-tetralone; (**c**) 3-chlorocyclopentene; (**d**) 2-carboethoxycyclopentanone; (**e**) norcarane; (**f**) tropylium fluroborate; (**g**) 1-methyl-2-tetralone; (**h**) adamantane; (**i**) buckminsterfullerene (buckyball).

4. Locate descriptions of procedures for the preparation or purification of the following reagents and solvents: (**a**) Raney nickel catalysts; (**b**) sodium borohydride; (**c**) dimethyl sulfoxide; (**d**) sodium amide; (**e**) diazomethane.

5. Find IR spectra for the following compounds: (**a**) *N*-cyclohexylbenzamide; (**b**) 4,5-dihydroxy-2-nitrobenzaldehyde; (**c**) benzyl acetate; (**d**) diisopropyl ether; (**e**) 3,6-diphenyl-2-cyclohexen-1-one; (**f**) 4-amino-1-butanol.

6. Find ^1H and ^{13}C NMR spectra of the following compounds: (**a**) benzyl acetate; (**b**) diisopropyl ether; (**c**) 4-amino-1-butanol; (**d**) 1-propanol; (**e**) indan.

7. *N*-Mesityl-*N*′-phenylformamidine (**3**) was first synthesized between 1950 and 1960. Find the primary research article in which this compound is described, and write an equation for the reaction used to prepare it.

3

8. *N*-Phenyl-*N'*-*p*-tolylformamidine (**4**) (the German name for this compound is the same as in English except that the final "e" is omitted) is reported in *Beilstein* to have a melting point of 86 °C. If you check the first reference given in *Beilstein*, however, you will find the surprising fact that the same chemist who reported this pure compound to have a melting point of 86 °C had described it as melting at 103.5–104.5 °C two years previously. The discrepancy between these reports was not explained until the ambiguity was reexamined during the period 1947–1956. Find the article that solved this mystery.

4

9. The *N*-benzoyl derivative of α-phenylethanamine (**5**) was first described in the form of the optically active (−)-isomer in 1905. Using the formula index of *Beilstein* and given the German name "benzosäure-1-α-phenyläthylamid," find the first reference to this compound in a primary research journal. If you can read the German, give (**a**) the method of preparation by writing the equation; (**b**) the melting point of the pure compound and the recrystallization solvent; and (**c**) the [α]$_D$. Find an article published in English between 1910 and 1920 for a description of the *N*-benzoyl derivative of **5**.

5

10. The name used in the formula indexes of *Chemical Abstracts* for the compound described in Exercise 9 is "benzamide, *N*-methylbenzyl" or "benzamide, *N*-α-methylbenzyl." (**a**) Find a second reference to the (−)-isomer that was published between 1930 and 1940, and compare the physical constants given there with the earlier data. (**b**) Find a reference to a paper published in a Czechoslovakian journal between 1950 and 1960 giving data on the racemic form of the compound. (**c**) Find a reference to data on the (+)-isomer of the compound in a paper published between 1960 and 1970.

11. Determine whether or not each of the compounds whose names are provided below has ever been synthesized, and if it has, give the reference to the first appearance of its synthesis in the literature.

(**a**) Vitamin A	(**f**) Penicillin V	(**k**) Morphine
(**b**) Strychnine	(**g**) Prostaglandin E$_2$	(**l**) Vitamin B$_{12}$
(**c**) Testosterone	(**h**) Lysergic acid diethylamide	(**m**) Cocaine
(**d**) Cephalosporin	(**i**) Erythromycin A	(**n**) Taxol®
(**e**) Reserpine	(**j**) Vinblastine	(**o**) Vernolepin

12. Determine whether or not each of the compounds **(a)–(l)** has ever been synthesized; if it has, give the reference to the first appearance of its synthesis in the literature.

(a) CH$_2$CH$_2$NMe$_2$

(e)

(i) H$_3$CO, OCH$_3$, C=C, H$_3$CO, OCH$_3$

(b) H$_3$C, CH$_3$, C=C=C=C, H$_3$C, CH$_3$

(f) H ... H

(j)

(c)

(g)

(k) Cl, O, Cl, Cl, O, Cl

(d)

(h)

(l) —CH$_2$OH, CH$_3$

13. Using *Science Citation Index*, list five research papers by complete title and journal citation that have cited the review by Martin, S. F. *Tetrahedron* **1980**, *36*, 419.

Page numbers in **boldface** designate locations where the structure of the named compound may be found.